Quantum Many-Particle Systems

ADVANCED BOOK CLASSICS
David Pines, Series Editor

Anderson, P.W., *Basic Notions of Condensed Matter Physics*
Bethe H. and Jackiw, R., *Intermediate Quantum Mechanics, Third Edition*
Feynman, R., *Photon-Hadron Interactions*
Feynman, R., *Quantum Electrodynamics*
Feynman, R., *Statistical Mechanics*
Feynman, R., *The Theory of Fundamental Processes*
Negele, J. W. and Orland, H., *Quantum Many-Particle Systems*
Nozières, P., *Theory of Interacting Fermi Systems*
Parisi, G., *Statistical Field Theory*
Pines, D., *The Many-Body Problem*
Quigg, C., *Gauge Theories of the Strong, Weak, and Electromagnetic Interactions*
Schwinger, J., *Particles, Sources, and Fields, Volume I*
Schwinger, J., *Particles, Sources, and Fields, Volume II*
Schwinger, J., *Particles, Sources, and Fields, Volume III*

Quantum Many-Particle Systems

John W. Negele
Massachusetts Institute of Technology

Henri Orland
Service de Physique Théorique, CEA Saclay, France

Advanced Book Program

CRC Press
Taylor & Francis Group
Boca Raton London New York

CRC Press is an imprint of the
Taylor & Francis Group, an **informa** business

First published 1988 by Westview Press

Published 2018 by CRC Press
Taylor & Francis Group
6000 Broken Sound Parkway NW, Suite 300
Boca Raton, FL 33487-2742

Visit the Taylor & Francis Web site at
http://www.taylorandfrancis.com

and the CRC Press Web site at
http://www.crcpress.com

Library of Congress Catalog Card Number: 98-88186

ISBN 13: 978-0-7382-0052-1 (pbk)

Cover design by Suzanne Heiser

For our parents, our wives, and Janette, Julia and Jonathan

Editor's Foreword

Perseus Books's *Frontiers in Physics* series has, since 1961, made it possible for leading physicists to communicate in coherent fashion their views of recent developments in the most exciting and active fields of physics—without having to devote the time and energy required to prepare a formal review or monograph. Indeed, throughout its nearly forty-year existence, the series has emphasized informality in both style and content, as well as pedagogical clarity. Over time, it was expected that these informal accounts would be replaced by more formal counterparts—textbooks or monographs—as the cutting-edge topics they treated gradually became integrated into the body of physics knowledge and reader interest dwindled. However, this has not proven to be the case for a number of the volumes in the series: Many works have remained in print on an on-demand basis, while others have such intrinsic value that the physics community has urged us to extend their life span.

The *Advanced Book Classics* series has been designed to meet this demand. It will keep in print those volumes in *Frontiers in Physics* or its sister series, *Lecture Notes and Supplements in Physics*, that continue to provide a unique account of a topic of lasting interest. And through a sizable printing, these classics will be made available at a comparatively modest cost to the reader.

The many-body problem, the study of quantum many-particle systems, is an essential part of the education of graduate students in both chemistry and physics. In the present informal text/monograph, John Negele and Henri Orland provide advanced undergraduates and beginning graduate students with a self-contained introduction to the key physical ideas and mathematical techniques currently employed in fields as diverse as nuclear physics, atomic physics, condensed matter physics, and theoretical chemistry. Written in lucid fashion, and containing homework problems at the end of each chapter, *Quantum Many-*

Particle Systems can serve as both a text for a one-semester introductory course or as a reference book for the novice and experienced researcher alike. I am very pleased that the *Advanced Book Classics* series will now make it readily accessible to new generations of readers.

David Pines
Aspen, Colorado
August 1998

PREFACE

The problem of understanding the properties of quantum systems possessing large or infinite numbers of degrees of freedom pervades all of theoretical physics. Hence, the theoretical methods and the physical insight which have been developed over the years for quantum many-particle systems comprise an essential part of the education of students in disciplines as diverse as solid state physics, field theory, atomic physics, condensed matter physics, quantum chemistry and nuclear physics. During the past decade, we have taught one- and two-semester courses on the quantum theory of many-particle systems to graduate students in these disciplines at the Massachusetts Institute of Technology, and this book is an outgrowth of these lectures.

Compared to the texts that appeared in the early 1970's, we have presented standard topics from a different perspective and included a number of new developments. Because of the physical appeal and utility of the Feynman path integral, we have used functional integrals as the foundation of our presentation. Functional integral techniques provide an economical formalism for deriving familiar results, such as perturbation expansions, and yield valuable new approximations and insight into such problems as quantum collective motion, tunneling decay, and phase transitions. Because of the power and physical insights provided by these techniques and their prevalence in the literature, we believe it is essential to teach them to students at this level.

Order parameters and broken symmetry play crucial roles in characterizing and understanding the phases in which matter exists and the transitions between these phases. These concepts, which are familiar from the Landau theory of phase transitions, arise quite naturally from our general development in terms of functional integrals, and are discussed in detail in this text.

Another new topic is the use of stochastic methods for many-body problems. Techniques have existed for a long time to use Markov random walks and Monte Carlo evaluation of integrals to calculate quantum mechanical observables of physical interest to any desired degree of accuracy. In the past, such techniques have received less attention than analytic methods involving summations of diagrams having undetermined convergence properties or other ultimately uncontrolled approximations. We believe that stochastic methods are intellectually interesting in their own right and that they provide a powerful tool to obtain definitive answers to certain classes of otherwise insolvable problems. Hence, we have included a pedagogical introduction to stochastic methods, showing how to calculate observables of interest, stressing the physical connection with path integrals, and demonstrating how to tailor the method to the physics of the problem under consideration.

The scope of this book is intended to be sufficiently broad to serve as a text for a one- or two-semester graduate course. Thus, in addition to these new topics, we have also included the basic body of methodology found in older texts, such as perturbation theory, Green's function techniques, and the Landau theory of Fermi liquids.

Our pedagogical objective is to convey the essential ideas and to prepare the student to read and understand the relevant research literature. We have attempted to present the formalism tersely, without undue emphasis on technical details and to show how it applies to a broad variety of interesting physical systems.

Homework problems are provided at the end of each chapter, and are crucial to a thorough mastery of the subject. Instructive alternative treatments of formal developments in the text are often presented as problems, as well as detailed calculations which are too lengthy for the text. One model system, particles in one dimension interacting via a δ−function two-body potential, is used extensively to illustrate methods presented in the text. For this system, both exact solutions and a multitude of common approximations can be worked out in detail analytically.

Finally, the organization of the book is as follows: We assume only an understanding of elementary quantum mechanics and statistical mechanics, so we begin in Chapter 1 with a thorough, self-contained treatment of second quantization and coherent states. Chapter 2 presents the general formalism of path integrals, perturbation theory and its resummations, and non-perturbative approximations in the formally simple case of the grand canonical ensemble at finite temperature. Specialization to zero temperature and the canonical ensemble is discussed in Chapter 3. Chapter 4 addresses the role of order parameters and broken symmetry in many-body theory and shows how mean field theory embodies the essential physical content of the Landau theory of order parameters and phase transitions. The next chapter develops the general properties of Green's functions, and their application in describing fundamental excitations and physical observables. The phenomenological description and microscopic foundation of the Landau theory of Fermi liquids are presented in Chapter 6. Chapter 7 describes a number of further developments of functional integral techniques, including alternative functional integral representations, the treatment of quantum mean field theory and tunneling decay, and the study of high orders of perturbation theory. The final chapter presents stochastic methods.

As in all such efforts, we are indebted to many people for their invaluable assistance in writing this book. This book was originally stimulated by David Pines and benefited from the editorial guidance of Rick Mixter and Allan Wylde. Although it is impossible to list all of the teachers, colleagues, and students whose insights have contributed to this work, we would particularly like to acknowledge the contributions of R. Balian, J. P. Blaizot, E. Brezin, C. De Dominicis, C. Itzykson, S. E. Koonin, S. Leibler, S. Levit, G. Ripka, and R. Schaeffer.

Portions of earlier drafts of the manuscript were typeset in TEX by Meredith Pollard, Karl Kowalski, and Dany Bunel. The majority of the TEX typesetting, as well as the final editing, improving the layout and appearance of formulas, and preparation of tables was performed by Roger L. Gilson, to whom the authors are particularly indebted for his outstanding work. We also wish to acknowledge the excellent technical art work by Don Souza and the work of the Addison-Wesley production department in providing the final copy.

JOHN NEGELE
HENRI ORLAND

CONTENTS

CHAPTER 1

SECOND QUANTIZATION AND COHERENT STATES

The quantum mechanics of a single particle is usually formulated in terms of the position operator $\hat{x}$ and the momentum operator $\hat{p}$. All other operators of physical interest may be expressed in terms of these operators, and a natural representation for quantum mechanics, the coordinate representation, is defined in terms of eigenfunctions of the position operator. In this chapter, an analogous formalism is developed for systems composed of many identical particles. For these systems it is useful to define operators which create or annihilate a particle in specified states. Operators of physical interest may be expressed in terms of these creation and annihilation operators, in which case they are said to be expressed in "second quantized" form. The eigenstates of the annihilation operators are coherent states. A natural representation for the quantum mechanics of many-particle systems, the holomorphic representation, is defined in terms of these coherent states.

As a prelude to the formalism for many-particle systems, it is useful to begin by reviewing some elementary aspects of quantum mechanics.

1.1 QUANTUM MECHANICS OF A SINGLE PARTICLE

Quantum mechanics describes the state of a particle by a state vector $|\phi\rangle$, which belongs to a Hilbert space $\mathcal{H}$. This Hilbert space $\mathcal{H}$ is the vector space of complex, square integrable functions, defined in configuration space. Using Dirac notation, the scalar product of vectors in $\mathcal{H}$ is:

$$\langle\phi|\psi\rangle = \int d^3r\phi^*(\vec{r})\psi(\vec{r}) \quad . \tag{1.1}$$

Then by definition, a vector $|\phi\rangle$ belongs to the Hilbert space $\mathcal{H}$ if the norm of $|\phi\rangle$ is finite:

$$\langle\phi|\phi\rangle = \int d^3r\,|\phi(\vec{r})|^2 < +\infty \quad . \tag{1.2}$$

Of particular importance are the vectors $|\vec{r}\rangle$ and $|\vec{p}\rangle$, eigenvectors of the quantum position operator $\hat{\vec{r}}$ and momentum operator $\hat{\vec{p}}$

$$\hat{\vec{r}}|\vec{r}\rangle = \vec{r}|\vec{r}\rangle \tag{1.3}$$

$$\hat{\vec{p}}|\vec{p}\rangle = \vec{p}|\vec{p}\rangle \tag{1.4}$$

Although these vectors do not belong to $\mathcal{H}$, because their norm is not finite, they span the whole Hilbert space $\mathcal{H}$. This is reflected by the following closure relations:

$$\int d^3r|\vec{r}\rangle\langle\vec{r}| = 1 \tag{1.5}$$

$$\int d^3p|\vec{p}\rangle\langle\vec{p}| = 1 \tag{1.6}$$

where 1 denotes the unit operator in $\mathcal{H}$.

A state vector $|\vec{r}\rangle$ represents a state in which the particle is localized at point $\vec{r}$, and a state vector $|\vec{p}\rangle$ represents a particle with a momentum $\vec{p}$. The overlap of these vectors is given by:

$$\langle\vec{r}|\vec{r}'\rangle = \delta^{(3)}(\vec{r}-\vec{r}') \tag{1.7}$$

$$\langle\vec{p}|\vec{p}'\rangle = \delta^{(3)}(\vec{p}-\vec{p}') \tag{1.8}$$

and

$$\langle\vec{r}\,|\vec{p}\rangle = \left(\frac{1}{2\pi\hbar}\right)^{\frac{3}{2}} \cdot e^{\frac{i\vec{p}\cdot\vec{r}}{\hbar}} \ . \tag{1.9}$$

The wave function of a particle in a state $|\phi\rangle$ is given in coordinate representation by:

$$\phi(\vec{r}) = \langle\vec{r}|\phi\rangle \tag{1.10}$$

and represents the probability amplitude for finding the particle at point $\vec{r}$.

In coordinate representation, the operators $\hat{\vec{r}}$ and $\hat{\vec{p}}$ act as follows:

$$\langle\vec{r}|\hat{\vec{r}}|\phi\rangle = \vec{r}\langle\vec{r}|\phi\rangle = \vec{r}\phi(\vec{r}) \tag{1.11}$$

and

$$\begin{aligned}\langle\vec{r}|\vec{p}|\phi\rangle &= \int d^3p\langle\vec{r}|\hat{\vec{p}}|\vec{p}\rangle\langle\vec{p}|\phi\rangle \\ &= \int d^3p\vec{p}\langle\vec{r}|\vec{p}\rangle\langle\vec{p}|\phi\rangle \\ &= \frac{\hbar}{i}\frac{\vec{\partial}}{\partial r}\int d^3p\langle\vec{r}|\vec{p}\rangle\langle\vec{p}|\phi\rangle \\ &= \frac{\hbar}{i}\frac{\vec{\partial}}{\partial r}\phi(\vec{r}) \ .\end{aligned} \tag{1.12}$$

Thus, in coordinate representation we may write:

$$\hat{\vec{r}} = \vec{r} \tag{1.13}$$

and

$$\hat{\vec{p}} = \frac{\hbar}{i}\frac{\partial}{\partial\vec{r}} \ . \tag{1.14}$$

For a particle of mass m in a local potential $V(\vec{r})$, the Hamiltonian is

$$\hat{H} = \frac{\hat{\vec{p}}^2}{2m} + V(\vec{r}) \ . \tag{1.15}$$

In this case, the time-independent Schrödinger equation

$$H|\phi_\alpha\rangle = e_\alpha|\phi_\alpha\rangle \tag{1.16}$$

has the familiar form in coordinate representation

$$\left[\frac{-\hbar^2}{2m}\nabla_r^2 + V(\vec{r})\right]\phi_\alpha(\vec{r}) = e_\alpha\phi_\alpha(\vec{r}) \ . \tag{1.17}$$

For particles having some internal degrees of freedom, such as spin $\vec{\sigma}$ or isospin $\vec{\tau}$, the state vector $|\phi\rangle$ has several components, each corresponding to a different set of values σ_z and τ_z of the internal degrees of freedom. Analogous to the definition of the states $|\vec{r}\rangle$ in Eq. (1.3), we may define a state $|\vec{r}\sigma\tau\rangle$, which is a state of the particle localized at point $\vec{r}$, with a projection of the spin σ and isospin τ. These states span the Hilbert space of the particle:

$$\sum_{\sigma=\pm\frac{1}{2}}\sum_{\tau=\pm\frac{1}{2}}\int d^3r\,|\vec{r}\sigma\tau\rangle\langle\vec{r}\sigma\tau| = 1 \tag{1.18}$$

and their overlap is given by:

$$\langle\vec{r}\sigma\tau|\vec{r}'\sigma'\tau'\rangle = \delta_{\sigma\sigma'}\delta_{\tau\tau'}\delta^{(3)}(\vec{r}-\vec{r}') \ . \tag{1.19}$$

Whenever no ambiguities will arise, we will use the notation:

$$|x\rangle \equiv |\vec{r}\sigma\tau\rangle \tag{1.20}$$

and

$$\int dx = \sum_{\sigma=\pm\frac{1}{2}}\sum_{\tau=\pm\frac{1}{2}}\int d^3r \tag{1.21}$$

and the convention:

$$\delta(x-x') = \delta_{\sigma\sigma'}\delta_{\tau\tau'}\delta^{(3)}(\vec{r}-\vec{r}') \ . \tag{1.22}$$

With this notation, equations (1.18) and (1.19) simplify to:

$$\int dx|x\rangle\langle x| = 1 \tag{1.23}$$

and

$$\langle x|x'\rangle = \delta(x-x') \ . \tag{1.24}$$

1.2 SYSTEMS OF IDENTICAL PARTICLES

The Hilbert space of states for a system of N identical particles is the space $\mathcal{H}_N$ of complex, square integrable functions, defined in the configuration space of the N particles. The wave function $\psi_N(\vec{r}_1, \vec{r}_2, ..., \vec{r}_N)$, which represents the probability amplitude for finding particles at the N positions $\vec{r}_1, \vec{r}_2, ..., \vec{r}_N$, must satisfy the condition:

$$\langle\psi_N|\psi_N\rangle = \int d^3r_1 \ldots d^3r_N |\psi_N(\vec{r}_1, \vec{r}_2, \ldots, \vec{r}_N)|^2 < +\infty \quad . \tag{1.25}$$

As we have defined it, the Hilbert space $\mathcal{H}_N$ is simply the Nth tensor product of the single-particle Hilbert space $\mathcal{H}$:

$$\mathcal{H}_N = \mathcal{H} \otimes \mathcal{H} \otimes \ldots \otimes \mathcal{H} \quad . \tag{1.26}$$

If $\{|\alpha\rangle\}$ is an orthonormal basis of $\mathcal{H}$, the canonical orthonormal basis of $\mathcal{H}_N$ is constructed from the tensor products:

$$|\alpha_1 \ldots \alpha_N) \equiv |\alpha_1\rangle \otimes |\alpha_2\rangle \otimes \ldots \otimes |\alpha_N\rangle \quad . \tag{1.27}$$

Note for future reference that the states defined in this way utilize a curved bracket in the ket symbol. These basis states have the wave functions:

$$\begin{aligned}\psi_{\alpha_1\alpha_2\ldots\alpha_N}(\vec{r}_1, \ldots, \vec{r}_N) &= (\vec{r}_1 \ldots \vec{r}_N|\alpha_1 \ldots \alpha_N) \\ &= (\langle\vec{r}_1| \otimes \langle\vec{r}_2| \otimes \ldots \otimes \langle\vec{r}_N|)\,(|\alpha_1\rangle \otimes |\alpha_2\rangle \otimes \ldots \otimes |\alpha_N\rangle) \\ &= \phi_{\alpha_1}(\vec{r}_1)\phi_{\alpha_2}(\vec{r}_2) \ldots \phi_{\alpha_N}(\vec{r}_N) \quad .\end{aligned} \tag{1.28}$$

The overlap of two vectors of the basis is given by:

$$\begin{aligned}(\alpha_1\alpha_2 \ldots \alpha_N|\alpha'_1\alpha'_2 \ldots \alpha'_N) &= (\langle\alpha_1| \otimes \langle\alpha_2| \otimes \ldots \otimes \langle\alpha_N|)\,(|\alpha'_1\rangle \otimes |\alpha'_2\rangle \otimes \ldots \otimes |\alpha'_N\rangle) \\ &= \langle\alpha_1|\alpha'_1\rangle\langle\alpha_2|\alpha'_2\rangle \ldots \langle\alpha_N|\alpha'_N\rangle\end{aligned} \tag{1.29}$$

and the completeness of the basis is obtained from the tensor product of the completeness relation for the basis $\{|\alpha\rangle\}$:

$$\sum_{\alpha_1, \alpha_2, \ldots, \alpha_N} |\alpha_1\alpha_2 \ldots \alpha_N)(\alpha_1\alpha_2 \ldots \alpha_N| = 1 \tag{1.30}$$

where 1 now represents the unit operator in $\mathcal{H}_N$. Physically, it is clear that the space $\mathcal{H}_N$ is generated by linear combinations of products of single-particle wave functions.

Thus far, in defining the Hilbert space $\mathcal{H}_N$, we have not taken into account the symmetry property of the wave function. In contrast to the multitude of pure and mixed symmetry states one could define mathematically, only totally symmetric and antisymmetric states are observed in nature. Particles occurring in symmetric or antisymmetric states are called Bosons and Fermions respectively.

The wave function of N Bosons is totally symmetric and thus satisfies

$$\psi\left(\vec{r}_{P1}, \vec{r}_{P2}, \ldots, \vec{r}_{PN}\right) = \psi\left(\vec{r}_1, \vec{r}_2, \ldots, \vec{r}_N\right) \tag{1.31a}$$

where $(P1, P2, \ldots, PN)$ represents any permutation, P, of the set $(1, 2, \ldots, N)$.

The wave function of N Fermions is antisymmetric under the exchange of any pair of particles and therefore satisfies:

$$\psi(\vec{r}_{P1}, \vec{r}_{P2}, \ldots, \vec{r}_{PN}) = (-1)^P \psi(\vec{r}_1, \vec{r}_2, \ldots, \vec{r}_N) \quad . \tag{1.31b}$$

Here, $(-1)^P$ denotes the sign, or parity, of the permutation P, and is defined as the parity of the number of transpositions of two elements which brings the permutation $(P1, P2, \ldots, PN)$ to its original form $(1, 2, \ldots, N)$.

Although the symmetry requirements for Bosons and Fermions are ultimately founded on experiment, it may be proven within the context of quantum field theory that given general assumptions of locality, causality and Lorentz invariance, particles with integer spin $(0, 1, 2, \ldots)$ are Bosons and particles with half-integer spin $(\frac{1}{2}, \frac{3}{2}, \ldots)$ are Fermions. Familiar examples of Bosons include photons, pions, mesons, gluons, and the ^4He atom. Examples of Fermions include protons, neutrons, electrons, muons, neutrinos, quarks, and the ^3He atom. Composite particles composed of any number of Bosons and an even or odd number of Fermions behave as Bosons or Fermions respectively at energies sufficiently low compared to their binding energy.

For convenience, we shall adopt the following unified notation for Bosons or Fermions:

$$\psi(\vec{r}_{P1}, \vec{r}_{P2}, \ldots, \vec{r}_{PN}) = \varsigma^P \psi(\vec{r}_1, \vec{r}_2, \ldots, \vec{r}_N) \tag{1.32}$$

where P is the parity of the permutation, and ς is $+1$ or -1 for Bosons or Fermions respectively.

These symmetry requirements imply corresponding restrictions of the Hilbert space $\mathcal{H}_N$ of N-particle systems. A wave function $\psi(\vec{r}_1, \ldots, \vec{r}_N)$ of $\mathcal{H}_N$ belongs to the Hilbert space of N Bosons, $\mathcal{B}_N$, or the Hilbert space of N Fermions, $\mathcal{F}_N$, if it is symmetric or antisymmetric respectively, under a permutation of the particles.

We will define the symmetrization operator $\mathcal{P}_B$ and the antisymmetrization operator $\mathcal{P}_F$ in $\mathcal{H}_N$ by their action on a wave function $\psi(\vec{r}_1, \ldots, \vec{r}_N)$:

$$\mathcal{P}_{\{{B \atop F}\}} \psi(\vec{r}_1, \ldots, \vec{r}_N) = \frac{1}{N!} \sum_P \varsigma^P \psi(\vec{r}_{P1}, \vec{r}_{P2}, \ldots \vec{r}_{PN}) \quad . \tag{1.33}$$

For example, for two Bosons:

$$\mathcal{P}_B \psi(\vec{r}_1, \vec{r}_2) = \frac{1}{2} \left(\psi(\vec{r}_1, \vec{r}_2) + \psi(\vec{r}_2, \vec{r}_1) \right) \tag{1.34a}$$

and for two Fermions:

$$\mathcal{P}_F \psi(\vec{r}_1, \vec{r}_2) = \frac{1}{2} \left(\psi(\vec{r}_1, \vec{r}_2) - \psi(\vec{r}_2, \vec{r}_1) \right) \quad . \tag{1.34b}$$

The manifestly hermitian operator $\mathcal{P}_{\{{B \atop F}\}}$ may be shown to be a projector as follows. For any wave function ψ of $\mathcal{H}_N$:

$$\mathcal{P}^2_{\{{B \atop F}\}} \psi(\vec{r}_1, \ldots, \vec{r}_N) = \frac{1}{N!} \frac{1}{N!} \sum_P \sum_{P'} \varsigma^{P'} \varsigma^P \psi(\vec{r}_{P'P1}, \vec{r}_{P'P2}, \ldots, \vec{r}_{P'PN}) \tag{1.35}$$

where $P'P$ denotes the group composition of P' and P. Since $\varsigma^{P+P'} = \varsigma^{P'P}$, the summation over P and P' can be replaced by a summation over $Q = P'P$ and P:

$$\begin{aligned} \mathcal{P}^2_{\{^B_F\}}\psi(\vec{r}_1,\ldots,\vec{r}_N) &= \frac{1}{N!}\sum_P \left(\frac{1}{N!}\sum_Q \varsigma^Q \psi(\vec{r}_{Q1},\ldots,\vec{r}_{QN})\right) \\ &= \frac{1}{N!}\sum_P \mathcal{P}_{\{^B_F\}}\psi(\vec{r}_1,\ldots,\vec{r}_N) \\ &= \mathcal{P}_{\{^B_F\}}\psi(\vec{r}_1,\ldots,\vec{r}_N) \ . \end{aligned} \tag{1.36}$$

Since this equality holds for any wave function ψ, the equality holds for the operator itself and the symmetrization and antisymmetrization operators are projectors. These operators project $\mathcal{H}_N$ onto the Hilbert space of Bosons $\mathcal{B}_N$ and the Hilbert space of Fermions $\mathcal{F}_N$:

$$\mathcal{B}_N = \mathcal{P}_B\mathcal{H}_N \tag{1.37a}$$

$$\mathcal{F}_N = \mathcal{P}_F\mathcal{H}_N \ . \tag{1.37b}$$

Using these projectors, a system of Bosons or Fermions, with one particle in state α_1, one particle in state $\alpha_2,\ldots,$ and one particle in state α_N, is represented as follows:

$$\begin{aligned} |\alpha_1\ldots\alpha_N\} &\equiv \sqrt{N!}\mathcal{P}_{\{^B_F\}}|\alpha_1\ldots\alpha_N) \\ &= \frac{1}{\sqrt{N!}}\sum_P \varsigma^P |\alpha_{P1}\rangle\otimes|\alpha_{P2}\rangle\otimes\ldots\otimes|\alpha_{PN}\rangle \ . \end{aligned} \tag{1.38}$$

Note that these symmetrized or antisymmetrized states utilize a curly bracket in the ket symbol.

The Pauli exclusion principle, that two Fermions cannot occupy the same state, is automatically satisfied for antisymmetric states. Suppose two states are identical, for example, $|\alpha_1\rangle = |\alpha_2\rangle = |\alpha\rangle$. Then:

$$\begin{aligned} |\alpha_1\alpha_2\alpha_3\ldots\alpha_N\} &= \sqrt{N!}\mathcal{P}_F|\alpha_1\alpha_2\alpha_3\ldots\alpha_N) \\ &= -\sqrt{N!}\mathcal{P}_F|\alpha_2\alpha_1\alpha_3\ldots\alpha_N) \\ &= 0 \end{aligned} \tag{1.39}$$

and no acceptable many-Fermion state exists in this case.

From Equations (1.37), it follows that if $|\alpha_1\ldots\alpha_N)$ is a basis of the Hilbert space $\mathcal{H}_N$, then $\mathcal{P}_{\{^B_F\}}|\alpha_1\ldots\alpha_N)$ is a basis of $\mathcal{B}_N$ or $\mathcal{F}_N$. The closure relation (1.30) in $\mathcal{H}_N$ becomes a closure relation in $\mathcal{B}_N$ or $\mathcal{F}_N$:

$$\sum_{\alpha_1\ldots\alpha_N} \mathcal{P}_{\{^B_F\}}|\alpha\ldots\alpha_N)(\alpha_1\ldots\alpha_N|\mathcal{P}_{\{^B_F\}} = \frac{1}{N!}\sum_{\alpha_1\ldots\alpha_N} |\alpha_1\ldots\alpha_N\}\{\alpha_1\ldots\alpha_N| = 1 \ . \tag{1.40}$$

Further, if the basis $|\alpha)$ is orthogonal in $\mathcal{H}$, then the basis $|\alpha_1\ldots\alpha_N)$ is orthogonal in $\mathcal{H}_N$, and the basis $|\alpha_1\ldots\alpha_N\}$ is orthogonal in $\mathcal{B}_N$ or $\mathcal{F}_N$.

The scalar product of two such vectors constructed from the same basis $|\alpha\rangle$ is:

$$\begin{aligned}\{\alpha'_1\dots\alpha'_N|\alpha_1\dots\alpha_N\} &= N!(\alpha'_1\dots\alpha'_N|\mathcal{P}^2_{\{{B \atop F}\}}|\alpha_1\dots\alpha_N) \\ &= N!(\alpha'_1\dots\alpha'_N|\mathcal{P}_{\{{B \atop F}\}}|\alpha_1\dots\alpha_N) \\ &= \sum_P \varsigma^P \langle\alpha'_1|\alpha_{P1}\rangle\dots\langle\alpha'_N|\alpha_{PN}\rangle \ .\end{aligned} \tag{1.41}$$

Because of the orthogonality of the basis $|\alpha\rangle$, the only non-vanishing terms in the right hand side of (1.41) are the permutations P such that:

$$\alpha'_1 = \alpha_{P1}, \alpha'_2 = \alpha_{P2}, \dots, \alpha'_N = \alpha_{PN} \ . \tag{1.42}$$

If $\alpha'_1, \dots, \alpha'_N$ is a permutation of $\alpha_1, \dots, \alpha_N$, the overlap may be evaluated straightforwardly. For Fermions, since there is at most one particle per state $|\alpha\rangle$, no two identical states can be present in the set $\{\alpha_1 \dots, \alpha_N\}$ and therefore, there exists only one permutation P which transforms $\alpha_1 \dots, \alpha_N$ into $\alpha'_1, \dots, \alpha'_N$. The sum in (1.41) thus reduces to one term, and if the states $|\alpha_i\rangle$ are normalized, we obtain

$$\{\alpha'_1\dots\alpha'_N|\alpha_1\dots\alpha_N\} = (-1)^P \quad (\text{ Fermions }) \ . \tag{1.43a}$$

For Bosons, many particles may be in the same state, and therefore, any permutation which does interchange particles in the same state contributes to the sum (1.41). The overlap (1.41) is thus equal to the total number of permutations which transform $\{\alpha_1 \dots \alpha_N\}$ into $\{\alpha'_1, \dots, \alpha'_N\}$. If the set of states $\{\alpha_1 \dots, \alpha_N\}$ represents a system with n_1 Bosons in state α_1, n_2 Bosons in state $\alpha_2, \dots,$ n_p Bosons in state α_p, where the states $\alpha_1, \dots, \alpha_p$ are distinct, the overlap is given by:

$$\{\alpha'_1\dots\alpha'_N|\alpha_1\dots\alpha_N\} = n_1!n_2!\dots n_p! \quad (\text{ Bosons }) \ . \tag{1.43b}$$

The results (1.43a) and (1.43b) may be combined efficiently by specifying a state with particles in states $\alpha_1, \dots, \alpha_N$ in terms of occupation numbers n_α for each state of the basis $|\alpha\rangle$. For Bosons, the occupation numbers are *a priori* not restricted, whereas for Fermions they can take only the value 0 or 1. In both cases, the sum of the occupation numbers, which counts the total number of occupied states, must be equal to the number of particles N:

$$N = \sum_\alpha n_\alpha \ . \tag{1.44}$$

For example, the three-Boson state with particles in states $\alpha_1, \alpha_1, \alpha_2$ corresponds to $n_1 = 2, n_2 = 1$, and $n_i = 0$ for $i \geq 3$. The three-Fermion state with particles in states $\alpha_1, \alpha_2, \alpha_3$ corresponds to $n_1 = 1, n_2 = 1, n_3 = 1$, and $n_i = 0$ for $i \geq 4$. Using the convention that $0! = 1$, the formulas (1.43) are equivalent to the single expression

$$\{\alpha'_1\dots\alpha'_N|\alpha_1\dots\alpha_N\} = \varsigma^P \prod_\alpha n_\alpha! \ . \tag{1.45}$$

Finally, an orthonormal basis for the Hilbert space $\mathcal{B}_N$ or $\mathcal{F}_N$ is obtained by utilizing (1.45) to normalize the states $|\alpha_1 \dots \alpha_N\}$:

$$|\alpha_1 \dots \alpha_N\rangle = \frac{1}{\sqrt{\prod_\alpha n_\alpha!}} |\alpha_1 \dots \alpha_N\}$$
$$= \frac{1}{\sqrt{N! \prod_\alpha n_\alpha!}} \sum_P \varsigma^P |\alpha_{P1}\rangle \otimes |\alpha_{P2}\rangle \otimes \dots \otimes |\alpha_{PN}\rangle \quad . \tag{1.46}$$

Note that in contrast to the states defined in (1.27) and (1.37), the normalized symmetric or antisymmetric states defined in (1.46) utilize an angular bracket in the ket symbol. Because orthonormality was used in the calculation of the normalization factor, it will be understood henceforth that whenever the symbol $|\alpha_1 \dots \alpha_N\rangle$ is utilized, the basis $\{|\alpha_i\rangle\}$ is orthonormal.

The overlap between a tensor product $|\beta_1 \dots \beta_N)$ constructed from an orthonormal basis $|\beta\rangle$ and the symmetrized or antisymmetrized state $|\alpha_1 \dots \alpha_N\rangle$ is

$$(\beta_1, \dots \beta_N | \alpha_1 \dots \alpha_N\rangle = \frac{1}{\sqrt{N! \prod_\alpha n_\alpha!}} \sum_P \varsigma^P \langle\beta_1|\alpha_{P1}\rangle \langle\beta_2|\alpha_{P2}\rangle \dots \langle\beta_N|\alpha_{PN}\rangle$$
$$\equiv \frac{1}{\sqrt{N! \prod_\alpha n_\alpha!}} S(\langle\beta_i|\alpha_j\rangle) \tag{1.47}$$

where $S(M_{ij})$ denotes a permanent for Bosons

$$Per(M_{ij}) \equiv \sum_P M_{1,P1} M_{2,P2} \dots M_{N,PN} \tag{1.48a}$$

and a determinant for Fermions

$$\det(M_{ij}) \equiv \sum_P (-1)^P M_{1,P1} M_{2,P2} \dots M_{N,PN} \quad . \tag{1.48b}$$

In coordinate representation, we thus obtain a basis of permanent wave functions for Bosons

$$\psi_{\beta_1 \dots \beta_N}(x_1 \dots x_n) = (x_1 \dots x_N | \beta_1 \dots \beta_N\rangle$$
$$= \frac{1}{\sqrt{N! \prod n_\alpha!}} Per(\phi_{\beta_i}(x_j)) \tag{1.49a}$$

and a basis of Slater determinants for Fermions

$$\psi_{\beta_1 \dots \beta_N}(x_1 \dots x_N) = (x_1 \dots x_N | \beta_1 \dots \beta_N\rangle$$
$$= \frac{1}{\sqrt{N!}} \det(\phi_{\beta_i}(x_j)) \quad . \tag{1.49b}$$

Similarly, the overlap of two normalized Boson or Fermion states is

$$\langle\beta_1 \dots \beta_N | \alpha_1 \dots \alpha_N\rangle = \frac{1}{\sqrt{\prod_\beta n'_\beta! \prod_\alpha n_\alpha!}} S(\langle\beta_i|\alpha_j\rangle) \quad . \tag{1.50}$$

Finally using (1.40) and the normalization in (1.46), the closure relation in $\mathcal{B}_N$ or $\mathcal{F}_N$ is given by:

$$\sum_{\alpha_1 \dots \alpha_N} \frac{\prod_\alpha n_\alpha!}{N!} |\alpha_1 \dots \alpha_N\rangle\langle\alpha_1 \dots \alpha_N| = 1 \quad . \tag{1.51}$$

1.3 MANY-BODY OPERATORS

We now consider matrix elements of many-body operators in the canonical basis of $\mathcal{H}_N$, Eq. (1.27). From these, the representation of operators in the spaces $\mathcal{B}_N$ and $\mathcal{F}_N$ follows straightforwardly using the symmetrization and antisymmetrization operators $\mathcal{P}_{\{{B \atop F}\}}$, Eq. (1.33).

Let $\mathcal{O}$ be an arbitrary operator in $\mathcal{B}_N$ or $\mathcal{F}_N$. Independent of whether the particles are Bosons or Fermions, their indistinguishability implies that $\mathcal{O}$ is invariant under any permutation of the particles. Thus, for any states, and any permutation P:

$$(\beta_{P1}\dots\beta_{PN}|\mathcal{O}|\beta'_{P1}\dots\beta'_{PN}) = (\beta_1\dots\beta_N|\mathcal{O}|\beta'_1\dots\beta'_N) \quad . \tag{1.52}$$

We begin by considering the case of one-body operators. An operator $\hat{U}$ is a one-body operator if the action of $\hat{U}$ on a state $|\alpha_1\dots\alpha_N)$ of N particles is the sum of the action of $\hat{U}$ on each particle:

$$\hat{U}|\alpha_1\dots\alpha_N) = \sum_{i=1}^{N}\hat{U}_i|\alpha_1\dots\alpha_N) \tag{1.53}$$

where the operator $\hat{U}_i$ operates only on the ith particle. For example, the kinetic energy operator $\hat{T}$ in the $\{\vec{p}\}$ basis, acts as:

$$\hat{T}|\vec{p}_1\dots\vec{p}_N) = \sum_{i=1}^{N}\frac{\hat{\vec{p}}_i^{\,2}}{2m}\cdot|\vec{p}_1\dots\vec{p}_N) \tag{1.54a}$$

and a local one-body potential in the $\{x\}$ basis acts as:

$$\hat{W}|x_1\dots x_N) = \sum_{i=1}^{N}W(x_i)|x_1\dots x_N) \quad . \tag{1.54b}$$

The matrix element of a one-body operator $\hat{U}$ between two states $|\alpha_1\dots\alpha_N)$ and $|\beta_1\dots\beta_N)$ is given by

$$\begin{aligned}(\alpha_1\dots\alpha_N|\hat{U}|\beta_1\dots\beta_N) &= \sum_{i=1}^{N}(\alpha_1\dots\alpha_N|\hat{U}_i|\beta_1\dots\beta_N)\\ &= \sum_{i=1}^{N}\prod_{k\neq i}\langle\alpha_k|\beta_k\rangle\cdot\langle\alpha_i|\hat{U}|\beta_i\rangle\end{aligned} \tag{1.55a}$$

and thus for two non-orthogonal states:

$$\frac{(\alpha_1\dots\alpha_N|\hat{U}|\beta_1\dots\beta_N)}{(\alpha_1\dots\alpha_N|\beta_1\dots\beta_N)} = \sum_{i=1}^{N}\frac{\langle\alpha_i|\hat{U}|\beta_i\rangle}{\langle\alpha_i|\beta_i\rangle} \; . \tag{1.55b}$$

The one-body operator U is entirely determined by its matrix elements $\langle\alpha|\hat{U}|\beta\rangle$ in the single-particle Hilbert space $\mathcal{H}$.

Similarly, an operator $\hat{V}$ is a two-body operator if the action of $\hat{V}$ on a state $|\alpha_1 ... \alpha_N)$ of N particles is the sum of the action of V on all distinct pairs of particles:

$$\hat{V}|\alpha_1 \ldots \alpha_N) = \sum_{1\le i<j\le N} \hat{V}_{ij}|\alpha_1 \ldots \alpha_N) \tag{1.56}$$

where V_{ij} operates only on particles i and j. The restriction $i < j$ yields a summation over distinct pairs. The symmetry requirement (1.52) on $\hat{V}$ implies that $\hat{V}_{ij} = \hat{V}_{ji}$, and the expression (1.56) can be rewritten as:

$$\hat{V}|\alpha_1 \ldots \alpha_N) = \frac{1}{2} \sum_{1\le i\neq j\le N} \hat{V}_{ij}|\alpha_1 \ldots \alpha_N) \ . \tag{1.57}$$

The matrix elements of $\hat{V}$ are given by:

$$\begin{aligned}(\alpha_1 \ldots \alpha_N|\hat{V}|\beta_1 \ldots \beta_N) &= \frac{1}{2}\sum_{i\neq j}(\alpha_1 \ldots \alpha_N|\hat{V}_{ij}|\beta_1 \ldots \beta_N) \\ &= \frac{1}{2}\sum_{i\neq j}\prod_{\substack{k\neq i\\ k\neq j}}\langle\alpha_k|\beta_k\rangle \cdot (\alpha_i\alpha_j|V|\beta_i\beta_j)\end{aligned} \tag{1.58a}$$

and for non-orthogonal states,

$$\frac{(\alpha_1 \ldots \alpha_N|\hat{V}|\beta_1 \ldots \beta_N)}{(\alpha_1 \ldots \alpha_N|\beta_1 \ldots \beta_N)} = \frac{1}{2}\sum_{i\neq j}\frac{(\alpha_i\alpha_j|\hat{V}|\beta_i\beta_j)}{\langle\alpha_i|\beta_i\rangle\langle\alpha_j|\beta_j\rangle} \ . \tag{1.58b}$$

Thus, a two-body operator V is entirely determined by its matrix elements $(\alpha\alpha'|\hat{V}|\beta\beta')$ in the Hilbert space $\mathcal{H}_2$ of two-particle systems.

Further, a two-body interaction $\hat{V}$ is said to be local, or velocity independent, when it is diagonal in configuration space (and spin and isospin if appropriate):

$$(\vec{r}_1\vec{r}_2|\hat{V}|\vec{r}_3\vec{r}_4) = \delta(\vec{r}_1 - \vec{r}_3)\delta(\vec{r}_2 - \vec{r}_4)v(\vec{r}_1 - \vec{r}_2) \ . \tag{1.59}$$

In this case, Equation (1.57) takes the simple form:

$$\hat{V}|\vec{r}_1 \ldots \vec{r}_N) = \frac{1}{2} \sum_{1\le i\neq j\le N} v(\vec{r}_i - \vec{r}_j)|\vec{r}_1 \ldots \vec{r}_N) \ . \tag{1.60}$$

In general, we define an n-body operator $\hat{R}$ as an operator which acts on a state $|\alpha_1 ... \alpha_N\rangle$ in the following way:

$$\hat{R}|\alpha_1 \ldots \alpha_N) = \frac{1}{n!} \sum_{1\le i_1\neq i_2\neq \ldots \neq i_n\le N} \hat{R}_{i_1 i_2 \ldots i_n}|\alpha_1 \ldots \alpha_N) \ . \tag{1.61}$$

That is, the action of $\hat{R}$ on a state of N particles is the sum of the action of $\hat{R}$ on all distinct subsets of n-particles. Analogous to the previous cases, the matrix elements of $\hat{R}$ satisfy:

$$\frac{(\alpha_1 \ldots \alpha_N|\hat{R}|\beta_1 \ldots \beta_N)}{(\alpha_1 \ldots \alpha_N|\beta_1 \ldots \beta_N)} = \frac{1}{n!}\sum_{i_1\neq i_2\neq \ldots \neq i_n}\frac{(\alpha_{i_1} \ldots \alpha_{i_n}|\hat{R}|\beta_{i_1} \ldots \beta_{i_n})}{\langle\alpha_{i_1}|\beta_{i_1}\rangle\langle\alpha_{i_2}|\beta_{i_2}\rangle \ldots \langle\alpha_{i_n}|\beta_{i_n}\rangle} \tag{1.62}$$

and an n-body operator is entirely determined by its matrix elements $(\alpha_1 \ldots \alpha_n|\hat{R}|\beta_1 \ldots \beta_n)$ in the Hilbert space $\mathcal{H}_n$ of n-particle systems.

1.4 CREATION AND ANNIHILATION OPERATORS

Creation and annihilation operators provide a convenient representation of the many-particle states and many-body operators introduced in the preceding sections. These operators generate the entire Hilbert space by their action on a single reference state and provide a basis for the algebra of operators of the Hilbert space.

For each single-particle state $|\lambda\rangle$ of the single-particle space $\mathcal{H}$, we define a Boson or Fermion creation operator $a^\dagger_\lambda$ by its action on any symmetrized or antisymmetrized state $|\lambda_1 \dots \lambda_N\}$ of $\mathcal{B}_N$ or $\mathcal{F}_N$ as follows:

$$a^\dagger_\lambda|\lambda_1 \dots \lambda_N\} \equiv |\lambda\lambda_1 \dots \lambda_N\} \ . \tag{1.63a}$$

For the present treatment, it is convenient to use an orthonormal basis $\{|\lambda_i\rangle\}$, for which the definition of $a^\dagger_\lambda$ may also be written

$$a^\dagger_\lambda|\lambda_1 \dots \lambda_N\rangle = \sqrt{n_\lambda + 1}|\lambda\lambda_1 \dots \lambda_N\rangle \tag{1.63b}$$

where n_λ is the occupation number of the state $|\lambda\rangle$ in $|\lambda_1 \dots \lambda_N\rangle$. Note that whereas we have previously denoted other operators by a circumflex, $\hat{O}$, since the symbols a and $a^\dagger$ will be reserved exclusively for creation and annihilation operators, no circumflex will be used. Physically the operator $a^\dagger_\lambda$ adds a particle in state $|\lambda\rangle$ to the state on which it operates, and symmetrizes or antisymmetrizes the new state. In the case of Fermions, since there can be at most one Fermion in a given state, Eq. (1.63b) takes the simpler form:

$$a^\dagger_\lambda|\lambda_1 \dots \lambda_N\rangle = \begin{cases} |\lambda\lambda_1 \dots \lambda_N\rangle & \text{if the state } |\lambda\rangle \text{ is not present in } |\lambda_1 \dots \lambda_N\rangle \\ 0 & \text{if the state } |\lambda\rangle \text{ is present in } |\lambda_1 \dots \lambda_N\rangle \ . \end{cases} \tag{1.64}$$

In the following, to avoid a distinction between Bosons and Fermions, we shall use Eq. (1.63).

In addition to the many-particle states we have used thus far, it is useful to define the vacuum state, denoted $|0\rangle$, which represents a state with zero particles. A natural extension of (1.63) is the definition that $a^\dagger_\lambda$ acting on the vacuum $|0\rangle$, creates a particle in state $|\lambda\rangle$

$$a^\dagger_\lambda|0\rangle = |\lambda\rangle \ . \tag{1.65}$$

The vacuum state $|0\rangle$ is a physical state with no particles, and must be distinguished from the zero of the Hilbert space. As a physical example, consider an atom in an excited state which can emit a photon. The initial state of the system is a direct product of the excited atomic state and the photon vacuum. After decay, the state is a product of the final atomic state and a one-photon state.

The creation operators $a^\dagger_\lambda$ do not operate within one space $\mathcal{B}_n$ or $\mathcal{F}_n$, but rather operate from any space $\mathcal{B}_n$ or $\mathcal{F}_n$ to $\mathcal{B}_{n+1}$ or $\mathcal{F}_{n+1}$. Hence, it is useful to define the Fock space as the direct sum of the Boson or Fermion spaces

$$\mathcal{B} = \mathcal{B}_0 \oplus \mathcal{B}_1 \oplus \mathcal{B}_2 \oplus \dots = \oplus_{n=0}^{\infty} \mathcal{B}_n \tag{1.66a}$$

$$\mathcal{F} = \mathcal{F}_0 \oplus \mathcal{F}_1 \oplus \mathcal{F}_2 \oplus \ldots = \oplus_{n=0}^{\infty} \mathcal{F}_n \tag{1.66b}$$

where by definition:

$$\mathcal{B}_0 = \mathcal{F}_0 = |0\rangle \tag{1.67a}$$

and

$$\mathcal{B}_1 = \mathcal{F}_1 = \mathcal{H} \ . \tag{1.67b}$$

A general state $|\phi\rangle$ of the Fock space is a linear combination of states with any number of particles. For instance, the state:

$$|\phi\rangle = \frac{1}{2}|0\rangle + \frac{1}{\sqrt{2}}|\lambda\rangle + \frac{1}{2}|\lambda\mu\rangle \tag{1.68}$$

belongs to the Fock space and represents a system of particles in which there is probability $\frac{1}{4}$ for having no particles, probability $\frac{1}{2}$ for having one particle in state $|\lambda\rangle$, and probability $\frac{1}{4}$ for having two particles in state $|\lambda\mu\rangle$.

Suitable bases for the Fock space utilize either unnormalized or normalized states of the proper symmetry, $\{|0\rangle, |\lambda_1\}, |\lambda_1\lambda_2\}, |\lambda_1\lambda_2 \ldots \lambda_n\}, \ldots\}$ or $\{|0\rangle, |\lambda_1\rangle, |\lambda_1\lambda_2\rangle, |\lambda_1\lambda_2 \ldots \lambda_3\rangle, \ldots, \}$ and in subsequent derivations we will choose the most convenient form. The closure relation in the Fock space may be written

$$\begin{aligned} 1 &= |0\rangle\langle 0| + \sum_{N=1}^{\infty} \frac{1}{N!} \sum_{\lambda_1 \ldots \lambda_N} |\lambda_1 \ldots \lambda_N\}\{\lambda_1 \ldots \lambda_N| \\ &= |0\rangle\langle 0| + \sum_{N=1}^{\infty} \frac{1}{N!} \sum_{\lambda_1 \ldots \lambda_N} \left(\prod_{\lambda} n_\lambda!\right) |\lambda_1 \ldots \lambda_N\rangle\langle\lambda_1 \ldots \lambda_N| \ . \end{aligned} \tag{1.69}$$

Any basis vector $|\lambda_1 \ldots \lambda_N\}$ or $|\lambda_1 \ldots \lambda_N\rangle$ may be generated by repeated action of the creation operators on the vacuum $|0\rangle$:

$$|\lambda_1 \ldots \lambda_N\} = a^\dagger_{\lambda_1} a^\dagger_{\lambda_2} \ldots a^\dagger_{\lambda_N} |0\rangle \tag{1.70a}$$

and

$$|\lambda_1 \ldots \lambda_N\rangle = \frac{1}{\sqrt{\prod_\lambda n_\lambda!}} a^\dagger_{\lambda_1} a^\dagger_{\lambda_2} \ldots a^\dagger_{\lambda_n} |0\rangle \ . \tag{1.70b}$$

Thus, the creation operators generate the entire Fock space by repeated action on the vacuum. The absence of factors of N or n_λ in Eq. (1.70a) is the reason for the definition of the states $|\ \}$ and their use throughout this section.

The symmetry or antisymmetry properties of the many-particle states impose commutation or anticommutation relations between the creation operators. For any state $|\lambda_1 \ldots \lambda_N\}$ and any single particle states $|\lambda\rangle$ and $|\mu\rangle$, we obtain:

$$\begin{aligned} a^\dagger_\lambda a^\dagger_\mu |\lambda_1\lambda_2 \ldots \lambda_N\} &= |\lambda\mu\lambda_1 \ldots \lambda_N\} \\ &= \varsigma|\mu\lambda\lambda_1 \ldots \lambda_N\} \\ &= \varsigma a^\dagger_\mu a^\dagger_\lambda |\lambda_1\lambda_2 \ldots \lambda_N\} \end{aligned} \tag{1.71}$$

where we have used the permutation properties of the states and, as usual, ς is $+1$ for Bosons and -1 for Fermions. Since Equation (1.71) holds for any state, the $a^\dagger$'s satisfy the operator equation

$$a_\lambda^\dagger a_\mu^\dagger - \varsigma a_\mu^\dagger a_\lambda^\dagger = 0 \ . \tag{1.72a}$$

Thus for Bosons, the creation operators commute:

$$[a_\lambda^\dagger, a_\mu^\dagger]_- \equiv [a_\lambda^\dagger, a_\mu^\dagger] = a_\lambda^\dagger a_\mu^\dagger - a_\mu^\dagger a_\lambda^\dagger = 0 \tag{1.72b}$$

whereas they anticommute for Fermions:

$$[a_\lambda^\dagger, a_\mu^\dagger]_+ = a_\lambda^\dagger a_\mu^\dagger + a_\mu^\dagger a_\lambda^\dagger = 0 \ . \tag{1.72c}$$

It is convenient to combine Eqs. (1.72) in a unified form with the following notation

$$[a_\lambda^\dagger, a_\mu^\dagger]_\varsigma = a_\lambda^\dagger a_\mu^\dagger - \varsigma a_\mu^\dagger a_\lambda^\dagger = 0 \ . \tag{1.73}$$

The operators $a_\lambda^\dagger$ are not self-adjoint, and therefore we define the annihilation operators a_λ as the adjoints of the creation operators $a_\lambda^\dagger$. The commutation relations of annihilation operators follow immediately from the adjoint of Eq. (1.73)

$$[a_\lambda, a_\mu]_{-\varsigma} = a_\lambda a_\mu - \varsigma a_\mu a_\lambda = 0 \tag{1.74}$$

and their action to the left on bras follows from the adjoints of Eqs. (1.64) and (1.65). The action of a_λ to the right is obtained by evaluating its matrix element between two arbitrary states. From the adjoint of Eq. (1.63a), we obtain

$$\{\alpha_1 \dots \alpha_m | a_\lambda | \beta_1 \dots \beta_n\} = \{\lambda \alpha_1 \dots \alpha_m | \beta_1 \dots \beta_n\} \ . \tag{1.75}$$

The r.h.s. of (1.75) is non-zero only if $m+1=n$, so one effect of a_λ acting to the right is to decrease the number of particles in the state on which it acts by one. In particular, when acting on the vacuum it yields 0 for any state $|\lambda\rangle$:

$$a_\lambda |0\rangle = 0 \ . \tag{1.76a}$$

Similarly

$$\langle 0 | a_\lambda^\dagger = 0 \ . \tag{1.76b}$$

From Eq. (1.76) it is evident that the vacuum is the kernel of the annihilation operators.

In order to calculate the action of a_λ on a many-particle state, consider an orthonormal single-particle basis $\{|\alpha\rangle\}$ such that $|\lambda\rangle$ is one of the basis states. Using the closure relation (1.69), we obtain:

$$\begin{aligned} a_\lambda |\beta_1 \dots \beta_n\} &= \sum_{P=0}^{\infty} \frac{1}{P!} \sum_{\alpha_1 \dots \alpha_P} \{\alpha_1 \dots \alpha_P | a_\lambda | \beta_1 \dots \beta_n\} |\alpha_1 \dots \alpha_P\} \\ &= \sum_{P=0}^{\infty} \frac{1}{P!} \sum_{\alpha_1 \dots \alpha_P} \{\lambda \alpha_1 \dots \alpha_P | \beta_1 \dots \beta_n\} |\alpha_1 \dots \alpha_P\} \ . \end{aligned} \tag{1.77}$$

Only the terms with $P=(n-1)$ and the set $\{\lambda, \alpha_1, \ldots, \alpha_P\}$ equal to a permutation of $\{\beta_1 \ldots \beta_n\}$ are non-vanishing and the scalar product is given by (1.41) with the result

$$\begin{aligned} a_\lambda |\beta_1 \ldots \beta_n\} &= \sum_{i=1}^{n} \varsigma^{i-1} \delta_{\lambda\beta_i} |\beta_1 \ldots \beta_{i-1}\beta_{i+1} \ldots \beta_n\} \\ &= \sum_{i=1}^{n} \varsigma^{i-1} \delta_{\lambda\beta_i} |\beta_1 \ldots \hat{\beta}_i \ldots \beta_n\} \end{aligned} \tag{1.78a}$$

where $\hat{\beta}_i$ indicates that the state β_i has been removed from the many-particle state $|\beta_1 \ldots \beta_n\}$ For a normalized state, $|\beta_1 \ldots \beta_n\rangle$ with occupation number n_λ for the state λ we may write

$$a_\lambda |\beta_1 \ldots \beta_n\rangle = \frac{1}{\sqrt{n_\lambda}} \sum_{i=1}^{n} \varsigma^{i-1} \delta_{\lambda\beta_i} |\beta_1 \ldots \hat{\beta}_i \ldots \beta_n\rangle \ . \tag{1.78b}$$

We observe from Eq. (1.78) that the effect of a_λ acting on any state is to annihilate one particle in the state λ from this state. For the case of Bosons, the general result is more conveniently expressed in occupation number representation. We denote by $|n_{\beta_1}, n_{\beta_2}, n_{\beta_3} \ldots\rangle$ a symmetrized state with n_{β_1} particles in $|\beta_1\rangle$, n_{β_2} particles in $|\beta_2\rangle$ and so on. Since n_λ non-vanishing equal terms arise in Eq. (1.78), the Boson result may be written

$$a_\lambda |n_{\beta_1} n_{\beta_2} \ldots n_\lambda \ldots\rangle = \sqrt{n_\lambda} |n_{\beta_1} n_{\beta_2} \ldots (n_\lambda - 1) \ldots\rangle \ . \tag{1.79a}$$

For Fermions, it is convenient to write Eq. (1.78) in the form

$$a_\lambda |\beta_1 \ldots \beta_n\rangle = \begin{cases} (-1)^{i-1} |\beta_1 \ldots \hat{\beta}_\lambda \ldots \beta_n\rangle & \text{if } |\lambda\rangle \text{ is occupied} \\ 0 & \text{if } |\lambda\rangle \text{ is unoccupied} \end{cases} \ . \tag{1.79b}$$

As expected physically, if there are no particles present in state $|\lambda\rangle$ to be annihilated, a_λ yields zero.

To close the algebra of the creation and annihilation operators, we need to evaluate the commutators of creation and annihilation operators. As in the previous calculation, we consider the case in which $|\lambda\rangle$ and $|\mu\rangle$ are two states belonging to the orthonormal basis $\{|\alpha\rangle\}$. Then, according to Eqs. (1.63) and (1.78)

$$\begin{aligned} a_\lambda a_\mu^\dagger |\alpha_1 \ldots \alpha_n\} &= a_\lambda |\mu\alpha_1 \ldots \alpha_n\} \\ &= \delta_{\lambda\mu} |\alpha_1 \ldots \alpha_n\} + \sum_{i=1}^{n} \varsigma^i \delta_{\lambda\alpha_i} |\mu\alpha_1 \ldots \hat{\alpha}_i \ldots \alpha_n\} \end{aligned} \tag{1.80}$$

and

$$a_\mu^\dagger a_\lambda |\alpha_1 \ldots \alpha_n\} = \sum_{i=1}^{n} \varsigma^{i-1} \delta_{\lambda\alpha_i} |\mu\alpha_1 \ldots \hat{\alpha}_i \ldots \alpha_n\} \ . \tag{1.81}$$

Hence

$$a_\lambda a^\dagger_\mu|\alpha_1\ldots\alpha_n\} = (\delta_{\lambda\mu} + \varsigma a^\dagger_\mu a_\lambda)|\alpha_1\ldots\alpha_n\} \tag{1.82}$$

for any state $|\alpha_1\ldots\alpha_n\}$, so the creation and annihilation operators satisfy the operator equations:

$$[a_\lambda, a^\dagger_\mu]_{-\varsigma} = a_\lambda a^\dagger_\mu - \varsigma a^\dagger_\mu a_\lambda = \delta_{\lambda\mu} \ . \tag{1.83a}$$

Thus, Bosons satisfy the commutation relations

$$[a_\lambda, a^\dagger_\mu]_- = \delta_{\lambda\mu} \quad (\text{ Bosons }) \tag{1.83b}$$

and Fermions satisfy the anticommutation relations

$$[a_\lambda, a^\dagger_\mu]_+ = \delta_{\lambda\mu} \quad (\text{ Fermions }) \ . \tag{1.83c}$$

Creation and annihilation operators in another basis may be obtained straightforwardly from the operators $\{a^\dagger_\alpha, a_\alpha\}$. Consider a transformation which transforms the orthonormal basis $\{|\alpha\rangle\}$ into another basis $\{|\tilde{\alpha}\rangle\}$ as follows

$$|\tilde{\alpha}\rangle = \sum_\alpha \langle\alpha|\tilde{\alpha}\rangle|\alpha\rangle \ . \tag{1.84}$$

By this equation and the definition of the creation operators $a^\dagger_{\tilde{\alpha}}$ and $a^\dagger_\alpha$,

$$\begin{aligned} a^\dagger_{\tilde{\alpha}}|\tilde{\alpha}_1\ldots\tilde{\alpha}_n\} &= |\tilde{\alpha}\tilde{\alpha}_1\ldots\tilde{\alpha}_n\} \\ &= \sum_\alpha \langle\alpha|\tilde{\alpha}\rangle|\alpha\tilde{\alpha}_1\ldots\tilde{\alpha}_n\} \\ &= \sum_\alpha \langle\alpha|\tilde{\alpha}\rangle a^\dagger_\alpha|\tilde{\alpha}_1\ldots\tilde{\alpha}_n\} \ . \end{aligned} \tag{1.85}$$

Since this result holds for a set of basis states $|\tilde{\alpha}_1\ldots\tilde{\alpha}_n\}$, the creation operators satisfy the operator equation

$$a^\dagger_{\tilde{\alpha}} = \sum_\alpha \langle\alpha|\tilde{\alpha}\rangle a^\dagger_\alpha \tag{1.86a}$$

and annihilation operators satisfy the adjoint equation

$$a_{\tilde{\alpha}} = \sum_\alpha \langle\tilde{\alpha}|\alpha\rangle a_\alpha \ . \tag{1.86b}$$

A convenient mnemonic for this result is the fact that the action of the operator Eq. (1.86a) on the vacuum reproduces the transformation for single particle states, Eq. (1.84).

Commutation and anticommutation relations for $a^\dagger_{\tilde{\alpha}}$ and $a_{\tilde{\beta}}$ are computed straightforwardly from (1.86). For example

$$\begin{aligned} [a_{\tilde{\beta}}, a^\dagger_{\tilde{\alpha}}]_{-\varsigma} &= \sum_{\alpha\beta} \langle\tilde{\beta}|\beta\rangle\langle\alpha|\tilde{\alpha}\rangle[a_\beta, a^\dagger_\alpha]_{-\varsigma} \\ &= \sum_\alpha \langle\tilde{\beta}|\alpha\rangle\langle\alpha|\tilde{\alpha}\rangle \\ &= \langle\tilde{\beta}|\tilde{\alpha}\rangle \end{aligned} \tag{1.87a}$$

and similarly

$$[a^\dagger_{\tilde{\alpha}}, a^\dagger_{\tilde{\beta}}]_{-\varsigma} = [a_{\tilde{\alpha}}, a_{\tilde{\beta}}]_{-\varsigma} = 0 \quad . \tag{1.87b}$$

In the event that the basis $\{|\tilde{\alpha}\rangle\}$ is orthonormal, Eqs. (1.87) produce the same commutation relations for $a^\dagger_{\tilde{\alpha}}$ and $a_{\tilde{\alpha}}$ as for $a^\dagger_\alpha$ and a_α, and the transformation (1.84) is canonical.

Of particular importance is the $\{|x\rangle\}$ basis defined in Eq. (1.20), where $|x\rangle$ represents $|\vec{r}\sigma\tau\rangle$ In this case, the creation and annihilation operators are traditionally denoted by $\hat{\psi}^\dagger(x)$ and $\hat{\psi}(x)$ and are called field operators. Their commutation relations follow from equation (1.87a).

$$[\hat{\psi}^\dagger(x), \hat{\psi}^\dagger(y)]_{-\varsigma} = [\hat{\psi}(x), \hat{\psi}(y)]_{-\varsigma} = 0 \tag{1.88a}$$

$$[\hat{\psi}(x), \hat{\psi}^\dagger(y)]_{-\varsigma} = \delta(x - y) \quad . \tag{1.88b}$$

The expansion of these operators on a basis $\{|\alpha\rangle\}$ follows from (1.86)

$$\hat{\psi}^\dagger(x) = \sum_\alpha \langle\alpha|x\rangle a^\dagger_\alpha = \sum_\alpha \phi^*_\alpha(x) a^\dagger_\alpha \tag{1.89a}$$

$$\hat{\psi}(x) = \sum_\alpha \langle x|\alpha\rangle a_\alpha = \sum_\alpha \phi_\alpha(x) a_\alpha \tag{1.89b}$$

where $\phi_\alpha(x)$ is the coordinate representation wave function of the state $|\alpha\rangle$

In addition to generating all the many-particle states in the Fock space by repeated action on the vacuum, a fundamental property of creation and annihilation operators is that they provide a basis for all operators in the Fock space. That is, any operator can be expressed as a linear combination of the set of all products of the operators $\{a^\dagger_\alpha, a_\alpha\}$.

A convenient technique for representing an operator in terms of creation and annihilation operators is to first use a basis in which it is diagonal and then transform to a general basis. To represent the operator in the diagonal basis, we define the number operator $\hat{n}_\alpha$:

$$\hat{n}_\alpha = a^\dagger_\alpha a_\alpha \quad . \tag{1.90}$$

Using Eqs. (1.63) and (1.78), we observe that the action of $\hat{n}_\alpha$ on a state $|\phi\rangle$ is to count the number of particles in state $|\alpha\rangle$ in the state $|\phi\rangle$:

$$\begin{aligned}\hat{n}_\alpha|\alpha_1 \ldots \alpha_N\} &= a^\dagger_\alpha a_\alpha |\alpha_1 \ldots \alpha_N\} \\ &= \sum_{i=1}^N \varsigma^{i-1} \delta_{\alpha\alpha_i} a^\dagger_\alpha |\alpha_1 \ldots \hat{\alpha}_i \ldots \alpha_N\} \\ &= \left(\sum_{i=1}^N \delta_{\alpha\alpha_i}\right) |\alpha_1 \ldots \alpha_i \ldots \alpha_N\}\end{aligned} \tag{1.91}$$

where $\sum_{i=1}^N \delta_{\alpha\alpha_i}$ yields the number of particles in state $|\alpha\rangle$ in $|\alpha_1 \ldots \alpha_N\}$. The operator $\hat{N}$ which counts the total number of particles in a state is given by:

$$\hat{N} = \sum_\alpha \hat{n}_\alpha = \sum_\alpha a^\dagger_\alpha a_\alpha \quad . \tag{1.92}$$

We now consider a one-body operator, $\hat{U}$, which is diagonal in the orthonormal basis $|\alpha\rangle$

$$\hat{U} = U_\alpha|\alpha\rangle \tag{1.93a}$$

$$U_\alpha = \langle\alpha|U|\alpha\rangle \ . \tag{1.93b}$$

Using (1.55), and (1.41) we obtain

$$\begin{aligned}\{\alpha_1'\ldots\alpha_N'|U|\alpha_1\ldots\alpha_N\} &= \sum_P \varsigma^P \sum_{i=1}^{N} \prod_{k\neq i} \langle\alpha_{Pk}'|\alpha_k\rangle\langle\alpha_{Pi}'|U|\alpha_i\rangle \\ &= \left(\sum_{i=1}^{N} U_{\alpha_i}\right)\{\alpha_1'\ldots\alpha_N'|\alpha_1\ldots\alpha_N\} \\ &= \{\alpha_1'\ldots\alpha_N'|\sum_\alpha U_\alpha \hat{n}_\alpha|\alpha_1\ldots\alpha_N\} \ .\end{aligned} \tag{1.94}$$

Since this equality holds for any states, we obtain the operator equation

$$\hat{U} = \sum_\alpha U_\alpha \hat{n}_\alpha = \sum_\alpha \langle\alpha|U|\alpha\rangle a_\alpha^\dagger a_\alpha \ . \tag{1.95}$$

Thus, in order to obtain the action of $\hat{U}$, we must sum over all states $|\alpha\rangle$ multiplying U_α by the number of particles in state $|\alpha\rangle$.

We may now transform from the diagonal representation to a general basis, using Eqs. (1.86). Then

$$\begin{aligned}\hat{U} &= \sum_{\alpha\lambda\mu} U_\alpha \langle\lambda|\alpha\rangle\langle\alpha|\mu\rangle a_\lambda^\dagger a_\mu \\ &= \sum_{\lambda\mu} \langle\lambda|U|\mu\rangle a_\lambda^\dagger a_\mu\end{aligned} \tag{1.96}$$

where

$$\begin{aligned}\langle\lambda|U|\mu\rangle &= \sum_\alpha \langle\lambda|\alpha\rangle U_\alpha \langle\alpha|\mu\rangle \\ &= \int dx dy\, \phi_\lambda^*(x)\langle x|U|y\rangle\phi_\mu(y) \ .\end{aligned} \tag{1.97}$$

For example, using field operators in the $\{\vec{x}\}$ representation, the kinetic energy operator $\hat{T}$, and a local one-body operator $\hat{U}$ may be written

$$\hat{T} = -\frac{\hbar^2}{2m}\int d^3x\, \hat{\psi}^\dagger(\vec{x})\nabla^2\hat{\psi}(\vec{x}) \tag{1.98}$$

$$\hat{U} = \int d^3x\, U(\vec{x})\hat{\psi}^\dagger(\vec{x})\hat{\psi}(\vec{x}) \tag{1.99}$$

and in the $\{\vec{p}\}$ representation, the kinetic energy is:

$$\hat{T} = \int d^3p \frac{\hat{\vec{p}}^2}{2m} \hat{\psi}^\dagger(\vec{p})\hat{\psi}(\vec{p}) \quad . \tag{1.100}$$

Similarly, a two-body operator $\hat{V}$ may be expressed in terms of creation and annihilation operators. Using the basis $V_{\alpha\beta}$ in which $\hat{V}$ is diagonal,

$$\hat{V}|\alpha\beta) = V_{\alpha\beta}|\alpha\beta) \tag{1.101a}$$

$$V_{\alpha\beta} = (\alpha\beta|V|\alpha\beta) \quad . \tag{1.101b}$$

We calculate a general matrix element as before:

$$\begin{aligned}\{\alpha'_1\dots\alpha'_N|V|\alpha_1\dots\alpha_N\} &= \sum_P \zeta^P \frac{1}{2}\sum_{i\neq j}\prod_{\substack{k\neq i\\ k\neq j}}(\alpha'_{Pk}|\alpha_k)(\alpha'_{Pi}\alpha'_{Pj}|\hat{V}|\alpha_i\alpha_j) \\ &= \Big(\frac{1}{2}\sum_{i\neq j}^{N} V_{\alpha_i\alpha_j}\Big)\{\alpha'_1\dots\alpha'_N|\alpha_1\dots\alpha_N\} \quad .\end{aligned} \tag{1.102}$$

The factor $\frac{1}{2}\sum_{i\neq j} V_{\alpha_i\alpha_j}$ is a sum over all distinct pairs of particles present in the state $|\alpha_1\dots\alpha_N\rangle$, so we need to construct an operator $\hat{P}_{\alpha\beta}$ which counts the number of pairs of particles in the states $|\alpha\rangle$ and $|\beta\rangle$. If $|\alpha\rangle$ and $|\beta\rangle$ are different, the number of pairs is $n_\alpha n_\beta$ whereas if $|\alpha\rangle = |\beta\rangle$, the number of pairs is $n_\alpha(n_\alpha - 1)$. Hence, the operator which counts pairs may be written

$$\hat{P}_{\alpha\beta} = \hat{n}_\alpha\hat{n}_\beta - \delta_{\alpha\beta}\hat{n}_\alpha \tag{1.103}$$

or in terms of creation and annihilation operators,

$$\begin{aligned}\hat{P}_{\alpha\beta} &= a^\dagger_\alpha a_\alpha a^\dagger_\beta a_\beta - \delta_{\alpha\beta}a^\dagger_\alpha a_\alpha \\ &= a^\dagger_\alpha \zeta a^\dagger_\beta a_\alpha a_\beta \\ &= a^\dagger_\alpha a^\dagger_\beta a_\beta a_\alpha \quad .\end{aligned} \tag{1.104}$$

Using this operator, Eq. (1.102) may be rewritten

$$\{\alpha'_1\dots\alpha'_n|\hat{V}|\alpha_1\dots\alpha_n\} = \{\alpha'_1\dots\alpha'_n|\frac{1}{2}\sum_{\alpha\beta}V_{\alpha\beta}\hat{P}_{\alpha\beta}|\alpha_1\dots\alpha_n\} \tag{1.105}$$

and hence, $\hat{V}$ satisfies the operator equation

$$\hat{V} = \frac{1}{2}\sum_{\alpha\beta}V_{\alpha\beta}\hat{P}_{\alpha\beta} = \frac{1}{2}\sum_{\alpha\beta}(\alpha\beta|V|\alpha\beta)a^\dagger_\alpha a^\dagger_\beta a_\beta a_\alpha \quad . \tag{1.106}$$

Analogous to the case of a one-body operator, the action of a two-body operator is obtained by summing over pairs of single-particle states $|\alpha\rangle$ and $|\beta\rangle$ and multiplying the matrix element $(\alpha\beta|V|\alpha\beta)$ by the number of pairs of such particles present in the physical state. Transforming from the diagonal representation to an arbitrary basis, the general expression for a two-body potential is

$$\hat{V} = \frac{1}{2}\sum_{\lambda\mu\nu\rho}(\lambda\mu|V|\nu\rho)a^\dagger_\lambda a^\dagger_\mu a_\rho a_\nu \quad . \tag{1.107a}$$

Note the order of annihilation operators in this expression which is required to remove the factor ς from Eq. (1.104), and thus yield a simple result for both Fermions and Bosons. Indeed, one of the great virtues of representing operators in terms of creation and annihilation operators is the simplicity and convenience of Eq. (1.107), which concisely handles all the bookkeeping for Fermi and Bose statistics. Occasionally, it is convenient to use the following variant of Eq. (1.107a) in which the matrix element is manifestly symmetrized or antisymmetrized:

$$\begin{aligned}\hat{V} &= \frac{1}{4}\sum_{\lambda\mu\nu\rho}\left[(\lambda\mu|V|\nu\rho)+\varsigma|\rho\nu)\right]a^\dagger_\lambda a^\dagger_\mu a_\rho a_\nu \\ &= \frac{1}{4}\sum_{\lambda\mu\nu\rho}\{\lambda\mu|V|\nu\rho\}a^\dagger_\lambda a^\dagger_\mu a_\rho a_\nu \quad .\end{aligned} \tag{1.107b}$$

As an example, if $\hat{V}$ is an interaction which is diagonal in the $\{\hat{x}\}$ representation, Eq. (1.107a) yields:

$$\hat{V} = \frac{1}{2}\int d^3x d^3y\, v(\vec{x}-\vec{y})\hat{\psi}^\dagger(\vec{x})\hat{\psi}^\dagger(\vec{y})\hat{\psi}(\vec{y})\hat{\psi}(\vec{x}) \quad . \tag{1.108}$$

In terms of its antisymmetrized matrix elements, Eq. (1.107b) may be used with:

$$\{\lambda\mu|V|\nu\rho\} = \int d^3x d^3y\, v(\vec{x}-\vec{y})\phi^*_\lambda(\vec{x})\phi^*_\mu(\vec{y})(\phi_\nu(\vec{x})\phi_\rho(\vec{y}) - \phi_\rho(\vec{x})\phi_\nu(\vec{y})) \quad . \tag{1.109}$$

The preceding derivations for one- and two-body operators may be straightforwardly generalized to n-body operators, with the result:

$$R = \frac{1}{n!}\sum_{\lambda_1\ldots\lambda_n}\sum_{\mu_1\ldots\mu_n}(\lambda_1\ldots\lambda_n|R|\mu_1\ldots\mu_n)a^\dagger_{\lambda_1}\ldots a^\dagger_{\lambda_n}a_{\mu_n}\ldots a_{\mu_1} \quad . \tag{1.110}$$

The only technical complication in the general derivation is to carefully count n-tuples of particles in states $|\alpha_1\rangle\ldots|\alpha_n\rangle$ when not all $\{\alpha_i\}$ are distinct (See Problem 1.1).

Finally, for future reference, it is useful to define normal ordering for a many-particle operator. By definition, an operator is in normal order if all the creation operators are to the left of all the annihilation operators. For example, the first line of Eq. (1.104) is not in normal order whereas the last line is in normal order. Clearly, any expression which is not in normal order may be brought into normal order by a sequence of applications of the commutation or anticommutation relations for creation and annihilation operators.

1.5 COHERENT STATES

Thus far we have used permanents or Slater determinants as a natural basis for the Fock space $\mathcal{B}$ or $\mathcal{F}$. Another extremely useful basis of the Fock space is the basis of coherent states which is analogous to the basis of position eigenstates in quantum mechanics. Although this is not an orthonormal basis, it spans the whole Fock space. Just as the position states $|\vec{r}\rangle$ are defined as eigenstates of $\hat{\vec{r}}$, the coherent states are defined as eigenstates of the annihilation operators.

To see why annihilation operators are selected rather than creation operators, it is useful to consider the existence of eigenstates of the creation and annihilation operators. If we denote by $|\phi\rangle$ a general vector of the Fock space, we can expand $|\phi\rangle$ as:

$$|\phi\rangle = \sum_{n=0}^{\infty} \sum_{\alpha_1 \ldots \alpha_n} \phi_{\alpha_1 \ldots \alpha_n} |\alpha_1 \ldots \alpha_n\rangle \quad . \tag{1.111}$$

Thus, $|\phi\rangle$ necessarily has a component with a minimum number of particles and if we apply any creation operator to $|\phi\rangle$, we see that the minimum number of particles in $|\phi\rangle$ is increased by one. Thus, the resulting state cannot be a multiple of the original state and therefore a creation operator cannot have an eigenstate. On the other hand, if we apply an annihilation operator to $|\phi\rangle$, this decreases the maximum number of particles in $|\phi\rangle$ by one, since $|\phi\rangle$ may contain components with all particle numbers, nothing *a priori* forbids $|\phi\rangle$ to have eigenstates.

Assuming we have succeeded in constructing an eigenstate $|\phi\rangle$ of the annihilation operators, then

$$a_\alpha |\phi\rangle = \phi_\alpha |\phi\rangle \quad . \tag{1.112}$$

When more than one annihilation operator acts upon such a coherent state, a significant difference arises between Bosons and Fermions. The commutation or anticommutation relations (1.74) imply corresponding relations for the eigenvalues

$$[\phi_\alpha, \phi_\beta]_{-\varsigma} = 0 \quad . \tag{1.113}$$

For Fermions, the eigenvalues anticommute, and in order to accommodate this unusual feature we subsequently will need to introduce anticommuting variables called Grassmann numbers. For Bosons, however, the eigenvalues commute and we will be able to proceed straightforwardly using ordinary numbers. Hence, we shall begin by considering Bosons.

BOSON COHERENT STATES

For Bosons, the eigenvalues ϕ_α of the annihilation operators may be real or complex numbers. It is convenient to expand a Boson coherent state in occupation number representation

$$|\phi\rangle = \sum_{n_{\alpha_1}, n_{\alpha_2} \ldots n_{\alpha_p}} \phi_{n_{\alpha_1} n_{\alpha_2} \ldots n_{\alpha_p} \ldots} |n_{\alpha_1} n_{\alpha_2} \ldots n_{\alpha_p} \ldots\rangle \tag{1.114}$$

where as usual $|n_{\alpha_1} n_{\alpha_2} \ldots n_{\alpha_p} \ldots\rangle$ denotes a normalized symmetrized state with n_{α_1} particles in state $|\alpha_1\rangle$, n_{α_2} particles in state $|\alpha_2\rangle, \ldots$ and $\{|\alpha_i\rangle\}$ is an orthonormal basis.

The eigenvalue condition, Eq. (1.112) for an annihilation operator a_{α_i} acting on $|\phi\rangle$ implies the following conditions on the coefficients for all $\{n_\alpha\}$

$$\phi_{\alpha_i}\phi_{n_{\alpha_1}n_{\alpha_2}\ldots(n_{\alpha_i}-1)\ldots} = \sqrt{n_{\alpha_i}}\,\phi_{n_{\alpha_1}n_{\alpha_2}\ldots n_{\alpha_i}\ldots} \quad . \tag{1.115}$$

Relating each coefficient by induction to the coefficient for the vacuum state which we arbitrarily set equal to 1, we obtain

$$\phi_{n_{\alpha_1}n_{\alpha_2}\ldots n_{\alpha_i}} = \frac{\phi_{\alpha_1}^{n_{\alpha_1}}\phi_{\alpha_2}^{n_{\alpha_2}}\ldots\phi_{\alpha_i}^{n_{\alpha_i}}\ldots}{\sqrt{n_{\alpha_1}!}\sqrt{n_{\alpha_2}!}\ldots\sqrt{n_{\alpha_i}!}\ldots} \quad . \tag{1.116}$$

Substituting (1.116) in (1.114) and using the fact that

$$|n_{\alpha_1}n_{\alpha_2}\ldots n_{\alpha_p}\ldots\rangle = \frac{(a^\dagger_{\alpha_1})^{n_{\alpha_1}}}{\sqrt{n_{\alpha_1}!}}\frac{(a^\dagger_{\alpha_2})^{n_{\alpha_2}}}{\sqrt{n_{\alpha_2}!}}\ldots\frac{(a^\dagger_{\alpha_p})^{n_{\alpha_p}}}{\sqrt{n_{\alpha_p}!}}\ldots|0\rangle \tag{1.117}$$

we finally obtain

$$\begin{aligned}
|\phi\rangle &= \sum_{n_{\alpha_1},n_{\alpha_2}\ldots n_{\alpha_p}} \frac{(\phi_{\alpha_1}a^\dagger_{\alpha_1})^{n_{\alpha_1}}}{n_{\alpha_1}!}\frac{(\phi_{\alpha_2}a^\dagger_{\alpha_2})^{n_{\alpha_2}}}{n_{\alpha_2}!}\ldots\frac{(\phi_{\alpha_p}a^\dagger_{\alpha_p})^{n_{\alpha_p}}}{n_{\alpha_p}!}|0\rangle \\
&= e^{\sum_\alpha \phi_\alpha a^\dagger_\alpha}|0\rangle && (1.118a)\\
\langle\phi| &= \langle 0|e^{\sum_\alpha \phi^*_\alpha a_\alpha} \quad . && (1.118b)
\end{aligned}$$

Note by taking the adjoint of (1.112), that the adjoint of a coherent state is a left eigenstate of the creation operators:

$$\langle\phi|a^\dagger_\alpha = \langle\phi|\phi^*_\alpha \quad . \tag{1.119}$$

The action of a creation operator $a^\dagger_\alpha$ on a coherent state is given by

$$\begin{aligned}
a^\dagger_\alpha|\phi\rangle &= a^\dagger_\alpha e^{\sum_{\alpha'}\phi_{\alpha'}a^\dagger_{\alpha'}}|0\rangle \\
&= \frac{\partial}{\partial\phi_\alpha}|\phi\rangle && (1.120a)
\end{aligned}$$

with the adjoint relation

$$\langle\phi|a_\alpha = \frac{\partial}{\partial\phi^*_\alpha}\langle\phi| \quad . \tag{1.120b}$$

The overlap of two coherent states is given by:

$$\begin{aligned}
\langle\phi|\phi'\rangle = &\sum_{n_{\alpha_1}\ldots n_{\alpha_p}\ldots}\ \sum_{n_{\alpha_1}\ldots n'_{\alpha_p}\ldots} \frac{\phi^{*n_{\alpha_1}}_{\alpha_1}}{\sqrt{n_{\alpha_1}!}}\ldots\frac{\phi^{*n_{\alpha_p}}_{\alpha_p}}{\sqrt{n_{\alpha_p}!}}\ldots\frac{\phi'^{n'_{\alpha_1}}_{\alpha_1}}{\sqrt{n'_{\alpha_1}!}}\ldots\frac{\phi'^{n'_{\alpha_p}}_{\alpha_p}}{\sqrt{n'_{\alpha_p}!}}\ldots \\
&\times\langle n_{\alpha_1}\ldots n_{\alpha_p}\ldots|n'_{\alpha_1}\ldots n'_{\alpha_p}\rangle \quad . && (1.121)
\end{aligned}$$

Since the basis $|\alpha\rangle$ is orthonormal, the scalar product $\langle n_{\alpha_1} \dots n_{\alpha_p} \dots | n'_{\alpha_1} \dots n'_{\alpha_p} \dots \rangle$ is equal to $\delta_{n_{\alpha_1} n'_{\alpha_1}} \dots \delta_{n_{\alpha_p} n'_{\alpha_p}} \dots$ which leads to:

$$\langle \phi | \phi' \rangle = e^{\sum_\alpha \phi^*_\alpha \phi'_\alpha} . \tag{1.122}$$

A crucial property of the coherent states is their overcompleteness in the Fock space, that is, the fact that any vector of the Fock space can be expanded in terms of coherent states. This is expressed by the closure relation

$$\int \prod_\alpha \frac{d\phi^*_\alpha d\phi_\alpha}{2i\pi} e^{-\sum_\alpha \phi^*_\alpha \phi_\alpha} |\phi\rangle\langle\phi| = 1 \tag{1.123}$$

where 1 is the unit operator in the Fock space, the measure is given by:

$$\frac{d\phi^*_\alpha d\phi_\alpha}{2i\pi} = \frac{d(\mathrm{Re}\,\phi_\alpha) d(\mathrm{Im}\,\phi_\alpha)}{\pi} \tag{1.124}$$

and the integration extends over all values of $\mathrm{Re}\phi_\alpha$ and $\mathrm{Im}\phi_\alpha$.

As explained in Problem 1.2, one may verify Eq. (1.123) straightforwardly by integrating the left hand side to obtain the familiar completeness relation Eq. (1.51). A more economical proof is provided by Schur's lemma, which in the present context states that if an operator commutes with all creation and annihilation operators, it is proportional to the unit operator in the Fock space. Using equations (1.120),

$$[a_\alpha, |\phi\rangle\langle\phi|] = (\phi_\alpha - \frac{\partial}{\partial\phi^*_\alpha})|\phi\rangle\langle\phi| \tag{1.125}$$

so that evaluating the commutator of Eq. (1.123) and integrating by parts,

$$\begin{aligned} &\left[a_\alpha, \int \prod_{\alpha'} \frac{d\phi^*_{\alpha'} d\phi_{\alpha'}}{2\pi i} e^{-\sum_{\alpha'} \phi^*_{\alpha'} \phi_{\alpha'}} |\phi\rangle\langle\phi|\right] \\ &= \int \prod_{\alpha'} \frac{d\phi^*_{\alpha'} d\phi_{\alpha'}}{2\pi i} e^{-\sum_{\alpha'} \phi^*_{\alpha'} \phi_{\alpha'}} (\phi_\alpha - \frac{\partial}{\partial\phi^*_\alpha})|\phi\rangle\langle\phi| \\ &= 0 \ . \end{aligned} \tag{1.126}$$

By taking the adjoint of Eq. (1.126), we observe that the left hand side of Eq. (1.123) commutes with all the creation operators as well as the annihilation operators so it must be proportional to the unit operator. The proportionality factor is calculated by taking the expectation value of the left hand side of (1.123) in the vacuum:

$$\begin{aligned} \int \prod_\alpha \frac{d\phi^*_\alpha d\phi_\alpha}{2i\pi} e^{-\sum_\alpha \phi^*_\alpha \phi_\alpha} \langle 0|\phi\rangle\langle\phi|0\rangle &= \int \prod_\alpha \frac{d\phi^*_\alpha d\phi_\alpha}{2i\pi} e^{-\sum_\alpha \phi^*_\alpha \phi_\alpha} \\ &= 1 \ . \end{aligned} \tag{1.127}$$

This proves Eq. (1.123).

This completeness relation provides a useful expression for the trace of an operator. Let A be any operator and let $\{|n>\}$ denote a complete set of states. Then

$$\begin{aligned}\mathrm{Tr}\, A &= \sum_n \langle n|A|n\rangle \\ &= \int \prod_\alpha \frac{d\phi_\alpha^* d\phi_\alpha}{2\pi i} e^{-\sum_\alpha \phi_\alpha^*\phi_\alpha} \sum_n \langle n|\phi\rangle\langle\phi|A|n\rangle \\ &= \int \prod_\alpha \frac{d\phi_\alpha^* d\phi_\alpha}{2\pi i} e^{-\sum_\alpha \phi_\alpha^*\phi_\alpha} \langle\phi|A\sum_n |n\rangle\langle n||\phi\rangle \\ &= \int \prod_\alpha \frac{d\phi_\alpha^* d\phi_\alpha}{2\pi i} e^{-\sum_\alpha \phi_\alpha^*\phi_\alpha} \langle\phi|A|\phi\rangle \ . \end{aligned} \tag{1.128}$$

In quantum mechanics, the completeness of the position eigenstates allows us to represent a state $|\psi\rangle = \int dx \psi(x)|x\rangle$ where $\psi(x) = \langle x|\psi\rangle$ is the coordinate representation of the state $|\psi\rangle$. Analogously, Equation (1.124) implies that any state $|\psi\rangle$ of the Fock space can be represented as:

$$|\psi\rangle = \int \prod_\alpha \frac{d\phi_\alpha^* d\phi_\alpha}{2\pi i} e^{-\sum_\alpha \phi_\alpha^*\phi_\alpha} \psi(\phi^*) \cdot |\phi\rangle \tag{1.129a}$$

where by definition:

$$\psi(\phi^*) = \langle\phi|\psi\rangle \tag{1.129b}$$

is the coherent state representation of the state $|\psi\rangle$. and ϕ denotes the set $\{\phi_\alpha^*\}$. The coherent state representation for Bosons is often referred to as the holomorphic representation, which arises from the fact that ψ is an analytic function of the variables ϕ_α^*. Physically, $\psi(\phi^*)$ is simply the wavefunction of the state $|\psi\rangle$ in the coherent state representation; that is, the probability amplitude to find the system in the coherent state $|\phi\rangle$.

Just as it is useful to know how the operators $\hat{x}$ and $\hat{p}$ act in coordinate representation, it is useful to exhibit how the operators $a_\alpha^\dagger$ and a_α act in the coherent state representation. Using Eqs. (1.119) and (1.120) we find:

$$\langle\phi|a_\alpha|f\rangle = \frac{\partial}{\partial\phi_\alpha^*} f(\phi^*) \tag{1.130a}$$

and

$$\langle\phi|a_\alpha^\dagger|f\rangle = \phi_\alpha^* f(\phi^*) \ . \tag{1.130b}$$

Thus we can write symbolically

$$a_\alpha = \frac{\partial}{\partial\phi_\alpha^*} \tag{1.131a}$$

and

$$a_\alpha^\dagger = \phi_\alpha^* \tag{1.131b}$$

which is consistent with the Boson commutation rules:

$$\left[\phi^*_\alpha, \phi^*_\beta\right] = \left[\frac{\partial}{\partial \phi^*_\alpha}, \frac{\partial}{\partial \phi^*_\beta}\right] = 0 \tag{1.132a}$$

$$\left[\frac{\partial}{\partial \phi^*_\alpha}, \phi^*_\beta\right] = \delta_{\alpha\beta} \ . \tag{1.132b}$$

Aside from factors of i, the behavior of a and $a^\dagger$ in the coherent state representation is thus analogous to that of $\hat{x}$ and $\hat{p}$ in coordinate representation.

The result, Eq. (1.130), yields a simple expression for the Schrödinger equation in the coherent state representation. If $H(a^\dagger_\alpha, a_\alpha)$ is the Hamiltonian in normal form, then projection of the Schrödinger equation

$$H(a^\dagger_\alpha, a_\alpha)|\psi\rangle = E|\psi\rangle \tag{1.133}$$

on the left by $\langle\phi|$ yields

$$H\left(\phi^*_\alpha, \frac{\partial}{\partial \phi^*_\alpha}\right) \psi(\phi^*) = E\psi(\phi^*) \ . \tag{1.134a}$$

For a standard Hamiltonian with one- and two-body operators, it reads:

$$\left(\sum_{\alpha,\beta} T_{\alpha\beta}\phi^*_\alpha \frac{\partial}{\partial \phi^*_\beta} + \frac{1}{2}\sum_{\alpha\beta\gamma\delta}(\alpha\beta|v|\gamma\delta)\phi^*_\alpha\phi^*_\beta \frac{\partial}{\partial \phi^*_\delta}\frac{\partial}{\partial \phi^*_\gamma}\right)\psi(\phi^*) = E\psi(\phi^*) \ . \tag{1.134b}$$

In the space of holomorphic functions $\psi(\phi^*)$, the unit operator is obtained by using (1.123):

$$\langle\phi|\psi\rangle = \int \prod_\alpha \frac{d\phi'^*_\alpha d\phi'_\alpha}{2\pi i} \cdot e^{-\sum_\alpha \phi'^*_\alpha \phi'_\alpha} \langle\phi|\phi'\rangle\langle\phi'|\psi\rangle \tag{1.135}$$

which implies:

$$\psi(\phi^*) = \int \prod_\alpha \frac{d\phi'^*_\alpha d\phi'_\alpha}{2\pi i} e^{-\sum_\alpha (\phi'^*_\alpha - \phi^*_\alpha)\phi'_\alpha} \cdot \psi(\phi'^*) \ . \tag{1.136}$$

Note that this is just a general form in the complex plane for the familiar representation of a δ-function $\delta(x - x') = \int \frac{dy}{2\pi} e^{iy(x-x')}$.

Another useful property of coherent states is the simple form of matrix elements of normal-ordered operators between coherent states. If we denote by $A(a^\dagger_\alpha, a_\alpha)$ an operator in normal form, the action of the a_α to the right and $a^\dagger_\alpha$ to the left on coherent states immediately yields

$$\langle\phi|A(a^\dagger_\alpha, a_\alpha)|\phi'\rangle = A(\phi^*_\alpha, \phi'_\alpha)\, e^{\sum_\alpha \phi^*_\alpha \phi'_\alpha} \tag{1.137}$$

where $A(\psi^*_\alpha, \psi'_\alpha)$ is the normal form of the operator where the creation operators $a^\dagger_\alpha$ have been replaced by ψ^*_α and the annihilation operators a_α have been replaced by ψ'_α. For example, a two-body potential is written:

$$\begin{aligned}\langle\phi|V|\phi'\rangle &= \frac{1}{2}\sum_{\lambda\mu\nu\rho}(\lambda\mu|v|\nu\rho)\langle\phi|a^\dagger_\lambda a^\dagger_\mu a_\rho a_\nu|\phi'\rangle \\ &= \frac{1}{2}\sum_{\lambda\mu\nu\rho}(\lambda\mu|v|\nu\rho)\phi^*_\lambda\phi^*_\mu\phi'_\rho\phi'_\nu e^{\sum_\alpha \phi^*_\alpha\phi'_\alpha} \ . \end{aligned} \tag{1.138}$$

From their definition, it is clear that coherent states do not have a fixed number of particles. Rather, the occupation number n_α for each state α is Poisson distributed with mean value $|\phi_\alpha|^2$

$$|\langle n_{\alpha_1} n_{\alpha_2} \ldots |\phi\rangle|^2 = \prod_\alpha \frac{|\phi_\alpha|^{2n_\alpha}}{n_\alpha!} \ . \tag{1.139}$$

Thus the distribution of particle numbers has the average value

$$\bar{N} = \frac{\langle\phi|\hat{N}|\phi\rangle}{\langle\phi|\phi\rangle} = \frac{\sum_\alpha \langle\phi|a^\dagger_\alpha a_\alpha|\phi\rangle}{\langle\phi|\phi\rangle} = \sum_\alpha \phi^*_\alpha\phi_\alpha \tag{1.140a}$$

and variance

$$\sigma^2 = \frac{\langle\phi|\hat{N}^2|\phi\rangle}{\langle\phi|\phi\rangle} - \bar{N}^2 = \sum_\alpha \phi^*_\alpha\phi_\alpha = \bar{N} \ . \tag{1.140b}$$

In the thermodynamic limit, where $\bar{N} \to +\infty$, the relative width $\frac{\sigma}{\bar{N}} = \frac{1}{\sqrt{\bar{N}}}$ goes to zero, and the coherent states become sharply peaked around $\overline{N}$, reflecting the fact that the product of Poisson distributions approaches a normal distribution.

GRASSMANN ALGEBRA

In order to construct coherent states for fermions which are eigenstates of annihilation operators, we have seen in Eq. (1.113) that it will be necessary to use anticommuting numbers. Algebras of anticommuting numbers are called Grassmann algebras, and in this section we briefly summarize their essential properties. For our present purposes, it is sufficient to view Grassmann algebra and the definitions of integration and differentiation as a clever mathematical construct which takes care of all the minus signs associated with antisymmetry without attempting to attach any physical significance to it. A complete treatment of Grassmann algebra is given in the treatise by Berezin (1965).

A Grassmann algebra is defined by a set of generators, which we denote by $\{\xi_\alpha\}, \alpha = 1, \ldots, n$. These generators anticommute:

$$\xi_\alpha\xi_\beta + \xi_\beta\xi_\alpha = 0 \tag{1.141a}$$

so that, in particular:

$$\xi^2_\alpha = 0 \ . \tag{1.141b}$$

The basis of the Grassmann algebra is made of all distinct products of the generators. Thus, a number in the Grassmann algebra is a linear combination with complex coefficients of the numbers $\{1, \xi_{\alpha_1}, \xi_{\alpha_1}\xi_{\alpha_2}, \ldots \xi_{\alpha_1}\xi_{\alpha_2}\ldots\xi_{\alpha_n}\}$ where by convention the indices α_i are ordered $\alpha_1 < \alpha_2 < \ldots \alpha_n$. The dimension of a Grassmann algebra with n generators is 2^n since distinct basis elements are produced by the two possibilities of including a generator 0 or 1 times for each of n generators. Hence, a matrix representation of Grassmann numbers requires matrices of dimension at least $2^n \times 2^n$.

In an algebra with an even number $n = 2p$ of generators, one can define a conjugation operation (called involution in some texts) in the following way. We select a set of p generators ξ_α, and to each generator ξ_α, we associate a generator which we denote ξ^*_α.

The following properties define conjugation in a Grassmann algebra:

$$(\xi_\alpha)^* = \xi^*_\alpha \tag{1.142a}$$

$$(\xi^*_\alpha)^* = \xi_\alpha \ . \tag{1.142b}$$

If λ is a complex number,

$$(\lambda\xi_\alpha)^* = \lambda^*\xi^*_\alpha \tag{1.143}$$

and for any product of generators:

$$(\xi_{\alpha_1}\ldots\xi_{\alpha_n})^* = \xi^*_{\alpha_n}\xi^*_{\alpha_{n-1}}\ldots\xi^*_{\alpha_1} \ . \tag{1.144}$$

To simplify notation, we now consider a Grassmann algebra with two generators. We can denote the generators by ξ and ξ^*, and the algebra is generated by the four numbers $\{1, \xi, \xi^*, \xi^*\xi\}$.

Because of property (1.141b), any analytic function f defined on this algebra is a linear function:

$$f(\xi) = f_0 + f_1\xi \tag{1.145}$$

and this is the form we will obtain for the coherent state representation of a wave function. Similarly, the coherent state representation of an operator in the Grassmann algebra will be a function of ξ^* and ξ and must have the form

$$A(\xi^*, \xi) = a_0 + a_1\xi + \overline{a}_1\xi^* + a_{12}\xi^*\xi \ . \tag{1.146}$$

As for ordinary complex functions, a derivative can be defined for Grassmann variable functions. It is defined to be identical to the complex derivative, except that in order for the derivative operator $\frac{\partial}{\partial\xi}$ to act on ξ, the variable ξ has to be anticommuted through until it is adjacent to $\frac{\partial}{\partial\xi}$. For instance:

$$\begin{aligned}\frac{\partial}{\partial\xi}(\xi^*\xi) &= \frac{\partial}{\partial\xi}(-\xi\xi^*) \\ &= -\xi^* \ .\end{aligned} \tag{1.147}$$

With these definitions:

$$\frac{\partial}{\partial \xi} A(\xi^*, \xi) = a_1 - a_{12}\xi^* \tag{1.148a}$$

$$\frac{\partial}{\partial \xi^*} A(\xi^*, \xi) = \bar{a}_1 + a_{12}\xi. \tag{1.148b}$$

$$\frac{\partial}{\partial \xi^*}\frac{\partial}{\partial \xi} A(\xi^*, \xi) = -a_{12}$$

$$= -\frac{\partial}{\partial \xi}\frac{\partial}{\partial \xi^*} A(\xi^* \xi) \ . \tag{1.148c}$$

Note from Eq. (1.148c) that the operators $\frac{\partial}{\partial \xi^*}$ and $\frac{\partial}{\partial \xi}$ anticommute.

In defining a definite integral, there is no analog of the familiar sum motivating the Riemann integral for ordinary variables. Hence, we define integration over Grassmann variables as a linear mapping which has the fundamental property of ordinary integrals over functions vanishing at infinity that the integral of an exact differential form is zero. This requirement implies that the integral of 1 is zero, since 1 is the derivative of ξ. The only non-vanishing integral is that of ξ, since ξ is not a derivative. Hence, the definite integral is defined as follows:

$$\int d\xi \, 1 = 0$$

$$\int d\xi \, \xi = 1 \tag{1.149}$$

and as in the case of a derivative, in order to apply (1.149b), one must first anticommute the variable ξ as required to bring it next to $d\xi$. A simple mnemonic for this definition is the fact that Grassmann integration is identical to Grassmann differentiation. Since half of the generators ξ^*_α have been defined arbitrarily to be conjugate variables but are otherwise equivalent to the generators ξ_α, it is natural to define integration for conjugate variables in the same way:

$$\int d\xi^* \, 1 = 0$$

$$\int d\xi^* \, \xi^* = 1 \ . \tag{1.150}$$

Note, however, that in contrast to a Riemann integral in which dx is an infinitesimal real variable, $d\xi^*$ is not a Grassmann number and it makes no sense to apply Eq. (1.144) to the quantity ($\int d\xi \xi^*$) to try to relate Eq. (1.149) to Eq. (1.150).

The following examples illustrate the application of these integration rules. Using (1.145), we obtain:

$$\int d\xi \, f(\xi) = f_1 \tag{1.151}$$

and using (1.146) we get:

$$\int d\xi A(\xi^*, \xi) = \int d\xi (a_0 + a_1\xi + \bar{a}_1\xi^* + a_{12}\xi^*\xi)$$

$$= a_1 - a_{12}\xi^* \tag{1.152a}$$

$$\int d\xi^* A(\xi^*,\xi) = \overline{a}_1 + a_{12}\xi \tag{1.152b}$$

and

$$\int d\xi^* d\xi A(\xi^*\xi) = -a_{12}$$

$$= -\int d\xi d\xi^* A(\xi^*\xi) \ . \tag{1.152c}$$

The motivation for this definition of integration is that with these conventions, many results look similar to those of complex integration. For instance, consider the definition of a Grassmann δ-function by:

$$\delta(\xi,\xi') \equiv \int d\eta e^{-\eta(\xi-\xi')}$$

$$= \int d\eta(1-\eta(\xi-\xi'))$$

$$= -(\xi-\xi') \ . \tag{1.153}$$

To verify that this definition has the desired behavior, we use Eq. (1.145) to obtain

$$\int d\xi'\delta(\xi,\xi')f(\xi') = -\int d\xi'(\xi-\xi')(f_0+f_1\xi')$$

$$= f_0 + f_1\xi$$

$$= f(\xi) \tag{1.154}$$

for any function $f(\xi)$.

Motivated by Eq. (1.135) for Boson coherent states, it will be useful to define a scalar product of Grassmann functions by:

$$\langle f|g\rangle = \int d\xi^* d\xi e^{-\xi^*\xi} f^*(\xi)g(\xi^*) \tag{1.155}$$

where $f(\xi)$ is defined by (1.145) and

$$g(\xi) = g_0 + g_1\xi \ . \tag{1.156}$$

With definition (1.155), we see that:

$$\langle f|g\rangle = \int d\xi^* d\xi(1-\xi^*\xi)(f_0^* + f_1^*\xi)(g_0+g_1\xi^*)$$

$$= -\int d\xi^* d\xi\, \xi^*\xi f_0^* g_0 + \int d\xi^* d\xi\, \xi\xi^* f_1^* g_1$$

$$= f_0^* g_0 + f_1^* g_1 \tag{1.157}$$

and it can be shown that Grassmann functions have the structure of a Hilbert space. The results we have presented for the case of two generators ξ and ξ^* generalize straightforwardly to 2p generators $\xi_1 \ldots \xi_p, \xi_1^* \ldots \xi_p^*$ as shown in Problem 1.3.

FERMION COHERENT STATES

If we try to construct coherent states for Fermions, we immediately encounter the difficulty that if we expand them according to (1.114), the coefficients must be Grassmann numbers. Therefore, in order to construct coherent states, we must enlarge the Fermion Fock space.

We first define a Grassmann algebra $\mathcal{G}$ by associating a generator ξ_α with each annihilation operator a_α, and a generator ξ^*_α with each creation operator $a^\dagger_\alpha$. We then construct the generalized Fock space as the set of linear combinations of states of the Fock space $\mathcal{F}$ with coefficients in the Grassmann algebra $\mathcal{G}$. Any vector $|\psi\rangle$ in the generalized Fock space can be expanded as:

$$|\psi\rangle = \sum_\alpha \chi_\alpha |\phi_\alpha\rangle \tag{1.158}$$

where the χ_α are Grassmann numbers and the $|\phi_\alpha\rangle$ vectors of the Fock space.

In order to treat expressions containing combinations of Grassmann variables and creation and annihilation operators, it is necessary to augment the definition of the Grassmann variables to specify the commutation relations between ξ's and a's and the adjoints of mixed expressions. To obtain results analogous to those obtained previously for Bosons, it is natural and convenient to require that

$$[\tilde{\xi}, \tilde{a}]_+ = 0 \tag{1.159a}$$

and

$$(\tilde{\xi}\tilde{a})^\dagger = \tilde{a}^\dagger \tilde{\xi}^* \tag{1.159b}$$

where $\tilde{\xi}$ denotes any Grassmann variable in $\{\xi_\alpha, \xi^*_\alpha\}$ and $\tilde{a}$ is any operator in $\{a^\dagger_\alpha, a_\alpha\}$.

We now define a Fermion coherent state $|\xi\rangle$ analogous to Boson coherent states by

$$\begin{aligned} |\xi\rangle &= e^{-\sum_\alpha \xi_\alpha a^\dagger_\alpha}|0\rangle \\ &= \prod_\alpha (1 - \xi_\alpha a^\dagger_\alpha)|0\rangle \ . \end{aligned} \tag{1.160}$$

Note that the combination $\xi_\alpha a^\dagger_\alpha$ commutes with $\xi_\beta a^\dagger_\beta$ so that the second line in Eq. (1.159) reproduces each non-vanishing term of the expansion of the exponential in the first line. Although the coherent state belongs to the generalized Fock space and not to $\mathcal{F}$, as we shall see, the crucial point is that any physical Fermion state of $\mathcal{F}$ can be expanded in terms of these coherent states.

We now verify that coherent states as defined in (1.159) are eigenstates of the annihilation operators. For a single state α, the anticommutation relations of $a_\alpha, a^\dagger_\alpha$, and ξ_α yield the relation.

$$\begin{aligned} a_\alpha(1 - \xi_\alpha a^\dagger_\alpha)|0\rangle &= +\xi_\alpha|0\rangle \\ &= \xi_\alpha(1 - \xi_\alpha a^\dagger_\alpha)|0\rangle \ . \end{aligned} \tag{1.161}$$

Using Eqs. (1.154) and (1.160) and the fact that a_α and ξ_α both commute with the combination $\xi_\beta a_\beta^\dagger$ for $\beta \neq \alpha$, we obtain the desired eigenvalue conditions.

$$\begin{aligned} a_\alpha|\xi\rangle &= a_\alpha \prod_\beta (1-\xi_\beta a_\beta^\dagger)|0\rangle \\ &= \prod_{\beta\neq\alpha}(1-\xi_\beta a_\beta^\dagger) a_\alpha (1-\xi_\alpha a_\alpha^\dagger)|0\rangle \\ &= \prod_{\beta\neq\alpha}(1-\xi_\beta a_\beta^\dagger)\xi_\alpha(1-\xi_\alpha a_\alpha^\dagger)|0\rangle \\ &= \xi_\alpha \prod_\beta (1-\xi_\beta a_\beta^\dagger)|0\rangle \\ &= \xi_\alpha|\xi\rangle \ . \end{aligned} \tag{1.162}$$

Similarly, the adjoint of the coherent state is

$$\langle\xi| = \langle 0|e^{-\sum_\alpha a_\alpha \xi_\alpha^*} = \langle 0|e^{\sum \xi_\alpha^* a_\alpha} \tag{1.163a}$$

and is a left-eigenfunction of $a_\alpha^\dagger$

$$\langle\xi|a_\alpha^\dagger = \langle\xi|\xi_\alpha^* \ . \tag{1.163b}$$

The action of $a_\alpha^\dagger$ on a coherent state is analogous to the Boson result, Eq.(1.120), and differs only in sign:

$$\begin{aligned} a_\alpha^\dagger|\xi\rangle &= a_\alpha^\dagger(1-\xi_\alpha a_\alpha^\dagger)\prod_{\beta\neq\alpha}(1-\xi_\beta a_\beta^\dagger)|0\rangle \\ &= a_\alpha^\dagger \prod_{\beta\neq\alpha}(1-\xi_\beta a_\beta^\dagger)|0\rangle \\ &= -\frac{\partial}{\partial\xi_\alpha}(1-\xi_\alpha a_\alpha^\dagger)\prod_{\beta\neq\alpha}(1-\xi_\beta a_\beta^\dagger)|0\rangle \\ &= -\frac{\partial}{\partial\xi_\alpha}|\xi\rangle \end{aligned} \tag{1.164a}$$

and similarly one may verify

$$\langle\xi|a_\alpha = +\frac{\partial}{\partial\xi^*}\langle\xi| \ . \tag{1.164b}$$

The overlap of two coherent states is easily calculated:

$$\begin{aligned} \langle\xi|\xi'\rangle &= \langle 0|\prod_\alpha(1+\xi_\alpha^* a_\alpha)(1-\xi_\alpha a_\alpha^\dagger)|0\rangle \\ &= \prod_\alpha(1+\xi_\alpha^*\xi_\alpha) \\ &= e^{\sum_\alpha \xi_\alpha^*\xi_\alpha} . \end{aligned} \tag{1.165}$$

As in the Boson case, the closure relation may be written

$$\int \prod_{\alpha} d\xi_{\alpha}^{*} d\xi_{\alpha} e^{-\sum_{\alpha} \xi_{\alpha}^{*}\xi_{\alpha}} |\xi\rangle\langle\xi| = 1 \tag{1.166}$$

where the 1 denotes the unit operator in the physical Fermion Fock space $\mathcal{F}$. This closure relation may be proved using Schur's lemma as we proved Eq. (1.123) once integration by parts has been derived for Grassmann variables (See Problem 1.4). Here, we present an alternative proof. We define the operator A to be the left hand side of Eq. (1.166).

$$A = \int \prod_{\alpha} d\xi_{\alpha}^{*} d\xi_{\alpha} e^{-\sum_{\alpha} \xi_{\alpha}^{*}\xi_{\alpha}} |\xi\rangle\langle\xi| \ . \tag{1.167}$$

To prove (1.167), it is sufficient to prove that for any vectors of the basis of the Fock space:

$$\langle\alpha_1 \dots \alpha_n|A|\beta_1 \dots \beta_m\rangle = \langle\alpha_1 \dots \alpha_n|\beta_1 \dots \beta_m\rangle \ . \tag{1.168}$$

Using the eigenvalue property of the coherent states (1.161) we obtain

$$\begin{aligned}\langle\alpha_1 \dots \alpha_n|\xi\rangle &= \langle 0|a_{\alpha_n} \dots a_{\alpha_1}|\xi\rangle \\ &= \xi_{\alpha_n} \dots \xi_{\alpha_1}\end{aligned} \tag{1.169}$$

and the analogous adjoint equations. Thus

$$\begin{aligned}\langle\alpha_1 \dots \alpha_n|A|\beta_1 \dots \beta_n\rangle &= \int \prod_{\alpha} d\xi_{\alpha}^{*} d\xi_{\alpha} e^{-\sum_{\alpha} \xi_{\alpha}^{*}\xi_{\alpha}} \langle\alpha_1 \dots \alpha_n|\xi\rangle\langle\xi|\beta_1 \dots \beta_n\rangle \\ &= \int \prod_{\alpha} d\xi_{\alpha}^{*} d\xi_{\alpha} \prod_{\alpha}(1 - \xi_{\alpha}^{*}\xi_{\alpha})\xi_{\alpha_n} \dots \xi_{\alpha_1}\xi_{\beta_1}^{*} \dots \xi_{\beta_n}^{*} \ .\end{aligned} \tag{1.170}$$

Now consider the integrals which may arise in Eq. (1.170) for a particular state γ:

$$\int d\xi_{\gamma}^{*} d\xi_{\gamma}(1 - \xi_{\gamma}^{*}\xi_{\gamma}) \begin{Bmatrix} \xi_{\gamma}\xi_{\gamma}^{*} \\ \xi_{\gamma}^{*} \\ \xi_{\gamma} \\ 1 \end{Bmatrix} = \begin{Bmatrix} 1 \\ 0 \\ 0 \\ 1 \end{Bmatrix} \ . \tag{1.171}$$

Thus, the integral in Eq.(1.170) is non-vanishing only if each state γ is either occupied in both $\langle\alpha_1 \dots \alpha_n|$ and $|\beta_1 \dots \beta_n\rangle$ or unoccupied in both states, which requires that m=n and $\{\alpha_1 \dots \alpha_n\}$ is some permutation P of $\{\beta_1 \dots \beta_n\}$. In this case, the integral is easily evaluated by writing $\xi_{\alpha_n} \cdots \xi_{\alpha_1}\xi_{\beta_1}^{*} \cdots \xi_{\beta_n}^{*} = (-1)^P \xi_{\alpha_n} \cdots \xi_{\alpha_1}\xi_{\alpha_1}^{*} \cdots \xi_{\alpha_n}^{*}$ and noting that an even number of anticommutations is required to bring the integral over each state into the form of Eq. (1.171), so that the value is just $(-1)^P$. The left hand side of Eq. (1.168) thus yields the result previously derived in Eq. (1.50) for the right hand side, so the equality is established for any vectors in the Fock space.

As in the case of Bosons, this completeness relation provides a useful expression for the trace of an operator. Because the matrix elements $\langle\psi_i|\xi\rangle$ and $\langle\xi|\psi_i\rangle$ between states $|\psi_i\rangle$ in the Fock space and coherent states contain Grassmann numbers, it follows from the anticommutation relations that

$$\langle\psi_i|\xi\rangle\langle\xi|\psi_j\rangle = \langle-\xi|\psi_j\rangle\langle\psi_i|\xi\rangle \ . \tag{1.172}$$

Hence, if we define a complete set of states $\{|n\rangle\}$ in the Fock space, the trace of an operator A may be written

$$\begin{aligned}\operatorname{Tr} A &= \sum_n \langle n|A|n\rangle \\ &= \int \prod_\alpha d\xi^*_\alpha d\xi_\alpha e^{-\sum_\alpha \xi^*_\alpha \xi_\alpha} \sum_n \langle n|\xi\rangle\langle\xi|A|n\rangle \\ &= \int \prod_\alpha d\xi^*_\alpha d\xi_\alpha e^{-\sum_\alpha \xi^*_\alpha \xi_\alpha} \langle-\xi|A \sum_n |n\rangle\langle n||\xi\rangle \\ &= \int \prod_\alpha d\xi^*_\alpha d\xi_\alpha e^{-\sum_\alpha \xi^*_\alpha \xi_\alpha} \langle-\xi|A|\xi\rangle \ .\end{aligned} \tag{1.173}$$

The overcompleteness of the Fermion coherent states allows us to define a Grassmann coherent state representation analogous to the coherent state representation for Bosons in Eq. (1.129a)

$$|\psi\rangle = \int \prod_\alpha d\xi^*_\alpha d\xi_\alpha e^{-\sum_\alpha \xi^*_\alpha \xi_\alpha} \psi(\xi^*)|\xi\rangle \tag{1.174a}$$

where

$$\langle\xi|\psi\rangle = \psi(\xi^*) \ . \tag{1.174b}$$

Within this representation, it follows from Eqs. (1.161) and (1.164) that the creation and annihilation operators satisfy:

$$\langle\xi|a_\alpha|\psi\rangle = \frac{\partial}{\partial\xi^*_\alpha}\psi(\xi^*) \tag{1.175a}$$

$$\langle\xi|a^\dagger_\alpha|\psi\rangle = \xi^*_\alpha\psi(\xi^*) \ . \tag{1.175b}$$

Thus, as in the Boson case, the operators a_α and $a^\dagger_\alpha$ are represented by the operators $\frac{\partial}{\partial\xi^*_\alpha}$ and ξ^*_α respectively. The anticommutation relation (1.83) is represented by

$$\left[\frac{\partial}{\partial\xi^*_\alpha}, \xi^*_\beta\right]_+ = \delta_{\alpha\beta} \ . \tag{1.176}$$

As in the Boson case, the matrix element of a normal-ordered operator $A(a^\dagger_\alpha a_\alpha)$ between two coherent states is very simple:

$$\langle\xi|A(a^\dagger_\alpha, a_\alpha)|\xi'\rangle = e^{\sum_\alpha \xi^*_\alpha \xi'_\alpha} A(\xi^*_\alpha, \xi'_\alpha) \ . \tag{1.177}$$

However, in contrast to the Boson case, the expectation value of the number operator is not a real number:

$$\frac{\langle\xi|N|\xi\rangle}{\langle\xi|\xi\rangle} = \sum_\alpha \xi^*_\alpha \xi_\alpha \tag{1.178}$$

and it is meaningless to speak about the average number of particles in a Fermion coherent state.

Finally, we conclude this section by contrasting the physical significance of Boson and Fermion coherent states. Boson coherent states are the physical states which emerge naturally when taking the classical limit of quantum mechanics or of a quantum field theory (See Problem 1.5). In the classical limit, when the field operators are assumed to commute, the definition of a classical field $\phi(x)$ at each point of space is identical to saying that the system is in the coherent state $|\phi\rangle = e^{\int d^3x \phi(x)\psi^\dagger(x)}|0\rangle$. For example, a classical electromagnetic field can be viewed as a coherent state of photons.

In contrast, Fermion coherent states are not contained in the Fermion Fock space, they are not physically observable, and there are no classical fields of Fermions. Nevertheless, Fermion coherent states are very useful in formally unifying many-Fermion and many-Boson problems, and we shall use this property extensively in the following chapters.

In the subsequent development, this physical difference will give rise to some significant differences in the treatment of Bosons and Fermions. For example, application of the stationary-phase approximation to an expression formulated in terms of Boson coherent states yields a useful expansion around a physical classical field configuration. For Fermions, no corresponding physical solution exists, and the Fermion degrees of freedom will have to be integrated out explicitly.

GAUSSIAN INTEGRALS

In the ensuing formal development, we will frequently evaluate matrix elements of the evolution operator in coherent states, leading to integrals of exponential functions which are polynomials in complex variables or Grassmann variables. In the case of quadratic forms, these are straightforward generalizations of the familiar Gaussian integral, and we present several useful integrals in this section for future reference. For brevity, we derive the identities for the special case of symmetric and Hermitian matrices, and refer the reader to the standard references for the general case.

We begin by proving the following identity for multi-dimensional integrals over real variables:

$$\int \frac{dx_1 \ldots dx_n}{(2\pi)^{\frac{n}{2}}} e^{-\frac{1}{2}x_i A_{ij} x_j + x_i J_i} = [\det A]^{-\frac{1}{2}} e^{\frac{1}{2}J_i A^{-1}_{ij} J_j} \tag{1.179}$$

where A is a real symmetric positive definite matrix and summation over repeated Latin indices is understood throughout this section. This identity is established straightforwardly by changing variables to reduce it to diagonal form and using the familiar Gaussian integral

$$\int_{-\infty}^{\infty} dx e^{-ax^2} = \sqrt{\frac{\pi}{a}} \,. \tag{1.180}$$

Performing the transformations $y_i = x_i - A^{-1}_{ij} J_j$ and $z_k = O^{-1}_{ki} y_i$, where O is the orthogonal transformation which diagonalizes A, we obtain

$$\begin{aligned}
\int dx_1 \ldots dx_n e^{-\frac{1}{2}x_i A_{ik} x_k + J_k x_k - \frac{1}{2} J_i A^{-1}_{ik} J_k} \\
&= \int dy_1 \ldots dy_n e^{-\frac{1}{2} y_i A_{ik} y_k} \\
&= \int dz_1 \ldots dz_n e^{\sum_m \frac{1}{2} a_m z_m^2} \\
&= \prod_{m=1}^{n} \sqrt{\frac{2\pi}{a_m}} \\
&= \frac{(2\pi)^{\frac{n}{2}}}{[\det A]^{\frac{1}{2}}}
\end{aligned} \tag{1.181}$$

which proves Eq. (1.179). Note that the positivity of A is essential for convergence of the Gaussian integral.

A similar identity for integrals over pairs of conjugate complex variables is

$$\int \prod_{i=1}^{n} \frac{dx_i^* dx_i}{2\pi i} e^{-x_i^* H_{ij} x_j + J_i^* x_i + J_i x_i^*} = [\det H]^{-1} e^{J_i^* H^{-1}_{ij} J_j} \tag{1.182}$$

which is valid for any matrix H with a positive Hermitian part. For the special case of a Hermitian matrix, it is proved in the same way as Eq. (1.179) by defining the transformation $y_i = x_i - H^{-1}_{ij} J_j$ and its complex conjugate, transforming H to diagonal form and evaluating the diagonal integral

$$\begin{aligned}
\int \frac{dz^* dz}{2\pi i} e^{-z^* a z} &= \int \frac{du dv}{\pi} e^{-a(u^2+v^2)} \\
&= \frac{1}{a} \ .
\end{aligned} \tag{1.183}$$

Finally, we wish to establish the analogous identity for Grassmann variables

$$\int \prod_{i=1}^{n} d\eta_i^* d\eta_i e^{-\eta_i^* H_{ij} \eta_j + \varsigma_i^* \eta_i + \varsigma_i \eta_i^*} = [\det H] e^{\varsigma_i^* H^{-1}_{ij} \varsigma_j} \tag{1.184}$$

where, for simplicity, H is again assumed to be Hermitian but not necessarily positive definite and $\{\eta_i, \eta_i^*, \varsigma_i, \varsigma_i^*\}$ are Grassmann variables. Note that Eq. (1.184) differs from Eq. (1.182) by the appearance of the determinant in the numerator instead of the denominator. To prove this identity, we need to derive two additional results for Grassmann variables; the transformation law for an integral under a change of variables and the formula for a Gaussian integral.

A Gaussian integral involving a single pair of conjugate Grassmann variables is easily evaluated as follows.

$$\int d\xi^* d\xi e^{-\xi^* a \xi} = \int d\xi^* d\xi (1 - \xi^* a \xi) = a \ . \tag{1.185}$$

Note that for a single variable, the Grassmann Gaussian integral yields a in contrast to $\frac{1}{a}$ in Eq. (1.183) for the ordinary Gaussian integral. Hence, if we can bring the multivariable Grassmann integral (1.184) into diagonal form, we expect to obtain the product of eigenvalues, and thus the determinant of H, in the numerator instead of in the denominator as for complex variables.

In order to transform Eq. (1.184) into diagonal form, we need to derive the law for linear transformations of Grassmann variables

$$\int d\varsigma_1^* d\varsigma_1 \ldots d\varsigma_n^* d\varsigma_n P(\varsigma^*, \varsigma) = \left|\frac{\partial(\eta^*, \eta)}{\partial(\varsigma^*, \varsigma)}\right| \int d\eta_1^* d\eta_1 \ldots d\eta_n^* d\eta_n \times P(\varsigma^*(\eta^*, \eta), \varsigma(\eta^*, \eta)) \tag{1.186}$$

which differs from the transformation law for complex variables by the appearance of the inverse of the Jacobian instead of the Jacobian. The derivation is facilitated by relabelling the variables as follows:

$$\begin{aligned}(\varsigma_1^* \varsigma_2^* \cdots \varsigma_n^* \varsigma_n \varsigma_{n-1} \cdots \varsigma_1) &\equiv (\tilde{\varsigma}_1 \tilde{\varsigma}_2 \ldots \tilde{\varsigma}_{2n}) \\ (\eta_1^* \eta_2^* \cdots \eta_n^* \eta_n \eta_{n-1} \cdots \eta_1) &\equiv (\tilde{\eta}_1 \tilde{\eta}_2 \ldots \tilde{\eta}_{2n})\end{aligned} \tag{1.187}$$

and writing

$$\tilde{\varsigma}_i = M_{ij} \tilde{\eta}_j \ . \tag{1.188}$$

The only non-vanishing contributions to Eq. (1.186) come from the term in the polynomial containing each $\tilde{\varsigma}_i$ as a factor, which we write as $p \prod_{i=1}^{2n} \tilde{\varsigma}_i$. Thus, we must evaluate J in the equation

$$\int d\varsigma_1^* d\varsigma_1 \ldots d\varsigma_n^* d\varsigma_n p \prod_{i=1}^{2n} \tilde{\varsigma}_i = J \int d\eta_1^* d\eta_1 \ldots d\eta_n^* d\eta_n p \prod_{i=1}^{2n} \left(\sum_j M_{ij} \tilde{\eta}_j \right) \ . \tag{1.189}$$

The left hand side yields $p(-1)^n$ and the right hand side is evaluated by noting that the only non-vanishing contributions arise from the $(2n)!$ distinct permutations P of the variables $\{\tilde{\eta}\}$ generated by the product. Thus

$$\begin{aligned}p(-1)^n &= Jp \int d\eta_1^* d\eta_1 \ldots d\eta_n^* d\eta_n \prod_i \left(\sum_j M_{ij} \tilde{\eta}_j \right) \\ &= Jp \int d\eta_1^* d\eta_1 \ldots d\eta_n^* d\eta_n \sum_P \prod_i M_{iPi} \tilde{\eta}_{Pi} \\ &= Jp \sum_P \prod_i M_{iPi} (-1)^P \int d\eta_1^* d\eta_1 \ldots d\eta_n^* d\eta_n \tilde{\eta}_1 \tilde{\eta}_2 \ldots \tilde{\eta}_{2n} \\ &= Jp \det M (-1)^n\end{aligned}$$

so that

$$J = (\det M)^{-1} = \left| \frac{\partial(\tilde{\eta})}{\partial(\tilde{\xi})} \right| = \left| \frac{\partial(\eta^*, \eta)}{\partial(\varsigma^*, \varsigma)} \right|$$

which proves Eq. (1.186) for a general linear transformation.

Finally, Eq. (1.184) may now be proved by defining the transformations $\rho_i = \eta_i - H^{-1}_{ij}\varsigma_j, \rho^*_i = \eta^*_i - H^{-1*}_{ij}\varsigma^*_j$, diagonalizing H with a unitary transformation U, defining $\xi_i = U^{-1}_{ij}\rho_j$ and $\xi^*_i = U^{-1*}_{ij}\rho^*_j$, noting that all the Jacobians are unity, and using the Gaussian integral Eq. (1.185). Thus

$$\begin{aligned}\int\prod_{i=1}^{n} d\eta^*_i\, d\eta_i e^{-\eta^*_i H_{ij}\eta_j + \varsigma^*_i\eta_i + \varsigma_i\eta^*_i - \varsigma^*_i H^{-1}_{ij}\varsigma_j} \\ &= \int\prod_{i=1}^{n} d\rho^*_i\, d\rho_i e^{-\rho^*_i H_{ij}\rho_j} \\ &= \int\prod_{i=1}^{n} d\xi^*_i\, d\xi_i e^{-\sum_i h_i \xi^*_i \xi_i} \\ &= \prod_{m=1}^{n} h_m = \det H \end{aligned} \tag{1.190}$$

which proves Eq. (1.184). Note, as in the previous case, that the derivation may be generalized to a non-Hermitian matrix H.

It may appear curious that the Gaussian integral for Grassmann variables requires no restrictions on the matrix H whereas for ordinary variables H must be positive definite for the integral to converge. Formally, the distinction arises because the expansion of the exponential $e^{a\eta^*\cdot\eta}$ terminates at first order, yielding a finite integral irrespective of the sign of a. This formal property, however, reflects a fundamental difference between Fermions and Bosons, and it is the Pauli principle restriction that occupation numbers be either 0 or 1 which guarantees finite results for Fermions irrespective of the eigenvalues of H.

This point is illustrated by the simple case of the partition function of non-interacting particles, which will be shown in the next chapter to yield Gaussian integrals of the form of Eq. (1.184). For non-interacting particles in the Grand Canonical ensemble, the partition function may be written

$$Z = \prod_\alpha \sum_{n_\alpha} e^{-\beta(\epsilon_\alpha - \mu)n_\alpha}$$

where n_α denotes the occupation number of the state α. For Fermions, the series for each α terminates after two terms

$$Z_F = \prod_\alpha \left(1 + e^{-\beta(\epsilon_\alpha - \mu)}\right)$$

so no restriction is imposed on$(\epsilon_\alpha - \mu)$. However, for Bosons,

$$Z_B = \prod_\alpha \left(1 + \sum_{n=1}^{\infty} \left(e^{-\beta(\epsilon_\alpha - \mu)}\right)^n\right)$$

so that the partition function is finite only if $e^{-\beta(\epsilon_\alpha - \mu)} < 1$ for all α which requires that $(\epsilon_\alpha - \mu)$ be positive for all α. Thus the operator $(H - \mu N)$ must be positive definite for Bosons, but no such requirement arises for Fermions.

The salient results from this introductory chapter are collected for subsequent reference in Table 1.1.

$$
\begin{aligned}
[a_\alpha, a^\dagger_\beta]_{-\zeta} &= \delta_{\alpha\beta} \\
|\xi\rangle &= e^{\zeta \sum_\alpha \xi_\alpha a^\dagger_\alpha}|0\rangle \\
a_\alpha|\xi\rangle &= \xi_\alpha|\xi\rangle \\
\langle\xi|a^\dagger_\alpha &= \langle\xi|\xi^* \\
a^\dagger_\alpha|\xi\rangle &= \zeta\frac{\partial}{\partial\xi_\alpha}|\xi\rangle \\
\langle\xi|a_\alpha &= \frac{\partial}{\partial\xi^*_\alpha}\langle\xi| \\
\langle\xi|A(a^\dagger_\alpha, a_\alpha)|\xi'\rangle &= e^{\sum_\alpha \xi^*_\alpha\xi'_\alpha} A(\xi^*_\alpha, \xi_\alpha) \\
1 &= \int d\mu(\xi) e^{-\sum_\alpha \xi^*_\alpha\xi_\alpha}|\xi\rangle\langle\xi| \\
trA &= \int d\mu(\xi) e^{-\sum_\alpha \xi^*_\alpha\xi_\alpha}\langle\zeta\xi|A|\xi\rangle \\
|\psi\rangle &= \int d\mu(\xi) e^{-\sum_\alpha \xi^*_\alpha\xi_\alpha}\psi(\xi^*_\alpha)|\xi\rangle \\
\psi(\xi^*) &= \langle\xi|\psi\rangle \\
\langle\xi|a^\dagger_\alpha|\psi\rangle &= \xi^*_\alpha\psi(\xi^*) \\
\langle\xi|a_\alpha|\psi\rangle &= \frac{\partial}{\partial\xi^*}\psi(\xi^*) \\
[\det H]^{-\zeta} e^{\sum_{\alpha\beta}\eta^*_\alpha H^{-1}_{\alpha\beta}\eta_\beta} &= \int d\mu(\xi) e^{-\sum_{\alpha\beta}\xi^*_\alpha H_{\alpha\beta}\xi_\beta + \sum_\alpha(\eta^*_\alpha\xi_\alpha + \eta_\alpha\xi^*_\alpha)} \\
d\mu(\xi) &= \frac{1}{\mathcal{N}}\prod_\alpha d\xi^*_\alpha d\xi_\alpha \\
\mathcal{N} &= \begin{cases} 2\pi i & \text{Bosons} \\ 1 & \text{Fermions} \end{cases} \\
\zeta &= \begin{cases} 1 & \text{Bosons} \\ -1 & \text{Fermions} \end{cases}
\end{aligned}
$$

Table 1.1 SUMMARY OF PRINCIPAL RESULTS OF CHAPTER 1. Formulas are written in a unified form for Fermions and Bosons using the conventions that for Bosons ξ and η denote complex variables and $\zeta = +1$ whereas for Fermions ξ and η denote Grassmann variables and $\zeta = -1$.

PROBLEMS FOR CHAPTER 1

The subject of quantum many-body theory is too vast to treat all the formalism and illustrative examples adequately in the text. Thus the problems are intended to be an integral part of the course, and in contrast to most texts, entirely new and separate topics will often be introduced. Readers are strongly encouraged to read through all the problems and solve those which appear appropriate. As a guide, introductory comments will be made concerning the problems in each chapter, and particularly crucial problems will be denoted by an *.

Chapter I was necessarily completely formal, and Problems 1–4 are intended to develop expertise with the formalism and fill in details omitted in the text. Problem 5 is intended to develop a physical appreciation for Boson coherent states.

Problems 6–9 deal with the physics of the Hartree Fock approximation. It is assumed that students have been exposed to the conventional treatment of the Hartree Fock and Fermi gas approximations in elementary quantum mechanics courses, so these topics are not repeated in the text. However, this mean-field physics is fundamental to all the subsequent physical applications, so it is optimal to have it clearly in mind before proceeding to the general formalism. For those unfamiliar with these approximations, sufficient details are provided to work out everything using the treatment of permanents, determinants, and matrix elements of one- and two-body operators provided in this chapter.

Problem 9 is particularly important because it introduces the problem of particles interacting in one spatial dimension with δ-function forces. Most of the methods developed in later chapters may be worked out analytically for this system, so it is essential to explore the elementary properties developed here as a foundation for subsequent applications.

PROBLEM 1.1 Show that the operator which counts the number of n-tuples in the states $|\alpha_1\rangle, |\alpha_2\rangle, \dots |\alpha_n\rangle$ may be written

$$\hat{P}_{\alpha_1\alpha_2\dots\alpha_n} = a^\dagger_{\alpha_1} a^\dagger_{\alpha_2} \dots a^\dagger_{\alpha_n} a_{\alpha_n} \dots a_{\alpha_2} a_{\alpha_1} \tag{1}$$

from which Eq. (1.110) immediately follows. One straightforward method is to repeat the argument leading to Eq. (1.103) for three particles and then generalize to obtain

$$\hat{P}_{\alpha_1\alpha_2\dots\alpha_n} = n_{\alpha_1}(n_{\alpha_2} - \delta_{\alpha_1\alpha_2})(n_{\alpha_3} - \delta_{\alpha_1\alpha_3})\dots(n_{\alpha_n} - \delta_{\alpha_1\alpha_n} - \dots - \delta_{\alpha_{n-1}\alpha_n})$$

from which (1) may be proved by induction. Note that the complications associated with non-distinct states only arise for Bosons.

PROBLEM 1.2 Derive the completeness relation Eq. (1.123) by integration. First consider one single-particle state α and let $|n\rangle$ denote the state with n particles in α. Show that by writing ϕ in polar form $\phi = \rho e^{i\theta}$ one obtains

$$\begin{aligned}\int \frac{d\phi^* d\phi}{2\pi i} e^{-\phi^*\phi} |\phi\rangle\langle\phi| &= \int \rho \frac{d\rho d\theta}{\pi} e^{-\rho^2} \sum_m \frac{(\rho e^{i\theta})^m}{\sqrt{m!}} |m\rangle \sum_n \frac{(\rho e^{-i\theta})^n}{\sqrt{n!}} \langle n| \\ &= \sum_n |n\rangle\langle n| \ .\end{aligned}$$

Now generalize to a set of single-particle states $\{|\alpha\rangle\}$ noting that the closure relation Eq. (1.89) may be written

$$\sum_{\{n_\alpha\}} |n_{\alpha_1} n_{\alpha_2} \dots\rangle\langle n_{\alpha_1} n_{\alpha_2} \dots| = 1$$

where $\{n_\alpha\}$ denotes a complete set of occupation numbers.

PROBLEM 1.3 Generalize the properties of Grassmann variables demonstrated in Section (1.5) for the pair of generators ξ, ξ^* to the case of 2p generators $\{\xi_1 \ldots \xi_p \xi^*, \ldots \xi_p^*\}$. In particular, determine the general form of a function $f(\xi_\alpha)$ and an operator $A(\xi_\alpha^*, \xi_\alpha)$, show that $\frac{\partial}{\partial \xi_\alpha}$, $\frac{\partial}{\partial \xi_\beta^*}$, ξ_γ, and ξ_δ^* anticommute, determine if an analogous property holds for integration, find and verify an expression for the p-dimensional δ–function $\delta^P(\vec{\xi} - \vec{\xi}')$, and generalize Eq. (1.157) for $\langle f|g\rangle$.

PROBLEM 1.4 Prove the closure relation Eq. (1.116) for Fermions using Schur's lemma. As in the Boson case, $[a_\alpha, |\phi\rangle\langle\phi|] = (\xi_\alpha - \frac{\partial}{\partial \xi_\alpha^*})|\phi\rangle\langle\phi|$, so one must show

$$\prod_\alpha d\xi_\alpha^* d\xi_\alpha e^{-\sum_\alpha \xi_\alpha^* \xi_\alpha}(\xi_\alpha - \frac{\partial}{\partial \xi_\alpha^*})A(\xi_\alpha, \xi_\alpha^*) = 0 \tag{1}$$

for any A.

First, prove Eq. (1) as it stands, establishing that the left-hand side of Eq. (1.116) must be proportional to unity, and evaluate the constant of proportionality by calculating the matrix element in the zero-particle state.

Then note that Eq. (1) is a special case of integration by parts for Grassmann variables. Show that the general rule for integration by parts is

$$\int \prod_\alpha d\xi_\alpha^* d\xi_\alpha A(\xi_\alpha^*, \xi_\alpha)\frac{\vec{\partial}}{\partial \xi_\alpha}B(\xi_\alpha^*, \xi_\alpha) = \int \prod_\alpha d\xi_\alpha^* d\xi_\alpha A(\xi_\alpha^*, \xi_\alpha)\frac{\overleftarrow{\partial}}{\partial \xi_\alpha}B(\xi_\alpha^*, \xi_\alpha) \tag{2}$$

where $\frac{\overleftarrow{\partial}}{\partial \xi_\alpha}$ acts to the left and the variable ξ_α must be anticommuted to the right to be adjacent to the derivative. Note in particular that the sign in Eq. (2) is reversed from the usual relation for complex variables and that expressions like $\int \prod_\alpha d\xi_\alpha^* d\xi_\alpha \frac{\partial}{\partial \xi_\alpha}[A(\xi_\alpha^*, \xi_\alpha)]B(\xi_\alpha^*, \xi_\alpha)$ do not reproduce the right-hand side.

PROBLEM 1.5 Coherent States and the Classical Limit of the Harmonic Oscillator Consider a simple harmonic oscillator

$$H = \frac{1}{2}\hat{p}^2 + \frac{1}{2}\omega^2\hat{x}^2 = \hbar\omega(a^\dagger a + \frac{1}{2})$$

where

$$\hat{x} = (\frac{\hbar}{2\omega})^{\frac{1}{2}}(a^\dagger + a)$$
$$\hat{p} = i(\frac{\hbar\omega}{2})^{\frac{1}{2}}(a^\dagger - a) \quad .$$

a) Show that a minimum uncertainty wave packet solves the time-dependent Schrödinger equation, giving a probability density

$$|\psi(x,t)|^2 = \sqrt{\frac{\omega}{\pi\hbar}}e^{-\frac{\omega}{\hbar}(x - x_0 \cos\omega t)^2} \quad .$$

b) Now consider the Heisenberg representation coherent state

$$|\phi\rangle = e^{a^\dagger \phi}|0\rangle = \sum_n \frac{\phi^n}{\sqrt{n!}}|n\rangle \ .$$

From the eigenvalue condition

$$\langle x|a|\phi\rangle = \frac{1}{\sqrt{2\hbar\omega}}\langle x|\omega\hat{x} + i\hat{p}|\phi\rangle = \phi\langle x|\phi\rangle$$

show that the coordinate space wave function $\phi(q) = \langle q|\phi\rangle$ satisfies

$$\frac{\partial}{\partial x}\phi(x) = \left(\sqrt{\frac{2\omega}{\hbar}}\phi - \frac{\omega}{\hbar}x\right)\phi(x)$$

so that

$$\phi(x) = Ce^{-\left[\left(\frac{\omega}{2\hbar}\right)^{\frac{1}{2}}x - \phi\right]^2} \ .$$

The Schrödinger representation wave function is obtained by noting

$$|\phi\rangle_{sch} = \sum_n \frac{\phi^n}{\sqrt{n!}} e^{-i\omega(n+\frac{1}{2})t}|n\rangle = e^{-i\frac{\omega}{2}t}|\phi e^{-i\omega t}\rangle \ .$$

Thus, neglecting the inconsequential phase factor $e^{-i\frac{\omega}{2}t}$, the Schrödinger wave function is obtained from the Heisenberg wave function by the substitution $\phi \to \phi e^{-i\omega t}$. Hence, show that the coherent state wave function has the properties

$$|\phi_{sch}(x,t)|^2 = \sqrt{\frac{\omega}{\pi\hbar}}e^{-2\left[\sqrt{\frac{\omega}{2\hbar}}x - |\phi|\cos\omega t\right]^2}$$

$$\langle\hat{x}\rangle = \left(\frac{2\hbar}{\omega}\right)^{\frac{1}{2}}|\phi|\cos\omega t$$

$$\langle\hat{p}\rangle = (2\hbar\omega)^{\frac{1}{2}}|\phi|\sin\omega t$$

$$\langle H\rangle = \frac{1}{2}[\langle\hat{p}^2\rangle + \omega^2\langle\hat{x}^2\rangle] = \hbar\omega(|\phi|^2 + \frac{1}{2})$$

where $\langle\ \rangle$ denotes the coherent state expectation value.

Thus, the coherent state $|\phi\rangle$ is the minimum uncertainty wave packet of part (a). The mean values of the coordinate and momentum satisfy the classical equations of motion and execute simple harmonic motion with amplitude proportional to $|\phi|$. The probability density associated with the oscillator ground state is simply translated by this classical motion, and the expectation value of the energy differs from the classical result $\frac{1}{2}[\langle p\rangle^2 + \omega^2\langle x\rangle^2]$ only by the zero point energy $\frac{1}{2}\hbar\omega$. Finally, use Stirling's approximation and the distribution of oscillator quanta $P(n) = C\frac{|\phi|^{2n}}{n!}$ to show that the most probable oscillator state in the coherent state has $n = |\phi|^2$ with $E_n = \hbar\omega(n + \frac{1}{2}) = \hbar\omega(|\phi|^2 + \frac{1}{2}) = \langle H\rangle$.

Further applications of coherent states may be found in the following two references. The preceding treatment of a single oscillator mode is generalized to treat the quantized electromagnetic field by R. Glauber (1963). The S-matrix for a harmonic oscillator coupled to an arbitrary time-dependent field is solved by P. Carruthers and M.M. Nieto (1965).

PROBLEM 1.6 The Hartree-Fock Approximation

a) Show that the expectation value of a Hamiltonian with a one-body kinetic energy operator Eq. (1.98) and local spin independent two-body potential Eq. (1.156) in a Slater determinant is

$$\langle H\rangle = \sum_{\alpha=1}^{N}\int dx\phi_\alpha^*(x)\left(\frac{-\nabla^2}{2m}\right)\phi_\alpha(x)$$
$$+\frac{1}{2}\sum_{\alpha,\beta=1}^{N}\int dxdy\phi_\alpha^*(x)\phi_\beta^*(y)v(x-y)\left(\phi_\alpha(x)\phi_\beta(y)-\phi_\alpha(y)\phi_\beta(x)\right) \quad . \quad (1)$$

By the Ritz variational principle, the best determinantal approximation to the exact wave function is obtained by minimizing $\langle H\rangle$ with respect to the single-particle wave function ϕ_α. Hence, introduce Lagrange multipliers with $\gamma_{\alpha\beta}^* = \gamma_{\beta\alpha}$ to maintain orthonormality and vary $[\langle H\rangle - \sum_{\alpha\beta}\gamma_{\alpha\beta}\langle\alpha|\beta\rangle]$ with respect to each wave function. The variation is simplified by noting that variation with respect to the real and imaginary parts of ϕ_α is equivalent to independent variation with respect to ϕ_α^* and ϕ_α. Finally, perform a unitary transformation on the wave functions ϕ_α (which cannot alter the determinant) to diagonalize $\gamma_{\alpha\beta}$ and obtain the Hartree-Fock equations

$$-\frac{1}{2m}\vec{\nabla}^2\phi_\alpha(x) + V_H(x)\phi_\alpha(x) - \int dyV_E(x,y)\phi_\alpha(y) = \epsilon_\alpha\phi_\alpha(x) \qquad (2)$$

where the local Hartree potential is

$$V_H(x) = \int dy\sum_\beta \phi_\beta^*(y)\phi_\beta(y)v(x-y)$$

the non-local exchange potential is

$$V_E(x,y) = \sum_\beta \phi_\beta(x)\phi_\beta^*(y)v(x-y)$$

and the diagonalized Lagrange multipliers are now denoted ϵ_α. Recall that x includes both space and spin variables, and write the spatial Hartree-Fock equations explicitly for the case of spin S, assuming the number of particles N is an integer multiple of the spin degeneracy $(2S+1)$.

b) Repeat the derivation for the case of Bosons with all particles in a single state ϕ. Compare the equations with the Fermion result in the special case $2S+1=N$. Explain this comparison physically.

c) Show that the total energy of the Hartree-Fock solution for Fermions may be written

$$E_{HF} = \frac{1}{2}\sum_\alpha(\epsilon_\alpha + T_\alpha) = \sum_\alpha \epsilon_\alpha - \frac{1}{2}\sum_{\alpha\beta}(\alpha\beta|V|\alpha\beta - \beta\alpha)$$

where $T_\alpha = \langle\alpha| -\frac{1}{2m}\nabla^2|\alpha\rangle$.

Also prove Koopmans' theorem which states that if one evaluates the Hartree-Fock energies of the N and $(N-1)$–particle system using the Hartree-Fock wave functions for the N–particle system,

$$E_{HF}(N) - E_{HF}(N-1) = \epsilon_\alpha$$

where ϵ_α is the eigenvalue of the Nth single particle state. Hence, to the extent to which single-particle wave functions only change to order δ when the particle number changes by 1, $-\epsilon_\alpha$ specifies the removal energy for the last particle to order δ^2.

d) Now, define a one-body potential U_{HF} such that diagonalizing $H_0 = T+U_{HF}$ reproduces the Hartree-Fock equations. In the subsequent study of perturbation theory, it will be useful to regard $T+U_{HF}$ as the unperturbed Hamiltonian and treat $\sum_{i<j} v(x_i - x_j) - U_{HF}$ as the perturbation.

PROBLEM 1.7 Stability and Equilibrium Conditions for Uniform Matter Consider the thermodynamic limit of a system of N particles at zero temperature confined in a box of volume V in a state with energy E having uniform density. Denote the density $\rho = \frac{N}{V}$ and the energy per particle $\epsilon(\rho) = \frac{E}{N}$.

a) Show that the pressure is given by $P = \rho^2 \frac{\partial \epsilon}{\partial \rho}$.

b) Show that the equilibrium density, often called the saturation density, for self-bound unconfined bulk matter occurs at the density which minimizes $\epsilon(\rho)$.

c) Show that the chemical potential $\mu = E(N+1) - E(N)$ is given by $\mu = \epsilon + \frac{P}{\rho}$ Thus, for a self-bound saturating system at equilibrium density $\mu = \epsilon$.

d) Examine the stability with respect to long wavelength density fluctuations as follows. Consider dividing the volume V in half, and calculate the total energy if one half has density $\rho + \delta$ and the other half has density $\rho - \delta$. Show that this energy exceeds that of the uniform configuration, and thus the system is stable, if $\frac{\partial^2}{\partial \rho^2}(\rho\epsilon(\rho)) > 0$ which is equivalent to $\frac{\partial P}{\partial \rho} > 0$. When we subsequently examine stability with respect to finite wavelength fluctuations using the random phase approximation, this result will be recovered in the long wavelength limit.

PROBLEM 1.8 Uniform Fermi Gas A particularly simple stationary solution to the Hartree-Fock equations for a translationally invariant system is a Slater determinant of plane waves. In a finite box with periodic boundary conditions, the normalized wave functions have the form

$$\psi_{k_n s} = \frac{1}{\sqrt{L_1 L_2 L_3}} e^{i(k_{n_1} x_1 + k_{n_2} x_2 + k_{n_3} x_3)} \chi_s$$

where $k_{n_i} = \frac{2\pi}{L_i} n_i$ and χ_s is a spinor representing spin, isospin, color, or any other non-spatial quantum number. If the degeneracy associated with the non-spatial quantum numbers is denoted n_s, the lowest $M = \frac{N}{n_s}$ momentum states are occupied. In the continuum limit in D dimensions, $\sum_{n=1}^{M} F(k_n) \to (\frac{L}{2\pi})^D \int^{k_F} d^D k F(k)$ and one simply integrates over all $\vec{k}$ less than the Fermi momentum k_f, which is specified by the density

$$\rho = \frac{N}{(L)^D} = n_s \int_0^{k_F} \frac{d^D k}{(2\pi)^D} \quad .$$

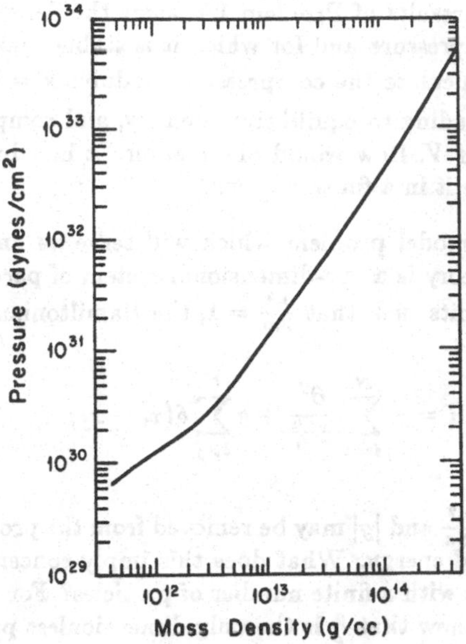

Fig. 1.1 Equation of state of neutron star matter

a) Consider a non-interacting gas, for which $\langle\psi|T|\psi\rangle$ is the exact energy. Find the energy per particle as a function of density, for non-relativistic and ultra-relativistic spin $\frac{1}{2}$ fermions in three dimensions.

b) As a concrete physical example, calculate the pressure (dynes/cm^2) as a function of mass density (gm/cm^3) for a non-relativistic neutron gas. Taking into account the fact that weak interactions allow a proton and electron to form a neutron (and a neutrino which escapes) and thus $E_F|_e + E_F|_p + m_e + m_p = E_F|_n + m_n$, calculate the pressure versus mass density for a neutral mixture of non-relativistic protons and neutrons and ultra-relativistic electrons. Assess the validity of this simple approximation for $10^{11}\frac{gm}{cc} < \rho < 10^{14}\frac{gm}{cc}$ and compare your results for the neutron gas and the neutron-proton-electron gas with the complete microscopic calculation including nuclear forces shown in the graph in Fig. 1.1.

c) Consider a D-dimensional system of Fermions with spin degeneracy $n_s = 2S+1$ and mass m interacting with the two-body potential $v(\vec{r}_i - \vec{r}_j) = -\alpha\delta(\vec{r}_i - \vec{r}_j)$, where $\alpha > 0$. Calculate the Fermi gas energy per particle $\frac{\langle H\rangle}{N} \equiv \epsilon(n)$ in the thermodynamic limit. Determine whether or not the system is stable in 1, 2, and 3 spatial dimensions.

d) Now, add a repulsive three-body potential $v(\vec{r}_i, \vec{r}_j, \vec{r}_k) = \beta\delta(\vec{r}_i - \vec{r}_j)\delta(\vec{r}_i - \vec{r}_k)$ with $\beta > 0$ and calculate $\epsilon(n)$ in three dimensions. How does this compare with the results in (c) when $n_s = 2$? Explain physically.

Explain why the case $n_s = 4$ is relevant to nuclear matter, that is, a system having equal densities of neutrons and protons interacting with nuclear forces but no Coulomb forces. Find values for α and β such that the minimum in $\epsilon(\rho)$ occurs at the experimental values $\rho_0 = 0.16 fm^{-3}$ and $\epsilon(\rho_0) = -16 MeV$. Sketch $\epsilon(\rho)$ for

these values. Using the results of Problem 1.7, show the density regions for which the system has positive pressure and for which it is stable against long wavelength density fluctuations. Calculate the compression modulus $\kappa = k_F^2 \frac{\partial^2 \epsilon}{\partial k_F^2}$ at the Fermi momentum k_F corresponding to equilibrium density, and compare with the experimental value $\kappa = 200 MeV$. How would one measure κ in a bucket of liquid ^{3}He? How would one measure it in a finite nucleus?

PROBLEM 1.9* **A model problem which will serve as an example for many topics in many-body theory is a one-dimensional system of particles interacting via δ−function forces. In units such that $\frac{\hbar^2}{2m} = 1$, the Hamiltonian is**

$$H = -\sum_{i=1}^{N} \frac{\partial^2}{\partial x_i^2} + g\sum_{i>j}^{N} \delta(x_i - x_j) \ . \tag{1}$$

a) Show that both $\frac{\hbar^2}{2m}$ and $|g|$ may be removed from the problem by appropriate scaling of the length and energy. What does this imply concerning a perturbation expansion in g for states with a finite number of particles? For uniform translationally invariant systems, show that $\frac{g}{\rho}$ is the only dimensionless parameter. Note that high density corresponds to the weak coupling limit.

b) Solve for the exact bound states for an attractive interaction as follows. Note that the totally symmetric function of the form $\exp\left(-\alpha \sum_{i<j} |x_i - x_j|\right)$ satisfies a free Schrödinger equation when the x's coincide. Thus obtain the following ground state wave functions and energies for any N bosons and for $N < n_s$ Fermions where n_s is the degeneracy $2S + 1$ for some additional quantum number we will call spin:

$$\psi_{\rm Boson}(x_1 \ldots x_N) = Ce^{\frac{-g}{4}\sum_{i<j}^{N} |x_i - x_j|},$$

$$\psi_{\rm Fermi}(x_1 \ldots x_N) = \psi_{\rm Bose}\chi,$$

and

$$E_N = \frac{-g^2}{48}(N^3 - N) \tag{2}$$

where χ denotes a totally antisymmetric spin wave function (spin singlet for $N = 2S + 1$) and C is a normalization constant. Note the behavior of the size and energy as $N \to \infty$. Do you expect there to be any Fermion bound states for $N > 2S + 1$? This Hamiltonian was first solved exactly by H.A. Bethe, (1931) and the wavefunction is often referred to as the Bethe *ansatz*.

c) For an attractive interaction and N Bosons or $N \leq n_s$ Fermions, solve for the approximate bound state in the Hartree-Fock approximation (see Problem 1.6). Note that the Hartree-Fock equation

$$\left[-\frac{\partial^2}{\partial x^2} - g(N-1)|\phi(x)|^2\right]\phi(x) = \epsilon\phi(x) \tag{3}$$

is the well-known cubic Schrödinger equation with localized solutions of the form $\frac{A}{\cosh(Bx)}$. Hence, show that

$$\phi(x) = \frac{(\frac{1}{8}g(N-1))^{\frac{1}{2}}}{\cosh(\frac{1}{4}g(N-1)x)} \tag{4}$$

$$E_{HF} = -\frac{g^2}{48}(N^3 - 2N^2 + N) \ .$$

Note that E_{HF} agrees with the exact E to leading order in N. For those who enjoy challenging integrals, it is interesting to calculate the exact one-body density distribution from the exact wave function (2) and show that it agrees with the Hartree-Fock density

$$\rho_{HF}(x) = \frac{\frac{1}{8}N(N-1)g}{\cosh^2(\frac{1}{4}g(N-1)x)} \tag{5}$$

to leading order in N. (The integral is done by F. Calogero and A. Degasperis (1975).) These two results which agree to leading order in N are special cases of the general result proved later that the mean field approximation gives the leading term in a $\frac{1}{N}$ expansion. Since the scaling argument in (a) shows that there is no expansion parameter in the Hamiltonian, $\frac{1}{N}$ is the only possible expansion parameter for the bound state problem.

d) Note that the energy in part (c) arises from the use of a variational wave function which is not translationally invariant, so that some spurious center of mass motion is included. A better approximation is to minimize the intrinsic Hamiltonian

$$H_{\text{intrinsic}} = H - H_{CM} \tag{6}$$

where, in units such that $\frac{\hbar^2}{2m} = 1$,

$$H_{CM} = \frac{1}{N}\left(\sum_{i=1}^{N} P_i\right)^2 = \frac{1}{N}\sum_i P_i^2 + \frac{1}{N}\sum_{i\neq j} P_i P_j \ . \tag{7}$$

Explain why the P_iP_j term vanishes for the permanents or determinants we are using and show that the Hartree-Fock energy obtained by minimizing the intrinsic Hamiltonian is $E_{HF}^{INT} = -\frac{g^2}{48}(N^3 - N^2)$, thereby eliminating one half of the $\mathcal{O}(\mathcal{N}^2)$ discrepancy.

e) Translationally invariant solutions. Using the result for a Fermi gas wave-function obtained in Problem 8 part (c), show that for Fermions with spin degeneracy $n_s = 2S + 1$ the energy per particle is

$$\frac{E}{N} = \frac{k_F^2}{3} + \frac{gk_F S}{\pi} \ . \tag{8}$$

Note that for an attractive g, $\frac{E}{N}$ has a minimum at $k_{min} = \frac{3gS}{2\pi}$ with $\frac{E}{N}|_{min} = -\frac{3|g|^2S^2}{4\pi^2}$. Compare this Hartree Fock energy per particle with that of a condensed

phase comprised of a dilute (and thus non-interacting) gas of Hartree Fock bound states, each containing $2S+1$ particles (see part (d) above). Show that the ratio is

$$\frac{\frac{E}{N}\big|_{\text{uniform}}}{\frac{E}{N}\big|_{\text{bound state}}} = \left(\frac{3}{\pi}\right)^2 \frac{1}{1+\frac{1}{2S}} < .91$$

so that in this approximation it is energetically favorable for the uniform phase to break up into clusters. Using the criterion in Problem 1.7, show that the uniform solution is stable against long wavelength fluctuations for all $k_F > \frac{gS}{\pi}$ which includes the equilibrium density. (We will see later that is unstable with respect to fluctuations of wavelength comparable to the size of the bound state.)

The following references treat the δ–function problem in greater detail. The exact bound states are discussed by J. McGuire (1965) and the complete S-matrix is calculated by C.N. Yang (1967). Time dependent mean field solutions and perturbative corrections to the Hartree-Fock approximation are given by B. Yoon and J.W. Negele (1977) and the model has been solved using functional integral techniques by C.R. Nohl (1976). Many of the salient results for the model and the fact that it is the non-relativistic limit of the sigma model are discussed in Section VI.A of the review by J.W. Negele (1982).

CHAPTER 2

GENERAL FORMALISM AT FINITE TEMPERATURE

2.1 INTRODUCTION

The ultimate objective of the quantum theory of many-particle systems is to understand experimentally observable properties of a diverse range of physical systems. Techniques will eventually be presented which are suitable for small finite systems as well as extensive macroscopic systems, for observables in isolated systems at zero temperature as well as finite-temperature systems. Once the appropriate theoretical groundwork has been established, specific experimental observables in a variety of systems will be treated in detail.

The intent of the present chapter, however, is to present the essence of the general theory in the particularly simple and convenient case of systems at finite temperature which may be treated in the grand canonical ensemble. To place the subsequent formalism in proper perspective, quantum statistical mechanics is reviewed, the observables and response functions accessible experimentally are summarized, and the general strategy of systematic approximations is discussed. Feynman path integrals are presented and functional integral techniques are developed for the many-body problem. Perturbation theory is derived and detailed rules are presented for Feynman diagrams. Finally, generating functions are derived for Green's Functions and irreducible diagrams, and the stationary-phase approximation is applied to the functional integral.

QUANTUM STATISTICAL MECHANICS

In quantum statistical mechanics, the themodynamic properties of a system in equilibrium are specified by the assumption of the equal occupation of accessible states. The physical problem is idealized by considering an appropriate ensemble of identical systems, and the probability of observing a particular state is given by the ratio of the number of systems in the ensemble in that state to the total number of systems in the ensemble.

Three ensembles are commonly used. The microcanonical ensemble describes an isolated system, which has fixed energy E and fixed particle number. The probability of observing a state of energy $E' \neq E$ is zero and that of observing a state of energy $E' = E$ is constant, equal to the inverse of the total number of states of energy E. The canonical ensemble describes a system of fixed particle number in equilibrium with a thermal reservoir with which it may exchange energy at a specified temperature. By enumerating the number of states in which the system and the reservoir combined have a fixed total energy, it follows that the probability of observing this system alone with energy E is proportional to $e^{-\frac{E}{kT}}$ where k is Boltzmann's constant. The average energy of the system is fixed and is controlled by the temperature T. The grand canonical ensemble describes a system in equilibrium with a particle reservoir, with which it may exchange particles, as well as a thermal reservoir. Enumeration of the states in which the system and reservoir have fixed total energy and particle number leads to the result that the probability of observing the system alone with energy E and particle number N is proportional to $e^{-\frac{(E-\mu N)}{kT}}$. In addition to the average energy of the system, the average particle number is now fixed and is controlled by the chemical potential μ.

The thermodynamic limit for a physical system of N particles contained in a volume V is defined by taking the limit as N and V go to infinity such that the ratio $\rho = \frac{N}{V}$ remains constant. An essential result of thermodynamics is that the three ensembles are equivalent for describing the macroscopic properties of most system in the thermodynamic limit. One is then free to select the ensemble on the basis of formal convenience, and as in the case of classical statistical mechanics, the grand canonical ensemble offers the greatest formal simplicity.

The principal case in which the three ensmbles yield nonequivalent results arises when some observable has divergent fluctuations in one ensemble and is constrained not to fluctuate in another ensemble. This situation occurs in physical systems composed of a mixture of phases such as a liquid-gas transition or Bose condensation. If one looks at a subsystem, the fact that droplets of one phase can move in the other phase is responsible for the divergence of the fluctuations of the particle number in the grand canonical ensemble, whereas these fluctuations are zero by definition in the canonical ensemble.

Given the fact that the probability of observing a state of energy E and particle number N is proportional to $e^{-\beta(E-\mu N)}$, where $\beta = \frac{1}{kT}$, the thermal average of an operator $\hat{R}$ may be expressed as

$$\langle \hat{R} \rangle = \frac{\sum_\alpha \langle \psi_\alpha | e^{-\beta(\hat{H}-\mu\hat{N})} \hat{R} | \psi_\alpha \rangle}{\sum_\alpha \langle \psi_\alpha | e^{-\beta(\hat{H}-\mu\hat{N})} | \psi_\alpha \rangle} \tag{2.1}$$

where $\{|\psi_\alpha\rangle\}$ denotes an orthonormal basis of the Fock space and $\hat{N}$ is the particle number operator, Eq. (1.92). It is convenient to define the partition function, Z,

$$Z = \mathrm{Tr}\, e^{-\beta(\hat{H}-\mu\hat{N})} \tag{2.2}$$

and the grand canonical potential, Ω

$$e^{-\beta\Omega} = Z\,. \tag{2.3}$$

By explicit differentiation of the function $\Omega(\mu, V, T)$ and use of Eqs. (2.1– 2.3) the familiar thermodynamic relations may be established:

$$\frac{\partial\Omega}{\partial\mu} = -N \tag{2.4a}$$

$$\frac{\partial\Omega}{\partial V} = -P \tag{2.4b}$$

$$\frac{\partial\Omega}{\partial T} = \frac{\Omega - \mathcal{U} - \mu N}{T} = -S \tag{2.4c}$$

where P is the pressure, $\mathcal{U}$ is the internal energy, and S is the entropy. The Gibbs-Duhem relation

$$\mathcal{U} = TS - PV + \mu N \tag{2.5}$$

follows from the fact that the internal energy $\mathcal{U}(N,V,S)$ is an extensive function of extensive variables and may be combined with Eq. (2.4c) to yield

$$\Omega = -PV \ . \tag{2.6}$$

Since the thermodynamic properties of a system at equilibrium are specified by Ω and derivatives thereof, one of the immediate tasks in this chapter will be to develop methods to calculate the grand potential, Ω. The techniques developed for Ω will also apply directly to thermal averages of operators, Eq. (2.1), allowing the calculation of any equilibrium properties of interest.

PHYSICAL RESPONSE FUNCTIONS AND GREEN'S FUNCTIONS

Before embarking on the development of the general formalism, it is valuable to specify conveniently calculable quantities which characterize the range of experimentally accessible observables to be addressed in physical systems. In addition to measuring the equilibrium properties discussed above, experimentalists learn about physical systems by measuring their response to a diversity of external probes. The results of such measurements are conveniently expressed in terms of response functions or Green's functions. The relation of specific experimental observables to these quantities will be derived in detail in later chapters, and in the present section we simply present general arguments to show how they arise naturally from experiments.

Consider a system which at initial time t_i is in an eigenstate $|\Psi_\alpha(t_i)\rangle$ of the hamiltonian $\hat{H}$. Subsequently, let it be subject to a time-dependent external field

$$\hat{H}_U(t) = \hat{H} + U(t)\hat{\mathcal{O}}_1 \tag{2.7}$$

where the field couples to the system through an operator denoted $\hat{\mathcal{O}}_1$ and the states and operators are in the Schrödinger representation.

A convenient representation of the evolution operator for a system with a time-dependent hamiltonian is given by a time-ordered exponential. The time-ordered product of a set of time-dependent creation and annihilation operators, denoted $\{\mathcal{O}_\alpha\}$, is defined

$$T[\mathcal{O}_{\alpha_1}(t_1)\mathcal{O}_{\alpha_2}(t_2)\ldots\mathcal{O}_{\alpha_n}(t_n)] = \varsigma^P \mathcal{O}_{\alpha_{P1}}(t_{P1})\mathcal{O}_{\alpha_{P2}}(t_{P2})\ldots\mathcal{O}_{\alpha_{Pn}}(t_{Pn}) \tag{2.8}$$

where according to our standard convention ς is -1 or 1 for Fermions or Bosons respectively and P is the permutation of $\{1,2,\ldots n\}$ which orders the times chronologically with the latest time to the left:

$$t_{P1} > t_{P2} > \ldots > t_{Pn}$$

and which orders creation operators to the left of annihilation operators (normal order) at equal times. The time-ordered exponential is defined by

$$T\, e^{-\int_{t_a}^{t_b} dt A(t)} = \lim_{M\to\infty} e^{-\epsilon A(t_M)} e^{-\epsilon A(t_{M-1})} \ldots e^{-\epsilon A(t_1)} e^{-\epsilon A(t_0)} \tag{2.9}$$

where $\epsilon = \frac{t_b - t_a}{M}$ and $t_n = t_a + n\epsilon$. As shown in Problem 2.3, it may be expanded in a Taylor series as follows:

$$T\, e^{-\int_{t_a}^{t_b} dt A(t)} = \sum_{n=0}^{\infty} \frac{(-1)^n}{n!} \int_{t_a}^{t_b} dt_1 \dots dt_n T[A(t_1)\dots A(t_n)] \ . \tag{2.10}$$

Using this time-ordered exponential, the evolution operator may be written

$$\hat{\mathcal{U}}(t, t_i) = T\, e^{-i\int_{t_i}^{t} \hat{H}_U(t')dt'} \tag{2.11}$$

since it satisfies the equation of motion

$$\begin{aligned} \frac{d}{dt}\hat{\mathcal{U}}(t, t_i) &= T(-i\hat{H}_U(t)\, e^{-i\int_{t_i}^{t} \hat{H}_U(t')dt'}) \\ &= -i\hat{H}_U(t)\hat{\mathcal{U}}(t, t_i) \end{aligned} \tag{2.12a}$$

and the boundary condition

$$\hat{\mathcal{U}}(t_i, t_i) = 1 \ . \tag{2.12b}$$

The response of a wave function originally in eigenstate $|\psi_\alpha\rangle$ at time t_i to an infinitesimal perturbation by an external field acting between time t_i and t is thus given by the functional derivative

$$\begin{aligned} \delta|\psi(t)\rangle &= \int_{t_i}^{t} dt_1 \delta U(t_1) \left.\frac{\delta\hat{\mathcal{U}}(t, t_i)}{\delta U(t_1)}\right|_{U=0} |\psi_\alpha(t_i)\rangle \\ &= -i\int_{t_i}^{t} dt_1 \delta U(t_1) T e^{-i\int_{t_1}^{t} \hat{H}_U(t')dt'} O_1 T\, e^{-i\int_{t_i}^{t_1} \hat{H}_U(t')dt'}\Big|_{U=0} |\psi_\alpha(t_i)\rangle \\ &= -i\int_{t_i}^{t} dt_1 \delta U(t_1) e^{-iH(t-t_1)} O_1 e^{-iH(t_1 - t_i)} |\psi_\alpha(t_i)\rangle \\ &= -i\int_{t_1}^{t} dt_1 \delta U(t_1) e^{-iHt} O_1^{(H)}(t_1) |\psi_\alpha^{(H)}\rangle \end{aligned} \tag{2.13}$$

where the operator $\hat{O}_1^{(H)}(t)$ in the Heisenberg representation is related to the operator $\hat{O}_1$ in the Schrödinger representation by

$$\hat{O}_1^{(H)}(t) \equiv e^{iHt}\hat{O}_1\, e^{-iHt} \tag{2.14a}$$

and the state $|\psi^{(H)}\rangle$ in the Heisenberg representation is related to the state $|\psi(t)\rangle$ in the Schrödinger representation by

$$|\psi^{(H)}\rangle \equiv e^{iHt}\, |\psi(t)\rangle \ . \tag{2.14b}$$

Finally, consider the expectation value of an operator O_2 evaluated at time t_2 in the state $|\psi(t_2)\rangle$. The response of this expectation value to an infinitesimal perturbation in the external field is given by

$$\begin{aligned}
\delta\langle\psi(t_2)|\hat{O}_2|\psi(t_2)\rangle &= -i\int_{t_i}^{t_2} dt_1 \delta U(t_1)\langle\psi_\alpha^{(H)}|\left[\hat{O}_2^{(H)}(t_2),\hat{O}_1^{(H)}(t_1)\right]|\psi_\alpha^{(H)}\rangle \\
&= -i\int_{-\infty}^{\infty} dt_1 \delta U(t_1)\theta(t_2-t_1)\langle\psi_\alpha^{(H)}|\left[\hat{O}_2^{(H)}(t_2),\hat{O}_1^{(H)}(t_1)\right]|\psi_\alpha^{(H)}\rangle
\end{aligned} \tag{2.15}$$

where a θ function has been inserted to extend the upper limit of the time integral and the lower limit t_i has been extended to $-\infty$ to include all possible variations of $U(t_1)$. This equation may be weighted by the Boltzmann factor $Z^{-1}\, e^{-(E_\alpha-\mu N_\alpha)}$ and summed over a complete set $\{\alpha\}$. Hence, the response of a measurement of $\langle\hat{O}_2(t_2)\rangle$ to a perturbation coupled to $\hat{O}_1$ is specified by the response function

$$\begin{aligned}
D(1,2) &\equiv \frac{\delta\langle\hat{O}_2(t_2)\rangle}{\delta U(t_1)} \\
&= -i\theta(t_2-t_1)\langle[\hat{O}_2^{(H)}(t_2),\hat{O}_1^{(H)}(t_1)]\rangle
\end{aligned} \tag{2.16}$$

where the brackets denote the thermal average, Eq. (2.1).

The physical response of a system to an external potential is thus characterized by correlation functions of the form $\langle\hat{O}_2^{(H)}(t_2)\hat{O}_1^{(H)}(t_1)\rangle$. This result that a transport coefficient characterizing the dissipation in a system is specified by a matrix element of the thermodynamic or ground state fluctuations of an operator is often called the fluctuation-dissipation theorem. For example consider measuring the magnetization of a spin system in the presence of a time and spatially varying magnetic field. Since the magnetic field couples to the spin through $\hat{\vec{\sigma}}(x)\cdot\vec{H}(x)$, the operators $\hat{O}_1$ and $\hat{O}_2$ are spin operators $\hat{\vec{\sigma}}$ and the response function is given in terms of the spin-spin correlation function $\langle\hat{\vec{\sigma}}(x_1,t_1)\hat{\vec{\sigma}}(x_2,t_2)\rangle$. Fourier transforming to momentum and frequency then directly specifies the dynamic magnetic susceptibility $\chi(k,\omega)$. Similarly, an electromagnetic field couples to a system of charged particles though the vector potential $\hat{\vec{j}}\cdot\vec{A}$, and in a gauge in which $\Phi=0, \vec{E}=-\frac{\partial}{\partial t}\vec{A}$. Thus, the response of the current to a variation in the electric field, that is the electrical conductivity, is given by the current-current correlation function $\langle\hat{\vec{j}}(x_1,t_1)\,\hat{\vec{j}}(x_2,t_2)\rangle$. This argument is evidently quite general, and a large class of experiments in which a system is subjected to a weak controllable external probe is characterized by simple correlation functions of operators at different times.

In addition to these experiments involving an external potential, there is an important class of inclusive scattering experiments used to study many-body systems which may be characterized similarly by correlation functions. Consider an external particle which interacts with the constituents of a system through a weak potential v. Let a particle with initial momentum $\vec{k}$ scatter from the system in initial state $|\psi_\alpha\rangle$ transferring energy ω and momentum $\vec{q}$ leading to a final state $|\psi_\beta\rangle$ and final particle momentum $\vec{k}-\vec{q}$. The matrix element for this transition is

$$
\begin{aligned}
T_{\alpha\beta} &= \int dx_1\dots dx_n dx \psi^*_\beta(x_1,\dots x_n)e^{-i(\vec{k}-\vec{q})\cdot\vec{x}}\sum_i v(x-x_i)e^{i\vec{k}\cdot\vec{x}}\psi_\alpha(x_1\dots x_n) \\
&= \int dx dy e^{i\vec{q}\cdot\vec{x}} v(x-y)|\langle\psi_\beta|\hat{\rho}(y)|\psi_\alpha\rangle \\
&= \tilde{v}(-q)\langle\psi_\beta|\hat{\rho}(-q)|\psi_\alpha\rangle
\end{aligned} \tag{2.17}
$$

where $\tilde{v}(q)=\tilde{v}(-q)$ is the Fourier transform of the potential $v(|x|)$ and the Fourier transform of the density operator is defined by

$$
\hat{\rho}(k)=\int e^{-i\vec{k}\cdot\vec{x}}\hat{\rho}(x)dx \ . \tag{2.18}
$$

If the interaction is weak, the inclusive cross section may be calculated in Born approximation by summing over the unobserved states $|\psi_\beta\rangle$ of the system having energy $E_\beta=E_\alpha+\omega$

$$
\sigma(q,\omega)=2\pi\sum_\beta\delta(E_\beta-E_\alpha-\omega)\tilde{v}(q)^2|\langle\psi_\beta|\rho(-q)|\psi_\alpha\rangle|^2 \ . \tag{2.19}
$$

Writing the δ-function in terms of a time integral and using completeness, this may be recast in the following form

$$
\begin{aligned}
\sigma(q,\omega) &= \tilde{v}(q)^2\int dt e^{i\omega t}\sum_\beta\langle\psi_\alpha|e^{iE_\alpha t}\hat{\rho}(q)e^{-iE_\beta t}|\psi_\beta\rangle\langle\psi_\beta|\hat{\rho}(-q)|\psi_\alpha\rangle \\
&= \tilde{v}(q)^2\int dt e^{i\omega t}\int dx dy e^{-iq(x-y)}\langle\psi_\alpha|e^{iHt}\hat{\rho}(x)e^{-iHt}\hat{\rho}(y)|\psi_\alpha\rangle \ .
\end{aligned} \tag{2.20}
$$

Summing over a complete set of states $|\psi_\alpha\rangle$ with thermal weighting factors yields the desired result that the inclusive cross section is specified by the correlation function $\langle\hat{\rho}^{(H)}(x,t)\hat{\rho}^{(H)}(y,o)\rangle$.

The inclusive cross section or structure function $\sigma(q,\omega)$ has been measured for a variety of systems of physical interest. Striking examples which will be discussed in Chapter 5 are liquid He^3 and He^4, for which the combination of neutron scattering and X-ray scattering provide high-precision results over a wide dynamical range. From Eq. (2.20), the integral $\int d\omega\sigma(q,\omega)$ yields the Fourier transform of the two-body correlation function $\langle\hat{\rho}(x)\ \hat{\rho}(y)\rangle$ which has now been measured to impressively high accuracy for liquid He. Equation (2.20) applies equally well to a finite system at zero temperature in its ground state, and extensive high-precision data are now available from inclusive electron scattering from finite nuclei.

From these considerations, we are led to the conclusion that a wide range of observables of direct experimental interest may be expressed in terms of the thermal average of products of operator at different times. This motivates the definition of Green's functions, which on one hand are the thermal averages of time-ordered products of operators that are the most convenient to calculate in perturbation theory and on the other hand can be related by suitable analytic continuation to the quantities arising from experimental observables.

The n-body real-time Green's function is defined by

$$\mathcal{G}^{(n)}(\alpha_1 t_1, \ldots \alpha_n t_n | \alpha'_1 t'_1, \ldots \alpha'_n t'_n) \qquad (2.21)$$
$$= (-i)^n \langle T[a^{(H)}_{\alpha_1}(t_1) \ldots a^{(H)}_{\alpha_n}(t_n) a^{(H)^\dagger}_{\alpha'_n}(t'_n) \ldots a^{(H)^\dagger}_{\alpha'_1}(t'_1)]\rangle$$

where $a^{(H)}_\alpha(t)$ and $a^{(H)^\dagger}_\alpha(t)$ denote annihilation and creation operators in the Heisenberg representation, Eq. (2.14), T denotes a time-ordered product, and the brackets denote a thermal average. Aside from θ-functions restricting relative times arising from the time-ordered product, all the response functions discussed above may be expressed in terms of two-particle Green's functions with suitable choice of the time arguments. We will demonstrate in Chapter 5 that when $\mathcal{G}$ is Fourier transformed from time to frequency ω, the effect of the θ function is merely to displace poles infinitesimally above or below the real axis and that physical observables may be obtained staightforwardly from Green's functions.

When one writes a real-time Green's function explicitly in terms of the thermal weighting factor $e^{-\beta(H-\mu N)}$ and Heisenberg operators $e^{iHt}a_\alpha e^{-iHt}$, the simultaneous appearance of real and imaginary times multiplying H is formally inconvenient. It is therefore advantageous to define thermal, or imaginary-time, Green's functions as follows:

$$\mathcal{G}^{(n)}(\alpha_1 \tau_1, \ldots \alpha_n \tau_n | \alpha'_1 \tau'_1, \ldots \alpha'_n \tau'_n) \qquad (2.22)$$
$$= \langle T[a^{(H)}_{\alpha_1}(\tau_1) \ldots a^{(H)}_{\alpha_n}(\tau_n) a^{(H)^\dagger}_{\alpha'_n}(\tau'_n) \ldots a^{(H)^\dagger}_{\alpha'_1}(\tau'_1)]$$

where $a^{(H)^\dagger}_\alpha(\tau)$ and $a^{(H)}_\alpha(\tau)$ are the imaginary-time Heisenberg representation of $a^\dagger_\alpha$ and a_α

$$a^{(H)^\dagger}_\alpha(\tau) = e^{\tau(\hat{H}-\mu\hat{N})} a^\dagger_\alpha e^{-\tau(\hat{H}-\mu\hat{N})} \qquad (2.23a)$$
$$a^{(H)}_\alpha(\tau) = e^{\tau(\hat{H}-\mu\hat{N})} a_\alpha e^{-\tau(\hat{H}-\mu\hat{N})} \qquad (2.23b)$$

and T now orders operators in imaginary time. Note that $a^\dagger_\alpha(\tau)$ and $a_\alpha(\tau)$ are not Hermitian adjoints. In Chapter 5 we will also show how real-time Green's functions can be obtained easily from imaginary-time Green's functions by analytic continuation.

The essential conclusion of this section is that the physical observables of interest may be obtained by analytic continuation of imaginary-time Green's functions. We will subsequently show in the remainder of this chapter how thermal averages of time-ordered products of Heisenberg operators in imaginary time emerge naturally from path integrals and how to evaluate them in perturbation theory.

APPROXIMATION STRATEGIES

We will presently develop a number of techniques which formally appear to provide a systematic sequence of successive approximations for many-particle systems. To maintain a realistic perspective, however, we should consider at the outset what level

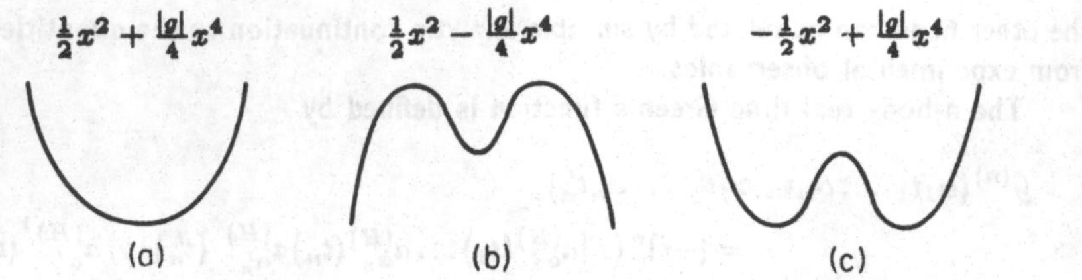

Fig. 2.1 Three qualitative cases for the quadratic plus quartic potential.

of mathematical rigor one can actually expect and how in practice one may select and assess an approximation scheme.

The most familiar systematic approximation scheme from elementary quantum mechanics is perturbation theory. It is based on the belief that the behavior of a physical system is continuous in some "small" parameter describing the difference between a solvable problem and the actual system. It is crucial to note, however, that in general perturbation theory yields an asymptotic rather than convergent series. Under appropriate circumstances, it may yield a useful physical approximation, but it cannot be applied systematically to abritrary precision.

The salient features of asymptotic expansions for our present purposes may be illustrated by the simple integral

$$Z(g) = \int \frac{dx}{\sqrt{2\pi}} e^{-\frac{x^2}{2} - \frac{g}{4}x^4} \tag{2.24}$$

corresponding to the classical partition function $Z = \int dx e^{-V(x)}$ for a particle in a quadratic plus quartic potential

$$V(x) = \frac{x^2}{2} + \frac{g}{4}x^4 \quad . \tag{2.25}$$

We will subsequently see that this integral is highly suggestive of the path integral describing the quantum mechanics of a particle in the same potential. Depending on the signs of the quadratic and quartic terms, the potential has three regimes of qualitatively different behavior sketched in Fig. 2.1

Physically, the classical or quantum behavior in this potential changes completely when g changes sign, since a particle is localized in (a) and not in (b). Thus, we expect that the theory is nonanalytic at $g = 0$ and that an expansion in powers of g has zero radius of convergence. This physical behavior is reflected in the integral $Z(g)$ which diverges for $g < 0$ and is thus non-analytic at $g = 0$.

The perturbation series for $Z(g)$ is easily obtained by expanding $e^{-\frac{g}{4}x^4}$, with the result

$$Z(g) = \sum_n g^n Z_n \tag{2.26a}$$

with

$$\begin{aligned} g^n Z_n &= \frac{(-g)^n}{n!} \frac{1}{4^n} \int \frac{dx}{\sqrt{2\pi}} e^{-\frac{x^2}{2}} x^{4n} \\ &= \frac{(-g)^n (4n-1)!!}{n!\, 4^n} \\ &= \frac{(-g)^n (4n)!}{n! 16^n (2n)!} \\ &\underset{n\to\infty}{\sim} \frac{1}{\sqrt{n\pi}} \left(\frac{4gn}{e}\right)^n \end{aligned} \tag{2.26b}$$

where the asymptotic behavior in the last line is obtained using Stirling's formula $n! = \sqrt{2\pi} n^{n+\frac{1}{2}} e^{-n}$. Since Z_n grows like $n!$, the series diverges as expected from the non-analyticity at $g = 0$. Later on when we think in terms of diagrams, we will understand the growth of Z_n in terms of the proliferation of the number of diagrams with increasing order and the factor $(4n-1)!!$ in the present example which overwhelms the factor $\frac{g^n}{n!}$ will be recognized as the number of diagrams generated in performing the Gaussian integral of x^{4n} using Wick's theorem.

The crucial point for our present discussion is the fact that, under appropriate circumstances, a finite number of terms of an asymptotic series may give an excellent approximation. The residual error after n terms is bounded by the $(n+1)^{st}$ term

$$\begin{aligned} R_n &\equiv \left| Z(g) - \sum_{m=0}^{n} g^m Z_m \right| \\ &= \int \frac{dx}{\sqrt{2\pi}} e^{-\frac{x^2}{2}} \left| e^{-\frac{g}{4}x^4} - \sum_{m=0}^{n} \frac{1}{m!} \left(-\frac{gx^4}{4}\right)^m \right| \\ &\leq \int \frac{dx}{\sqrt{2\pi}} e^{-\frac{x^2}{2}} \frac{1}{(n+1)!} \left(\frac{gx^4}{4}\right)^{n+1} = g^{n+1} |Z_{n+1}| \end{aligned} \tag{2.27}$$

so the approximation continues to improve as long as $g^n Z_n$ decreases. Figure 2.2 shows R_n and $g^n |Z_n|$ as a function of n for several values of the coupling constant g. To assess the precision, one should compare the minimum error with the total shift due to g, $|Z(g) - Z(0)| = R_0$. As g decreases, the minimum error decreases and occurs at larger values of n so that at $g = .01$, for example, one attains the extremely high precision of $R_{25}/R_0 = 1.1 \times 10^{-10}$ after 25 terms. From the asymptotic expression for $g^n Z_n$, Eq. (2.26b), we note that the minimum occurs for $n \sim \frac{1}{4g}$, so that the minimum error is $g^n Z_n\big|_{\min} \sim \sqrt{4g/\pi} e^{-\frac{1}{4g}}$. This exponential dependence on the inverse coupling constant is characteristic of perturbation theory and indicates the exceedingly high precision obtainable from a divergent theory for weak enough coupling. Quantum electrodynamics is a good example of a theory in which such weak coupling is realized. As in our physical argument for the quartic potential, it is clear physically that electrodynamics is non-analytic at $\alpha = 0$ since negative $\alpha = \frac{e^2}{4\pi}$ would change the signs of all electromagnetic forces. Nevertheless, $\frac{1}{137}$ is so small that perturbation theory is adequate for all conceivable purposes.

The real problem comes at large coupling constant where the series begins to diverge in very low order. At $g = 0.1$, for example, one must stop at the third term

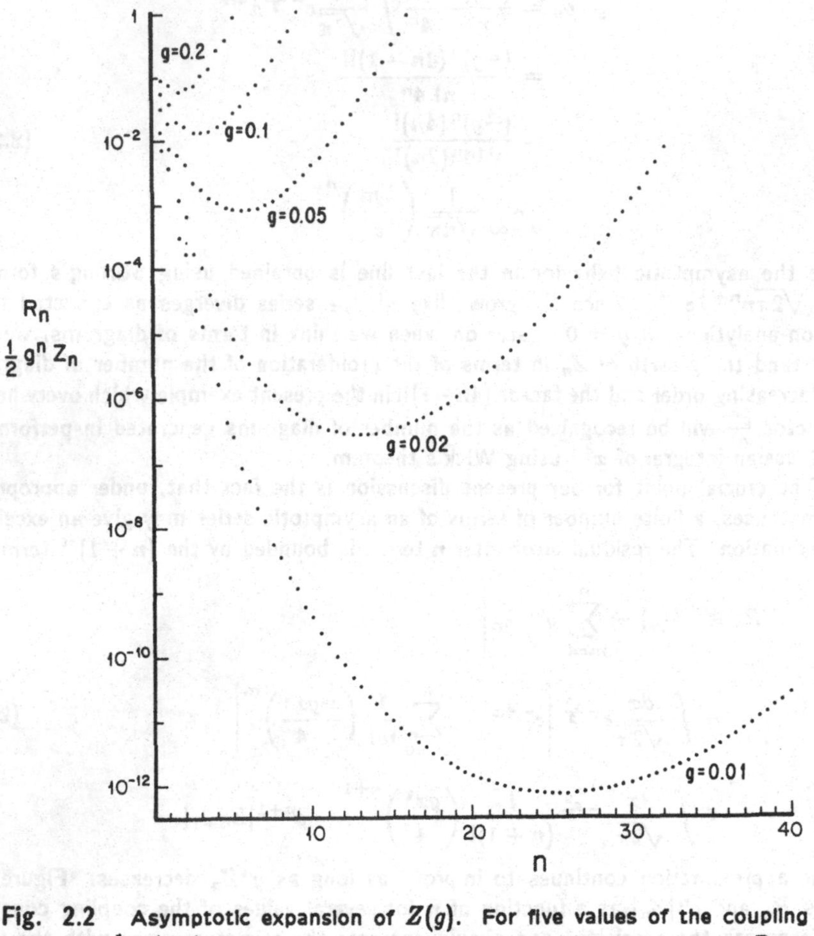

Fig. 2.2 Asymptotic expansion of $Z(g)$. For five values of the coupling constant, $\frac{1}{2}g^n|Z_n|$ is plotted at each integer n and the residual error R_n is plotted at each half integer $n+\frac{1}{2}$.

after only attaining an accuracy of 25%. If one wishes to do better, one must either extract the non-analytic part of $Z(g)$ using the Borel summation discussed in Section 7.5 or find an alternative expansion.

For other regimes of the potentials sketched in Fig. 2.1, other expansions are more appropriate than (2.26). For large positive g, it is more reasonable to make a strong coupling expansion by starting with the solution to the quartic potential and treating the quadratic potential as the perturbation. For case (c), the natural approach is the stationary phase approximation which will be discussed at length in Section 2.5. In this case, the partition function is written $\tilde{Z}(g) = \int \frac{dx}{\sqrt{2\pi}} e^{\frac{x^2}{2} - \frac{g}{4}x^4} = \int \frac{dy}{\sqrt{2\pi g}} e^{-\frac{1}{g}\left[-\frac{y^2}{2}+\frac{y^4}{4}\right]}$ and the exponent is expanded around its two maxima $y = \pm 1$ corresponding to the minimum of the potential.

The general features illustrated by the quadratic plus quartic potential are pertinent to the richer and more complicated case of the many-body problem. We will find that different approximations will address different aspects of the physics. Perturbation theory in the two-body interaction, while at best an asymptotic expansion, will serve to organize our thinking and elucidate much of the general structure of the problem. Various resummations will focus on different parts of the physics such as short-range or long-range correlations. The stationary phase approximation will address still other aspects of the problem such as large amplitude collective motion and tunnelling. In all of these expansions, there will be some formal expansion parameter which provides no real mathematical control on the problem. The most we can ask in practice, as in the example in Fig. 2.2, is that successive terms in the series of approximations decrease.

2.2 FUNCTIONAL INTEGRAL FORMULATION

Functional integrals provide a powerful tool for the study of many-particle systems. The partition function is represented by an integral over field configurations which provides both a physically intuitive description of the system and a useful starting point for approximations. Approximations which arise naturally from functional integrals include perturbation expansions, loop expansions around stationary solutions, approximations in terms of solitons or instantons, and stochastic approximations. Before proceeding to the general case of many-particle systems, it is instructive to illustrate the essential idea with Feynman path integrals. The essence of the path integral was introduced in a germinal paper by P.A.M. Dirac (1933) and developed extensively by R.P. Feynman (1948,1949,1950). All these historic papers are published in the reprint volume edited by J. Schwinger(1953).

THE FEYNMAN PATH INTEGRAL

For physical clarity, we will first introduce the Feynman path integral in real time. Subsequently, in order to represent the partition function, we shall perform an analytic continuation to imaginary time where this path integral is closely related to the Wiener integral and is mathematically well-defined.

Consider a matrix element of the evolution operator for a particle governed by the Hamiltonian $H(\hat{p},\hat{x})$

$$\mathcal{U}(x_f t_f, x_i t_i) = \langle x_f | e^{-\frac{i}{\hbar}\hat{H}(t_f - t_i)} | x_i \rangle \quad . \tag{2.28}$$

Whereas the matrix elements of the evolution operator cannot be evaluated exactly for finite time intervals, for infinitesimal time intervals they may be calculated to any desired degree of accuracy. Thus, the basic idea of the Feynman path integral is to break a finite time interval into infinitesimal steps, evaluate the evolution operator for each step, and chain the matrix elements together to obtain the result for the finite interval.

Let the time interval $t_f - t_i$ be divided into M equal steps of size ϵ

$$\epsilon = \frac{t_f - t_i}{M} \tag{2.29a}$$

with intermediate times denoted

$$t_n = t_i + (n-1)\epsilon \ . \tag{2.29b}$$

With this notation,

$$t_0 \equiv t_i \quad \text{and} \quad t_M \equiv t_f \tag{2.29c}$$

and it will be convenient to use the same convention for initial and final coordinates

$$x_0 \equiv x_i \quad \text{and} \quad x_M \equiv x_f \ . \tag{2.29d}$$

By inserting the closure relation Eq. (1.23) $M-1$ times, the matrix element of the evolution operator may be written:

$$\begin{aligned} \mathcal{U}(x_f t_f, x_i t_i) &= \langle x_f | \left(e^{-i\frac{\epsilon}{\hbar}\hat{H}}\right)^M | x_i \rangle \\ &= \int \prod_{k=1}^{M-1} dx_k \langle x_f | e^{-i\frac{\epsilon}{\hbar}\hat{H}} | x_{M-1}\rangle \langle x_{M-1} | e^{-i\frac{\epsilon}{\hbar}\hat{H}} | x_{M-2}\rangle \\ &\quad \times \langle x_{M-2} | \ldots e^{-i\frac{\epsilon}{\hbar}\hat{H}} | x_1 \rangle \langle x_1 | e^{-i\frac{\epsilon}{\hbar}\hat{H}} | x_i \rangle \ . \end{aligned} \tag{2.30}$$

The key step is to find an appropriate approximation for the matrix element of the infinitesimal evolution operator, which may be written

$$\langle x_n | e^{-i\frac{\epsilon}{\hbar}H(\hat{p},\hat{x})} | x_{n-1}\rangle = \int d^3p_n \langle x_n | p_n \rangle \langle p_n | e^{-i\frac{\epsilon}{\hbar}H(\hat{p},\hat{x})} | x_{n-1} \rangle \ . \tag{2.31}$$

For our purposes in obtaining a practical functional integral, we desire an approximation to the infinitesimal evolution operator $e^{-\frac{\epsilon}{\hbar}H(\hat{p},\hat{x})}$ which not only reproduces the exact evolution of a wave function in the limit $\epsilon \to 0$, but also yields acceptable results when acting on position and momentum eigenstates $|x\rangle$ and $|p\rangle$ as in Eq. (2.31). We obtain such an approximation, which generalizes directly to the subsequent treatment of coherent state functional integrals, by considering a form of normal-ordered exponential.

For operators expressed in terms of $\hat{p}$ and $\hat{x}$, we will define an operator to be in normal form when all the $\hat{p}$'s appear to the left of all the $\hat{x}$'s, and the result of reordering an operator $\mathcal{O}(\hat{p},\hat{x})$ into normal form will be denoted $:\mathcal{O}(\hat{p},\hat{x}):$. For example, the Hamiltonian for a single particle in a potential

$$H_v(\hat{p},\hat{x}) = \frac{\hat{p}^2}{2m} + V(\hat{x}) \tag{2.32}$$

is in normal form and

$$:e^{-i\frac{\epsilon}{\hbar}H_v(\hat{p},\hat{x})}: = \sum_{n=0}^{\infty} \left(-i\frac{\epsilon}{\hbar}\right)^n \sum_{k=0}^{n} \frac{1}{k!(n-k)!} \left(\frac{\hat{p}^2}{2m}\right)^k (V(\hat{x}))^{n-k} \ . \tag{2.33}$$

The Hamiltonian for a particle in a magnetic field described by a vector potential $A(x)$ may be rewritten in normal form as follows

$$\begin{aligned} H_A(\hat{p},\hat{x}) &= \frac{1}{2m}\left(\hat{\vec{p}} - \frac{e}{c}\vec{A}(\hat{x})\right)^2 \\ &= \frac{1}{2m}\left(\hat{\vec{p}}^2 - \frac{e}{c}2\hat{\vec{p}}\hat{A}(\hat{x}) - \frac{e}{c}i\vec{\nabla}\cdot\vec{A}(\hat{x}) + \left(\frac{e}{c}A(\hat{x})\right)^2\right) . \end{aligned} \tag{2.34}$$

For any $H(\hat{p},\hat{x})$ in normal form,

$$e^{-i\frac{\epsilon}{\hbar}H(\hat{p},\hat{x})} = :e^{-i\frac{\epsilon}{\hbar}H(\hat{p},\hat{x})}: - \left(\frac{\epsilon}{\hbar}\right)^2 \sum_{n=0}^{\infty}\frac{(-i\frac{\epsilon}{\hbar})^n}{(n+2)!}\left(H(\hat{p},\hat{x})^{n+2} - :[H(\hat{p},\hat{x})]^{n+2}:\right) \tag{2.35}$$

and for the special case Eq. (2.32), the leading correction is

$$-\frac{\epsilon^2}{2\hbar^2}\left[V,\frac{\hat{p}^2}{2m}\right] = -\frac{\epsilon^2}{4m\hbar^2}(V'' + 2iV'\hat{p}) . \tag{2.36}$$

Thus, if the infinitesimal evolution operator is approximated by the normal-ordered evolution operator, the error is of order ϵ^2 times an operator which may be expressed in terms of multiple commutators of the operators comprising the Hamiltonan. When acting upon a normalizable, differentiable wave function $\psi(x)$, the error term is ϵ^2 times a finite number, so that in the limit $\epsilon \rightarrow 0$, we are assured that $:e^{-\frac{\epsilon}{\hbar}H(\hat{p},\hat{x})}:$ yields the correct evolution of the wave function. Furthermore, in contrast to other approximations which are valid to first order in ϵ, the normal-ordered evolution operator may be used in the integral in Eq. (2.31).

$$\begin{aligned} \langle x_n|:e^{-i\frac{\epsilon}{\hbar}H(\hat{p},\hat{x})}:|x_{n-1}\rangle &= \int d^3p_n\langle x_n|p_n\rangle\langle p_n|:e^{-i\frac{\epsilon}{\hbar}H(\hat{p},\hat{x})}:|x_{n-1}\rangle \\ &= \int\frac{d^3p_n}{(2\pi\hbar)^3}e^{ip_n(x_n-x_{n-1})}e^{-i\frac{\epsilon}{\hbar}H(p_n,x_{n-1})} . \end{aligned} \tag{2.37}$$

For the case of a particle in a potential, Eq. (2.32), this integral over p is a Gaussian integral yielding

$$\begin{aligned} \langle x_n|e^{-\frac{\epsilon}{\hbar}\left(\frac{\hat{p}^2}{2m}+V(\hat{x})\right)}|x_{n-1}\rangle &= \langle x_n|:e^{-\frac{\epsilon}{\hbar}\left(\frac{\hat{p}^2}{2m}+V(\hat{x})\right)}:|x_{n-1}\rangle + O(\epsilon^2) \\ &= \int\frac{d^3p}{(2\pi\hbar)^3}e^{i\frac{p}{\hbar}(x_n-x_{n-1})-i\frac{\epsilon}{\hbar}\frac{p^2}{2m}-i\frac{\epsilon}{\hbar}V(x_{n-1})} + O(\epsilon^2) \\ &= \left(\frac{m}{2\pi i\epsilon\hbar}\right)^{\frac{3}{2}}e^{\frac{i}{\hbar}\left(\frac{m}{2\epsilon}(x_n-x_{n-1})^2-\epsilon V(x_{n-1})\right)} + O(\epsilon^2) . \end{aligned} \tag{2.38}$$

At this point, approximating the infinitesimal evolution operator by $:e^{-i\frac{\epsilon}{\hbar}\left(\frac{\hat{p}^2}{2m}+V(\hat{x})\right)}:$ instead of $1 - i\frac{\epsilon}{\hbar}\left(\frac{\hat{p}^2}{2m}+V(\hat{x})\right)$ or some other expression valid to order ϵ may well appear artificial. The essential issue is to obtain an approximation

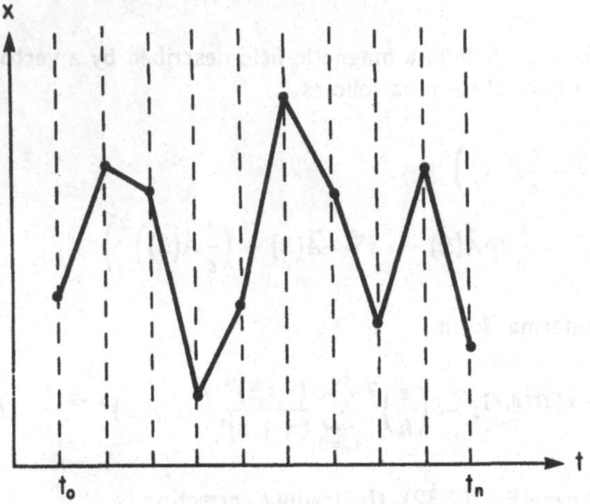

Fig. 2.3 A typical trajectory contributing to a path integral.

yielding convergent momentum integrals both in the present case of real time and in subsequent expressions continued to imaginary time. Whereas individual terms in the Taylor series expansion yield powers of p which diverge, since p^2 is bounded from below, both $e^{-\frac{\epsilon}{\hbar}\frac{p^2}{2m}}$ and $e^{-i\frac{\epsilon}{\hbar}\frac{p^2}{2m}}$ are bounded. Mathematically precise derivations of successive levels of generality are provided by Simon (1979), Trotter (1959), and Kato (1978).

Although it is crucial to use a form in which p^2 has been exponentiated, there still remains some arbitrariness in the approximation Eq. (2.37). As shown in Problem 2.4, one may replace $V(x_{n-1})$ by $V(x_n)$ or $\frac{1}{2}(V(x_{n-1})+V(x_n))$ and in applications such as in Chapter 8 where ϵ remains finite, this freedom may be exploited to improve the approximation.

The problems arising when the Hamiltonian contains terms in which $\hat{p}$ and $\hat{x}$ are combined are exhibited in Problem 2.5 for a particle in a magnetic field. The ultimate justification for any expression such as Eq. (2.38) used to approximate $\langle x_n|e^{-\frac{\epsilon}{\hbar}\left(\frac{\hat{p}^2}{2m}+V(\hat{x})\right)}|x_{n-1}\rangle$ is that it reproduces the correct evolution of the wave function, and this may be verified straightforwardly (see Problems 2.4 and 2.5).

Using Eq. (2.38) and the notation of Eq. (2.29), the matrix element of the evolution operator Eq. (2.30) may be written

$$\mathcal{U}(x_f t_f, x_i t_i) = \lim_{M\to\infty}\int\prod_{k=1}^{M-1} dx_k \left(\frac{m}{2\pi i\epsilon\hbar}\right)^{\frac{M}{2}} e^{\frac{i}{\hbar}\epsilon\sum_{k=1}^{M}\left(\frac{m}{2}\left(\frac{x_k-x_{k-1}}{\epsilon}\right)^2-V(x_{k-1})\right)} \,. \tag{2.39}$$

The set of points $\{x_o, x_1, \ldots x_M\}$ defines a trajectory as sketched in Fig. 2.3. For notational convenience, in the limit $M\to\infty$ we will often denote this trajectory by $x(t)$ with starting point $x(t_i) = x_i$ and endpoint $x(t_f) = x_f$, but it is crucial to note that this notation does not imply continuity or differentiability. Rather, the trajectory should always be thought of as a set of M points $x(t_k)$ indexed by the discrete times

t_k. In the same spirit, it is convenient to represent $\frac{x_k - x_{k-1}}{\epsilon}$ by the symbol $\frac{dx}{dt}$. Again, no differentiability is implied and the precise definition of $\frac{dx}{dt}$ is given by the finite-difference expression. With this notation, the Riemann sums in the exponent may be indicated symbolically

$$\epsilon \sum_{k=1}^{M} \frac{m}{2}\left(\frac{x_k - x_{k-1}}{\epsilon}\right)^2 \to \int_{t_i}^{t_f} dt \frac{m}{2}\left[\frac{dx}{dt}\right]^2 \tag{2.40a}$$

and

$$\epsilon \sum_{k=1}^{M} V(x_{k-1}) \to \int_{t_i}^{t_f} dt\, V(x(t)) \ . \tag{2.40b}$$

The Feynman path integral, which is defined as the limit of Eq. (2.39) as $M \to \infty$, is denoted

$$\mathcal{U}(x_f t_f, x_i t_i) = \int_{(x_i,t_i)}^{(x_f,t_f)} \mathcal{D}[x(t)] e^{\frac{i}{\hbar}\int_{t_i}^{t_f} dt\left(\frac{m}{2}\left(\frac{dx}{dt}\right)^2 - V(x(t))\right)} = \int_{(x_i,t_i)}^{(x_f,t_f)} \mathcal{D}[x(t)] e^{\frac{i}{\hbar}S[x(t)]} \tag{2.41}$$

where

$$\int_{(x_i,t_i)}^{(x_f,t_f)} \mathcal{D}[x(t)] = \lim_{M\to\infty} \int \prod_{k=1}^{M-1} dx_k \left(\frac{m}{2\pi i\epsilon\hbar}\right)^{\frac{3M}{2}} \tag{2.42}$$

represents a sum over all trajectories starting at position x_i at time t_i and ending at position x_f at time t_f, the action $S[x(t)]$ is

$$S[x(t)] = \int_{t_i}^{t_f} dt L[x(t)] \tag{2.43}$$

and the Lagrangian $L[x(t)]$ is

$$L[x(t)] = \frac{1}{2} m \left(\frac{dx}{dt}\right)^2 - V(x(t)) \ . \tag{2.44}$$

The matrix element of the evolution operator between states $|x_i\rangle$ and $|x_f\rangle$ is thus the sum over all trajectories beginning at x_i at time t_i and ending at x_f at time t_f of the exponential $\frac{i}{\hbar}$ times the action along the trajectory.

Several remarks concerning the Feynman path integral, Eq. (2.41) are germane at this point. Because the path integral is an exact representation of the evolution operator, it may be used as the starting point for the formulation of quantum mechanics (see Feynman and Hibbs (1965)). The superposition principle, which may be written at any time t in the form

$$\mathcal{U}(x_f t_f, x_i t_i) = \int dx\, \mathcal{U}(x_f t_f, xt)\,\mathcal{U}(xt, x_i t_i) \tag{2.45}$$

is expressed in terms of path integrals as

$$\int_{(x_i t_i)}^{(x_f t_f)} \mathcal{D}[x(t)] e^{\frac{i}{\hbar}\int_{t_i}^{t_f} dt' L[x(t')]} = \int dx \int_{(xt)}^{(x_f t_f)} \mathcal{D}[x(t)] e^{\frac{i}{\hbar}\int_{t}^{t_f} dt' L[x(t')]} \times \int_{(x_i t_i)}^{(x,t)} \mathcal{D}[x(t)] e^{\frac{i}{\hbar}\int_{t_i}^{t} dt' L[x(t')]} \tag{2.46}$$

and quantum mechanical interference arises directly from the sums over trajectories. A natural approximation to the path integral, Eq. (2.41) in the limit as $\hbar \rightarrow 0$ is the stationary-phase approximation. As will be shown in detail in Section (2.5), the dominant contribution to the transition amplitude in this limit comes from trajectories surrounding the classical trajectory joining x_i to x_f. Finally, since the measure in Eq. (2.42) is still ill-defined when ϵ goes to zero, it is useful to note that the functional integral may be normalized by solutions of an analytically solvable reference problem. For example, one may require that when the potential V is set to zero, the transition amplitude is

$$\begin{aligned}\mathcal{U}_0(x_f t_f, x_i t_i) &= \langle x_f | e^{\frac{i}{\hbar}\frac{\hat{p}^2}{2m}(t_f - t_i)} | x_i \rangle \\ &= \left[\frac{m}{2\pi i\hbar(t_f - t_i)}\right]^{\frac{3}{2}} e^{i\frac{m}{2\hbar}\frac{(x_f - x_i)^2}{t_f - t_i}} \quad .\end{aligned} \tag{2.47}$$

The functional integral in Eq. (2.41) is called the Lagrangian form and requires that the Hamiltonian have quadratic momentum dependence as in Eq. (2.32). The Hamiltonian form of the functional integral is obtained by substituting Eq. (2.37) in Eq. (2.30) without performing the p integration, in which case the matrix element of the evolution operator becomes

$$\begin{aligned}\mathcal{U}(x_f t_f, x_i t_i) &= \lim_{M\rightarrow\infty} \int \prod_{k=1}^{M-1} dx_k \prod_{k=1}^{M} \frac{dp_k}{(2\pi\hbar)^3} e^{\frac{i}{\hbar}\sum_{k=1}^{M}[p_k(x_k - x_{k-1}) - \epsilon\frac{p_k^2}{2m} - \epsilon V(x_{k-1})]} \\ &\rightarrow \int_{(x_i t_i)}^{(x_f, t_f)} \mathcal{D}[x(t)]\mathcal{D}[p(t)] e^{\frac{i}{\hbar}\int_{t_i}^{t_f} dt[p(t)\frac{\partial}{\partial t}x(t) - H(p(t), x(t))]} \quad .\end{aligned} \tag{2.48}$$

The trajectories $x(t)$ obey the same boundary conditions as in the Lagrangian form and the trajectories $p(t)$ have no boundary conditions. The Hamiltonian form of this functional integral is *a priori* more general than the Lagrangian form, but requires care in the ordering of the non-commuting operators $\hat{x}$ and $\hat{p}$ when $\mathcal{H}$ contains mixed terms in $\hat{x}$ and $\hat{p}$ (see Problem 2.5).

At this point, it is useful to note that path integrals automatically represent time-ordered products. Let $\mathcal{O}_1(\hat{x}, t_1)$ and $\mathcal{O}_2(\hat{x}, t_2)$ be operators acting at times t_1 and t_2 with $t_1 \geq t_2$ and let t_m denote the discrete time in Eq. (2.30) closest to t_1 and t_n

denote the discrete time closest to t_2. Then

$$\begin{aligned}
&\langle x_f|TO_1(\hat{x},t_1)O_2(\hat{x},t_2)e^{-\frac{i}{\hbar}\int_{t_i}^{t_f}\hat{H}(t)dt}|x_i\rangle \\
&=\langle x_f|Te^{-\frac{i}{\hbar}\int_{t_1}^{t_f}\hat{H}(t)dt}O(\hat{x},t_1)Te^{-\frac{i}{\hbar}\int_{t_2}^{t_1}\hat{H}(t)dt}O(\hat{x},t_2)Te^{-\frac{i}{\hbar}\int_{t_i}^{t_2}\hat{H}(t)dt}|x_i\rangle \\
&=\lim\int\prod_{k=1}^{M-1}dx_k\langle x_f|e^{-i\frac{\epsilon}{\hbar}H}\dots|x_m\rangle\langle x_m|O_1(\hat{x})e^{-i\frac{\epsilon}{\hbar}H}|x_{m-1}\rangle\langle x_{m-1}|\dots \\
&\quad\times e^{-i\frac{\epsilon}{\hbar}H}|x_n\rangle\langle x_n|O_2(\hat{x})e^{-i\frac{\epsilon}{\hbar}H}|x_{n-1}\rangle\langle x_{n-1}|\dots e^{-i\frac{\epsilon}{\hbar}H}|x_i\rangle \\
&=\int_{(x_i,t_i)}^{(x_f,t_f)}\mathcal{D}[x(t)]O_1(x(t_1))O_2(x(t_2))e^{\frac{i}{\hbar}\int_{t_i}^{t_f}dtL[x(t)]} \qquad (2.49) \\
&=\int_{(x_i,t_i)}^{(x_f,t_f)}\mathcal{D}[x(t)]\mathcal{D}[p(t)]O_1(x(t_1))O(x(t_2))e^{\frac{i}{\hbar}\int_{t_i}^{t_f}dt(p(t)\dot{x}(t)-H[p(t),x(t)])} \quad .
\end{aligned}$$

Thus, although there is no explicit indication in the notation that operators have been time-ordered, in order for the operators $O_1(\hat{x},t_1)$, $O_2(\hat{x},t_2)$, and $\hat{H}(t)$ to be replaced by the c-numbers $O_1(x(t_1))$, $O_2(x(t_2))$, and $H[p(t),x(t)]$, it is implicit in the construction of the functional integral that each operator had to act on the complete set of states introduced at the corresponding discrete time. Operators depending upon the momentum $\hat{p}$ are treated in the same way by letting them act on the complete set of momentum states introduced at the corresponding time. The fact that functional integrals necessarily yield time-ordered products is the reason for the assertion in Section 2.1 that time-ordered products are the quantities which arise naturally in the formalism and that physical response functions should ultimately be evaluated in terms of them. From the definition of the time-ordered exponential, Eq. (2.9), and the path integral, it is evident that both quantities deal with the non-commutativity of operators in quantum mechanics in the same way. In both cases the time interval is divided into sufficiently small subintervals that the commutator terms $\frac{\epsilon^2}{2}[\frac{p^2}{2m},V]$ become negligible. There is no reason why the continuous parameter must necessarily be the physical time, and we will see that it is also useful to use temperature (or imaginary time) as the formal parameter for developing path integrals.

IMAGINARY-TIME PATH INTEGRAL AND THE PARTITION FUNCTION

The partition function for a single particle may be written

$$\begin{aligned}
Z &= \operatorname{Tr} e^{-\beta\hat{H}} \\
&= \int dx\langle x|e^{-\beta\hat{H}}|x\rangle
\end{aligned} \qquad (2.50)$$

and may be thought of as a sum over diagonal matrix elements of the imaginary time evolution operator

$$\mathcal{U}(x_f\tau_f, x_i\tau_i) = \langle x_f|e^{-(\tau_f-\tau_i)\frac{\hat{H}}{\hbar}}|x_i\rangle \tag{2.51}$$

evaluated for the interval $\tau_f - \tau_i = \beta\hbar$. Note that for the one-particle problem, we work in the canonical ensemble and there is no chemical potential. Given the observation that the essence of a path integral or time-ordered exponential is the subdivision of the interval into sufficiently small intervals that commutators of the quantum operators appearing in H may be neglected, all the steps in the derivation of the real-time path integral may be repeated for the case of imaginary time. For the Hamiltonian, Eq. (2.32), we obtain

$$\begin{aligned}
\mathcal{U}(x_f t_f, x_i t_i) &= \lim_{M\to\infty}\int \prod_{k=1}^{M-1} d^3x_k \prod_{k=1}^{M} \langle x_k|e^{-\frac{\epsilon}{\hbar}H(\hat{p},\hat{x})}|x_{k-1}\rangle \\
&= \lim_{M\to\infty}\int \prod_{k=1}^{M-1} d^3x_k \prod_{k=1}^{M} d^3p_k \langle x_k|p_k\rangle\langle p_k| :e^{-\frac{\epsilon}{\hbar}H(\hat{p},\hat{x})}: + \mathcal{O}\epsilon^2|x_{k-1}\rangle \\
&= \lim_{M\to\infty}\int \prod_{k=1}^{M-1} d^3x_k \prod_{k=1}^{M} \frac{d^3p_k}{(2\pi\hbar)^3} e^{\sum_{k=1}^{M}\left[\frac{ip_k}{\hbar}(x_k-x_{k-1})-\frac{\epsilon}{\hbar}\left(\frac{p_k^2}{2m}+V(x_{k-1})\right)\right]} \\
&= \lim_{M\to\infty}\int \prod_{k=1}^{M-1} d^3x_k \left(\frac{m}{2\pi\epsilon\hbar}\right)^{\frac{3M}{2}} e^{-\frac{\epsilon}{\hbar}\sum_{k=1}^{M}\left[\frac{m}{2}\frac{(x_k-x_{k-1})^2}{\epsilon^2}+V(x_{k-1})\right]} \\
&= \int_{(x_i,\tau_i)}^{(x_f,\tau_f)} \mathcal{D}[x(\tau)]e^{-\frac{1}{\hbar}\int_{\tau_i}^{\tau_f} d\tau\left(\frac{m}{2}\left(\frac{dx(\tau)}{d\tau}\right)^2+V(x(\tau))\right)} \\
&= \int_{(x_i,\tau_i)}^{(x_f,\tau_f)} \mathcal{D}[x(\tau)]e^{-\frac{1}{\hbar}\int_{\tau_i}^{\tau_f} d\tau H[x(\tau)]}
\end{aligned} \tag{2.52}$$

where $\epsilon = \frac{1}{M}(\tau_f - \tau_i)$. Thus, the imaginary-time path integral is a sum over trajectories starting at (x_i, τ_i) and ending at (x_f, τ_f) of the exponential of a modified action in which a change in sign of the kinetic term yields the Hamiltonian instead of the Lagrangian.

An alternative derivation, which shows explicitly how the Lagrangian in the real-time case is transformed into the Hamiltonian in the imaginary time case, is to preform an analytic continuation of Eq. (2.41) to imaginary time. This continuation, known as a Wick rotation because it may be viewed as a rotation of the integration contour in the complex t-plane, is effected by the variable transformation

$$t = -i\tau \quad . \tag{2.53a}$$

Thus,

$$\frac{dx}{d\tau} = \frac{dt}{d\tau}\frac{dx}{dt} = -i\frac{dx}{dt} \tag{2.53b}$$

and the action, which is called the Euclidean action, becomes

$$\frac{i}{\hbar}\int_{t_1}^{t_2} dt\left[\frac{m}{2}\left(\frac{dx}{dt}\right)^2 - V(x(t))\right] = -\frac{1}{\hbar}\int_{\tau_1}^{\tau_2} d\tau\left[\frac{m}{2}\left(\frac{dx}{d\tau}\right)^2 + V(x(\tau))\right] \quad . \tag{2.53c}$$

The kinetic energy thus changes signs because each time derivative acquires a factor of i. The same sign reversal arises in the classical equations of motion in imaginary time, and the interpretation of a particle moving in an inverted potential will subsequently provide a picturesque way to visualize the stationary solutions to path integrals in classically forbidden regimes.

For the imaginary-time path integral, the measure appearing in Eq. (2.52) is equivalent to the Wiener measure defined in the study of continuous stochastic processes (Wiener, (1924,1932)) and the functional integral can be given a rigorous mathematical definition. This path integral will provide the foundation of the stochastic method presented in Chapter 8, and the nature of the trajectories which contribute to it will be studied more thoroughly in that context.

Using Eqs. (2.50) – (2.52), the partition function may be expressed

$$Z = \int dx \int_{x(0)=x}^{x(\beta\hbar)=x} \mathcal{D}[x(\tau)]e^{-\frac{1}{\hbar}\int_0^{\beta\hbar} d\tau\left(\frac{m}{2}\left(\frac{dx(\tau)}{d\tau}\right)^2 + V(x(\tau))\right)}$$

$$\equiv \int_{x(\beta\hbar)=x(0)} \mathcal{D}[x(\tau)]e^{-\frac{1}{\hbar}\int_0^{\beta\hbar} d\tau\left(\frac{m}{2}\left(\frac{dx(\tau)}{d\tau}\right)^2 + V(x(\tau))\right)} \quad . \tag{2.54}$$

The partition function is thus a sum over all periodic trajectories of period $\beta\hbar$ and the shorthand notation in the last line emphasizes the fact that the integral over x_m at the endpoint of the interval is equivalent to the integral over each of the internal x_k's within the interval. For notational clarity, except when we are specifically concerned with the classical limit in which $\hbar \to 0$, it will be convenient to use units in which $\hbar = 1$.

Finally, the Feynman path integral in real or imaginary time may be straightforwardly extended to many-particle systems. For example, using the symmetrized or antisymmetrized states defined in Eq. (1.38) where we used $\varsigma = \pm 1$ for Bosons or Fermions, the partition function for an N-particle system may be written

$$Z = \frac{1}{N!}\int \prod_{i=1}^{N} dx_i \, \{x_1 \ldots x_n|e^{-\beta H}|x_1 \ldots x_n\}$$

$$= \frac{1}{N!}\sum_P \varsigma^P \int \prod_{i=1}^{N} dx_i (x_{P1} \ldots x_{PN}|e^{-\beta H}|x_1 \ldots x_n) \quad . \tag{2.55}$$

As in the case of a simple variable, the time interval β may be divided into infinitesimal steps. However, now we have the additional choice of inserting at each step the closure relation Eq. (1.30) with product states or Eq. (1.40) using symmetrized or antisymmetrized states. Either choice yields exact evolution and the symmetry or antisymmetry of the final states in the trace suffices to impose the proper statistics. The use of product states yields the simplest formal result completely analogous to Eq. (2.52), and for the case of a Hamiltonian of the form

$$H = \sum_{i=1}^{N} \frac{\hat{p}_i^2}{2m} + \frac{1}{2} \sum_{i \neq j} v(\hat{x}_i - \hat{x}_j) \tag{2.56}$$

the partition function may be written

$$Z = \frac{1}{N!} \sum_P \varsigma^P \int_{\substack{x_1(\beta) = x_{P1}(0) \\ x_N(\beta) = x_{PN}(0)}} \mathcal{D}[x_1(\tau)] \ldots \mathcal{D}[x_N(\tau)] \times e^{-\int_0^\beta d\tau \left[\sum_{i=1}^N \frac{m}{2} \left(\frac{dx_i(\tau)}{d(\tau)} \right)^2 + \frac{1}{2} \sum_{i \neq j} v(x_i(\tau) - x_j(\tau)) \right]} . \tag{2.57}$$

In the case of stochastic evolution of path integrals for Fermions, the alternative choice of using antisymmetrized states at intermediate steps will prove advantageous in certain applications, as discussed in Chapter 8.

COHERENT STATE FUNCTIONAL INTEGRAL

For a general many-particle Hamiltonian expressed in second quantized form, a functional integral representation for the many-body evolution operator may be obtained using the coherent states $|\phi\rangle$, Eq. (1.118) and (1.160) instead of the position and momentum eigenstates used for the Feynman path integral. Recall that the relations for Fermions and Bosons have the identical form tabulated in Table 1.1 at the end of Chapter 1 where the integration variables are understood to be complex variables for Bosons and Grassman variables for Fermions. We will evaluate the matrix element of the evolution operator between an initial coherent state $|\phi_i\rangle$ having components $\phi_{\alpha,i}$ and a final state $\langle \phi_f|$ with components $\phi^*_{\alpha,f}$. As before, the integral $[t_i, t_f]$ is broken into M times steps of size $\epsilon = \frac{t_f - t_i}{M}$, a closure relation in the notation of Table 1.1

$$1 = \frac{1}{\mathcal{N}} \prod_\alpha d\phi^*_{\alpha,k} d\phi_{\alpha,k} e^{-\sum_\alpha \phi^*_{\alpha,k}\phi_{\alpha,k}} |\phi_{\alpha,k}\rangle\langle\phi_{\alpha,k}|$$

is inserted at the kth time step, and we use the notation at the end points:

$$\begin{aligned} \phi_{\alpha,0} &\equiv \phi_{\alpha,i} \\ \phi^*_{\alpha,M} &\equiv \phi^*_{\alpha,f} \end{aligned} \quad . \tag{2.58}$$

For second quantized operators, the appropriate form of normal ordering is that defined in Section (1.4) with all creation operators to the left of annihilation operators, and we will assume that $H(a^\dagger_\alpha, a_\alpha)$ is written in normal form. As in the path integral case,

$$e^{\epsilon H(a^\dagger a)} = :e^{\epsilon H(a^\dagger a)}: + \mathcal{O}(\epsilon^2) \tag{2.59}$$

where the term of order ϵ^2 is ϵ^2 times an operator which is finite when acting on a normalized, differential wave function $\psi(\phi^*_\alpha)$. Thus, using Eqs. (1.137) or (1.177) to evaluate coherent state matrix elements of the normal ordered exponentials, the matrix element of the evolution operator may be written

$$\begin{aligned}
\mathcal{U}(\phi^*_{\alpha,f}t_f;\phi_{\alpha,i}t_i) &= \lim_{M\to\infty}\langle\phi_f|e^{-\frac{i}{\hbar}H(t_f-t_i)}|\phi_i\rangle \\
&= \lim_{M\to\infty}\int\prod_{k=1}^{M-1}\prod_\alpha\frac{1}{\mathcal{N}}d\phi^*_{\alpha,k}d\phi_{\alpha,k}e^{-\sum_{k=1}^{M-1}\sum_\alpha\phi^*_{\alpha,k}\phi_{\alpha,k}} \\
&\quad\times\prod_{k=1}^{M}\langle\phi_k|:e^{-\frac{i\epsilon}{\hbar}H(a^\dagger_\alpha,a_\alpha)}:+O(\epsilon^2)|\phi_{k-1}\rangle \qquad (2.60)\\
&= \lim_{M\to\infty}\int\prod_{k=1}^{M-1}\prod_\alpha\frac{1}{\mathcal{N}}d\phi^*_{\alpha,k}d\phi_{\alpha,k}e^{-\sum_{k=1}^{M-1}\sum_\alpha\phi^*_{\alpha,k}\phi_{\alpha,k}} \\
&\quad\times e^{\sum_{k=1}^{M}\left(\sum_\alpha\phi^*_{\alpha,k}\phi_{\alpha,k-1}-\frac{i\epsilon}{\hbar}H(\phi^*_{\alpha,k}\phi_{\alpha,k-1})\right)}\ .
\end{aligned}$$

Note that in the case of Fermions, since there is no metric in the Grassman algebra, all the integrals indicated in Eq. (2.60) are finite. For Bosons, the argument is analogous to that for the path integral case. In real time $e^{-\frac{i\epsilon}{\hbar}H}$ is oscillatory and the factor $e^{\phi^*_{\alpha,k}\phi_{\alpha,k}}$ arising from the measure produces convergence. In imaginary time we again rely upon the physical fact that the Hamiltonian is bounded from below, which implies that $H(\phi^*_{\alpha,k},\phi_{\alpha,k-1})\equiv\frac{\langle\phi_k|H|\phi_{k-1}\rangle}{\langle\phi_k|\phi_{k-1}\rangle}$ is also bounded from below. Hence $e^{-\frac{\epsilon}{\hbar}H(\phi^*_{\alpha,k},\phi_{\alpha,k-1})}$ is bounded and the Gaussian factor from the measure again ensures convergence.

As in the case of the path integral, it is convenient to introduce a trajectory $\phi_\alpha(t)$ to represent the set $\{\phi_{\alpha,1}\phi_{\alpha,2}\ldots\phi_{\alpha,M}\}$ and to introduce the notation

$$\phi^*_{\alpha,k}\frac{(\phi_{\alpha,k}-\phi_{\alpha,k-1})}{\epsilon}\equiv\phi^*_\alpha(t)\frac{\partial}{\partial t}\phi_\alpha(t) \qquad (2.61a)$$

and

$$H(\phi^*_{\alpha,k};\phi_{\alpha,k-1})\equiv H(\phi^*_\alpha(t),\phi_\alpha(t)) \qquad (2.61b)$$

in which case the exponent in Eq. (2.60) may be rewritten symbolically

$$\begin{aligned}
&\sum_\alpha\phi^*_{\alpha,M}\phi_{\alpha,M-1}-i\frac{\epsilon}{\hbar}H(\phi^*_{\alpha,M};\phi_{\alpha,M-1}) \\
&\quad+i\epsilon\sum_{k=1}^{M-1}\left[i\sum_\alpha\phi^*_{\alpha,k}\left(\frac{\phi_{\alpha,k}-\phi_{\alpha,k-1}}{\epsilon}\right)-\frac{1}{\hbar}H(\phi^*_{\alpha,k};\phi_{\alpha,k-1})\right] \\
&=\sum_\alpha\phi^*_\alpha(t_f)\phi_\alpha(t_f)+\frac{i}{\hbar}\int_{t_i}^{t_f}dt\left[\sum_\alpha i\hbar\phi^*_\alpha(t)\frac{\partial\phi_\alpha(t)}{\partial t}-H(\phi^*_\alpha(t),\phi_\alpha(t))\right] \\
&=\sum_\alpha\phi^*_\alpha(t_f)\phi_\alpha(t_f)+\frac{i}{\hbar}\int_{t_i}^{t_f}dt\,L(\phi^*_\alpha(t),\phi_\alpha(t)) \qquad (2.61c)
\end{aligned}$$

where the Schrödinger Lagrangian operator is $i\hbar\frac{\partial}{\partial t} - H$. As in the Feynman path integral, the trajectory and derivative notation is purely symbolic, and for any case in which ambiguity may arise, the correct physical quantity is calculated by performing the integral over the discrete action in Eq. (2.61c) and then taking the limit $M \to \infty$. With this notation,

$$\mathcal{U}(\phi^*_{\alpha,f}, t_f; \phi_{\alpha,i} t_i) = \int_{\phi_\alpha(t_i)=\phi_{\alpha,i}}^{\phi^*_\alpha(t_f)=\phi^*_{\alpha,f}} \mathcal{D}[\phi^*_\alpha(t)\phi_\alpha(t)] e^{\sum_\alpha \phi^*_\alpha(t_f)\phi_\alpha(t_f)} \times e^{\frac{i}{\hbar}\int_{t_i}^{t_f} dt \left[\sum_\alpha i\hbar\phi^*_\alpha(t)\frac{\partial\phi_\alpha(t)}{\partial t} - H(\phi^*_\alpha(t),\phi_\alpha(t))\right]} \tag{2.62a}$$

where

$$\int_{\phi_\alpha(t_i)}^{\phi^*_\alpha(t_f)} \mathcal{D}[\phi^*_\alpha(t)\phi_\alpha(t)] = \lim_{M\to\infty} \int \prod_{k=1}^{M-1} \prod_\alpha \frac{1}{\mathcal{N}} d\phi^*_{\alpha,k} d\phi_{\alpha,k} \ . \tag{2.62b}$$

Note in the discrete expression, that the boundary conditions specified $\phi_{\alpha,0}$ and $\phi^*_{\alpha,M}$, that there were no variables $\phi^*_{\alpha,0}$ or $\phi_{\alpha,M}$, and that all the internal conjugate variables $\phi^*_{\alpha,k}$ and $\phi_{\alpha,k}$ for $k = 1, M-1$ are integrated. In the trajectory notation $\phi^*_\alpha(t)$ and $\phi_\alpha(t)$ are associated with variables displaced by one time step, $\phi^*_{\alpha,k}$ and $\phi_{\alpha,k-1}$, respectively, so that $\phi^*_\alpha(t_f)$ and $\phi_\alpha(t_i)$ are specified by the boundary conditions but $\phi_\alpha(t_f)$ and $\phi^*_\alpha(t_f)$ correspond to internal variables of integration not subject to boundary conditions. The boundary term $\phi^*_{\alpha,M}\phi_{\alpha,M-1}$ appearing in the exponent, Eq. (2.61c) represents a term left over at the end of the path from our grouping of terms in defining the derivative Eq. (2.61a). Had we chosen the alternative convention $\frac{1}{\epsilon}(-\phi^*_{\alpha,k+1} + \phi^*_{\alpha,k})\phi_{\alpha,k} = \left(-\frac{\partial}{\partial t}\phi^*_\alpha(t)\right)\phi(t)$, the remaining boundary term would have been $\phi^*_1\phi_0$. Both results correspond to the same fundamental discrete expression and are thus equivalent. If a symmetric formal expression is desired, one may use the average of the two.

One significant difference between the coherent state functional integral, Eq. (2.62) and the Feynman path integral is the dependence upon $\hbar$. In the Feynman case, $\frac{1}{\hbar}$ appears as a constant multiplying the entire exponent, so that the stationary-phase expansion immediately yields the classical limit. In the present case, the action contains $\hbar$ within the Lagrangian as well as a multiplicative factor, so that the stationary-phase method yields a result quite distinct from the classical limit.

THE PARTITION FUNCTION FOR MANY-PARTICLE SYSTEMS

As in Eq. (2.57), the partition function for a many-particle system may be expressed as the trace of an imaginary-time evolution operator. Using Eqs. (1.128) or (1.173) for the trace with Boson or Fermion coherent states and units such that $\hbar = 1$, the partition function may be written

$$\begin{aligned} Z &= \mathrm{Tr}\, e^{-\beta(\hat{H}-\mu\hat{N})} \\ &= \int \prod_\alpha d\phi^*_\alpha d\phi_\alpha e^{-\sum_\alpha \phi^*_\alpha\phi_\alpha} \langle \zeta\phi | e^{-\beta(\hat{H}-\mu\hat{N})} | \phi \rangle \ . \end{aligned} \tag{2.63}$$

When the continuation of Eq. (2.50) to imaginary time is substituted in this expression, the trace imposes the periodic or antiperiodic boundary conditions

$$\begin{aligned} \phi_{\alpha,0} &= \phi_\alpha \\ \phi^*_{\alpha,M} &= \varsigma\phi^*_\alpha \ . \end{aligned} \tag{2.64}$$

The equivalence of the interior and exterior coherent state integrals is emphasized by relabelling $\phi_\alpha \equiv \varsigma\phi_{\alpha,M}$ and the resulting partition function is

$$Z = \lim_{M\to\infty} \int \prod_{k=1}^{M} \prod_{\alpha} \frac{1}{\mathcal{N}} d\phi^*_{\alpha,k} d\phi_{\alpha,k} e^{-S(\phi^*,\phi)} \tag{2.65a}$$

where

$$\begin{aligned} S(\phi^*,\phi) = \epsilon \sum_{k=2}^{M} &\left[\sum_\alpha \phi^*_{\alpha k} \left\{ \frac{(\phi_{\alpha,k} - \phi_{\alpha,k-1})}{\epsilon} - \mu\phi_{\alpha,k-1} \right\} + H(\phi^*_{\alpha,k}, \phi_{\alpha,k-1}) \right] \\ + \epsilon &\left[\sum_\alpha \phi^*_{\alpha 1} \left\{ \frac{\phi_{\alpha,1} - \varsigma\phi_{\alpha,M}}{\epsilon} - \mu\varsigma\phi_{\alpha,M} \right\} + H(\phi^*_{\alpha 1}, \varsigma\phi_{\alpha,M}) \right] \ . \end{aligned} \tag{2.65b}$$

Using the trajectory notation, this may be rewritten

$$Z = \int_{\phi_\alpha(\beta)=\varsigma\phi_\alpha(0)} \mathcal{D}(\phi^*_\alpha(\tau)\phi_\alpha(\tau)) e^{-\int_0^\beta d\tau \left\{ \sum_\alpha \phi^*_\alpha(\tau)(\frac{\partial}{\partial\tau} - \mu)\phi_\alpha(\tau) + H(\phi^*_\alpha(\tau),\phi_\alpha(\tau)) \right\}} \tag{2.66}$$

with the usual understanding that the derivatives and integrals are defined in terms of the discrete expression, Eq. (2.62). Note that the integration is over complex variables satisfying periodic boundary conditions for Bosons and Grassman variables satisfying antiperiodic boundary conditions for Fermions.

In a formal sense, the problem has now been reduced to quadrature and we only need to develop techniques to evaluate the integral in Eq. (2.62). Our overall approach will be to group the one-body part of $H(\phi^*, \psi)$ together with the other quadratic terms in the exponent and to develop a perturbation series in which the exponential of the many-body part of $H(\phi^*, \phi)$ is expanded in a Taylor series. This will give rise to a series of integrals of the products of a Gaussian times polynomials which may be evaluated straightforwardly using the techniques developed in the next section.

The thermal Green's function, defined in Eq(2.22), has a simple form expressed in terms of a coherent state path integral. It is useful to give the creation and annihilation operators a formal τ label, $\{a^\dagger_\alpha(\tau), a_\alpha(\tau)\}$, denoting the time slice τ upon which they are defined. This purely formal τ label is introduced to allow the time-ordering operator to appropriately interlace operators with no explicit τ-dependence, and when the evolution operator is represented by a functional integral, the operators $\{a^\dagger_\alpha(\tau), a_\alpha(\tau)\}$ on the time slice τ will be replaced by the coherent state variables $\{\psi^*_\alpha(\tau), \psi_\alpha(\tau)\}$. To facilitate manipulation of the time-ordered product, it is convenient to write the thermal Green's function, Eq. (2.22), as follows:

$$
\begin{aligned}
\mathcal{G}^{(n)}&(\alpha_1\tau_1\ldots\alpha_n\tau_n|\alpha_{2n}\tau_{2n}\ldots\alpha_{n+1}\tau_{n+1}) \qquad (2.67a)\\
&= \frac{1}{Z}\mathrm{Tr}\left[e^{-\beta(\hat{H}-\mu\hat{N})}T\,a^{(H)}_{\alpha_1}(\tau_1)\ldots a^{(H)}_{\alpha_n}(\tau_n)a^{(H)}_{\alpha_{n+1}}{}^{\dagger}(\tau_{n+1})a^{(H)\dagger}_{\alpha_{2n}}(\tau_{2n})\right]\\
&= \frac{1}{Z}\mathrm{Tr}\left[e^{-\beta(\hat{H}-\mu\hat{N})}\zeta^P\tilde{a}^{(H)}_{\alpha_{P1}}(\tau_{P1})\tilde{a}^{(H)}_{\alpha_{P2}}(\tau_{P2})\ldots\tilde{a}^{(H)}_{\alpha_{P2N}}(\tau_{P2N})\right]
\end{aligned}
$$

where the permutation P arranges the times in chronological order and $\tilde{a}_{\alpha_i}$ is an annihilation operator for $i \leq n$ and a creation operator for $i > n$. Using the definition of the Heisenberg operator, Eq(2.23), and the fact that a functional integral corresponds to a time-ordered product,Eq(2.49), the Green's function may be written as follows:

$$
\begin{aligned}
\mathcal{G}^{(n)}&(\alpha_1\tau_1\ldots\alpha_n\tau_n|\alpha_{2n}\tau_{2n}\ldots\alpha_{n+1}\tau_{n+1}) \qquad (2.67b)\\
&= \frac{1}{Z}\zeta\mathrm{Tr}\Big[e^{-\beta(\hat{H}-\mu\hat{N})}e^{\tau_{P1}(\hat{H}-\mu\hat{N})}\tilde{a}_{\alpha_{P1}}e^{-\tau_{P1}(\hat{H}-\mu\hat{N})}\\
&\qquad \times e^{\tau_{P2}(\hat{H}-\mu\hat{N})}\tilde{a}_{\alpha_{P2}}e^{-\tau_{P2}(\hat{H}-\mu\hat{N})}\ldots e^{\tau_{P2n}(\hat{H}-\mu\hat{N})}\tilde{a}_{\alpha_{P2n}}e^{-\tau_{P2n}(\hat{H}-\mu\hat{N})}\Big]\\
&= \frac{1}{Z}\zeta^P\mathrm{Tr}\left[e^{-\int_{\tau_{P1}}^{\beta}(\hat{H}-\mu\hat{N})}\tilde{a}_{\alpha_{P1}}e^{-\int_{\tau_{P2}}^{\tau_{P1}}(\hat{H}-\mu\hat{N})}\tilde{a}_{\alpha_{P2}}\ldots\tilde{a}_{\alpha_{P2n}}e^{-\int_{0}^{\tau_{P2n}}(\hat{H}-\mu\hat{N})}\right]\\
&= \frac{1}{Z}\mathrm{Tr}\left[Te^{-\int_0^\beta(\hat{H}-\mu\hat{N})}a_{\alpha_1}(\tau_1)\ldots a_{\alpha_N}(\tau_N)a^\dagger_{\alpha_{N+1}}(\tau_{N+1})\ldots a^\dagger_{\alpha_{2n}}(\tau_{2n})\right]\\
&= \frac{1}{Z}\int\mathcal{D}[\phi^*_\alpha(\tau)\phi_\alpha(\tau)]\Big[e^{-\int_0^\beta d\tau\left[\sum_\alpha\phi^*_\alpha(\tau)(\frac{\partial}{\partial\tau}-\mu)\phi_\alpha(\tau)+H[\phi^*_\alpha(\tau),\phi_\alpha(\tau)]\right]}\\
&\qquad \times\phi_{\alpha_1}(\tau_1)\ldots\phi_{\alpha_n}(\tau_n)\phi^*_{\alpha_{n+1}}(\tau_{n+1})\ldots\phi^*_{\alpha_{2n}}(\tau_{2n})\Big]\;.
\end{aligned}
$$

As preparation for the general case of a many-body Hamiltonian, it is useful to evaluate the partition function for a system of non-interacting particles described by a one-body Hamiltonian. For convenience, we choose a basis in which H_0 is diagonal:

$$
H_0 = \sum_\alpha \epsilon_\alpha a^\dagger_\alpha a_\alpha\;. \qquad (2.68)
$$

The discrete expression for the partition function, Eq. (2.65), may be written

$$
\begin{aligned}
Z_0 &= \lim_{M\to\infty}\prod_\alpha\left[\prod_{k=1}^{n}\int\frac{1}{\mathcal{N}}d\phi^*_k d\phi_k e^{-\sum_{j,k=1}^{M}\phi^*_j S^{(\alpha)}_{jk}\phi_k}\right] \qquad (2.69)\\
&= \lim_{M\to\infty}\prod_\alpha[\det S^{(\alpha)}]^{-\zeta}
\end{aligned}
$$

where, with the convention that the time index increases with increasing row and column index

$$S^{(\alpha)} = \begin{bmatrix} 1 & 0 & & \cdots & 0 & -\varsigma a \\ -a & 1 & 0 & & & 0 \\ 0 & -a & 1 & \ddots & & \vdots \\ & 0 & -a & \ddots & 0 & \\ \vdots & & 0 & \ddots & 1 & 0 \\ 0 & & & \cdots & -a & 1 \end{bmatrix} \qquad \phi = \begin{bmatrix} \phi_1 \\ \phi_2 \\ \\ \vdots \\ \\ \phi_M \end{bmatrix} \tag{2.70a}$$

and where

$$a = 1 - \frac{\beta}{M}(\epsilon_\alpha - \mu) \ . \tag{2.70b}$$

The determinant of $S^{(\alpha)}$ may be evaluated by expanding by minors along the first row

$$\begin{aligned} \lim_{M\to\infty} \det S^{(\alpha)} &= \lim_{M\to\infty} \left[1 + (-1)^{M-1}\varsigma(-a)^M\right] \\ &= \lim_{M\to\infty} \left[1 - \varsigma\left(1 - \frac{\beta(\epsilon_\alpha - \mu)}{M}\right)^M\right] \\ &= 1 - \varsigma e^{-\beta(\epsilon_\alpha - \mu)} \ . \end{aligned} \tag{2.71}$$

Substitution in Eq. (2.69) yields the familiar result for non-interacting particles

$$Z_0 = \prod_\alpha \left(1 - \varsigma e^{-\beta(\epsilon_\alpha - \mu)}\right)^{-\varsigma} \ . \tag{2.72}$$

Note that $\frac{1}{\epsilon}S$ arising from the definition of the functional integral corresponds to a specific discrete approximation to the continuum operator $\frac{\partial}{\partial\tau} + \epsilon_\alpha - \mu$ with periodic or antiperiodic boundary conditions. As seen in Problem 2.6, other discrete approximations to $\frac{\partial}{\partial\tau} + \epsilon_\alpha - \mu$ give inequivalent results emphasizing the fact that Eq. (2.65) rather than the continuum shorthand, Eq. (2.66) is the defining expression.

Finally, we evaluate the single-particle Green's function for non-interacting particles, $\mathcal{G}_0$. Let τ_q correspond to the time $q\frac{\beta}{M}$ and τ_r correspond to the time $r\frac{\beta}{M}$ for integers q and r. Using Eq(2.67) for the non-interacting Hamiltonian, H_0, we obtain:

$$
\begin{aligned}
\mathcal{G}_0(\alpha\tau_q|\gamma\tau_r) &= \frac{1}{Z_0}\mathrm{Tr}\left[Te^{-\int_0^\beta d\tau(H_0-\mu N)}a_\alpha(\tau_q)a_\gamma^\dagger(\tau_r)\right] \\
&= \frac{1}{Z_0}\lim_{M\to\infty}\int\prod_\delta\prod_{k=1}^{M}\frac{1}{\mathcal{N}}d\phi_{\delta,k}^*d\phi_{\delta,k}e^{-\sum\limits_{j,k=1}^{M}\phi_{\delta,j}^*S_{jk}^{(\delta)}\phi_{\delta,k}}\phi_{\alpha,q}\phi_{\gamma,r}^* \\
&= \delta_{\alpha\gamma}\frac{\int\prod_k d\phi_k^*d\phi_k e^{-\sum\limits_{j,k=1}^{M}\phi_j^*S_{jk}^{(\alpha)}\phi_k}\phi_q\phi_r^*}{\int\prod_k d\phi_k^*d\phi_k e^{-\sum\limits_{j,k=1}^{M}\phi_j^*S_{jk}^{(\alpha)}\phi_k}} \\
&= \delta_{\alpha\gamma}\frac{\varsigma\partial^2}{\partial J_q^*\partial J_r}\left.\frac{\int\prod_k d\phi_k^*d\phi_k e^{-\sum_{j,k}\phi_j^*S_{jk}^{(\alpha)}\phi_k+\sum_i(J_i^*\phi_i+\phi_i^*J_i)}}{\int\prod_k d\phi_k^*d\phi_k e^{-\sum_{j,k}\phi_j^*S_{jk}^{(\alpha)}\phi_k}}\right|_{J=J^*=0} \\
&= \delta_{\alpha\gamma}\frac{\varsigma\partial^2}{\partial J_q^*\partial J_r}e^{\sum_{j,k}J_j^*S_{jk}^{(\alpha)-1}J_k}\Big|_{J=J^*=0} \\
&= \delta_{\alpha\gamma}S_{qr}^{(\alpha)-1}\ .
\end{aligned}
\tag{2.73}
$$

The inverse of S in Eq. (2.70a), with a defined by Eq. (2.70b) is

$$
S^{(\alpha)-1} = \frac{1}{1-\varsigma a^M}\begin{bmatrix}
1 & \varsigma a^{M-1} & \varsigma a^{M-2} & \dots & \varsigma a \\
a & 1 & \varsigma a^{M-1} & & \varsigma a^2 \\
a^2 & a & 1 & & \\
\vdots & a^2 & a & & \vdots \\
 & & a^2 & & \\
a^{M-3} & & & & \\
a^{M-2} & a^{M-3} & & & \varsigma a^{M-1} \\
a^{M-1} & a^{M-2} & a^{M-3} & \dots & 1
\end{bmatrix}\ . \tag{2.74}
$$

Hence, for $q \geq r$

$$
\begin{aligned}
\lim_{M\to\infty}S_{q,r}^{(\alpha)-1} &= \lim_{M\to\infty}\frac{a^{q-r}}{1-\varsigma a^M} \\
&= \lim_{M\to\infty}\left(1-\frac{\beta}{M}(\epsilon_\alpha-\mu)\right)^{q-r}\left(1+\frac{\varsigma}{(1-\frac{\beta}{M}(\epsilon_\alpha-\mu))^{-M}-\varsigma}\right) \\
&= e^{-(\epsilon_\alpha-\mu)(\tau_q-\tau_r)}\left(1+\frac{\varsigma}{e^{\beta(\epsilon_\alpha-\mu)}-\varsigma}\right) \\
&= e^{-(\epsilon_\alpha-\mu)(\tau_q-\tau_r)}(1+\varsigma n_\alpha)
\end{aligned}
\tag{2.75a}
$$

where n_α is the familiar Boson or Fermion occupation probability

$$
n_\alpha = \frac{1}{e^{\beta(\epsilon_\alpha-\mu)}-\varsigma}\ . \tag{2.75b}
$$

Similarly, for $q \leq r$

$$\begin{aligned}\lim_{M\to\infty} S^{-1}_{q,r} &= \lim_{M\to\infty} \frac{\varsigma a^{M+q-r}}{1-\varsigma a^M}\\ &= \lim_{M\to\infty} \left(1-\frac{\beta}{M}(\epsilon_\alpha-\mu)\right)^{q-r} \frac{\varsigma}{(1-\frac{\beta}{M}(\epsilon_\alpha-\mu))^{-M}-\varsigma}\\ &= e^{-(\epsilon_\alpha-\mu)(\tau_q-\tau_r)}\varsigma n_\alpha \ .\end{aligned} \tag{2.76}$$

The two results specify the single-particle Green's function when $\tau_q \leq \tau_r$ and when $\tau_q \geq \tau_r$ respectively, so there only remains the case in which creation and annihilation operators act at equal physical times, as occurs for example whenever a second-quantized operator is evaluated at a specific time. Using the fact that the time-ordered product is defined to be equal to a normal-ordered product at equal time, the equal-time propagator may be obtained two equivalent ways. If the operator $a^\dagger_\alpha a_\beta$ is surrounded by evolution operators, subdivision of the interval in the usual way yields

$$\begin{aligned} e^{-\epsilon H}|\phi_k\rangle\langle\phi_k|a^\dagger_\alpha a_\beta e^{-\epsilon H}|\phi_{k-1}\rangle\langle\phi_{k-1}|\ldots \\ = e^{-\epsilon H}|\phi_k\rangle\left(\phi^*_{\alpha,k}\phi_{\beta,k-1}e^{-\epsilon H(\phi^*_k,\phi_{k-1})}+O(\epsilon)\right)\langle\phi_{k-1}|\ldots \ .\end{aligned} \tag{2.77a}$$

Thus, like the operators in H, the creation operator is evaluated one time step later than the annihilation operator, and the appropriate expression for the Green's function corresponding to equal times is Eq. (2.76) rather than Eq. (2.75). Alternatively, the time-ordered product may be written $T[a_\beta(\tau)a^\dagger_\alpha(\tau)] = \varsigma a^\dagger_\alpha(\tau)a_\beta(\tau) = a_\beta(\tau)a^\dagger_\alpha(\tau) - \delta_{\alpha\beta}$ in which case the evolution operator is expressed

$$\begin{aligned} |\phi_{k+1}\rangle\langle\phi_{k+1}|e^{-\epsilon H}a_\alpha|\phi_k\rangle\langle\phi_k|a^\dagger_\beta e^{-\epsilon H}|\phi_{k-1}\rangle\langle\phi_{k-1}|\ldots \\ = |\phi_{k+1}\rangle e^{-\epsilon H(\phi^*_{k+1},\phi_k)}\phi_{\alpha,k}\phi^*_{\beta,k}e^{-\epsilon H(\phi^*_k,\phi_{k-1})}\langle\phi_{k-1}|\ldots\end{aligned} \tag{2.77b}$$

and ϕ_α and ϕ^*_β are evaluated at equal times. Thus, $\langle Ta_\alpha(\tau)a^\dagger_\alpha(\tau)\rangle = S^{-1}_{r,r} - 1 = \varsigma n_\alpha$ as before.

Combining these results, the single-particle Green's function may be written

$$\begin{aligned}\mathcal{G}_0(\alpha\tau|\alpha'\tau') &= \langle Ta_\alpha(\tau)a^\dagger_{\alpha'}(\tau')\rangle\\ &= \delta_{\alpha\alpha'}e^{-(\epsilon_\alpha-\mu)(\tau-\tau')}\left\{\theta(\tau-\tau'-\eta)(1+\varsigma n_\alpha)+\varsigma\theta(\tau'-\tau+\eta)n_\alpha\right\}\\ &= \delta_{\alpha\alpha'}g_\alpha(\tau-\tau'-\eta)\end{aligned} \tag{2.78}$$

where the infinitesimal η serves as a reminder that the second term contributes at equal times. (A convenient mnemonic for the η is the fact that the time τ' associated with the creation operator is always shifted one time step later).

Although we have derived $\mathcal{G}$ carefully as the inverse of the discrete expression appearing in the exponent of the partition function, it may be obtained directly as the inverse of $(\partial_\tau + \epsilon_\alpha - \mu)$ by solving the differential equation

$$(\partial_\tau + \epsilon_\alpha - \mu)\mathcal{G}(\alpha\tau|\alpha\tau') = \delta(\tau-\tau') \tag{2.79a}$$

subject to the boundary condition

$$\mathcal{G}(\alpha\beta|\alpha\tau') = \varsigma\mathcal{G}(\alpha 0|\alpha\tau') \quad . \tag{2.79b}$$

The only ambiguity in the continuum derivation is the result at equal times. Whereas the discrete expression defining the functional integral produces the physical result at equal time, other discrete approximations to the continuum expression may be incorrect (see Problem 2.6).

2.3 PERTURBATION THEORY

In this section, we consider the case of a Hamiltonian which has been decomposed into the sum of a one-body operator H_0 and the residual Hamiltonian V, which in general may contain a one-body interaction as well as many-body interactions, and develop a systematic perturbation expansion in powers of V. The basis will be chosen to diagonalize $H_0 = \sum_\alpha \epsilon_\alpha a^\dagger_\alpha a_\alpha$ and we will write the normal-ordered many-body part as $V(a^\dagger_\alpha a^\dagger_\beta \ldots a_\gamma a_\delta \ldots)$ The starting point is to express the Grand partition function in terms of thermal averages defined with respect to H_0. To establish notation which will be used subsequently, we will define equivalent expressions in terms of operators and functional integrals. As in Eq. (2.67), creation and annihilation operators will be given a formal τ label denoting the time slice upon which they are defined. These operators $\{a^\dagger_\alpha(\tau), a_\alpha(\tau)\}$ are replaced by the coherent state variables $\{\psi^*_\alpha(\tau), \psi_\alpha(\tau)\}$ on the corresponding time slice in the functional integral and should not be confused with the Heisenberg operators $\{a^{(H)\dagger}_\alpha(\tau), a^{(H)}_\alpha\}$ defined in Eq. (2.23). The operator form of the partition function is written

$$\begin{aligned} Z &= \mathrm{Tr}\left[Te^{-\int_0^\beta d\tau\left(\sum_\alpha(\epsilon_\alpha-\mu)a^\dagger_\alpha(\tau)a_\alpha(\tau)+V(a^\dagger_\alpha(\tau)a^\dagger_\beta(\tau)\ldots a_\gamma(\tau)a_\delta(\tau))\right)}\right] \\ &= Z_0\langle e^{-\int_0^\beta d\tau V(a^\dagger_\alpha(\tau)a^\dagger_\beta(\tau)\ldots a_\gamma(\tau)a_\delta(\tau))}\rangle_0 \end{aligned} \tag{2.80a}$$

where the thermal average of an operator F is written

$$\begin{aligned} &\langle F(a^\dagger_\alpha(\tau_i)a^\dagger_\beta(\tau_j)\ldots a_\gamma(\tau_k)a_\delta(\tau_l)\ldots)\rangle_0 \\ &\quad = \frac{1}{Z_0}\mathrm{Tr}\left[Te^{-\int_0^\beta d\tau\sum_\alpha(\epsilon_\alpha-\mu)a^\dagger_\alpha(\tau)a_\alpha(\tau)}F(a^\dagger_\alpha(\tau_i)a^\dagger_\beta(\tau_j)\ldots a_\gamma(\tau_k)a_\delta(\tau_l)\ldots)\right] \quad . \end{aligned} \tag{2.80b}$$

It is crucial to note that all the operators in F are subject to the time-ordering operator included in the definition of $\langle F\rangle_0$.

Equivalently, in terms of functional integrals, the partition function may be written

$$\begin{aligned} Z &= \int\limits_{\psi(\beta)=\varsigma\psi(0)} \mathcal{D}(\psi^*_\alpha\psi_\alpha)e^{-\int_0^\beta dt\left(\sum_\alpha\psi^*_\alpha(\tau)(\partial_t+\epsilon_\alpha-\mu)\psi_\alpha(\tau)+V(\psi^*_\alpha(\tau),\psi^*_\beta(\tau)\ldots,\psi_\gamma(\tau),\psi_\delta(\tau)\ldots)\right)} \\ &= Z_0\langle e^{-\int_0^\beta d\tau V(\psi^*_\alpha(\tau),\psi^*_\beta(\tau)\ldots,\psi_\gamma(\tau),\psi_\delta(\tau)\ldots)}\rangle_0 \end{aligned} \tag{2.81a}$$

where the thermal average is defined

$$\langle F(\psi^*_\alpha(\tau_i),\psi^*_\beta(\tau_j)\ldots\psi_\gamma(\tau_k)\psi_\delta(\tau_l))\rangle_0$$

$$= \frac{1}{Z_0}\int\limits_{\psi(\beta)=\zeta\psi(0)} D(\psi^*_\alpha\psi_\alpha)e^{-\sum\limits_\alpha \int_0^\beta dt\psi^*_\alpha(\partial_t+\epsilon_\alpha-\mu)\psi_\alpha} \tag{2.81b}$$

$$\times F(\psi^*_\alpha(\tau_i)\psi^*_\beta(\tau_j)\ldots\psi_\gamma(\tau_k)\psi_\delta(\tau_l)\ldots)\ .$$

Note that the time-ordering which was explicit for operators is implicit because the functional integral always represents time-ordered products.

The partition function of the non-interacting system, Z_0, appearing in Eqs. (2.58) and (2.81) may be written

$$Z_0 = \mathrm{Tr}\left[Te^{-\int_0^\beta d\tau\sum_\alpha(\epsilon_\alpha-\mu)a^\dagger_\alpha(\tau)a_\alpha(\tau)}\right] \tag{2.82a}$$

$$= \int\limits_{\psi(\beta)=\zeta\psi(0)} D(\psi^*_\alpha\psi_\alpha)e^{-\sum_\alpha\int_0^\beta d\tau\psi^*_\alpha(\tau)(\partial_\tau+\epsilon_\alpha-\mu)\psi_\alpha(\tau)}\ . \tag{2.82b}$$

Because of the equivalence of Eqs. (2.80) and (2.81), we will henceforth pass freely between thermal averages $\langle\ \rangle_0$ of operators and of complex or Grassman variables. Note that analogous expression will also be introduced later for thermal averages $\langle\ \rangle$ defined with respect to the full Hamiltonian by replacing H_0 by H in Eqs. (2.80b) and (2.81b).

The perturbation expansion is obtained by expanding Eq. (2.81a) in a power series

$$\frac{Z}{Z_0} = \sum_{n=0}^\infty \frac{(-1)^n}{n!}\int_0^\beta d\tau_1\ldots d\tau_n \tag{2.83}$$

$$\langle V(\psi^*_\alpha(\tau_1)\ldots,\psi_\gamma(\tau_1)\ldots)\ldots V(\psi^*_\alpha(\tau_n)\ldots,\psi_\gamma(\tau_n)\ldots)\rangle_0\ .$$

We will proceed by deriving a form of Wick's theorem to evaluate the thermal averages of the products of ψ^* and ψ which arise in Eq. (2.83) and then develop a systematic set of rules for constructing Feynman diagrams.

WICK'S THEOREM

In the form we will use it, Wick's theorem corresponds to the following identity for the integral of a product of a polynomial with a Gaussian

$$\frac{\int D(\psi^*\psi)\psi_{i_1}\psi_{i_2}\ldots\psi_{i_n}\psi^*_{j_n}\ldots\psi^*_{j_2}\psi^*_{j_1}e^{-\sum\limits_{ij}\psi^*_iM_{ij}\psi_j}}{\int D(\psi^*\psi)e^{-\sum\limits_{ij}\psi^*_iM_{ij}\psi_j}} = \sum_P \zeta^P M^{-1}_{i_{Pn},j_n}\ldots M^{-1}_{i_{P1},j_{P1}} \tag{2.84}$$

where, as usual (ψ^*,ψ) denote complex or Grassman variables, $D(\psi^*\psi)$ is the appropriate measure, and we have simplified the notation by letting j denote the state and time labels. We will first prove this identity and then relate it to evaluation of Green's functions, thermal averages, and the traditional statement of Wick's theorem.

The identity Eq. (2.84) is a generalization of the result Eq. (2.73) and may be derived in the same way using the generating function

$$G(J^*,J) = \frac{\int D(\psi^*\psi)e^{-\sum_{ij}\psi_i^* M_{ij}\psi_j + \sum_i (J_i^*\psi_i + \psi_i^* J_i)}}{\int D(\psi^*\psi)e^{-\sum_{ij}\psi_i^* M_{ij}\psi_j}} \tag{2.85}$$
$$= e^{\sum_{ij} J_i^* M_{ij}^{-1} J_j} .$$

Differentiation of the first line of Eq. (2.85) with respect to the sources J and J^* yields

$$\left.\frac{\delta^{2n} G}{\delta J_{i_1}^* \ldots \delta J_{i_n}^* \delta J_{j_n} \ldots \delta J_{j_1}}\right|_{\substack{J=0\\ J^*=0}} = (\varsigma)^n \frac{\int D(\psi^*\psi)e^{-\sum \psi_i^* M_{ij}\psi_j}\psi_{i_1}\ldots\psi_{i_n}\psi_{j_n}^*\ldots\psi_{j_1}^*}{\int D(\psi^*\psi)e^{-\sum\psi_i^* M_{ij}\psi_j}} . \tag{2.86}$$

Note that in deriving this result, we used the fact that all the terms in the exponent are even in the J's and ψ's and thus commute with ψ, ψ^*, J and J^* and the fact that an odd number of interchanges is required for each differentiation with respect to J. Differentiation of the second line of Eq. (2.85) yields

$$\left.\frac{\delta^{2n}}{\delta J_{i_1}^* \ldots \delta J_{i_n}^* \delta J_{j_n} \ldots \delta J_{j_1}}\left(e^{\sum_{ij} J_i^* M_{ij} J_i}\right)\right|_{\substack{J^*=0\\ J=0}}$$
$$= \varsigma^n \left.\frac{\delta^n}{\delta J_{i_1}^* \ldots \delta J_{i_n}^*}\left(\sum_{k_n} J_{k_n}^* M_{k_n j_n}^{-1}\right)\ldots\left(\sum_{k_1} J_{k_1}^* M_{k_1 j_1}^{-1}\right)e^{\sum_{ij} J_i^* M_{ij}^{-1} J_j}\right|_{\substack{J^*=0\\ J=0}}$$
$$= \varsigma^n \sum_P \varsigma^P M_{i_{Pn},j_n}^{-1} \ldots M_{i_{P1},j_1}^{-1} . \tag{2.87}$$

Equating the two expressions Eqs. (2.86) and (2.87) proves Eq. (2.84).

We now apply this identity to the case of physical interest by defining M_{ij} to be the discrete matrix representing $(\partial_t + H_0 - \mu)$, Eq. (2.69), replacing ψ_j by $\psi_{\alpha,k}$, where α denotes the basis states in the diagonal representation and k denotes the time point on a mesh of M points with $\Delta\tau = \frac{\beta}{M}$. As shown in the preceding section, M_{ij}^{-1} is then the single particle Green's function, Eq. (2.78):

$$\mathcal{G}_0(\alpha_1\tau_1|\alpha_2\tau_2) = \frac{\int D(\psi^*\psi)\psi_{\alpha_1}(\tau_1)\psi_{\alpha_2}^*(\tau_2)e^{-\int dt \sum_\alpha \psi_\alpha^*(\partial_\tau + \epsilon_\alpha - \mu)\psi_\alpha}}{\int D(\psi^*\psi)e^{-\int dt\sum_\alpha \psi_\alpha(\partial_\tau+\epsilon_\alpha-\mu)\psi_\alpha}}$$
$$= (\partial_\tau + \epsilon_\alpha - \mu)^{-1}_{\alpha_1\tau_1;\alpha_2\tau_2} \tag{2.88}$$
$$\equiv \delta_{\alpha_1\alpha_2}\mathcal{G}_{\alpha_1}(\tau_1 - \tau_2 - \eta) .$$

The identity (2.84) then states that the n-particle Green's function for a non-interacting system is the sum of all permutations of the products of one-particle Green's functions

$$\mathcal{G}^{(n)}(\alpha_1\tau_1\ldots\alpha_n\tau_n|\alpha_1'\tau_1'\ldots\alpha_n'\tau_n')$$
$$= \sum_P \varsigma^P \delta_{\alpha_{P1},\alpha_1'}\ldots\delta_{\alpha_{Pn}\alpha_n'}\mathcal{G}_{\alpha_1'}(\tau_{P1}-\tau_1')\ldots\mathcal{G}_{\alpha_n'}(\tau_{Pn}-\tau_n') . \tag{2.89}$$

We may establish contact with the traditional statement of Wick's theorem by defining contractions of time-dependent operators. Let $\tilde{a}_\alpha(\tau)$ denote any creation operator $a^\dagger_\alpha(\tau)$ or annihilation operator $a_\alpha(\tau)$ and let $\tilde{\psi}_\alpha(\tau)$ denote the corresponding complex or Grassman variable $\psi^*_\alpha(\tau)$ or $\psi_\alpha(\tau)$. A contraction is then defined as

$$\overbrace{\tilde{a}_\alpha(\tau)\tilde{a}_{\alpha'}(\tau')} = \langle T[\tilde{a}_\alpha(\tau)\tilde{a}_{\alpha'}(\tau')]\rangle_0 = \langle \tilde{a}_\alpha(\tau)\tilde{a}_{\alpha'}(\tau')\rangle_0 \tag{2.90}$$

where the thermal average is defined by Eq. (2.80) and the explicit T-product may be omitted because the operators are necessarily time-ordered by the definition of the thermal average. An equivalent definition is given by

$$\overbrace{\tilde{\psi}_\alpha(\tau)\tilde{\psi}_{\alpha'}(\tau')} = \langle \tilde{\psi}_\alpha(\tau)\tilde{\psi}_{\alpha'}(\tau')\rangle_0 \tag{2.91}$$

where the thermal average is given by Eq. (2.81). From Eq. (2.88) we obtain

$$\begin{aligned}\overbrace{a_\alpha(\tau)a^\dagger_{\alpha'}(\tau')} &= \overbrace{\psi_\alpha(\tau)\psi^*_{\alpha'}(\tau')}\\ &= \delta_{\alpha\alpha'}g_\alpha(\tau-\tau')\end{aligned} \tag{2.92a}$$

and

$$\begin{aligned}\overbrace{a^\dagger_{\alpha'}(\tau')a_\alpha(\tau)} &= \overbrace{\psi^*_{\alpha'}(\tau')\psi_\alpha(\tau)}\\ &= \zeta\delta_{\alpha\alpha'}g_\alpha(\tau-\tau')\ .\end{aligned} \tag{2.92b}$$

Because the expectation value of two creation or annihilation operators is zero in any state of definite particle number, or by explicit integration of the corresponding Gaussian integral, the following contractions vanish:

$$\overbrace{a^\dagger_\alpha(\tau)a^\dagger_{\alpha'}(\tau')} = \overbrace{\psi^*_\alpha(\tau)\psi^*_{\alpha'}(\tau')} = 0 \tag{2.92c}$$

and

$$\overbrace{a_\alpha(\tau)a_{\alpha'}(\tau')} = \overbrace{\psi_\alpha(\tau)\psi_{\alpha'}(\tau')} = 0\ . \tag{2.92d}$$

Given these definitions, note that with $M \equiv (\partial_t + H_0 - \mu)$ the left hand side of the identity Eq. (2.84) is $\langle \psi_{i_1}\psi_{i_2}\ldots\psi_{i_n}\psi^*_{j_n}\ldots\psi^*_{j_2}\psi^*_{j_1}\rangle_0$ and each factor in the right hand side corresponds to a contraction $\overbrace{\psi_i\psi^*_j} = (\partial_\tau + H_0 - \mu)^{-1}_{ij}$. Thus, in this case, the thermal average is given by the sum over all complete sets of contractions, where a complete contraction is a configuration in which each ψ is contracted with a ψ^* and the overall sign is specified by ζ^P where P is the permutation such that $\psi_{i_{P_n}}$ is contracted with $\psi^*_{j_n}$. If one considered the expectation value of a product of an unequal number of ψ's and ψ^*'s, it would still be equal to the sum of all contractions since the complete expectation value would vanish and at least one contraction in each complete set of contractions would also vanish. Thus, the general statement of Wick's theorem, using the notation of Eqs. (2.90) and (2.91) to denote creation or annihilation operators, is the following:

$$\begin{aligned}\langle T[\tilde{a}_{\alpha_1}(\tau_1)\ldots\tilde{a}_{\alpha_n}(\tau_n)]\rangle_0 &= \langle \tilde{\psi}_{\alpha_1}(\tau_1)\ldots\tilde{\psi}_{\alpha_n}(\tau_n)\rangle_0\\ &= \sum \text{all complete contractions}\ .\end{aligned} \tag{2.93}$$

For example,

$$\begin{aligned}\langle T[a_{\alpha_1}(\tau_1)a_{\alpha_2}(\tau_2)a^\dagger_{\alpha'_2}(\tau'_2)a^\dagger_{\alpha'_1}(\tau'_1)]\rangle &= a_{\alpha_1}(\tau_1)a_{\alpha_2}(\tau_2)a^\dagger_{\alpha'_2}(\tau'_2)a^\dagger_{\alpha'_1}(\tau'_1) \\ &\quad + a_{\alpha_1}(\tau_1)a_{\alpha_2}(\tau_2)a^\dagger_{\alpha'_2}(\tau'_2)a^\dagger_{\alpha'_1}(\tau'_1) \\ &= \delta_{\alpha_1\alpha'_1}\delta_{\alpha_2\alpha'_2}g_{\alpha_1}(\tau_1-\tau'_1)g_{\alpha_2}(\tau_2-\tau'_2) \\ &\quad + \zeta\delta_{\alpha_1\alpha'_2}\delta_{\alpha_2\alpha'_1}g_{\alpha_1}(\tau_1-\tau'_2)g_{\alpha_2}(\tau_2-\tau'_1) \ .\end{aligned} \tag{2.94}$$

The utility of Wick's theorem is now evident, since the power series for $\frac{Z}{Z_0}$ in Eq. (2.83) may be represented as the sum of all contractions in which single-particle propagators $g_\alpha(\tau-\tau')$ connect the products of the potential V in all possible ways. Although we have used a diagonal basis for notational convenience, Wick's theorem holds in a general basis $|\alpha\rangle$ which is not an eigenbasis of H_0, in which case the only difference is that the single-particle propagators are no longer diagonal. An alternative derivation of Wick's theorem as an operator identity is presented in Problem 2.8.

LABELED FEYNMAN DIAGRAMS

Using Wick's theorem, the perturbation expansion for $\frac{Z}{Z_0}$ is obtained from the series Eq. (2.83) by enumerating all the complete sets of contractions contributing to the thermal average of products of the potential V. It is convenient to represent this expansion in terms of diagrams, and in this section we will develop simple rules for constructing diagrams which provide an economical representation of the enormous number of contractions. For simplicity, we will consider the special case in which V is a two-body interaction

$$V(\psi^*(\tau),\psi(\tau)) = \frac{1}{2}\sum_{\alpha\beta\gamma\delta}(\alpha\beta|v|\gamma\delta)\psi^*_\alpha(\tau)\psi^*_\beta(\tau)\psi_\delta(\tau)\psi_\gamma(\tau) \tag{2.95}$$

and subsequently address the generalization of the rules to n-body interactions. The term of order n in V in Eq. (2.83) for $\frac{Z}{Z_0}$ in this case is

$$\begin{aligned}\left(\frac{Z}{Z_0}\right)_n &= \frac{(-1)^n}{n!2^n}\sum_{\substack{\alpha_1\beta_1\\ \gamma_1\delta_1}}\cdots\sum_{\substack{\alpha_n\beta_n\\ \gamma_n\delta_n}}(\alpha_1\beta_1|v|\gamma_1\delta_1)\ldots(\alpha_n\beta_n|v|\gamma_n\delta_n)\int_0^\beta d\tau_1\ldots d\tau_n \\ &\quad\times\langle\psi^*_{\alpha_1}(\tau_1)\psi^*_{\beta_1}(\tau_1)\psi_{\delta_1}(\tau_1)\psi_{\gamma_1}(\tau_1)\ldots\psi^*_{\alpha_n}(\tau_n)\psi^*_{\beta_n}(\tau_n)\psi_{\delta_n}(\tau_n)\psi_{\gamma_n}(\tau_n)\rangle_0 \ .\end{aligned} \tag{2.96}$$

A faithful representation of all the complete sets of contractions contributing to $(\frac{Z}{Z_0})_n$ is given by labeled diagrams defined to reproduce each of the contributions to Eq. (2.96). Each contraction will join some $\psi^*_{\eta_i}(\tau_i)$ to some $\psi_{\eta_j}(\tau_j)$ yielding a propagator $\delta_{\eta_i\eta_j}g_{\eta_i}(\tau_j-\tau_i)$ which will be represented by a directed line originating at $\psi^*_{\eta_i}(\tau_i)$ and terminating at $\psi_{\eta_j}(\tau_i)$. Each interaction will yield a vertex having two incoming lines, corresponding to $\psi_{\delta_i}(\tau_i)\psi_{\gamma_i}(\tau_i)$ and two outgoing lines, corresponding to $\psi^*_{\alpha_i}(\tau_i)\psi^*_{\beta_i}(\tau_i)$. The n interactions in Eq. (2.96) will thus be represented by n vertices with two outgoing lines α_i, β_i and two incoming lines γ_i, δ_i acting at time τ_i

corresponding to the factor

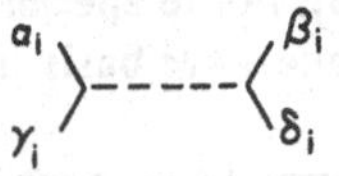

$$= (\alpha_i\beta_i|v|\gamma_i\delta_i) \; . \tag{2.97}$$

Each outgoing line will be connected to an ingoing line with a directed line corresponding to the propagator

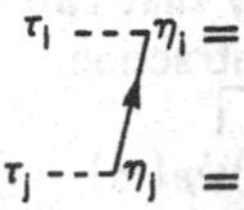

$$\begin{aligned} &= \delta_{\eta_i\eta_j} g_{\eta_i}(\tau_j - \tau_i) \\ &= \delta_{\eta_i\eta_j} e^{-(\epsilon_{\eta_i}-\mu)(\tau_j-\tau_i)}[(1+\zeta n_{\eta_i})\theta(\tau_j-\tau_i) + \zeta n_{\eta_i}\theta(\tau_j-\tau_i)] \; . \end{aligned} \tag{2.98}$$

The set of all possible ways of connecting interactions with propagators corresponds precisely to the set of all the contractions arising from Wick's theorem, so summation over a complete set of distinct diagrams will faithfully reproduce each of the desired contractions. At order $n = 1$, there are two diagrams corresponding to the two contractions contributing to Eq. (2.96).

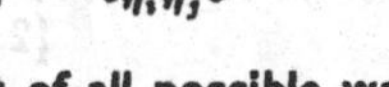

$$\begin{aligned} &= -\frac{1}{2}\int_0^\beta d\tau \sum_{\alpha\beta\gamma\delta} \psi_\alpha^*(\tau)\psi_\beta^*(\tau)\psi_\delta(\tau)\psi_\gamma(\tau)(\alpha\beta|v|\gamma\delta) \\ &= -\frac{1}{2}\int_0^\beta d\tau \sum_{\gamma\delta} g_\gamma(0)g_\delta(0)(\gamma\delta|v|\gamma\delta) \end{aligned} \tag{2.99a}$$

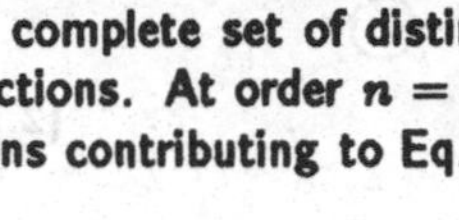

$$\begin{aligned} &= -\frac{1}{2}\int_0^\beta d\tau \sum_{\alpha\beta\gamma\delta} \psi_\alpha^*(\tau)\psi_\beta^*(\tau)\psi_\delta(\tau)\psi_\gamma(\tau)(\alpha\beta|v|\gamma\delta) \\ &= -\frac{1}{2}\zeta \int_0^\beta d\tau \sum_{\gamma\delta} g_\gamma(0)g_\delta(0)(\delta\gamma|v|\gamma\delta) \; . \end{aligned} \tag{2.99b}$$

Note that because the propagator is diagonal in the single-particle indices, each directed line in a diagram is labeled by a single index which is summed over all states. Recalling the form of the propagator at equal times, we obtain

$$\frac{Z}{Z_0} = 1 - \frac{\beta}{2}\sum_{\gamma\delta} n_\gamma n_\delta[(\gamma\delta|v|\gamma\delta) + \zeta(\delta\gamma|v|\gamma\delta)] \; . \tag{2.100a}$$

For Fermions, this is just the finite-temperature generalization of the expectation value of the energy in a Slater determinant composed of eigenfunctions of H_0. Expanding the partition function to first order in the grand canonical potential yields

$$\Omega = \Omega_0 + \frac{1}{2}\sum_{\gamma\delta} n_\gamma n_\delta[(\gamma\delta|v|\gamma\delta) + \zeta(\delta\gamma|v|\gamma\delta)] \; . \tag{2.100b}$$

For Fermions in the low temperature limit, the occupation numbers simply sum over the occupied states reproducing the interaction energy derived in Problem 1.6. Although in general the single-particle eigenfunctions $|\alpha\rangle$ may not coincide with the self-consistent Hartree-Fock wave functions (see Problem 1.6), in the special case of uniform matter, any translationally-invariant H_0 will yield a plane wave basis in which case Eq. (2.100) corresponds to the Hartree-Fock energy.

The diagram rules presented thus far correctly account for all the contractions, propagators, and matrix elements, and we have only to augment them by the rules for the overall sign and factor and a careful general definition of summation over all distinct diagrams.

In higher orders, one could imagine deforming a diagram in such a way that two different drawings correspond to the same contraction. For example the contraction

$$\psi^*_\rho(\tau_2)\psi^*_\sigma(\tau_2)\psi_\sigma(\tau_2)\psi_\delta(\tau_2)\psi^*_\beta(\tau_1)\psi^*_\delta(\tau_1)\psi_\gamma(\tau_1)\psi_\alpha(\tau_1)\psi^*_\alpha(\tau_3)\psi^*_\gamma(\tau_3)\psi_\rho(\tau_3)\psi_\beta(\tau_3) \tag{2.101a}$$

may be drawn the following ways

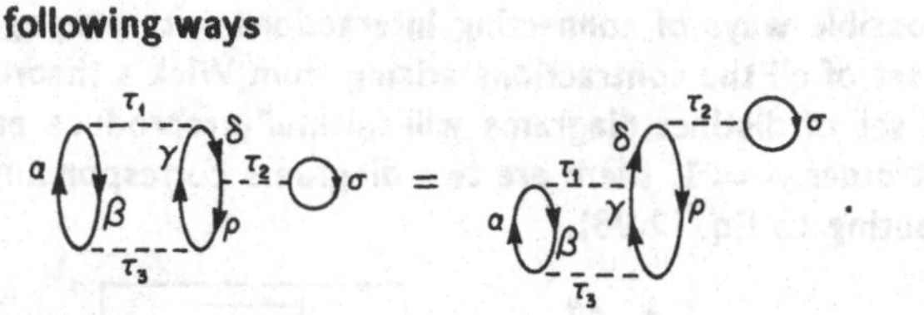

(2.101b)

Although these drawings look different, they are not distinct diagrams; rather they are merely deformations of the same diagram. The essential feature of the propagator line labeled δ is that it begins at the right side of the interaction at τ_1, and ends at the left side of the interaction at τ_2 and whether the diagram is drawn so that the arrow is pointing upward or downward is immaterial.

For subsequent developments, it will be useful to introduce a precise general definition of distinct diagrams. Imagine that the interactions and propagators comprising a diagram are arbitrarily flexible and may be lifted off the plane upon which they are drawn and twisted and deformed in all possible ways. All labels and arrows associated with interactions and propagators, however, must be retained as the diagram is deformed. Two diagrams are defined to be distinct if they cannot be made to coincide with respect to topological structure, direction of arrows, and labels by some deformation. By this definition, the two diagrams drawn above are clearly not distinct, because deforming the left diagram by raising the interaction τ_2 above τ_1 will make it coincide precisely with the right diagram. The summation over all distinct labeled diagrams defined in this way then correctly counts each contraction once and only once.

Since the state label on each propagator is summed over all single-particle states, it is superfluous to include these labels on each diagram. The only essential role these labels have played thus far is to distinguish the variables ψ^*_α and ψ_γ associated with the left-hand side of the interaction $(\alpha\beta|v|\gamma\delta)$ from the variables ψ^*_β and ψ_δ associated with the right. Thus, we may make the replacement

(2.102)

and a labeled diagram is now specified by the times τ_i, the arrows on propagators, and the left (L) and right (R) designation on interactions.

Finally, we must consider the overall factor associated with each diagram. Each term of nth order in Eq. (2.96) contains the overall factor $\frac{(-1)^n}{2^n n!}$ multiplied by a sign associated with the specific contractions (which is always positive for Bosons). Each contraction will result in a diagram in which the single-particle propagators form some number of closed loops, n_c. For example the first-order direct diagram, hboxEq. (2.99a) has $n_c = 2$, the first-order exchange diagram, Eq. (2.99b) has $n_c = 1$ and the contraction in Eq. (2.101) has $n_c = 3$. Each closed loop corresponds to a cycle of interactions which begins at the left or right side of some vertex, connects to one side of another vertex, connects to some side of yet another vertex and so on until it finally returns to the original side of the original vertex. For example, the contraction in Eq. (2.101) generating the closed loop comprised of the γ, δ, ρ propagators may be rewritten schematically:

$$\psi_L^*\psi_R^*\psi_R\psi_L \quad \psi_L^*\psi_R^*\psi_R\psi_L \quad \psi_L^*\psi_R^*\psi_R\psi_L \tag{2.103a}$$

where we have specifically indicated the variables corresponding to the left and right sides of each vertex and suppresed all other labels since we are only concerned with the signs. Since at each vertex ψ_L is separated from ψ_L^* by two variables, $\psi_L^*\psi_R^*\psi_R\psi_L$ may be rewritten $\psi_L^*\psi_L\psi_R^*\psi_R$ and the closed loop has the form

$$\psi_L^*\psi_L\psi_R^*\psi_R \quad \psi_L^*\psi_L\psi_R^*\psi_R \quad \psi_L^*\psi_L\psi_R^*\psi_R \ . \tag{2.103b}$$

The pairs included in the cycle corresponding to the closed loop under consideration may be reordered without affecting the sign until the cycle of contractions has the following form:

$$\psi^*\psi \quad \psi^*\psi \quad \psi^*\psi \ldots \psi^*\psi \ . \tag{2.103c}$$

The sign associated with this closed loop is now obvious. By Eq. (2.92) the interior contractions $\psi\psi^*$ each yield $+g$ whereas the outer contraction $\psi^*\psi$ produces the factor ςg. Any closed loop may be expressed in a similar form, in which the pairs $\psi^*\psi$ involved in the cycle of contractions are separated by pairs of variables not involved in the contraction. The sign is not altered by commuting all the pairs included in the cycle to the left of all the other pairs. Thus the contribution of a single cycle has been separated from the rest of the contractions, which may subsequently be analyzed by similar decomposition into cycles. Hence, each closed loop contributes a factor ς in addition to the propagators already included in the diagram rules and a diagram with n_L closed loops thus acquires a factor ς^{n_L}.

In summary, we have now derived the rules for constructing labeled diagrams which provide a faithful representation of the complete set of contractions contributing to the nth order perturbation expansion of $\frac{Z}{Z_0}$. The rules stated below are distinguished from subsequent variants by the subscript L indicating that the diagrams are labeled.

1_L Draw all distinct labeled diagrams composed of n vertices 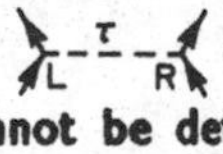

connected by directed lines . Two diagrams are distinct if they cannot be deformed so as to coincide completely, including all times labels τ_i, left-right labels L-R and the direction of arrows on propagators. For each distinct diagram, evaluate the contribution as follows.

2_L Assign a single-particle index to each directed line and include the corresponding factor

$$= g_\alpha(\tau - \tau') = e^{-(\epsilon_\alpha - \mu)(\tau - \tau')}[(1 + \varsigma n_\alpha)\theta(\tau - \tau' - \eta) + \varsigma n_\alpha \theta(\tau' - \tau + \eta)]$$

where the infinitesimal η is included in the θ functions to indicate that the second term is to be used at equal times.

3_L For each vertex, include the factor

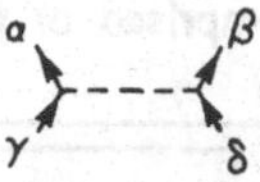

$$= (\alpha\beta|v|\gamma\delta) \ .$$

4_L Sum over all single-particle indices and integrate all times over the interval $[0, \beta]$.

5_L Multiply the result by the factor $\frac{(-1)^n}{n!2^n}\varsigma^{n_L}$ where n_L is the number of closed loops of single-particle propagators in the diagram.

UNLABELED FEYNMAN DIAGRAMS

The number of labeled diagrams proliferates dramatically with order. At order n in the interaction there are $2n$ variables ψ^* to contract with $2n$ variables ψ, yielding $(2n)!$ diagrams. At third order there are thus 720 diagrams and at fourth order 40320 diagrams so that combinations quickly become unmanageable. Since many distinct diagrams have the same numerical contribution, it is worthwhile investing some additional effort in developing a systematic method to account efficiently for all diagrams which have the same contribution.

For a general interaction v which has no special properties other than the symmetry

$$(\alpha\beta|v|\gamma\delta) = (\beta\alpha|v|\delta\gamma) \tag{2.104}$$

there are two types of transformations which leave the value of a diagram invariant: permutation of the time labels and exchange of the extremities of each vertex. Since all the time labels $\tau_1\tau_2\ldots\tau_n$ are integrated over the same interval $[0,\beta]$, any permutation of the time labels leaves the value of a diagram invariant. For n interactions, there are $n!$ permutations which yield the same value. Any vertex in a diagram is connected to the rest of the diagram by four propagators, and we may schematically represent the dependence of a diagram Γ upon one particular vertex as follows

$\Gamma =$ 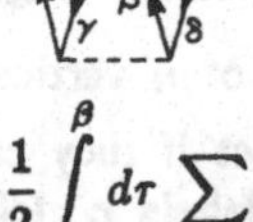

(2.105a)

$$= \frac{1}{2}\int_0^\beta d\tau \sum_{\alpha\beta\gamma\delta} (\alpha\beta|v|\gamma\delta) g_\alpha(\tau_\alpha - \tau) g_\gamma(\tau - \tau_\gamma)$$
$$\times\, g_\beta(\tau_\beta - \tau) g_\delta(\tau - \tau_\delta) F_{\alpha\beta\gamma\delta}(\tau_\alpha, \tau_\beta, \tau_\gamma, \tau_\delta) \ .$$

The graph Γ' obtained by exchanging the extremities of the vertex is

$\Gamma' =$ 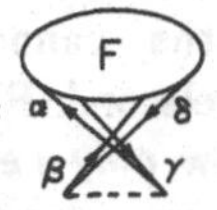

(2.105b)

$$= \frac{1}{2}\int_0^\beta d\tau \sum_{\alpha\beta\gamma\delta} (\beta\alpha|v|\delta\gamma) g_\alpha(\tau_\alpha - \tau) g_\gamma(\tau - \tau_\gamma)$$
$$\times\, g_\beta(\tau_\beta - \tau) g_\delta(\tau - \tau_\delta) F_{\alpha\beta\gamma\delta}(\tau_\alpha, \tau_\beta, \tau_\gamma, \tau_\delta) \ .$$

By the symmetry of the matrix element, Eq. (2.104) the value of the diagram Γ' equals that of Γ. For a diagram with n interactions there are 2^n exchanges which yield the same value. Note that extremity exchange symmetry corresponds to interchanging the spatial integration variables in a potential matrix element such as Eq. (1.109), so that all the transformations leaving a labeled Feynman diagram invariant correspond to permutations of spatial or time variables of integration.

The most general transformation which leaves the value of a labeled diagram invariant is a combination of a permutation of the time labels and an exchange of vertex extremities. Some such transformations acting on a given diagram generate distinct diagrams while others simply generate deformations of the original diagram which must not be counted as distinct diagrams. Thus, the combinatorial task facing us is to calculate the number of distinct diagrams which have the same value so that we can multiply this number by the value of any one such diagram.

To this end, we note that the set of transformations of an nth order diagram generated by any permutation of the time labels and any exchange of extremities of vertices is a group, which we will denote by G, and has $2^n n!$ elements. Consider the action of this group of transformations, G, on a labeled diagram Γ. Some set of transformations G_Γ will transform Γ into a deformation of itself and the rest of the transformations will yield diagrams which are distinct from Γ. The set G_Γ defined in this way is a subgroup of G.

We define the symmetry factor S of the diagram Γ as the number of deformations of Γ generated by the action of G. By the definition of G_Γ, S is the number of elements of G_Γ and since G_Γ is a subgroup of G, S is a divisor of $2^n n!$. Now consider a graph Γ' distinct from Γ which is obtained from Γ by a transformation in G. Since Γ' corresponds to some permutation of the τ and LR labels on Γ, it transforms into a deformation of itself under G_Γ and S of the remaining graphs are therefore deformations of Γ'. Continuing in this way by selecting a graph distinct from all the previous graphs and identifying its S deformations, the $2^n n!$ diagrams generated by G acting on Γ can be grouped into $\frac{2^n n!}{S}$ sets of S diagrams, such that the diagrams in each set are deformations of each other and diagrams in different sets are distinct. We therefore conclude that the group G of transformations which leave the value of a labeled diagram invariant generates exactly $\frac{2^n n!}{S}$ distinct diagrams with the same value. All distinct diagrams will thus be correctly counted if we take the value of one diagram and multiply by $\frac{2^n n!}{S}$.

Finally, a single diagram from the set of $2^n n!$ diagrams generated by the group of transformations G is specified by an unlabeled diagram. An unlabeled diagram is obtained from a labeled diagram by removing the time labels and the L-R labels on the vertices. Thus it is composed of completely unlabeled vertices connected by directed lines. As before, two diagrams are distinct if they cannot be deformed so as to coincide. However, since there are now no time labels or L-R labels, the condition that two diagrams coincide is less stringent: they now only need to have the same topological structure and the same direction of arrows on propagators. With the time and L-R labels removed, all the permutation and extremity exchange transformations in G transform an unlabeled diagram into a deformation of itself. Thus the contribution of all contractions, that is of all distinct labeled diagrams, is obtained by calculating all distinct unlabeled diagrams and multiplying by $\frac{2^n n!}{S}$. Note that the factor $2^n n!$ cancels the factor $\frac{1}{2^n n!}$ in Eq. (2.96).

The rules for calculating the nth order contribution to the perturbation expansion of $\frac{Z}{Z_0}$ using unlabelled Feynman diagrams are summarized below. These rules are designated by the subscript F for Feynman and we will subsequently assume Feynman diagrams to be unlabled unless explicitly stated to the contrary.

1_F Draw all distinct unlabeled diagrams composed of n vertices  connected by directed lines . Two diagrams are distinct if they cannot be deformed so as to coincide completely including the direction of arrows on propagators. For each distinct unlabeled diagram, evaluate the contribution as follows.

2_F Calculate the symmetry factor S for the diagram. This may be accomplished by adding times and L-R labels to the diagram to make it a labeled diagram. Then S is equal to the number of transformations composed of time permutations and vertex extremity exchanges which transform the labeled diagram into a deformation of itself.

3_F Assign a time label τ_i to each vertex, and a single-particle index to each directed line. For each directed line include the factor

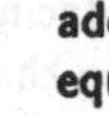

$$= g_\alpha(\tau-\tau') = e^{-(\epsilon_\alpha-\mu)(\tau-\tau')}[(1+\zeta n_\alpha)\theta(\tau-\tau'-\eta)+\zeta n_q\theta(\tau'-\tau+\eta)] \, .$$

4_F For each vertex, include the factor

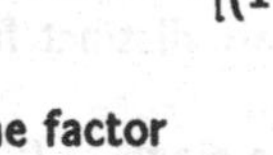

$$= (\alpha\beta|v|\gamma\delta) \, .$$

5_F Sum over all single-particle indices and integrate all times over the interval $[0,\beta]$

6_F Multiply the result by the factor $\frac{(-1)^n}{S}\zeta^{n_L}$ where n_L is the number of closed loops and S is the symmetry factor.

At this point, some detailed examples may be useful in clarifying the definitions and rules. First we consider the definition of distinct diagrams. Considered as labeled diagrams, the following sets of diagrams are distinct

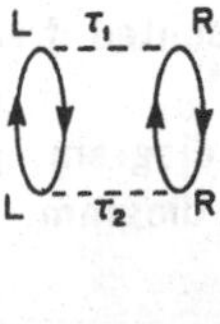

$\neq$

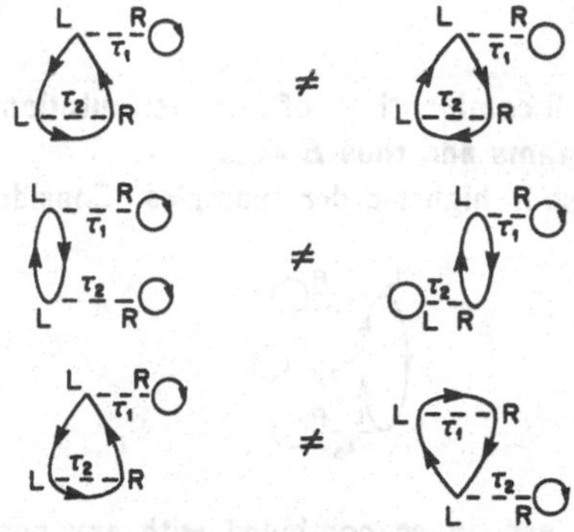

(2.106)

In the first three cases, any deformation which superimposes the directed propagators interchanges a set of L-R labels and in the last case it interchanges the $\tau_1 - \tau_2$ labels. However, when these diagrams are considered as unlabeled diagrams, in each case the left diagram is a deformation of the right diagram and thus not distinct.

The simplest examples of symmetry factors for unlabeled Feynman diagrams are the first-order diagrams which we have already considered in Eq. (2.99). Since there is only one time, the only symmetry operations are vertex exchanges. There are two transformations, unity and the exchange of the vertex extremities, which transform the labeled direct and exchange diagrams

(2.107)

into themselves. Hence the symmetry factor is 2. According to rule 6_F, the overall factor for the direct term is $-\frac{1}{2}$ and for the exchange term is $-\frac{\xi}{2}$, which agree with Eqs. (2.99).

In second order, both time permutations and extremity exchanges are possible. For example, the diagram

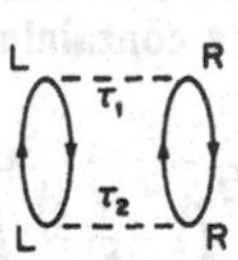

(2.108)

is transformed into a deformation of itself by permuting τ_1 and τ_2, by simultaneously interchanging the extremities of both vertices, and by a combination of both transformations. Thus, counting unity, we obtain $S = 4$. In contrast, the diagram

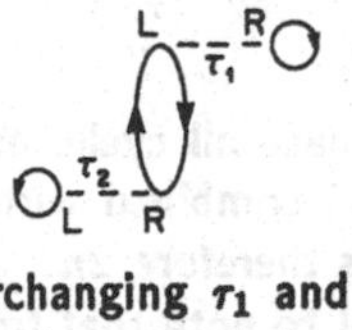

(2.109)

has lower symmetry. Simply interchanging τ_1 and τ_2 yields a distinct diagram, since the τ_2 vertex is then connected to the central closed loop by the right end and the τ_1 vertex is connected by the left end. Hence, the only non-trivial transformation which yields a deformation of the original diagram is the simultaneous interchange of the times and exchange of the extremeties of both vertices, yielding $S = 2$. The diagram

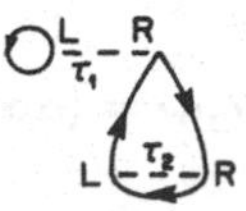

(2.110)

has even lower symmetry: all combinations of time permutations and vertex extremity exchange yield distinct diagrams and thus $S = 1$.

Additional features arise in higher-order examples. Consider the diagram

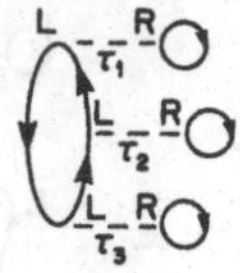

(2.111)

for which vertex extremity exchanges combined with any permutation always yield distinct diagrams. Any cyclic permutation of (τ_1, τ_2, τ_3) yields a deformation. However, although any cyclic permutation of (τ_3, τ_2, τ_1) yields a diagram of the same topological structure, the direction of the arrows on the large closed loop is reversed and the diagram is distinct. Hence, we obtain $S = 3$. For the third-order direct ring diagram

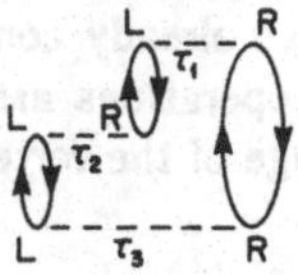

(2.112)

deformations are obtained for each permutation of (τ_1, τ_2, τ_3) when it is combined with the appropriate combination of exchanges of extremities. For example, one may verify that the permutation (τ_2, τ_1, τ_3) must be combined with the simultaneous exchange of the extremities of all three vertices and that (τ_2, τ_3, τ_1) requires the exchange of the extremities of the vertices labeled by τ_1 and τ_3. The symmetry factor is therefore $3! = 6$. Higher-order direct ring diagrams containing n interactions, such as,

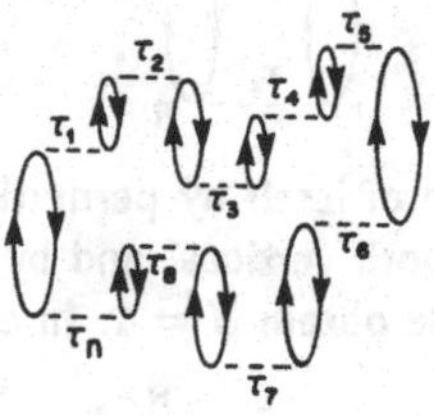

(2.113)

may be analyzed similarly. In this case all cyclic permutations of $(\tau_1 \tau_2 \tau_3 \ldots \tau_n)$ and cyclic permutations of $(\tau_n \ldots \tau_3 \tau_2 \tau_1)$ combined with appropriate vertex extremity exchanges yield deformations and S is therefore $2n$.

For future reference, it is useful to note that from the symmetry factors we have derived, the factors from rule 6_F for the first-order exchange graph and all direct ring diagrams are $\frac{(-\varsigma)^n}{2n}$. If we regard the product of an interaction and two propagators as a matrix in the time and single-particle labels

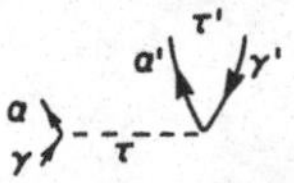

$$= [vgg]_{\alpha\gamma\tau,\alpha'\gamma'\tau'} = (\alpha\alpha'|v|\gamma\gamma')g_{\alpha'}(\tau' - \tau)g_{\gamma'}(\tau - \tau') \quad (2.114)$$

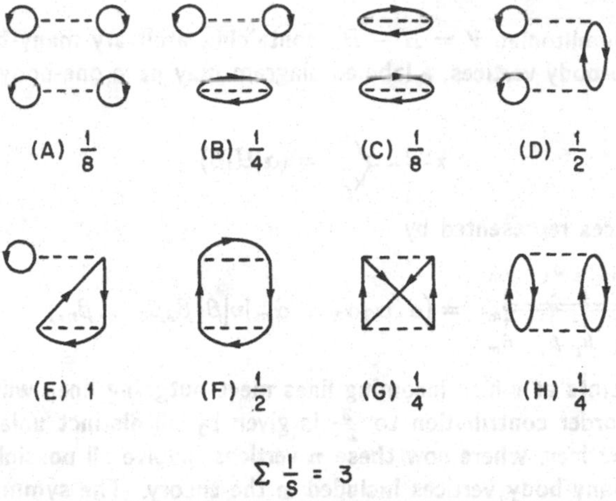

Fig. 2.4 Second order unlabeled Feynman diagrams with symmetry factor $\frac{1}{S}$.

then the sum of the first-order diagrams and all direct ring diagrams is

$$-\frac{1}{2}\mathrm{Tr}[\ln(1-\varsigma vgg)] = \sum_{n=1}^{\infty}\frac{(-\varsigma)^n}{2n}\mathrm{Tr}[vgg]^n \tag{2.115}$$

= 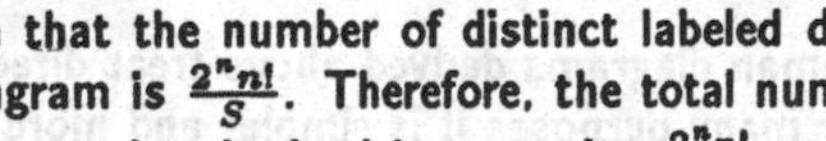 .

By now it may have become evident that the enumeration of distinct unlabeled diagrams and the calculation of symmetry factors are susceptible to human fallibility. Hence it is useful to derive a sum rule which provides a consistency check on all the diagrams and symmetry factors contributing to a given order of perturbation theory. We have already shown that the number of distinct labeled diagrams corresponding to a given unlabeled diagram is $\frac{2^n n!}{S}$. Therefore, the total number of distinct labeled diagrams, which is $2n!$, must be obtained by summing $\frac{2^n n!}{S}$ over all distinct unlabeled diagrams. Hence

$$\sum \frac{1}{S} = \frac{(2n)!}{n!2^n} = (2n-1)!! = (2n-1)(2n-3)\ldots 5\cdot 3\cdot 1 \tag{2.116}$$

where the sum is over all distinct unlabeled diagrams. This sum rule is trivially satisfied in first order where the direct and exchange diagrams each have $S = 2$. The case of second order perturbation theory is more interesting. Whereas there are 24 labeled diagrams in second order, there are only 8 unlabeled diagrams which are shown in Fig. 2.4 together with $\frac{1}{S}$.

The symmetry factors indeed satisfy the condition $\sum \frac{1}{S} = 3!!$, and it is a useful exercise to verify each symmetry factor and to check that there are no other distinct unlabeled second order diagrams.

Thus far, we have restricted our attention to the case in which V is a two-body interaction. The preceding diagram analysis may be generalized straightforwardly to treat a residual Hamiltonian $V = H - H_0$ containing arbitrary many-body operators. In addition to two-body vertices, a labeled diagram may have one-body vertices represented by

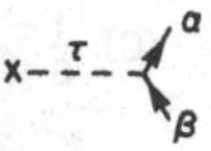

$$= \langle\alpha|U|\beta\rangle \tag{2.117a}$$

and m-body vertices represented by

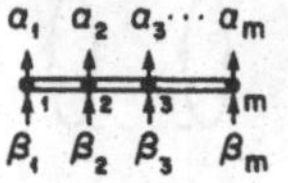

$$= (\alpha_1\alpha_2\alpha_3\ldots\alpha_m|v|\beta_1\beta_2\beta_3\ldots\beta_m) \ . \tag{2.117b}$$

The numbered points at which incoming lines meet outgoing lines will be denoted as joints. The nth order contribution to $\frac{Z}{Z_0}$ is given by all distinct unlabeled diagrams composed of n vertices, where now these n vertices involve all possible combinations of the one and many-body vertices included in the theory. The symmetry factor S is now equal to the number of transformations composed of time permutations among identical vertices and permutations of the numbered joints of each many-body vertex which yield a deformation.

As an example, for a residual Hamiltonian, V, containing a one-body potential U and a three-body potential W, one possible graph is

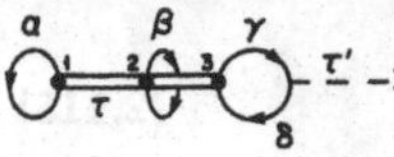

$$\begin{aligned} &= (-1)\varsigma^3 \frac{1}{2} \sum_{\alpha\beta\gamma\delta} (\alpha\beta\gamma|W|\alpha\beta\delta)\langle\delta|U|\gamma\rangle g_\alpha(0) g_\beta(0) \\ &\quad \times \int_0^\beta d\tau d\tau' g_\gamma(\tau' - \tau) g_\delta(\tau - \tau') \ . \end{aligned} \tag{2.118}$$

The symmetry factor is 2 since the only transformation which produces a deformation of the original diagram is the exchange of joints 1 and 3.

HUGENHOLTZ DIAGRAMS

The unlabeled Feynman diagrams derived above treat direct and exchange matrix elements separately. For many purposes it is simpler and more convenient to combine them in a single symmetrized or antisymmetrized matrix element. The resulting diagrams, called Hugenholtz diagrams, are readily derived starting from the symmetrized or antisymmetrized version of the residual Hamiltonian, V. For simplicity, we again consider the case of a two-body potential.

$$\begin{aligned} V &= \frac{1}{4} \sum_{\alpha\beta\gamma\delta} \{\alpha\beta|v|\gamma\delta\} a^\dagger_\alpha a^\dagger_\beta a_\delta a_\gamma \\ &= \frac{1}{4} \sum_{\alpha\beta\gamma\delta} [(\alpha\beta|v|\gamma\delta) + \varsigma(\alpha\beta|v|\delta\gamma)] a^\dagger_\alpha a^\dagger_\beta a_\delta a_\gamma \ . \end{aligned} \tag{2.119}$$

Since we no longer wish to distinguish direct and exchange diagrams, this vertex will now be represented graphically by a dot with two incoming and two outgoing lines

$$= \{\alpha\beta|v|\gamma\delta\} \ . \tag{2.120}$$

According to Wick's theorem, we again construct diagrams by drawing n vertices and connecting all incoming lines with outgoing lines. Note, however, that by the symmetry or antisymmetry of the vertex, two contractions corresponding to the exchange of the incoming or outgoing lines associated with a given vertex are equal:

$$\sum_{\gamma\delta}\{\alpha\beta|v|\gamma\delta\}a^\dagger_\alpha a^\dagger_\beta a_\delta a_\gamma \ldots a^\dagger_{\alpha'} a^\dagger_{\beta'} = \sum_{\gamma\delta}\{\alpha\beta|v|\gamma\delta\}a^\dagger_\alpha a^\dagger_\beta a_\delta a_\gamma \ldots a^\dagger_{\alpha'} a^\dagger_{\beta'} \ . \tag{2.121}$$

Thus, in contrast to the previous Feynman diagrams, we will no longer need to distinguish the two incoming or outgoing lines of a vertex.

A diagram which does not distinguish the two incoming lines or two outgoing lines at each vertex represents many sets of contractions. To count the number of contractions associated with a given Hugenholtz diagram it is useful to define an equivalent pair of lines as two directed propagators which begin at the same vertex, end at the same vertex (possibly the original vertex), and point in the same direction.

First, consider the two outgoing propagators from a given vertex v_i, and assume that they are not an equivalent pair. Then they terminate at different vertices, say v_j and v_k, and in addition other propagators terminating at v_j and v_k cannot belong to equivalent pairs. Now associate the factor $\sqrt{2}$ with each end of every line which does not belong to an equivalent pair. Then the product of the $\sqrt{2}$'s accounts for the two choices for contracting non-equivalent outgoing lines at any vertex, such as v_i and the two choices for contracting non-equivalent incoming lines at any vertex, such as v_j and v_k. It is convenient to use the four factors of $\sqrt{2}$ associated with the non-equivalent pair of lines originating from a vertex to cancel the factor $\frac{1}{4}$ associated with that vertex from Eq. (2.119).

Secondly, consider the case in which the two outgoing propagators from a given vertex v_i are an equivalent pair. This pair of lines must terminate at the same vertex and therefore has only 2 equal contractions. When combined with the factor $\frac{1}{4}$ associated with the vertex at which the pair originated, we see that each equivalent pair produces an overall factor $\frac{1}{2}$. Hence all the contractions arising from Wick's theorem are correctly counted if for each Hugenholtz diagram one eliminates the factor $\frac{1}{4^n}$ arising from the n interactions and includes a factor of $\frac{1}{2}$ for each equivalent pair of propagators.

Since we do not distinguish pairs of lines entering or leaving vertices, labeled Hugenholtz diagrams only possess time labels and arrows denoting the direction of propagators. The symmetry factor S is then the number of permutations of the time labels which yield a deformation of the original labeled Hugenholtz diagram, and the number of distinct labeled Hugenholtz diagrams corresponding to a given unlabeled diagram is $\frac{n!}{S}$.

Finally, the sign of a Hugenholtz diagram is most conveniently determined by selecting any order of assigning the propagators to the left and right sides of a conventional vertex in a corresponding Feynman diagram, and applying the previous sign rule $(-1)^n\varsigma^{n_L}$ to the resulting diagram. By the equivalence with respect to interchange at a vertex, although different assignments may produce different Feynman diagrams, they must yield the same contribution. For example

 $\rightarrow$ 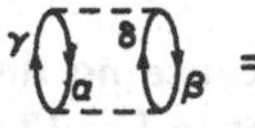

$$= \frac{1}{8}\varsigma^2\{\alpha\beta|v|\gamma\delta\}\{\gamma\delta|v|\alpha\beta\} \tag{2.122a}$$

or

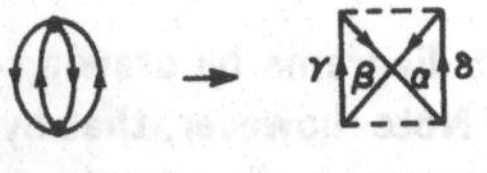

$$= \frac{1}{8}\varsigma\{\alpha\beta|v|\gamma\delta\}\{\gamma\delta|v|\beta\alpha\}$$
$$= \frac{1}{8}\varsigma^2\{\alpha\beta|v|\gamma\delta\}\{\gamma\delta|v|\alpha\beta\} \; . \qquad (2.122b)$$

These arguments give rise to the following rules, designated by the subscript H, for calculating the nth order contribution to the perturbation expansion of $\frac{Z}{Z_o}$ using Hugenholtz diagrams:

1_H Draw all distinct unlabeled diagrams composed of n vertices  connected by directed lines . Two diagrams are distinct if they cannot be deformed so as to coincide completely, including the direction of arrows on propagators. For each distinct unlabeled Hugenholtz diagram, evaluate the contribution as follows.

2_H Calculate the symmetry factor S for the diagram. This may be done by adding time labels to each vertex in the diagram. Then S is equal to the number of time permutations which transform the labeled diagram into a deformation of itself.

3_H Assign a time label τ_i to each vertex and a single-particle index to each directed line. For each directed line include the factor

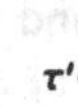

$$= g_\alpha(\tau-\tau') = e^{-(\epsilon_\alpha-\mu)(\tau-\tau')}[(1+\varsigma n_\alpha\theta(\tau-\tau'-\eta)+\varsigma n_\alpha\theta(\tau'-\tau+\eta)] \; .$$

4_H For each vertex, include the factor

$$= \{\alpha\beta|v|\gamma\delta\} = (\alpha\beta|v|\gamma\delta) + \varsigma(\alpha\beta|v|\delta\gamma) \; .$$

5_H Sum over all single-particle indices and integrate all times over the interval $[0,\beta]$.

6_H Multiply the result by the factor $\frac{(-1)^n\varsigma^{n_L}}{2^{n_e}S}$. The symmetry factor S is defined above in (2_H). The number of equivalent pairs of lines, n_e, is the number of pairs of lines beginning at the same vertex, terminating at the same vertex, and oriented in the same direction. The number of closed loops n_L is obtained by replacing each Hugenholtz vertex 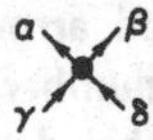 by a conventional vertex 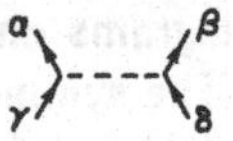 and counting the number of closed loops as for a Feynman diagram. Note that the order of the labels on the conventional vertex must agree with the matrix elements in (4_H).

It is an instructive exercise to generalize these Hugenholtz diagram rules to the case of n-body interactions (see Problem 2.9).

At this point, it is useful to recalculate in terms of Hugenholtz diagrams some of the contributions we previously evaluated using Feynman diagrams. The only diagram for $n = 1$ is

 . (2.123)

There is one equivalent pair of lines and because no time permutations are possible, $S = 1$. Thus the contribution is equal to that in Eq. (2.99).

In second order, whereas there are 24 labeled diagrams and eight unlabeled Feynman diagrams, there are three Hugenholtz diagrams. The diagram

$$= \frac{1}{8}\int_0^\beta d\tau_1 \int_0^\beta d\tau_2 g_\alpha(\tau_1-\tau_2)g_\beta(\tau_1-\tau_2)g_\gamma(\tau_2-\tau_1)g_\delta(\tau_2-\tau_1) \times [(\alpha\beta|v|\gamma\delta)+\varsigma(\alpha\beta|v|\delta\gamma)][(\gamma\delta|v|\alpha\beta)+\varsigma(\gamma\delta|v|\beta\alpha)] \tag{2.124a}$$

corresponds to the two unlabeled Feynman diagrams G and H of Fig. (2.4). It has two equivalent pairs and the symmetry factor is 2 since interchange of the upper and lower interactions is a deformation of the original diagram. Note that the two contributions from $(\alpha\beta|v|\gamma\delta)(\gamma\delta|v|\alpha\beta)$ and $(\alpha\beta|v|\delta\gamma)(\gamma\delta|v|\beta\alpha)$ are equal, reproducing the direct graph (H) of Fig. (2.4) with the correct weight $\frac{1}{4}$ and that the contributions from $(\alpha\beta|v|\gamma\delta)(\gamma\delta|v|\beta\alpha)$ and $(\alpha\beta|v|\delta\gamma)(\gamma\delta|v|\alpha\beta)$ are equal, reproducing the exchange graph (G) of Fig. (2.4) with weight $\frac{1}{4}$. Similarly, the graph

(2.124b)

has no equivalent pairs and symmetry factor 2 so that when all the combinations of direct and exchange matrix elements are enumerated, the three Feynman graphs (D), (E), and (F) of Fig. (2.4) are reproduced with the proper weight.

The counting for higher-order ring diagrams proceeds as for the case of Feynman diagrams. The generic n^{th} order ring diagram

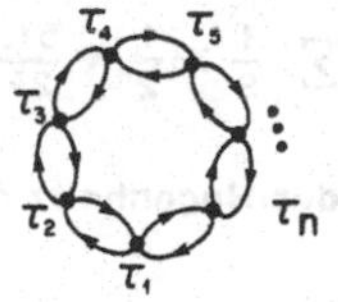

(2.125)

has no equivalent pairs and a symmetry factor $S = 2n$ to account for the cyclic permutations of $(\tau_1\tau_2\ldots\tau_n)$ and $(\tau_n\ldots\tau_2\tau_1)$ which yield deformations of the original diagram. Thus, for $n = 1$ and $n > 2$, the factors $\frac{1}{2n}$ for Hugenholtz ring diagrams are identical to those we derived before for Feynman direct ring diagrams. However, for $n = 2$, because of the presence of two pairs of equivalent lines, the correct factor of $\frac{1}{8}$ is one half of the general ring factor $\frac{1}{2n}$. Hence, with antisymmetrized matrix elements, the formula $-\frac{1}{2}\mathrm{Tr}[\ln(1-\varsigma vgg)]$ overcounts the second-order diagram, and a correction term must therefore be subtracted explicitly.

As in the case of unlabeled Feynman diagrams, it is valuable to derive a sum rule which provides a consistency check on the enumeration of Hugenholtz diagrams and the calculation of symmetry factors. We have already seen that the number of distinct labeled diagrams corresponding to a given unlabeled diagram is $\frac{n!}{S}$. For each distinct labeled diagram, there correspond $\frac{4^n}{2^{n_e}}$ contractions, where a factor of 4 arises for each vertex from which a non-equivalent pair of lines originates and a factor of 2 occurs for each vertex from which an equivalent pair originates. Therefore, the total number of

(a) $\frac{1}{2} \times \frac{1}{2^2}$ (b) $\frac{1}{2} \times 1$ (c) $\frac{1}{2} \times \frac{1}{2^2}$

$$\sum \frac{1}{S} \times \frac{1}{2^{n_e}} = \frac{3}{2^2}$$

(A) $\frac{1}{6} \times \frac{1}{2^3}$ (B) $\frac{1}{2} \times \frac{1}{2^3}$ (C) $\frac{1}{2} \times \frac{1}{2}$ (D) $\frac{1}{2} \times 1$

(E) $\frac{1}{3} \times 1$ (F) $1 \times \frac{1}{2}$ (G) $\frac{1}{3} \times \frac{1}{2^3}$ (H) $\frac{1}{6} \times 1$

$$\sum \frac{1}{S} \times \frac{1}{2^{n_e}} = \frac{5!!}{2^3}$$

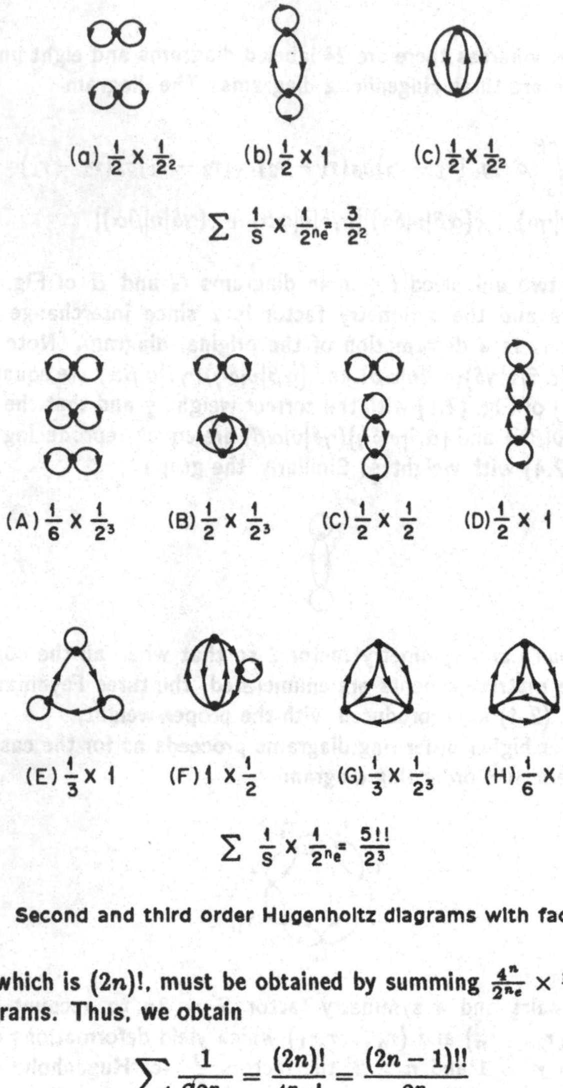

Fig. 2.5   Second and third order Hugenholtz diagrams with factors $\frac{1}{S} \times \frac{1}{2^{n_e}}$.

contractions, which is $(2n)!$, must be obtained by summing $\frac{4^n}{2^{n_e}} \times \frac{n!}{S}$ over all distinct unlabeled diagrams. Thus, we obtain

$$\sum \frac{1}{S 2^{n_e}} = \frac{(2n)!}{4^n n!} = \frac{(2n-1)!!}{2^n} \tag{2.126}$$

where the sum is over all distinct unlabeled Hugenholtz diagrams.

In Fig. 2.5 all the second-order and third-order Hugenholtz diagrams are shown, together with the factor $\frac{1}{S} \times \frac{1}{2^{n_e}}$. Again, it is an instructive exercise to verify the diagrams, symmetry factors and sum rule. In comparison with the 720 original contractions, the 8 third-order Hugenholtz diagrams represent a significant simplification.

FREQUENCY AND MOMENTUM REPRESENTATION

For systems which are homogeneous in space such as a liquid or gas, it is frequently advantageous to perform the perturbation expansion in momentum representation. Similarly, for problems which are homogeneous in time, it is often useful to Fourier transform from time to frequency.

The following conventions will be used in defining Fourier series and Fourier transforms. In one dimension for a periodic function $F(x)$ in a box of length L, we write the Fourier series as

$$\tilde{f}(k_n) = \int_0^L dx e^{-ik_n x} f(x) \tag{2.127a}$$

$$f(x) = \frac{1}{L}\sum_{k_n} e^{ik_n x}\tilde{f}(k_n) \tag{2.127b}$$

$$k_n = \frac{2\pi n}{L} \tag{2.127c}$$

and the corresponding Fourier transform for an infinite system as

$$\tilde{f}(k) = \int_{-\infty}^{\infty} dx e^{-ikx} f(x) \tag{2.128a}$$

$$f(x) = \frac{1}{2\pi}\int_{-\infty}^{\infty} dk e^{ikx}\tilde{f}(k) \ . \tag{2.128b}$$

For a function of time $g(\tau)$ which is periodic or antiperiodic in the interval $(0,\beta)$, we write the Fourier series as

$$\tilde{g}(\omega_n) = \int_0^\beta d\tau e^{i\omega_n \tau} g(\tau) \tag{2.129a}$$

$$g(\tau) = \frac{1}{\beta}\sum_{\omega_n} e^{-i\omega_n \tau}\tilde{g}(\omega_n) \tag{2.129b}$$

where for periodic functions

$$\omega_n = \frac{2\pi n}{\beta} \tag{2.129c}$$

and for antiperiodic functions

$$\omega_n = \frac{(2n+1)\pi}{\beta} \ . \tag{2.129d}$$

Consider the evaluation of a diagram for a spatially homogeneous system. In this case H_0 will be translationally invariant so that the eigenstates will be plane waves, and it is most convenient to formulate the problem in a finite box with periodic boundary conditions. Denoting the discrete momentum $\vec{k}_n \equiv (k^x_{n_x}, k^y_{n_y}, k^z_{n_z})$ and volume $\mathcal{V} = L_x L_y L_z$, the matrix element of the potential evaluated with normalized eigenstates is

$$\begin{aligned}(\vec{k}_{n_1}\vec{k}_{n_2}|v|\vec{k}_{n_3}\vec{k}_{n_4}) &= \int \frac{d^3x_1 d^3x_2}{\mathcal{V}^2} v(\vec{x}_1 - \vec{x}_2) e^{i[(\vec{k}_{n_1}-\vec{k}_{n_3})\cdot\vec{x}_1 + (\vec{k}_{n_2}-\vec{k}_{n_4})\cdot\vec{x}_2]} \\ &= \int \frac{d^3r d^3R}{\mathcal{V}^2} v(\vec{r}) e^{i\vec{R}(\vec{k}_{n_1}+\vec{k}_{n_2}-\vec{k}_{n_3}-\vec{k}_{n_4}) + \frac{i\vec{r}}{2}(\vec{k}_{n_1}-\vec{k}_{n_3}+\vec{k}_{n_4}-\vec{k}_{n_2})} \\ &= \frac{1}{\mathcal{V}}\delta^{(3)}_{\vec{k}_{n_1}+\vec{k}_{n_2},\vec{k}_{n_3}+\vec{k}_{n_4}} \tilde{v}(\vec{k}_{n_1} - \vec{k}_{n_3}) \ . \end{aligned} \tag{2.130}$$

Thus, each interaction conserves momentum, with the Kronecker δ requiring that the sum of the momenta of propagators entering the vertex equal the sum of the momenta leaving the vertex, and carries a factor of $\frac{1}{\mathcal{V}}$.

Every diagram can be decomposed into connected parts, that is parts in which an uninterrupted chain of propagators connects any vertex to any other vertex. For example in Fig. 2.5, diagram A has three connected parts, diagrams B and C have two connected parts, and diagrams D through H have one connected part.

The diagram rules for a connected diagram with m interactions give rise to the following contributions. There will be $2m$ propagators $g_{k_n}(\tau_i - \tau_j)$, each summed independently over a complete set of momentum states k_n. The m interactions will give rise to the product of m momemtum-conserving Kronecker δ's and a factor of $\mathcal{V}^{-m}$. Since the diagram is linked and momentum is conserved at each vertex, one of the m Kronecker δ's is redundant. Thus, there are $(m-1)$ constraints on the $2m$ momenta for the propagator, leading to a sum over $(m+1)$ independent momenta. In the continuum limit, each sum over an independent mometum is replaced by an integral $\sum_{k_n} F(k_n) \rightarrow \frac{\mathcal{V}}{(2\pi)^3} \int d^3k F(k)$. Combining the factor $\mathcal{V}^{m+1}$ from these independent momentum integrals with the factor $\mathcal{V}^{-m}$ from the m matrix elements thus yields an overall factor of the volume $\mathcal{V}$, for each connected diagram. A diagram having n_c connected parts yields a contribution proportional to $\mathcal{V}^{n_c}$ and it is evident that only completely connected diagrams, having $n_c = 1$, produce extensive contributions.

The diagram rules presented previously for Feynman and Hugenholtz diagrams are thus modified as follows in mometum representation:

$3_{F,H}$ Assign momentum labels to each directed line as follows. Within each connected part of the diagram containing m interactions, select $m+1$ propagators to label independently and use conservation of momentum at each vertex to label the remaining propagators in terms of combinations of the independent momenta. Assign time labels as before and for each directed line include the factor

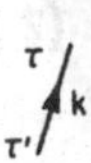

$$= g_k(\tau - \tau') = e^{-(\epsilon_k - \mu)(\tau - \tau')}[(1 + \varsigma n_k)\theta(\tau - \tau' - \eta) + \varsigma n_k \theta(\tau' - \tau + \eta)] \; .$$

The single-particle energy ϵ_k may include the contribution of a general non-local one-body potential in addition to the kinetic energy $\frac{k^2}{2m}$.

4_F For each vertex include the factor

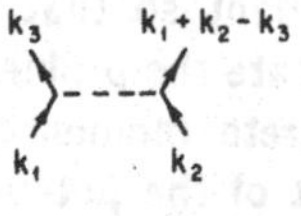

$$= \tilde{v}(\vec{k}_1 - \vec{k}_3) \; .$$

4_H For each vertex include the factor

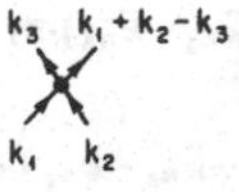

$$= \tilde{v}(\vec{k}_1 - \vec{k}_3) + \varsigma \tilde{v}(\vec{k}_2 - \vec{k}_3) \; .$$

$5_{F,H}$ For each independent momentum, perform the integral $\int \frac{d^3k}{(2\pi)^3}$ and integrate all times over the interval $[0, \beta]$.

$6_{F,H}$ **Multiply the previously defined factor by $\mathcal{V}^{n_c}$ where n_c is the number of connected parts in the diagram.**

Fourier transformation from τ to ω for a problem which is homogeneous in time proceeds analogously, except that periodic or antiperiodic functions of τ are integrated over the finite interval $[0,\beta]$. The propagator $g(\tau_1,\tau_2)$ defined in Eq. (2.98) is periodic or antiperiodic with period β in each of the times and therefore also in the relative time. Although the relative time ranges from $-\beta$ to β, it is sufficient to calculate the Fourier coefficients by integrating over the range $[0,\beta]$. Using Eq. (2.74b) for n_α, the definitions Eqs. (2.129) with $\omega_n = \frac{2\pi n}{\beta}$ for Bosons and $\omega_n = \frac{(2n+1)\pi}{\beta}$ for Fermions, and noting that $e^{i\beta\omega_n} = \varsigma$, we obtain

$$\begin{aligned}
\tilde{g}_\alpha &= \int_0^\beta d\tau e^{(i\omega_n-(\epsilon_\alpha-\mu))\tau}[\theta(\tau)(1+\varsigma n_\alpha)+\varsigma\theta(-\tau)n_\alpha] \\
&= \frac{e^{\beta(i\omega_n-(\epsilon_\alpha-\mu))}-1}{i\omega_n-(\epsilon_\alpha-\mu)}\,\frac{e^{\beta(\epsilon_\alpha-\mu)}}{e^{\beta(\epsilon_\alpha-\mu)}-\varsigma} \\
&= \frac{\varsigma-e^{\beta(\epsilon_\alpha-\mu)}}{[(i\omega_n-\epsilon_\alpha-\mu)](e^{\beta(\epsilon_\alpha-\mu)}-\varsigma)} \\
&= \frac{-1}{i\omega_n-(\epsilon_\alpha-\mu)}
\end{aligned} \tag{2.131a}$$

and hence

$$g_\alpha(\tau) = \sum_{\omega_n}\frac{-1}{\beta}e^{-i\omega_n\tau}\frac{1}{i\omega_n-(\epsilon_\alpha-\mu)}\ . \tag{2.131b}$$

One additional consideration is required to treat the special case of a propagator which begins and ends at the same vertex. In order to invoke the proper θ-function in $g_\alpha(\tau)$ corresponding to the occupation probability n_k, the argument is always shifted by an infinitesimal η to $g_\alpha(\tau-\eta)$. (See, for example, Eq. (2.78)). This same result may be obtained by multiplying $\tilde{g}_\alpha(\omega_n)$ by the factor $e^{i\omega_n\eta}$, in which case

$$\begin{aligned}
\frac{1}{\beta}\sum_{\omega_n}e^{-i\omega_n(\tau-\eta)}\tilde{g}_\alpha(\omega_n) &= \frac{-1}{\beta}\sum_{\omega_n}e^{-i\omega_n(\tau-\eta)}\frac{1}{i\omega_n-(\epsilon_\alpha-\mu)} \\
&= g_\alpha(\tau-\eta)\ .
\end{aligned} \tag{2.131c}$$

The time integration associated with the vertex

yields the integral

$$\begin{aligned}
&\int d\tau g_{\alpha_1}(\tau-\tau_1)g_{\alpha_2}(\tau-\tau_2)g_{\alpha_3}(\tau_3-\tau)g_{\alpha_4}(\tau_4-\tau) \\
&= \int_0^\beta d\tau e^{-i(\omega_{n_1}+\omega_{n_2}-\omega_{n_3}-\omega_{n_4})\tau}\prod_{i=1}^4\frac{\tilde{g}(\omega_{n_i})}{\beta}e^{i(\omega_{n_1}\tau_1+\omega_{n_2}\tau_2-\omega_{n_3}\tau_3-\omega_{n_4}\tau_4)} \\
&= \beta\delta_{\omega_{n_1}+\omega_{n_2},\omega_{n_3}+\omega_{n_4}}\prod_{i=1}^4\frac{\tilde{g}(\omega_{n_i})}{\beta}e^{i(\omega_{n_1}\tau_1+\omega_{n_2}\tau_2-\omega_{n_3}\tau_3-\omega_{n_4}\tau_4)}\ .
\end{aligned} \tag{2.132}$$

Thus, at every vertex, we obtain a factor β and a Kronecker δ requiring that the sum of the frequencies associated with the propagators entering the vertex equal the sum of the frequencies of propagators leaving the vertex. As was the case with momentum conservation, a connected diagram with m interactions will have $m+1$ independent frequencies. Since in a general n^{th} order diagram each of the $2n$ propagators has factor of $\frac{1}{\beta}$ and each of the n time integrals yields a factor β, the overall contribution is proportional to β^{-n}. The diagram rules presented previously for Feynman and Hugenholtz diagrams are then modified as follows in the frequency representation.

$3_{F,H}$ Assign frequency labels to each directed line as follows. Within each connected part of a diagram containing m interactions, select $m+1$ propagators to label independently and use frequency conservation at each vertex to label the remaining propagators in terms of combinations of independent momenta. Assign single-particle indices as before. For each directed line include the factor

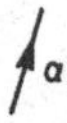

$$\alpha, \omega_n \quad = \tilde{g}_\alpha(\omega_n) = \frac{-1}{i\omega_n - (\epsilon_\alpha - \mu)} \; .$$

For propagators beginning and ending at the same vertex, include an additional factor $e^{i\omega_n\eta}$.

$5_{F,H}$ Sum over all single-particle labels and sum over all independent frequencies ω_n.

$6_{F,H}$ Multiply the previously defined factor by $\frac{1}{\beta^m}$, where m is the number of interactions.

The generalization of these rules using momentum and frequency representation to n-body interactions is straightforward. It is often convenient to use momentum and frequency representation simultaneously in which case one associates with each propagator a four-momentum (ω_n, k) which is conserved at each vertex. In the zero temperature limit which will be discussed in detail in Chapter 3, $\beta \to \infty$, the Fourier series in ω_n becomes a Fourier integral, and the diagram rules can be expressed in the covariant form familiar in relativistic field theory.

THE LINKED CLUSTER THEOREM

We have seen that the expansion for Z contains all powers of the volume $\mathcal{V}$, with an individual diagram with n_c connected parts being proportional to $\mathcal{V}^{n_c}$. In contrast, the ground potential $\Omega = -\frac{1}{\beta}\ln Z = -P\mathcal{V}$ is an extensive quantity so it must be possible to regroup the expansion to obtain an extensive expansion for Ω. Whereas it is plausible that Ω may be expressed in terms of connected diagrams, the combinatorial factors are not *a priori* obvious. The linked cluster theorem states that $\ln Z$ is in fact given by the sum of all connected diagrams.

We will derive this theorem using the replica technique, both for the sake of brevity and to introduce this useful method. A standard derivation is given in Problem 2.10. The basic idea of the replica method is to evaluate Z^n for integer n by replicating the system n times and expand the result as follows.

$$Z^n = e^{n\ln Z} = 1 + n\ln Z + \sum_{m=2}^{\infty} \frac{(n\ln Z)^m}{m!} \; . \tag{2.133}$$

Thus, if we evaluate Z^n for integer n in perturbation theory, lnZ will be given by the coefficients of the terms proportional to n. A more general statement of the method is to calculate Z^n for integer n, continue the function to $n = 0$ (which is valid by Carlson's theorem) and evaluate an appropriate expression involving the continued function to calculate the observable of interest. In the present case, we calculate

$$\lim_{n\to 0} \frac{d}{dn} Z^n = \lim_{n\to 0} \frac{d}{dn}(e^{n \ln Z}) \\ = \ln Z \; . \tag{2.134}$$

We will first evaluate Z^n for integer n by perturbation theory, and according to equation (2.133), Log Z will be given by the coefficient of the graphs proportional to n. Since

$$\frac{Z}{Z_0} = \frac{1}{Z_0} \int_{\psi_\alpha(\beta)=\varsigma\psi_\alpha(0)} \mathcal{D}(\psi^*_\alpha(\tau), \psi_\alpha(\tau)) e^{-\int_0^\beta d\tau \left(\sum_\alpha \psi_\alpha(\tau)^*(\partial_\tau+\epsilon_\alpha-\mu)\psi_\alpha(\tau) - V(\psi^*_\alpha(\tau),\psi_\alpha(\tau))\right)} \; . \tag{2.135}$$

We may write Z^n as a functional integral over n sets of fields $\{\psi^{*\sigma}_\alpha(\tau), \psi^\sigma_\alpha(\tau)\}$ where the index σ runs from 1 to n.

$$\left(\frac{Z}{Z_0}\right)^n = \frac{1}{Z_0^n} \int_{\psi^\sigma_\alpha(\beta)=\varsigma\psi^\sigma_\alpha(0)} \prod_{\sigma=1}^{n} \mathcal{D}(\psi^{\sigma*}_\alpha(\tau), \psi^\sigma_\alpha(\tau)) \\ \times e^{-\int_0^\beta d\tau \sum_{\sigma=1}^n \left(\sum_\alpha \psi^{\sigma*}_\alpha(\partial_\tau+\epsilon_\alpha-\mu)\psi^\sigma_\alpha - V(\psi^{\sigma*}_\alpha,\psi^\sigma_\alpha)\right)} \; . \tag{2.136}$$

The Feynman rules for $(\frac{Z}{Z_0})^n$ are the same as those for $(\frac{Z}{Z_0})$, except that each propagator now carries an index σ, all propagators entering or leaving a given vertex have the same index σ, and all σ's are summed from 1 to n. It is evident that each connected part of a diagram must carry a single index σ, which when summed from 1 to n, yields a factor n. Thus, a graph with n_c connected parts is proportional to n^{n_c} and the graphs proportional to n are those with only one connected part, that is the connected graphs. As a consequence, we obtain the linked cluster theorem:

$$\Omega - \Omega_0 = -\frac{1}{\beta} \sum (\text{all connected graphs}) \tag{2.137a}$$

where Ω_0 is the grand potential of the unperturbed system

$$\Omega_0 = \frac{\varsigma}{\beta} \sum_\alpha \ln(1 - \varsigma e^{-\beta(\epsilon_\alpha-\mu)}) \; . \tag{2.137b}$$

CALCULATION OF OBSERVABLES AND GREENS FUNCTIONS

The linked expansion for the expectation value of any n-body operator R is analogous to that for Ω. One way to proceed, as outlined in Problem 2.11, is to calculate $\Omega(\lambda)$ for the Hamiltonian $\hat{H}_\lambda = \hat{H} + \lambda\hat{R}$ and evaluate $\langle R \rangle = \frac{d}{d\lambda}\Omega(\lambda)$. An alternative method, which has the advantage of showing how symmetry factors are simplified, is to use the replica technique again. Using the definition for the unperturbed thermal average, Eq. (2.81b), the expectation value of R may be written

$$\begin{aligned}\langle R\rangle &= \frac{\int \mathcal{D}(\psi^*_\alpha,\psi_\alpha)e^{-\int dt \sum_\alpha \psi^*_\alpha(\tau)(\partial_\tau+\epsilon_\alpha-\mu)\psi_\alpha(\tau)+V(\psi^*_\alpha(\tau),\psi_\alpha(\tau))}R(\psi^*_\alpha(0),\psi_\alpha(0))}{\int \mathcal{D}(\psi^*_\alpha,\psi_\alpha)e^{-\int dt \sum_\alpha \psi^*_\alpha(\tau)(\partial_\tau+\epsilon_\alpha-\mu)\psi_\alpha(\tau)+V(\psi^*_\alpha(\tau),\psi_\alpha(\tau))}} \\ &= \frac{\langle e^{-\int dt V(\psi^*_\alpha(\tau),\psi_\alpha(\tau))}R(\psi^*_\alpha(0)\psi_\alpha(0))\rangle_0}{\langle e^{-\int dt V(\psi^*_\alpha(\tau),\psi_\alpha(\tau))}\rangle_0} \ . \end{aligned} \tag{2.138}$$

Let us again introduce n fields $\{\psi^{*\sigma}_\alpha(\tau),\psi^\sigma_\alpha(\tau)\}$ where the index σ runs from 1 to n and define

$$\begin{aligned}R_n = Z_0^{-n}\int \prod_{\sigma=1}^{n} \mathcal{D}(\psi^{\sigma*}_\alpha(\tau),\psi^\sigma_\alpha(\tau))R(\psi^{1*}_\alpha(0)),\psi^1_\alpha(0)) \\ \times e^{-\sum_{\sigma=1}^{n}\int_0^\beta dt(\sum_\alpha \psi^{\sigma*}_\alpha(\tau)(\partial_\tau+\epsilon_\alpha-\mu)\psi^\sigma_\alpha(\tau)+V(\psi^{*\sigma}_\alpha(\tau),\psi^\sigma_\alpha(\tau))}\ .\end{aligned} \tag{2.139a}$$

Note that the operator R is calculated with the field $\psi^{1*}_\alpha,\psi^1_\alpha$ associated with $\sigma=1$ and is evaluated at $\tau=0$. By separating the $\sigma=1$ component from the $(n-1)$ other components we observe that

$$R_n = \left\langle e^{-\int_0^\beta dt V(\psi^*_\alpha(\tau),\psi_\alpha(\tau))} R(\psi^*_\alpha(0),\psi_\alpha(0))\right\rangle_0 \left\langle e^{-\int_0^\beta dt V(\psi^*_\alpha(\tau)\psi_\alpha(\tau))}\right\rangle_0^{n-1} \tag{2.139b}$$

and thus the desired expectation value is obtained for $n=0$:

$$R_0 = \langle R\rangle \ . \tag{2.140}$$

The perturbation expansion for R_n is obtained by expanding Eq. (2.139a) in powers of V

$$\begin{aligned}R_n = \sum_{p=0}^{\infty}\frac{(-1)^p}{p!}\sum_{\sigma_1=1}^{n}\cdots\sum_{\sigma_p=1}^{n}\int_0^\beta d\tau_1\ldots d\tau_p \\ \times\langle V(\psi^{*\sigma_1}_\alpha(\tau_1),\psi^{*\sigma_1}_\alpha(\tau_1))\ldots V(\psi^{*\sigma_p}_\alpha(\tau_p),\psi^{\sigma_p}_\alpha(\tau_p))R(\psi^{*1}_\alpha(0)\psi^1_\alpha(0))\rangle_0\end{aligned} \tag{2.141}$$

and diagram rules for unlabeled diagrams, labeled Feynman diagrams, Hugenholtz diagrams, or the momentum frequency transform of any of these diagrams are developed as a straightforward generalization of the rules in the preceding sections. We will describe here the case of a two-body operator R and unlabeled Feynman diagrams.

The expansion of Eq. (2.141) consists of all distinct diagrams containing one vertex $(\alpha\beta|R|\gamma\delta)$ at time $\tau=0$ and any number of vertices $(\alpha\beta|v|\gamma\delta)$ at other times which are integrated from 0 to β. As in the case of the linked cluster proof, all propagators now carry an index σ, all propagators entering or leaving an interaction vertex have the same index, and all indices are summed from 1 to n. In addition, the index associated with R is one so that all the propagators entering and leaving R are restricted to have $\sigma = 1$. Now, consider a diagram composed of a set of interaction vertices v connected to R and one or more additional unconnected parts. Since R is constrained to carry $\sigma = 1$ and all propagators and v vertices conserve σ, σ must be 1 everywhere in the portion of the diagram linked to R. In the additional unconnected parts, however, there is at least one free summation over σ leading to an overall factor of at least one power of n. Hence all diagrams with disconnected parts vanish when n is set equal to zero and $\langle R \rangle$ is given by the sum of all connected diagrams linked to R. Having played their role in the proof that only connected diagrams contribute to $\langle R \rangle$, the σ indices which are constrained to be 1 in all components linked to the vertex R are superfluous and may be omitted from the final diagram rules.

The symmetry factors for diagrams contributing to $\langle R \rangle$ are much simpler than for corresponding contributions to the grand potential. Since the symmetry of a diagram is reduced by singling out a particular vertex to be labeled R, it is clear that S, which specifies the number of combinations of time perturbations and vertex extremity exchanges which transform a labeled diagram into a deformation of itself, will be reduced. In fact, as we shall now demonstrate, the symmetry is reduced so much that the only possible values of S are 1 or 2.

To see how the symmetry factors are constrained, consider the typical labeled diagram in which nine v-vertices at times $\tau_1 \ldots \tau_9$ are denoted by 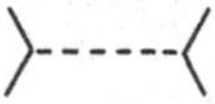 and the R-vertex at time 0 is denoted by 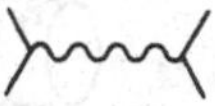

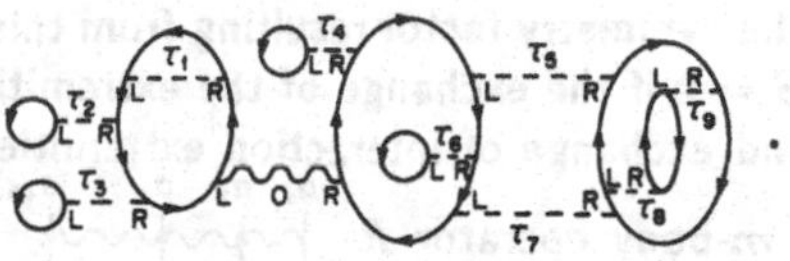

$$. \qquad (2.142a)$$

One may see that there are no permutations of time labels which produce a deformation of this diagram by the following argument. Each diagram may be considered as a series of closed Fermion loops connected by one or more vertices, and in this example there are eight such loops. Consider any loop which has one time fixed. Since the propagators comprising the loop are directed, there are no permutations of the vertex time labels within the loop which yield a deformation. Therefore, all the time labels in the loop are fixed. If another loop is connected to the first loop by at least one interaction, then at least one of its time labels is fixed and there is no freedom to permute time labels within that loop. By induction, one observes that no permutations of time labels within closed loops are possible. In this example, the time 0 fixes the labels $\tau_4, \tau_5, \tau_6, \tau_7$ which in turn fix the labels τ_8 and τ_9 in the next loop. The only remaining possibility, then, is to permute times between loops. However, since in the case of the diagram (2.142a) the sub-diagrams connected to the left and right sides of the R-vertex are topologically inequivalent, none of these permutations produce a deformation.

The following diagram is an example in which exchanging times between different

loops will produce a deformation.

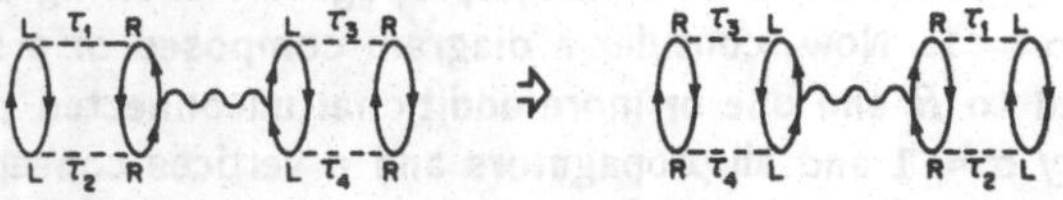

(2.142*b*)

Note that the subdiagram connected to the left side of the R-vertex is topologically equivalent to that connected to the right side. Interchanging τ_1 with τ_3 and τ_2 with τ_4 and simultaneously interchanging the extremities of all the vertices thus yields a deformation of the original diagrams as shown at the right. This simultaneous extremity exchange and exchange of all the times in the topologically equivalent subdiagrams connected to the R-vertex is in fact the only operation which can produce a deformation of the diagram. We have seen that no other operations which change time labels produce deformations, and by analyzing the L-R labels in each closed propagator loop of a general diagram, one can also see that pure extremity exchange without time permuations can never produce a deformation. Thus, the symmetry factor for diagram (2.142b) is $S = 2$. Another example is the exchange diagram shown in below which has symmetry factor $S = 2$ because of the time and extremity permutations shown at the right

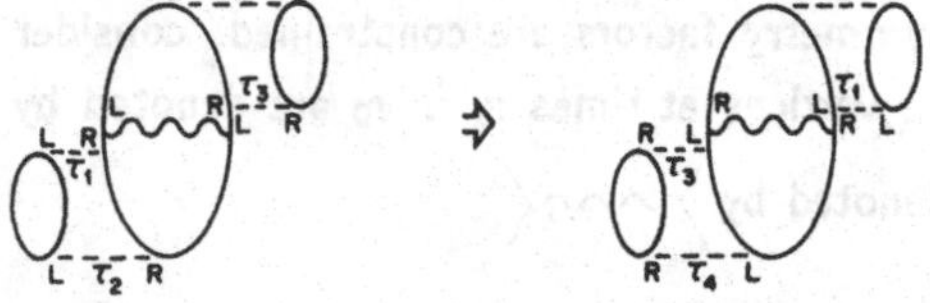

(2.142*c*)

The general rule for the symmetry factor resulting from this analysis is simple. For a two-body operator R, $S = 2$ if the exchange of the extremities of R combined with some time permutation and exchange of interaction extremities yields a deformation; otherwise $S = 1$. For an m-body operator R, 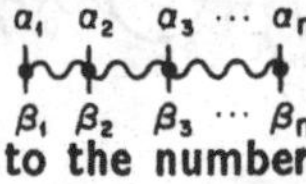 **the argument is easily generalized. The symmetry factor S is equal to the number of permutations of the m joints of R, which, when combined with time and interaction extremity exchanges, yields a deformation.**

The rules for calculating the pth order contribution to the perturbation expansion of the expectation value of a two-body operator $\langle R \rangle$ using unlabeled Feynman diagrams may be summarized as follows:

1. **Draw all distinct unlabeled connected diagrams composed of one R-vertex** 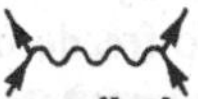 **and p v-vertices** 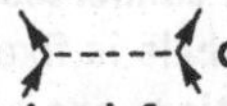 **connected by directed lines** **. Two diagrams are distinct if they cannot be deformed so as to coincide completely including the direction of arrows on propagators. For each distinct unlabeled diagram, evaluate the contribution as follows.**
2. **Calculate the symmetry factor S for the diagram. If the exchange of the extremities of R combined with some time permutation and exchange of interaction extremities yields a deformation of the original diagram, $S = 2$. Otherwise, $S = 1$.**
3. **Assign a time label τ_i to each of the p v-vertices, associate the time $\tau = 0$ with the R-vertex, and assign a single-particle index to each directed line. For each**

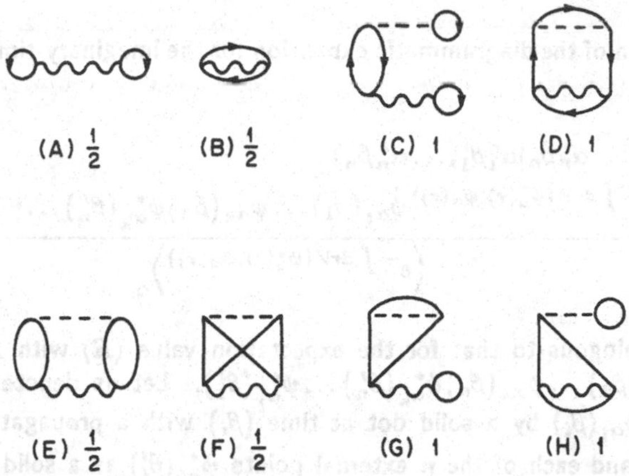

Fig. 2.6 Unlabeled Feynman diagrams for $\langle R\rangle$ with factor $\frac{1}{S}$

directed line include the factor

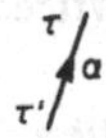

$$= g_\alpha(\tau-\tau') = e^{-(\epsilon_\alpha-\mu)(\tau-\tau')}[(1+\varsigma n_\alpha)\theta(\tau-\tau'-\eta)+\varsigma n_\alpha\theta(\tau'-\tau-\eta)] \ .$$

4. For each v-vertex include the factor

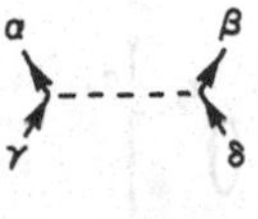

$$= (\alpha\beta|v|\gamma\delta)$$

and for the R-vertex include the factor

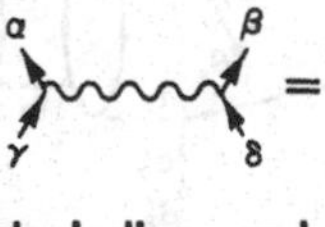

$$= (\alpha\beta|R|\gamma\delta) \ .$$

5. Sum over all single particle indices and integrate the p times over the interval $[0,\beta]$.
6. Multiply the result by the factor $\frac{(-1)^p}{S}\varsigma^{n_L}$ where n_L is the number of closed loops and S is the symmetry factor.

The eight unlabeled diagrams contributing to $\langle R\rangle$ in orders $p=0$ and 1 are shown in Fig. 2.6, together with the factors $\frac{1}{S}$.

The contribution at order 0 is given by diagrams A and B

$$\langle R\rangle^{(0)} = \frac{1}{2}\sum_{\alpha,\beta}[(\alpha\beta|R|\alpha\beta)+\varsigma(\alpha\beta|R|\beta\alpha)]n_\alpha n_\beta \tag{2.143}$$

and a typical contribution at order 1 is that of diagram E

$$\langle R\rangle^{(E)} = -\frac{1}{2}\sum_{\alpha\beta\gamma\delta}\int_0^\beta d\tau(\alpha\beta|R|\gamma\delta)(\gamma\delta|v|\alpha\beta)g_\alpha(-\tau)g_\beta(-\tau)g_\gamma(\tau)g_\delta(\tau) \ . \tag{2.144}$$

The derivation of the diagrammatic expansion for the imaginary-time Green's function

$$\mathcal{G}^{(n)}(\alpha_1\beta_1, \ldots \alpha_n\beta_n|\alpha_1'\beta_1', \ldots \alpha_n'\beta_n') \tag{2.145}$$
$$= \frac{\left\langle e^{-\int d\tau V(\psi_\alpha^*(\tau),\psi_\alpha(\tau))}\psi_{\alpha_1}(\beta_1)\ldots\psi_{\alpha_n}(\beta_n)\psi_{\alpha_n'}^*(\beta_n')\ldots\psi_{\alpha_1'}^*(\beta_1')\right\rangle_0}{\left\langle e^{-\int d\tau V(\psi_\alpha^*(\tau),\psi_\alpha(\tau))}\right\rangle_0}$$

is completely analogous to that for the expectation value $\langle R\rangle$ with $R(\psi_\alpha^*(0)\psi_\alpha(0))$ replaced by $\psi_{\alpha_1}(\beta_1)\ldots\psi_{\alpha_n}(\beta_n)\psi_{\alpha_n'}^*(\beta_n')\ldots\psi_{\alpha_1'}^*(\beta_1')$. Let us denote each of the n external points $\psi_{\alpha_i}(\beta_i)$ by a solid dot at time (β_i) with a propagator for state α_i entering it, , and each of the n external points $\psi_{\alpha_i'}^*(\beta_i')$ as a solid dot at time β_i' with a propagator for state α_i' leaving it, 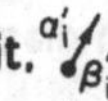. These external points are to be treated analogous to the R-vertex in the previous derivation. Introduction of m replicas labeled by indices σ_i with the constraint that $\psi_{\alpha_i}(\beta_i)$ and $\psi_{\alpha_i'}^*(\beta_i')$ be associated with σ_1, yields, in the limit $m \to 0$, the sum of all linked diagrams in which the interaction vertices, v, are linked to the $2n$ external points.

An example of a diagram contributing to the three-particle Green's function $G_3^{(I)}$ is the following.

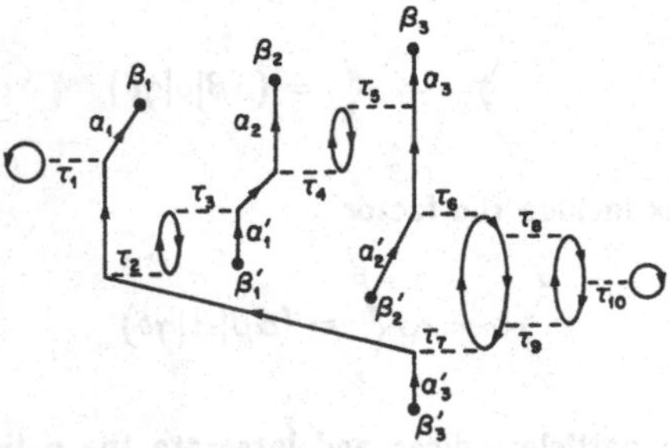

(2.146)

It consists of three directed lines beginning at the external points β_1', β_2' and β_3' and terminating at the external points β_1, β_2 and β_3 as well as an arbitrary number, in this case six, of closed loops.

In contrast to a diagram for a three-body operator, where the propagators entering and leaving each of the three joints may be exchanged, each of the external points in (2.146) is fixed. Hence, its symmetry factor is even lower than that of the corresponding contribution to a three-body operator, and as we shall now demonstrate, $S = 1$. The analysis is similar to that for the expectation value $\langle R\rangle$. In the present case, because there is no freedom to exchange propagators at the external points, there are no permutations involving the times along the three propagators starting and terminating at these external points which can produce a deformation. Since these times are now uniquely fixed, propagator loops connected to these lines by interaction vertices in turn have all their times fixed and by induction all times are fixed. For example, in the diagram (2.146), propagators beginning at β_3' and β_2' fix times τ_6 and τ_7 which in turn fix τ_8 and τ_9 which then fix τ_{10}. Similarly, the fact that propagators cannot be exchanged at external points prevents any exchange of vertex extremities from producing a deformation. Since no time permutations, extremity exchanges, or combinations thereof can yield a deformation of the original diagram, $S = 1$. The difference between the case of the expectation value $\langle R\rangle$ and a Green's function is illustrated by the diagram below which shows the contribution to $G_2^{(I)}$ analogous to diagram (2.142b).

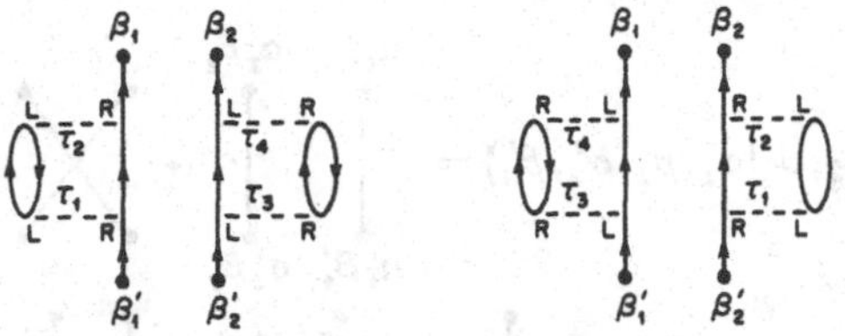

(2.147)

The simultaneous interchange of τ_1 with τ_3 and τ_2 with τ_4 and the exchange of all interaction extremities which produced a deformation in diagram (2.142b) yields the distinct labeled diagram in the right of (2.147).

Whereas the symmetry factor is simpler for Green's functions than for any other quantity we have evaluated, the sign of the contractions contributing to a diagram is slightly more complicated. First, consider the sign associated with the contractions of $\langle\psi_{\alpha_1}(\beta_1)\ldots\psi_{\alpha_n}(\beta_n)\psi^*_{\alpha'_n}(\beta'_n)\cdots\psi^*_{\alpha'_1}(\beta'_1)\rangle_0$ without any interaction vertices. The contraction $\langle\psi_{\alpha_1}(\beta_1)\psi_{\alpha_2}(\beta_2)\ldots\psi_{\alpha_n}(\beta_n)\psi^*_{\alpha_{n'}}(\beta'_n)\ldots\psi^*_{\alpha_{2'}}(\beta'_2)\psi_{\alpha'_1}(\beta'_1)\rangle$ in which $\psi^*_{\alpha'_m}(\beta'_m)$ is contracted with $\psi_{\alpha_m}(\beta_m)$ for all m has sign $+1$, and for any permutation P such that $\psi^*_{\alpha'_m}(\beta'_m)$ is contracted with $\psi_{\alpha_{Pm}}(\beta_{Pm})$ the overall sign is $(\varsigma)^P$. Adding any number of interaction vertices to produce a diagram with n_L closed loops modifies the overall sign by the factor $(\varsigma)^{n_L}$ by the same argument as presented previously for diagrams for the grand potential.

An equivalent sign rule is to close up a Green's functions diagram by deforming it so that each external point β'_m coincides with β_m. The diagram then looks like the expectation value of $\varsigma^n\langle\psi^*_{\alpha'_n}\ldots\psi^*_{\alpha'_1}\psi_{\alpha_1}\ldots\psi_{\alpha_n}\rangle$ where the factor ς^n arises from placing the $\psi^*\psi$ in normal order. The sign is then $\varsigma^{n+\tilde{n}_L}$ where $\tilde{n}_L$ is the number of closed loops in the deformed diagram.

Application of these rules to the diagram (2.146), yields $+1$. Using the first method, we note that (123) is an even permutation of (312) and the six closed loops yield an additional factor $\varsigma^6 = 1$. According to the second rule, joining the external points β'_m with β_m produces a seventh closed loop, so that $\varsigma^{n+\tilde{n}_L} = \varsigma^{3+7} = 1$

In summary, the rules for calculating the rth order contribution to the perturbation expansion of $\mathcal{G}^{(n)}(\alpha_1\beta_1,\ldots\alpha_n,\beta_n|\alpha'_1\beta'_1,\ldots\alpha'_n\beta'_n)$ using unlabeled Feynman diagrams are as follows:

1. Draw all distinct unlabeled connected diagrams composed of n external points $\psi_{\alpha_i}(\beta_i)$, n external points $\psi^*_{\alpha'_i}(\beta'_i)$, and r interaction vertices connected by directed lines. Two diagrams are distinct if, holding the external points fixed, the vertices and internal propagators cannot be deformed so as to coincide completely including the direction of arrows on propagators. The contribution for each distinct unlabeled diagram is evaluated as follows:

$G_1^{(I)}(\alpha_1,\beta_1|\alpha_2',\beta_2') =$ 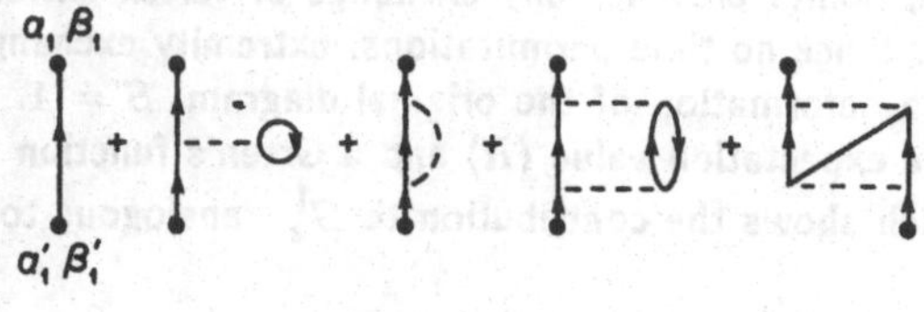

$G_2^{(I)}(\alpha_1,\beta_1;\alpha_2,\beta_2|\alpha_1',\beta_1';\alpha_2',\beta_2') =$ 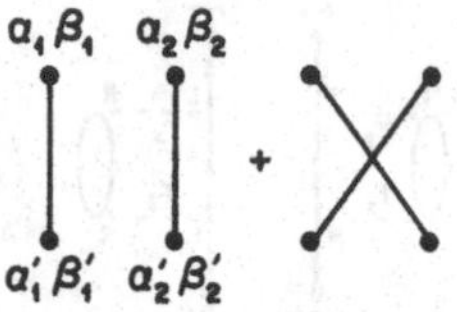

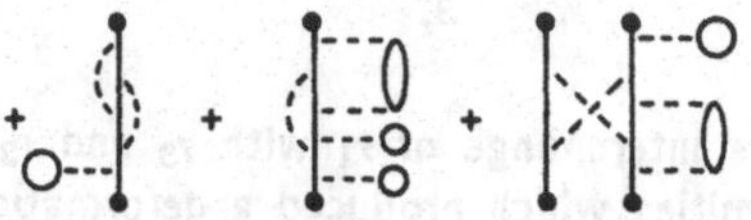

Fig. 2.7 Diagrams contributing to one- and two-particle Green's functions with overall sign factors.

2. **Each external point β_i, α_i corresponds to a specified state α_i and time β_i. Assign an internal time label τ_i to each of the r interaction vertices and for any propagator which is not connected to an external point assign an internal single-particle index. For each directed line include the factor**

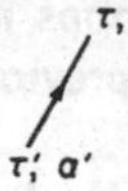

$$= \delta_{\alpha'\alpha} g_\alpha(\tau-\tau')$$

$$= \delta_{\alpha'\alpha} e^{-(\epsilon_\alpha-\mu)(\tau-\tau')}[(1+\varsigma n_\alpha)\theta(\tau-\tau'-\eta)+\varsigma n_\alpha\theta(\tau'-\tau+\eta)]$$

where τ and τ' denote either internal times τ_i or external times β_i. Propagators connected to an interaction vertex will only have one single-particle index α, in which case the factor $\delta_{\alpha'\alpha}$ is superfluous.

3. **For each interaction vertex include the factor**

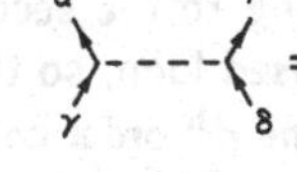

$$= (\alpha\beta|v|\gamma\delta) \ .$$

4. **Sum over all internal single-particle indices and integrate the r internal times τ_i over the interval $[0,\beta]$.**
5. **Multiply the result by the factor $(-1)^r\varsigma^P\varsigma^{n_L}$ where n_L is the number of closed propagator loops and ς^P is the sign of the permutation P such that each propagator line originating at the external point $\psi^*_{\alpha'_m}(\beta'_m)$ terminates at the external point $\psi_{\alpha_{Pm}}(\beta_{Pm})$.**

Examples of graphs contributing to the one- and two-particle Green's functions, together with the overall sign factor are given in Fig. (2.7).

2.4 IRREDUCIBLE DIAGRAMS AND INTEGRAL EQUATIONS

With the foundations of diagrammatic perturbation theory presented in the preceding section, it is now possible to derive exact integral equations relating connected Green's functions and irreducible vertex functions. Since these equations include contributions of all orders of perturbation theory, they are useful in defining consistent approximations involving infinite resummations of diagrams and in treating the renormalization of divergent field theories. They also lead naturally to the effective potential, Γ, which will be useful in understanding spontaneous symmetry breaking in Chapter 3, and to the self energy, Σ, which governs the propagation of a single particle in a many-body medium.

GENERATING FUNCTION FOR CONNECTED GREEN'S FUNCTIONS

As seen in the preceding sections, it is often useful to define a generating function by adding to the physical Hamiltonian under consideration additional terms in which field operators are coupled to external sources. The simplest possibility is to couple the field operators to an external source J by adding the term

$$S = \sum_{\alpha} \int d\tau [(J^*_\alpha(\tau) a_\alpha(\tau) + a^\dagger_\alpha(\tau) J_\alpha(\tau))] \tag{2.148}$$

where the sources $J_\alpha(\tau)$ are complex or Grassman variables for Bosons and Fermions respectively. Other more general sources, such as the bilinear form

$$\begin{aligned} S = \sum_{\alpha\beta} \int d\tau d\tau' [& a^\dagger_\alpha(\tau_1) a^\dagger_\beta(\tau_2) \eta_{\alpha\beta}(\tau_1,\tau_2) \\ &+ a^\dagger_\alpha(\tau_1) a_\beta(\tau_2) \bar{\eta}_{\alpha\beta}(\tau_1,\tau_2) + a_\alpha(\tau_1) a_\beta(\tau_2) \eta^*_{\alpha\beta}(\tau_1,\tau_2)] \end{aligned} \tag{2.149}$$

are useful for specific applications.

Similar to the generating function, Eq. (2.85) used to derive Wick's theorem, the generating function for imaginary-time Green's Function is defined as the partition function for the full Hamiltonian plus the source term Eq. (2.148). Using the functional integral representation for the partition function, Eq. (2.64), the generating function may be written

$$\begin{aligned} \mathcal{G}(J^*_\alpha(\tau), J_\alpha(\tau)) &\equiv \frac{1}{Z} \int \mathcal{D}[\psi^*_\alpha(\tau)\psi_\alpha(\tau)] e^{-\int_0^\beta d\tau [\sum_\alpha \psi_\alpha(\tau)(\partial_\tau - \mu)\psi_\alpha(\tau) + H[\psi^*_\alpha(\tau),\psi_\alpha(\tau)]} \\ &\quad \times e^{-\int_0^\beta d\tau \sum_\alpha [J^*_\alpha(\tau)\psi_\alpha(\tau) + \psi^*_\alpha(\tau) J_\alpha(\tau)]} \\ &= \langle e^{-\int_0^\beta d\tau \sum_\alpha [J^*_\alpha(\tau)\psi_\alpha(\tau) + \psi^*_\alpha(\tau) J_\alpha(\tau)]} \rangle \end{aligned} \tag{2.150a}$$

where, analagous to the thermal average with respect to H_0 in Eq. (2.81), we define the thermal average with respect to the full Hamitonian H as

$$\langle F(\psi^*,\psi)\rangle = \frac{1}{Z}\int \mathcal{D}[\psi_\alpha^*(\tau)\psi_\alpha(\tau)]F(\psi_\alpha^*,\psi_\alpha)e^{-\int_0^\beta d\tau[\sum_\alpha \psi_\alpha(\tau)(\partial_\tau-\mu)\psi_\alpha(\tau)+H[\psi_\alpha^*(\tau),\psi_\alpha(\tau)]]}\,. \tag{2.150b}$$

Differentiation with respect to the sources $J_\alpha^*(\tau)$ and $J_\alpha(\tau)$ yields

$$\frac{\delta \mathcal{G}(J_\alpha^*(\tau),J_\alpha(\tau))}{\delta J_{\alpha_1}^*(\tau_1)} = -\langle\psi_{\alpha_1}(\tau_1)e^{-\int_0^\beta d\tau\sum_\alpha[J_\alpha^*(\tau)\psi_\alpha(\tau)+\psi_\alpha^*(\tau)J_\alpha(\tau)]}\rangle \tag{2.151a}$$

and

$$\frac{\delta \mathcal{G}(J_\alpha^*(\tau),J_\alpha(\tau))}{\delta J_{\alpha_1}(\tau_1)} = -\zeta\langle\psi_{\alpha_1}^*(\tau_1)e^{-\int_0^\beta d\tau\sum_\alpha[J_\alpha^*(\tau)\psi_\alpha(\tau)+\psi_\alpha^*(\tau)J_\alpha(\tau)]}\rangle \tag{2.151b}$$

so that the n-particle imaginary-time Green's function may be written

$$\mathcal{G}^{(n)}(\alpha_1,\tau_1;\ldots\alpha_n,\tau_n|\alpha_1',\tau_1';\ldots;\alpha_n',\tau_n') = \frac{1}{\zeta^n}\left.\frac{\delta^{2n}\mathcal{G}(J_\alpha^*(\tau),J_\alpha(\tau))}{\delta J_{\alpha_1}^*(\tau_1)\ldots\delta J_{\alpha_n}^*(\tau_n)\delta J_{\alpha_n'}(\tau_n')\ldots\delta J_{\alpha_1'}(\tau_1')}\right|_{J=J^*=0}. \tag{2.152}$$

Although the diagrams for Green's functions derived in Section 2.3 have all the interaction vertices linked to the external legs, the diagrams are not all connected. For example, the first four diagrams in Fig. (2.7) for $\mathcal{G}^{(2)}$ are disconnected; that is, they may be separated into distinct subdiagrams in which the external points and vertices of one subdiagram are not connected by any interactions or propagators to any other subdiagrams. Since the sum of all disconnected diagrams simply corresponds to combinations of products of fewer-particle Green's functions, it is useful to deal with connected Green's functions, where $\mathcal{G}_c^{(n)}(\alpha_1,\tau_1;\ldots\alpha_n,\tau_n|\alpha_1',\tau_1';\ldots\alpha_n',\tau_n')$ is defined as the sum of all connected diagrams linked to the external points $(\alpha_1,\tau_1;\ldots\alpha_n,\tau_n)$ and $(\alpha_1',\tau_1';\ldots\alpha_n',\tau_n')$. The last diagram in Fig. (2.7) is an example of a contribution to $\mathcal{G}_c^{(2)}$.

The generating function for connected Green's Functions, which we shall denote $W(J_\alpha^*(\tau),J_\alpha(\tau))$, may be obtained from the generating function $\mathcal{G}(J_\alpha^*(\tau),J_\alpha(\tau))$ by using the replica technique once again. The functional $[\mathcal{G}(J_\alpha^*(\tau),J_\alpha(\tau))]^p$ may be written as a functional integral over p distinct fields $\{\psi_\alpha^{*\sigma},\psi_\alpha^\sigma\}$ and the resulting Green's function diagrams will have the property that all connected diagrams will be proportional to p and all disconnected diagrams will contain at least two factors of p. The terms proportional to p are singled out by

$$W(J_\alpha^*(\tau),J_\alpha(\tau)) = \lim_{p\to 0}\frac{\partial}{\partial p}(\mathcal{G}(J_\alpha^*(\tau),J_\alpha(\tau)))^p$$

so that

$$W(J^*_\alpha(\tau), J_\alpha(\tau)) = \ln \mathcal{G}(J^*_\alpha(\tau), J_\alpha(\tau)) \tag{2.153}$$

and

$$\mathcal{G}^{(n)}_c(\alpha_1, \tau_1; \ldots \alpha_n, \tau_n | \alpha'_1, \tau'_1; \ldots \alpha'_n, \tau'_n) = \zeta^n \left. \frac{\delta^{2n} W(J^*_\alpha(\tau), J_\alpha(\tau))}{\delta_\alpha J^*_{\alpha_1}(\tau_1) \ldots \delta J^*_{\alpha_n}(\tau_n) \delta J_{\alpha'_n}(\tau'_n) \ldots \delta J_{\alpha'_1}(\tau'_1)} \right|_{J^*=J=0} . \tag{2.154}$$

Note that by Eqs. (2.150) and (2.153), $W(J^*, J) = -\beta(\Omega(J^*, J) - \Omega(0,0))$ so that physically, W represents the difference between the grand canonical potential in the presence and absence of sources.

The structure of the terms produced by this generating function is illustrated by the examples of one- and two-particle connected Green's functions. Abbreviating $\sum_\alpha \int d\tau [J^*_\alpha(\tau)\psi_\alpha(\tau) + \psi^*_\alpha(\tau) J_\alpha(\tau)]$ by $J^*\psi + \psi^* J$ and $\{J^*_{\alpha_1}(\tau_1), \psi^*_{\alpha_1}(\tau_1), J_{\alpha'_1}(\tau'_1), \psi_{\alpha'_1}(\tau'_1)\}$ by $\{J^*_1, \psi^*_1, J_{1'}, \psi_{1'}\}$ in obvious notation, we obtain:

$$\begin{aligned}
\mathcal{G}^{(n)}_c(1|1') &= \zeta \frac{\delta^2}{\delta J^*_1 \delta J_{1'}} \left[\ln \langle e^{-(J^*\psi + \psi^* J)} \rangle \right]_{J=J^*=0} \\
&= -\frac{\delta}{\delta J^*_1} \left[\langle e^{-(J^*\psi+\psi^* J)} \rangle^{-1} \langle \psi^*_{1'} e^{-(J^*\psi+\psi^* J)} \rangle \right]_{J=J^*=0} \\
&= \Big[\langle e^{-(J^*\psi+\psi^* J)} \rangle^{-1} \langle \psi_1 \psi^*_{1'} e^{-(J^*\psi+\psi^* J)} \rangle \\
&\quad - \langle e^{-(J^*\psi+\psi^* J)} \rangle^{-2} \langle \psi_1 e^{-(J^*\psi+\psi^* J)} \rangle \langle \psi^*_{1'} e^{-(J^*\psi+\psi^* J)} \rangle \Big]_{J=J^*=0} .
\end{aligned} \tag{2.155}$$

Sucessive derivatives act on both the numerator and denominator of each term, yielding the familiar structure of a cumulant expansion in which all possible linked and unlinked combinations of $n\psi$'s and $n\psi^*$'s are generated. Note that because of the presence of Grassman variables, the order of the factors is important in calculating derivatives. Except in the presence of spontaneous symmetry breaking, as arises for example in the case of Bose condensation treated in Section 2.5, all expectation values involving unequal numbers of creation and annihilation operators vanish, greatly simplifying the general result. For the one-body connected Green's Function in the absence of symmetry breaking we thus obtain:

$$\mathcal{G}^{(1)}_c(1|1') = \langle \psi_1 \psi^*_{1'} \rangle = \mathcal{G}^{(1)}(1, 1') . \tag{2.156}$$

The two-body connected Green's Function is evaluated similarly:

$$\begin{aligned}
\mathcal{G}^{(2)}_c(1,2|1',2') &= \frac{\delta^4}{\delta J^*_1 \delta J^*_2 \delta J_{2'} \delta J_{1'}} \left[\ln \langle e^{-(J^*\psi+\psi^* J)} \rangle \right]_{J=J^*=0} \\
&= \frac{\delta^2}{\delta J^*_1 \delta J^*_2} \Big[\langle e^{-(J^*\psi+\psi^* J)} \rangle^{-1} \langle \psi^*_{2'} \psi^*_{1'} e^{-(J^*\psi+\psi^* J)} \rangle \\
&\quad - \langle e^{-(J^*\psi+\psi^* J)} \rangle^{-2} \langle \psi^*_{2'} e^{-(J^*\psi+\psi^* J)} \rangle \langle \psi^*_{1'} e^{-(J^*\psi+\psi^* J)} \rangle \Big]_{J=J^*=0} .
\end{aligned} \tag{2.157}$$

The only non-vanishing terms generated by the two remaining derivatives, assuming no symmetry breaking, are those in which derivatives acting on numerator terms add two ψ's to $\langle\psi^*\psi^*\rangle$ and one ψ to each of the factors $\langle\psi^*\rangle$, yielding

$$\begin{aligned} \mathcal{G}_c^{(2)}(1,2|1',2') &= \langle\psi_1\psi_2\psi^*_{2'}\psi^*_{1'}\rangle - \langle\psi_2\psi^*_{2'}\rangle\langle\psi_1\psi^*_{1'}\rangle - \varsigma\langle\psi_1\psi^*_{2'}\rangle\langle\psi_2\psi^*_{1'}\rangle \qquad (2.158a)\\ &= \mathcal{G}^{(2)}(1,2|1'2') - \left[\mathcal{G}^{(1)}(1|1')\mathcal{G}^{(1)}(2|2') + \varsigma\mathcal{G}^{(1)}(1|2')\mathcal{G}^{(1)}(2|1')\right] \;. \end{aligned}$$

Note that the factor ς in the last term arises from the fact that $\frac{\delta}{\delta J_2^*}$ had to be commutted through an odd number of Grassman variables to act on $\langle\psi_1^* e^{-(J^*\psi+\psi^*J)}\rangle$. A compact graphical representation of Eq. (2.158a) is

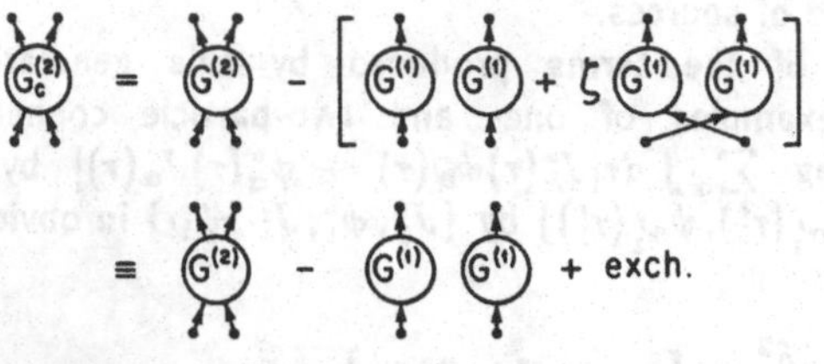

(2.158b)

where  denotes an internal propagator connected to an external point $\psi_\alpha(\tau)$ and the abbreviation *exch.* for exchange indicates the sum of all possible permutations, P, of the external points, in this case 2, with associated factor ς^P. Higher order connected Green's functions are evaluated similarly, and it is straightforward to verify that the three-body connected Green's function satisfies the following equation, which is also evident from the diagram rules for Green's functions;

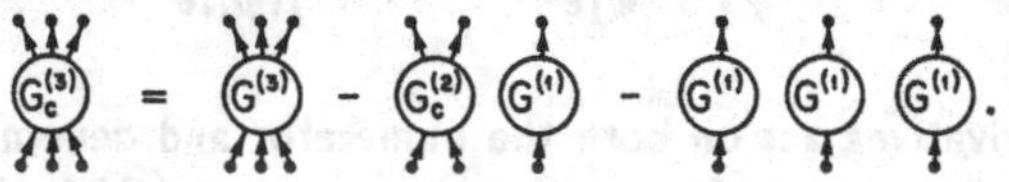

(2.159)

Since the Green's functions only involve equal numbers of ψ and ψ^*, a generating function with a bilinear source, Eq. (2.149), may also be used, and it is an instructive exercise to reproduce the preceding results with a bilinear source (see Problem 2.12).

THE EFFECTIVE POTENTIAL

In the presence of sources, the operators $\{a_\alpha^\dagger(\tau), a_\alpha(\tau)\}$ acquire non-zero expectation values. Let us define the average field

$$\begin{aligned} \phi_\alpha &= \langle a_\alpha\rangle_{J^*,J}\\ &= \langle\psi_\alpha\rangle_{J^*,J}\\ &= \frac{\int D[\psi_\alpha^*\psi_\alpha]\psi_\alpha e^{-\int_0^\beta d\tau\left[\sum_\alpha \psi_\alpha(\partial_\tau-\mu)\psi_\alpha + H[\psi_\alpha^*\psi_\alpha] + \sum_\alpha [J_\alpha^*\psi_\alpha+\psi_\alpha^* J_\alpha]\right]}}{\int D[\psi_\alpha^*\psi_\alpha] e^{-\int_0^\beta d\tau\left[\sum_\alpha \psi_\alpha(\partial_\mu-\mu)\psi_\alpha + H[\psi_\alpha^*\psi_\alpha] + \sum_\alpha [J_\alpha^*\psi_\alpha+\psi_\alpha^* J_\alpha]\right]}} \qquad (2.160a)\\ &= -\frac{\delta}{\delta J_\alpha^*(\tau)} W[J_\alpha^*(\tau), J_\alpha(\tau)] \end{aligned}$$

and its complex conjugate field

$$\begin{aligned}\phi^*_\alpha(\tau) &= \langle a^\dagger_\alpha(\tau)\rangle_{J^*,J} \\ &= -\varsigma\frac{\delta}{\delta J_\alpha(\tau)}W[J^*_\alpha(\tau), J_\alpha(\tau)] \ .\end{aligned} \tag{2.160b}$$

Here, in an obvious notation $\langle\ \ \rangle_{J^*,J}$ denotes a thermal average with respect to H plus the source term of the form introduced in Eqs. (2.80) - (2.82) for H_0 and in Eq. (2.150b) for H and the τ arguments of ψ_α and J_α have been suppressed for compactness.

Instead of dealing with the generating function W as a function of the sources J^*, J, it is useful to perform a Legendre transformation to obtain a function of the fields ϕ^*, ϕ. The motivation for performing this transformation is evident from considering the familiar example of a spin system in a magnetic field, which will be treated in detail in Chapter 4. Denoting the Hamiltonian of the spin variable as $\mathcal{H}(s)$, the free energy as a function of an external magnetic field $\vec{H}$ is given by

$$\mathrm{Tr}\, e^{-\beta(\mathcal{H}(s) - \vec{H}\cdot\sum_i \vec{s}_i)} = e^{-\beta F(H)} \tag{2.161a}$$

from which it follows that the magnetization is given by

$$M = -\frac{\partial F(H)}{\partial H} \ . \tag{2.161b}$$

A state function which depends upon the magnetization instead of the external magnetic field is obtained by inverting the relation (2.161b) to obtain $H(M)$ and defining the Legendre transform

$$G(M) = F(H(M)) + MH(M) \ . \tag{2.162}$$

It follows that G satisfies the reciprocity relation

$$\frac{\partial G}{\partial M} = \frac{\partial F}{\partial H}\frac{\partial H}{\partial M} + H + M\frac{\partial H}{\partial M} = H \ . \tag{2.163}$$

Whereas both $F(H)$ and $G(M)$ contain the same physical information, in the case of the first-order ferromagnetic phase transition, $G(M)$ has better analytic properties and will thus be preferable to approximate. Below the critical temperature, since M is non-vanishing and has the same sign as H, equation (2.161b) implies that $F(H)$ has non-zero positive slope for positive H and a non-zero negative slope for negative H so that it has a cusp at the origin. In contrast, $G(M)$ is a smooth double well, and the discontinuity associated with the first-order phase transition arises from moving from the solution to Eq. (2.163) in one well to the solution in the second well.

In the case of the generating function $W[J^*_\alpha(\tau), J_\alpha(\tau)]$, the equations (2.160) for $\phi_\alpha(J^*_\alpha, J_\alpha)$ and $\phi^*_\alpha(J^*_\alpha, J_\alpha)$ are inverted to obtain the sources as functions of the fields $J^*_\alpha(\phi^*_\alpha, \phi_\alpha)$ and $J_\alpha(\phi^*_\alpha, \phi_\alpha)$ and the effective potential (or effective action) is defined as the Legendre transform

$$\Gamma[\phi^*_\alpha(\tau), \phi_\alpha(\tau)] = -W[J^*_\alpha(\tau), J_\alpha(\tau)] - \sum_\gamma \int_0^\beta d\tau' [\phi^*_\gamma(\tau')J_\gamma(\tau') + J^*_\gamma(\tau')\phi_\gamma(\tau')] \ . \tag{2.164}$$

As in the example $G(M)$, the effective potential satisfies the reciprocity relation

$$\frac{\partial}{\partial\phi_\alpha^*(\tau)}\Gamma[\phi_\alpha^*(\tau),\phi_\alpha(\tau)] = \sum_\gamma \int_0^\beta d\tau' \left[-\frac{\partial W}{\partial J_\gamma^*(\tau')}\frac{\partial J_\gamma^*(\tau')}{\partial\phi_\alpha^*(\tau)} - \frac{\partial W}{\partial J_\gamma(\tau')}\frac{\partial J_\gamma(\tau')}{\partial\phi_\alpha^*(\tau)}\right.$$
$$\left. -\delta_{\alpha\gamma}\delta(\tau-\tau')J_\gamma(\tau') - \zeta\phi_\gamma^*(\tau')\frac{\partial J_\gamma(\tau')}{\partial\phi_\alpha^*(\tau)} - \frac{\partial J_\gamma^*(\tau')}{\partial\phi_\alpha^*(\tau)}\phi_\gamma(\tau')\right]$$
$$= -J_\alpha(\tau) \qquad (2.165a)$$

and the companion equation

$$\frac{\partial}{\partial\phi_\alpha(\tau)}\Gamma[\phi_\alpha^*(\tau),\phi_\alpha(\tau)] = -\zeta J_\alpha^*(\tau) \ . \qquad (2.165b)$$

When the sources are set equal to zero, Eqs. (2.165) show that the effective potential is stationary. That is, if we denote the fields in the absence of sources by $\tilde{\phi}_\alpha^*(\tau)$ and $\tilde{\phi}_\alpha(\tau)$, then

$$\frac{\delta\Gamma(\tilde{\phi}_\alpha^*(\tau),\tilde{\phi}_\alpha(\tau))}{\delta\tilde{\phi}_\alpha^*(\tau)} = \frac{\delta\Gamma(\tilde{\phi}_\alpha^*(\tau),\tilde{\phi}_\alpha(\tau))}{\delta\tilde{\phi}_\alpha(\tau)} = 0 \ . \qquad (2.166)$$

As we have already mentioned, in the case of Bose condensation Eqs. (2.166) have non-zero solutions $\{\tilde{\phi}_\alpha^*(\tau),\tilde{\phi}_\alpha(\tau)\}$ and these solutions will be studied in Section 2.5. In the absence of symmetry breaking, the fields $\{\tilde{\phi}_\alpha^*(\tau),\tilde{\phi}_\alpha(\tau)\}$ are zero and all Green's functions which do not have equal numbers of creation and annihilation operators vanish.

The effective potential is a generating function for vertex functions. These vertex functions are generated by differentiating the effective potential $\Gamma[\phi_\alpha^*(\tau)\phi_\alpha(\tau)]$ in the same way as connected Green's functions are generated from $W[J_\alpha^*(\tau)J_\alpha(\tau)]$:

$$\Gamma_{m\phi^*,n\phi}(\alpha_1\tau_1,\ldots\alpha_m\tau_m|\alpha_1'\tau_1',\ldots\alpha_n'\tau_n') \qquad (2.167)$$
$$= \left.\frac{\delta^{m+n}}{\delta\phi_{\alpha_1}^*(\tau_1)\ldots\delta\phi_{\alpha_m}^*(\tau_m)\delta\phi_{\alpha_n'}(\tau_n')\ldots\delta\phi_{\alpha_1'}(\tau_1')}\Gamma[\phi_\alpha^*(\tau),\phi_\alpha(\tau)]\right|_{J_\alpha^*=J_\alpha=0} .$$

Note that evaluation at $J_\alpha^* = J_\alpha = 0$ is equivalent to evaluation at the stationary solutions $\{\tilde{\phi}_\alpha^*,\tilde{\phi}_\alpha\}$ of Eq. (2.166).

The vertex functions $\Gamma_{m\phi^*,n\phi}$ defined in this way have several important properties. One feature is that they are one-particle irreducible and thus cannot be disconnected by removing a single internal propagator. Another significant property is the fact that connected Green's functions may be constructed from vertex functions using only tree diagrams, that is, diagrams containing no closed propagator loops. This property is extremely useful in renormalization of field theories, since all the divergences arise from loop integrals which are isolated in the vertex functions Γ, and in the definition of consistent truncated expansions. These properties are most easily seen by deriving a hierarchy of integral equations satisfied by the vertex functions and Green's functions.

THE SELF-ENERGY AND DYSON'S EQUATION

Before proceeding to the general case, it is useful to study the vertex function $\Gamma_{\phi^*\phi}$. An economical method to derive the desired integral equations for vertex functions is to take successive derivatives of the generating function $W[J^*,J]$ with respect to $\phi_{\alpha'}(\tau')$ and $\phi^*_{\alpha'}(\tau')$ using the chain rule and Eqs. (2.165) to write functional derivatives of a functional F

$$\begin{aligned}\frac{\delta F(J^*,J)}{\delta\phi_{\alpha_1}(\tau_1)} &= \sum_{\alpha_2}\int_0^\beta d\tau_2\left[\frac{\delta F}{\delta J^*_{\alpha_2}(\tau_2)}\frac{\delta J^*_{\alpha_2}(\tau_2)}{\delta\phi_{\alpha_1}(\tau_1)} + \frac{\delta F}{\delta J_{\alpha_2}(\tau_2)}\frac{\delta J_{\alpha_2}(\tau_2)}{\delta\phi_{\alpha_1}(\tau_1)}\right]\\
&= \sum_{\alpha_2}\int_0^\beta d\tau_2\left[-\varsigma\frac{\delta F}{\delta J^*_{\alpha_2}(\tau_2)}\frac{\delta^2\Gamma}{\delta\phi_{\alpha_1}(\tau_1)\delta\phi_{\alpha_2}(\tau_2)}\right.\\
&\qquad\left.-\frac{\delta F}{\delta J_{\alpha_2}(\tau_2)}\frac{\delta^2\Gamma}{\delta\phi_{\alpha_1}(\tau_1)\delta\phi^*_{\alpha_2}(\tau_2)}\right]\end{aligned}\tag{2.168a}$$

and similarly

$$\begin{aligned}\frac{\delta F(J^*,J)}{\delta\phi^*_{\alpha_1}(\tau_1')} &= \sum_{\alpha_2}\int_0^\beta d\tau_2\left[-\varsigma\frac{\delta F}{\delta J^*_{\alpha_2}(\tau_2)}\frac{\delta^2\Gamma}{\delta\phi^*_{\alpha_1}(\tau_1)\delta\phi_{\alpha_2}(\tau_2)}\right.\\
&\qquad\left.-\frac{\delta F}{\delta J_{\alpha_2}(\tau_2)}\frac{\delta^2\Gamma}{\delta\phi^*_{\alpha_1}(\tau_1)\delta\phi^*_{\alpha_2}(\tau_2)}\right] .\end{aligned}\tag{2.168b}$$

The lowest order equation, a general matrix form of Dyson's equation, is obtained by differentiating each of the quantities $\phi_{\alpha_3}(\tau_3) = -\frac{\delta W}{\delta J^*_{\alpha_3}(\tau_3)}$ and $\phi^*_{\alpha_3}(\tau_3) = -\varsigma\frac{\delta W}{\delta J_{\alpha_3}(\tau_3)}$ with respect to $\phi_{\alpha_1}(\tau_1)$ and $\phi^*_{\alpha_1}(\tau_1)$. Calculating $\frac{\delta\phi_{\alpha_3}(\tau_3)}{\delta\phi_{\alpha_1}(\tau_1)}$ in detail, we obtain

$$\begin{aligned}\delta_{\alpha_3\alpha_1}\delta(\tau_3,\tau_1) &= \frac{\delta\phi_{\alpha_3}(\tau_3)}{\delta\phi_{\alpha_1}(\tau_1)} = \frac{\delta}{\delta\phi_{\alpha_1}(\tau_1)}\left[-\frac{\delta W}{\delta J^*_{\alpha_3}(\tau_3)}\right]\\
&= \sum_{\alpha_2}\int_0^\beta d\tau_2\left[\varsigma\frac{\delta^2 W}{\delta J^*_{\alpha_2}(\tau_2)\delta J^*_{\alpha_3}(\tau_3)}\frac{\delta^2\Gamma}{\delta\phi_{\alpha_1}(\tau_1)\delta\phi_{\alpha_2}(\tau_2)}\right.\\
&\qquad\left.+\frac{\delta^2 W}{\delta J_{\alpha_2}(\tau_2)\delta J^*_{\alpha_3}(\tau_3)}\frac{\delta^2\Gamma}{\delta\phi_{\alpha_1}(\tau_1)\delta\phi^*_{\alpha_2}(\tau_2)}\right]\end{aligned}\tag{2.169}$$

which may be rewritten in the more compact notation

$$\delta(31) = \int d2\left[\varsigma\frac{\delta^2 W}{\delta J^*(2)\delta J^*(3)}\frac{\delta^2\Gamma}{\delta\phi^*(1)\delta\phi(2)} + \frac{\delta^2 W}{\delta J(2)\delta J^*(3)}\frac{\delta^2\Gamma}{\delta\phi(1)\delta\phi^*(2)}\right]\tag{2.170a}$$

where 1 denotes the variables $\{\alpha_1,\tau_1\}$ and $\int d2$ implies a sum over α_2 and an integral over τ_2. The remaining three derivatives yield the equations

$$\begin{aligned}\delta(31) &= \frac{\delta}{\delta\phi^*(1)}\phi^*(3) = \frac{\delta}{\delta\phi^*(1)}\left[-\varsigma\frac{\delta W}{\delta J(3)}\right]\\
&= \int d2\left[\frac{\delta^2 W}{\delta J^*(2)\delta J(3)}\frac{\delta^2\Gamma}{\delta\phi^*(1)\delta\phi(2)} + \varsigma\frac{\delta^2 W}{\delta J(2)\delta J(3)}\frac{\delta^2\Gamma}{\delta\phi^*(1)\delta\phi^*(2)}\right]\end{aligned}\tag{2.170b}$$

$$0 = \frac{\delta}{\delta\phi^*(1)}\phi(3) = \frac{\delta}{\delta\phi^*(1)}[-\frac{\delta W}{\delta J^*(3)}] \tag{2.170c}$$
$$= \int d2 \left[\varsigma \frac{\delta^2 W}{\delta J^*(2)\delta J^*(3)} \frac{\delta^2\Gamma}{\delta\phi^*(1)\delta\phi(2)} + \frac{\delta^2 W}{\delta J(2)\delta J^*(3)} \frac{\delta^2\Gamma}{\delta\phi^*(1)\delta\phi^*(2)}\right]$$

and

$$0 = \frac{\delta}{\delta\phi(1)}\phi^*(3) = \frac{\delta}{\delta\phi(1)}\left[-\varsigma\frac{\delta W}{\delta J(3)}\right] \tag{2.170d}$$
$$= \int d2 \left[\frac{\delta^2 W}{\delta J^*(2)\delta J(3)} \frac{\delta^2\Gamma}{\delta\phi(1)\delta\phi(2)} + \varsigma \frac{\delta^2 W}{\delta J(2)\delta J(3)} \frac{\delta^2\Gamma}{\delta\phi(1)\delta\phi^*(2)}\right] .$$

The equations (2.170) may be expressed in the matrix form

$$\int d2 \begin{pmatrix} \frac{\delta^2 W}{\delta J^*(3)\delta J(2)} & \varsigma\frac{\delta^2 W}{\delta J^*(3)\delta J^*(2)} \\ \varsigma\frac{\delta^2 W}{\delta J(3)\delta J(2)} & \frac{\delta^2 W}{\delta J(3)\delta J^*(2)} \end{pmatrix} \begin{pmatrix} \frac{\delta^2\Gamma}{\delta\phi^*(2)\delta\phi(1)} & \frac{\delta^2\Gamma}{\delta\phi^*(2)\delta\phi^*(1)} \\ \frac{\delta^2\Gamma}{\delta\phi(2)\delta\phi(1)} & \frac{\delta^2\Gamma}{\delta\phi(2)\delta\phi^*(1)} \end{pmatrix} = \delta(31)\begin{pmatrix} 1 & 0 \\ 0 & 1 \end{pmatrix} \tag{2.171}$$

which shows that the matrix of second derivatives of Γ is the inverse of the matrix of second derivatives of W. Thus, the matrix composed of $\Gamma_{\phi^*\phi}, \Gamma_{\phi^*\phi^*}, \Gamma_{\phi\phi}$ and $\Gamma_{\phi\phi}$ is the inverse of the matrix of connected Green's functions

$$\begin{pmatrix} \Gamma_{\phi^*\phi} & \Gamma_{\phi^*\phi^*} \\ \Gamma_{\phi\phi} & \Gamma_{\phi\phi^*} \end{pmatrix} = \varsigma \begin{pmatrix} \langle\psi\psi^*\rangle & \langle\psi\psi\rangle \\ \langle\psi^*\psi^*\rangle & \langle\psi^*\psi\rangle \end{pmatrix}^{-1} . \tag{2.172}$$

In order to understand the properties and physical significance of $\Gamma_{\phi^*\phi}$, we now consider a system with no symmetry breaking, in which case Green's functions with unequal numbers of ψ^*'s and ψ's vanish and Eqs. (2.170– 2.172) simplify to

$$\int d2\, \mathcal{G}_c^{(1)}(1,2)\Gamma_{\phi^*\phi}(2,3) = \int d2\, \Gamma_{\phi^*\phi}(1,2)\mathcal{G}_c^{(1)}(2,3) = \delta(1,3) \tag{2.173}$$

and

$$\Gamma_{\phi^*\phi}(1,2) = [\mathcal{G}_c^{(1)}]^{-1}(1,2) . \tag{2.174}$$

For a non-interacting system, rewriting Eq. (2.78) for the single-particle Green's functions with explicit $\{\alpha,\tau\}$ dependence yields

$$\sum_{\alpha_2}\left(\delta_{\alpha_1\alpha_2}\left(\frac{\partial}{\partial\tau_1} - \mu\right) + \langle\alpha_1|H_0|\alpha_2\rangle\right)\mathcal{G}_{0,c}^{(1)}(\alpha_2,\tau_1|\alpha_3,\tau_3) = \delta_{\alpha_1\alpha_3}\delta(\tau_1-\tau_3) \tag{2.175}$$

where H_0 is the single-particle Hamiltonian. Thus, for a non-interacting system

$$\Gamma^{(0)}_{\phi^*\phi}(\alpha_1,\tau_1|\alpha_2,\tau_2) = [\mathcal{G}_{0,c}^{(1)}]^{-1}(\alpha_1\tau_1|\alpha_2\tau_2) \tag{2.176}$$
$$= \left(\delta_{\alpha_1\alpha_2}\left(\frac{\partial}{\partial\tau_1} - \mu\right) + \langle\alpha_1|H_0|\alpha_2\rangle\right)\delta(\tau_1-\tau_2) .$$

The one-particle vertex function is a special case, and rather than deal with it directly, it is conventional to define the self energy Σ as the difference between the vertex function for the interacting and non-interacting systems

$$\Gamma_{\phi^*\phi}(1,2) \equiv \Gamma^{(0)}_{\phi^*\phi}(1,2) + \Sigma(1,2) \ . \tag{2.177}$$

Simplifying the notation by writing $\mathcal{G}^{(1)}_c(1,2)$ as the matrix $\mathcal{G}$, with the arguments and labels suppressed, equation (2.177) may be expressed

$$\mathcal{G}^{-1} = [\mathcal{G}_0]^{-1} + \Sigma \tag{2.178}$$

which when multiplied on the left by $\mathcal{G}_0$ and on the right by $\mathcal{G}$ yields the Dyson equation

$$\begin{aligned}\mathcal{G} &= \mathcal{G}_0 - \mathcal{G}_0\Sigma\mathcal{G} \\ &= \mathcal{G}_0 - \mathcal{G}_0\Sigma\mathcal{G}_0 + \mathcal{G}_0\Sigma\mathcal{G}_0\Sigma\mathcal{G}_0 \ldots\end{aligned} \tag{2.179a}$$

or, exhibiting the explicit $\{\alpha, \tau\}$ dependence,

$$\mathcal{G}^{(1)}_c(\alpha_1\tau_1|\alpha_4\tau_4) = \mathcal{G}^{(1)}_{0,c}(\alpha_1, \tau_1|\alpha_4, \tau_4) \tag{2.179b}$$

$$- \sum_{\alpha_2\alpha_3} \int_0^\beta d\tau_2 \, d\tau_3 \, \mathcal{G}^{(1)}_{0,c}(\alpha_1, \tau_1|\alpha_2\tau_2)\Sigma(\alpha_2\tau_2, \alpha_3\tau_3)\mathcal{G}^{(1)}_c(\alpha_3, \tau_3|\alpha_4, \tau_4) \ .$$

The graphical expansion of the self-energy Σ is evident from expressing the Dyson equation and its series expansion in diagrams. With the following graphical notation for the one-particle Green's functions non-interacting Green's functions, and self energy.

$$\overset{2}{\underset{1}{\big\Uparrow}} = \mathcal{G}^{(1)}_c(2,1) \qquad \overset{2}{\underset{1}{\big\uparrow}} = \mathcal{G}^{(1)}_{0,c}(2,1) \qquad \overset{2}{\underset{1}{\textcircled{Σ}}} = \Sigma(2,1)$$

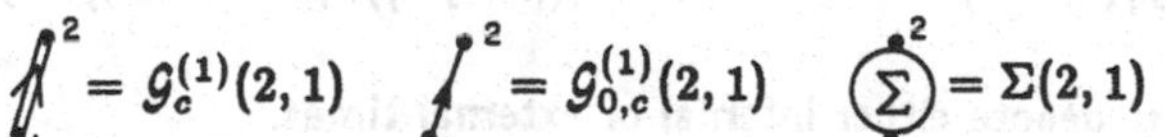

equation (2.179a) yields

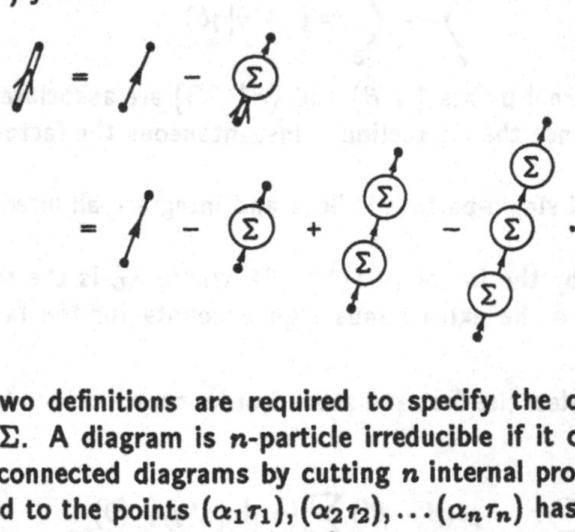

$$\tag{2.179c}$$

The following two definitions are required to specify the diagrams contributing to the self energy Σ. A diagram is n-particle irreducible if it cannot be separated into two or more disconnected diagrams by cutting n internal propagators. An amputated diagram attached to the points $(\alpha_1\tau_1), (\alpha_2\tau_2), \ldots (\alpha_n\tau_n)$ has no free propagator $G^{(0)}$ attached to these points; hence each point must connect directly to an interaction vertex. With these definitions, it is evident from (2.179c) that $-\Sigma(\alpha_2\tau_2, \alpha_1\tau_1)$ is the sum of all one-particle irreducible amputated diagrams connecting the points (α_1, τ_1) and (α_2, τ_2). To see this, imagine enumerating all the diagrams for the one-particle Green's function according to the rules given in Section 2.3. Amputate the external propagators, and let the set of one-particle irreducible diagrams define $-\Sigma$. Finally, consider the set of graphs generated by substituting this set of diagrams into the last line of Eq. (2.179c). By construction, each original Green's function diagram which could have been disconnected by cutting n different propagators will be generated once and only once by the nth term in this expansion (2.179c)

The rules for calculating the self energy $\Sigma(\alpha\beta, \alpha'\beta')$ thus follow directly from thoses enumerated in Section 2.4 for the one-particle Green's functions. The rules for the nth order contribution using unlabeled Feynman diagrams are summarized as follows:

1. Draw all distinct, unlabeled, one-particle irreducible, amputated diagrams composed of n interaction vertices 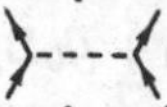 with the label $\{\alpha, \beta\}$ assigned to one outgoing arrow of an interaction vertex, the label $\{\alpha', \beta'\}$ assigned to one ingoing arrow of an interaction vertex, and all other arrows of the interaction vertices connected by directed lines . Two diagrams are distinct if, holding the points $\{\alpha, \beta\}$ and $\{\alpha', \beta'\}$ fixed, the lines and propagators cannot be deformed to coincide completely including the direction of arrows on propagators. The contribution for each distinct unlabeled diagram is evaluated as follows:
2. Assign an internal time label τ_i to each interaction vertex which is not assigned to one of the external time values β or β'. For every directed line, assign a single-particle index γ and include the factor

$$\gamma = g_\gamma(\tau - \tau') = e^{-(\epsilon_\gamma - \mu)(\tau - \tau')}[(1 + \varsigma n_\gamma)\theta(\tau - \tau' - \eta) + \varsigma n_\gamma \theta(\tau' - \tau + \eta)]$$

   where τ, τ' denote either internal or external times.
3. For each interaction vertex include the factor

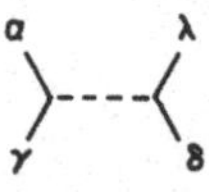

$$= (\alpha\lambda|v|\gamma\delta) \ .$$

   Note that if the external points $\{\alpha, \beta\}$ and $\{\alpha', \beta'\}$) are associated with the same interaction vertex, since the interaction is instantaneous the factor $\delta(\beta - \beta')$ must also be included.
4. Sum over all internal single-particle indices and integrate all internal times τ_i over the interval $[0, \beta]$.
5. Multiply the result by the factor $(-1)^{n-1}\varsigma^{n_L}$ where n_L is the number of closed propagator loops and the extra minus sign accounts for the fact that (2.179c) specifies $-\Sigma$.

To illustrate these rules, the first and second order contributions to the self-energy are the following:

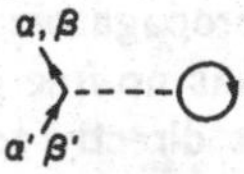

$$= \varsigma\delta(\beta - \beta')\sum_{\alpha}(\alpha\gamma|v|\alpha'\gamma)g_\gamma(0)$$

$$= \delta(\beta - \beta') \sum_{\alpha} (\alpha\gamma|v|\alpha'\gamma) n_\gamma \tag{2.180a}$$

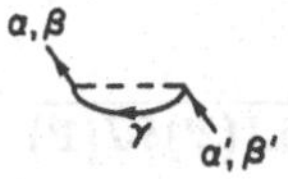

$$= \delta(\beta - \beta') \varsigma \sum_{\alpha} (\alpha\gamma|v|\gamma\alpha') n_\gamma \tag{2.180b}$$

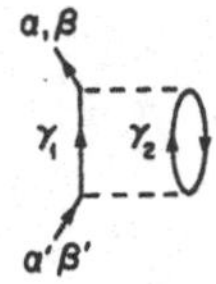

$$= -\varsigma \sum_{\gamma_1\gamma_2\gamma_3} (\alpha_1\gamma_3|v|\gamma_1\gamma_2)(\gamma_1\gamma_2|v|\alpha'\gamma_3)$$
$$\times\, g_{\gamma_1}(\beta - \beta') g_{\gamma_2}(\beta - \beta') g_{\gamma_3}(\beta' - \beta) \tag{2.180c}$$

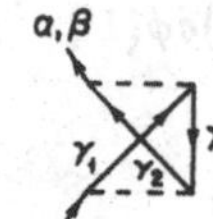

$$= - \sum_{\gamma_1\gamma_2\gamma_3} (\alpha_1\gamma_3|v|\gamma_2\gamma_1)(\gamma_1\gamma_2|v|\alpha'\gamma_3)$$
$$\times\, g_{\gamma_1}(\beta - \beta') g_{\gamma_2}(\beta - \beta') g_{\gamma_3}(\beta' - \beta) \ . \tag{2.180d}$$

Note that amputated diagrams, such as (2.180) are drawn with incoming and outgoing lines which do not begin or end with solid dots, indicating that the external labels are assigned to the vertex and no propagator is present. In contrast, diagrams having propagators in the external legs such as the Green's function diagrams in Fig. 2.7 are drawn with solid dots for the external points and it is understood that a propagator is to be included between each external point and the interaction to which it is connected.

The physical interpretation of Σ as a self energy or effective one-body potential is evident from using Eq. (2.171) to rewrite (2.178) in the form

$$[\mathcal{G}_c^{(1)}]^{-1}(\alpha_1\tau_1|\alpha_2\tau_2) = \Big(\delta_{\alpha_1\alpha_2}\big(\frac{\partial}{\partial\tau_1} - \mu\big) + \langle\alpha_1|H_0|\alpha_2\rangle\Big)\delta(\tau_1 - \tau_2) + \Sigma(\alpha_1\tau_1|\alpha_2\tau_2) \ . \tag{2.181}$$

The full Green's functions specifying the propagation of a particle in the many-body medium is obtained from the propagator in a non-interacting system by adding to $\langle\alpha_1|H_0|\alpha_2\rangle$ all the irreducible graphs describing the particle's interaction with the rest of the system. The contributions (2.180a and b) are instantaneous like H_0 and specify the shift in energy due to the Hartree Fock mean field. These and the higher order contributions will be discussed in detail in Chapter 5.

HIGHER-ORDER VERTEX FUNCTIONS

The essential features of Σ, which differs from the one-particle vertex function $\Gamma_{\phi^*\phi}$ only by the trivial term $\Gamma^0_{\phi^*\phi}$, are that it is one-particle irreducible and that the full one-particle Green's function is obtained from it by Eq. (2.179) which involves no loop integrals. We will now demonstate that these two features generalize to n-particle vertex functions.

We begin by introducing an economical graphical representation for differentiations

of the form which led to (2.170). Let

$$\equiv \zeta^n \frac{\delta^{m+n} W(J^*, J)}{\delta J^*(1)\delta J^*(2)\ldots\delta J^*(m)\delta J(n')\ldots\delta J(2')\delta J(1')} \tag{2.182a}$$

and

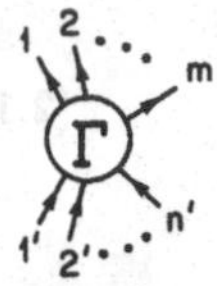

$$\equiv \frac{\delta^{m+n}\Gamma(\phi^*, \phi)}{\delta\phi^*(1)\delta\phi(2)\ldots\delta\phi^*(m)\delta\phi(n')\ldots\delta\phi(2')\delta\phi(1')} \ . \tag{2.182b}$$

With this notation, Eq. (2.170a) may be written

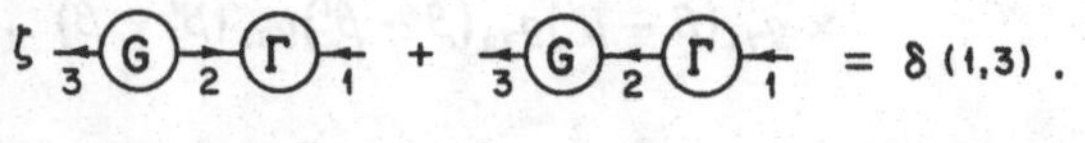

(2.183)

To emphasize the essential structure, we shall condense the notation still further by omitting signs, disregarding the direction of the arrows, letting $\frac{\delta}{\delta\phi}$ represent either $\frac{\delta}{\delta\phi(i)}$ or $\frac{\delta}{\delta\phi^*(i)}$ and letting $\frac{\delta}{\delta J}$ represent either $\frac{\delta}{\delta J(i)}$ or $\frac{\delta}{\delta J^*(i)}$. Then a functional derivative $\frac{\delta}{\delta\phi}$ applied to $\frac{\delta^n}{\delta\phi^n}\Gamma$ increases the number of legs by one:

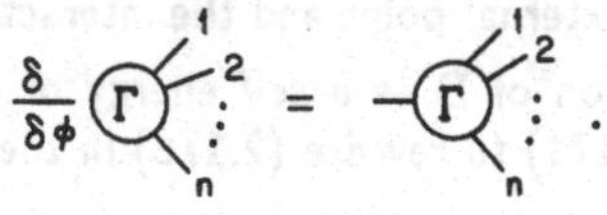

(2.184)

Using the chain rule $\frac{\delta}{\delta\phi} = \frac{\delta J}{\delta\phi}\frac{\delta}{\delta J}$, Eq. (2.168), the functional derivative $\frac{\delta}{\delta\phi}$ applied to $\frac{\delta^n}{\delta J^n}W$ adds a leg containing $\frac{\delta J}{\delta\phi} = \frac{\delta^2\Gamma}{\delta\phi^2}$:

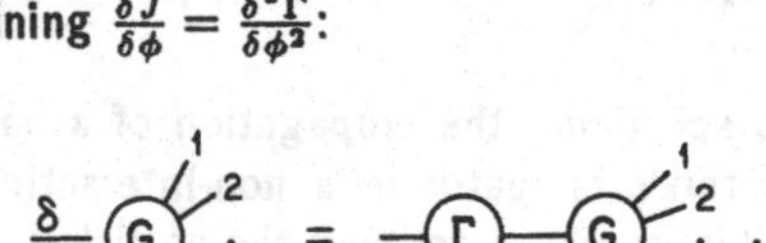

(2.185)

With this compact notation, evaluation of $\frac{\delta^n}{\delta\phi^n}[\phi] = \frac{\delta^n}{\delta\phi^n}[\frac{\partial W}{\partial J}]$ for successive values of n yields the desired hierarchy of equations. For $n = 1$, we recover the abbreviated form of (2.183)

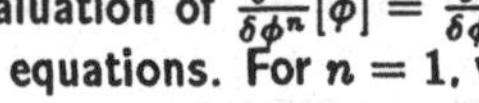

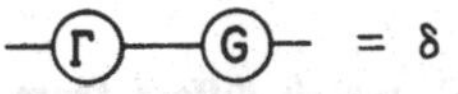

(2.186a)

Evaluating successive derivatives by letting $\frac{\delta}{\partial\phi}$ act on Γ's using (2.184) or on G's using (2.185) yields

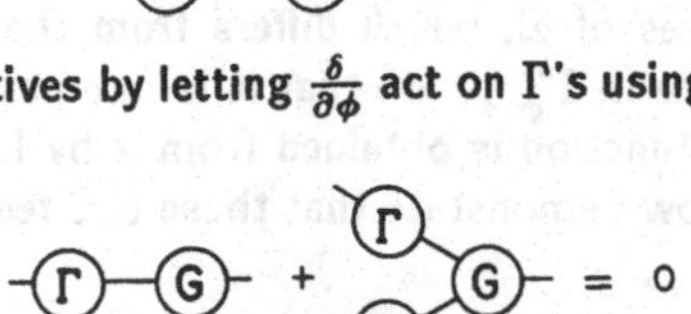

(2.186b)

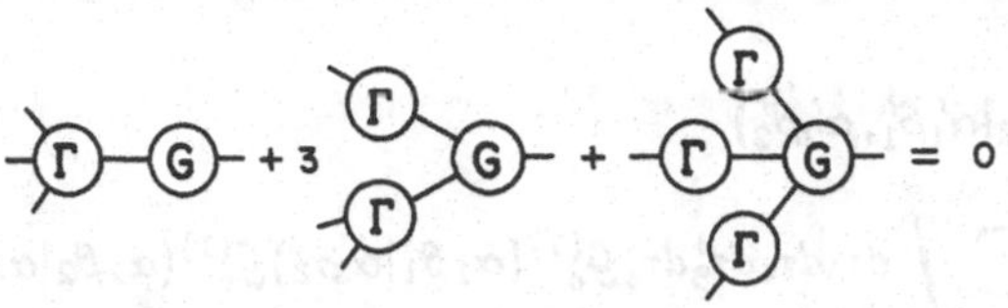

(2.186c)

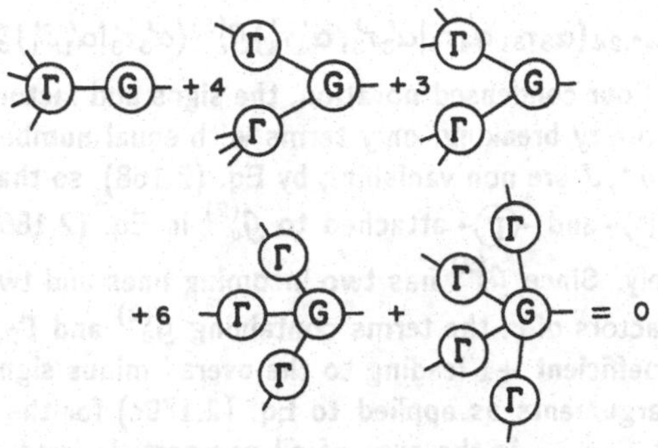

(2.186d)

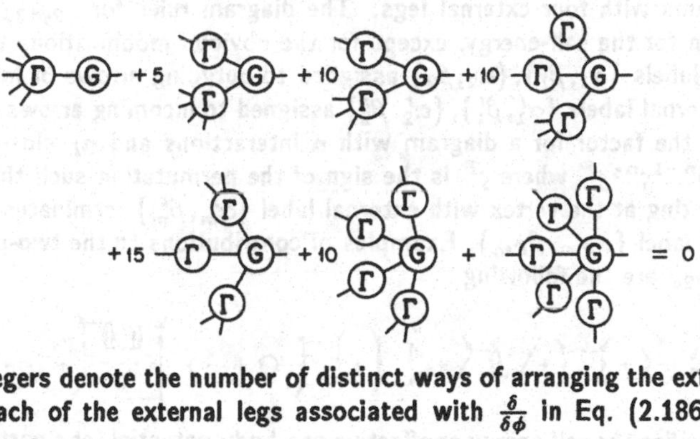

(2.186e)

where the integers denote the number of distinct ways of arranging the external labels. Multiplying each of the external legs associated with $\frac{\delta}{\delta\phi}$ in Eq. (2.186c) by $G_1 = (\Gamma_1)^{-1}$, we obtain

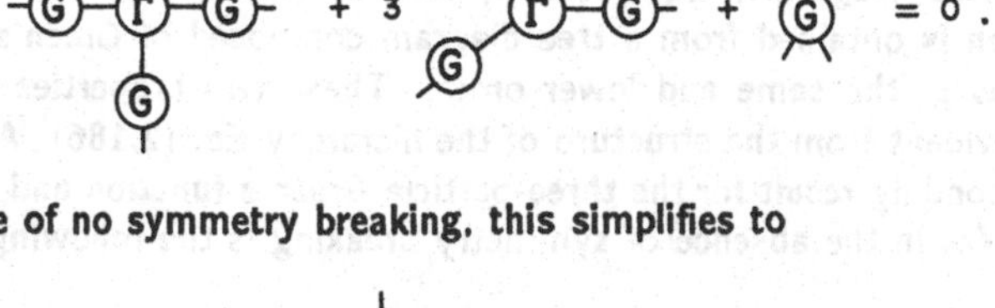

(2.187)

In the case of no symmetry breaking, this simplifies to

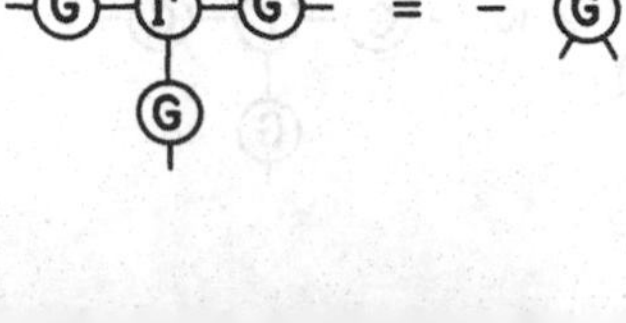

(2.188a)

or explicitly

$$\mathcal{G}_c^{(2)}(\alpha_1\beta_1, \alpha_2\beta_2|\alpha_1'\beta_1', \alpha_2'\beta_2') \tag{2.188b}$$

$$= -\sum_{\substack{\alpha_3\alpha_4\\ \alpha_3'\alpha_4'}} \int_0^\beta d\tau_3 d\tau_4 d\tau_3' d\tau_4' \mathcal{G}_c^{(1)}(\alpha_1\beta_1|\alpha_3\tau_3)\mathcal{G}_c^{(1)}(\alpha_2\beta_2|\alpha_4\tau_4)$$

$$\times \Gamma_{2\phi^*,2\phi}(\alpha_3\tau_3, \alpha_4\tau_4|\alpha_3'\tau_3', \alpha_4'\tau_4')\mathcal{G}_c^{(1)}(\alpha_3'\tau_3'|\alpha_1'\beta_1')\mathcal{G}_c^{(1)}(\alpha_4'\tau_4'|\alpha_2'\beta_2') \ .$$

Note that in spite of our condensed notation, the signs and factors of ς are obvious: In the absence of symmetry breaking, only terms with equal numbers of derivatives with respect to ϕ^*, ϕ or J^*, J are non vanishing by Eq. (2.168), so that ingoing or outgoing legs of the form $\rightarrow\!\!(\Gamma)\!\!-$ and $\leftarrow\!\!(\Gamma)\!\!-$ attached to $\mathcal{G}_c^{(2)}$ in Eq. (2.186d) have factors (-1) and $(-\varsigma)$ respectively. Since $\mathcal{G}_c^{(2)}$ has two incoming lines and two outgoing lines and, by Eq. (2.154), 2 factors of ς, the terms containing $\mathcal{G}_c^{(2)}$ and $\Gamma_{2\phi^*2\phi}$ in Eqs. (2.186d) and (2.187) have coefficient $+1$ leading to the overall minus sign in (2.188a).

By the same arguments as applied to Eq. (2.179c) for the self energy, equation (2.188) shows that $\Gamma_{2\phi^*,2\phi}$ is the sum of all one-particle irreducible amputated connected diagrams with four external legs. The diagram rules for $\Gamma_{2\phi^*2\phi}$ are identical to those given for the self-energy, except for the obvious modifications that there are two external labels $\{\alpha_1, \beta_1\}, \{\alpha_2, \beta_2\}$ assigned to outgoing arrows of interaction vertices, two external labels $\{\alpha_1', \beta_1'\}, \{\alpha_2', \beta_2'\}$ assigned to incoming arrows of interaction vertices, and the factor for a diagram with n interactions and n_L closed propagator loops is $(-1)^{n-1}\varsigma^{n_L}\varsigma^P$ where ς^P is the sign of the permutation such that each propagator originating at the vertex with external label $\{\alpha_m', \beta_m'\}$ terminates at the vertex with external label $\{\alpha_{Pm}, \beta_{Pm}\}$. Examples of contributions to the two-particle vertex function $\Gamma_{2\phi^*2\phi}$ are the following

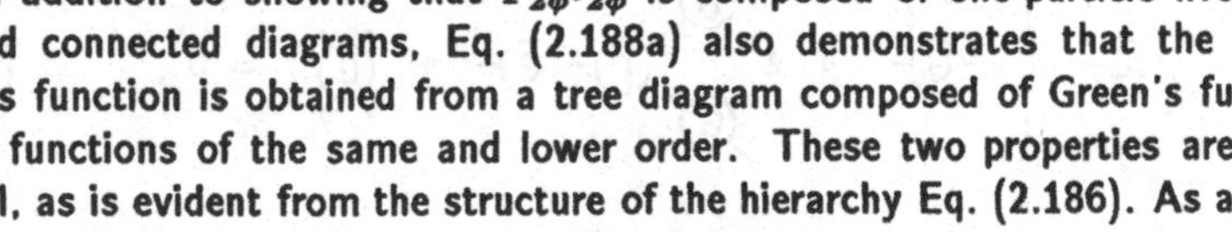

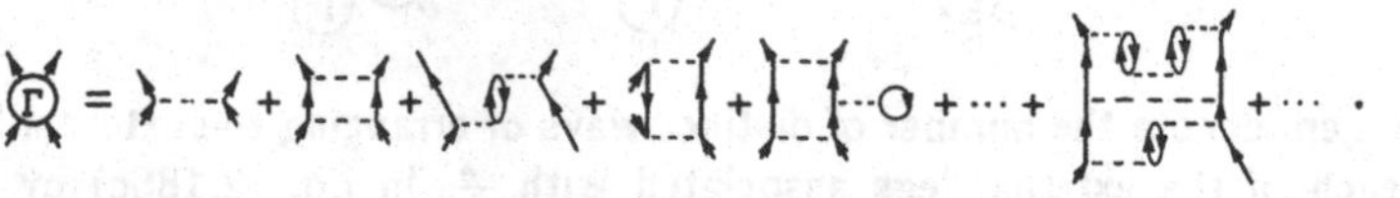

(2.189)

Just as Σ specifies the self energy or effective one-body potential for a particle propagating in a many-particle system, $\Gamma_{2\phi^*2\phi}$ corresponds to the effective two-body interaction between two particles propagating in a many-particle medium.

In addition to showing that $\Gamma_{2\phi^*2\phi}$ is composed of one-particle irreducible amputated connected diagrams, Eq. (2.188a) also demonstrates that the two-particle Green's function is obtained from a tree diagram composed of Green's functions and vertex functions of the same and lower order. These two properties are completely general, as is evident from the structure of the hierarchy Eq. (2.186). As a final example, the corresponding result for the three-particle Green's function and vertex function from Eq. (2.184e) in the absence of symmetry breaking is the following:

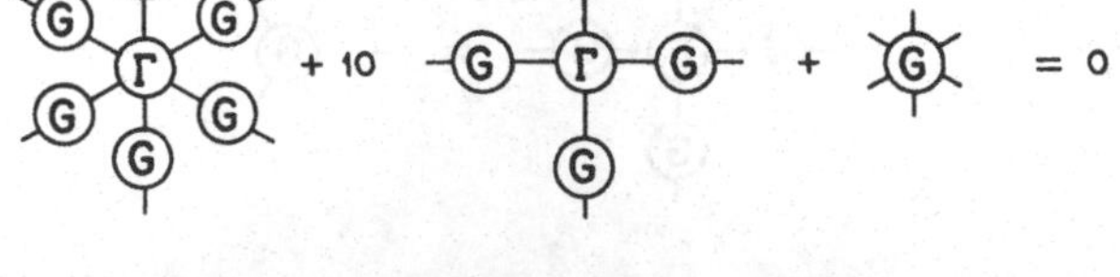

(2.190)

again demonstrating the one-particle irreducibility and tree-diagram structure.

Analogous equations in which vertex functions are n-particle irreducible amputated connected diagrams may be derived straightforwardly by the same approach we have used starting with sources coupled to all combinations of n creation and annihiltion operators. For example using the bilinear source, (2.149), the principal steps in deriving the two-particle irreducible theory are the following. The generating function for connected Green's functions is

$$W(\eta,\bar{\eta},\eta^*) = ln\langle e^{-\int_0^\beta d\tau \sum_{\alpha,\beta} \eta_{\alpha\beta}\psi^*_\alpha\psi^*_\beta + \bar{\eta}_{\alpha\beta}\psi^*_\alpha\psi_\beta + \eta^*_{\alpha\beta}\psi_\alpha\psi_\beta}\rangle \tag{2.191}$$

from which we define the expectation values

$$\begin{aligned} \xi_{\alpha\beta} &= -\frac{\delta W}{\delta\eta^*_{\alpha\beta}} = \langle\psi_\alpha\psi_\beta\rangle \\ \xi^*_{\alpha\beta} &= -\frac{\delta W}{\delta\eta_{\alpha\beta}} = \langle\psi^*_\alpha\psi^*_\beta\rangle \\ \bar{\xi}_{\alpha\beta} &= -\frac{\delta W}{\delta\bar{\eta}_{\alpha\beta}} = \langle\psi^*_\alpha\psi_\beta\rangle \end{aligned} \tag{2.192}$$

and Legendre transform

$$\Gamma(\xi,\bar{\xi},\xi^*) = -W(\eta,\bar{\eta},\eta^*) - \sum_{\alpha\beta}\int_0^\beta d\tau[\xi^*_{\alpha\beta}\eta_{\alpha\beta} + \bar{\xi}_{\alpha\beta}\bar{\eta}_{\alpha\beta} + \xi_{\alpha\beta}\eta^*_{\alpha\beta}] \tag{2.193}$$

with the reciprocity relations

$$\frac{\partial\Gamma}{\delta\xi^*_{\alpha\beta}} = -\eta_{\alpha\beta} \quad \frac{\delta\Gamma}{\delta\bar{\xi}_{\alpha\beta}} = -\bar{\eta}_{\alpha\beta} \quad \frac{\delta\Gamma}{\delta\xi_{\alpha\beta}} = -\eta^*_{\alpha\beta}\ . \tag{2.194}$$

Using the schematic notation of Eq. (2.184 – 2.188) in which $\frac{\delta}{\delta\xi}$ represents $\frac{\delta}{\delta\xi_{\alpha\beta}}$, $\frac{\delta}{\delta\xi^*_{\alpha\beta}}$, or $\frac{\delta}{\delta\bar{\xi}_{\alpha\beta}}$, $\frac{\delta}{\delta\eta}$ represents $\frac{\delta}{\delta\eta_{\alpha\beta}}$, $\frac{\delta}{\delta\eta^*_{\alpha\beta}}$, or $\frac{\delta}{\delta\bar{\eta}_{\alpha\beta}}$ and indices and directions of propagators are suppressed, a functional derivative $\frac{\delta}{\delta\xi}$ applied to $\frac{\delta^m}{\delta\xi^m}\Gamma$ increases the number of legs by two

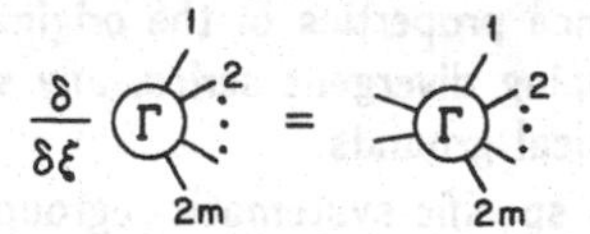

(2.195a)

and applied to $\frac{\delta^m}{\delta\eta^m}W$ adds a leg containing $\frac{\delta\eta}{\delta\xi} = \frac{\delta^2\Gamma}{\delta\xi^2}$:

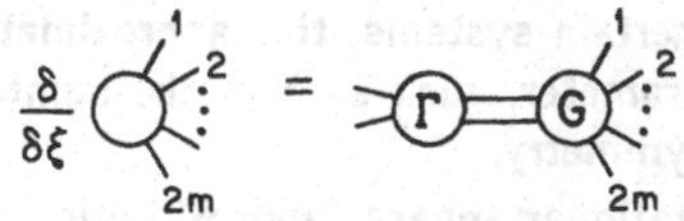

(2.195b)

Evaluation of successive deriatives $\frac{\delta^n}{\delta\xi}[\xi] = \frac{\delta^n}{\delta\xi}[\frac{\delta W}{\partial\eta}]$ yields a hierarchy of the form of Eq. (2.186)

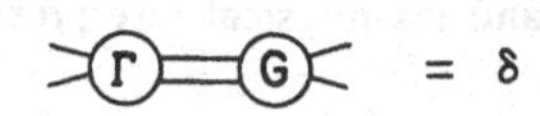

(2.196a)

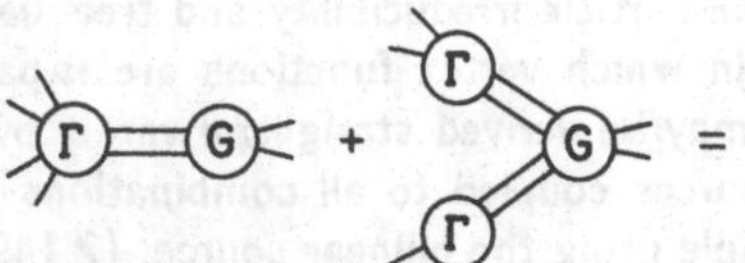

(2.196*b*)

The first equation shows that the matrix of four-point vertex functions is the inverse of the matrix of two-particle Green's functions. In the absence of symmetry-breaking, all $\mathcal{G}$'s and Γ's must have equal numbers of incoming and outgoing lines and the pairs of lines in these equations represent all combinations of aligned or anti-aligned propagators consistent with this restriction. By the same arguments used for Eqs. (2.186), Eqs. (2.196) show that the Γ's are composed of two-particle irreducible amputated connected diagrams and that the n-particle Green's functions is obtained from tree graphs involving fewer particle Green's functions and $2n$-point and fewer point vertex functions. It is an instructive exercise to derive these equations in detail (see Problem 2.12).

2.5 STATIONARY-PHASE APPROXIMATION AND LOOP EXPANSION

Whereas perturbation theory is valuable for the formal developments of Section 2.3 and is directly applicable to a limited class of physical problems with weak interactions characterized by a small expansion parameter, many problems of physical interest involve strong many-body interactions for which a perturbation expansion in the interaction strength is inappropriate. A natural approach for such problems is to reorganize the perturbation expansion into a series in powers of a new small parameter. We have already seen examples of such infinite resummations of perturbation theory in Section 2.4, where solution of Dyson's equations with some finite set of self energy diagrams or, in general, calculation of Green's functions with any finite set of diagrams for n-particle irreducible vertex functions Γ resums an infinite number of terms of the original perturbation series. Other physically motivated resummations will be presented subsequently for specific systems. In view of the general lack of mathematical control on the convergence properties of the original series and the obvious ambiguity associated with regrouping divergent series, any such resummations must be understood ultimately on physical grounds.

In this section, we consider a specific systematic regrouping of terms obtained by applying the stationary-phase approximation to the functional integral for the partition function. This regrouping will be seen to be ordered in the number of loops occuring in the Feynman diagrams. For certain systems, this approximation generates an asymptotic expansion in a small parameter, such as $\hbar$ or the number of degrees of freedom associated with an internal symmetry.

We first review the stationary-phase approximation in the case of a one-dimensional integral, and then generalize to the cases of the Feynman path integral and the partition function for many-particle systems. As usual, the primary emphasis will be on the essence of the method and its physical interpretation, rather than mathematical rigor.

ONE-DIMENSIONAL INTEGRAL

The stationary-phase approximation, also referred to as the saddle point approximation or method of steepest descent, is a method for developing an asymptotic expansion in powers of $\frac{1}{\ell}$ for an integral of the form

$$I(\ell) = \int_{-\infty}^{\infty} dt e^{-\ell f(t)} \tag{2.197}$$

where ℓ is a real parameter and in general $f(t)$ is an analytic function in the complex t-plane.

For simplicity, we will first consider the special case of a real function $f(t)$ with an absolute minimum at $t = t_0$. As ℓ increases, the integral becomes sharply peaked around the point t_0, and the dominant contribution to the integral arises from the vicinity of t_0. Expanding $f(t)$ around t_0, recognizing the fact that $f'(t_0) = 0$ and $f''(t_0) > 0$ since t_0 is the minimum, the integral may be rewritten:

$$\begin{aligned} I(\ell) &= e^{-\ell f_0} \int_{-\infty}^{\infty} dt\, e^{-\frac{\ell}{2} f_0''(t-t_0)^2 - \ell \sum_{n=3}^{\infty} \frac{(t-t_0)^n}{n!} f_0^{(n)}} \\ &= \sqrt{\frac{2\pi}{\ell f_0''}}\, e^{-\ell f_0} \int_{-\infty}^{\infty} \frac{d\tau}{\sqrt{2\pi}} e^{-\frac{\tau^2}{2} - \sum_{n=3}^{\infty} \frac{\tau^n}{n!} \frac{f_0^{(n)}}{\ell^{(\frac{n}{2}-1)} (f_0'')^{\frac{n}{2}}}} \end{aligned} \tag{2.198}$$

where derivatives evaluated at t_0 are denoted f_0'' and $f_0^{(n)}$ and the change of variables $\tau = (t - t_0)\sqrt{\ell f_0''}$ has been introduced to rescale the Gaussian to unit width. As $\ell \to \infty$, the terms with $n \geq 3$ go to zero and we may expand $I(\ell)$ in powers of $\frac{1}{\ell}$. For the problems of physical interest, we will usually be interested in expanding the logarithm of $I(\ell)$, and in the present example, it is straightforward to expand the exponential in Eq. (2.149), perform the Gaussian integrals, and exponentiate the result to the desired power of $\frac{1}{\ell}$ (see Problem 2.13).

A more economical derivation, however, may be obtained by utilizing our knowledge of Wick's theorem and the linked cluster expansion. Just as we defined contractions in connection with Eq. (2.84), we may define the contraction of τ as

$$\overbrace{\tau \cdot \tau} = \int_{-\infty}^{\infty} \frac{d\tau}{\sqrt{2\pi}} \tau \cdot \tau e^{-\frac{\tau^2}{2}} = 1 \ . \tag{2.199}$$

The coefficients of τ^n for $n \geq 3$ in Eq. (2.198) are regarded as vertices with n lines

$$V_n = \frac{1}{n!} \frac{f_0^{(n)}}{\ell^{(\frac{n}{2}-1)} (f_0'')^{\frac{n}{2}}} = \tag{2.200}$$

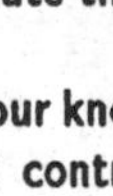

Diagrams representing all possible contractions contributing to Eq. (2.198) are obtained by drawing any number of vertices $V_{n_1} V_{n_2} \ldots V_{n_N}$ and connecting them with propagators equal to 1. In addition to the vertices, V_n, a diagram of order N has the overall factor $\frac{(-1)^N}{N!}$. By the linked cluster theorem,

$$I(\ell) = e^{-\ell f_0} \sqrt{\frac{2\pi}{\ell f_0''}}\, e^{(\text{sum of all linked diagrams})} \ . \tag{2.201}$$

The following diagrams contribute to lowest order in $\frac{1}{\ell}$

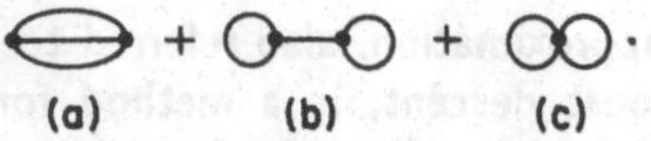

$$\ominus + \bigcirc\!\!-\!\!\bigcirc + \bigcirc\!\bigcirc \ . \tag{2.202}$$
$$(a) \qquad\qquad (b) \qquad\qquad (c)$$

Counting the 3×2 ways of contracting the three lines of each vertex in diagram (a) and the 3×3 ways of picking one line from each vertex to contract in diagram (b), the contribution of these two diagrams is

$$\frac{(-1)^2}{2!}(3 \times 2 + 3 \times 3)(\frac{f_0^{(3)}}{3!})^2 \frac{1}{\ell(f_0'')^3} \ .$$

Similarly, counting the three distinct contractions, diagram (c) contributes $(-1)\frac{3}{4!}\frac{f_0^{(4)}}{\ell(f_0'')^2}$, so that the asymptotic expansion of $I(\ell)$ is thus

$$I(\ell) = e^{-\ell f_0 + \frac{1}{2}ln(\frac{2\pi}{\ell f_0''}) + \frac{1}{\ell}(\frac{5}{24}\frac{(f_0^{(3)})^2}{(f_0'')^3} - \frac{1}{8}\frac{f_0^{(4)}}{(f_0'')^2}) + O(\frac{1}{\ell^2})} \ . \tag{2.203}$$

The contribution of order $\frac{1}{\ell^2}$ involves diagrams containing $V_6, V_5V_3, (V_4)^2, V_4(V_3)^2$, and $(V_3)^4$ and is treated in Problem 2.13.

The previous discussion generalizes straighforwardly to a complex function $f(t)$ which is analytic in some region of the complex t plane. Consider the case in which $f(t)$ has a single stationary point t_0 such that $f'(t_0) = 0$. Since by the Cauchy integral formula $\mathrm{Re}\, f(t_0)$ is equal to the average of $\mathrm{Re}\, f(t)$ on a circle centered on t_0, t_0 must be a saddle point for the function $\mathrm{Re}\, f(t)$. For an arbitrary contour passing through t_0, $\mathrm{Im}\, f(t)$ will vary along the contour giving rise to arbitrarily rapid oscillations in the integrand as $\ell \to \infty$. However, by selecting a contour such that $\mathrm{Im}\, f(t)$ has the constant value $\mathrm{Im}\, f(t_0)$ in the vicinity of t_0, the integral assumes the form

$$I(\ell) = e^{-\ell\,\mathrm{Im}\, f(t_0)} \int_c e^{-\ell\,\mathrm{Re}\, f(t)} dt \tag{2.204}$$

in the region of t_0. Furthermore, writing $f''(t_0) = \rho_0 e^{i\phi_0}$ in polar form and expanding to second order around t_0

$$f(t_0 + \rho e^{i\phi}) = f(t_0) + \frac{1}{2}\rho_0 e^{i\phi_0}\rho^2 e^{2i\phi} \tag{2.205}$$

it is evident that the two directions $\phi = -\frac{\phi_0}{2}$ and $\phi = -\frac{\phi_0}{2} + \frac{\pi}{2}$ which keep $\mathrm{Im}\, f$ constant correspond to the directions of maximum positive and negative curvature for $\mathrm{Re}\, f(t)$. Hence, the countour c is deformed from the real axis such that in the vicinity of t_0 it coincides with the curve having $\mathrm{Im}\, f(t) = \mathrm{Im}\, f(t_0)$ which passes through the saddle point of the function $e^{\ell\,\mathrm{Re}\, f(t)}$ in the direction of steepest descent, and the resulting real integral, Eq. (2.204) is evaluated as in the previous case.

In the case of multiple stationary points, the analysis is complicated in two respects. The first, essentially technical, complication, is the necessity of globally analyzing the surface of $e^{-\ell\,\mathrm{Re}\, f(t)}$ in order to connect the positive and negative t axes at infinity with a contour traversing a sequence of intermediate saddle points. The more substantive problem is the fact that integration of an infinite expansion around each of two separate stationary points has the potential for double counting contributions to the integral. For well-separated stationary points, it is often assumed (without justification) that low-order contributions from each stationary point may simply be added. When two stationary points come sufficiently close together, the combined contribution of both stationary points is treated by an appropriate form of uniform approximation (Berry (1966), Miller (1970), Connor and Marcus, (1971)).

FEYNMAN PATH INTEGRAL

The Feynman path integral for the evolution operator of a particle in a potential $V(x)$ is the limit of a product of integrals over the variables x_k at each time slice k, Eq. (2.39), so that the stationary-phase approximation may be applied straightforwardly to each integral in the product. For convenience, we will use here the continuum notation of Eq. (2.41)

$$\mathcal{U}(x_f t_f; x_i t_i) = \int_{(x_i,t_i)}^{(x_f,t_f)} \mathcal{D}[x(t)] e^{\frac{i}{\hbar}\int_{t_i}^{t_f} dt\left[\frac{1}{2}m\left(\frac{\partial x}{\partial t}\right)^2 - V(x(t))\right]} = \int_{(x_i,t_i)}^{(x_f,t_f)} \mathcal{D}[x(t)] e^{\frac{i}{\hbar}S[x(t)]} \tag{2.206}$$

with the understanding that $\int dt$ and $\left(\frac{\partial x}{\partial t}\right)$ are defined by the discrete expressions (2.40). Due to the multiplicative factor $\frac{1}{\hbar}$ in the exponent, it is evident that the stationary-phase approximation will generate the semi-classical expansion of the evolution operator in powers of $\hbar$.

Since $S[x(t)]$ is the classical action, stationarity of $S[x(t)]$ yields the Euler-Lagrange equation of motion for the classical trajectory $x_c(t)$

$$m\frac{d^2 x_c}{dt^2} = -\nabla V(x_c) \tag{2.207a}$$

with the boundary conditions

$$x_c(t_i) = x_i \qquad x_c(t_f) = x_f \ . \tag{2.207b}$$

Expanding the action around the classical trajectory $x_c(t)$ and introducing the change of variables

$$\eta(t) = \frac{1}{\sqrt{\hbar}}(x(t) - x_c(t)) \tag{2.208}$$

we obtain

$$\mathcal{U}(x_f t_f, x_i t_i) = e^{\frac{i}{\hbar}S(x_c(t))} \int_{(0,t_i)}^{(0,t_f)} \mathcal{D}[\eta(t)]$$
$$\times\, e^{i\int_{t_i}^{t_f} dt\left\{\frac{1}{2}\eta(t)\left[-m\frac{d^2}{dt^2} - V''(x_c(t))\right]\eta(t) + \sum_{n=3}^{\infty}\frac{\hbar^{\left(\frac{n}{2}-1\right)}}{n!}V^{(n)}(x_c(t))\eta^n\right\}} \tag{2.209}$$

where $\eta(t)$ is required to vanish at the end points because $x(t)$ and $x_c(t)$ satisfy the same boundary conditions. Introducing an infinitesimal real term,

ϵ, to render the integral well-defined and noting that the quadratic form $\int dt \frac{1}{2}\eta(t)\left[-\epsilon - im\frac{d^2}{dt^2} - iV''(x_c(t))\right]\eta(t)$ actually represents a discrete sum $\eta_k A_{k\ell}\eta_\ell$, the Gaussian integral may be performed using Eq. (1.179) with the result

$$\mathcal{U}(x_f t_f, x_i t_i) = e^{\frac{i}{\hbar}S(x_c(t)) - \frac{1}{2}\ln\det\left(m\frac{d^2}{dt^2} + V''(x_c(t))\right) + O(\hbar)} \tag{2.210}$$

where the determinant may either be calculated directly from the discrete expression for the quadratic form and the measure Eq. (2.42) or from the appropriately normalized product of eigenvalues E_n of the equation

$$\left(m\frac{d^2}{dt^2} + V''(x_c(t)) - E_n\right)\psi_n(t) = 0 \tag{2.211a}$$

with

$$\psi_n(t_i) = \psi_n(t_f) = 0 \ . \tag{2.211b}$$

It is an instructive exercise to explicitly evaluate the determinant for the case of the harmonic oscillator to obtain the exact propagator (see Problem 2.14).

Physically, the leading term in Eq. (2.210) of order $\frac{1}{\hbar}$ is given by the classical trajectory, with the next term of order unity corresponding to the sum of all possible quadratic fluctuations around the classical trajectory. Higher order terms in $\hbar$ may be obtained by summing linked diagrams in which the vertices $V^{(n)}(x_c)$ are connected by propagators $[m\frac{d^2}{dt^2} + V''(x_c)]^{-1}$ as in the previous section.

It is interesting to note at this point that application of the stationary-phase approximation to the imaginary-time path integral, Eq. (2.54) corresponds to an analogous expansion around the stationary solution

$$m\frac{d^2 x_c}{dt^2} = -\nabla(-V(x_c)) \ . \tag{2.212}$$

Here the minus sign associated with the transformation (2.53) has been grouped with the potential to indicate that the stationary trajectory corresponds to the classical solution in the inverted potential. Thus, in tunneling and barrier penetration problems for which the classically forbidden region does not support appropriate classical solutions, these stationary solutions in the inverted potential will serve as the starting point for a quantum mechanical expansion.

MANY-PARTICLE PARTITION FUNCTION

The coherent state functional integral for the many-particle evolution operator, Eq. (2.62), may be written in coordinate representation as

$$\mathcal{U}(\phi_f^*(x), t_f; \phi_i(x), t_i) = \int_{\phi(x,t_i)=\phi_i(x)}^{\phi^*(x,t_f)=\phi_f^*(x)} \mathcal{D}[\phi^*(x,t), \phi(x,t)] e^{\int dx \phi^*(x,t_f)\phi(x,t_f)}$$

$$\times e^{\frac{i}{\hbar}\int_{t_i}^{t_f} dt \left[\int dx \phi^*(x,t)(i\hbar\frac{\partial}{\partial t} + \frac{\hbar^2}{2m}\nabla^2)\phi(x,t) - \frac{1}{2}\int dx dy \phi^*(x,t)\phi^*(y,t)v(x-y)\phi(y,t)\phi(x,t)\right]} \tag{2.213}$$

Whereas an explicit factor $\frac{1}{\hbar}$ multiplies the integral $\int dt$ in the exponent, in contrast to the Feynman path integral in the previous section, $\hbar$ also appears in the integrand of the exponent. Thus application of the stationary-phase approximation to Eq. (2.213) does not strictly yield a semi-classical expansion in powers of $\hbar$. A similar situation arises in field theory where, for example, the action associated with a scalar field has the form

$$\frac{1}{\hbar}S(\phi) = \frac{1}{\hbar}\int dtd^3x\left[\frac{1}{2}\partial_\mu\phi\partial^\mu\phi - \frac{1}{2}(\frac{m}{\hbar})^2\phi^2 + \lambda\phi^4\right] \tag{2.214}$$

and $\hbar$ appears explicitly in the mass term. In either case, to interpret the stationary-phase approximation as an expansion in $\hbar$, one must imagine two separate $\hbar$'s, with the $\hbar$ appearing within the integrand as a fixed constant and only the multiplicative factor as the expansion parameter.

In the absence of a strict expansion in an explicit small parameter, our present treatment of the stationary-phase approximation will be analogous to that of perturbation theory. We will introduce a parameter ℓ multiplying the action, which for the sake of the derivation will be assumed to be large, just as the formal parameter λ which is often introduced in the potential in perturbation theory is assumed to be small. In this way, it is straightforward to develop the stationary-phase expansion and demonstrate its correspondence to perturbation theory. From subsequent application to specific examples and the treatment of the closely related $\frac{1}{N}$ expansion, the physical conditions will become apparent under which a suitable expansion parameter arises and the method becomes preferable to perturbation theory.

With the introduction of the parameter ℓ and suppressing factors of $\hbar$, the partition function Eq. (2.66) may be written

$$Z(\ell) = \int_{\phi(x,\beta)=\varsigma\phi(x,0)} \mathcal{D}(\phi^*(x,\tau)\phi(x,\tau))e^{-\ell\int_0^\beta d\tau\int dx\phi^*(x,\tau)\left(\frac{\partial}{\partial\tau}-\frac{\nabla^2}{2m}-\mu\right)\phi(x,\tau)}$$
$$\times e^{-\frac{\ell}{2}\int_0^\beta d\tau\int dxdy\phi^*(x,\tau)\phi^*(y,\tau)v(x-y)\phi(y,\tau)\phi(x,\tau)} \, . \tag{2.215a}$$

Variation of the exponent yields the following equations for the stationary solutions ϕ_c^* and ϕ_c:

$$\frac{\delta S}{\delta\phi_c^*(x,\tau)} = \left[\frac{\partial}{\partial\tau} - \frac{\nabla^2}{2m} - \mu + \int dy\, v(x-y)\phi_c^*(y,\tau)\phi_c(y,\tau)\right]\phi_c(x,\tau) = 0 \tag{2.215b}$$

$$\frac{\delta S}{\delta\phi_c(x,\tau)} = \left[-\frac{\partial}{\partial\tau} - \frac{\nabla^2}{2m} - \mu + \int dy\, v(x-y)\phi_c^*(y,\tau)\phi_c(y,\tau)\right]\phi_c^*(x,t) = 0 \tag{2.215c}$$

with the boundary conditions

$$\phi_c(x,\beta) = \varsigma\phi_c(x,0) \quad . \tag{2.215d}$$

One trivial solution about which to expand for either Fermions or Bosons is the solution $\phi_c(x,\tau) = \phi_c^*(x,\tau) = 0$. In this case, $Z(\ell)$ is evaluated in the usual way using perturbation theory with propagators $\frac{1}{\ell}(\frac{\partial}{\partial\tau} - \frac{\nabla^2}{2m} - \mu)^{-1}$ and interaction $\ell v(x-y)$. The ℓ-dependence of each linked diagram for the grand potential is then $\ell^{n_V - n_I}$ where n_V denotes the number of vertices and n_I indicates the number of internal propagator lines. Since each vertex is connected to four lines and each line is connected to two vertices, $n_I = 2n_V$. Finally for a translationally invariant system, let n_M denote the number of momentum loops, defined as the number of independent momentum integrals performed in the diagram. Note that the number of momentum loops differs from n_L, the number of closed propagator loops defined earlier to determine the sign of a Feynman diagram. Recalling from Section 2.3 that there are $n_V - 1$ momentum conserving δ-functions, the number of independent momenta is $n_M = n_I - n_V + 1$. The ℓ-dependence of a general diagram may therefore be expressed as ℓ^{-n_V} or ℓ^{-n_M+1}. The former result may also be obtained trivially by rescaling the fields ϕ in Eq. (2.215) by a factor $\frac{1}{\sqrt{\ell}}$ and observing that only the ratio $\frac{v}{\ell}$ occurs in the rescaled action.

Thus ordinary perturbation theory may be regarded as an expansion in $\frac{1}{\ell}$ and is equivalent to a loop expansion in the number of independent momentum integrals. Note, however, that when expanding around $\phi_c = \phi_c^* = 0$, the leading contribution to Ω is of order $\ln \ell$.

In addition to the trivial solution $\phi_c = \phi_c^* = 0$, Eqs. (2.215) admit non-trivial static and time-dependent solutions. In the case of Fermions, the stationary solutions in terms of Grassman variables lie outside the space of physical observables and the functional integrals over the shifted variables must be performed to obtain a physical result. The most efficient way to proceed for Fermions is thus to introduce an auxiliary field to enable the integrations over Grassman variables to be done exactly, and then apply the stationary-phase approximation to the integral over the auxiliary field. We defer this treatment of Fermions until Chapter 7, and address the simpler case of Bosons here.

For Bosons, application of the stationary-phase approximation to $\int \mathcal{D}(\phi^*\phi)\, e^{F(\phi^*\phi)}$ may be regarded as approximating a double real integral $\int du dv\, e^{F(u,v)}$ where u and v are real variables representing the real and imaginary parts of ϕ. Since the general method of steepest descent requires consideration of all complex stationary points of u and v, in principle we should consider solutions for which ϕ_c^* is not necessarily the complex conjugate of ϕ_c. Again, the most general case will be deferred and we consider here the special case that ϕ_c^* is the complex conjugate of ϕ_c. Because of the opposite signs of the time derivatives in (2.215b) and (2.215c), such solutions must necessarily be time-independent and satisfy the static Hartree equation:

$$\left(-\frac{\nabla^2}{2m} - \mu + \int dy\, v(x-y)|\phi_c(y)|^2\right)\phi_c(x) = 0 \quad . \tag{2.216}$$

The general solutions to Eq. (2.216) for translationally invariant systems are plane waves. We will assume for the present treatment that $\tilde{v}(0)$, which is the zero-momentum Fourier transform or volume integral of $V(x-y)$, is positive. In this case the solution which minimizes the action is the zero-momentum Bose condensate

$$\psi_c = \sqrt{\frac{\mu}{\tilde{v}(0)}} \tag{2.217}$$

with action

$$\ell S_c = -\frac{\ell\beta}{2}\mathcal{V}\frac{\mu^2}{\tilde{v}(0)}$$

where $\mathcal{V}$ is the volume of the system, and we have chosen the arbitrary phase to make ψ_c real. The significance of the phase of ψ_c will be addressed in Chapter 4 in the general discussion of order parameters. With the change of variables

$$\begin{aligned} \psi(x,\tau) &= \sqrt{\ell}(\phi(x,\tau) - \psi_c) \\ \psi^*(x,\tau) &= \sqrt{\ell}(\phi^*(x,\tau) - \psi_c) \end{aligned} \tag{2.218}$$

the action becomes

$$\ell S\left(\psi^*(x,\tau),\psi(x,\tau)\right) = \ell S_c + S^{(2)} + \frac{1}{\sqrt{\ell}}S^{(3)} + \frac{1}{\ell}S^{(4)} \tag{2.219a}$$

where

$$S^{(2)} = \frac{1}{2}\int_0^\beta d\tau \int dxdy[\psi^*(x,\tau)\psi(x,\tau)]D\begin{bmatrix}\psi(y,\tau)\\ \psi^*(y,\tau)\end{bmatrix} \tag{2.219b}$$

$$D = \begin{bmatrix} \delta(x-y)\left(\frac{\partial}{\partial\tau} - \frac{\nabla^2}{2m}\right) + \psi_c^2 v(x-y) & \psi_c^2 v(x-y) \\ \psi_c^2 v(x-y) & \delta(x-y)\left(-\frac{\partial}{\partial\tau} - \frac{\nabla^2}{2m}\right) + \psi_c^2 v(x-y) \end{bmatrix} \tag{2.219c}$$

$$S^{(3)} = \psi_c\int_0^\beta d\tau \int dxdy\, v(x-y)|\psi(x,\tau)|^2(\psi^*(y,\tau) + \psi(y,\tau)) \tag{2.219d}$$

and

$$S^{(4)} = \int_0^\beta d\tau \int dxdy\, v(x-y)|\psi(x,\tau)|^2|\psi(y,\tau)|^2 \ . \tag{2.219e}$$

It is convenient to simplify the notation by suppressing (x,τ) arguments, using the matrix D, Eq. (2.219c), and abbreviating the vertices in $S^{(3)}$ and $S^{(4)}$ as g_3 and g_4, so that the partition function is written as

$$Z(\ell) = e^{-\ell S_c}\int \mathcal{D}(\psi^*\psi)e^{-\frac{1}{2}(\psi^*\psi)D\binom{\psi}{\psi^*} - \frac{1}{\sqrt{\ell}}|\psi|^2 g_3(\psi+\psi^*) - \frac{1}{\ell}|\psi|^2 g_4|\psi|^2} \ . \tag{2.220}$$

Although the quadratic form appearing in this functional integral differs slightly from that in Eq. (2.84) used in establishing Wick's theorem due to the presence of the $\psi^*\psi^*$ and $\psi\psi$ terms, it is shown in Problem 2.15 that a straightforward generalization of the usual linked diagram expansion is obtained. Writing the inverse of the matrix D in the block form

$$(D)\begin{pmatrix} G_{\psi\psi^*} & G_{\psi\psi} \\ G_{\psi^*\psi^*} & G_{\psi^*\psi}\end{pmatrix} = \begin{pmatrix}1 & 0\\ 0 & 1\end{pmatrix} \tag{2.221}$$

the contractions of the fields $\{\psi^*,\psi\}$ are given by

$$\begin{aligned} \int \mathcal{D}(\psi^*\psi)e^{-\frac{1}{2}(\psi^*\psi)D\binom{\psi}{\psi^*}}&\begin{Bmatrix}\psi(1)\psi^*(2) & \psi(1)\psi(2)\\ \psi^*(1)\psi^*(2) & \psi^*(1)\psi(2)\end{Bmatrix} \\ &= \begin{Bmatrix} G_{\psi\psi^*}(1,2) & G_{\psi\psi}(1,2) \\ G_{\psi^*\psi^*}(1,2) & G_{\psi^*\psi}(1,2)\end{Bmatrix} \ . \end{aligned} \tag{2.222}$$

Note that the positivity of D requires that ψ_c correspond to a minimum of the action.

The propagators corresponding to these contractions are written as follows:

 $= G_{\psi\psi^*}(1,2) = \langle\psi(1)\psi^*(2)\rangle$

 $= G_{\psi\psi}(1,2) = \langle\psi(1)\psi(2)\rangle$

 $= G_{\psi^*\psi^*}(1,2) = \langle\psi^*(1)\psi^*(2)\rangle$ (2.223)

 $= G_{\psi^*\psi}(1,2) = \langle\psi^*(1)\psi(2)\rangle$.

As usual, arrows entering or leaving a dot indicate a ψ or ψ^* respectively and the new feature arising from the shifted fields ψ is the introduction of lines with two arrows pointing in opposite directions. The vertices corresponding to the cubic and quartic terms in Eq. (2.219) are denoted

 $= \frac{1}{\sqrt{\ell}}\psi_c v(1,2)$

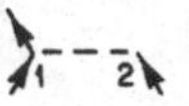 $= \frac{1}{\sqrt{\ell}}\psi_c v(1,2)$ (2.224)

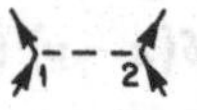 $= \frac{1}{\ell} v(1,2)$.

By calculating the linked diagrams composed of these vertices connected by the propagators (2.223), the partition function may be written

$Z(\ell) = e^{-\ell S_c - \frac{1}{2}\ln(\det D) + \frac{1}{\ell}\{$ 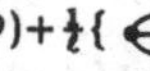 $+$  $+$  $\} + O\left(\frac{1}{\ell^2}\right)}$ (2.225a)

where the diagrams denote the sum of all contractions of the indicated topology, for example

 $=$  $+$ 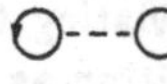 $+$  (2.225b)

and similarly for the remaining diagrams.

Note that in contrast to the expansion of $Z(\ell)$ about $\phi = 0$, which corresponded to ordinary perturbation theory in powers of v or $\frac{1}{\ell}$, the expansion about ψ_c in Eq. (2.205a) yields contributions to Ω of order ℓ and order unity. Such terms cannot be obtained by expansion in powers of ℓ, so the stationary-phase expansion corresponds to an infinite resummation of perturbation theory.

This resummation may be understood physically in terms of Bose condensation. The zero-momentum mode is macroscopically occupied, with amplitude $\psi_c = \sqrt{\frac{\mu}{\tilde{v}(0)}}$ controlled by the chemical potential. The shift in variables (2.218) then gives rise to vertices in which one or more of the extremities has a zero-momentum condensate factor ψ_c which we will denote by the symbol  rather than a field variable ψ or ψ^* indicated in the usual way by or . We will write the quadratic martix D from Eq. (2.219b) and its inverse G from Eq. (2.221) in the following schematic form:

$$D = G_0^{-1} + V = G^{-1} \tag{2.226a}$$

where

$$G_0^{-1} = \begin{pmatrix} \frac{\partial}{\partial\tau} - \frac{\nabla^2}{2m} & 0 \\ 0 & -\frac{\partial}{\partial\tau} - \frac{\nabla^2}{2m} \end{pmatrix} \qquad V =$$ 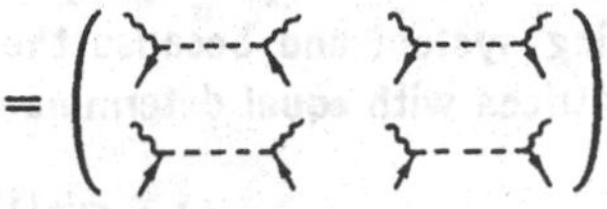 **(2.226b)**

and will denote G_0 and G by broken and solid propagators:

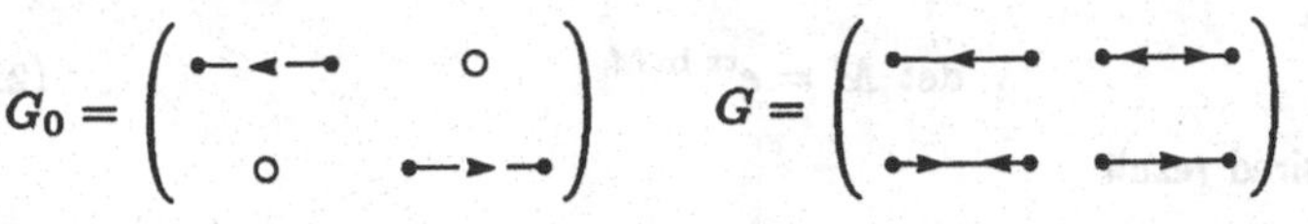

(2.226c)

The Dyson equation $G = G_0 - G_0 V G$ following from (2.226a) then has the diagrammatic form:

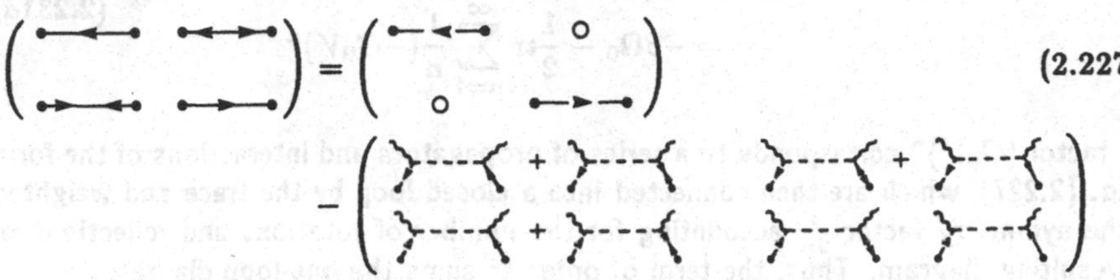

(2.227a)

Iteration yields the series:

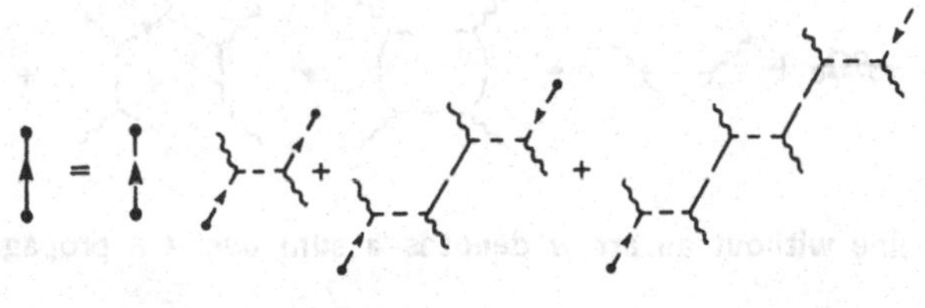

(2.227b)

and

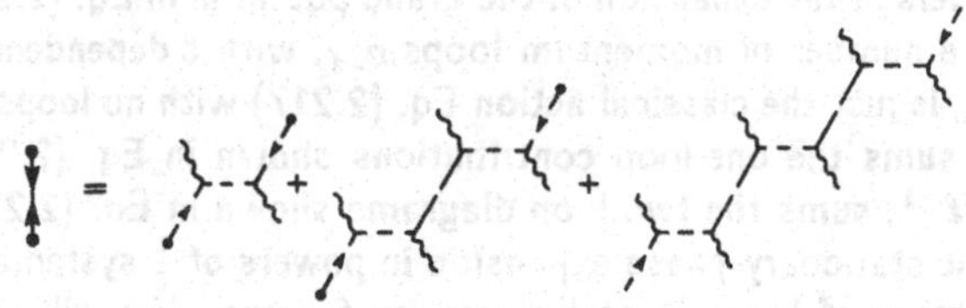

(2.227c)

where internal lines are summed over G_0 propagators with arrows in each direction. The propagator G thus sums self-energy insertions analogous to the exchange term  of Eq. (2.180b) where the factor ψ_c^2 plays the role of the occupation number n_γ for the zero momentum state. Note that in deriving Eq. (2.199), direct terms analogous to the Hartree self-energy of the form $\tilde{v}(0)\psi_c^2$ exactly cancelled the chemical potential, so in fact both direct and exchange terms have been resummed to all orders.

The term of order $\ell^{(0)}$ in Eq. (2.225a), $\frac{1}{2}\ln(\det D)$, also has a simple diagrammatic interpretation. We first write

$$\det D = \det\left(G_0^{-1} + V\right) = \det G_0^{-1} \det\left(1 + G_0 V\right) . \qquad (2.228)$$

Since from Eq. (2.68), $\det\left(\frac{\partial}{\partial\tau}-\frac{\nabla^2}{2m}\right)^{-1}$ is the partition function $Z_0 = e^{-\beta\Omega_0}$ for the non-interacting system and because the block form for G_0^{-1}, Eq. (2.226b) contains two such matrices with equal determinants, it follows that

$$[\det G_0^{-1}]^{-\frac{1}{2}} = e^{-\beta\Omega_0} \quad . \tag{2.229}$$

Using the relation

$$\det M = e^{\mathrm{tr}\,\ln M} \tag{2.230}$$

we obtain the desired result

$$\begin{aligned} -\frac{1}{2}\ln\det D &= -\beta\Omega_0 - \frac{1}{2}\mathrm{tr}\,\ln(1+G_0V) \\ &= -\beta\Omega_0 - \frac{1}{2}\mathrm{tr}\sum_{n=1}^{\infty}\frac{1}{n}(-G_0V)^n \quad . \end{aligned} \tag{2.231a}$$

The factor $(G_0V)^n$ corresponds to a series of propagators and interactions of the form in Eq. (2.227), which are then connected into a closed loop by the trace and weighted by the symmetry factor $\frac{1}{2n}$ accounting for the number of rotations and reflections of the resulting diagram. Thus, the term of order ℓ^0 sums the one-loop diagrams

$-\frac{1}{2}\ln\det D = -\beta\Omega_0 +$  $+$ 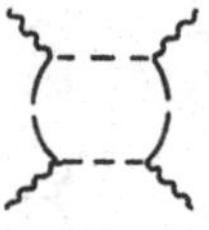 $+$ 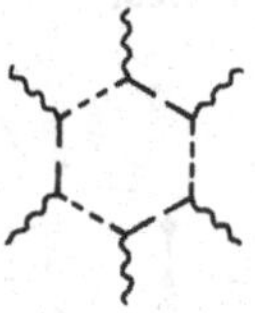 $+ \cdots$ (2.231b)

where again the line without an arrow denotes a sum over G_0 propagators in each direction.

The first few orders of the expansion of the grand potential in Eq. (2.225a) thus are clearly ordered by the number of momentum loops, n_M, with ℓ dependence $\ell^{-(n_M-1)}$. The leading term ℓS_c is just the classical action Eq. (2.217) with no loops. The second term, of order $\ell^{(0)}$, sums the one-loop contributions shown in Eq. (2.231b) and the third term, of order ℓ^{-1}, sums the two loop diagrams shown in Eq. (2.225).

The fact that the stationary-phase expansion in powers of $\frac{1}{\ell}$ systematically orders diagrams by the number of loops is easily seen as follows. We will explicitly draw the condensate factor  on the vertices in (2.224) as an external line so that the vertices

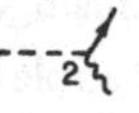 $= \frac{1}{\sqrt{\ell}}\psi_c v(1,2) =$ 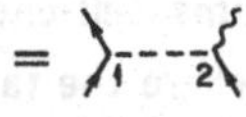 (2.232a)

connect to three internal lines and one external line and the vertex

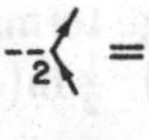 $= \frac{1}{\ell}v(1,2)$ (2.232b)

connects to four internal lines. Now, consider a general diagram obtained by contracting n_V vertices with propagators. Of these n_V vertices, assume n_{VE} are of the form of

(2.232a) containing an external condensate line and the remaining $n_V - n_{VE}$ are of the form (2.232b) with no external lines. Since the propagators are independent of ℓ, the overall ℓ-dependence of the diagram from (2.232) will be $\ell^{-n_V+\frac{1}{2}n_{VE}}$. If one counts the lines coming out of all vertices, the internal lines connecting to two vertices will be counted twice and the external condensate lines will be counted once, giving the topological constraint

$$2n_I + n_{VE} = 4n_V \tag{2.233a}$$

where n_I denotes the number of internal lines. As in the analysis with no condensate, the number of independent momentum loops, n_M, after accounting for the $n_V - 1$ momentum conservation constraints is

$$n_M = n_I - n_V + 1 \tag{2.233b}$$

or using (2.233a)

$$n_M = n_V - \frac{1}{2}n_{VE} + 1 \; . \tag{2.233c}$$

Thus, for an arbitrary diagram, the overall ℓ-dependence may be written $\ell^{-(n_M-1)}$ establishing the loop expansion. An analogous analysis shows that the series for the effective action is also a loop expansion. In contrast to the expansion around $\psi = 0$, for which the loop expansion coincided with a perturbation expansion, in the presence of a condensate diagrams with n_M loops contain from $n_M - 1$ to $2(n_M - 1)$ vertices of the form of Eq. (2.232) as well as the infinite resummation of interactions Eq. (2.227) contained in the propagators.

PROBLEMS FOR CHAPTER 2

The first two problems review elementary aspects of quantum statistical mechanics used in this chapter. Problem 3 indicates the derivation of the time-ordered exponential used in the text, for those unfamiliar with it. The two most crucial problems for understanding path integrals are Problems 4 and 5, which are emphasized by an *. The next two problems treat discrete and continuum derivations of the propagators. For those who wish to broaden their understanding of fundamental derivations in the text and establish contact with traditional derivations in the original literature, alternative derivations of Wick's theorem, the linked-cluster theorem, the evaluation of expectation values, and the stationary phase expansion are presented in Problems 8, 10, 11, and 13. Generalizations of topics introduced in the text are treated in Problems 9, 12, and 15 and a specific example of the evaluation of the determinant arising from quadratic fluctuations is given in Problem 14.

PROBLEM 2.1 Show that Ω defined in quantum statistical mechanics by Eq. (2.40) is equal to the grand canonical potential defined in thermodynamics by the Legendre transformation

$$\Omega_T(T,V,\mu) = U - TS - \mu N$$

where $U(S,V,N)$ is the state function specifying the internal energy. First, define $\phi(T,V,\mu) = \Omega - \Omega_T$. Derive Eqs. (2.4) and use these results to show that ϕ must be a constant times T, introducing at most an additive constant in the entropy. Then, take the zero temperature limit of $\frac{\partial \Omega}{\partial T}$ to show that this additive constant is zero.

PROBLEM 2.2 Show that the probability for a system in contact with thermal and particle reservoirs to be in a state with energy E_α and containing N_α particles is proportional to $e^{-\frac{1}{kT}(E_\alpha-\mu N_\alpha)}$. Let the combination of the reservoir and system have total energy E and particle number N. Use the definition of the entropy to show that the density of states for the reservoir to have energy $E-E_\alpha$ and particle number $N-N_\alpha$ is proportional to $e^{\frac{1}{k}S(E-E_\alpha,N-N_\alpha)}$ from which the final result is obtained by expanding in E_α and N_α.

PROBLEM 2.3 Show that the two forms of the time-ordered exponential, Eqs. (2.9) and (2.10), are equivalent. Consider an interval which has been subdivided into a discrete number of time slices M and show that the terms of order ϵ^n obtained from $\frac{(-1)^n}{n!}T\left[\sum_{i=1}^{M}\epsilon A(t_i)\right]^n$ agree with the terms of order ϵ^n obtained from $\prod_i\left[\sum_{n=0}^{\infty}\frac{(-\epsilon A(t_i))^n}{n!}\right]$.

PROBLEM 2.4* The validity of an approximation to the matrix element of an infinitesimal evolution operator may be checked by showing that the wave function evolves properly to leading order in ϵ, that is

$$\int dy\langle x|e^{-\epsilon H}|y\rangle\psi(y)=\left(1-i\epsilon H+O(\epsilon^2)\right)\psi(x)\ .$$

Begin with the simple case in which the Hamiltonian is the kinetic energy operator (with $\hbar=m=1$ for convenience) and show

$$\begin{aligned}\frac{1}{\sqrt{2\pi i\epsilon}}\int dy\, e^{i\frac{(x-y)^2}{2\epsilon}}\psi(y)&=\frac{1}{\sqrt{2\pi i\epsilon}}\int dy\, e^{i\frac{(x-y)^2}{2\epsilon}}\sum_{n=0}^{\infty}\frac{(y-x)^n}{n!}\frac{d^n}{dx^n}\psi(x)\\&=\sum_{n=0}^{\infty}\left(\frac{i\epsilon}{2}\right)^n\frac{d^{2n}}{dx^{2n}}\psi(x)\\&=\left[1-i\epsilon\left(-\frac{1}{2}\frac{d^2}{dx^2}\right)+O(\epsilon^2)\right]\psi(x)\ .\end{aligned}$$

Now, include a potential $V(\hat{x})$ and show

$$\begin{aligned}&\frac{1}{\sqrt{2\pi i\epsilon}}\int dy\, e^{i\frac{(x-y)^2}{2\epsilon}-i\epsilon V(y)}\psi(y)\\&\quad=\frac{1}{\sqrt{2\pi i\epsilon}}\int dy\, e^{i\frac{(x-y)^2}{2\epsilon}}\sum_n\frac{(-i\epsilon V(y))^n}{n!}\psi(y)\\&\quad=\left[1-i\epsilon\left(-\frac{1}{2}\frac{d^2}{dx^2}+V(x)\right)+O(\epsilon^2)\right]\psi(x)\ .\end{aligned}$$

Demonstrate that the same answer is obtained when $V(y)$ is replaced by $V(x)$ or by any linear combination $\eta V(y)+(1-\eta)V(x)$. Show that when the symmetric combination $\frac{1}{2}V(y)+\frac{1}{2}V(x)$ is used, the evolution is actually correct through order ϵ^2. The same result may be obtained more economically by noting $e^{-i\epsilon(T+V)}=e^{-i\epsilon\frac{V}{2}}e^{-i\epsilon T}e^{-i\epsilon\frac{V}{2}+O(\epsilon^3)}$ and evaluating $\int\frac{dp}{2\pi}\langle x|e^{-i\epsilon\frac{V(\hat{x})}{2}}|p\rangle\langle p|e^{-i\epsilon\frac{\hat{p}^2}{2}}e^{-i\epsilon\frac{V(\hat{x})}{2}}|y\rangle$.

PROBLEM 2.5* The formulation of path integrals for a particle in a magnetic field exhibits the general problems arising when the Hamiltonian contains terms in which $\hat{x}$ and $\hat{p}$ are combined.

Recall that the Lagrangian for a particle in a magnetic field is

$$L = \frac{1}{2}\left(\frac{dx}{dt}\right)^2 + \frac{dx}{dt}A$$

leading to the classical Hamiltonian

$$H = \frac{1}{2}(P - A)^2$$

where $\hbar = m = 1$ and the factor $\frac{e}{c}$ has been absorbed in the vector potential A.

Begin with the normal-ordered form of the quantum Hamiltonian, Eq. (2.34), and as in Problem 2.4, show that the evolution operator

$$\int d^3p\langle x|p\rangle\langle p| : e^{-i\frac{\epsilon}{2}\sum_k(\hat{p}_k^2 - 2\hat{p}_kA_k(\hat{x}) - i\partial_kA_k(\hat{x}) + A_k^2(\hat{x}))} : |y\rangle$$

$$= \int \frac{d^3p}{(2\pi)^3} e^{\sum_k[ip_k(x_k-y_k) - i\frac{\epsilon}{2}(p_k^2 - 2p_kA_k(y) - i\partial_kA_k(y) + A_k^2(y))]}$$

$$= \left(\frac{1}{2\pi i\epsilon}\right)^{\frac{3}{2}} e^{\sum_k\left[\frac{1}{2}\frac{(x_k-y_k)^2}{\epsilon} + i(x_k-y_k)A_k(y) - \frac{\epsilon}{2}\partial_kA_k(y)\right]}$$

yields the proper equation of motion to leading order in ϵ. Now, show that the following alternative expressions also give the correct evolution:

$$\langle x|e^{-\epsilon H}|y\rangle \approx \int \frac{d^3p}{(2\pi)^3} e^{\sum_k[ip_k(x_k-y_k) - i\frac{\epsilon}{2}(p_k^2 - p_kA_k(y) - A_k(x)p_k + A_k^2(y))]}$$

$$\langle x|e^{-\epsilon H}|y\rangle \approx \left(\frac{1}{2\pi i\epsilon}\right)^{\frac{3}{2}} e^{\sum_k\left[\frac{i}{2}\frac{(x_k-y_k)^2}{2} + i(x_k-y_k)\left(\frac{A_k(x)+A_k(y)}{2}\right)\right]} .$$

Note that this last result looks like a path integral containing the Lagrangian with a definite discrete approximation:

$$\frac{dx}{dt}A \approx \frac{(x_k - y_k)}{\epsilon}\left(\frac{A_k(x) + A_k(y)}{2}\right) .$$

By comparing with the previous result, it is clear that the alternative discrete approximation

$$\frac{dx}{dt}A \not\approx \frac{x_k - y_k}{\epsilon}A_k(y)$$

yields a different answer, which is only repaired by the inclusion of the $\partial_kA_k(y)$ term. Contrast this dependence on the point at which $A(\hat{x})$ is evaluated in the term $\hat{p}A(\hat{x})$ with the independence of the point at which $V(\hat{x})$ is evaluated demonstrated in Problem 2.4 and thereby explain how the ordering of operators in ordinary quantum mechanics is reflected in path integrals.

PROBLEM 2.6 Contrast the discrete matrix S, Eq. (2.69), arising from the defining expression of the functional integral for the partition function for non-interacting particles with the alternative discrete approximation to the continuum expression $(\frac{\partial}{\partial t} - \mu + \epsilon_\alpha)$

$$\tilde{S} = \begin{bmatrix} -a & 1 & & \cdots & & 0 \\ 0 & -a & 1 & & & \\ & 0 & -a & \ddots & & \\ & & 0 & \ddots & 1 & \\ & & & \ddots & -a & 1 \\ \varsigma & & & & 0 & -a \end{bmatrix} .$$

Show that whereas S yields the correct partition function, has Boson eigenvalues with positive real parts (so that Gaussian integrals converge) and yields the correct propagator for $\phi\phi^*$ at equal times, $\tilde{S}$ yields an incorrect partition function, has eigenvalues with negative real parts, and evaluation of the equal-time propagator using Eq. (2.76b) gives the wrong result. This exercise emphasizes the fact that one must take the limit as $M \to \infty$ of the discrete result obtained from the defining expression for the functional integral.

PROBLEM 2.7 Obtain the propagator for the non-interacting system Eq. (2.77) by solving the differential Eq. (2.78a) with periodic or antiperiodic boundary conditions. One straightforward way to procede is to find the periodic or antiperiodic eigenfunctions $(\frac{\partial}{\partial \tau} + \epsilon_\alpha)\psi_n(\tau) = \lambda_n \psi_n(\tau)$ and write $G(\tau, \tau') = \sum_n \frac{1}{\lambda_n}\psi_n(\tau)\psi_n^*(\tau')$, thereby obtaining the sum in Eq. (2.131).

PROBLEM 2.8 The Operator Form of Wick's Theorem. Although we will only use Wick's theorem to express thermal averages or ground state expectation values as sums over products of propagators, the original theorem by Wick (1950) is an operator identity. Let T and N denote the time-ordering and normal-ordering operators respectively. Recall that both operators include a factor ς if an odd number of interchanges take place and that T normal-orders operators at equal times. Define the contraction of two operators $\hat{A}_1$ and $\hat{A}_2$ as the difference

$$\overbrace{\hat{A}_1 \hat{A}_2} = T(\hat{A}_1, \hat{A}_2) - N(\hat{A}_1, \hat{A}_2) \ . \tag{1}$$

With these definitions, Wick's theorem states that the time-ordered product of n operators is equal to the sum of the normal-ordered products of all possible contractions.

$$\begin{aligned} T(\hat{A}_1\hat{A}_2\ldots\hat{A}_n) = {} & N(\hat{A}_1\hat{A}_2\ldots\hat{A}_n) \\ & + \sum_{\substack{\text{single} \\ \text{contractions}}} N(\overbrace{\hat{A}_1\hat{A}_2}\ldots\hat{A}_n) + \sum_{\substack{\text{pairs of} \\ \text{contractions}}} N(\overbrace{\hat{A}_1\hat{A}_2}\ldots\overbrace{\ldots\hat{A}_n}) + \ldots \\ & + \sum_{\substack{\text{complete} \\ \text{contractions}}} N(\overbrace{\hat{A}_1\hat{A}_2}\overbrace{\hat{A}_3\ldots}\overbrace{\ldots\hat{A}_n}) \ . \end{aligned} \tag{2}$$

The proof is by induction, and the theorem holds for $n = 2$ by Eq. (1). Since N and T are distributive, it is sufficient to consider the $\hat{A}_i$'s to be creation or annihilation operators. Consider $n+1$ operators and denote the operator with the earliest time as $\hat{B}$. Assuming the theorem holds for n operators,

$$\begin{aligned} T(\hat{A}_1\hat{A}_2\ldots\hat{A}_n\hat{B}) &= T(\hat{A}_1\hat{A}_2\ldots\hat{A}_n)\hat{B} \\ &= N(\hat{A}_1\hat{A}_2\ldots\hat{A}_n)\hat{B} \\ &\quad + \sum_{\substack{\text{contractions}\\ \text{of A's}}} N(\overline{\hat{A}_1\hat{A}_1}\ldots\overline{\ldots\hat{A}_n})\hat{B} \end{aligned} \tag{3}$$

show that Eq. (3) yields Wick's theorem for $n+1$ particles as follows: If $\hat{B}$ is an annihilation operator, explain how the right-hand side yields the normal product of all contractions except pairings with $\hat{B}$ and show that all pairings with $\hat{B}$ vanish. If $\hat{B}$ is a creation operator, commute it to the left of all the creation operators in each normal product, and show that this commutation generates the normal product of all contractions.

Finally, let $a_\alpha|\phi\rangle = 0$ for all annihilation operators. Show that $\overline{\hat{A}_i\hat{A}_j}$ is the propagator $\langle\phi|T\hat{A}_i\hat{A}_j|\phi\rangle$ and that

$$\langle\phi|T\hat{A}_1\hat{A}_2\cdots\hat{A}_n|\phi\rangle = \sum_{\substack{\text{all complete}\\ \text{contractions}}} \overline{\hat{A}_1\hat{A}_2}\ldots\overline{\ldots\hat{A}_n}$$

where

$$\overline{\hat{A}_1\hat{A}_2} = \langle\phi|T\hat{A}_1\hat{A}_2|\phi\rangle \ .$$

PROBLEM 2.9 Generalize the Hugenholtz diagram rules to the case in which V contains one-body, two-body, ... up to n-body interactions. The principal issue is what to do about m-tuples of equivalent lines. Note that in general an m-tuple of equivalent lines could start at an n_1-body vertex and terminate at an n_2-body vertex and that multiple m-tuples could originate or terminate at the same vertex.

PROBLEM 2.10 The Linked-Cluster Theorem. Show that the exponential of the sum of all linked graphs yields the sum of all diagrams with the appropriate factors using unlabeled Feynman diagrams. First, let a_c denote the contribution of some linked cluster and consider a graph composed of n of these clusters. Explain why the symmetry factor is $n!s_c^n$ where s_c is the symmetry factor in a_c. Hence, show that the contribution of any number (including 0) of clusters a_c is $\sum_{n=0}^{\infty} \frac{a_c^n}{n!} e^{a_c}$. Generalize to show that $e^{\sum_c a_c}$ yields the sum of all diagrams with the correct symmetry factors.

PROBLEM 2.11 Expectation Values of Operators. Consider the Hamiltonian $\mathcal{H}(\lambda) = \mathcal{H} + \lambda R$ and define the grand potential $e^{-\beta\Omega(\lambda)} = \mathrm{Tr}[e^{-\beta(\mathcal{H}+\lambda R-\mu N)}]$. Show that $\frac{d}{d\lambda}\Omega(\lambda)|_{\lambda=0} = \mathrm{Tr}[e^{-\beta(\mathcal{H}-\mu N)}R] = \langle R\rangle$. Hence explain why $\langle R\rangle$ is given by the sum of all linked diagrams containing any number of interaction terms from $\mathcal{H}$ and a single operator R. Show that the rules and symmetry factors are those presented in Section 2.3.

PROBLEM 2.12 In Section 2.4, the generating function, effective potential, vertex functions, and hierarchy of integral equations were derived in detail for the linear source, Eq.(2.148). Repeat the steps for the bilinear source, Eq.(2.31) and derive the equations indicated schematically in Eqs.(2.191– 2.196).

PROBLEM 2.13 The Stationary Phase Approximation. Starting from Eq. (2.198), expand $e^{-\sum_{n=3}^{\infty} \frac{z^n}{n!} \frac{f_0^{(n)}}{\ell^{(\frac{n}{2}-1)}(f_0'')^{\frac{n}{2}}}}$ to order $\frac{1}{\ell^2}$, perform the resulting Gaussian integrals, and exponentiate the result to obtain Eq. (2.203) complete with $\mathcal{O}(\frac{1}{\ell^2})$ terms.

As in Eq.(2.202), obtain the same result by evaluating the diagrams:

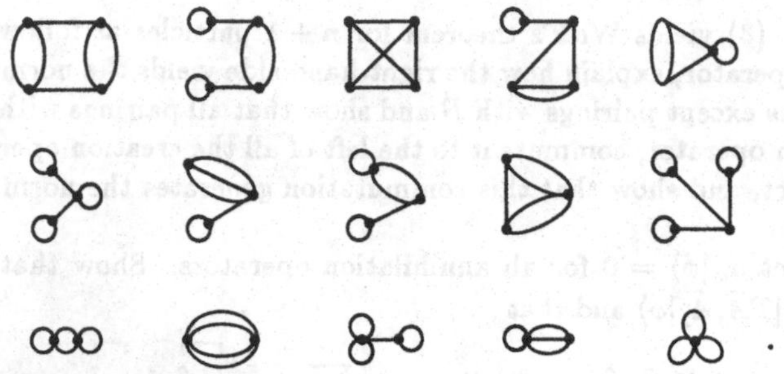

PROBLEM 2.14 Propagator for the Harmonic Oscillator. Since the action in the Feynman path integral for the harmonic oscillator is a quadratic form in the coordinates, the stationary-phase result, Eq.(2.211) is exact. Solve for the propagator for a particle in one spatial dimension moving in the potential $V(x) = \frac{m\omega^2}{2}x^2$ by explicitly evaluating a discrete path integral and show:

$$\mathcal{U}(x_f t_f, x_i, t_i) = \left(\frac{m\omega e^{-i\frac{\pi}{2}}}{2\pi\sin(\omega t)}\right)^{\frac{1}{2}} e^{i\frac{m\omega}{2}\left[\left(x_f^2+x_i^2\right)\cot(\omega t)-\frac{2x_f x_i}{\sin(\omega t)}\right]} .$$

It is convenient to use the one-dimensional analog of Eq.(2.39) with $V(x_{k-1})$ replaced by the symmetrized form $\frac{1}{2}(V(x_k)+V(x_{k-1}))$. Note that the determinant of the p-dimensional tridiagonal matrix occurring in the exponent

$$D(p) = \begin{vmatrix} 1 & -a & & & \\ -a & 1 & \ddots & & \\ & -a & \ddots & -a & \\ & & \ddots & 1 & -a \\ & & & -a & 1 \end{vmatrix}$$

can be evaluated by expanding in minors to obtain the two-term recursion relation

$$D(p) = D(p-1) - a^2 D(p-2)$$

or

$$\begin{bmatrix} D(p) \\ D(p-1) \end{bmatrix} = \begin{bmatrix} 1 & -a^2 \\ 1 & 0 \end{bmatrix} \begin{bmatrix} D(p-1) \\ D(p-2) \end{bmatrix}$$

which may be solved by diagonalizing a 2×2 matrix.

PROBLEM 2.15 Generalize Wick's theorem as presented in Section 2.3 to the case of the functional integral Eq. (2.220) encountered in Bose condensation. First introduce sources and show

$$\int \mathcal{D}(\psi^*\psi)e^{-\frac{1}{2}(\psi^*\psi)D\binom{\psi}{\psi^*}+\frac{1}{2}(J^*J)\binom{\psi}{\psi^*}+\frac{1}{2}(\psi^*\psi)\binom{J}{J^*}} = (Det\, D)^{-\frac{1}{2}}e^{\frac{1}{2}(J^*J)G\binom{J}{J^*}}$$

where D and G are defined in Eqs. (2.219b) and (2.221). Note that although there are half as many fields ψ as the dimension of D, the integral over real and imaginary parts yields a real quadratic form of the proper dimension. By differentiation with respect to $\{J, J^*\}$ and using the properties $G_{\psi\psi^*}(12) = G_{\psi^*\psi}(21)$ and $G_{\psi\psi}(12) = G_{\psi\psi}(21)$ prove (2.223). Finally, show that the integral of any even number of ψ's and ψ^*'s weighted with $e^{-\frac{1}{2}(\psi^*\psi)D\binom{\psi}{\psi^*}}$ is given by the sum of all contractions.

CHAPTER 3

PERTURBATION THEORY AT ZERO TEMPERATURE

In this chapter, the general formalism introduced in Chapter 2 for the grand canonical ensemble at finite temperature is specialized to systems of fixed particle number at zero temperature. The coherent state functional integral formalism may be applied directly in this special case by replacing the trace occurring in the partition function by the expectation value in a non-interacting ground state. The methodology of enumerating all contractions with Feynman diagrams is unchanged, and the diagram rules are modified only by simple changes in the propagators and associated factors.

We begin by expressing physical observables in terms of matrix elements in the non-interacting ground state which are strictly analogous to corresponding finite temperature traces in Chapter 2 and which may then be evaluated in perturbation theory using Wick's theorem. As a preliminary step, propagators for the zero-temperature expansion are first evaluated in the zero-particle vacuum, and the result is then applied to a many-Fermion system by transforming to particle and hole operators defined relative to the non-interacting ground state. Given the perturbation theory for Fermions, Bosons may be treated directly by introducing a formal spin variable with spin degeneracy equal to the particle number, so that restricting the non-interacting ground state to be an antisymmetric singlet state in this variable renders the overall wave function symmetric in the physical variables. In subsequent sections, time-ordered Goldstone diagrams will be introduced and the zero-temperature limit will be discussed.

3.1 FEYNMAN DIAGRAMS

The finite temperature diagram expansion was obtained by expressing the observables of interest in terms of thermal averages of operators, Eq. (2.80b),

$$\langle F(a^\dagger_\alpha(\tau_i)\ldots a_\delta(\tau_\ell))\rangle_0 = \frac{\mathrm{Tr}[Te^{-\int_0^\beta d\tau(\hat{H}_0-\mu\hat{N})}F(a^\dagger_\alpha(\tau_i)\ldots a_\delta(\tau_\ell))\,]}{\mathrm{Tr}[Te^{-\int_0^\beta d\tau(\hat{H}_0-\mu\hat{H})}\,]} \tag{3.1}$$

where all operators including those in $\hat{H}_0$ and $\hat{N}$ have a formal label τ denoting the time slice upon which they are defined. Expressing the thermal average by the equivalent functional integral, Eq. (2.81b), and expanding $F(\psi^*_\alpha(\tau_i)\ldots\psi_\delta(\tau_\ell))$ in a series of polynomials in $\{\psi^*_\alpha, \psi_\alpha\}$ then yields a series of Gaussian integrals which produce all the contractions represented by Feynman diagrams.

OBSERVABLES

The first task is to derive expressions for the ground state energy, expectation values of operators and Green's functions analogous to Eqs. (2.81a), (2.138) and (2.145) where the thermal average has been replaced by an appropriate expectation value. Let us begin with the ground state energy and denote the eigenvalues and

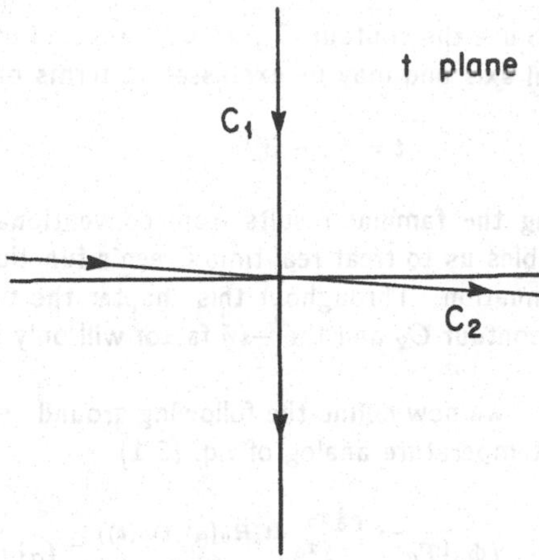

Fig. 3.1 Contours in the complex t-plane

normalized N-particle eigenfunctions of the one-body Hamiltonian H_0 and complete Hamiltonian $\hat{H} = \hat{H}_0 + \hat{V}$ as follows:

$$\hat{H}_0|\Phi_n\rangle = W_n|\Phi_n\rangle \tag{3.2a}$$

and

$$(\hat{H}_0 + \hat{V})|\Psi_n\rangle = E_n|\Psi_n\rangle \quad . \tag{3.2b}$$

The basic idea is seen by considering the following quantity

$$\frac{\langle\Phi_0|e^{-i(\hat{H}_0+\hat{V})T_0}|\Phi_0\rangle}{\langle\Phi_0|e^{-i\hat{H}_0T_0}|\Phi_0\rangle} = \sum_n |\langle\Phi_0|\Psi_n\rangle|^2 e^{-i(E_n-W_0)T_0} \underset{Im T_0\to-\infty}{\longrightarrow} e^{2\ln\langle\Phi_0|\Psi_0\rangle - i(E_0-W_0)T_0} \tag{3.3}$$

where the right-hand side is obtained by inserting a complete set of states $|\psi_n\rangle\langle\psi_n|$ in the numerator. As long as $|\Phi_0\rangle$ is not orthogonal to $|\Psi_0\rangle$ and $|\Psi_0\rangle$ is non-degenerate, in the limit $Im\ T_0 \to -\infty$ the contributions of all excited states are exponentially damped and the desired ground state energy may be obtained.

To recast Eq. (3.3) into a form analogous to Eq. (3.1), and to prepare for the subsequent treatment of Green's functions, it is useful to select a contour from $-\frac{1}{2}T_0$ to $\frac{1}{2}T_0$ in the complex t-plane. Let the distance from $-\frac{1}{2}T_0$ to $\frac{1}{2}T_0$ along this contour be specified by a real variable $\tilde{t}$. The time-ordered product for a set of operators having times on the contour is then defined by ordering the operator in terms of increasing $\tilde{t}$ and a path integral is defined by breaking $\tilde{t}$ into arbitrarily small intervals and introducing a complete set of states at each $\tilde{t}$ slice in the usual way. There is considerable freedom in choosing the contour, since by analyticity the result will be independent of the shape of the contour and the only requirement on T_0 is that its imaginary part must ultimately go to $-\infty$. One possible contour is C_1, shown in Fig. 3.1, corresponding to the imaginary time used in Chapter 2. For our present purposes, however, it is preferable to use the contour C_2, which is rotated by an arbitrarily small angle $\tilde{\eta}$ relative to the real axis and may be expressed in terms of a real variable $\tilde{t}$ as

$$t = (1 - i\tilde{\eta})\tilde{t} \ . \tag{3.4}$$

In addition to reproducing the familiar results from conventional derivations in real time, this choice also enables us to treat real-time Green's functions without the need for explicit analytic continuation. Throughout this chapter the time variable t will be understood to be on the contour C_2 and the $-i\tilde{\eta}$ factor will only be included explicitly when necessary.

With this preparation, we now define the following ground state average specified by H_0 which is the zero-temperature analog of Eq. (3.1)

$$\langle F(a^\dagger_\alpha(t_i)\ldots a_\delta(t_\ell))\rangle_{H_0} = \frac{\langle\Phi_0|Te^{-i\int_{-\frac{1}{2}T_0}^{\frac{1}{2}T_0} dt\, H_0(a^\dagger(t)a(t))} F\,(a^\dagger_\alpha(t_i)\ldots a_\delta(t_\ell))|\Phi_0\rangle}{\langle\Phi_0|Te^{-i\int_{-\frac{1}{2}T_0}^{\frac{1}{2}T_0} dt\, \hat{H}_0}|\Phi_0\rangle} \ . \tag{3.5}$$

As before, it is crucial to note that T time orders along C_2 all the operators in F as well as interleaving them with the operators in $\hat{H}_0(a^\dagger_\alpha(t),\, a_\alpha(t))$. Note that H_0 enters the definition implicitly through the fact that H_0 defines $|\Phi_0\rangle$, as well as explicitly.

Using this definition, Eq. (3.3) may be rewritten in the following form

$$\begin{aligned} &\lim_{T_0\to\infty} \frac{i}{T_0}\ln\left[\left\langle e^{-i\int_{-\frac{1}{2}T_0}^{\frac{1}{2}T_0} dt\, V(a^\dagger(t)a(t))}\right\rangle_{H_0}\right] \\ &\qquad = \lim_{T_0\to\infty} \frac{i}{T_0}\ln\frac{\langle\Phi_0|e^{-i(\hat{H}_0+\hat{V})T_0}|\Phi_0\rangle}{\langle\Phi_0|e^{-i\hat{H}_0T_0}|\Phi_0\rangle} \\ &\qquad = \lim_{T_0\to\infty} \frac{i}{T_0}\left(2\ln|\langle\Phi_0|\Psi_0\rangle| - i(E_0 - W_0)T_0\right) \\ &\qquad = E_0 - W_0 \end{aligned} \tag{3.6}$$

which is analogous to the finite temperature result following from Eq. (2.80a) that $\Omega - \Omega_0 = \frac{1}{\beta}\ln\langle e^{-\int_0^\beta dt\,\hat{V}}\rangle_0$. As before, expansion of $e^{-i\int dt\,V}$ and application of Wick's theorem yields the sum of all diagrams which, when exponentiated, yields the sum of all linked diagrams. In order for the limit $T_0 \to \infty$ to exist, it is crucial that each linked diagram be proportional to T_0 for large T_0. This T_0 dependence follows from the fact that the time integrals in the nth order term $\int_{-\frac{1}{2}T_0}^{\frac{1}{2}T_0} dt_1\ldots dt_n\,\langle V(t_1)\ldots V(t_n)\rangle_{H_0}$ may be expressed as $(n-1)$ integrals over relative time, each of which converges because of the choice of contour and positive definite excitation energies in propagators, and one integral over absolute time which yields a factor T_0. Physically, ground state Feynman diagrams are always of finite time extent because they represent time histories of virtual excitations which by the energy-time uncertainty principle, can only be sustained for limited times.

In the same way, expectation values of operator and Green's functions may be expressed in forms strictly analogous to the finite temperature results in Chapter 2.

Let $R(0)$ denote an operator $\hat{R}$ acting at time $t = 0$. Then, the ground state expectation value of R may be expressed in the same form as Eq. (2.138) as follows

$$
\begin{aligned}
&\lim_{T_0\to\infty} \frac{\left\langle e^{-i\int_{-\frac{1}{2}T_0}^{\frac{1}{2}T_0} dt\, V(a^\dagger(t)a(t))} R(0)\right\rangle_{H_0}}{\left\langle e^{-i\int_{-\frac{1}{2}T_0}^{\frac{1}{2}T_0} dt\, V(a^\dagger(t)a(t))}\right\rangle_{H_0}} \\
&= \lim_{T_0\to\infty} \frac{\langle\Phi_0|e^{-i\int_{-\frac{1}{2}T_0}^{\frac{1}{2}T_0} dt\,\left(H_0(a^\dagger(t)a(t))+V(a^\dagger(t)a(t))\right)} R(0)|\Phi_0\rangle}{\langle\Phi_0\left|e^{-iT_0(\hat{H}_0+\hat{V})}\right|\Phi_0\rangle} \\
&= \lim_{T_0\to\infty} \frac{\sum_{m,n}\langle\Phi_0|\Psi_m\rangle\langle\Psi_m\left|e^{-\frac{1}{2}T_0H}\,\hat{R}\,e^{-\frac{1}{2}T_0H}\right|\Psi_n\rangle\langle\Psi_n|\Phi_0\rangle}{\sum_n\langle\Phi_0|\Psi_n\rangle\, e^{-iT_0H}\,\langle\Psi_n|\Phi_0\rangle} \\
&= \lim_{T_0\to\infty} \frac{\sum_{m,n}\langle\Phi_0|\Psi_m\rangle\,\langle\Psi_m|\hat{R}|\Psi_n\rangle\langle\Psi_n|\Phi_0\rangle\, e^{-iT_0\frac{(E_m+E_n)}{2}}}{\sum_n|\langle\Phi_0|\Psi_n\rangle|^2\, e^{-iT_0E_n}} \\
&= \langle\Psi_0|\hat{R}|\Psi_0\rangle\ .
\end{aligned}
\tag{3.7}
$$

Similarly, by repeating the steps in Eq. (2.67), the n-particle Green's function defined in Eq. (2.21) may be expressed in the same form as Eq. (2.145). As in Eq. (2.67), let $\tilde{a}_{\alpha_i}$ denote an annihilation operator for $i \leq n$ and a creation operator for $i > n$, let P denote the permutation which arranges the times t_{Pi} in chronological order and let $\{a^{(H)\dagger}, a^{(H)}\}$ denote the Heisenberg operators defined in Eq. (2.14a). Then

$$
\begin{aligned}
&\lim_{T_0\to\infty}(-i)^n \frac{\left\langle e^{-i\int_{-\frac{1}{2}T_0}^{\frac{1}{2}T_0} dt V(a^\dagger(t)a(t))}\, a_{\alpha_1}(t_1)\ldots a_{\alpha_n}(t_n)a^\dagger_{\alpha_{n+1}}(t_{n+1})\ldots a^\dagger_{\alpha_{2n}}(t_{2n})\right\rangle_{H_0}}{\left\langle e^{-i\int_{-\frac{1}{2}T_0}^{\frac{1}{2}T_0} dt\, V(a^\dagger(t)a(t))}\right\rangle_{H_0}} \\
&= \lim_{T_0\to\infty}(-i)^n \frac{\langle\Phi_0|Te^{-i\int_{-\frac{1}{2}T_0}^{\frac{1}{2}T_0} dt\, H(a^\dagger(t)a(t))}\prod_{i=1}^{2n}\tilde{a}_{\alpha_i}(t_i)|\Phi_0\rangle}{\langle\Phi_0|e^{-iT_0H}|\Phi_0\rangle} \\
&= \lim_{T_0\to\infty} \frac{(-i)^n\ \ \varsigma^P}{|\langle\Phi_0|\Psi_0\rangle|^2 e^{-iE_0T_0}} \sum_{\ell,m}\langle\Phi_0|\Psi_\ell\rangle\langle\Psi_\ell|e^{-i(\frac{1}{2}T_0-t_{P1})H}\,\tilde{a}_{\alpha_{P1}}(t_{P1}) \\
&\quad\times e^{-i(t_{P1}-t_{P2})H}\,\tilde{a}_{\alpha_{P2}}(t_{P2})\ldots\tilde{a}_{\alpha_{P2n}}(t_{P2n})e^{-i(t_{P2n}-\frac{1}{2}T_0)H}|\Psi_m\rangle\langle\Psi_m|\Phi_0\rangle \\
&= \lim_{T_0\to\infty} \frac{(-i)^n\ \ \varsigma^P}{|\langle\phi_0|\Psi_0\rangle|^2\ e^{-iE_0T_0}} \sum_{\ell,m}\langle\Phi_0|\Psi_\ell\rangle\langle\Psi_m|\Phi_0\rangle \\
&\quad\times e^{-iT_0\frac{(E_\ell+E_m)}{2}}\langle\Psi_\ell|\prod_{i=1}^{2n}\tilde{a}^{(H)}_{\alpha_{Pi}}(t_{Pi})|\Psi_m\rangle \\
&= (-i)^n\langle\psi_0|Ta^{(H)}_{\alpha_1}(t_1)\ldots a^{(H)}_{\alpha_n}a^{(H)\dagger}_{\alpha_{n+1}}(t_{n+1})\ldots a^{(H)\dagger}_{\alpha_{2n}}(t_{2n})|\psi_0\rangle \\
&= G_n(\alpha_1t_1\ldots\alpha_nt_n|\alpha_{2n}t_{2n}\ldots\alpha_{n+1}t_{n+1})\ .
\end{aligned}
\tag{3.8}
$$

The left-hand sides of Eqs. (3.7) and (3.8) are of the form of Eqs. (2.138) and (2.145), respectively, and by the standard replica argument are given by the sum of all linked diagrams containing the operators R or $a_{\alpha_1}(t_1)\ldots a^\dagger_{\alpha_{2n}}(t_{2n})$. Thus, all zero-temperature observables of interest may be obtained by evaluating expressions of the form $\langle\, e^{-i\int_{-\infty}^{\infty} dt\, V(a^\dagger(t)a(t))}\, \mathcal{O}(t)\, \rangle_{H_0}$ using Wick's theorem and the appropriate zero-temperature propagators derived in the next section.

It is useful to note in passing that there exists an alternative method to obtain the same diagram expansions in real time by considering the evolution operator for a Hamiltonian $H_\epsilon(t) = H_0 + e^{-\epsilon|t|}V$ in which the interaction V is adiabatically turned off at large positive and negative times. As outlined in Problem (3.1), a theorem by Gell-Mann and Low (1951) shows how to obtain an eigenstate of the interacting system by applying the evolution operator for $H_\epsilon(t)$ to an eigenstate of the non-interacting system and taking the limit as $\epsilon \to 0$, from which observables may be evaluated straightforwardly.

ZERO-TEMPERATURE FERMION PROPAGATORS

The generating function for a non-interacting N-Fermion system with one-body Hamiltonian H_0 at zero-temperature is defined

$$\begin{aligned} \mathcal{M}_{H_0}(J^*, J) &= \langle e^{\int_{-\frac{1}{2}T_0}^{\frac{1}{2}T_0} dt \sum_\alpha [J^*_\alpha(t) a_\alpha + a^\dagger_\alpha(t) J_\alpha(t)]} \rangle_{H_0} \\ &= \frac{\langle \Phi_0 | T e^{\int_{-\frac{1}{2}T_0}^{\frac{1}{2}T_0} dt \sum_\alpha [-i\epsilon_\alpha a^\dagger_\alpha(t) a_\alpha(t) + J^\dagger_\alpha(t) a_\alpha(t) + a^\dagger_\alpha(t) J_\alpha(t)]} | \Phi_0 \rangle}{\langle \Phi_0 | T e^{-i\int_{-T_0/2}^{T_0/2} dt \sum_\alpha \epsilon_\alpha a^\dagger_\alpha(t) a_\alpha(t)} | \Phi_0 \rangle} \end{aligned} \tag{3.9}$$

where we have used a single-particle basis which diagonalizes H_0 with eigenvalues ϵ_α. $J^*_\alpha(t)$ and $J_\alpha(t)$ denote Grassmann sources and $|\Phi_0\rangle$ denotes the N-Fermion ground state of H_0. In the usual way, the average defined in Eq. (3.5) for any product of creation and annihilation operators may be obtained from appropriate derivatives of $\mathcal{M}_{H_0}$ with respect to the sources evaluated at $\tau^* = \tau = 0$.

As a first step toward the evaluation of $\mathcal{M}_{H_0}$, it is useful to calculate the related quantity

$$\mathcal{M}_0(J^*, J) = \langle 0 | T e^{\int_{-T_0/2}^{T_0/2} dt \sum_\alpha [-i\epsilon_\alpha a^\dagger_\alpha(t) a_\alpha(t) + J^*_\alpha(t) a_\alpha(t) + a^\dagger_\alpha(t) J_\alpha(t)]} | 0 \rangle \tag{3.10}$$

where the N-particle state $|\Phi_0\rangle$ has been replaced by the zero particle state $|0\rangle$ and the denominator is temporarily ignored. It is also convenient to pick the zero of the energy scale such that all $\epsilon_\alpha > 0$. Since the Fermion zero-particle state $|0\rangle$ is the Grassmann coherent state $|\psi\rangle$ with $\psi = 0$, the discrete functional integral for $\mathcal{M}_0(J^*, J)$ may be written

$$\begin{aligned} \mathcal{M}_0(J^* J) = \lim_{M\to\infty} \int \prod_{k=1}^{M-1} D(\psi^*_{\alpha,k}\psi_{\alpha,k}) e^{-\sum_\alpha \sum_{k=1}^{M-1} \psi^*_{\alpha,k}\psi_{\alpha k}} \\ \times e^{-\sum_\alpha \sum_{k=1}^{M} \left[\psi^*_{\alpha,k}\left(-1 + i\frac{T}{M}\epsilon_\alpha\right)\psi_{\alpha,k-1} - J^*_{\alpha,k}\psi_{\alpha,k} - \psi^*_{\alpha,k} J_{\alpha,k}\right]_{\psi_0=0,\, \psi^*_M=0}} \end{aligned}$$

$$= \lim_{M\to\infty} \prod_{\alpha} \left[\int \prod_{k=1}^{M} D(\psi_k^* \psi_k) e^{-\sum_{jk=1}^{M-1} \psi_j^* S(\epsilon_\alpha)_{jk} \psi_k + \sum_{k=1}^{M-1} (J_{\alpha,k}^* \psi_k + \psi_k^* J_{\alpha,k})} \right]$$
$$= \lim_{M\to\infty} \prod_{\alpha} \left[\mathrm{Det} S^{(\alpha)} e^{\sum_{j,k=1}^{M-1} J_{\alpha,j}^* S(\epsilon_\alpha)_{jk}^{-1} J_{\alpha,k}} \right] \tag{3.11}$$

where, using the notation of Eq. (2.70)

$$S(\epsilon_\alpha) \begin{bmatrix} 1 & 0 & \dots & 0 & 0 \\ -a & 1 & 0 & & 0 \\ 0 & -a & 1 & \ddots & \vdots \\ & 0 & -a & \ddots & 0 \\ \vdots & & 0 & \ddots & 1 & 0 \\ 0 & & \dots & -a & 1 \end{bmatrix} \qquad \psi = \begin{bmatrix} \psi_1 \\ \psi_2 \\ \\ \vdots \\ \\ \psi_{M-1} \end{bmatrix} \tag{3.12a}$$

and

$$a = 1 - i\frac{T_0}{M}\epsilon_\alpha \ . \tag{3.12b}$$

The essential difference between the matrix $S(\epsilon_\alpha)$ in the present case and Eq. (2.70) for the partition function is the boundary condition. Expanding by minors, $\mathrm{Det}S = 1$ and the inverse is

$$S^{-1}(\epsilon_\alpha) = \begin{bmatrix} 1 & 0 & 0 & \dots & 0 \\ a & 1 & 0 & & \\ a^2 & a & 1 & & \\ \vdots & a^2 & a & & \vdots \\ a^{M-3} & \vdots & & 1 & 0 \\ a^{M-2} & a^{M-3} & & \dots\ a & 1 \end{bmatrix} \ . \tag{3.13}$$

Thus, $S^{-1}(\epsilon_\alpha)_{q,r}$ vanishes for $q < r$ and for $q \geq r$, we obtain

$$\begin{aligned} S^{-1}(\epsilon_\alpha)_{q,r} &= \lim_{M\to\infty} (a)^{q-r} \\ &= \lim_{M\to\infty} (1 - i\tfrac{T_0}{M}\epsilon_\alpha)^{(\frac{t_q - t_r}{T_0})M} \\ &= e^{-i\epsilon_\alpha(t_q - t_r)} \ . \end{aligned} \tag{3.14}$$

Hence, the non-interacting single-particle Green's function, or contraction, is given by

$$\begin{aligned} iG_0(\alpha, t|\alpha', t') &= \lim_{T_0\to\infty} \frac{\langle 0|Te^{-i\int_{-\frac{1}{2}T_0}^{\frac{1}{2}T_0} dt \sum_\alpha \epsilon_\alpha a_\alpha^\dagger(t) a_\alpha(t)} a_\alpha(t) a_{\alpha'}^\dagger(t')|0\rangle}{\langle 0|Te^{-i\int_{-\frac{1}{2}T_0}^{\frac{1}{2}T_0} dt \sum_\alpha \epsilon_\alpha a_\alpha^\dagger(t) a_\alpha(t)}|0\rangle} \\ &= \lim_{T_0\to\infty} \frac{1}{\mathcal{M}_0(0,0)} \left[\frac{-\delta^2}{\delta J_\alpha^*(t) \delta J_{\alpha'}(t')} \mathcal{M}_0(J^* J) \right]_{J^* = J = 0} \\ &= \delta_{\alpha\alpha'} e^{-i\epsilon_\alpha(t - t')} \theta(t - t' - \eta') \ . \end{aligned} \tag{3.15}$$

Note that since normal-ordered operators have creation operators represented by a coherent state one time step later than the annihilation operator, as in Section 2.2, an infinitesimal η' has been introduced in Eq. (3.15) to emphasize that the result vanishes at equal time. The fact that the propagator (3.15) is non-vanishing only for positive times $(t-t')$ whereas the thermodynamic propagator (2.78) has components for both positive and negative times arises directly from the difference between the boundary conditions $\psi^*(\frac{T_0}{2}) = \psi(\frac{T}{2}) = 0$ in the present case and the antiperiodic boundary conditions $\psi(\beta) = -\psi(0)$ in the finite temperature case. One also observes in Eq. (3.15) an example of the general feature that the imaginary part in Eq. (3.2b) defining the contour C_2 makes the propagator converge at large values of $|t-t'|$. Physically, Eq. (3.15) for the non-interacting Green's function is obvious. For $t > t'$, $a^\dagger_\alpha$ creates a one-particle state in eigenstate α of time t' which propagates with the evolution operator $e^{-i\epsilon_\alpha(t''-t')}$ until it is annihilated at time $t'' = t$. For $t \le t'$, $a_\alpha(t)$ acts on $|0\rangle$ giving zero.

The generating function $\mathcal{M}_{H_0}$ defined relative to the N-Fermion non-interacting ground state $|\Phi_0\rangle$ can be reduced to the simple form of $\mathcal{M}_0$ by introducing a particle-hole transformation. Let the basis states of H_0 be labeled in order of increasing eigenvalues ϵ_α. Then, the ground state Slater determinant is obtained by occupying the lowest N states

$$|\Phi_0\rangle = \prod_{\alpha=1}^{N} a^\dagger_\alpha |0\rangle \tag{3.16}$$

and occupation numbers may be defined

$$n_\alpha = \begin{cases} 1 & \alpha \le N \\ 0 & \alpha > N \end{cases} . \tag{3.17}$$

In the familiar case of a translationally invariant system, the states α are plane waves, the Fermi gas ground state includes all momenta up to the Fermi momentum k_F, and the occupation numbers are specified by $n_k = \theta(k_F - |\vec{k}|)$.

It is crucial to the present treatment that $|\Phi_0\rangle$ be non-degenerate. This requires that there be a finite energy gap $\epsilon_{N+1} - \epsilon_N$ between the last occupied state and the first unoccupied state, since otherwise if $\epsilon_{N+1} = \epsilon_N$ the state $a^\dagger_{N+1} \prod_{\alpha=1}^{N-1} a^\dagger_\alpha |0\rangle$ would be degenerate with $|\Phi_0\rangle$. (Degenerate cases, such as arise in partially-filled shells in atomic and nuclear physics, require an appropriate form of degenerate perturbation theory such as the Bloch-Horowitz (1958) expansion.) Here, we will assume the existence of a gap, and shift the energies such that:

$$\begin{aligned} \epsilon_\alpha &< 0 \quad \alpha \le N \\ \epsilon_\alpha &> 0 \quad \alpha > N . \end{aligned} \tag{3.18}$$

This shift is useful in defining propagators which converge as $|t-t'| \to \infty$.

Finally, we define particle-hole operators $\{b^\dagger_\alpha, b_\alpha\}$ by the canonical transformation

$$b^\dagger_\alpha = \begin{cases} a^\dagger_\alpha & \alpha > N \\ a_\alpha & a \le N \end{cases} \quad b_\alpha = \begin{cases} a_\alpha & a > N \\ a^\dagger_\alpha & a \le N \end{cases} . \tag{3.19}$$

For an unoccupied state $\alpha > N$, $b^\dagger_\alpha$ creates a particle whereas for an occupied state $\alpha \leq N$, $b^\dagger_\alpha$ destroys a particle creating a hole so $\{b^\dagger_\alpha\}$ corresponds to the set of particle-hole creation operators relative to $|\Phi_0\rangle$. For both $\alpha > N$ and $\alpha \leq N$, Eqs. (3.16) and (3.19) imply

$$b_\alpha|\Phi_0\rangle = 0 \quad . \tag{3.20}$$

so that $\{b_\alpha\}$ is the set of annihilation operators relative to $|\Phi_0\rangle$ and $|\Phi_0\rangle$ is the zero-particle, zero-hole vacuum. Thus, when coherent states are constructed using the operators $\{b^\dagger_\alpha\}$, the state $|\psi\rangle$ with $\psi_\alpha = 0$ will be $|\Phi_0\rangle$.

In terms of particle-hole operators, H_0 may be written

$$\begin{aligned} H_0 &= \sum_{\alpha=1}^{N} \epsilon_\alpha b_\alpha b^\dagger_\alpha + \sum_{\alpha=N+1}^{\infty} \epsilon_\alpha b^\dagger_\alpha b_\alpha \\ &= \sum_{\alpha=1}^{N} \epsilon_\alpha (1 - b^\dagger_\alpha b_\alpha) + \sum_{\alpha=N+1}^{\infty} \epsilon_\alpha b^\dagger_\alpha b_\alpha \end{aligned} \tag{3.21}$$

and the generating function is

$$\begin{aligned} \mathcal{M}_{H_0}(J^*, J) = &\left[\langle\Phi_0|Te^{\int_{-\frac{1}{2}\tau_0}^{\frac{1}{2}\tau_0} dt [\sum_{\alpha=1}^{N} i\epsilon_\alpha b^\dagger_\alpha(t) b_\alpha(t) + \sum_{\alpha=N+1}^{\infty} -i\epsilon_\alpha b^\dagger_\alpha(t) b_\alpha(t)\,]}|\Phi_0\,\rangle\right]^{-1} \\ &\times \langle\Phi_0|Te^{\int_{-\frac{1}{2}\tau_0}^{\frac{1}{2}\tau_0} dt \sum_{\alpha=1}^{N} [i\epsilon_\alpha b^\dagger_\alpha(t) b_\alpha(t) - b^\dagger_\alpha(t) J^*_\alpha(t) - J_\alpha(t) b_\alpha(t)\,]} \\ &\times e^{\int_{-\frac{1}{2}\tau_0}^{\frac{1}{2}\tau_0} dt \sum_{\alpha=N+1}^{\infty} [-i\epsilon_\alpha b^\dagger_\alpha(t) b_\alpha(t) + J^*_\alpha(t) b_\alpha(t) + b^\dagger_\alpha(t) J_\alpha(t)\,]}|\Phi_0\rangle \end{aligned} \tag{3.22}$$

where the sum $\sum_{\alpha=1}^{N} \epsilon_\alpha$ has been cancelled out of the numerator and denominator. Because $|\Phi_0\rangle$ is the zero-particle, zero-hole vacuum for all the operators $\{b_\alpha\}$, Eq. (3.20), the evaluation of $\mathcal{M}_{H_0}(J^*, J)$ by a discrete functional integral is identical to that for $\mathcal{M}_0$ Eq. (3.11).

Note that since Det $S^{(\alpha)} = 1$, the denominator is unity. As before, the numerator factorizes into products of integrals for each α, so that the Green's function is diagonal. Thus, the generating function may be written

$$\mathcal{M}_{H_0}(J^*, J) = \lim_{M\to\infty} \prod_{\alpha\leq N} [e^{\sum_{j,k=1}^{M-1} J_{\alpha,j} S(-\epsilon_\alpha)^{-1}_{jk} J^*_{\alpha,k}}\,] \prod_{\alpha>N} [e^{\sum_{j,k=1}^{M-1} J^*_{\alpha,j} S(\epsilon_\alpha)^{-1}_{jk} J_{\alpha,k}}\,] \tag{3.23}$$

where the notation $S(-\epsilon_\alpha)$ arises because the hole state energies enter with a minus sign in Eq. (3.22). For particle states, $\alpha > N$, the Green's function is calculated as before in Eq. (3.15)

$$\begin{aligned} iG_{0\,\alpha\alpha'}(t_q, t_r)(1 - n_\alpha) &= \frac{-\delta^2}{\delta J^*_{\alpha,q} \delta J_{\alpha',r}} \mathcal{M}_{H_0}(J^*, J)(1 - n_\alpha)\Big|_{J^*=J=0} \\ &= \delta_{\alpha\alpha'}(1 - n_\alpha) S^{-1}(\epsilon_\alpha)_{qr} \\ &= \delta_{\alpha\alpha'}(1 - n_\alpha) e^{-i\epsilon_\alpha(t_q - t_r)} \theta(q \geq r) \\ &= \delta_{\alpha\alpha'}(1 - n_\alpha) e^{-i\epsilon_\alpha(t_q - t_r)} \theta(t_q - t_r - \eta') \quad . \end{aligned} \tag{3.24a}$$

Recall that the $-\eta'$ arises because at equal physical times the creation operator (r) is evaluated one time slice later than the annihilation operator (q) so that the θ function yields zero.

For hole states, $\alpha \leq N$, we obtain the new result

$$\begin{aligned}
iG_{0_{\alpha\alpha'}}(t_q, t_r)n_\alpha &= \frac{-\delta^2}{\delta J^*_{\alpha q}\delta J_{\alpha' r}} \mathcal{M}_{H_0}(J^*, J)n_\alpha \\
&= \delta_{\alpha\alpha'} n_\alpha \frac{-\delta^2}{\delta J^*_{\alpha,q}\delta J_{\alpha,r}} e^{\sum_{j,k} J_{\alpha,j} S^{-1}(-\epsilon_\alpha)_{j,k} J^*_{\alpha,k}}\Big|_{J^*=J=0} \\
&= -\delta_{\alpha\alpha'} n_\alpha \; S^{-1}(-\epsilon_\alpha)_{rq} \\
&= -\delta_{\alpha\alpha'} n_\alpha e^{-i(-\epsilon_\alpha)(t_r - t_q)} \theta(r \geq q) \\
&= -\delta_{\alpha\alpha'} n_\alpha e^{-i\epsilon_\alpha(t_q - t_r)} \theta(t_r - t_q + \eta') \;.
\end{aligned} \tag{3.24b}$$

The case of equal times for hole states is different than for particle states. Since at equal times, the time-ordered product of $a^\dagger a$ is defined to be the normal-ordered product, for $\alpha \leq N$, it follows that the time-ordered product of $b^\dagger_\alpha b_\alpha$ at equal times is $-b_\alpha b^\dagger_\alpha$. Hence, when a path integral is evaluated, $e^{-\epsilon H}\; b_\alpha b^\dagger_\alpha \; e^{-\epsilon H}$ is broken up as $e^{-\epsilon H}\; b_\alpha \int D(\psi^*\psi)e^{-\psi^*\psi} \;|\psi\rangle\langle\psi| b^\dagger_\alpha \; e^{-\epsilon H}$ so that b_α and $b^\dagger_\alpha$ must be evaluated on the same time slice and the θ function in Eq. (3.24b) is satisfied at equal times.

Combining the two results for particles and holes, Eq. (3.24), the complete Green's function for the non-interacting system may be written

$$iG_0(\alpha t|\alpha' t') = \delta_{\alpha\alpha'} iG_{0_\alpha}(t - t') \tag{3.25a}$$

where

$$iG_{0_\alpha}(t - t') = e^{-i\epsilon_\alpha(t-t')}[\theta(t - t' - \eta')(1 - n_\alpha) - \theta(t' - t + \eta')n_\alpha] \;. \tag{3.25b}$$

Note that this result is precisely of the form of the thermal propagator, Eq. (2.78), with the occupation numbers n_α replacing the thermal weights $(e^{\beta(\epsilon_\alpha - \mu)} - 1)^{-1}$, real time t replacing $-i\tau$, and no chemical potential. Physically, the first term represents the propagation of particles in the system. When α is an occupied state and $t > t'$, $a^\dagger_\alpha$ acting on $|\Phi_0\rangle$ creates an additional particle which propagates with energy ϵ_α until it is later annihilated at time τ. Similarly, when α is an occupied state and $t < t'$, a_α acting on $|\Phi_0\rangle$ destroys a particle in state α leaving a hole which propagates with energy ϵ_α until is is later filled at time t' by the re-creation of a particle in state α, so that the second term represents hole propagation. Finally, note that the sign of ϵ_α from (3.18) ensures that along the contour C_2, Eq. (3.4), both the particle and hole terms go to zero as $|t - t'| \to \infty$.

FERMION DIAGRAM RULES

Feynman diagrams for expectation values of operators and Green's functions are obtained for many-Fermion systems in the same as in Section 2.3. Multiple differentiation of $\mathcal{M}_{H_0}(J^*, J)$ with respect to J^* and J yields Wick's theorem as in Eq. (2.84) with the contractions defined by iG_0, Eq. (3.25). Thus, all the results in Chapter 2.3

may be applied at zero temperature with the replacements $\tau \to it$, $\int_0^\beta d\tau \to \int_{-T/2}^{T/2} dt$, $(-1)^n \to (-i)^n$ and propagators $g_\alpha(\tau-\tau')$, Eq. (2.98), replaced by $iG_{0_\alpha}(t-t')$, Eq. (3.25b).

The result, Eq. (2.137a), that $\Omega - \Omega_0$ is given by $-\frac{1}{\beta}$ times the sum of all linked diagrams where each unlabelled Feynman diagram contains the factor $\frac{(-1)^n}{S}\,\zeta^{n_L}$ has the zero-temperature counterpart for the Fermion ground state energy:

$$E_0 - W_0 = \lim_{t_0\to\infty} \frac{i}{T_0} \sum \text{all linked diagrams} \tag{3.26}$$

where each unlabelled Feynman diagram with n interactions, n_L closed loops and symmetry factor S contains the factor $\frac{(-i)^n}{S}$ $(-1)^{n_L}$ and propagators iG_{0_α}. For example, the Hartree-Fock diagrams yield

$$E_0 - W_0\,|_{HF} = \lim_{T_0\to\infty} \frac{i}{T_0}\Big[\ \text{(diagrams)}\ \Big]$$

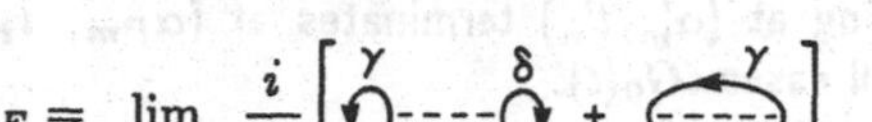

$$= \lim_{T_0\to\infty} \frac{i}{T_0} \int_{-\frac{T_0}{2}}^{\frac{T_0}{2}} \frac{-i}{2} \sum_{\gamma\delta} iG^0_\gamma(0) iG^0_\delta(0) [(-1)^2 (\gamma\delta|v|\gamma\delta) + (-1)(\gamma\delta|v|\delta\gamma)\,]$$

$$= \frac{1}{2} \sum_{\gamma,\delta=1}^{N} [(\gamma\delta|v|\gamma\delta) - (\gamma\delta|v|\delta\gamma)\,] \tag{3.27}$$

in agreement with Eq. (2.100b) and Problem (1.6). Similarly, a direct second-order contribution to $E_0 - W_0$ is

$$E_0 - W_0\,|_D^{(2)} = \lim_{t_0\to\infty} \frac{i}{T_0}\ \text{(diagram)}$$

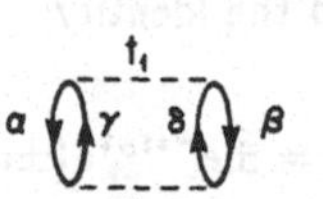

$$= \lim_{T_0\to\infty} \frac{i}{T_0} \frac{(-i)^2}{4} (-1)^2 \sum_{\alpha\beta\gamma\delta} (\alpha\beta|v|\gamma\delta)(\gamma\delta|v|\alpha\beta)$$

$$\times \int_{-\frac{T_0}{2}}^{\frac{T_0}{2}} dt_1 \int_{-\frac{T_0}{2}}^{\frac{T_0}{2}} dt_2\, iG_{0_\alpha}(t_1-t_2) iG_{0_\beta}(t_1-t_2) iG_{0_\gamma}(t_2-t_1) iG_{0_\delta}(t_2-t_1)$$

$$= \frac{-i}{2} \sum_{\substack{a,b>N\\ A,B\le N}} |\,(ab|v|AB)\,|^2 \int_0^\infty dt\; e^{-i(\epsilon_a+\epsilon_b-\epsilon_A-\epsilon_B)t} \tag{3.28}$$

$$= -\frac{1}{2} \sum_{\substack{a,b>n\\ A,B\le N}} \frac{|\,(ab|v|AB)\,|^2}{\epsilon_a+\epsilon_b-\epsilon_A-\epsilon_B}\ .$$

Note that, because the t integral is evaluated along the contour C_2, the contribution at the upper limit is zero and that the sum over $\alpha\beta\gamma\delta$ yields two equal contributions when either of the pairs $\{\alpha\beta\}$, $\{\gamma\delta\}$ is occupied and the other is unoccupied. It will be useful henceforth to distinguish sums over states by the convention that upper case Roman labels denote occupied states, lower case Roman labels denote unoccupied states, and Greek labels run over all states:

$$\sum_A : \text{occupied states}, \quad \sum_a : \text{unoccupied states}, \quad \sum_\alpha : \text{all states} . \tag{3.29}$$

Thus, the restrictions a, $b > N$ and A, $B \le N$ indicated explicitly in Eq. (3.28) will be implied by this notation.

Diagrams for expectation values of operators and Green's functions are obtained in the same way. A diagram for $\langle R \rangle$ with p interactions, n_L Fermion loops and symmetry factor S has the factor $\frac{(-i)^p}{S}(-1)^{n_L}$. A contribution to an n-particle Green's function $iG_n(\alpha_1 t_1, \ldots \alpha_n t_n \mid \alpha_1' t_1', \ldots \alpha_n' t_n')$ with r interactions and n_L Fermion loops has the factor $(-i)^r\,(-1)^P\,(-1)^{n_L}$ where $(-1)^P$ is the sign of the permutation P such that each propagator line originating at (α_m', t_m') terminates at (α_{Pm}, t_{Pm}). The single-particle propagators are in all cases $iG_0(t)$.

The frequency representation of a zero-temperature propagator is given by a Fourier integral rather than the Fourier series obtained at finite temperature, Eq. (2.131b) because the time integral extends over the infinite domain $(-\infty, \infty)$ instead of the finite interval $(0, \beta)$. The frequency transform of the non-interacting zero-temperature Green's function is defined by the relations

$$i\tilde{G}_{0_\alpha}(\omega) = \int_{-\infty}^{\infty} dt\, e^{i\omega t}\, iG_{0_\alpha}(t) \tag{3.30a}$$

$$iG_{0_\alpha}(t) = \frac{1}{2\pi}\int_{-\infty}^{\infty} d\omega\, e^{-i\omega t} i\tilde{G}_{0_\alpha}(\omega) . \tag{3.30b}$$

To Fourier transform propagators, we need the identity

$$\frac{1}{2\pi}\int_{-\infty}^{\infty} d\omega \frac{ie^{-i\omega t}}{\omega - \epsilon_\alpha \pm i\eta} = \pm e^{-i\epsilon_\alpha t}\theta(\pm t) \tag{3.31a}$$

with inverse

$$\int_{-\infty}^{\infty} dt\, e^{i\omega t}\, e^{-i\epsilon_\alpha t}\theta(\pm t) = \pm \frac{i}{\omega - \epsilon_\alpha \pm i\eta} . \tag{3.31b}$$

Equation (3.31a) is established by contour integration as follows. Consider the case $+i\eta$ for which there is a single pole below the real axis. For $t < 0$, $e^{-i\omega t}$ vanishes in the upper half ω plane as $|\omega| \to \infty$ so that the contour may be closed in the upper half plane without enclosing any poles and thus yields 0. For $t > 0$, the contour must be closed in the lower half plane encircling the pole at $\omega = \epsilon_\alpha - i\eta$ and yielding $-2\pi i$ times the residue $ie^{-i\epsilon_\alpha t}$. Hence, the integral is given by Eq. (3.30) for all t. The proof for $-i\eta$ is analogous. Using Eqs. (3.31) and (3.25b), the propagator is

$$\begin{aligned}\tilde{G}_{0_\alpha}(\omega) &= \frac{1 - n_\alpha}{\omega - \epsilon_\alpha + i\eta} + \frac{n_\alpha}{\omega - \epsilon_\alpha - i\eta} \\ &= \frac{1}{\omega - \epsilon_\alpha + i\eta \,\mathrm{sgn}\, \epsilon_\alpha}\end{aligned} \tag{3.32a}$$

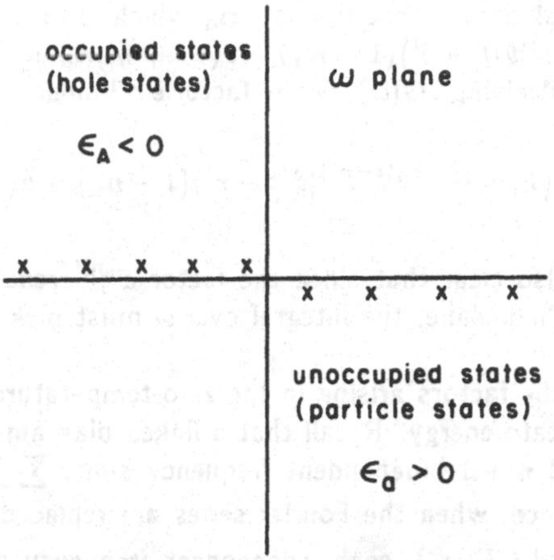

Fig. 3.2 The poles of $\tilde{G}_{0_\alpha}(\omega)$ in the complex ω plane.

where we recall that ϵ_α is positive for unoccupied states and negative for occupied states. An alternative derivation in which the integral for $\tilde{G}_{0_\alpha}(\omega)$ manifestly converges is obtained by using the contour C_2 specified by $t = \tilde{t}(1 - i\tilde{\eta})$, Eq. (3.4), along which the propagator $G_{0_\alpha}(t)$ converges as the real variable $\tilde{t} \to \pm\infty$. Then, $G^{(0)}(\omega)$ should be defined as the Fourier transform with respect to the real variable $\tilde{t}$ from which we obtain

$$
\begin{aligned}
i\tilde{G}_{0_\alpha}(\omega) &= \int_{-\infty}^{\infty} d\tilde{t}\; e^{i\omega\tilde{t}} iG_{0_\alpha}(\tilde{t}(1 - i\tilde{\eta})) \\
&= (1 - n_\alpha)\int_0^\infty d\tilde{t}\; e^{i(\omega - \epsilon_\alpha(1 - i\tilde{\eta}))\tilde{t}} - n_\alpha \int_{-\infty}^0 d\tilde{t}\; e^{i(\omega - \epsilon_\alpha(1 - i\tilde{\eta}))\tilde{t}} \\
&= (1 - n_\alpha)\frac{e^{i(\omega - \epsilon_\alpha + i\epsilon_\alpha\tilde{\eta})\tilde{t}}\big|_0^\infty}{i(\omega - \epsilon_\alpha + i\epsilon_\alpha\tilde{\eta})} - n_\alpha \frac{e^{i(\omega - \epsilon_\alpha + i\epsilon_\alpha\tilde{\eta})\tilde{t}}\big|_{-\infty}^0}{i(\omega - \epsilon_\alpha - i\epsilon_\alpha\tilde{\eta})} \, .
\end{aligned} \tag{3.32b}
$$

Noting that $\epsilon_\alpha\tilde{\eta}$ is a positive infinitesimal for unoccupied states in the first term and a negative infinitesimal for occupied states in the second term, Eq. (3.32b) reproduces the previous result (3.32a).

The location of the poles of $\tilde{G}_{0_\alpha}(\omega)$ in the complex ω plane is shown in Fig. 3.2 and is characteristic of the general structure of one-particle Green's functions established subsequently in Chapter 5. The essential features are that the poles for unoccupied, or particle, states are displaced below the real axis, occupied, or hole, states are above the real axis, and because of the restriction to a non-degenerate ground state, there is a gap between the lowest particle state and highest hole state. Our convention of shifting the zero of the energy scale such that $\epsilon_A < 0$ and $\epsilon_a > 0$ corresponds to measuring energies and ω relative to a Fermi energy or chemical potential located in the gap.

As at finite temperature, an additional factor is required to treat the special case of a propagator which begins and ends at the same physical time. We have seen previously that the equal time propagator is $-n_\alpha$ which is assured in Eq. (3.25b) by writing $iG^{(0)}(t) = e^{-i\epsilon_\alpha t}[\theta(t-\eta')(1-n_\alpha) - \theta(-t+\eta')n_\alpha]$. As before, this result may be enforced by multiplying $i\tilde{G}(\omega)$ by the factor $e^{i\omega\eta'}$ since

$$\int \frac{d\omega}{2\pi} e^{-i\omega(t-\eta')} i\tilde{G}_{0_\alpha}(\omega) = e^{-i\epsilon_\alpha(t-\eta')}[\theta(t-\eta')(1-n_\alpha) - \theta(-t+\eta')n_\alpha] \, . \quad (3.33)$$

From Fig. (3.2), it is also clear that since the factor $e^{i\omega\eta'}$ requires that the contour be closed in the upper half plane, the integral over ω must pick up only the occupied state contributions.

We now consider the factors arising in the zero-temperature frequency representation for the ground state energy. Recall that a linked diagram at finite temperature with n interactions had $n+1$ independent frequency sums $\sum_{\omega_n}$ and an overall factor for $\frac{Z}{Z_0}$ of β^{-n}. Hence, when the Fourier series are replaced by Fourier integrals, $\sum_{\omega_n} F(\omega_n) \to \frac{\beta}{2\pi} \int d\omega \, F(\omega_n)$, each independent frequency sum is replaced by an integral $\int \frac{d\omega}{2\pi}$ and the diagram has an overall factor β. This is completely analogous to the overall volume factor $\mathcal{V}$ obtained for each linked diagram in momentum representation. Similarly, at zero temperature, each linked diagram acquires an overall factor T_0 which cancels $\frac{1}{T_0}$ in Eq. (3.26) yielding a constant in the limit $T_0 \to \infty$.

The contribution to $E_0 - W_0$ of a linked diagram with n interactions and n_L closed Fermion loops in calculated as follows. Interaction vertices are included as at finite temperature. The propagator $i\tilde{G}^{(0)}_\alpha(\omega)$, Eq. (3.32), is included for each Fermion line with an additional factor $e^{i\omega\eta'}$ for propagators beginning and ending at the same interaction. After utilizing the $(n-1)$ frequency conservation conditions at interaction vertices, frequency integrals $\int \frac{d\omega}{2\pi}$ are performed for the $(n+1)$ remaining independent frequencies. From Eq. (3.26), the result should be multiplied by the factor $\frac{(-i)^{n-1}}{S}(-1)^{n_L}$, where S is the symmetry factor. A slightly more convenient rule is to use propagators $\tilde{G}_{0_\alpha}(\omega)$, integrals $\int d\omega$ and to include the factor $(i)^{2n}$ for $2n$ propagators and $(\frac{1}{2\pi})^{n+1}$ for $n+1$ independent momenta in a new overall factor $(\frac{i}{2\pi})^{n+1} \frac{(-1)^{n_L}}{S}$.

As an example, let us evaluate once again the second-order direct contribution considered in Eq. (3.28)

$$E_0 - W_0 \,|^{(2)}_D = \lim \frac{i}{T_0} \quad (3.34a)$$

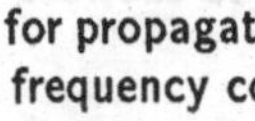

$$= \frac{1}{4}\left(\frac{i}{2\pi}\right)^3 \sum_{\alpha\beta\gamma\delta} \int d\omega_\alpha d\omega_\beta d\omega_\gamma \left[\frac{1-n_\alpha}{\omega_\alpha - \epsilon_\alpha + i\eta} + \frac{n_\alpha}{\omega_\alpha - \epsilon_\alpha - i\eta}\right]$$

$$\times \left[\frac{1-n_\beta}{\omega_\beta - \epsilon_\beta + i\eta} + \frac{n_\beta}{\omega_\beta - \epsilon_\beta - i\eta}\right]\left[\frac{1-n_\alpha}{\omega_\gamma - \epsilon_\gamma + i\eta} + \frac{n_\alpha}{\omega_\gamma - \epsilon_\gamma - i\eta}\right]$$

$$\times \left[\frac{1-n_\delta}{\omega_\alpha + \omega_\beta - \omega_\gamma - \epsilon_\gamma + i\eta} + \frac{n_\delta}{\omega_\alpha + \omega_\beta - \omega_\gamma - \epsilon_\delta - i\eta}\right] | (\alpha\beta|v|\gamma\delta) |^2 \, .$$

We already know from the time representation that $\{\alpha\beta\}$ must be particles and $\{\gamma\delta\}$ holes or vice versa. This is reflected in the fact that the only non-vanishing contributions in Eq. (3.34a) arise from terms in which the poles associated with n_α and n_β

are on one side of the real axis and the poles associated with n_γ and n_δ are on the other side as may be verified by considering the other alternatives: If the n_α and n_γ poles are in the same half plane, the integral $\int d\omega_\alpha \frac{1}{\omega_\alpha-\epsilon_\alpha\pm i\eta}\frac{1}{\omega_\alpha+\omega_\beta-\omega_\gamma-\epsilon_\delta\pm i\eta}$ may be closed in the other half plane yielding zero. The same argument applies to n_β and n_δ, so that the n_α and n_β poles must be in the opposite plane from the n_γ pole. Finally, the n_γ and n_δ poles must be in the same half plane, since otherwise $\int d\omega_\gamma \frac{1}{\omega_\gamma-\epsilon_\alpha\pm i\eta}\ \frac{1}{\omega_\alpha+\omega_\beta-\omega_\gamma-\epsilon_\gamma\pm i\eta}$ yields zero. The contribution in which $\{\alpha\beta\}$ are particles and $\{\gamma\delta\}$ are holes is equal to that in which particles and holes are interchanged, so we may calculate twice the contribution of the former case. Using the convention that $\{ab\}$ denote particles and $\{AB\}$ holes, we obtain the following result

$$E_0 - W_0\,|_D^{(2)} = \frac{1}{2}\left(\frac{i}{2\pi}\right)^3 \sum_{\substack{ab\\AB}} \int d\omega_a d\omega_b d\omega_A \frac{1}{\omega_a-\epsilon_a+i\eta}\ \frac{1}{\omega_b-\epsilon_b+i\eta}$$
$$\times\frac{1}{\omega_A-\epsilon_A-i\eta}\ \frac{1}{\omega_a+\omega_b-\omega_A-\epsilon_B-i\eta}\,|\,(AB|v|ab)\,|^2\ . \tag{3.34b}$$

Closing the ω_a and ω_b contours in the lower half plane and the ω_A contour in the upper half plane yields

$$\begin{aligned}
&\int d\omega_a d\omega_b d\omega_A \frac{1}{\omega_a-\epsilon_a+i\eta}\ \frac{1}{\omega_b-\epsilon_b+i\eta}\ \frac{1}{\omega_A-\epsilon_A-i\eta}\ \frac{1}{\omega_a+\omega_b-\omega_A-\epsilon_B-i\eta}\\
&\quad=(-2\pi i)\int d\omega_b d\omega_A \frac{1}{\omega_b-\epsilon_b+i\eta}\ \frac{1}{\omega_A-\epsilon_A-i\eta}\ \frac{1}{\epsilon_a+\omega_b-\omega_A-\epsilon_B-i\eta}\\
&\quad=(-2\pi i)^2\int d\omega_A \frac{1}{\omega_A-\epsilon_A-i\eta}\ \frac{1}{\epsilon_a+\epsilon_b-\omega_A-\epsilon_B-i\eta}\\
&\quad=(2\pi i)^3\frac{1}{\epsilon_a+\epsilon_b-\epsilon_A-\epsilon_B-i\eta}\ .
\end{aligned}\tag{3.34c}$$

Since $\epsilon_a-\epsilon_b-\epsilon_A-\epsilon_B$ is positive definite, the $i\eta$ is irrelevant and Eqs. (3.34b,c) reproduce the result (3.28). The general structure of calculations in the frequency representation is of this form, with one contour integral being required for each independent frequency. For many calculations, the time representation or the time ordered diagrams introduced in a later section are therefore more convenient.

The rules for the contribution to the ground state expectation value of operator R from a linked diagram with p interactions and n_L closed Fermion loops are obtained analogously. Since the operator R is evaluated at time $t=0$, there is no overall factor of T_0. The finite temperature rules are applied with the modification that each independent frequency corresponds to the integral $\int \frac{d\omega}{2\pi}$, each Fermion propagator corresponds to $i\tilde{G}_{0\alpha}(\omega)$ with an additional factor $e^{-i\omega\eta'}$ if it begins and ends at the same vertex, and the overall factor is $\frac{(-i)^p}{S}\ (-1)^{n_L}$. For example, the lowest-order contribution for a two-body operator R is

$$\langle R\rangle\,|_H = $$ 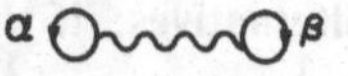

$$
\begin{aligned}
&= \frac{(-i)^0}{2}(-i)^2 \sum_{\alpha\beta} \int \frac{d\omega_\alpha}{2\pi}\frac{d\omega_\beta}{2\pi} i\tilde{G}_{0_\alpha}(\omega_\alpha)e^{i\omega_\alpha\eta'} i\tilde{G}_{0_\beta}(\omega_\beta)e^{i\omega_\beta\eta'}(\alpha\beta|v|\alpha\beta)\\
&= \frac{1}{2}\sum_{\alpha\beta} n_\alpha n_\beta (\alpha\beta|v|\alpha\beta)\\
&= \frac{1}{2}\sum_{AB}(AB|v|AB)
\end{aligned}
\tag{3.35}
$$

where Eq. (3.33) was used to evaluate the frequency integrals.

The frequency representation of the many-particle Green's function

$$
\begin{aligned}
i\tilde{G}^{(n)}(\alpha_1\omega_1,\ldots\alpha_n\omega_n \mid \alpha'_1\omega'_1,\ldots\alpha'_n\omega'_n) &= \int dt_1\ldots dt_n dt'_1\ldots dt'_n\\
&\times e^{i\sum_{j=1}^{n}(\omega_j t_j - \omega'_j t'_j)} iG^{(n)}(\alpha_1 t_1,\ldots\alpha_n t_n \mid \alpha'_1 t'_1,\ldots\alpha'_n t'_n)
\end{aligned}
\tag{3.36}
$$

is evaluated by the same modification of the finite temperature rules. For a diagram with r interactions, n_L closed Fermion loops and permutation P such that the Fermion lines originating at (α_i,ω_i) terminate at $(\alpha'_{Pi},\omega'_{Pi})$, the contribution to $i^n\tilde{G}_n$ contains independent frequency integrals $\int\frac{d\omega}{2\pi}$, Fermion propagators $i\tilde{G}^0_\alpha(\omega)$ and overall factor $(-i)^r(-1)^P(-1)^{n_L}$. For example, the Hartree contribution to the one-particle Green's function yields

$$i\tilde{G}(\beta,\omega \mid \alpha,\omega)\,|_H = $$ 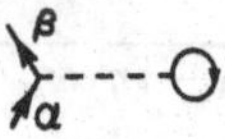

$$
\begin{aligned}
&= (-i)(-1)^1\sum_\delta \int \frac{d\omega_\delta}{2\pi} iG_{0_\alpha}(\omega) iG_{0_\beta}(\omega) iG_{0_\delta}(\omega_\delta)e^{i\omega_\delta\eta}(\beta\delta|v|\alpha\delta)\\
&= G_{0_\alpha}(\omega)G_{0_\beta}(\omega)\sum_D \langle\beta D|v|\alpha D\rangle\ .
\end{aligned}
\tag{3.37}
$$

To relate the self-energy to irreducible contributions to the one-body Green's function, we need to recognize a sign difference arising between the finite and zero-temperature cases as a result of our other sign conventions. Physically, the self-energy Σ represents the energy associated with propagation in the medium and should be defined with a sign such that the combination $H_0+\Sigma$ enters into the total Green's functions. Since at finite temperature $G_0{}^{-1}=\partial_\tau+H_0$, the self-energy was defined in Eqs. (2.178) and (2.181) so that $G^{-1}(\tau)=G_0^{-1}(\tau)+\Sigma=\partial_\tau+H_0+\Sigma$ from which it followed that $G(\tau)=G_0(\tau)-G_0(\tau)\Sigma G(\tau)$. (Note that since we are only interested in an overall sign, the matrix indices on G and Σ are suppressed and the labels (τ) and (t) are only used to distinguish the finite and zero temperature cases, respectively.) At zero temperature, $G_{0_\alpha}^{-1}(t)=i\frac{\partial}{\partial t}-\epsilon_\alpha$ with ϵ_α entering with the opposite sign so that the appropriate definition of the self energy is

$$
\begin{aligned}
G^{-1}(t) &= G_0^{-1}(t)-\Sigma(t)\\
&= i\frac{\partial}{\partial t}-H_0-\Sigma
\end{aligned}
\tag{3.38a}
$$

and the Dyson equation is

$$G(t) = G_0(t) + G_0(t)\Sigma(t)G(t) \quad . \tag{3.38b}$$

The self energy, Σ, is thus defined in terms of one-particle irreducible amputated Green's function diagrams as before, but without the overall minus sign. The result for the Hartree insertion, Eq. (3.37) is consistent with the general argument, since after amputating G_α and G_β, the contribution to the self-energy $\Sigma_{\alpha\beta}$ is the correct Hartree expression $\sum_D (\beta D|v|\alpha D)$.

For future reference, it is also useful to calculate the second-order direct contribution to the self-energy. The second-order contribution to the Green's function is

$$i\tilde{G}(\alpha',\omega \mid \alpha,\omega)\,\Big|_D^{(2)} =$$

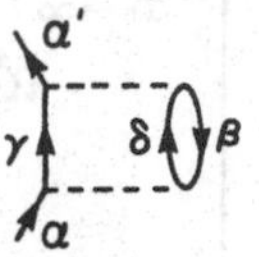

$$\begin{aligned} &= iG_{0'_\alpha}(\omega)iG_{0_\alpha}(\omega)(-i)^{r=2}(-1)^{n_L=1}\sum_{\beta\gamma\delta}\int\frac{d\omega_\gamma}{2\pi}\,\frac{d\omega_\delta}{2\pi}iG_{0_\gamma}(\omega_\gamma) \\ &\quad\times iG_{0_\delta}(\omega_\delta)iG_{0_\beta}(\omega_\gamma+\omega_\delta-\omega)(\alpha'\beta|v|\gamma\delta)(\gamma\delta|v|\alpha\beta) \qquad (3.39a) \\ &= iG_{0_{\alpha'}}(\omega)iG_{0_\alpha}(\omega)\Sigma(\alpha',\omega \mid \alpha,\omega)\,\big|_{2D} \quad . \end{aligned}$$

Separating the two non-vanishing contributions $\{\alpha\gamma\} = \{AB\}$, $\beta = B$ and $\{\gamma\delta\} = \{AB\}$, $\beta = b$ and evaluating the ω integrals as in Eq. (3.34) yields the result

$$\Sigma(\alpha'\omega \mid \alpha,\omega)\,\Big|_D^{(2)} = \sum_{\substack{ab \\ B}}\frac{(\alpha'B|v|ab)(ab|v|\alpha B)}{\omega+\epsilon_B-\epsilon_a-\epsilon_b+i\eta} - \sum_{\substack{AB \\ b}}\frac{(\alpha'b|v|AB)(AB|v|\alpha b)}{-\omega-\epsilon_b+\epsilon_A+\epsilon_B+i\eta} \quad . \tag{3.39b}$$

The correspondence between finite and zero temperature rules is summarized in Table 3.1. The finite temperature rules enumerated in Chapter 2 showed in detail how to evaluate observables using the diagram elements and factors displayed in the left column. When these elements and factors are replaced by the entries on the right, the corresponding zero temperature observables are obtained.

BOSONS

The ground state of a non-interacting N-Boson system is a Bose condensate

$$|\Phi_B(N)\rangle = \frac{(a_0^\dagger)^N}{\sqrt{N!}}|0\rangle \tag{3.40}$$

where $a_0^\dagger$ is a creation operator for the lowest single-particle state of the non-interacting Hamiltonian H_0. By Eqs. (1.63b) and (1.79a),

$$\begin{aligned} a_0|\Phi_B(N)\rangle &= \sqrt{N}|\Phi_B(N-1)\rangle \\ a_0^\dagger|\Phi_B(N)\rangle &= \sqrt{N+1}|\Phi_B(N+1)\rangle \end{aligned} \tag{3.41}$$

$T \neq 0$	$T = 0$ Fermions
Elements of Diagrams	
$G^0_\alpha(\tau) = e^{-(\epsilon_\alpha-\mu)\tau}[\theta(\tau-\eta)(1+\varsigma\tilde{n}_\alpha)$ $+\varsigma\theta(-\tau+\eta)\tilde{n}_\alpha]$	$iG^0_\alpha(t) = e^{-i\epsilon_\alpha t}[\theta(t-\eta)(1-n_\alpha)$ $-\theta(-t+\eta)n_\alpha]$
$\tilde{n}_\alpha = \frac{1}{e^{\beta(\epsilon_\alpha-\mu)}-\varsigma}$	$n_\alpha = \begin{cases} 1 & \alpha \leq N \\ 0 & \alpha > N \end{cases}$
$\int_0^\beta d\tau$	$\lim_{T_0\to\infty} \int_{-T_0/2}^{T_0/2} dt \qquad [t = (1-i\eta)\tilde{t}]$
$\tilde{G}^0_\alpha(\omega_n) = \frac{-1}{i\omega_n-(\epsilon_\alpha-\mu)}$	$i\tilde{G}^0_\alpha(\omega) = \frac{i}{\omega-\epsilon_\alpha+i\eta\,\mathrm{sgn}\,\epsilon_\alpha}$
$e^{i\omega_n\eta}$ for equal times	$e^{i\omega\eta}$ for equal times
$\sum_{\omega_n}\left[\omega_n = \begin{Bmatrix} 2n \\ 2n+1 \end{Bmatrix}\frac{\pi}{\beta}\right]$	$\int \frac{d\omega}{2\pi}$
Factors for Observables	
$\Omega - \Omega_0 : \frac{1}{\beta}\frac{(-1)^{m-1}}{S}\varsigma^{n_L}$	$E_0 - W_0 : \lim_{T_0\to\infty}\frac{1}{T_0}\frac{(-i)^{m-1}}{S}(-1)^{n_L}$
$\langle R\rangle : \frac{(-1)^m}{S}\varsigma^{n_L}$	$\langle R\rangle : \frac{(-i)^m}{S}(-1)^{n_L}$
$G_n : (-1)^m\varsigma^P\varsigma^{n_L}$	$iG_n : (-i)^m(-1)^P(-1)^{n_L}$
Definitions	
$m =$ number of interactions $n_L =$ number of closed propagator loops $S =$ symmetry factor $\varsigma^P =$ sign of permutation connecting incoming and outcoming external lines	

Table 3.1 Summary of Correspondence Between Finite and Zero Temperature Diagram Rules

so that a_0 and $a_0^\dagger$ yield factors depending on the occupation of the ground state. Thus, the expectation value of an arbitrary product of operators in the Bose ground state cannot be separated into independent products of contractions and there is no zero-temperature Wick's theorem as in the Fermion case. Technically, the point at which our previous derivations for Fermions cannot be generalized to Bosons is the construction of a canonical transformation analogous to Eq. (3.19) such that all the new annihilation operators annihilate the ground state.

There are two options for treating Bosons. The first method treats the Bose condensate as a classical c-number field (Bogoliubov, 1947; Hugenholtz and Pines, 1959). For large N, there is negligible difference between $\sqrt{N}$ and $\sqrt{N-m} = \sqrt{N} - \frac{m}{2}N^{-1/2} + 0(N^{-3/2})$ so in the thermodynamic limit one may define new operators

$$b_0 = \frac{1}{\sqrt{\mathcal{V}}}\, a_0 \qquad b_0^\dagger = \frac{1}{\sqrt{\mathcal{V}}}\, a_0^\dagger \tag{3.42}$$

which eliminate the problem of factors depending on the occupation of the condensate. Since

$$\lim_{\mathcal{V}\to\infty}\left[b_0, b_0^\dagger\right] = \lim_{\mathcal{V}\to\infty}\frac{1}{\mathcal{V}} = 0 \tag{3.43}$$

and

$$\begin{aligned} b_0|\Phi_B(N-m)\rangle &= \sqrt{\frac{N-m}{\mathcal{V}}}|\Phi_B(N-m-1)\rangle \\ &\to \sqrt{\rho}|\Phi_B(N-m-1)\rangle \\ b_0^\dagger|\Phi_B(N-m)\rangle &= \sqrt{\frac{N-m+1}{\mathcal{V}}}|\Phi_B(N-m+1)\rangle \\ &\to \sqrt{\rho}|\Phi_B(N-m-1)\rangle \end{aligned} \tag{3.44}$$

b_0 and $b_0^\dagger$ may be replaced by the c-number $\sqrt{\rho}$. All the remaining operators $a_i^\dagger$ $i > 0$ annihilate on the state $|\Phi_B\rangle$ so that Wick's theorem may be obtained as before. The resulting diagram expansion is quite similar to that obtained in Chapter 2 for a Bose condensate and contains four kinds of vertices with all possible combinations of a's and b's. For example, $a^\dagger a^\dagger aa$ yields a conventional vertex connecting to four propagators 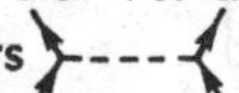 whereas $a^\dagger b^\dagger bb$ has only one propagator and three "condensate" lines 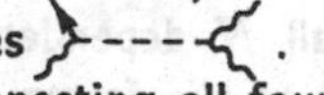. The general expansion is obtained by enumerating the diagrams connecting all four kinds of vertices and a simple special case is treated in Problem (3.3).

A second alternative which is simpler and more general is to make the Boson problem into an equivalent Fermion problem by introducing a formal spin variable with spin degeneracy equal to the particle number (Gentile, 1940, 1942; Brandow, 1971). This method has the advantage that it is valid for any particle number, does not involve the bookkeeping associated with four different interaction vertices, and involves only trivial modifications of the Fermion rules derived in the previous section.

To avoid confusion with the physical spin and to make the language more colorful, let us call the formal spin variable color. Each physical single-particle state now has N degenerate substates with different color projections. We now treat this augmented problem, with both physical and color variables, as a many-Fermion problem and use zero-temperature perturbation theory to find the totally antisymmetric ground state. The non-interacting ground state, Φ_0, is a determinant with all particles in the lowest physical single-particle state and having each color projection occupied once. Thus, Φ_0 is the product of an antisymmetric color-singlet wave function times a wave function which is totally symmetric in the physical variables. The Hamiltonian in the enlarged

Fermion space is the product of the physical Hamiltonian and the unit operator in color space, $H = H(x)1_{\text{color}}$. Since the unit operator 1_{color} cannot change the color wave function, the interacting ground state $|\Phi\rangle = \lim_{Im\, T_0 \to -\infty} e^{-iHT}|\Phi_0\rangle$ must also factorize into the product of an antisymmetric color singlet times a totally symmetric wave function in the physical space corresponding to the Boson ground state. The only assumption, that $e^{-iHT}|\Phi_0\rangle$ yields the ground state instead of an excited eigenstate, is a reflection of the ever present assumption that we have selected a non-interacting ground state which is not orthogonal to the true ground state.

The Boson diagram rules are extremely simple. Each propagator is summed over physical single-particle labels and over color projections. Since H contains the diagonal unit operator in color space, there can be no color flips and the color projections must be the same for all Fermion propagators comprising a closed loop. When propagators are summed over all physical quantum numbers and color projections, each closed propagator thus acquires a factor of N and there is no other remnant of the artificial color variables. Since the non-interacting Boson ground state has only a single physical orbital occupied, the sum over N occupied orbitals $\{A\}$ for Fermions is replaced by a single occupied orbital which we shall denote by $\{0\}$. The sum over unoccupied orbitals is denoted as a sum over an infinite set of labels $\{a\}$ as for Fermions. Thus, there are only two changes relative to Fermions: each closed loop contributes a factor N and the sum over occupied states $\{A\}$ is replaced by a single state $\{0\}$.

Several comments are useful concerning the N-dependence and signs. Direct and exchange graphs now differ by a factor of N. For example, the Hartree-Fock diagrams, Eq. (3.27) now yield

$$E_0 - W_0\,|_{HF} = \lim \frac{i}{T_0}\Big[\quad\Big]$$

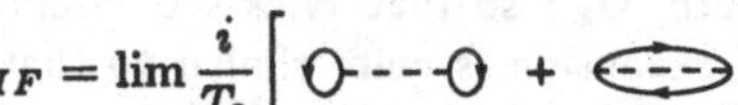

$$= \frac{1}{2}\Big[N^2\langle 00|v|00\rangle - N\langle 00|v|00\rangle\Big] \quad . \tag{3.45}$$

For an infinite system in the thermodynamic limit, the overall N dependence must be determined by considering the volume dependence as well as the formal N dependence. Each interaction vertex contributes a factor $\frac{1}{\mathcal{V}}$ as before, so the Hartree energy goes as $\frac{N^2}{\mathcal{V}} = N\rho$. Counting independent momenta is now different than for Fermions, however, since occupied states now correspond to momentum $k = 0$. For example, the second order diagram 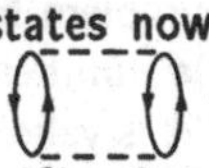 will have momenta $0, 0, \vec{k}$, and $-\vec{k}$ corresponding to a single sum $\sum_k \to \frac{\mathcal{V}}{(2\pi)^3}\int dk$ so that the two interactions, one momentum sum, and two closed loops yield $\frac{\mathcal{V}}{\mathcal{V}^2}N^2 = N\rho$.

Whereas the minus signs in the Boson rules at first appear unintuitive, they play a very simple role in generating the factors $\sqrt{N-m}$ arising from successive depletion of the condensate. Consider, for example, evaluating the matrix element arising from four potential interactions

$$M = \langle\Phi(N)\,|\,a_0^\dagger a_0^\dagger a_1 a_2\;\; a_0^\dagger a_0^\dagger a_3 a_4\;\; a_4^\dagger a_3^\dagger a_0 a_0\;\; a_2^\dagger a_1^\dagger a_0 a_0\,|\,\Phi(N)\rangle \tag{3.46a}$$

by summing contractions. By successive application of Eq. (3.41), the result must be $M = N(N-1)(N-2)(N-3)$. Since the non-vanishing contractions of the operators

a_i $i \neq 0$ are unique, we may represent M by the graph

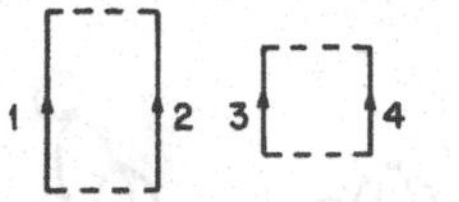

(3.46b)

and enumerate all the ways of joining the ends of the upper vertices and lower vertices with downgoing condensate propagators. The 4! contractions generate the following topologies the indicated number of times:

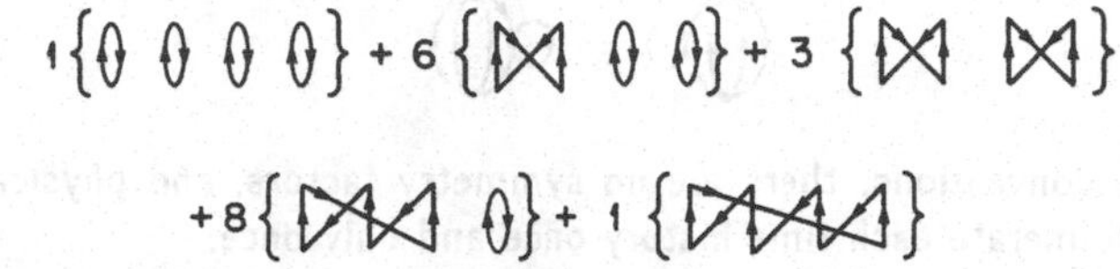

(3.46c)

Note that in (3.46c) the dashed interaction lines are omitted and all distinct permutations of the labels are counted. Including a factor of (-1) and N for each closed loop, as required by the Boson diagram rules, we obtain $M = N^4 - 6N^3 + 11N^2 - 6N = N(N-1)(N-2)(N-3)$ as required. Thus, the Fermion minus signs combined with the factor of N from the color sums do just what is required to generate the correct Bose factors. A detailed application of these rules to an interacting Bose gas is given in Problem 3.3.

3.2 TIME-ORDERED DIAGRAMS

It is often useful to explicitly separate the particle components $e^{-i\epsilon_\alpha(t-t')}\theta(t-t')$ and hole components $-e^{-\epsilon_A(t-t')}\theta(t'-t)$ of the zero temperature propagator $G_{0_\alpha}(t-t')$ which are distinguished by the relative time order of t and t'. This may be done by rewriting each term in the expansion of the exponential of the interaction in the form

$$\frac{1}{n!}\int_{\frac{-T_0}{2}}^{\frac{T_0}{2}} dt_n \int_{\frac{-T_0}{2}}^{\frac{T_0}{2}} dt_{n-1} \ldots \int_{\frac{-T_0}{2}}^{\frac{T_0}{2}} dt_1 \langle v(t_n)v(t_{n-1})\ldots v(t_1)\rangle$$

$$= \int_{\frac{-T_0}{2}}^{\frac{T_0}{2}} dt_n \int_{\frac{-T_0}{2}}^{t_n} dt_{n-1} \ldots \int_{\frac{-T_0}{2}}^{t_2} dt_1 \langle v(t_n)v(t_{n-1})\ldots v(t_1)\rangle \quad . \tag{3.47}$$

Thus, we enumerate time-ordered diagrams representing all contractions with a specific relative time order and omit the factor of $\frac{1}{n!}$.

Adopting the convention that time increases in the upward direction, an upgoing line will always denote a particle state  $= e^{-i\epsilon_\alpha(t'-t)}$ and a downgoing line will denote a hole state 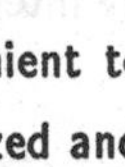 $= e^{-i\epsilon_A(t'-t)}$. It will be convenient to use the Hugenholtz convention, in which all matrix elements are antisymmetrized and all $L-R$ exchanges are accounted for by $\frac{1}{2^{n_e}}$ where n_e is the number of equivalent pairs of lines. Two diagrams are distinct if they cannot be deformed to coincide completely in topology and direction of arrows while maintaining the same relative time ordering. Thus, in contrast to the Feynman diagrams (2.101b), the following diagrams are distinct

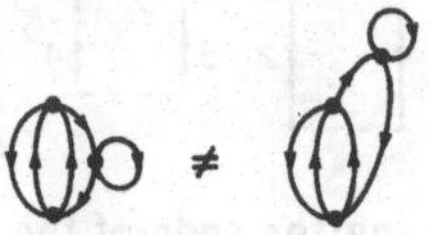

(3.48a)

whereas in both cases the following diagrams are equivalent

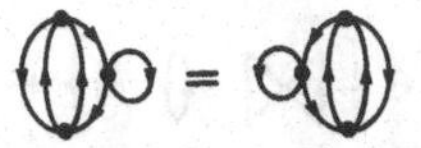

(3.48b)

With these conventions, there are no symmetry factors, and physically the diagrams simply enumerate each time history once and only once.

The integrals over relative time may be performed by breaking each propagator into a product of factors involving all intermediate times. For example, if t_n is later than t_m with the same number of intervening times, particle and hole propagators are written as follows:

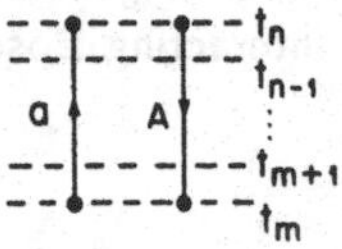

$$e^{-i\epsilon_a(t_n-t_m)} = \prod_{k=m}^{n-1} e^{-i\epsilon_a(t_{k+1}-t_k)}$$

$$e^{-i\epsilon_A(t_m-t_n)} = \prod_{k=m}^{n-1} e^{-i(-\epsilon_A)(t_{k+1}-t_k)} \quad . \tag{3.49}$$

Note that the hole energies ϵ_A enter into Eq. (3.49) with the opposite sign from the particle energies because the relative time entering the propagator is the time at which the arrow terminates minus the time at which it begins. The segment of the total propagator for an arbitrary diagram between times t_m and t_{m+1} then contributes the factor

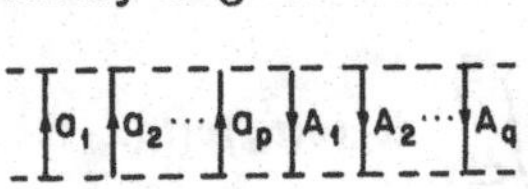

$$= e^{-iS_m(t_{m+1}-t_m)} \tag{3.50a}$$

where

$$S_m = \sum_{i=1}^{p} \epsilon_{a_i} - \sum_{i=1}^{q} \epsilon_{A_i} \quad . \tag{3.50b}$$

Physically, S_m is just the excitation energy for a state with p particles and q holes relative to the non-interacting ground state between times t_m and t_{m+1}. For ground state observables $p = q$ so that the intermediate state is a p-particle, p-hole excitation and with our sign convention that $\epsilon_a > 0$ and $\epsilon_A < 0$, S_m is positive.

Consider now the integral over all the ordered times in an arbitrary linked diagram with n interactions

$$I = \int_{-\frac{T_0}{2}}^{\frac{T_0}{2}} dt_n \prod_{m=1}^{n-1} \int_{-\frac{T_0}{2}}^{t_{m+1}} dt_m \, e^{-iS_m(t_{m+1}-t_m)} \quad . \tag{3.51a}$$

Performing the integral over each t_m, $1 \leq m \leq n-1$, along the contour C_2 in the limit $T_0 \to \infty$ yields

$$\lim_{T_0\to\infty} \int_{-\frac{T_0}{2}}^{\tilde{t}_{m+1}} d\tilde{t}_m \; e^{-iS_m(\tilde{t}_{m+1}-\tilde{t}_m)(1-i\tilde{\eta})} = \lim_{T_0\to\infty} \frac{1-e^{-iS_m(\tilde{t}_{m+1}+\frac{T_0}{2})(1-i\tilde{\eta})}}{iS_m(1-i\tilde{\eta})}$$

$$\to \frac{-i}{S_m - i\eta} \qquad (3.51b)$$

where $\eta > 0$ since S_m and $\tilde{\eta}$ are positive. Hence, the time integrals of the propagator arising in the ground state energy yield

$$\lim_{T_0\to\infty} \frac{1}{T_0} I = \prod_{m=1}^{n-1} \frac{-i}{S_m - i\eta} \quad . \qquad (3.51c)$$

In addition to antisymmetrized Hugenholtz matrix elements for each interaction vertex $\{\alpha\beta|v|\gamma\delta\} = (\alpha\beta|v|(|\gamma\delta) - |\delta\gamma)$ and Eq. (3.51c), the expansion for the ground state energy $E_0 - W_0$ contains the usual factor $(-i)^{n-1}(-1)^{n_L}$ from Table 4.1, the factor $\frac{1}{2^{n_e}}$ from the Hugenholtz rules which together with the explicit enumeration of all time orders completely replaces the symmetry factor S, and a factor $(-1)^{n_h}$ to account for the minus sign associated with each of the n_h hole line propagators. Each energy denominator is positive definite, so that the $i\eta$ is superfluous. Thus, the final formula for the ground state energy is

$$E_0 - W_0 = \sum_{\substack{\text{all linked} \\ \text{diagrams}}} \frac{(-1)^{n_L+n_h}}{2^{n_e}} \prod \frac{1}{-(\sum_a \epsilon_a - \sum_A \epsilon_A)} \prod \{\alpha\beta \mid v \mid \gamma\delta\} \quad . \qquad (3.52)$$

A global energy denominator $-(\sum_a \epsilon_a - \sum_A \epsilon_A)$ is computed between each successive interaction including all the particle lines and hole lines present in that time interval and n_h and n_e denote the number of hole lines and equivalent pairs, respectively. An antisymmetrized matrix element is included for each interaction, and n_L is obtained by counting the number of closed loops in the corresponding diagram with direct matrix elements at each vertex.

Historically, the expansion for the ground state energy in terms of linked time-ordered diagrams, Eq. (3.52), was first obtained by Goldstone (1957) using the Dyson expansion, and is known as the Goldstone expansion. Time-ordered diagrams are therefore also frequently called Goldstone diagrams. It is instructive to compare this result with Brillouin-Wigner perturbation theory, which has the wrong N dependence in higher orders (see Problem 3.6). As shown in Problem 3.7, Rayleigh-Schrödinger perturbation theory may be obtained from Brillouin-Wigner perturbation theory by systematically expanding energy demoninators $\frac{1}{E_0-H_0} = \frac{1}{W_0-H_0}\sum_m \left(\frac{W_0-E_0}{W_0-H_0}\right)^m$ and has the proper N-dependence in each order. As observed by Brueckner (1959), the terms in Rayleigh-Schrödinger perturbation theory which would correspond to unlinked diagrams cancel in low orders of perturbation theory (see Problem 3.8) and it is the generality of this cancellation which is established by Goldstone's theorem or Eq. (3.52).

As an example of the evaluation of a Goldstone diagram for the ground state energy we will evaluate the second-order contribution which includes both direct and exchange terms

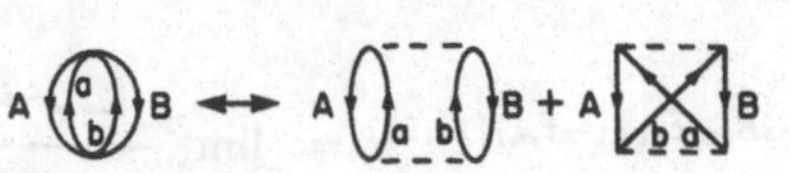

(3.53)

There are two equivalent pairs, two hole lines, and the associated direct diagram has two closed loops. Hence,

$$E_0 - W_0|_2 = \frac{1}{4}\sum_{\substack{ab\\AB}} \frac{(AB|v[|ab) - |ba)](ab|v|[|AB) - |BA)]}{-(\epsilon_a + \epsilon_b - \epsilon_A - \epsilon_B)}$$
$$= -\frac{1}{2}\sum_{\substack{ab\\AB}} \frac{|(AB|v|ab)|^2 - |(AB|v|ba)|^2}{\epsilon_a + \epsilon_b - \epsilon_A - \epsilon_B} . \tag{3.54}$$

The direct term agrees with the previous, more lengthy calculations in Eq. (3.28) and (3.34).

The rules for the expansion of the ground state expectation value of an operator are analogous and are left as an exercise. One new feature arises in the evaluation of Green's functions using time-ordered graphs. In addition to r interactions, a one-particle Green's function contains a creation operator at some time t_i and an annihilation operator at some time t_f. For the case $t_f > t_i$ the general form of the relative times and a specific example are shown below.

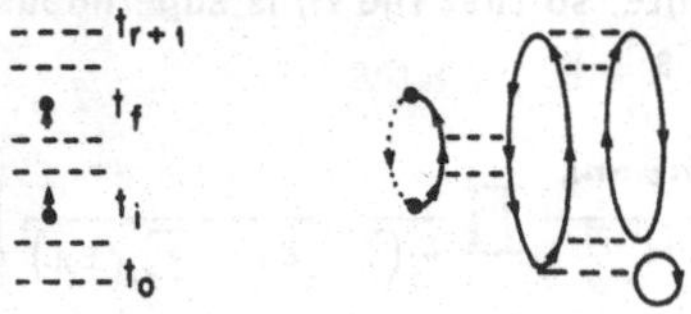

(3.55)

Including all factors as in the case of $E_0 - W_0$ and Fourier transforming from relative times $t_f - t_i$ to frequency ω in the usual way, the one-particle Green's function may be written

$$iG_i(\omega) = \frac{(-i)^r(-)^{n_L+n_h}}{2^{n_e}} \int_{-\frac{T_0}{2}}^{\frac{T_0}{2}} dt_{r+1} \prod_{m=0}^{r} \int_{-\frac{T_0}{2}}^{t_{m+1}} dt_m$$
$$\times e^{i\omega(t_f - t_i)} e^{-iS_m(t_{m+1}-t_m)} \prod\{\alpha\beta|v|\gamma\delta\} . \tag{3.56}$$

Breaking $e^{-i\omega(t_f - t_i)}$ into a product of factors for each time interval between t_f and t_i as in Eq. (3.49), we see that ω may be grouped with S_m for all $t_i \le m \le t_f$ and that it enters precisely as a hole line. Similarly, if $t_f < t_i$, ω combines with S_m at intermediate times as a particle line. Thus, performing the time integrals as in Eq. (3.51b) and retaining the $i\eta$ factors, the general result for the one-particle Green's function for either time order is

$$G(\omega) = \sum_{\substack{\text{all linked}\\\text{diagrams}}} \frac{(-1)^{n_L+n_h}}{2^{n_e}} \prod_{m=0}^{r} \frac{1}{-\tilde{S}_m + i\eta} \prod\{\alpha\beta|v|\gamma\delta\} \tag{3.57a}$$

where

$$\tilde{S}_m = \begin{cases} \sum_a \epsilon_a - \sum_A \epsilon_A - \operatorname{sgn}(t_f - t_i)\omega & \text{intervals between } t_f \text{ and } t_i \\ \sum_a \epsilon_a - \sum_A \epsilon_A & \text{intervals outside } t_f \text{ and } t_i \end{cases} . \tag{3.57b}$$

A simple mnemonic to correctly include ω in the appropriate demoninator is to add a ficticious propagator, denoted by the dotted line in (3.55), running from the annihilation operator at t_f to the creation operator at t_i. Depending upon the direction of the arrow, it is treated like a particle or hole with single-particle energy ω in computing energy denominators. (It must, however, be ignored when calculating n_L, n_h and n_e).

To illustrate these rules and to show how several time-ordered graphs reproduce a single Feynman graph, we will calculate the second-order one-particle irreducible diagrams contributing to the one-particle Green's function. Consider first the following time order which includes a two-particle – one-hole intermediate state:

$G(d'\omega|d\omega)\,|_{2p1h} =$ 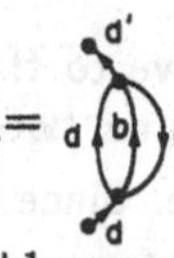

$$= \frac{(-1)^{1+1}}{2^1} \sum_{abB} \frac{\{d'B \mid v \mid ab\}\{ab \mid v \mid dB\}}{(-\epsilon_{d'} + \omega + i\eta)\,(-\epsilon_a - \epsilon_b + \epsilon_B + \omega + i\eta)\,(-\epsilon_d + \omega + i\eta)}$$

$$= G_{0d'}(\omega) G_{0d}(\omega) \frac{1}{2} \sum_{abB} \frac{\{d'B \mid v \mid ab\}\{ab \mid v \mid dB\}}{\omega + \epsilon_B - \epsilon_a - \epsilon_b + i\eta} \tag{3.58a}$$

where the sign has been determined by considering the direct term which has one closed loop. Separating the direct and exchange terms, the self energy is

$$\Sigma(d'\omega|d\omega)\,|_{2p1h} = \sum_{abB} \frac{(d'B \mid v \mid ab)(ab \mid v \mid dB) - (d'B \mid v \mid ab)(ab \mid v \mid Bd)}{\omega + \epsilon_B - \epsilon_a - \epsilon_b + i\eta} \tag{3.58b}$$

the direct term of which agrees with (3.39b). In this case, the time-ordered rules again provides the result more economically than the sequence of frequency integrals required for Feynman diagrams. In contrast, there are five different time orders involving an intermediate state with two holes and one particle (in addition to d and d') represented by the following diagrams:

$G(d'\omega)|d\omega|_{1p2h} =$ 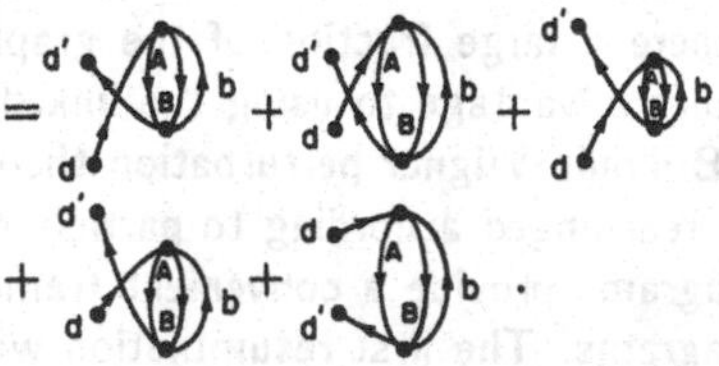

(3.59a)

As shown in Problem 3.10, each term is evaluated as in Eq. (3.12a) and the sum of the five resulting terms yields the self-energy.

$$\Sigma(d'\omega|d\omega)\,|_{1p2h} = -\sum_{aAB} \frac{(d'a \mid v \mid AB)(AB \mid v \mid da) - (d'a \mid v \mid AB)(AB \mid v \mid ad)}{\epsilon_A + \epsilon_B - \epsilon_a - \omega + i\eta} . \tag{3.59b}$$

The overall minus sign relative to Eq. (3.58b) results from the additional hole line. The direct term again agrees with Eq. (3.39b) and this case illustrates how a general class of topologically identical time-ordered diagrams combine to give a simple result.

Whereas most of the diagrams appearing in the Goldstone expansion have the direct physical interpretation of enumerating a time history of the N-particle system, several special cases arise which merit comment. One case is the presence of diagrams which appear to violate the Pauli principle. Although there is a temptation to eliminate such graphs on physical grounds, our derivation shows that they must be included. Their role is easily understood by considering the following pair of linked and unlinked graphs which have a propagator for state a appearing twice at the same time:

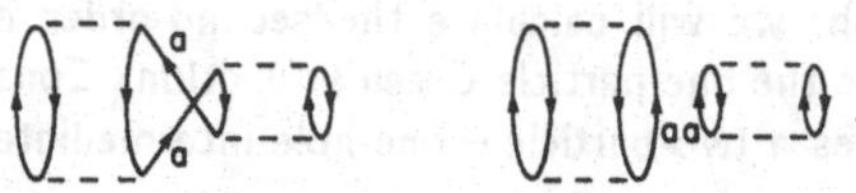

(3.60)

Because the right-hand graph has an additional loop relative to the left-hand graph and all matrix elements and denominators are identical, the two contributions to $\frac{\langle\Phi_0|e^{-i(\hat{H}_0+\hat{V})T_0}|\Phi_0\rangle}{\langle\Phi_0|e^{-i\hat{H}_0 T_0}|\Phi_0\rangle}$ cancel identically and there is no mistake. Since the exponential of the sum of all linked diagrams necessarily generates all possible unlinked diagrams irrespective of labels, including the right-hand graph in (3.60), it is essential to retain the left-hand graph to preserve the physical cancellation. Similarly, there exist canceling pairs of graphs such as the following

(3.61)

for which the linked partner has no physical meaning. Once a particle has been scattered out of an occupied state A into an unoccupied state a, it is impossible to have a subsequent interaction involving state A corresponding to the left-hand diagram. However, this unphysical contribution exactly cancels the unlinked Pauli violating diagram on the right, and by the preceding argument, must be retained. Note that no corresponding cancellation arises if A in the left-hand graph is replaced by a particle line, since closed loops beginning and ending at the same vertex as in the right-hand graph are necessarily hole lines. These special cases only arise for graphs involving duplicate labels which occur with relative probability $\frac{1}{N}$ for large systems. For very small systems, where a large fraction of the graphs are in some sense unphysical, there is no particular advantage to using the linked cluster expansion, and it may be preferable to use Brillouin Wigner perturbation theory or some other formalism which is not deliberately rearranged according to particle number.

Goldstone diagrams provide a convenient framework for physically motivated resummations of diagrams. The first resummation we consider here corresponds to the independent particle or mean-field approximation. The use of the Hartree-Fock basis to include the average effect of all the other particles has already been discussed in Problem 1.6 and in Chapter 2 in connection with the use of the Hartree-Fock self-energy to define single particle propagators. In terms of diagrams, it is convenient to decompose H into $H_0 = T + U_{HF}$ and $V = \sum_{ij} v(x_i - x_j) - U_{HF}$ and to enumerate all diagrams containing both the two-body potential v and the one-body potential $-U_{HF}$. Then for every diagram one can draw with the potential $-U_{HF}$, there is a corresponding diagram with a Hartree-Fock self-energy

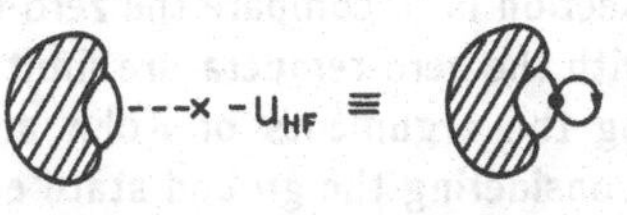

(3.62a)

and U_{HF} may be defined as in Problem 1.6 to identically cancel all Hartree-Fock self-energy insertions

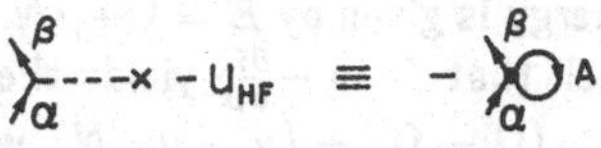

$$\langle\beta|U_{HF}|\alpha\rangle \equiv \sum_A \{BA|v|\alpha A|\} \ . \tag{3.62b}$$

Thus, use of the Hartree Fock potential performs an infinite summation of all possible insertions of the Hartree-Fock self-energy in propagators

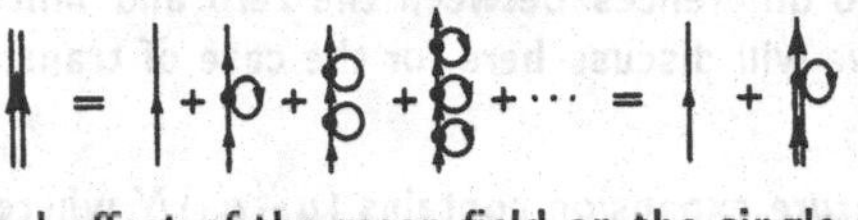

(3.63)

and includes the physical effect of the mean field on the single particle wave functions and propagators.

Just as the independent-particle approximation describes the propagation of a single particle by replacing its interaction with all the other particles by the mean field, the next resummation corresponds to the independent pair approximation which treats the two-body interaction between a pair of particles in the medium while replacing the interaction of the pair with all other particles by the mean field. In terms of diagrams, one sums all orders of ladder diagrams in which two interacting particles rescatter any number of times in intermediate particle states (using Hartree-Fock propagators from the previous resummation). Thus, Goldstone diagrams are reexpressed in terms of the Brueckner reaction matrix

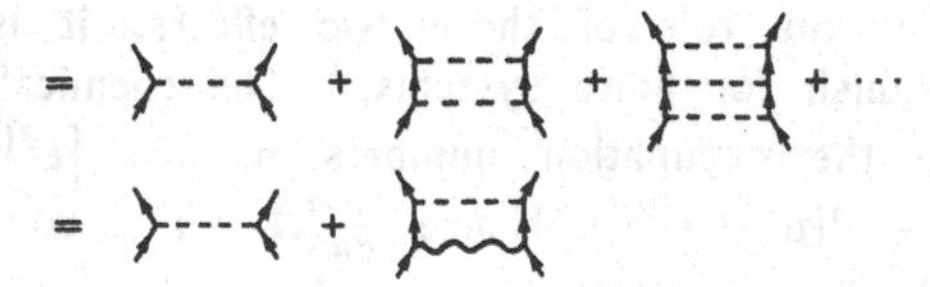

(3.64a)

which satisfies the integral equation

$$(\alpha\beta \mid G(W) \mid \gamma\delta) = (\alpha\beta \mid v \mid \gamma\delta) - \sum_{ab} \frac{(\alpha\beta \mid v \mid ab)(ab \mid G(W) \mid \gamma\delta)}{\epsilon_a + \epsilon_b - W} \ . \tag{3.64b}$$

The hierarchy of resummations may be continued with the solution of the three-body Bethe-Faddeev equation which, using single-particle Hartree-Fock propagators and two-body reaction matrix elements, sums all successive rescatterings of three particles in intermediate particle states. Since time-ordered diagrams clearly separate particles and holes, this sequence of resummations gives rise to an expansion ordered in the number of hole lines. The hole-line expansion is useful in treating dense quantum liquids, such as liquid Helium and nuclear matter.

3.3 THE ZERO-TEMPERATURE LIMIT

The objective of this final section is to compare the zero-temperature perturbation theory derived in this chapter with the zero-temperature limit of the finite temperature theory of Chapter 2. Following the arguments of Kohn and Luttinger (1960), the essential point is evident from considering the ground state energy for Fermions in the lowest order of perturbation theory in which the two theories can differ. A general formal proof may be found in the work of Luttinger and Ward (1960).

At finite temperature, the energy is given by $E = \Omega + \mu N + TS$, where the chemical potential μ must be adjusted such that $N = -\frac{\partial\Omega}{\partial\mu}$ yields the desired particle number. Thus, we must compare $\lim_{T\to\infty}(\Omega - \Omega_0 + (\mu - \mu_0)N)$ with the zero-temperature result $E_0 - W_0$. Since finite temperature propagators with $n_\alpha = \frac{1}{e^{\beta(\epsilon_\alpha - \mu)}+1}$ approach zero-temperature propagators with $n_\alpha = 1$ or 0 and each zero-temperature graph for $E_0 - W_0$ corresponds to the zero-temperature limit of some finite-temperature graph for $\Omega - \Omega_0$, the two expansions agree in most respects.

There are just two differences between the zero and finite temperature theories which, for simplicity, we will discuss here for the case of translationally invariant systems:

1) The finite temperature expansion contains $(\mu - \mu_0)N$ where the chemical potential for the non-interacting system, defined such that $\frac{\partial\Omega_0}{\partial\mu}\big|_{\mu_0} = N$, need not agree with the chemical potential of the interacting system defined by $\frac{\partial\Omega}{\partial\mu}\big|_{\mu} = N$.

2) The finite temperature expansion admits anomalous graphs such as 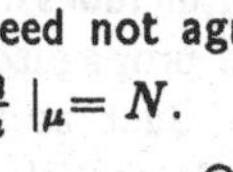 in which the same state appears as both an upgoing and downgoing line. Note that since we have assumed translational invariance throughout this section we will label states by single-particle momenta. At finite temperature, $n_k(1 - n_k)$ does not vanish at the Fermi surface whereas such anomalous graphs are never present in the zero temperature theory.

To understand the role of these two effects, it is useful to note first that they both vanish for finite systems. The chemical potential enters $\Omega - \Omega_0$ only through the occupation numbers $n_k = \left[e^{\beta(\epsilon_k - \mu)} + 1\right]^{-1}$ and as $T \to \infty$, $n_k \to \theta(\mu - \epsilon_\alpha)$. Hence $\frac{\partial}{\partial\mu}(\Omega - \Omega_0)$ contains μ only through $\frac{\partial n_k}{\partial\mu} \to \delta(\mu - \epsilon_k)$. For discrete energies ϵ_k in a finite periodic box, the δ-function is never satisfied so that $\frac{\partial\Omega}{\partial\mu} = \frac{\partial\Omega_0}{\partial\mu}$ and hence $\mu = \mu_0$. Similarly, consider the lowest order (second order in the potential) anomalous diagram

$\Omega_A^{(2)} =$ 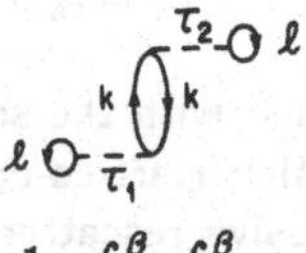

$$= \frac{1}{2\beta}\int_0^\beta\int_0^\beta d\tau_1 d\tau_2 (1 - n_k)(-n_k)\left(\sum_\ell n_\ell\{k\ell \mid v \mid k\ell\}\right)^2$$

$$= -\frac{\beta}{2}(1 - n_k)n_k\, U_k^2 \tag{3.65a}$$

where U_k denotes the Hartree Fock potential

$$U_k = \sum_\ell \{k\ell \mid v \mid k\ell\} n_\ell \ . \tag{3.65b}$$

The feature that an anomalous diagram yields $\beta(1 - n_k)n_k$ is general since oppositely directed propagators of the same k yield the factors $(1 - n_k)(-n_k)$ with no exponential so that integration over relative time yields an additional factor β. In the zero-temperature limit, this factor yields

$$\begin{aligned}\beta n_k(1 - n_k) &= \beta \frac{e^{\beta(\epsilon_k - \mu)}}{(1 + e^{\beta(\epsilon_k - \mu)})^2} \\ &= -\frac{\partial}{\partial \epsilon_k} n_k \\ &\underset{T \to 0}{=\!\!\longrightarrow} \delta(\epsilon_k - \mu) \ .\end{aligned} \tag{3.66}$$

Hence, like the term $N(\mu - \mu_0)$, anomalous diagrams produce δ-functions which are never satisfied for discrete ϵ_k. Therefore, for finite systems the $T \to 0$ limit agrees with the zero-temperature expansion.

For infinite systems, we wish to compare finite and zero-temperature theories in the thermodynamic limit, so we consider the double limit $\lim_{T\to\infty}\{\lim_{V\to\infty}\}$. In this case both the anomalous diagrams and $N(\mu - \mu_0)$ are non-vanishing, but as suggested by the fact that both terms yield the same δ-function, it is possible that they may cancel. The criterion for cancellation is easy to see in lowest order. Let us separate the Grand potential into the following terms:

$$\Omega = \Omega_0 + \Omega_Z + \Omega_A \tag{3.67}$$

where Ω_Z denotes all the diagrams which have zero-temperature counterparts and Ω_A denotes all the remaining anomalous diagrams. Note that the ground state energy calculated in zero-temperature theory may be written

$$E_0 = (\Omega_0 + \Omega_Z - N\mu)_{\substack{\mu=\mu_0 \\ T=0}} \ . \tag{3.68}$$

This follows because $W_0 = \Omega_0 + N\mu \mid_{\substack{\mu=\mu_0 \\ T=0}}$ and each of the diagrams in Ω_Z goes over to its zero temperature counterpart when evaluated with propagators evaluated at $T = 0$ and μ_0. Thus, we may write the difference between the zero-temperature limit of finite-temperature theory and E_0 as

$$\begin{aligned}\Delta &= \lim_{T\to 0}(\Omega + \mu N) - E_0 \\ &= \lim_{T\to 0}\left[\Omega_A \mid_\mu + (\Omega_0 + \Omega_Z + N\mu) \mid_\mu - (\Omega_0 + \Omega_Z + N\mu) \mid_{\mu_0}\right]\end{aligned} \tag{3.69a}$$

where μ and μ_0 satisfy

$$\frac{\partial}{\partial \mu}(\Omega_0 + \Omega_Z + \Omega_A) \mid_\mu = N = \frac{\partial}{\partial \mu}\Omega_0 \mid_{\mu_0} \ . \tag{3.69b}$$

Since Ω_Z is first-order in the potential and Ω_A is second order, it is straightforward to use (3.69b) to solve for $\mu - \mu_0$ to first order and to use this result to obtain Δ to second order. The details are explained in Problem 3.2 and denoting the order in the potential to which quantities are evaluated by superscripts, the result is

$$\Delta^{(2)} = \lim_{T\to 0}\left\{\Omega_A^{(2)}(\mu_0) - \frac{1}{2}\frac{\left(\frac{\partial\Omega_z^{(1)}(\mu_0)}{\partial\mu_0}\right)^2}{\frac{\partial^2\Omega_0}{\partial\mu_0^2}}\right\}$$

$$= -\frac{1}{2}\sum_k \delta(\epsilon_k - \mu_0)\langle(U - \langle U\rangle)^2\rangle \tag{3.70a}$$

where the average of any function of k at the Fermi surface is defined by

$$\langle F(k)\rangle = \frac{\sum_k \delta(\epsilon_k - \mu_0)U(k)}{\sum_k \delta(\epsilon_k - \mu_0)} \tag{3.70b}$$

and U denotes the Hartree-Fock potential, Eq. (3.65b).

The physical essence of the result, Eq. (3.70) is the relation between the symmetry of the non-interacting zero-temperature ground state and the exact ground state. When $\Delta = 0$, both theories yield the physical ground state. When $\Delta < 0$, the $T \to 0$ limit of the finite-temperature theory yields the ground state whereas the zero temperature theory produces an excited eigenstate with some incorrect symmetry it has inherited from the non-interacting ground state. Since Δ measures the fluctuations in U_k at the unperturbed Fermi surface, it will vanish for a spherical Fermi surface and a Hartree-Fock potential U_k which depends only on $|k|$. This will occur, for example, when ϵ_k is the kinetic energy or any single-particle energy depending on $|k|$ and when $v(r_i - r_j)$ is a central potential. The cancellation has also been demonstrated for spin 1/2 particles interacting with tensor forces.

The feature which can destroy the cancellation and produce negative Δ is a difference in the symmetry of the non-interacting problem, as reflected in the Fermi surface, and the interacting problem, as reflected in U. A simple example is a system of spin-1/2 particles in an external magnetic field. Taking H_0 as the kinetic energy, the Fermi surface is two equal Fermi spheres for spin up and spin down. Repeating the derivation of Eq. (3.70) for a one-body perturbation $-\mu B\sigma_Z$ simply replaces $U(k)$ by the matrix element of the one-body operator, $-\mu Bm$ where m is the spin projection. In this case, the variance of μBm evaluated over the two Fermi spheres with $m = \pm 1$ is non-vanishing. This corresponds to the fact that the perturbation cannot flip spins so there is no way to get from the non-interacting state with two equal Fermi spheres to the true interacting ground state with unequal Fermi spheres. We will return to this example again in Chapter 5.

In conclusion, it is evident that the decomposition $H_0 = T + U$, $V = v - U$ plays a different role in the finite and zero temperature theories. In both cases, V must be small in some sense for the perturbation expansion to be useful. In addition, at zero-temperature H_0 defines $|\Phi_0\rangle$ from which we determine $|\Psi\rangle = \lim_{Im\, T_0 \to -\infty} e^{-iT_0(H_0+V)}|\Phi_0\rangle$. Thus $|\Psi\rangle$ may be trapped in the space of some symmetry selected by $|\Phi_0\rangle$ and simply produce the lowest eigenstate in that space. In contrast, the finite temperature theory evaluates a trace which necessarily has access to all possible symmetries and will not become trapped. The ultimate conclusion, then, is that there is nothing fundamentally wrong with the zero-temperature theory. Rather, to do sensible physics, one must pick an intelligent choice for H_0 such that $|\Phi_0\rangle$ has the right symmetries and corresponds to the correct physical phase of the system, and such that $V = v - U$ is sufficiently small to obtain reasonable convergence.

PROBLEMS FOR CHAPTER 3

Several problems are designed to establish connections between the formalism we have presented and traditional treatments, both to deepen understanding of the physics and to facilitate reading the original literature. Problem 1 shows how to obtain the linked-cluster expansion using the Dyson expansion and the Gell-Mann Low theorem. Problems 6-8 demonstrate how Brillouin-Wigner perturbation theory, which has inappropriate dependence on the particle number N, can be rearranged to obtain Rayleigh-Schrödinger perturbation theory which, after the cancellation of unlinked terms, has the correct N-dependence. A little known but interesting property of Brillouin-Wigner perturbation theory is derived in Problem 9. Problems 2 and 10 work out details of results cited in the text associated with the cancellation of anomalous diagrams and combining topologically equivalent time-ordered diagrams. Zero temperature theory is applied to the dilute Bose gas in Problem 3 and to Fermions in one dimension interacting via δ-function forces in Problems 4 and 5.

PROBLEM 3.1 Zero temperature perturbation theory may also be obtained by the following traditional derivation. Let ψ and $H = H_0 + H_1$ denote the wave function and Hamiltonian in the Schrödinger representation, define the evolution operator such that $\psi(t) \equiv U(t,t_0)\psi(t_0)$ and allow H_1 to be time dependent. The corresponding quantities in the interaction representation, denoted in this problem by a circumflex, are defined as follows:

$$\hat{\psi}(t) = e^{iH_0t}\psi(t)$$
$$\hat{U}(t-t_0) = e^{iH_0t}U(t-t_0)e^{-iH_0t}$$
$$\hat{H} = e^{iH_0t}H\,e^{-iH_0t} \quad .$$

a) From the time-dependent Schrödinger equation, show that

$$\frac{d}{dt}\hat{U}(t-t_0) = -i\hat{H}_1(t)\hat{U}(t-t_0)$$

with the boundary condition $\hat{U}(0) = 1$ which may be integrated to obtain

$$\hat{U}(t-t_0) = 1 - i\int_{t_0}^{t} dt'\,\hat{H}_1(t')\hat{U}(t'-t_0) \quad .$$

By iterating, obtain the Dyson expansion

$$\hat{U}(t-t_0) = \sum_{n=0}^{\infty}\frac{(-i)^n}{n!}\int_{t_0}^{t} dt_1 \ldots \int_{t_0}^{t} dt_n\; T(\hat{H}_1(t_1)\hat{H}_1(t_2)\ldots\hat{H}_1(t_n))$$

where T denotes time-ordering.

b) The essential idea is to start with an eigenstate $|\phi_0\rangle$ of H_0 with energy W_0 and turn on the interaction V very slowly so that $|\phi_0\rangle$ adiabatically evolves into an eigenstate of the fully interacting Hamiltonian. It is convenient to parameterize this switching-on by defining $\hat{H}_1 = \lambda e^{-\epsilon|t|}\hat{V}$, and to consider the limit $\epsilon \to 0$. The corresponding evolution operator is defined by $U_\epsilon(t, t_0)$. The Gell-Mann Low theorem (1951) then states that

$$\frac{|\psi_0\rangle}{\langle\phi_0|\psi_0\rangle} = \lim_{\epsilon\to 0}\frac{|\psi_\epsilon\rangle}{\langle\phi_0|\psi_\epsilon\rangle} \equiv \lim_{\epsilon\to 0}\frac{\hat{U}_\epsilon(0,-\infty)|\phi_0\rangle}{\langle\phi_0|\hat{U}_\epsilon(0,-\infty)|\phi_0\rangle}$$

is an eigenstate of $H_0 + \lambda V$, assuming that the right-hand side exists in all orders of perturbation theory. The limit does not exist separately for the numerator, so the denominator is essential to remove an otherwise infinite phase associated with unlinked diagrams. To prove the theorem, show

$$\begin{aligned}(\hat{H}_0 - W_0)\hat{U}_\epsilon(0,-\infty)|\phi_0\rangle &= [\hat{H}_0, \hat{U}_\epsilon(0,-\infty)\,]|\phi_0\rangle \\ &= \sum_{n=1}^{\infty}\frac{(-i)^n}{n!}\int_{-\infty}^{0} dt_1\ldots dt_n\, e^{\epsilon(t_1+t_2+\ldots+t_n)}(-i)\sum_{i=1}^{n}\frac{\partial}{\partial t_i}T[\lambda\hat{V}(t_1)\ldots\lambda\hat{V}(t_n)\,]|\phi_0\rangle \\ &= -\lambda\hat{V}\,\hat{U}_\epsilon(0,-\infty)|\phi_0\rangle + i\epsilon\lambda\frac{d}{d\lambda}\hat{U}_\epsilon(0,-\infty)|\phi_0\rangle\ .\end{aligned}$$

Note that the interaction representation operators are evaluated at $t = 0$ and so are equal to the corresponding Schrödinger operator. Thus, show

$$i\epsilon\lambda\frac{d}{d\lambda}\left[\frac{|\psi_\epsilon\rangle}{\langle\phi_0|\psi_\epsilon\rangle}\right] = [H_0 + \lambda V - E(\epsilon)]\frac{|\psi_\epsilon\rangle}{\langle\phi_0|\psi_\epsilon\rangle}$$

where $E(\epsilon) = W_0 + \frac{\langle\phi_0|\lambda V|\psi_\epsilon\rangle}{\langle\phi_0|\psi_\epsilon\rangle}$. Finally, use the fact that $\lim_{\epsilon\to 0}\frac{|\psi_\epsilon\rangle}{\langle\phi_0|\psi_\epsilon\rangle}$ exists in perturbation theory at $\lambda = 1$ to argue that $\frac{|\psi_0\rangle}{\langle\phi_0|\psi_0\rangle}$ is an eigenstate of $H_0 + V$ with $E = W_0 + \frac{\langle\phi_0|V|\psi_0\rangle}{\langle\phi_0|\phi_0\rangle}$.

(c) Obtain the familiar diagram expansion by substituting the Dyson expansion for $U_\epsilon(0,-\infty)$ in the numerator and denominator of

$$E - E_0 = \lim_{\epsilon\to 0}\frac{\langle\phi_0|\hat{H}_1(0)U_\epsilon(0,-\infty)|\phi_0\rangle}{\langle\phi_0|U_\epsilon(0,-\infty)|\phi\rangle_0\rangle}\ .$$

Use the operator form of Wick's theorem in Problem 2.8 and perform the time integrals over relative times to show that the numerator yields the sum of all diagrams linked to $\hat{H}_1(0)$ times the product of all possible linked diagrams. Similarly, show that the denominator just cancels off the product of all possible linked diagrams and that the final expansion may be written:

$$E - E_0 = \langle\phi_0|V\sum_n\left(\frac{1}{W_0 - H_0}H_1\right)^n|\phi_0\rangle_{\text{linked}}\ .$$

Note that in contrast to the derivations in the text, this derivation is not restricted to the ground state, but in principle may be applied to any eigenstate.

PROBLEM 3.2 Cancellation of Anomalous Diagrams.

a) Derive the first line of Eq. (3.70a) as follows. Note that Ω_0 is independent of the potential, Ω_Z is first order in the interaction, and Ω_A is second order. Show by expanding the left hand side of Eq. (3.69b) around μ_0 that to first order in the potential

$$\mu - \mu_0 = -\frac{\frac{\partial \Omega_Z^{(1)}}{\partial \mu}}{\frac{\partial^2 \Omega_0}{\partial \mu^2}} \ .$$

Similarly, expand Eq. (3.69a) around μ_0,

$$\Delta = \Omega_A(\mu) + \frac{\partial}{\partial \mu}(\Omega_0 + \Omega_Z + \mu N)[\mu - \mu_0] + \frac{1}{2}\frac{\partial^2}{\partial \mu^2}(\Omega + \Omega_Z + \mu N)[\mu - \mu_0]^2 + \ldots$$

and show that the first line of Eq. (3.70a) is obtained when all second order terms are retained.

b) Now evaluate all the quantities appearing in $\Delta^{(2)}$. Using Eq. (2.72) show

$$\frac{d^2\Omega_0}{\partial \mu^2} = \frac{\partial}{\partial \mu}\left[\sum_k \frac{-1}{e^{\beta(\epsilon_k - \mu)} + 1}\right] \underset{T\to 0}{\longrightarrow} -\sum_k \delta(\epsilon_k - \mu) \ .$$

Using Eq. (2.100b) which evaluates the Hartree-Fock contribution $\Omega_z^{(1)}$, show

$$\frac{d\Omega_Z^{(1)}}{\partial \mu} = \frac{\partial}{\partial \mu}\frac{1}{2}\sum_{k\ell} n_k n_\ell \{k\ell \mid v \mid k\ell\} \underset{T\to 0}{\longrightarrow} \sum_k \delta(\epsilon_k - \mu) U_k$$

where U is defined in Eq. (3.65b). Finally, using Eq. (3.65a) for $\Omega_A^{(2)}$ show $\Omega_A^{(2)} \underset{T\to 0}{\longrightarrow} -\frac{1}{2}\sum_k \delta(\epsilon_k - \mu) U_k^2$. Substitution of these quantities in the result of part (a) yields the final result, Eq. (3.70).

PROBLEM 3.3* Bose Gas. Consider the binding energy per particle of a dilute Bose gas with a repulsive interaction. Dilute means that the scattering length a is much less than the average particle spacing $(n)^{-1/3}$. In this limit, S-wave scattering dominates and the relevant low momentum matrix elements of the potential are independent of the momentum. To first order in na^3, the answer is

$$\frac{E}{V} = \frac{2\pi n^2 a \hbar^2}{m}\left[1 + \frac{128}{15}\left(\frac{na^3}{\pi}\right)^{1/2}\right] \ .$$

It is instructive to solve the problem two different ways:

a) Make use of the fact that the number of particles excited out of the condensate is much smaller than the number left in, so that interactions between excited particles may be neglected compared to interactions between condensed and excited particles and between condensed particles. Hence, obtain the effective Hamiltonian:

$$H_{\text{eff}} = \sum_{P} \frac{p^2}{2m} \frac{1}{2} (a_p^\dagger a_p + a_{-p}^\dagger a_{-p}) + \frac{V}{2\Omega} [N_0^2 + 2N_0 \sum_{p\neq 0} (a_{-p}^\dagger a_{-p} + a_p^\dagger a_p)$$
$$+ N_0 \sum_{p\neq 0} (a_p^\dagger a_{-p}^\dagger + a_p a_{-p})] \ .$$

N_0 may be removed by writing

$$N = N_0 + \frac{1}{2} \sum_{p\neq 0} (a_p^\dagger a_p + a_{-p}^\dagger a_{-p}) \ .$$

The resulting H_{eff} may be diagonalized by the transformation:

$$a_p = \frac{1}{\sqrt{1 - A_p^2}} (\alpha_p + A_p \alpha_{-p}^\dagger)$$

and its adjoint, where A_p is specified by the requirement that H_{eff} be diagonal in α_p. Finally, write out the second order perturbation expression for the scattering length in terms of V and eliminate V from the ground state energy. The resulting integrals expressed in terms of a converge and yield the desired result.

The excitation spectrum is given by the coefficient of $\alpha_p^\dagger \alpha_p$ in the diagonalized H_{eff}. Note that there is no gap and that the energy is linear in p as $p \to 0$. Compare this slope with the sound velocity calculated from the appropriate derivative of the total energy as a function of density.

b) Now, approach the problem diagramatically, using number-conserving zero-temperature Boson diagrams.

1) First, note that the scattering length pertains to the T-matrix which is the ladder sum of all possible successive interactions between two particles. Since S-wave scattering dominates in the low density limit, define a pseudo-potential $V = U\delta^3(x_1 - x_2)$ which reproduces the scattering length, and use V to represent all ladders of the true interaction. Calculate U.

2) For a diagram with m interactions, find the topologies which have the leading N dependence, where N is the number of Bosons. Show that any Hartree insertion 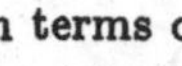 increases the order of V and N by one. Hence, remove all such insertions by appropriate definition of H_0. Of the remaining diagrams, show that the ring diagrams dominate. Typical ring diagrams containing only forward-going chains and containing a single backward-going segment are shown below:

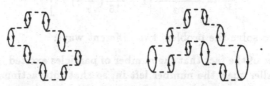

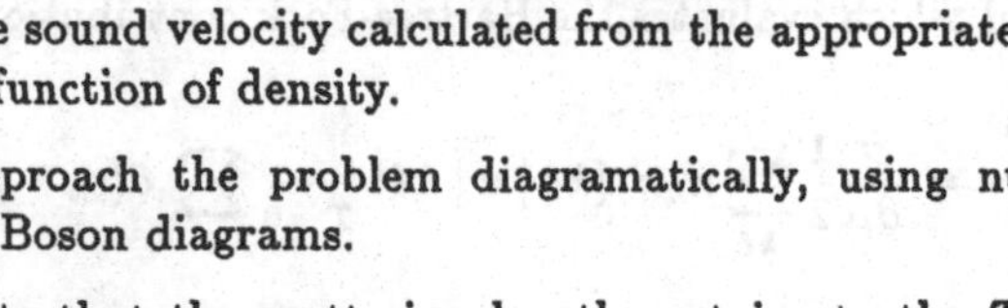

3) The easiest set to sum is all ring diagrams with forward-going chains of the form shown at the left. Note that whereas each ring diagram individually diverges, the sum is well-behaved. Show that the result is

$$\frac{E}{\Omega} = \frac{2\pi n^2 a\hbar^2}{m}[1 + \sqrt{8\pi}(na^3)^{1/2}] \; .$$

This result was obtained by Brueckner and Sawada (1957) and the $(na^3)^{1/2}$ term agrees with the result in part 1 to within 4%.

4) Sum all ring diagrams with one backward-going segment and determine the remaining fractional error.

PROBLEM 3.4 One Dimensional Interacting Fermi Gas. Consider a translationally invariant system of Fermions in one dimension interacting with the attractive δ-function potential of Problem 1.9 and having spin degeneracy $2S + 1$. Recall that g/ρ is the expansion parameter for a uniform density.

a) Evaluate the second-order energy and compare with the lowest order result calculated previously. Sketch the energy as a function of density for both cases and determine how the minimum changes in second order.

b) Calculate the sum of all ladder graphs with intermediate states above the Fermi sea:

$$\langle MN \mid G(W) \mid PQ \rangle = \langle MN \mid \left[v - \sum_{\substack{a>k_F \\ b>k_F}} \frac{v \mid ab\rangle\langle ab \mid G(W)}{E_a + E_b - W} \right] \mid PQ \rangle \; .$$

Then, evaluate

$$\frac{1}{2} \sum_{M,\, N<k_F} \langle MN \mid G(E_M + E_N) \mid MN \rangle - \langle MN \mid G(E_M + E_N) \mid NM \rangle$$

to obtain the contribution to the ground state energy. You may be left with one integral which cannot be done analytically, in which case expand in the low density limit, and compare with the first and second order results. How does this result affect the comparison with the energy of localized solutions in Problem 1.9?

PROBLEM 3.5* 1/N Expansion of Ground State Expectation Values for the One-Dimensional δ-Function Problem. It was shown in Problem 1.9 that the Hartree-Fock approximation yields the correct N^3 term in the energy for a bound state of N Fermions with spin degeneracy $2S + 1 = N$.

a) You have already found the lowest eigenfunction and eigenvalue of $-\frac{N-1}{N}\phi''(x) + U_{\rm HF}(x)\phi(x) = \epsilon\, \phi(x)$ for the HF potential

$$U_{\rm HF}(x) = -\frac{N(N-1)g^2}{8\cosh^2(\frac{Ngx}{4})} \; .$$

Now obtain the complete set of all higher eigenfunctions and eigenvalues. (Hint: Look in Morse and Feshbach (1953) or play around with exponentials and hyperbolic tangents.) Note that all eigenfunctions are functions of $(\frac{Ngx}{4})$.

b) Now consider enumerating all time-ordered graphs for the energy in this HF basis. (i.e., $H_0 = \sum_i (\frac{N-1}{N} p_i^2 + U_{\rm HF}(x_i))$). Prove that the N-dependence of an arbitrary diagram is N^{C-I+2} where C is the number of closed loops and I is the number of interactions. Hence, enumerate all the graphs of order N^3. Does this explain why the HF energy was correct to order N^3?

c) Define the class of graphs of order N^2. One of these should already be included in the HF energy. Write the expression for the next lowest order graph in this class. (Note evaluation of this expression yields $-\frac{.996\, N^2 g^2}{48}$, thereby removing almost all of the $\mathcal{O}(N^2)$ error.)

d) Now consider the expectation value of the one-body density operator, $\hat\rho(x) = \sum_i \delta(x - x_i)$. Derive a formula for the N-dependence of an arbitrary density diagram and evaluate the leading order approximation. For what other operators is it possible to derive general expressions for the N-dependence of individual diagrams?

PROBLEM 3.6 Brillouin-Wigner Perturbation Theory. **Let $H_0|\phi_0\rangle = W_0|\phi_0\rangle$ and $(H_0 + H_1)|\psi_0\rangle = E_0|\psi_0\rangle$ with the normalization condition $\langle\phi_0|\phi_0\rangle = \langle\phi_0|\psi_0\rangle = 1$ and define the projector $Q = 1 - |\phi_0\rangle\langle\phi_0|$**

a) Show that

$$\begin{aligned} E_0 - W_0 &= \langle\phi_0|H_1|\psi_0\rangle \\ |\psi_0\rangle &= \frac{1}{E_0 - H_0} H_1|\psi_0\rangle \\ &= \frac{|\phi_0\rangle\langle\phi_0|}{E_0 - H_0} H_1|\psi_0\rangle + \frac{Q}{E_0 - W_0} H_1|\psi_0\rangle \\ &= \sum_{n=0}^{\infty}\left(\frac{Q}{E_0 - H_0} H_1\right)^n |\phi_0\rangle \end{aligned}$$

and thus

$$E_0 - W_0 = \langle\phi_0|H_1 \sum_{n=0}^{\infty}\left(\frac{Q}{E_0 - H_0} H_1\right)^n |\phi_0\rangle \ .$$

b) Now, consider N Fermions in a periodic box of volume $\mathcal{V}$ in the limit $N \to \infty$ at fixed ρ, and let H_0 be the kinetic energy operator and H_1 a two-body potential $v(r_i - r_j)$. Show that whereas $\langle\phi_0|H_1|\phi_0\rangle$ is the usual Hartree-Fock energy which is proportional to N, the next order correction $\sum_{m\neq 0} \frac{|\langle\phi_0|H_1|\phi_m\rangle|^2}{E_0 - W_m}$ is independent of N. In analyzing the N-dependence, it is useful to use momentum conservation to enumerate all states $|\phi_m\rangle$ to which H_1 can connect $|\phi_0\rangle$ and to examine the factors of $\mathcal{V}$ and N arising from matrix elements, momentum sums, and energy denominators. Consider whether this problem persists in higher order and explain the implications for convergence of the perturbation series for large N.

PROBLEM 3.7 Rayleigh-Schrödinger Perturbation Theory. The form of perturbation theory used in most quantum mechanics textbooks is Rayleigh-Schrödinger perturbation theory. Let $H_0|\phi_n\rangle = W_n|\phi_n\rangle$ and $(H_0 + \lambda H_1)|\psi_0\rangle = E_0|\psi_0\rangle$, where we now associate a formal expansion parameter λ as follows: $E_0 = \sum_{n=0}^{\infty} \lambda^n \epsilon_n$ and $|\psi_0\rangle = \sum_{n=0}^{\infty} \lambda^n |n\rangle$. We use the same normalization as before $\langle\phi_0|\psi_0\rangle = 1$, which implies $\langle 0|n\rangle = \delta_{n0}$, denote $Q = 1 - |\phi_0\rangle\langle\phi_0|$, and note that $\epsilon_0 = W_0$ and $|0\rangle = |\phi_0\rangle$.

a) By equating terms of each power of λ in the Schrödinger equation, obtain the recursion relations

$$\epsilon_n = \langle 0|H_1|n-1\rangle$$
$$|n\rangle = \frac{Q}{W_0 - H_0}\left[(H_1 - \epsilon_1)|n-1\rangle - \epsilon_2|n-2\rangle - \epsilon_3|n-3\rangle \ldots \epsilon_{n-1}|1\rangle\right] .$$

Write out the explicit expressions for $|n\rangle$ and ϵ_n through ϵ_3.

b) This result may also be obtained from Brillouin-Wigner perturbation theory by expanding the energy denominator

$$\frac{1}{E_0 - H_0} = \frac{1}{(W_0 - H_0) - (W_0 - E_0)} = \frac{1}{W_0 - H_0}\sum_{n=0}^{\infty}\left[\frac{W_0 - E_0}{W_0 - H_0}\right]^n$$

and substituting the series for $(W_0 - E_0)$ each time it arises. Carry out this substitution through ϵ_3.

c) Compare the N dependence of ϵ_2 for N Fermions in a box interacting via a two-body potential with the result obtained in Problem 3.6. Note that it was the replacement of $\frac{1}{E_0 - H_0}$ by $\frac{1}{W_0 - H_0}$ which solved this problem.

PROBLEM 3.8 Cancellation of Unlinked Terms. Although, as shown in the preceding problem, ϵ_2 has the correct N-dependence, Rayleigh Schrödinger perturbation theory also contains terms with higher powers of N. For the theory to make sense, one must prove that these terms cancel.

a) Show that the term $\langle\phi_0|H_1\left(\frac{Q}{E_0 - H_0}H_1\right)^2|\phi_0\rangle$ contributing to ϵ_3 has contributions proportional to N^2 as well as contributions proportional to N. Associate diagrams with these contributions and relate the connectedness to the N-dependence.

b) Show that the term $-\langle\phi_0|H_1\left(\frac{Q}{E_0 - H_0}\right)^2 H_1|\phi_0\rangle\langle\phi_0|H_1|\phi_0\rangle$ in ϵ_3 exactly cancels the N^2 contribution in part (a). With sufficient effort, similar cancellation of unlinked diagrams can be verified in the next few orders of perturbation theory, and it was on this basis that the linked cluster theorem was first conjectured.

PROBLEM 3.9 Variational Property of Brillouin-Wigner Perturbation Theory Consider the following trial wave function:

$$|\chi_N(z)\rangle = \sum_{n=0}^{N}\left(\frac{Q}{z - H_0}H_1\right)^n |\phi_0\rangle$$

where z is to be regarded as a parameter.

a) Show that

$$\langle\chi_N(z)|H_1 - (z - H_0)|\chi_N(z)\rangle = \sum_{n=0}^{2N} \Delta E_n(z) + W_0 - z$$

where $\Delta E_n(z) \equiv \langle\phi_0|H_1\left(\frac{Q}{z-H_0}H_1\right)^n|\phi_0\rangle$. Note that $\Delta E_n(E_0)$ corresponds to the nth term in the Brillouin-Wigner expansion for the energy in Problem 3.6.

b) Show $\langle\chi_N(z)|\chi_N(z)\rangle \geq 1$ and use the variational principle to obtain a bound on E_0. State explicitly any conditions z must satisfy for this bound to be valid.

c) Thus, show $E_0 \leq W_0 + \sum_{n=0}^{2n}\Delta E_n(z)$ for $z = W_0 + \sum_{n=0}^{2n}\Delta E_n(z)$. Verify that this z satisfies the conditions established in (b). Hence, show that all odd orders of Brillouin-Wigner perturbation theory yield upper bounds on the ground state energy.

PROBLEM 3.10 Summation of Topologically Equivalent Time-Ordered Diagrams.

a) Consider the energy denominators $\frac{1}{-\tilde{S}+i\eta}$ contributing to each of the Green's function diagrams in Eq. (3.59a). Show that the sum of the first two diagrams yields

$S_{1+2} \equiv$ 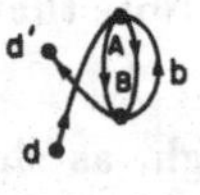

$+$ 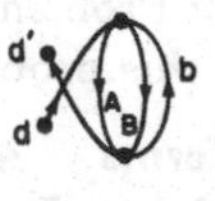

$$= \frac{1}{(-\epsilon_d - \epsilon_b + \epsilon_A + \epsilon_B + i\eta)(\omega - \epsilon_d - \epsilon_{d'} - \epsilon_B + \epsilon_A + \epsilon_B + i\eta)}$$

$$\times\left[\frac{1}{\omega - \epsilon_d + i\eta} + \frac{1}{-\epsilon_{d'} - \epsilon_b + \epsilon_A + \epsilon_B + i\eta}\right]$$

$$= \frac{1}{(-\epsilon_d - \epsilon_b + \epsilon_A + \epsilon_B + i\eta)(\omega - \epsilon_d + i\eta)(-\epsilon_{d'} - \epsilon_b + \epsilon_A + \epsilon_B + i\eta)}\ .$$

Similarly, combine the third and fourth diagrams in Eq. (3.59a) to obtain

$$S_{3+4} = \frac{1}{(\omega - \epsilon_{d'} + i\eta)(\omega - \epsilon_d + i\eta)(-\epsilon_{d'} - \epsilon_b + \epsilon_A + \epsilon_B + i\eta)}\ .$$

Finally, add the fifth diagram to S_{1+2} and add the result to S_{3+4} to obtain

$$S_{1+2} + S_{3+4} + S_5 = \frac{1}{(\omega - \epsilon_{d'} + i\eta)(-\omega - \epsilon_b + \epsilon_A + \epsilon_B + i\eta)(\omega - \epsilon_d + i\eta)}\ .$$

Hence, by including the proper signs and factors, obtain the self-energy Eq. (3.59b).

b) From the algebra in (a), it is evident that individual time-ordered diagrams having global energy denominators depending on the entire state of the system may be combined such that the final result is a product of components each of which has energy denominators depending upon propagators within that component. In the case of the Green's function, we obtained $G_{d'}(\omega)\ G_d(\omega)\ \Sigma(\omega)$, where G_d and $G_{d'}$ were independent of propagators in Σ and energy denominators in Σ were independent of d and d'. Bethe, Brandow, and Petschek (1963) showed that this generalized time ordering is quite general. As a simple example, combine the energy diagrams

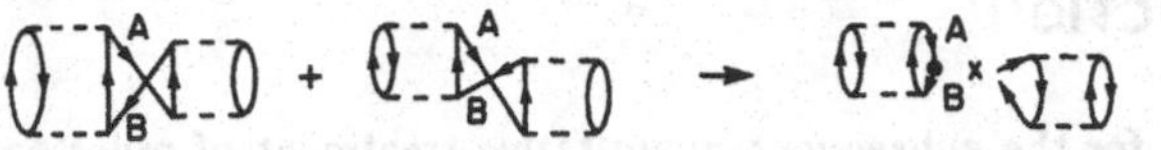

to obtain a product in which the energy denominator of each unit only involves its own propagators. This result may be generalized to the sum of diagrams of arbitrary complexity in which the position of the top interaction in the right hand insertion is held fixed and the lower interactions assume all time-orderings consistent with the topology.

CHAPTER 4

ORDER PARAMETERS AND BROKEN SYMMETRY

One of the goals of many-body theory is to understand the phases in which matter exists and the transitions between these phases. This chapter addresses the fundamental role of order parameters and symmetry breaking in characterizing phases and understanding phase transitions.

4.1 INTRODUCTION

To prepare for the subsequent quantitative treatment of order parameters, mean field theory, and fluctuations, it is useful to review briefly several elementary aspects of phase transitions.

PHASES OF TWO FAMILIAR SYSTEMS

A ferromagnetic material, such as iron, and a typical substance having solid, liquid, and gas phases illustrate the essential features of how the phase of a thermodynamic system changes as a function of the temperature and an external thermodynamic field. In addition to the temperature and external field, it will be useful to characterize the state of the system by a thermodynamic variable conjugate to the external field and to study the surface of the equation of state in the space of three variables: the temperature, the external field, and the conjugate variable.

The surface of the equation of state and phase diagrams for a ferromagnet are shown in Fig. 4.1 The magnetic field $\vec{H}$ is the external field which together with the temperature determine the state of the system, and the phase diagram in the $H-T$ plane is shown in the upper left of Fig. 4.1. Below the critical temperature T_c, the system is spontaneously magnetized and the state of the system is characterized by the magnetization $\vec{M}$, the thermal average of the microscopic spins. If we choose $\vec{H}$ to be along the $\hat{z}$ axis, these will be two ferromagnetic states at low temperature, the $\uparrow$ state for positive H and the $\downarrow$ state for negative H. These two states are separated by a phase boundary, indicated by the solid line. This phase boundary terminates at a critical point C at temperature T_c, and above the critical point, the system has a single paramagnetic phase. The system may either be brought from the ferromagnetic $\downarrow$ phase to the ferromagnetic $\uparrow$ phase discontinuously, by selecting a path which crosses the phase boundary, or continuously by choosing a path which travels around the critical point into the paramagnetic phase.

The magnetization M is the thermodynamic variable conjugate to H, and a more complete characterization of the phases of the ferromagnet is provided by the surface of the equation of state in the space of M, H, and T sketched in the upper right of Fig. 4.1. The vertical shaded surface denotes the region of phase coexistence separating the $\uparrow$ and $\downarrow$ ferromagnetic phases. Isotherms below the critical temperature $T_<$, at the critical temperature T_c, and above the critical temperature $T_>$, are shown on the surface of the equation of state. The top view of this surface along the negative M axis corresponds to the phase diagram in the $H-T$ plane we have already discussed.

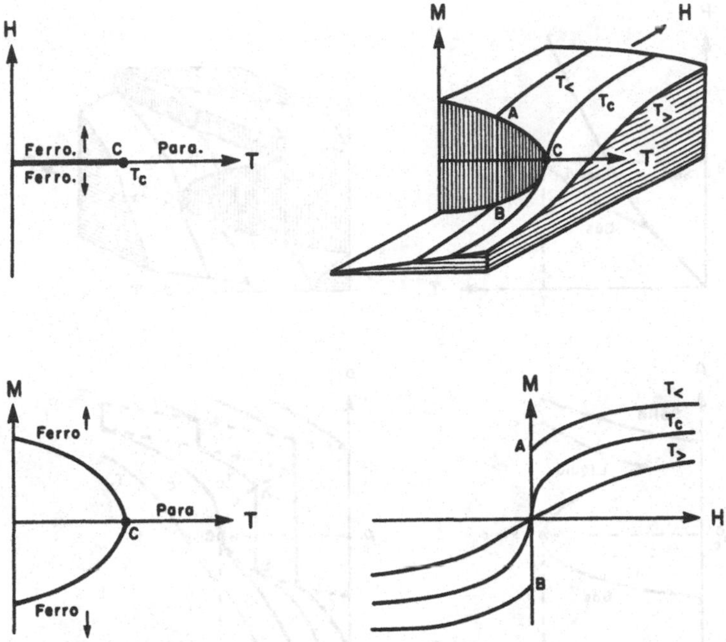

Fig. 4.1 Phase diagrams for a ferromagnet. The surface of the equation of state in the space of magnetization M, external magnetic field H, and temperature T is shown in the upper right. Projections of this surface are shown in the $H-T$, $M-T$, and $M-H$ planes. Isotherms are shown at the critical temperature T_c, at $T_<$ below T_c, and at $T_>$ above T_c.

Viewed from the right, along the negative T axis, this surface corresponds to the set of isotherms plotted in the $M-H$ plane in the lower right. For $T_>$ above T_c, the curves are continuous and represent the single paramagnetic phase. For $T_<$ below T_c, the curved segments of the isotherm represent the ferromagnetic ↑ and ↓ phases and the vertical segment joining points A and B represents the region of coexistence of the two phases at zero field. Note that the magnetization is non-analytic at A and B and that these points correspond to a single point on the phase boundary in the $H-T$ plane. Along the critical isotherm, the coexistence region is reduced to a single point, the critical point, and the two phases coincide.

The final view of the ferromagnet phase surface, the front view of the H axis, yields the boundary of the coexistence region in the $M-T$ plane shown at the lower left. We observe that at zero H field, M is non-zero below T_c, approaches zero at T_c, and is zero above T_c and we will subsequently use this behavior to identify M as the order parameter.

The analogous phase surface and projections for a typical* substance having solid,

* Note that water is atypical in the sense that ice is less dense than water at the freezing point and is not considered in the present discussion.

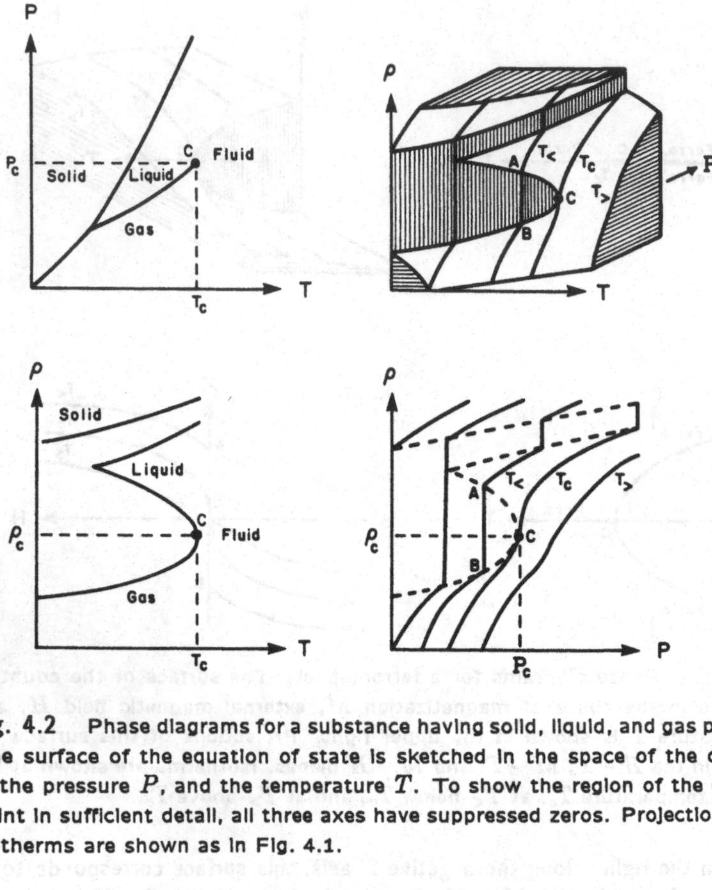

Fig. 4.2 **Phase diagrams for a substance having solid, liquid, and gas phases. The surface of the equation of state is sketched in the space of the density ρ, the pressure P, and the temperature T. To show the region of the critical point in sufficient detail, all three axes have suppressed zeros. Projections and isotherms are shown as in Fig. 4.1.**

liquid and gas phases are shown in Fig. 4.2. The pressure is the external thermodynamic field, and the $P-T$ phase diagram is shown at the upper left. At low temperature, depending upon the pressure, the material can exist as a gas, liquid, or solid. At high temperature, the liquid and gas phases become indistinguishable and the material has a unique fluid phase. The liquid-gas phase boundary terminates at the critical point marked C, having critical temperature T_c and pressure P_c. In the present discussion, we will be particularly interested in the behavior in the vicinity of the critical point, and its similarity to the preceding example of the ferromagnet. The liquid, gas, and fluid phases are analogous to the ferromagnetic ↑ ferromagnetic ↓ and the paramagnetic phases respectively, and as in the previous case, it is possible to go continuously from the gas to liquid phases by passing around the critical point through the fluid phase.

Although the thermodynamic variable conjugate to the field P is often chosen to be the volume, the analogy to the magnetic case is most evident if we use the density ρ as the conjugate variable ($\rho = \frac{1}{v}$ where v is the specific volume). The surface of the equation of state in the space of ρ, P, and T is shown in the upper right and isotherms in the $\rho - P$ plane are plotted in the lower right. Again, the vertical segment

$A-B$ of the isotherm $T_<$ below the critical temperature represents the region of phase coexistence, this time between gas and liquid. The projection onto the $\rho - T$ plane in the lower left yields a phase boundary near the critical point c analogous to that already observed for the magnetization as a function of temperature, and will subsequently lead us to identify $\rho_{\text{liquid}} - \rho_{\text{gas}}$ as the order parameter.

Thus, the two surfaces of the equations of state in Figs. 4.1 and 4.2 show the same essential behavior in the vicinity of the critical point. The magnetic case is somewhat simpler to visualize because the coexistence region lies precisely in the $M - T$ plane whereas the liquid gas coexistence region is a curved surface. In both cases, below the critical temperature a thermodynamic variable, M or ρ, varies discontinuously across the coexistence surface as a function of the conjugate field, H or P and the discontinuity approaches zero at the critical point.

Since non-analytic behavior in thermodynamic variables such as $M(H,T)$ is characteristic of phase transitions, it is useful to note the possible reasons for which the sum of analytic functions comprising the partition function $Z = \sum_n e^{-\frac{E_n}{T}}$ could become non-analytic. The energies E_n could be non-analytic functions of the parameters of the system as in the case of the quartic oscillator in Section 2.1. The finite sum over n could diverge as in the case of the hydrogen atom. Alternatively, non-analyticity could arise from taking the zero temperature limit or from taking the thermodynamic limit in which the number of degrees of freedom goes to infinity. Except in special cases, we shall see that the origin of non-analyticity is in fact the thermodynamic limit.

PHENOMENOLOGICAL LANDAU THEORY

Considerable insight into the structure of the general theory developed subsequently is provided by a purely phenomenological theory by Landau (Landau and Lifshitz, 1958). In this theory, an order parameter for a phase transition is defined as a quantity which vanishes in one phase, called the disordered phase, and is non-zero in the other phase, called the ordered phase. An order parameter may not exist for some transitions, such as the Kosterlitz-Thouless transition in a two dimensional classical spin system, and when it does exist, it is not unique (since any power also satisfies the definition). For our present purpose, it will always be possible to choose the order parameter to be the thermodynamic average of an observable. We will also use the Landau definition that a transition is first order if the order parameter is discontinuous at the transition point and second order if it is continuous.*

In the case of a ferromagnet, we observe from Fig. 4.1 that the magnetization per spin is an order parameter. We will use the convention that the total magnetization $\vec{M}$ is defined as the thermodynamic average of the spin and that $\vec{m}$ denotes the magnetization per spin

$$\vec{M} = \langle \sum_i \vec{S}_i \rangle \qquad \vec{m} = \langle \frac{1}{N} \sum_i \vec{S}_i \rangle \ . \tag{4.1}$$

The magnetization vanishes in the high-temperature paramagnetic phase, and the phase transition is second order since m is continuous. The physical order in this case is the

* Note that for $n > 2$, this differs from the Ehrenfest definition of an nth order transition as one in which $\frac{\partial^n F}{\partial T^n}$ is the lowest discontinuous derivative.

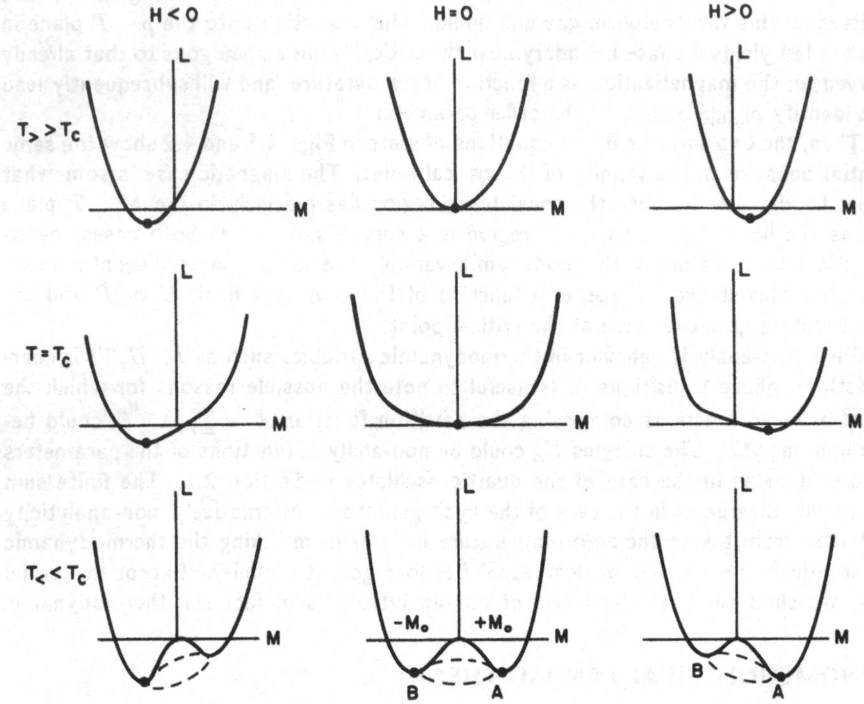

Fig. 4.3 The Landau function $\mathcal{L}(m, H, T)$ for a ferromagnet.

alignment of the microscopic spins. For the case of the liquid-gas phase transition, it is evident that $\rho_{\text{liquid}} - \rho_{\text{gas}}$ is a possible order parameter and that the transition is also second order.

The phenomenological theory is based on the Landau function $\mathcal{L}(m, H, T)$, a function of the order parameter, which we will denote as m in this discussion, the conjugate field, which we will denote as H, and the temperature T. The Landau function has the property that at any fixed H and T, the state of the system is specified by the absolute minimum of $\mathcal{L}$ with respect to m.

The only physical constraint on the form of $\mathcal{L}$ is that it be consistent with the known symmetries of the system. One of the great accomplishments of this theory is its description of the non-analytic behavior at phase transitions in terms of the discontinuous jumps in the position of the absolute minimum of a function which is itself varying continuously with T and H. Although we will eventually see a clear relation between the Landau function and an energy functional appearing in a functional integral for the exact partition function, for the present let us simply see how the observed ferromagnetic phase transition is described by the function $\mathcal{L}(m, H, T)$ sketched in Fig. 4.3

First, consider the case $T > T_c$, shown in the top row of Fig. 4.3. As an external field is applied tending to align $\vec{m}$ with $\vec{H}$, the minimum of $\mathcal{L}$ is continuously displaced from negative to positive m consistent with the continuous behavior in the paramagnetic region of Fig. 4.1. Below T_c, however, the absolute minimum jumps discontinuously from $-m_0$ to m_0 when H passes through zero, reproducing the discontinuous behavior in the ferromagnetic region. Thus, the double minimum provides a very simple description of a first-order phase transition. Furthermore, if we look ahead and anticipate the fact that the minimum of the Landau function will correspond to the stationary phase approximation to the exact theory, which must be corrected by fluctuations around the minimum, we see that since there is always a barrier between the two minima for $T < T_c$, there is no reason for the fluctuations to change appreciably as H passes through zero.

Now consider the dependence of $\mathcal{L}$ at zero field as the temperature decreases through T_c as shown by the middle column of Fig. 4.3. As T approaches T_c, the minimum becomes flatter and flatter, and one might expect the fluctuations to grow without bound. Infinitesimally below T_c, the single minimum bifurcates into two infinitesimally separated minima, which continue to separate with decreasing T, generating the structure observed in the $m-T$ phase diagram of Fig. 4.1. This generic behavior of a single minimum flattening out and finally forming two degenerate minima thus describes a second-order phase transition.

In discussing classical or quantum spin systems, there is a conventional nomenclature designating the number of spin components, or equivalently, the dimensionality of the space in which the spins can move. Ising spins have a single component which is restricted to be parallel or antiparallel to a fixed axis and is specified by the values ± 1. Spins in the $x-y$ model have two components in a plane and Heisenberg spins have three components in three dimensions. A generic name for a spin model with n components is the $O(n)$ model.

In the case of Ising spins, which can only have the values ± 1, the magnetization would only point in one of two distinct directions, and the Landau function would lie in a plane as denoted by the solid curves in Fig. 4.3. For $x-y$ spins, the Landau function is a surface in three dimensions, and this three-dimensional character is indicated by the dotted lines in the bottom representing the local minimum for each angle. Similarly, the Heisenberg model is described by a four-dimensional surface which we will not attempt to sketch.

Because the order parameter m changes continuously at a second order phase transition, the essential physics may often be seen by expanding $\mathcal{L}$ in the immediate neighborhood of the transition as a power series in m. To the extent to which only a few terms are relevant* and the form of these terms is highly constrained by symmetry, one may identify essentially universal behavior shared by classes of diverse physical systems. Such universality is already evident in the basic similarity of the phase surfaces in the vicinity of the critical points in Figs. 4.1 and 4.2. For the case of a magnetic system, let us expand $\mathcal{L}(m, H, T)$ through fourth order (the sixth order case is considered in Problem 4.1).

$$\mathcal{L}(m, H, T) = \sum_{n=0}^{4} a_n(H, T) m^n \quad . \tag{4.2}$$

* There is no guarantee that the series may be truncated, and we shall subsequently see by dimensional analysis alone that all orders are relevant in two dimensions.

As discussed in connection with the middle column of Fig. 4.3, the conditions for a second order transition are

$$\frac{\partial \mathcal{L}}{\partial m} = \frac{\partial^2 \mathcal{L}}{\partial m} = \frac{\partial^3 \mathcal{L}}{\partial m} = 0 \ , \qquad \frac{\partial^4 \mathcal{L}}{\partial m} > 0 \ \text{(second order)} \ . \tag{4.3a}$$

The second derivative must vanish because the curve changes from concave upward to concave downward and the third derivative must vanish to insure that the critical point is in fact a minimum. Similarly, from the bottom row, the conditions for a first order transition are

$$\left.\frac{\partial \mathcal{L}}{\partial m}\right|_{m_A} = \left.\frac{\partial \mathcal{L}}{\partial m}\right|_{m_B} = 0 \qquad \mathcal{L}(m_A) = \mathcal{L}(m_B) \quad \text{(first order)} \ .$$

The form of $\mathcal{L}$ may be further restricted by using the symmetry of the problem and expanding each of the coefficients $a_n(H,T)$ in the vicinity of the critical point in terms of the reduced parameters $t \equiv T - T_c$ and $h \equiv H - H_c = H$

$$a_n(H,T) = b_n + c_n h + d_n t \ . \tag{4.3b}$$

Since, by inversion symmetry, $\mathcal{L}(m,H,T) = \mathcal{L}(-m,-H,T)$, the only invariants are even powers of m, even powers of h or combinations of odd powers of m and odd powers of H so that $c_n = 0$ for even n and $b_n = d_n = 0$ for odd n. Furthermore, since a_2 must change sign at the critical point, $b_2 = 0$ and $d_2 > 0$ and since $a_4 > 0$ in the vicinity of the critical point, $b_4 > 0$ and d_2 is irrelevant. Hence, the most general form for the Landau function near the critical point is

$$\mathcal{L}(m,h,t) = c_1 hm + d_2 tm^2 + c_3 hm^3 + b_4 m^4 \qquad d_2 > 0 \ , \quad b_4 > 0 \ . \tag{4.4}$$

In the present case, this expansion succinctly summarizes the essential features of Fig. 4.3. In the case of crystals, with point group and translational symmetries and more complicated order parameters, the Landau rule that the symmetry group of the lower symmetry state must be a subgroup of the higher symmetry state for a second order transition provides a powerful tool for identifying the invariants which may appear in $\mathcal{L}$ and thus characterize the possible phase transitions.

To avoid misunderstanding, it is important to distinguish the Landau function from another function which also has the property that its minimum specifies the state of the system. Recall from the discussion in connection with Eqs. (2.162 – 2.163) that if the free energy is defined $e^{-\beta F(H)} = Tr\, e^{-\beta(\mathcal{H} - \vec{H}\cdot\sum_i \vec{S}_i)}$, then

$$\frac{\partial F}{\partial H} = -M \tag{4.5a}$$

and the Gibbs free energy is the Legendre transform with the property

$$\Gamma(M) \equiv F(H(M)) + MH(M) \tag{4.5b}$$

$$\frac{\partial \Gamma}{\partial M} = H \ . \tag{4.5c}$$

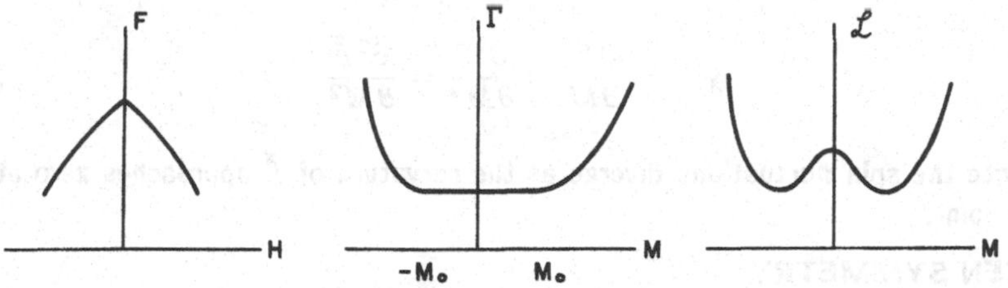

Fig. 4.4 The Helmholtz Free energy $F(H)$, the Gibbs free energy $\Gamma(M)$, and the Landau function $\mathcal{L}(M)$ for the ferromagnet below T_c.

At zero field, then, the state of the system is specified by the condition $\frac{\partial\Gamma}{\partial M} = 0$ so that when M is an order parameter, $\Gamma(M)$ appears to satisfy the definition of the Landau function. In general, at finite field, one is tempted to consider the following function as the Landau function

$$\begin{aligned}\tilde{\mathcal{L}}(M,H) &\equiv \Gamma(M) - MH \\ &= F(H(M)) + M(H(M) - H)\end{aligned} \tag{4.5d}$$

since $\frac{\partial\tilde{\mathcal{L}}(M,H)}{\partial M} = 0$ and its minimum specifies the state. (Note that the parameter H is distinct from the function $H(M)$.) Above T_c, where the isotherms in Fig. 4.1 are smooth curves, $\tilde{\mathcal{L}}$ becomes qualitatively like the top row of Fig. 4.3 and there is no problem with this association. Below the critical point, however, there is an essential difference between the behavior of $\tilde{\mathcal{L}}$ and $\mathcal{L}$ which is shown in Fig. 4.4 for the zero field case (where $\tilde{\mathcal{L}} = \Gamma$). Note that by Eq. (4.5c) $\frac{\partial\Gamma}{\partial M} = 0$ at the points $M = \pm M_0$ corresponding to values of M in the ferromagnetic phase and the curve bends continuously upwards outside these points. However, because the isotherm has $H = 0$ for all values of $-M_0 < M < M_0$, $\Gamma = F(H = 0)$ is a constant between these points. This non-analytic behavior of Γ is thus qualitatively different from the continuous double well of the phenomenological Landau function. Although, by definition, it describes the phase transition exactly, it does not embody the essential feature of the Landau function of describing a phase transition in terms of competing minima of a continuously varying smooth function.

One useful result of the correspondence between $\tilde{\mathcal{L}}$ and $\mathcal{L}$ outside the region of phase coexistence is a clarification of the earlier discussion of fluctuations. The magnetic susceptibility directly measures the spin fluctuations, since

$$\begin{aligned}\chi = \frac{\partial m}{\partial H} &= \frac{\partial}{\partial H}\frac{tr\left(\frac{1}{N}\sum_i S_i e^{-\beta(\mathcal{H}_0 - \sum_j \vec{S}_j\cdot\vec{H})}\right)}{\text{tr}\left(e^{-\beta(\mathcal{H}_0 - \sum_j \vec{S}_j\cdot\vec{H}}\right)} \\ &= \frac{\beta}{N}\left(\left\langle\left(\sum_i \vec{S}_i\right)^2\right\rangle - \left(\sum_i \langle S_i\rangle\right)^2\right)\ .\end{aligned} \tag{4.6}$$

By Eqs. (4.5) the susceptibility χ is also directly related to the second derivative of $\tilde{\mathcal{L}}$

$$\chi^{-1} = \frac{\partial H}{\partial M} = \frac{\partial^2 \Gamma}{\partial M^2} = \frac{\partial^2 \tilde{\mathcal{L}}}{\partial M^2} \tag{4.7}$$

and hence the spin fluctuations diverge as the curvature of $\tilde{\mathcal{L}}$ approaches zero at the critical point.

BROKEN SYMMETRY

The Landau function sketched in Fig. 4.3 clearly displays the phenomenon of symmetry breaking which we will observe in mean field theory.

Although the Landau function for $H = 0$, $T < T_c$ displays all the symmetries of the magnetic Hamiltonian, its minima break the symmetry. For the case of Ising spins, the system is invariant under global spin reflections $S_i \longrightarrow -S_i$ and the Landau function is a symmetric double well possessing the same symmetry. However, each of the two minima of the double well break the symmetry and transform into each other under spin reflection. For $x-y$ and Heisenberg spins, the Landau function is a surface with a continuous set of degenerate minima reflecting the symmetry with respect to continuous global spin rotations. The configurations corresponding to each of these minima break the spin rotation symmetry and transform into one another under spin rotation. This structure is quite general. For a Hamiltonian with a symmetry group $\mathcal{G}$, above T_c the disordered phase will be invariant under $\mathcal{G}$ whereas below T_c, the degenerate disordered phases will not be invariant and will transform among each other under the action of $\mathcal{G}$.

Since the thermodynamic properties of a macroscopic magnetic system below T_c are non-analytic at $H = 0$, some care is required in discussing the zero field properties. On the one hand, physically we know that the magnetic field of a ferromagnet can point in some specific direction. On the other hand, the strict definition of a thermodynamic average as the sum over all states with the appropriate Boltzmann weights would yield zero net magnetization since all degenerate orientations of the spins are weighted equally. To reconcile these two facts, one must examine carefully whether all the states of a magnetic system in zero field actually are accessed with equal probability.

To avoid ambiguity, we shall define the order parameter at zero field as the limit of the order parameter in the presence of a weak external conjugate field as the field goes to zero. For a magnetic system, the zero field magnetization is defined

$$M = \lim_{H \to 0} M(H) \quad . \tag{4.8}$$

We first consider the case of Ising spins. For an arbitrarily weak positive H, the symmetry of the Landau function is broken and the minimum at positive M, denoted by A in the lower right of Fig. 4.3, is slightly lower than B. Using the fact that H couples to the spins through the interaction $-H\sum_i S_i$, the Boltzmann weight then yields the ratio of probabilities for observing the system in state B and state A as

$$\frac{P_B}{P_A} = e^{-2N\beta m H} \tag{4.9}$$

where, by Eq. (4.1), m is the magnetization per spin. In the thermodynamic limit, $N \to \infty$ so that $P_B \to 0$ for any H and as $H \to 0$, the system is in state A with $M = M_0$. The zero field state thus depends upon the history by which it is prepared, and if instead we had taken a negative field, it would have approached state B with $M = -M_0$. The crucial role of the thermodynamic limit in this argument is clear. If N remained finite as $H \to 0$, then $\frac{P_B}{P_A} \to 1$ and the two minima would be equally populated. In fact, for a large but finite system, all the thermodynamic functions would be analytic. All the sharp edges of the phase surface in Fig. 4.1 would be smoothed and the isotherms below T_c would be very steep but not infinite at $H = 0$, yielding a unique solution $M = 0$. For infinite N, however, the behavior is non-analytic at $H = 0$ and one must define the state of the system by a specific path on the surface of the equation of state as $H \to 0$. Heuristically, if we interpret the Landau function as the free energy for the moment, because there is in some sense an infinite barrier between A and B, the system is not ergodic at $H = 0$ and may be trapped in either state A or B depending on the history.

The situation can be slightly more complicated in the case of continuous symmetry. For $x-y$ spins, the Landau function is a surface with a minimum corresponding to the circle denoted by dashed lines in Fig. 4.3. As $H \to 0$, the minimum becomes flat, and all the degenerate configurations can be accessed without having to cross an infinite barrier. Similarly, for Heisenberg spins in zero field, there is a sphere of degenerate configurations corresponding to all the orientations of fixed $|\vec{m}|$ in three dimensions. We will see that thermal fluctuations can excite collective Goldstone modes in which the system samples degenerate configurations at the bottom of the valley. Depending on the dimensionality of the space and tensorial character of the order parameter, these collective modes can either destroy the order by equally mixing all the degenerate phases, or have no effect on the order because their effect is suppressed by phase space factors.

4.2 GENERAL FORMULATION WITH ORDER PARAMETERS

We now seek to formulate a general theory which in leading approximation embodies the physics of the Landau function of the order parameter. Since the order parameter is singled out as the collective variable representing the essential degrees of freedom governing phase transition, it is natural to cast the partition function in the form of a functional integral over the order parameter, and in the stationary phase approximation, the theory has precisely the Landau form. The corrections to the SPA systematically correct the leading Landau contribution. To introduce the basic ideas, we first present a simple illustrative model and then discuss the general case.

INFINITE RANGE ISING MODEL

In order to relate the preceding discussion to the general theory involving a functional integral over an order parameter, it is instructive to consider an admittedly artificial model, which has the important property that the stationary phase approximation is exact. In this model, each of the Ising spins S_i ($S_i = \pm 1$) interacts with every other spin S_i with an exchange energy $-\frac{J}{N}$. The scaling of the exchange energy as $\frac{1}{N}$ in this model is required in order to have a thermodynamic limit. In the presence of a magnetic field H, the partition function Z is

$$Z(\beta, H, N) = \sum_{\{S_i\}} e^{\frac{\beta}{2}\frac{J}{N}\sum_{i,j} S_i S_j + \beta H \sum_i S_i} \tag{4.10}$$

where $\sum_{\{ij\}}$ means a summation over all the 2^N spin configurations in which each spin may have $S = \pm 1$.

We now introduce an extremely useful technique which we will use frequently and replace the inconvenient spin sum $\sum_{\{S_i\}} e^{\sum S_i S_i}$ by the integral over an auxiliary field of the easily evaluated spin sum $\sum_{\{S_i\}} e^{S_i}$. Using the Gaussian identity,

$$e^{\frac{\beta J}{2N}(\sum_i S_i)^2} = \int_{-\infty}^{+\infty} \frac{d\mu}{\sqrt{\frac{2\pi}{N\beta J}}} e^{-\frac{N\beta J}{2}\mu^2 + \beta J \mu \sum_{i=1}^{N} S_i} \tag{4.11}$$

and noting that

$$\sum_{\{S_i\}} e^{\beta(J\mu + H)\sum_i S_i} = \prod_{i=1}^{N} \left(e^{\beta(J\mu+H)} + e^{-\beta(J\mu+H)} \right) \tag{4.12}$$

we obtain

$$Z(\beta, H, N) = \int_{-\infty}^{+\infty} \frac{d\mu}{\sqrt{\frac{2\pi}{N\beta J}}} e^{-\frac{N\beta J}{2}\mu^2 + N \log 2ch\beta(H+\mu J)} \quad . \tag{4.13}$$

The magnetization m and susceptibility χ are obtained by taking derivatives of (4.13) with respect to H:

$$m = \frac{1}{N}\frac{1}{\beta}\frac{\partial \log Z}{\partial H} = \langle th\beta(H + \mu J)\rangle \tag{4.14a}$$

$$\chi = \frac{1}{\beta N}\frac{\partial m}{\partial H} \underset{N\to+\infty}{\simeq} \langle th^2\beta(H+\mu J)\rangle - \langle th\beta(H+\mu J)\rangle^2 \quad . \tag{4.14b}$$

As mentioned in the introduction, we see that as long as β and N are finite, $Z(\beta, H, N)$ is an analytic function. Singular behavior occurs only if $N \to +\infty$ or $\beta \to +\infty$.

It is useful to define a function $\mathcal{L}(\mu, H)$, which we shall later identify with the Landau function, as the exponent in the expression for Z:

$$\mathcal{L}(\mu, H) = \frac{J}{2}\mu^2 - \frac{1}{\beta}\log 2\,ch\,\beta(H + \mu J) \tag{4.15a}$$

so that:

$$Z = \int_{-\infty}^{+\infty} \frac{d\mu}{\sqrt{\frac{2\pi}{N\beta J}}} \cdot e^{-\beta N \mathcal{L}(\mu, H)} \quad . \tag{4.15b}$$

In the thermodynamic limit, $N \to \infty$, the stationary phase method applied to Z, Eq. (4.15b) becomes exact. The stationarity condition:

$$\frac{\partial}{\partial\mu}\mathcal{L}(\mu, H) = 0 \tag{4.16a}$$

yields

$$\mu = th\,\beta(H + \mu J) \ . \tag{4.16b}$$

Let us denote by μ_S the solutions for Eq. (4.16) which are separated local minima of the function $\mathcal{L}(\mu, H)$. Then we know that in the thermodynamic limit:

$$Z = \sum_S e^{-\beta N \mathcal{L}(\mu_S, H)} \tag{4.17a}$$

and

$$m = \frac{\sum_S \mu_S e^{-\beta N \mathcal{L}(\mu_S, H)}}{\sum_S e^{\beta H \mathcal{L}(\mu_S, H)}} \ . \tag{4.17b}$$

If there are several minima μ_S for $\mathcal{L}(\mu, H)$, they correspond to the different stable or metastable phases in which the system can exist. The probability to find the system in the phase μ_S is given by the Boltzmann factor:

$$P(\mu_S) = \frac{e^{-\beta N \mathcal{L}(\mu_S, H)}}{\sum_S e^{-\beta N \mathcal{L}(\mu_S, H)}} \ . \tag{4.18a}$$

If one minimum, μ_0, is an absolute minimum of $\mathcal{L}(\mu, H)$, then in the limit $N \to \infty$ $P(\mu_0) \to 1$ and $P(\mu_S) \to 0$ for all $S \neq 0$. Thus, in the thermodynamic limit, by Eq. (4.17b) $\mu = \mu_0$ and we observe that the stationary value of this auxiliary field corresponds to the mean magnetization. Let us study Eq. (4.16) in more detail. First consider the case $H = 0$ for which

$$\mathcal{L}(\mu, 0) = \frac{J}{2}\mu^2 - \frac{1}{\beta}\log 2ch\beta J\mu \ , \tag{4.19a}$$

and the stationary condition is

$$\mu = th\,\beta J\mu \ . \tag{4.19b}$$

The solutions of Eq. (4.19b) are the intersection of the straight line μ and the hyperbolic tangent graphed in Fig. 4.5a.

There are two qualitatively different types of solutions, depending on the slope of the hyperbolic tangent curve at the origin μ. If $\beta J < 1$, that is $T > J$, there is only one solution $\mu = 0$, corresponding to zero magnetization, and this solution is identified with the paramagnetic phase. The shape of $\mathcal{L}(\mu, 0)$ is shown in Fig. 4.5b for $T > T_c$ and has a single minimum at $\mu = 0$, with free energy $F/N = \mathcal{L}(\mu, 0) = -T \ln 2$. The susceptibility is obtained by differentiating Eq. (4.16b) with respect to H, $H = 0$.

$$\chi_0 = \frac{1}{\beta}\frac{\partial\mu}{\partial H}\bigg|_{H=0} = \frac{1 - m^2}{1 - \beta J(1 - m^2)}\bigg|_{H=0} \tag{4.20a}$$

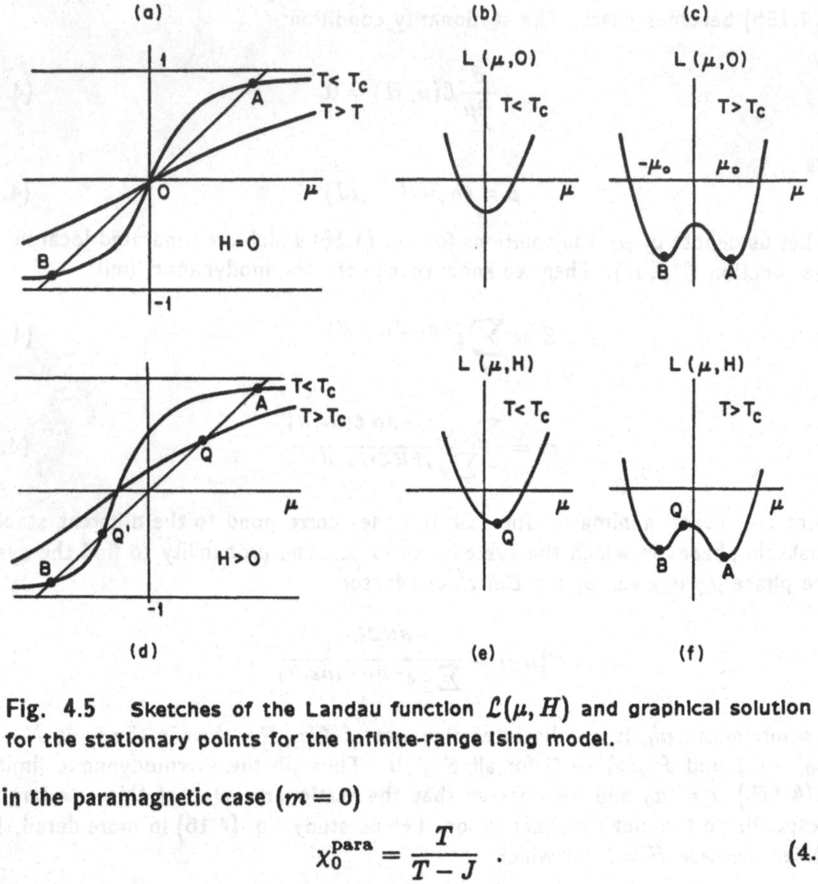

Fig. 4.5 Sketches of the Landau function $\mathcal{L}(\mu, H)$ and graphical solution for the stationary points for the infinite-range Ising model.

and in the paramagnetic case ($m = 0$)

$$\chi_0^{\text{para}} = \frac{T}{T - J} \; . \tag{4.20b}$$

The susceptibility diverges at $T_c = J$, which we thus identify as a critical point.

If $\beta J > 1$, that is $T < T_c$, Eq. (4.19b) has three solutions, which are the extrema of $\mathcal{L}(\mu, 0)$ graphed in Fig. 4.5c. Since the stationary solutions at the minima determine the physical solution, we ignore the solution $\mu = 0$ corresponding to a maximum.

The two physical solutions are the minimum A, corresponding to the positive magnetization $\uparrow$ phase and the minimum B corresponding to the negative magnetization $\downarrow$ phase. Both $\uparrow$ and $\downarrow$ phases have the same free energy, and thus both phases can be present in the system.

Close to T_c, the magnetization μ_0 of one phase can be calculated by expanding Eq. (4.19b) to third order in μ around $\mu = 0$,

$$\mu_0 \simeq \beta J \mu_0 - \frac{1}{3}(\beta J)^3 \mu_0^3 \tag{4.21a}$$

which yields:

$$\mu_0 \simeq \sqrt{\frac{3T^2}{T_c^3}(T_c - T)} \; . \tag{4.21b}$$

The model provides a concrete example of the need to define the zero field magnetization carefully as the limit in which an external field is turned off. Simply arguing in terms of equal Boltzmann weights, one would naively calculate the zero field magnetization to be

$$m = \frac{1}{2}(\mu_A + \mu_B) = \frac{1}{2}(\mu_0 - \mu_0) = 0 \quad . \tag{4.22}$$

However, as we shall now see, introduction of an infinitesimal field H, which we will assume to be positive, lifts the degeneracy between the two phases and leads to a unique absolute minimum of the Landau function. The stationarity condition $\mu = th\,\beta(H + \mu J)$, Eq. (4.16b), is solved graphically for a small magnetic field in Fig., 4.5d.

Away from the critical point, using the definition of the zero field susceptibility, the shift $\delta\mu$ away from the zero field solution μ_0 induced by a small field H is

$$\begin{aligned} \delta\mu &= \left.\frac{\partial\mu}{\partial H}\right|_{H=0} H \\ &= \beta\chi_0 H \ . \end{aligned} \tag{4.23}$$

So, as shown in Fig. 4.5, for $T > T_c$, there is only one solution, point Q, with magnetization $m = \chi_0 H$, which goes to zero as H goes to zero. The minimum of the Landau functional is displaced in the direction of H. If $T < T_c$ as shown in Fig. 4.5 there are three solutions: points A, B and Q'.

The solution Q', corresponding to $\mu_{Q'} = \frac{\beta H}{1-\beta J}$ is again a maximum of the Landau function and thus we must only consider the two phases A, corresponding to ↑, and B, corresponding to ↓. The magnetization of these phases is given by:

$$\mu_{\{{A \atop B}\}} = \pm\mu_0 + \beta\chi_0 H \ . \tag{4.24}$$

Their corresponding Landau function and Boltzman factors are given by

$$\mathcal{L}\left(\mu_{\{{A \atop B}\}}, H\right) = \mathcal{L}(\mu_0, 0) \mp \mu_0 H \tag{4.25}$$

$$P_{\{{A \atop B}\}} = P(\mu_0)e^{\pm N\beta\mu_0 H} \quad . \tag{4.26}$$

Thus, we see that the introduction of a small magnetic field is sufficient to lift the degeneracy between the ↑ and ↓ phases, and that:

$$\lim_{N\to\infty} \frac{P_B}{P_A} = \lim_{N\to\infty} e^{-2N\beta\mu_0 H} \quad . \tag{4.27}$$

So, the probability for the system to be in a phase opposite to the magnetic field is zero in the thermodynamic limit (as long as H does not go to zero as $\frac{1}{N}$ or faster) and the magnetization is

$$\mu_a = (\mu_0 + \beta\chi_0 H)\,P_A + (-\mu_0 + \beta\chi_0 H)\,P_B \tag{4.28}$$

which becomes equal to μ_0 when H goes to zero. Thus, an infinitesimal field selects the phase with magnetization parallel to the field and the system acquires a finite magnetization.

The non-analyticity of the free energy at $H = 0$ comes from the fact that at $H = 0$, the magnetization of the system flips. The free energy thus has the form sketched in Fig. 4.4 and the magnetization as a function of H behaves as the $M - H$ phase diagram in Fig. 4.1.

Because the stationary-phase approximation is exact for this model, we see clearly how the system locks onto one or the other of the two degenerate free field solutions as H is turned off, depending on the way the system is prepared. In practice, we will see that the process of turning off an external field means that whenever we encounter discretely separated degenerate minima in the Landau function, we calculate observables by taking into account only one of the minima.

GENERALIZATIONS

The features we have just seen for the infinite-range Ising model are very suggestive of the structure which will emerge in the next section when we address more general systems. In preparation, it is useful to discuss more generally some of the salient concepts.

Order Parameters. In many cases of physical interest, we will be able to express the partition function as an integral $Z = \int d\mu e^{-\beta \mathcal{L}(\mu,H)}$ such that at the stationary point, μ_S corresponds to the thermal average of a microscopic observable and satisfies the Landau definition of an order parameter. In the infinite-range Ising model, $\mu_s = m = \langle \frac{1}{N} \sum_i S_i \rangle$ and we found that m was zero above $T_c = T$, and non-zero below T_c. The essential issue is to identify the physical order parameter. Once it is known, it is straightforward to formulate a suitable integral for Z.

Thus far, we have restricted our attention to global order parameters which specify the value of some physical observable averaged over the entire system. It will also prove useful subsequently to consider a local order parameter specifying the average value of the observable at a specific position. Thus, for the magnetic example, the local order parameter is the local magnetization $m_i = \langle S_i \rangle$ and the global order parameter is the total magnetization $m = \frac{1}{N} \sum_i m_i$ representing the integral over all space. In Fourier space, m is just the $k = 0$ mode. We shall see that it is easy to generalize the Gaussian transformation for a single global order parameter μ to a general Gaussian integral over a local order parameter μ_i defined on all the lattice sites.

Landau Energy. The exponent $\mathcal{L}(\mu, H)$ appearing in the integral for the partition function naturally satisfies the definition of the Landau function. The absolute minimum of $\mathcal{L}(\mu, H)$ dominates Z in the stationary phase approximation, so the minimum of $\mathcal{L}$ specifies the state of the system.

Phase Boundary and Critical Point. The order parameter varies discontinuously at a first-order phase transition and continuously at a second-order phase transition. In the infinite-range Ising model, m varied discontinuously as H crossed zero below T_c and the transition was thus first order. This discontinuity in H clearly arose from the $N \to \infty$ limit since the ratio of the thermodynamic weights of two nearly degenerate wells in Eq. (4.10), $\frac{P_A}{P_B} = e^{-2N\beta\mu_0 H}$, would go to one as $H \to 0$ at finite N. The magnetization approached zero continuously at zero field at the second order critical point T_c. For second-order transitions, the behavior near the critical point is characterized by the critical exponent β, defined by

$$m \underset{T \to T_c}{\sim} \left(\frac{T_c - T}{T}\right)^{\beta} \qquad T < T_c \, . \tag{4.29}$$

For the infinite range Ising model, by Eq. (4.21b) we observe $\beta = \frac{1}{2}$. The definitions of critical exponents introduced in this and subsequent sections are summarized in Table 4.3 at the end of the chapter for reference.

Conjugate Field. A field which couples linearly to the microscopic variable whose expectation value is the order parameter is called the field conjugate to the order parameter. In the Ising example, H is the conjugate field because of the $H \cdot \sum_i S_i$ coupling in Eq. (4.10). The conjugate field, or symmetry breaking field, breaks the degeneracy of the degenerate double well and drives the discontinuous transition below T_c from finite positive m at positive H to finite negative m at negative H. A second order transition occurs for some specific value H_c of the conjugate field at the critical temperature. The behavior of m at T_c for H near H_c is described by the critical exponent δ defined

$$m \underset{H \to H_c}{\sim} |H - H_c|^{\frac{1}{\delta}} \, . \tag{4.30a}$$

For our Ising example, $H_c = 0$, and expanding Eq. (4.16b) to third order in m at T_c yields

$$m^3 = \frac{3}{J} H \tag{4.30b}$$

so that $\delta = 3$.

Susceptibility. For the Ising example we have seen in Eq. (4.6) that the susceptibility for the order parameter, defined as the linear response of the order parameter to an infinitesimal conjugate field, is also the fluctuation of the order parameter. This relation holds in general for all cases in which the conjugate field couples linearly to the order parameter. Thus, in general, the susceptibility diverges at a second order phase transition, and the critical exponent γ characterizing this divergence is defined by

$$\chi \underset{T \to T_c}{\sim} |T - T_c|^{-\gamma} \, . \tag{4.31}$$

For the infinite range example, according to Eq. (4.20) $\gamma = 1$. In the case of a first order transition, there is no divergence in the susceptibility.

Universal Behavior. In the present formulation, the critical behavior of a system is described by an integral over an order parameter of an action which is a functional of the order parameter. To the extent to which the functionals of the order parameter are equivalent for different systems, the critical behavior of these systems will coincide. The discussion of the power series expansion of the Landau function is highly suggestive that only a finite number of terms may be relevant to the critical behavior, and the form of these terms may in fact be highly constrained by symmetry considerations. Thus, it is not surprising that a large number of physically distinct systems may belong to the same universality class in the sense that they have identical critical exponents. For example, the ferromagnet $YFeO_3$, the antiferromagnet FeF_2, the fluid Xe, the alloy

β-brass, the molecular crystal $NH_4C\ell$, and the three-dimensional Ising model are all characterized by a spin-like order parameter and belong to the same universality class.

Symmetry Breaking. Very often, a phase transition occurs through the mechanism of a symmetry breaking. In the magnetic example, the Hamiltonian $H = \frac{J}{2N}\sum_{ij} S_i S_j$ is invariant with respect to reversal of all the spins $S_i \to -S_i$. Normally, we would expect that $\langle S_i \rangle = \langle -S_i \rangle = 0$. The appearance of a finite magnetization is the manifestation of a broken symmetry, in which $\langle S_i \rangle \neq \langle -S_i \rangle$.

More generally, let us consider a Hamiltonian $\mathcal{H}(q_i)$ invariant under a group $\mathcal{G}$ of transformations on its degrees of freedom $\{q_i\}$:

$$\forall g \in \mathcal{G},\ \forall q_i : \quad \mathcal{H}(gq_i) = \mathcal{H}(q_i) \quad . \tag{4.32}$$

Then, all states $\{q_i'\}$ obtained from $\{q_i\}$ by the action of the group G have the same energy, and thus the same Boltzman factor. The conjugate field must have the property that when added linearly to the Hamiltonian, it lifts (at least partially) the degeneracy of $\mathcal{H}$. For example, in the magnetic case, $H \cdot \sum S_i$ breaks the invariance with respect to spin reversal. Hence, the symmetry group $\mathcal{G}'$ of the perturbed Hamiltonian

$$\mathcal{H}'(q_i) = \mathcal{H}(q_i) + H \cdot m(q_i) \tag{4.33}$$

must be a subgroup of the group $\mathcal{G}$ ($\mathcal{G}'$ strictly included in $\mathcal{G}$). The perturbed Hamiltonian $\mathcal{H}'$ then satisfies

$$\forall g \in \mathcal{G}' : \quad \mathcal{H}'(gq_i) = \mathcal{H}'(q_i) \tag{4.34a}$$

and

$$\forall g \in \mathcal{G}, g \notin \mathcal{G}', \quad \mathcal{H}'(gq_i) \neq \mathcal{H}'(q_i) \quad . \tag{4.34b}$$

We will see in Section 4 that the nature of the residual subgroup $\mathcal{G}'$ controls the stability and excitation spectrum of the ordered low-temperature phase. In general, the knowledge of the order parameter is guided by general symmetry considerations, and by some physical intuition of the nature of the phase transition and of the conjugate field. In some cases, however, the order parameter can be a non-measurable quantity, not uniquely defined, or its physical nature can be very complex (as in the case of spin glasses).

PHYSICAL EXAMPLES

Table 4.1 summarizes the order parameter governing the phase transitions for a variety of physical systems, as well as their coupling to conjugate fields. The antiferromagnet is similar to the ferromagnet we have discussed at length with the exception that the interaction $J\sum_{\langle ij \rangle} \vec{S}_i \vec{S}_j$ favors antialignment of adjacent spins. For bipartite lattices, such that every spin on one sublattice A is surrounded by nearest neighbors on the other sublattice B and *vice versa*, m and H for the ferromagnet are simply replaced by staggered fields m_S and H_S defined with opposite signs on the two sublattices. The ferroelectric transition in polar crystals is analogous to the ferromagnet with the spins S_i and magnetic field H replaced by the polarization d_i and electric field E. Similarly, the binary alloy transition is analogous to the liquid-gas transition we have described.

Phase Transition	Order Parameter	Conjugate Field Coupling	N
Ferromagnetic	magnetization $\vec{m}_i = \langle \vec{S}_i \rangle$	magnetic field $\vec{H} \cdot \sum_i \vec{S}_i$	1,2,3
Antiferromagnetic	staggered magnetization $\vec{m}_i = \begin{cases} \vec{S}_i & i \in A \\ -\vec{S}_i & i \in B \end{cases}$	staggered field $\vec{H}_S \cdot \left(\sum_{i\in A} \vec{S}_i - \sum_{i \in B} \vec{S}_i \right)$	1,2,3
Ferroelectric (polar crystals)	polarization $\vec{d}_i$	electric field $\vec{E} \cdot \sum_i \vec{d}_i$	1,2,3
Liquid-gas	density difference $\rho_{\text{liquid}} - \rho_{\text{gas}}$	pressure P	1
Binary alloy	sublattice concentration	chemical potential μ	1
Superfluid λ transition in ^{4}He	condensate amplitude $\phi(x) = \langle \hat{\psi}(x) \rangle$	condensate source $J\hat{\psi}^\dagger + \text{h.c.}$	2
Superconducting (metals)	electron pair amplitude $\Delta(x,x') = \langle \hat{\psi}_\uparrow(x) \hat{\psi}_\downarrow x' \rangle$	electron pair source $J \hat{\psi}^\dagger_{k\uparrow} \hat{\psi}^\dagger_{k\downarrow} + \text{h.c.}$	2

Table 4.1 Order parameters, the coupling to conjugate fields, and the number of components N for a variety of physical phase transitions.

The new form of transition included in Table 4.1 is that associated with Bose condensation, which underlies the superconducting and superfluid transitions. For orientation, recall the simple case of Bose condensation of a non-interacting gas. The density of particles in a box of volume $\mathcal{V}$ is given by the sum over the discrete momentum states in the box

$$n = \frac{1}{\mathcal{V}} \sum_p \frac{1}{e^{\beta\left(\frac{p^2}{2m} - \mu\right)} - 1} . \tag{4.35}$$

The density increases monotonically as μ approaches zero, at which point the $p = 0$ mode becomes macroscopically occupied and must be separated from the rest of the sum which may be replaced by an integral

$$\begin{aligned} n &= n_0 + \int \frac{d^3p}{(2\pi)^3} \frac{1}{e^{\beta \frac{p^2}{2m}} - 1} \\ &= n_0 + \left(\frac{mkT}{2\pi} \right)^{3/2} \varsigma\left(\frac{3}{2}\right) \end{aligned} \tag{4.36}$$

where $\varsigma(3/2) = 2.612\ldots$ is the Riemann ς-function. Bose condensation, that is, macroscopic occupation of the $p = 0$ mode, thus sets in at the critical temperature

$$T_c = \frac{2\pi}{km} \left(\frac{n}{\varsigma(3/2)} \right)^{2/3} \tag{4.37}$$

below which the condensate density may be written

$$n_0 = n \left(1 - \left(\frac{T}{T_c} \right)^{3/2} \right) . \tag{4.38}$$

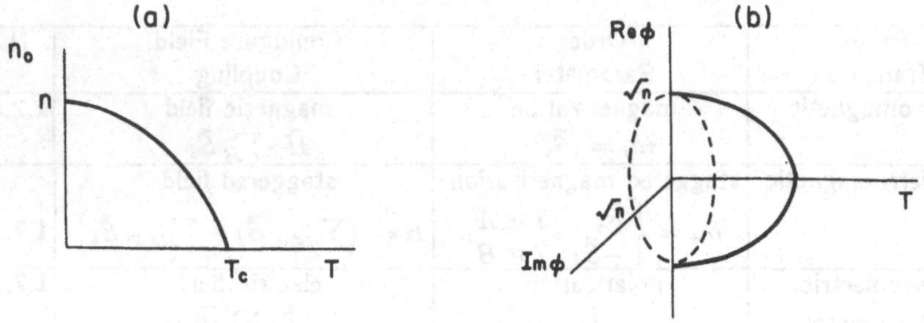

Fig. 4.6 Bose condensate density as a function of temperature (a) and sketch of the phase surface of the complex order parameter ϕ (b).

The condensate density thus has the behavior sketched in Fig. 4.6a characteristic of an order parameter or a function of an order parameter.

To identify the local order parameter and understand the similarity to the magnetic example, it is helpful to recall the stationary-phase treatment of Bose condensation in Section 2.5. We wrote the path integral in the space of coherent states $|\phi\rangle = e^{\int dx\, \phi(x)\, \psi^\dagger(x)}|0\rangle$ which have the property $\frac{\langle\phi|\psi(x)|\phi\rangle}{\langle\phi|\phi\rangle} = \phi(x)$. The action in the resulting path integral, Eq. (2.215a), depended upon the fields only through bilinear combinations $\phi^*(x,t)\ \phi(x',t)$, so the action was invariant with respect to the global gauge transformation $\phi(x,t) \rightarrow e^{i\alpha}\phi(x,t)$. In the stationary-phase approximation, the action was dominated by a constant solution, with $\phi(x) = 0$ (appropriate for the high temperature phase) or $\phi(x) = \phi_c$ where ϕ_c is a complex constant with magnitude $\frac{\mu}{\bar{v}(0)}$ and arbitrary phase reflecting the gauge symmetry. Thus, in this approximation, the condensed phase is characterized by a state in which $\langle\hat{\psi}(x)\rangle = \phi_c$ which is analogous to the magnetic case characterized by the state with $\langle\vec{S}(x_i)\rangle = \vec{m}$. We therefore identify $\langle\hat{\psi}(x)\rangle$ as the local order parameter and $\int dx \langle\hat{\psi}(x)\rangle$ as the global order parameter. Note that because the solution is spatially uniform, only the zero momentum mode is non-vanishing. Also, by the properties of coherent states $\frac{\langle\phi_c|\hat{\psi}^\dagger(x)\hat{\psi}(x)|\phi_c\rangle}{\langle\phi_c|\phi_c\rangle} = |\phi_c|^2$ so that the condensate density is $n_0 = |\phi_c|^2$ and, as expected, is a function of the order parameter. Because the order parameter ϕ_c is complex with an arbitrary phase, the phase diagram has the structure sketched in Fig. 4.6b. The two degrees of freedom in the complex order parameter ϕ are analogous to two components of $\vec{m}$ in the $x - y$ model for which the magnitude $|\vec{m}|$ is fixed but the orientation is arbitrary. As in the spin case, there is a Goldstone mode associated with the degenerate phase angle.

The analogy with spin may be extended further by considering the addition of external sources $\int dx[\hat{\psi}^\dagger(x) \cdot J(x) + J^*(x)\hat{\psi}(x)]$ as in Section 2.4 analogous to the coupling to an external magnetic field $\int dx_i\, \vec{H}(x_i) \cdot \vec{S}(x_i)$. For finite sources J^*, J, the Hamiltonian is no longer number-conserving, so for the exact eigenstate we know $\langle\hat{\psi}(x)\rangle_{J^*,J} \neq 0$. It is only in the case of identically vanishing sources that number conservation implies $\langle\hat{\psi}(x)\rangle_{J^*=J=0} = 0$, just as for the zero field ferromagnet, equal Boltzmann weighting of all states would imply $\langle\vec{S}(x_i)\rangle = 0$. Thus, in the zero field limit, the Bose condensate exhibits broken symmetry, and depending upon the path along which the sources are turned off, the condensate has a definite phase.

If one wants to construct a solution of fixed particle number, one must project by taking the appropriate linear combination of the degenerate solutions. For example, if we write $\phi_c = \rho e^{i\phi}$, a state of definite particle number may be projected from a coherent state as follows

$$\begin{aligned}\int_0^{2\pi} \frac{d\phi}{2\pi} e^{-i\phi N} |\phi_c\rangle &= \int_0^{2\pi} \frac{d\phi}{2\pi} e^{-i\phi N} e^{\phi_c\, a^\dagger(k=0)} |0\rangle \\ &= \int_0^{2\pi} \frac{d\phi}{2\pi} e^{-i\phi N} \sum_m \left(\rho e^{i\phi}\right)^m \frac{1}{\sqrt{m}} |m\rangle \\ &= \frac{\rho^N}{\sqrt{N}} |N\rangle \,.\end{aligned} \tag{4.39}$$

Once the order parameter for Bose condensation is recognized, the order parameter for superconducting or superfluid transitions follow naturally. For superconducting metals, pairs of electrons at the Fermi surface form bound states (Cooper pairs) which behave effectively as Bosons and thus undergo Bose condensation. Since the bound states are spin singlets, and the annihilation operator for a pair is $\hat{\psi}_\uparrow(x)\hat{\psi}_\downarrow(x')$, there is just one complex order parameter $\Delta = \langle \hat{\psi}_\uparrow(x)\hat{\psi}_\downarrow(x')\rangle$. In liquid ^{3}He, the short range repulsion is so strong that pairs in relative S-waves are much less bound than in relative p-waves, and by antisymmetry the p-wave pair wave function must be a spin triplet. Coupling spin 1 and angular momentum 1 involves 9 different complex amplitudes, and the resulting 18 real components of the order parameter give rise to an exceedingly rich physical phase structure.

4.3 MEAN FIELD THEORY

The mean field theory, first introduced by P. Weiss in the study of the ferromagnetic transition, provides a simple yet powerful theory of phase transitions. We will see that its validity depends crucially on the spatial dimension d. For d sufficiently large, greater than an upper critical dimension d_c, the mean field theory is very good at all temperatures. It yields the exact critical exponents, which means that it correctly takes into account the divergences which occur at the critical point, and it provides a starting point for systematic corrections. Below d_c but above a lower critical dimension d_ℓ, the mean field still works well except close to the critical point, where it yields the incorrect critical behavior. Below the lower critical dimension, d_ℓ, the mean field theory is invalid and its predictions are qualitatively wrong.

LEGENDRE TRANSFORM

To formulate the theory in terms of the order parameter $\vec{m}_i$ rather than the external field $\vec{H}_i$, it is natural to use the Gibbs free energy, $\Gamma(\vec{m}_i)$, which is the Legendre transform of the free energy $F(\vec{H}_j)$. Hence, let us review several basic ideas from Section 2.4, where the source J_α corresponded to the external field H_i, the field ϕ_α corresponded to the order parameter m_i, and the generating function for connected

Green's functions $-W(J_\alpha)$ corresponded to the free energy $F(H_i)$. The order parameter is given by

$$\vec{m}_i(\vec{H}_j) = -\frac{\partial F}{\partial \vec{H}_i} \tag{4.40}$$

and the Gibbs free energy is defined

$$\Gamma(\vec{m}_i) = F\left(\vec{H}_i(\vec{m}_j)\right) + \sum_i \vec{H}_i(\vec{m}_j)\cdot\vec{m}_i \tag{4.41}$$

where $\vec{H}_i(\vec{m}_j)$ is obtained by inverting Eq. (4.40).

Below the transition temperature, we know that the zero field order parameter depends on the path by which $\vec{H}$ approaches zero, so the equation of state

$$\lim_{H_j\to 0}\frac{\partial F}{\partial \vec{H}_i} = \lim_{\vec{H}_j\to 0} -\vec{m}_i(\vec{H}_j) \tag{4.42}$$

also depends upon the path. In contrast, the reciprocity relation

$$\vec{H}_i = \frac{\partial \Gamma}{\partial \vec{m}_i} \tag{4.43}$$

leads to the unique zero-field equation of state

$$\frac{\partial \Gamma}{\partial \vec{m}_i} = 0 \tag{4.44}$$

which admits a set of broken symmetry solutions which correspond to all the different orientations assumed by the magnetic field when it is taken to zero.

Since $F(\vec{H}_i)$ is the generating function for connected Green's functions, the Green's function specifying the spin-spin correlation function and response of $\vec{m}_i$ to $\vec{H}_j$ is calculated in the usual way:

$$\begin{aligned}
-\frac{1}{\beta}\frac{\partial^2 F}{\partial H_i \partial H_j} &= \frac{1}{\beta}\frac{\partial m_j}{\partial H_i} \\
&= \frac{1}{\beta}\frac{\partial}{\partial H_i}\frac{tr\left(S_j e^{-\beta(\mathcal{H}-\sum_i \vec{S}_i\cdot\vec{H}_i)}\right)}{tr\left(e^{-\beta(\mathcal{H}-\sum_i \vec{S}_i\cdot\vec{H}_i)}\right)} \\
&= \langle S_i S_j\rangle - \langle S_i\rangle\langle S_j\rangle \\
&\equiv G_{ij}\ .
\end{aligned} \tag{4.45}$$

For a translationally invariant system, G depends only on the relative distance $(\vec{r}_i - \vec{r}_j)$ and the Fourier transform is defined

$$G(\vec{q}) = \sum_j e^{-i\vec{q}\cdot(\vec{r}_i-\vec{r}_j)} G(r_i - r_j) \tag{4.46}$$

and in the long wavelength limit yields the static susceptibility, Eq. (4.6)

$$\begin{aligned}\lim_{\vec{q}\to 0} \beta G(\vec{q}) &= \sum_j \beta G_{ij} \\ &= \frac{\beta}{N}\sum_{ij}(\langle S_i S_j\rangle - \langle S_i S_j\rangle) \\ &= \chi \ .\end{aligned} \tag{4.47}$$

By the same argument used in Eq. (2.169), the Green's function may be expressed in terms of the Gibbs free energy using the relation

$$\begin{aligned}\delta_{ij} &= \frac{\partial m_i}{\partial m_j} = \frac{\partial m_i}{\partial H_k}\frac{\partial H_k}{\partial m_j} \\ &= -\frac{\partial^2 F}{\partial H_i \partial H_k}\frac{\partial^2 \Gamma}{\partial m_k \partial m_j}\end{aligned} \tag{4.48a}$$

so that

$$G_{ij} = -\frac{1}{\beta}\frac{\partial^2 F}{\partial H_i \partial H_j} = \frac{1}{\beta}\left[\frac{\partial^2 \Gamma}{\partial m_i \partial m_j}\right]^{-1} \ . \tag{4.48b}$$

Finally, as discussed in Sections 2.4 and 2.5, we recall that the perturbative expansion of the Legendre transform is identical to the loop expansion for the free energy. Hence, by using the stationary phase approximation to obtain the loop expansion of the free energy around the mean field, we will obtain a systematic perturbative solution to the equation of state, Eq. (4.44).

FERROMAGNETIC TRANSITION FOR CLASSICAL SPINS

To illustrate the mean field theory, we shall study the classical $O(n)$ spin model on a D-dimensional cubic lattice with lattice spacing a. The Hamiltonian is

$$\mathcal{H}(\vec{S}_i) = -\frac{1}{2}\sum_{ij} J_{ij}\vec{S}_i\cdot\vec{S}_j - \sum_j \vec{H}_i\cdot\vec{S}_i \tag{4.49a}$$

with the nearest-neighbor interaction

$$J_{ij} = \begin{cases} J & \text{if i and j are nearest neighbors} \\ 0 & \text{otherwise}\end{cases} \tag{4.49b}$$

and the constraint

$$\vec{S}_i^2 \equiv \sum_{\alpha=1}^{n}(S_i^\alpha)^2 = 1 \tag{4.49c}$$

where $\vec{S}_i$ denotes the n components $\{S_i^\alpha\}$ of a unit spin on the ith site of a cubic lattice, and $\vec{H}_i$ denotes the n components of a spatially varying magnetic field on the ith site. Recall that the cases $n = 1, 2, 3$ correspond to the Ising, $x-y$, and Heisenberg models, respectively. Note that in the absence of a magnetic field, the scalar products $\vec{S}_i\cdot\vec{S}_j$ are invariant if all the spins are rotated by the same angle so the Hamiltonian has $O(n)$ symmetry. In the presence of a constant magnetic field $\vec{H}_i = \vec{H}$, the symmetry group is reduced to $O(n-1)$, the global rotations of spins around the direction of $\vec{H}$.

To evaluate the partition function, as in the case of the infinite-range Ising model, we introduce an integral over an auxiliary field using the Gaussian identity Eq. (1.179)

$$\begin{aligned}
Z &= \int \prod_i d\vec{S}_i \mathcal{N}^{-1} \delta(\vec{S}_i^2 - 1) e^{\frac{\beta}{2}\sum_{ij} J_{ij} \vec{S}_i \cdot \vec{S}_j + \beta \sum_i \vec{S}_i \cdot \vec{H}_i} \\
&= C \int \prod_i d\vec{\phi}_i e^{-\frac{\beta}{2}\sum_{ij} J_{ij}^{-1} (\vec{\phi}_i - \vec{H}_i)\cdot(\vec{\phi}_j - \vec{H}_j)} \int \prod_i d\vec{S}_i \mathcal{N}^{-1} \delta(\vec{S}_i^2 - 1) e^{\beta \sum_i \vec{\phi}_i \cdot \vec{S}_i} \\
&= C \int \prod_i d\vec{\phi}_i e^{-\beta S(\vec{\phi}_i, \vec{H}_i)}
\end{aligned} \tag{4.50a}$$

where

$$S(\vec{\phi}_i, \vec{H}_i) = \frac{1}{2}\sum_{ij} J_{ij}^{-1} (\vec{\phi}_i - \vec{H}_i)\cdot(\vec{\phi}_j - \vec{H}_j) - \frac{1}{\beta}\sum_i \ln\left(\int \frac{d\vec{S}}{\mathcal{N}} \delta(\vec{S}^2 - 1) e^{\beta \vec{\phi}_i \cdot \vec{S}}\right) . \tag{4.50b}$$

The constant $C = \left(\frac{2\pi}{\beta}\right)^{-N/2} (\det J_{ij})^{-1/2}$ and the factor $\mathcal{N}$ normalizing the spin sum correspond to additive constants in the free energy which will not affect our results.

In the case of one-component Ising spins, the spin sum on a single site appearing in (4.41) is simple

$$\begin{aligned}
\mathcal{N} \int dS \delta(S^2 - 1) e^{\beta \phi_i S} &= \sum_{S=\pm 1} e^{\beta \phi_i S} \\
&= 2ch\, \beta \phi_i \ .
\end{aligned} \tag{4.51a}$$

For higher n, it may be evaluated in polar coordinates yielding

$$\mathcal{N} \int d\vec{S} \delta(\vec{S}^2 - 1) e^{\beta \vec{\phi}_i \cdot \vec{S}_i} \propto \int_0^\pi d\theta \sin^{n-2}\theta\, e^{\beta |\phi_i| \cos\theta} \tag{4.51b}$$

which may be expressed in terms of Bessel functions. Alternatively, the spin sum may be expanded in powers of ϕ as in Problem 4.3.

For simplicity, we shall first consider the single component Ising model, for which

$$S(\phi_i, H_i) = \frac{1}{2}\sum_{ij} (\phi_i - H_i) J_{ij}^{-1} (\phi_j - H_j) - \frac{1}{\beta}\sum_i \ln\left[2ch(\beta\phi_i)\right] \ . \tag{4.52}$$

The mean-field approximation is obtained by applying the stationary-phase approximation

$$\begin{aligned}
e^{-\beta F} &= \int \prod_i d\phi_i e^{-\beta S(\phi_i, H_i)} \\
&\approx e^{-\beta S(\bar{\phi}_i, H_i)}
\end{aligned} \tag{4.53}$$

where the stationary solutions $\overline{\phi}_i$ satisfy the mean-field equations

$$\left.\frac{\partial S}{\partial \phi_i}\right|_{\overline{\phi}} = 0 = \sum_j J_{ij}^{-1}(\overline{\phi}_j - H_j) - th(\beta\overline{\phi}_i) \tag{4.54a}$$

or

$$\overline{\phi}_i = H_i + \sum_j J_{ij}\, th(\beta\overline{\phi}_j) \ . \tag{4.54b}$$

Since $F(H_i) = S(\overline{\phi}_i(H_k), H_i)$ in this approximation and $\frac{\partial S}{\partial \phi_i} = 0$, the magnetization is given by

$$\begin{aligned} m_i &= -\frac{dS}{dH_i} = -\frac{\partial S}{\partial H_i} \\ &= \sum_j J_{ij}^{-1}(\overline{\phi}_j - H_j) \\ &= th(\beta\overline{\phi}_i) \ . \end{aligned} \tag{4.55}$$

Using Eqs. (4.54b) and (4.55) to determine $H_i(m_j)$

$$H_i = -\frac{1}{\beta}\sum_j J_{ij}m_j + \frac{1}{\beta}th^{-1}m_i \tag{4.56}$$

and using the relations $ch(th^{-1}x) = (1-x^2)^{-1/2}$ and $th^{-1}(x) = \frac{1}{2}\ln\left(\frac{1+x}{1-x}\right)$, the Legendre transform is

$$\begin{aligned} \Gamma(m_i) &= S\left(\overline{\phi}_j(m_i), H_j(m_i)\right) + \sum_i H_i(m_j)m_i \\ &= \frac{1}{2}\sum_{ij} m_iJ_{ij}m_j - \frac{1}{\beta}\sum_i \ln\left[2(1-m_i^2)^{-1/2}\right] - \sum_{ij} m_iJ_{ij}m_j + \frac{1}{\beta}\sum_i m_i th^{-1}m_i \\ &= -\frac{1}{2}\sum_{ij} m_iJ_{ij}m_j - \frac{N}{\beta}\ln 2 + \frac{1}{2\beta}\sum_i \{(1+m_i)\ln(1+m_i) + (1-m_i)\ln(1-m_i)\} \, . \end{aligned} \tag{4.57}$$

It is straightforward to verify that the equation of state given by $H_i = \frac{\partial \Gamma(m_j)}{\partial m_i}$ is identical to Eq. (4.55).

Specializing to the case of uniform magnetization $m_i = m$, noting that the number of nearest neighbor bonds on a cubic lattice in D dimensions with N sites is DN, and expanding to fourth order in m, we obtain

$$\begin{aligned} \frac{1}{N}\Gamma(m) &= -JDm^2 - \frac{1}{\beta}\ln 2 + \frac{1}{2\beta}\{(1+m)\ln(1+m) + (1-m)\ln(1-m)\} \\ &\approx -T\ln 2 + \frac{m^2}{2}(T - 2JD) + \frac{T}{12}m^4 + O(m^6) \ . \end{aligned} \tag{4.58}$$

We observe that $\Gamma(m)$ has the form of the zero field Landau function sketched in Fig. 4.3 and given in Eq. (4.4) and thus, the system has a second-order phase transition at

$$T_c = 2JD \ . \tag{4.59}$$

For a uniform field, the equation of state $H_i = \frac{\partial \Gamma}{\partial m_i}$ yields

$$H = m(T - T_c) + \frac{T}{3} m^3 \qquad T < T_c \tag{4.60}$$

which implies the zero-field magnetization below T_c

$$m = \pm\sqrt{\frac{3}{T}}(T_c - T)^{1/2} \qquad T < T_c \tag{4.61}$$

corresponding to the critical exponent $\beta = \frac{1}{2}$. At T_c, Eq. (4.59) yields

$$H = \frac{T_c}{3} m^3 \tag{4.62}$$

so that $\delta = 3$.

The spin-spin correlation function and susceptibility near the critical point are obtained using Eqs. (4.48b) and (4.56):

$$[\beta G_{ij}]^{-1} = \frac{\partial^2 \Gamma}{\partial m_i \partial m_j} = -J_{ij} + T\delta_{ij}(1 + m_i^2) + O(m^4) \ . \tag{4.63}$$

Since J_{ij} is a function of the distance between sites and vanishes except for nearest neighbors, its Fourier transform is

$$\begin{aligned} J(\vec{q}) &= a^D \sum_i e^{-i\vec{q}\cdot(\vec{r}_i - \vec{r}_j)} J(\vec{r}_i - \vec{r}_j) \\ &= 2Ja^D \sum_{\alpha=1}^{D} \cos q_\alpha a \\ &= 2Ja^D \left(D - \frac{a^2}{2}\vec{q}^2 + O\left((\vec{q}^2)^2\right) \right) \ . \end{aligned} \tag{4.64}$$

It is customary to simplify the formulas by using $a = 1$ and measuring all lengths in units of the lattice spacing. The Fourier transform of Eq. (4.62) yields the low-q behavior of the Green's function near the critical point

$$\begin{aligned} \beta G(\vec{q}) &= \frac{1}{T - 2J \sum_{\alpha=1}^{D} \cos q_\alpha + Tm^2} \\ &\underset{q\to 0}{\sim} \frac{1}{T - T_c + J\vec{q}^2 + Tm^2} \ . \end{aligned} \tag{4.65}$$

For $T > T_c$, the magnetization is zero, so

$$\beta G(\vec{q}) \underset{q\to 0}{\sim} \frac{1}{T - T_c + T\vec{q}^2} \qquad T > T_c \qquad (4.66a)$$

and, by Eq. (4.47)

$$\chi_+ = \frac{1}{T - T_c} \qquad T > T_c \ . \qquad (4.66b)$$

Note that we will use + and − subscripts to distinguish quantities above and below the critical point.

The Green's function (4.66a) implies an exponential decay in coordinate space

$$G(r) \sim e^{-|\vec{r}|\sqrt{\frac{T-T_c}{J}}} \qquad T > T_c \qquad (4.66c)$$

from which we define the correlation length

$$\xi_+ = \sqrt{\frac{J}{T - T_c}} \qquad T > T_c \ . \qquad (4.66d)$$

As expected physically at a second order transition, the correlation length diverges, and this divergence is measured in general by the critical exponent ν

$$\xi \underset{T\to T_c}{\sim} |T - T_c|^{-\nu} \ . \qquad (4.67)$$

In the mean field approximation, we thus find $\nu = \frac{1}{2}$ and $\gamma = 1$ from Eqs. (4.62d and b), respectively.

Below the critical point, the magnetization is given by Eqs. (4.60) so that we obtain

$$\beta G(\vec{q}) \underset{q\to 0}{\sim} \frac{1}{2(T_c - T) + J\vec{q}^2} \qquad T < T_c \qquad (4.68a)$$

$$\chi_- = \frac{1}{2(T_c - T)} \qquad T < T_c \qquad (4.68b)$$

$$G(r) = e^{-|\vec{r}|\sqrt{\frac{2(T_c-T)}{J}}} \qquad T < T_c \qquad (4.68c)$$

$$\xi_- = \sqrt{\frac{J}{2(T_c - T)}} \qquad T < T_c \qquad (4.68d)$$

with the same critical exponents $v = \frac{1}{2}$ and $\gamma = 1$.

At T_c, the decay of the correlation function is no longer exponential. Rather, it is given by

$$G(r) = \frac{T}{J}\int \frac{d^D q}{(2\pi)^D}\, \frac{e^{i\vec{q}\cdot\vec{r}}}{\vec{q}^2} \propto \frac{1}{|\vec{r}|^{D-2}} \qquad (4.69)$$

where the r dependence is easily recognized by changing variables $\vec{h} \equiv \vec{q}\cdot|\vec{r}|$. The general behavior of the correlation function near a critical point is parameterized by the critical exponent η

$$G(r) \underset{T\to T_c}{\sim} \frac{e^{-r/\xi}}{|\vec{r}|^{D-2+\eta}} \qquad (4.70)$$

so that in the mean field approximation, $\eta = 0$.

The final thermodynamic quantity we consider is the specific heat,

$$C_H = \left.\frac{\partial U}{\partial T}\right|_H = \left.\frac{\partial U}{\partial S}\right|_H \left.\frac{\partial S}{\partial T}\right|_H = T\left.\frac{\partial S}{\partial T}\right|_H = -T\left.\frac{\partial^2 F}{\partial T^2}\right|_H . \tag{4.71}$$

Using (4.58) and (4.61) to evaluate $F(H=0,T) = \Gamma(M(H=0),T)$ yields

$$F(H=0,T) = \begin{cases} -TN \ln 2 - \frac{3N}{4T}(T_c - T)^2 & T < T_c \\ -TN \ln 2 & T > T_c \end{cases} \tag{4.72}$$

so that

$$C_H = \begin{cases} \frac{3N}{2} & T < T_c \\ 0 & T > T_c \end{cases} \tag{4.73}$$

and the mean field theory yields a discontinuity in the specific heat but no singularity. Because of the possibility of an analytic background term in the specific heat, the critical behavior is defined by the relation

$$C_H \underset{T \to T_c}{\sim} |T - T_c|^{-\alpha} + c \tag{4.74}$$

and the mean-field value of the critical exponent is $\alpha = 0$. All the preceding definitions of critical exponents and the mean-field values are tabulated in Table 4.3 at the end of this chapter for reference.

All these mean field results for Ising spins are straightforwardly generalized to an n-component $O(n)$ spin system near the critical point where $S(\vec{\phi}, \vec{H}_i)$, Eq. (4.50b), may be expanded in powers of ϕ. As shown in Problem 4.3, the Legendre transform expanded to fourth order with an irrelevant additive constant omitted is

$$\Gamma(\vec{m}_i) = -\frac{1}{2}\sum_{ij} \vec{m}_i J_{ij} \vec{m}_j + \frac{A}{2} T \sum_i \vec{m}_i^2 + \frac{B}{4}\sum_i (\vec{m}^2)^2 . \tag{4.75a}$$

For uniform magnetization $\vec{m}_i = \vec{m}$ we obtain:

$$\frac{\Gamma(\vec{m})}{N} = \frac{A}{2}\vec{m}^2 \left(T - \frac{2JD}{A}\right) + \frac{B}{4}(\vec{m}^2)^2 \tag{4.75b}$$

and in zero magnetic field, the equation of state is

$$\left(T - \frac{2JD}{A}\right)\vec{m} + \frac{B}{A}(\vec{m}^2)\vec{m} = 0 . \tag{4.76}$$

Again, we see that there is a second-order transition at

$$T_c = \frac{2DJ}{A} . \tag{4.77}$$

The shape of $\Gamma(\vec{m})$ for a two-component spin system is a surface of revolution suggested by the dashed line in Fig. 4.3. For $T > T_c$, $\Gamma(\vec{m})$ has a single paraboloid minimum at $\vec{m} = 0$ whereas for $T < T_c$ the minimum is a circular valley with radius

$$\vec{m}^2 = \frac{A}{B}(T_c - T) \ . \tag{4.78}$$

For larger numbers of components, the minimum is a sphere in higher dimension with radius given by Eq. (4.77). Since $(\vec{m}) \sim (T_c - T)^{1/2}$, we again obtain the critical exponent $B = \frac{1}{2}$.

The correlation function now depends on the lattice sites and spin components and is given by

$$\begin{aligned} \left[\beta G_{ij}^{\alpha\beta}\right]^{-1} &= \frac{\delta^2 \Gamma}{\delta m_i^\alpha \delta m_j^\beta} \\ &= -J_{ij}\delta_{\alpha\beta} + \delta_{ij}\left\{\left(AT + B\vec{m}_i^2\right)\delta_{\alpha\beta} + 2Bm_i^\alpha m_i^\beta\right\} \end{aligned} \tag{4.79}$$

and the Fourier transform for a translationally invariant state with uniform magnetization $\vec{m}$ is

$$\left[\beta G^{\alpha\beta}(\vec{q})\right]^{-1} \underset{q\to 0}{\sim} \delta_{\alpha\beta}\left\{Jq^2 + A(T - T_c) + B\vec{m}^2\right\} + 2Bm^\alpha m^\beta \ . \tag{4.80}$$

The direction of $\vec{m}$ is defined as the longitudinal axis, and all other perpendicular directions are called transverse directions. Above T_c, $\vec{m} = 0$ and the correlation function and susceptibility are isotropic and equal to the Ising result:

$$\beta G^{\alpha\beta}(\vec{q}) = \frac{\delta_{\alpha\beta}}{A(T - T_c) + J\vec{q}^2} \qquad T > T_c \tag{4.81a}$$

$$\chi^{\alpha\beta} = \frac{\delta_{\alpha\beta}}{A(T - T_c)} \qquad T > T_c \tag{4.81b}$$

$$\xi_+ = \sqrt{\frac{J}{A(T - T_c)}} \qquad T > T_c. \tag{4.81c}$$

Below T_c, using Eq. (4.69), we obtain qualitatively different results in the longitudinal and transverse directions. In the longitudinal direction, where $m^\alpha = |\vec{m}|$,

$$\beta G_L(\vec{q}) = \frac{1}{2A(T_c - T) + J\vec{q}^2} \qquad T < T_c \tag{4.82a}$$

$$\chi_L = \frac{1}{2A(T_c - T)} \qquad T < T_c \tag{4.82b}$$

$$\xi_-^L = \sqrt{\frac{J}{2A(T_c - T)}} \qquad T < T_c \tag{4.82c}$$

whereas in the transverse direction, where $m^\alpha = 0$,

$$\beta G_T^{\alpha\beta}(\vec{q}) = \frac{\delta_{\alpha\beta}}{Jq^2} \qquad T < T_c \tag{4.83a}$$

$$\chi_T^{\alpha\beta} = \infty \qquad T < T_c \tag{4.83b}$$

$$\xi^T = \infty \qquad T < T_c. \tag{4.83c}$$

Thus, for any number n of spin components, including $n = 1$, the behavior in the direction of the magnetization is identical. In addition, however, for $n \geq 2$, there are components perpendicular to the direction of $\vec{m}$ which behave totally differently. Since the minimum of the Gibbs free energy Γ is degenerate in these other directions, it costs negligible energy to create excitations in these directions, and the resulting proliferation of spin waves is responsible for the divergence of the susceptibility and the correlation length.

APPLICATION TO GENERAL SYSTEMS

The mean field theory may be applied quite generally to any system for which the local order parameter has been identified. Let us consider the general partition function

$$Z = e^{-\beta F} = \int \Pi_i dq_i e^{-\beta\left[\mathcal{H}(q_i)+\int dx\, \mu(x,q_i)U(x)\right]} \tag{4.84}$$

where $\{q_i\}$ denotes the microscopic degrees of freedom, $\mu(x, q_i)$ is the microscopic local order parameter, $U(x)$ is the external field which couples to the order parameter, and $\int dx$ represents a spatial integral for continuous systems or a sum over lattice sites for crystals.

The basic idea is to express Z as a functional integral over a field which is an order parameter

$$Z = \int \mathcal{D}\left(f(x)\right) e^{-\beta\left[\mathcal{F}(f(x))+\int dx\, f(x)\, U(x)\right]} \ . \tag{4.85}$$

The integration variable $f(x)$ in this equation is an order parameter, since by virtue of its linear coupling to $U(x)$, its thermal average is equal to that of the microscopic order parameter

$$\left.\frac{\partial F}{\partial U(x)}\right|_{U=0} = \langle\mu(x, q_i)\rangle = \langle f(x)\rangle \ , \tag{4.86}$$

where in an obvious notation, the thermal averages $\langle\mu(x, q_i)\rangle$ and $\langle f(x)\rangle$ are defined by Eqs. (4.84) and (4.85), respectively. Hence, application of the stationary phase approximation to Eq. (4.85) directly yields the desired form of a free energy as a function of an order parameter.

Any linear transform of the microscopic order parameter is also an order parameter in the Landau sense, because it vanishes in one phase and is non-zero in the other. Thus, the action in Eq. (4.85) may have a slightly more general coupling term $\int dx\, dy\, U(x)M(x,y)g(y)$. Recall, for example, the functional integral for Ising spins, Eqs. (4.52, 4.53)

$$Z_I = \int d\phi_i\, e^{-\beta\left[\mathcal{F}[\phi_i]-\sum_{ij} H_i J_{ij}^{-1}\phi_j+\frac{1}{2}\sum_{ij} H_i J_{ij}^{-1} H_j\right]} \tag{4.87}$$

in which case the order parameter ϕ was linearly related to the microscopic order parameter by

$$\phi_i = \sum_j J_{ij} \langle S_j \rangle \quad . \tag{4.88}$$

Note that in addition to being an acceptable order parameter in the Landau sense, ϕ_i in this case has the physical interpretation of a mean molecular field. That is, ϕ_i gives the potential seen at site i as a result of interactions J_{ij} with the mean value $\langle S_j \rangle$ of each of the surrounding spins. Clearly, changing the variable of integration in Eq. (4.85) by an arbitrary linear transformation would not change the physical result, and we could just as well use the variable $\chi_i \equiv J_{ij}^{-1} \phi_i$, in which case $\chi_i = \langle S_i \rangle$.

Thus far, in the case of spin systems, we have seen an example of a specific technique, the use of an auxiliary field, to obtain a functional integral for the partition function. However, in general, it is straightforward to obtain the functional integral Eq. (4.85) from Eq. (4.84) by introducing the constraint $f(x) = \mu(x, q_i)$ by use of the identity

$$1 = \int \mathcal{D} f(x) \, \delta\left(f(x) - \mu(x, q_i)\right) \tag{4.89}$$

with the result

$$\begin{aligned} Z &= \int \Pi_i dq_i \int \mathcal{D} f(x) \, \delta\left(f(x) - \mu(x, q_i)\right) e^{-\beta\left[\mathcal{H}(q_i) + \int dx \, \mu(x, q_i) U(x)\right]} \\ &= \int \mathcal{D} f(x) e^{-\beta\left[\mathcal{F}[f(x)] + \int dx f(x) \, U(x)\right]} \end{aligned} \tag{4.90a}$$

where

$$\mathcal{F}[f(x)] = -\frac{1}{\beta} \ln \int \Pi_i dq_i e^{-\beta \mathcal{H}(q_i)} \delta\left(f(x) - \mu(x, q_i)\right) \quad . \tag{4.90b}$$

Even in cases for which $\mathcal{F}[f(x)]$ is too complicated to evaluate exactly, it may be expanded in powers of $f(x)$ and used to study the behavior in the region of the critical point where $f(x)$ is small.

Let us illustrate this method for the liquid-gas transition. The grand partition function for a classical gas, with two-body potential $v(\vec{r})$ in an external one-body potential $U(\vec{r})$ is

$$Z = \sum_{N=0}^{\infty} \frac{e^{\beta\mu N}}{N!} \int \prod_{i=1}^{N} \frac{d^D r_i d^D p_i}{(2\pi\hbar)^D} e^{-\sum_{i=1}^N \beta \frac{p_i^2}{2m} - \frac{\beta}{2} \sum_{i,j=1}^N v(\vec{r}_i - \vec{r}_j) - \beta \sum_{i=1}^N U(\vec{r}_i)} \quad . \tag{4.91a}$$

The momenta $\vec{p}_i$ can be integrated out, yielding:

$$Z = \sum_{N=0}^{\infty} \frac{\lambda^N}{N!} \int \prod_{i=1}^{N} d^D r_i e^{-\frac{\beta}{2} \sum_{i,j} v(\vec{r}_i - \vec{r}_j) - \beta \sum_i U(\vec{r}_i)} \tag{4.91b}$$

where λ is the fugacity, given by:

$$\lambda = e^{\beta\mu} \cdot \left(\frac{m}{2\pi\beta\hbar^2}\right)^{D/2} \quad . \tag{4.91c}$$

As discussed in the introduction, a natural order parameter in the liquid gas phase transition is the density of the fluid (or rather the difference $\rho_L - \rho_G$). Defining the density as

$$\rho(\vec{r}) = \sum_{i=1}^{N} \delta(\vec{r} - \vec{r}_i) \tag{4.92}$$

and using the integral representation of the δ-function

$$\delta\left(\rho(\vec{r}) - \sum_i \delta(\vec{r} - \vec{r}_i)\right) = \int \mathcal{D}\phi(r) e^{i \int d^D r \phi(r)[\rho(\vec{r}) - \sum_i \delta(\vec{r} - \vec{r}_i)]} \tag{4.93}$$

we write the partition function of the system as:

$$\begin{aligned} Z = \sum_{N=0}^{\infty} \frac{\lambda^N}{N!} \int \mathcal{D}\rho(\vec{r}) \mathcal{D}\phi(\vec{r}) e^{i \int d^D r \phi(\vec{r})\rho(\vec{r}) - \frac{\beta}{2} \int d^D r\, d^D r' \rho(\vec{r}) v(\vec{r} - \vec{r}') \rho(\vec{r}')} \\ \times e^{-\beta \int d^D r\, U(\vec{r})\, \rho(\vec{r})} \int \prod_{i=1}^{N} d^D r_i e^{-i \sum_i \phi(r_i)} \\ = \int \mathcal{D}\rho(\vec{r}) \mathcal{D}\phi(\vec{r}) e^{i \int d^D r \phi(\vec{r})\rho(\vec{r}) - \frac{\beta}{2} \int d^D r d^D r' \rho(\vec{r}) v(\vec{r} - \vec{r}') \rho(\vec{r}') + \lambda \int d^D r e^{-i\phi(\vec{r})}} \\ \times e^{-\beta \int d^D r\; U(\vec{r}) \rho(\vec{r})} \quad . \end{aligned} \tag{4.94}$$

Regarded as an integral over $\rho(\vec{r})$, this is precisely of the desired form, Eq. (4.85). However, in the present case, because the action is quadratic in $\rho(r)$, one may proceed even further by integrating out the field $\rho(r)$ to obtain

$$Z = \frac{1}{\mathcal{N}} \int \mathcal{D}\phi(\vec{r}) e^{-\frac{\beta}{2} \int d^D r\, d^D r' (\phi(\vec{r}) - U(\vec{r})) v^{-1}(\vec{r} - \vec{r}') (\phi(\vec{r}') - U(\vec{r}'))} e^{\lambda \int d^D r\; e^{-\beta\phi(\vec{r})}} \tag{4.95}$$

where $v^{-1}(\vec{r})$ is the inverse of $v(\vec{r})$ and we have redefined $\phi(r)$ to be equal to $\frac{i}{\beta}\phi(r)$. We observe that the result is completely analogous to the Ising spin case, written in Eqs. (4.53) and (4.87). The field $\phi(\vec{r})$ plays the role of ϕ_i, $U(\vec{r})$ corresponds to the magnetic field, the potential term $e^{\lambda \int d^D r\, e^{-\beta\phi(\vec{r})}}$ replaces $2\, ch\, \beta\phi_i$, and $\phi(\vec{r}) = \int d^D r'\, v(\vec{r} - \vec{r}') \langle \sum_{i=1}^{N} \delta(\vec{r}' - \vec{r}_i) \rangle = \sum_i \langle v(\vec{r} - \vec{r}_i) \rangle$ is the mean field at point r generated by the equilibrium distribution of the surrounding particles. The liquid gas transition is studied in more detail in Problem 4.4.

Before proceeding beyond the mean field theory, several observations should be made. The first is that the Landau theory discussed in Section 4.1, which was originally introduced on a purely phenomenological basis, follows naturally from mean field theory. Recall that the Landau function was defined as a function of the global order parameter with the property that its minimum specifies the state of the system. When the partition function is written in the form Eq. (4.85) as a functional integral over an order parameter $f(\vec{r})$ and the stationary phase approximation produces a spatially uniform solution, the minimum of the exponent $\left[\mathcal{F}(f(\vec{r})) + \int d\vec{x}\, f(\vec{x})\, U(\vec{x})\right]$ also specifies the state of the system. Hence, the exponent of the functional integral provides a microscopic definition of the Landau function, and the criterion for the validity of the Landau theory is the validity of the stationary phase approximation.

Note that since an analytic function of the order parameter is also an order parameter, there is considerable freedom to define alternative Landau functions. For example, in the case of Ising spins, both $S(\phi_i)$, Eq. (4.52), and $\Gamma(m_i)$: Eq. (4.57) are equivalent Landau functions and are related by the transformation $m_i = th(\beta\phi_i) = \beta\phi_i + \theta(\phi_i^3)$. Since these two order parameters are proportional near the critical point, they necessarily have the same critical behavior.

The previous discussion of the symmetry properties of the Landau function applies both to the Legendre transform and to the exponent of the functional integral, and is frequently useful in identifying the allowed forms of their series expansions. For example, the $O(n)$ symmetry of the n-component spin model implies that only powers of $\vec{m}^2$ may appear as in Eq. (4.75). In contrast, there are no exact symmetries for the order parameter $\phi = \rho_{\text{liquid}} - \rho_{\text{gas}}$ in the liquid gas phase transition, and in fact odd terms in ϕ enter the expansion after fourth order.

The second observation about the mean field theory is the remarkable absence of any dependence of the critical indices of the spin models we considered on the spatial dimension D, the number of spin components n, or even the details of the coupling matrix J_{ij}. These features only affect the actual value of the critical temperature. The critical exponents for the nearest neighbor $O(n)$ model with any n or D are identical to those of the infinite-range model for which the stationary phase approximation is exact. When we examine corrections to the stationary phase approximation, that is fluctuations around the mean field, we will see that n, D, and J_{ij} may play an important role. In some cases, such as the infinite range model or high enough D, we will see that the fluctuations do not affect the critical behavior and that the mean field result is exact. At the other extreme, for cases such as the Ising model in 1 dimension or the $x-y$ model in 2 dimensions, we will see that fluctuations completely alter the mean-field result. Hence, we now proceed to calculate the contributions of fluctuations.

4.4 FLUCTUATIONS

LANDAU GINZBURG THEORY AND DIMENSIONAL ANALYSIS

We have just seen that the Landau function corresponds to the expansion of the exponent of a functional integral over an order parameter for the special case of a spatially uniform order parameter. In studying fluctuations, we will be interested in the more general case of a spatially dependent local order parameter, and the corresponding expansion of the exponent in terms consistent with the symmetry of the system yields the Landau-Ginzburg functional. Suppose we have a problem with local order parameter $\phi(r)$, possessing symmetry with respect to $\phi(r) \to -\phi(r)$, and with translation and rotation invariance. In addition to the even powers of $\phi(r)$ discussed previously in connection with the Landau function, the action may contain invariant derivative couplings of the fields. The lowest-order derivative term consistent with the symmetries is $(\vec{\nabla}\phi(r))^2$, in which case the Landau-Ginzburg functional in zero field would have the form

$$\mathcal{L}[\phi] = \int d^D r \left[\frac{1}{2}(\nabla\phi(r))^2 + \frac{r_0}{2}\phi^2(r) + \frac{u_0}{4}\phi^4(r)\right] \quad . \tag{4.96}$$

For a uniform order parameter $\phi(r) = m$, we recover the familiar Landau form and the $(\nabla\phi(r))^2$ term accounts for the additional energy required if the order parameter is non-uniform. Just as only a limited number of powers of ϕ are required to study the critical behavior, it is shown in Problem 4.2 that terms involving higher derivatives of ϕ or higher powers of $(\nabla\phi)$ are not necessary to study the critical behavior.

In general, the Landau Ginzburg functional can always be obtained by expanding the exponent of a functional integral over an order parameter, and we will now illustrate how the form Eq. (4.96) is obtained for the Ising model in D dimensions. Expanding the Ising action, Eq. (4.52) to fourth order in zero field yields

$$\beta S(\phi_i) = \frac{\beta}{2}\sum_{ij}\phi_i J_{ij}^{-1}\phi_j - \frac{\beta^2}{2}\sum_i \phi_i^2 + \frac{\beta^4}{4}\sum_i \phi_i^4 \ . \tag{4.97}$$

Since we will be interested in dimensions, note that in units such that $\hbar = c = k_B = 1$, $\beta \sim L$, $J \sim \frac{1}{L}$ and $\phi_i \sim H_i \sim \frac{1}{L}$ so that βS_i is dimensionless as required. Defining a continuum field $\tilde{\phi}(\vec{n}a) = \phi_{\vec{n}}$, replacing sums by integrals according to $\frac{1}{a^D}\sum_{\vec{n}} a^D \phi_{\vec{n}} = \frac{1}{a^D}\int d^D r\, \tilde{\phi}(\vec{r})$, and calculating the first term in a Taylor series expansion of J^{-1} by noting

$$\begin{aligned}\sum_n J_{mn}\chi_n &= \sum_{\alpha=1}^{D} J\left[e^{a\frac{\partial}{\partial x_\alpha}} + e^{-a\frac{\partial}{\partial x_\alpha}}\right]\chi_m \\ &\approx 2JD\left[1 + \frac{\nabla^2}{2D}a^2\right]\chi \end{aligned} \tag{4.98a}$$

so that

$$J^{-1} \approx \frac{1}{2JD}\left[1 - \frac{\nabla^2}{2D}a^2 + \ldots\right] \tag{4.98b}$$

we obtain the result

$$\beta S(\tilde{\phi}) \approx \int d^D r \left\{\frac{\beta}{8JD^2 a^{D-2}}\left[-\tilde{\phi}\nabla^2\tilde{\phi}\right] + \frac{\beta}{2a^D}\left(\frac{1}{2JD} - \beta\right)\tilde{\phi}^2 + \frac{\beta^4}{12a^D}\tilde{\phi}^4\right\} \ . \tag{4.99}$$

It is conventional to rescale the fields such that the coefficient of the gradient term is $\frac{1}{2}$, so we define $\phi = \left[\frac{\beta}{4JD^2a^{D-2}}\right]^{1/2}\tilde{\phi}$ and thus obtain the form Eq. (4.96)

$$\beta S(\phi) \approx \int d^D r\left\{\frac{1}{2}\left(\vec{\nabla}\phi(r)\right)^2 + \frac{r_0}{2}\phi^2(r) + \frac{u_0}{4}\phi^4(r)\right\} \tag{4.100a}$$

where

$$r_0 = \frac{T_c}{Ja^2T}(T - T_c) \tag{4.100b}$$

$$u_0 = \frac{T_c^4}{3J^2T^2}a^{D-4} \tag{4.100c}$$

and we have used $T_c = 2JD$.

It is important to appreciate the physical motivation for taking the continuum limit. Although the Ising problem on a lattice has a fundamental microscopic length scale, the lattice spacing a, we have seen that the behavior near the critical point is characterized by fluctuations and a correlation length which become arbitrarily large. Thus, for the long-wavelength excitations of the system which dominate the physics, the microscopic length a is irrelevant, and the correlation length determined by the coefficients in Eq. (4.100) is the physically relevant length scale. At the critical point, where the correlation length diverges, the system has no characteristic length scale and is expected to behave as a scale invariant system. The dominance of the long wavelength, or infrared, behavior of the action near the critical point, illustrated here for the Ising model, is completely general. For example, in the liquid-gas phase transition, the correlation length becomes arbitrarily large compared to the microscopic length scale associated with the molecular potential $v(r)$, and the potential only affects the magnitude of the coefficients in the expansion of the exponent of Eq. (4.94). By the same argument, the critical behavior of a quantum system is also dominated by length scales arbitrarily large relative to the microscopic quantum scale, and quantum effects only contribute to the numerical values of coefficients.

Having established that the critical behavior is governed by a Landau Ginzburg functional of the form Eq. (4.96), we now examine the role of the spatial dimension D by dimensional analysis. Denoting the dimensions of a quantity by square brackets, we find from the term $\int d^D r \left(\vec{\nabla}\phi(r)\right)^2$ that

$$L^D L^{-2}[\phi]^2 = 1 \tag{4.101a}$$

or

$$[\phi] = L^{-\frac{D-2}{2}} \quad . \tag{4.101b}$$

Similarly, the term $r_0 \int d^D r\, \phi^2(r)$ yields

$$[r_0] = L^{-2} \tag{4.101c}$$

and the term $\mu_0 \int d^D r\, \phi^4(r)$ gives

$$[\mu_0] = L^{D-4} \tag{4.101d}$$

consistent with the specific Ising results, Eq. (4.100). Recasting the action in dimensionless form by the transformation

$$\begin{aligned} \vec{r} &= \vec{\xi} \cdot \vec{r}_0^{\,-1/2} \\ \phi &= \chi r_0^{\frac{D-2}{4}} \end{aligned} \tag{4.102}$$

the partition function for the system may be written

$$Z = \int \mathcal{D}\chi(\vec{r})\, e^{-\int d^D\xi \left\{ \frac{1}{2}\left(\frac{\partial}{\partial \vec{\xi}}\chi(\vec{\xi})\right)^2 + \frac{1}{2}\chi^2(\vec{\xi}) + \frac{1}{4}g\chi^4(\vec{\xi}) \right\}} \tag{4.103a}$$

where

$$g = u_0\, r_0^{\frac{D-4}{2}} \quad . \tag{4.103b}$$

Recall that the critical point occurs when the quadratic term $r_0(T)\phi^2$ changes sign, so that as seen in Eq. (4.4) for the general case and Eq. (4.100b) for the Ising model, we may write

$$r_0 \equiv \tilde{a}(T - T_c) \tag{4.104}$$

so that the effective perturbation parameter for the quartic term in Eq. (4.103) is

$$g = u_0\, \tilde{a}^{\frac{D-4}{2}}\, |T - T_c|^{\frac{D-4}{2}} \quad . \tag{4.105}$$

This result clearly displays the role of the spatial dimension D in governing the behavior of the functional integral Eq. (4.103a) for the partition function. We distinguish two qualitatively different cases. For $D > 4$, $g \to 0$ as $T \to T_c$ and a perturbative expansion in g around the mean field result becomes increasingly accurate as T approaches T_c. Hence, the mean field accurately describes the critical behavior. For $D < 4$, $g \to \infty$ as $T \to T_c$ and perturbation theory breaks down within some range of T_c. Although mean field theory may still be accurate away from the critical point, it will give the wrong critical behavior. The case $D = 4$ is marginal and requires special care. In general, however, one should not expect mean field theory to apply for $D = 4$.

Thus, we have identified the upper critical dimension, d_c, discussed at the beginning of Section 4.3, which specifies the dimension above which mean field theory is valid around T_c. For the action (4.96), we have just shown that $d_c = 4$. For other systems which admit other invariants in the action, dimensional analysis will yield different results. For example, as shown in Problem 4.2, a cubic term in the action implies $d_c = 6$.

For $D < d_c$, one can determine the temperature region, called the Ginzburg region, in which one expects the mean field theory to begin to break down by the condition $g \geq 1$. Using Eq. (4.105), we write $1 = u_0\, \tilde{a}^{\frac{D-2}{2}} (\Delta T)^{\frac{D-4}{2}}$ which yields

$$\Delta T = \frac{u_0^{\frac{2}{4-D}}}{\tilde{a}} \quad . \tag{4.106a}$$

For subsequent reference, using the explicit parameters (4.100) for the Ising model we find

$$\Delta T_I = \left(\frac{4D^2}{3} \right)^{\frac{2}{4-D}} J \quad . \tag{4.106b}$$

The size of the Ginzburg region may be estimated for physical systems by finding a combination of observables which yield Eq. (4.106a). First, note that using $\beta F = \mathcal{L}[\phi]$ given in Eq. (4.96), the mean field approximation to the free energy below T_c is $F = \mathcal{V} T \tilde{a}^2 (T - T_c)^2 / (2u_0)$ so that the discontinuity in the specific heat per unit volume at T_c analogous to Eq. (4.73) is

$$\Delta C = \left[-\frac{T}{\mathcal{V}} \frac{\partial^2 F}{\partial T^2} \right]_{T_c^+}^{T_c^-} = \frac{\tilde{a}^2 T_c^2}{u_0} \quad . \tag{4.107a}$$

The correlation length for $\mathcal{L}[\phi]$ is calculated in mean field theory as in Eqs. (4.63 – 4.68) using $G^{-1}(x,y) \propto \frac{\partial^2 \mathcal{L}}{\partial\phi(x)\partial\phi(y)}$, from which it follows that $G^{-1}(g) \propto q^2 + \tilde{a}(T-T_c)$ and $\xi = [\tilde{a}(T-T_c)]^{-1/2}$. Since mean field theory is only valid away from the critical point, we will estimate the correlation length well away from the critical point $|T_c - T| = \mathcal{O}(T_c)$ as follows

$$\xi_0 \sim [\tilde{a}T_c]^{-1/2} \ . \tag{4.107b}$$

Combining Eqs. (4.106a, 4.107a and 4.107b), we obtain the result of Ginzburg (1960)

$$\Delta T = \left[\frac{\xi_0^{-D}}{\Delta C}\right]^{\frac{2}{4-D}} \cdot T_c \xrightarrow[D=3]{} \frac{T_c}{\xi_0^6 (\Delta C)^2} \ . \tag{4.107c}$$

This dependence on the correlation length scale accounts for the dramatic difference between the size of the Ginzburg region for the superfluid transition in liquid Helium and for the superconducting transition in metals. In both cases, ΔC is of the order of several joule/cm^3 $^\circ K$ and T_c is of the order of several $^\circ K$. However, the characteristic range of intermolecular correlations in liquid Helium is of the order of several Å yielding $\Delta T \sim 1^\circ K$ whereas for a superconductor the correlation length is the size of a Cooper pair $\sim 10^3$ Å so that $\Delta T \sim 10^{-15}\,^\circ K$.

ONE-LOOP CORRECTIONS

We will now explicitly calculate the leading order correction to the mean field theory for the simple case of the Ising model, and defer until the next section the additional complications arising from continuous symmetry. Using the Ising action S given in Eq. (4.52), for which

$$\frac{\delta^2 S}{\delta\phi_i \delta\phi_j} = J_{ij}^{-1} - \beta\left(1 - th^2 \beta\phi_i\right)\delta_{ij} \tag{4.108}$$

and including an explicit expansion parameter ℓ to keep track of orders in the stationary-phase approximation, the stationary phase result including quadratic fluctuations is

$$\begin{aligned} Z_I(H_i) &= \int \Pi_i d\phi_i \, e^{-\ell\beta\, S(\phi_i, H_i)} \\ &\approx \det\left[\ell\beta \frac{\delta^2 S}{\delta\phi_i \delta\phi_j}\right]^{-1/2} e^{-\ell\beta\, S}\Big|_{\overline{\phi}} \\ &\equiv C\, e^{-\beta F(\overline{\phi}, H_i)} \ . \end{aligned} \tag{4.109a}$$

Here

$$F(\overline{\phi}, H_i) = \frac{1}{2}\sum_{ij}\left(\overline{\phi}_i - H_i\right) J_{ij}^{-1}\left(\overline{\phi}_j - H_j\right) - \frac{1}{\beta}\sum_i \ln ch\,\beta\overline{\phi}_i + \frac{1}{2\beta\ell}A(\overline{\phi}) \tag{4.109b}$$

with

$$A(\overline{\phi}) = \ln\det\left[\delta_{ij} - \beta(1 - th^2\beta\overline{\phi}_i)J_{ij}\right] \ , \tag{4.109c}$$

the stationary solution satisfies the mean field equations

$$\overline{\phi}_i = H_i + \sum_j J_{ij}\, th\, \beta\overline{\phi}_j \tag{4.109d}$$

and we will ignore the overall multiplicative constant $C = \left(\frac{2\pi}{\beta\ell}\right)^{n/2}\left[\det J^{-1}\right]^{-1/2} 2^{\ell N}$. To leading order in $\frac{1}{\ell}$, the magnetization is given by

$$m_i = -\frac{\partial F}{\partial H_i} = \sum_j J_{ij}^{-1}(\overline{\phi}_j - H_j) - \frac{1}{2\beta\ell}\,\frac{\partial A(\overline{\phi})}{\partial H_i} \tag{4.110a}$$

and as shown in Problem 4.5 the Legendre transform is

$$\Gamma[m_i] = -\frac{1}{2}\sum_{ij} m_i J_{ij} m_j + \frac{1}{2\beta}\sum_i \left[(1-m_i)\ln(1-m_i) + (1+m_i)\ln(1+m_i)\right]$$
$$+ \frac{1}{2\ell\beta}\ln\det\left[\delta_{ij} - \beta(1-m_i^2)J_{ij}\right] \quad . \tag{4.110b}$$

The equation of state is:

$$H_i = \frac{\partial\Gamma}{\partial m_i} = -\sum_j J_{ij}m_j + \frac{1}{\beta} th^{-1}\, m_i + \frac{1}{\ell}\sum_\ell J_{i\ell}\left(\delta_{rs} - \beta(1-m_r^2)J_{rs}\right)^{-1}_{\ell i} m_i \tag{4.111}$$

and the susceptibility in the high temperature region, where $m_i = 0$, is given by

$$\chi^{-1} = \frac{\partial^2\Gamma}{\partial m^2} = -2DJ + \frac{1}{\beta} + \frac{1}{\ell}\sum_i J_{i\ell}(\delta_{rs} - \beta J_{rs})^{-1}_{\ell i} \quad . \tag{4.112a}$$

The sum is simplified by taking the Fourier transform

$$\chi^{-1} = -2DJ + \frac{1}{\beta} + \frac{1}{\ell}\int \frac{d^D q}{(2\pi)^D}\,\frac{J(\vec{q})}{1-\beta J(\vec{q})} \tag{4.112b}$$

where $J(\vec{q})$ is the Fourier transform of J_{ij} defined by Eq. (4.63) and with the convention that the lattice spacing $a = 1$, the integral $\int d^D q$ is over the Brillouin zone $\pi < q_\alpha < \pi$.

The critical temperature is defined by the divergence of the susceptibility so

$$0 = -2DJ + T_c + \frac{1}{\ell}\int \frac{d^D q}{(2\pi)^D}\cdot\frac{J(\vec{q})}{1-\frac{J(\vec{q})}{T_c}} \quad . \tag{4.113a}$$

Hence, to leading order in $\frac{1}{\ell}$

$$T_c = 2DJ - \frac{1}{\ell}\int \frac{d^D q}{(2\pi)^D}\cdot\frac{J(\vec{q})}{1-\frac{J(\vec{q})}{2JD}} \quad . \tag{4.113b}$$

We thus observe that the effect of fluctuations is to decrease the critical temperature below the mean field result, $T_c^{MF} = 2DJ$. This downward shift in T_c is reasonable physically because fluctuations tend to disorder the system.

In order to expand χ^{-1} around T_c, we subtract Eq. (4.113a) from Eq. (4.112b) with the result

$$\chi^{-1} = (T - T_c) + \frac{(T_c - T)}{\ell} \int \frac{d^D q}{(2\pi)^D} \frac{\tilde{J}^2(q)}{(T - \tilde{J}(q))(T_c - \tilde{J}(q))} \quad . \tag{4.114}$$

Since the critical region is dominated by the low $\vec{q}$ fluctuations, we expand $\tilde{J}(\vec{q})$ to second order in $\vec{q}$ as in Eq. (4.64). To order $\frac{1}{\ell}$, we obtain the low q behavior

$$\chi^{-1} \underset{T\to T_c}{\sim} (T - T_c)\left(1 - \frac{1}{\ell}\int \frac{d^D q}{(2\pi)^D} \cdot \frac{T_c^2}{J\vec{q}^2(J\vec{q}^2 + T - T_c)}\right) \quad . \tag{4.115a}$$

From this result we observe that the $\frac{1}{\ell}$ correction to χ^{-1} at T_c for $D = 4$ contains $\int \frac{q^3 dq}{q^4}$ and is thus logarithmically divergent. For any other dimension, we make the change of variable

$$\vec{k} = \sqrt{\frac{J}{T - T_c}} \cdot \vec{q} \tag{4.115b}$$

and the susceptibility becomes:

$$\chi^{-1} \underset{T\to T_c}{\sim} (T - T_c)\left(1 - \frac{1}{\ell} \cdot \frac{(T - T_c)^{\frac{D-4}{2}} T_c^2}{J^{D/2}} \int \frac{d^D k}{(2\pi)^D} \frac{1}{\vec{k}^2(\vec{k}^2 + 1)}\right) \quad . \tag{4.115c}$$

This result displays the role of the dimension D observed previously from dimensional analysis. If $D > 4$, then the $\frac{1}{\ell}$ corrections to χ^{-1} are finite, and do not modify the dominant singular behavior of χ. In particular, the critical exponent γ remains 1. It can be shown that this result persists to all orders in $\frac{1}{\ell}$.

For $D \leq 4$, the $\frac{1}{\ell}$ corrections to χ^{-1} diverge around T_c, and therefore dominate the critical behavior. The critical regime is said to be fluctuation dominated. The higher order terms in $\frac{1}{\ell}$ become more and more divergent around T_c, and the mean field expansion is inapplicable. The critical behavior in this region may be understood using the renormalization group. An introduction is presented by Huang (1986) and more extensive treatments are given by Ma (1976) and by Pfeuty and Toulouse (1975).

From Eq. (4.115c), we can estimate the size of the Ginzburg region by calculating the value of $T - T_c$ such that the $\frac{1}{\ell}$ term equals 1, with the result

$$\Delta T = J\left[4D^2 \int \frac{d^D k}{(2\pi)^D} \frac{1}{\vec{k}^2(1 + \vec{k}^2)}\right]^{\frac{2}{4-D}} \tag{4.116}$$

which is consistent with the rough estimate (4.106b).

Before leaving the Ising model, several additional comments are in order. In two dimensions, it may be solved exactly, and as an alternative to the lengthy original derivation of Onsager (1944), a simple derivation using Grassmann variables is presented in

Problem 4.8. One should also note that there are physical systems such as $K_2\,Co\,F_4$ which are accurately described by the two-dimensional Ising model: the anisotropic ionic field allows only two orientations for the magnetic Co ions and the anisotropic exchange interactions between Co ions via intermediate ions are essentially confined within planes. In three dimensions, the critical behavior is approximately known numerically, and in Table 4.3 at the end of the chapter, the critical exponents for $D = 2$ and 3 are tabulated to show how they approach the mean field results as D increases.

Calculation of the leading order fluctuation contributions for n-component spins is similar to that we have just presented for Ising spins, except for the existence of a zero eigenvalue in the fluctuation matrix associated with symmetry with respect to continuous global spin rotations. Hence, we now consider the general problem of continuous symmetry.

CONTINUOUS SYMMETRY

We now consider the fluctuations around the mean field solution of a functional integral of the form:

$$Z = \int \mathcal{D}\phi(x)\; e^{-S(\phi(x))} \tag{4.117a}$$

where the action has a continuous invariance group $\mathcal{G}$:

$$\forall g \in \mathcal{G}: \quad S\left(g \cdot \phi(x)\right) = S\left(\phi(x)\right) \quad . \tag{4.117b}$$

The mean field equation:

$$\frac{\delta S}{\delta \phi_c(x)} = 0 \tag{4.117c}$$

has degenerate solutions, since if $\tilde{\phi}_c(x)$ is a particular solution, $g\tilde{\phi}_c(x)$ is also a solution. Since $\mathcal{G}$ is a continuous group, its elements may be parameterized by a continuous variable, θ, which may be real, complex, or a multicomponent vector, and we may write

$$g = g(\theta) \tag{4.118a}$$

and

$$g \cdot \tilde{\phi}_c(x) = \phi_c(x, \theta) \quad . \tag{4.118b}$$

To compute the quadratic fluctuations, we must diagonalize the operator:

$$A(x, y) = \frac{\delta^2 S}{\delta \phi_c(x) \delta \phi_c(y)} \quad . \tag{4.119}$$

In particular, we must identify the zero eigenvalues, and treat these modes explicitly.

Note that if $\phi_c(x, \theta_0)$ is a particular mean field solution to Eq. (4.117), then $\frac{\partial \phi_c(x,\theta_0)}{\partial \theta_0}$ is an eigenvector of Eq. (4.119) with zero eigenvalue since:

$$\begin{aligned}
\int dy\, A(x, y) \frac{\partial \phi_c(y, \theta_0)}{\partial \theta_0} &= \int dy \frac{\delta^2 S}{\delta \phi_c(x, \theta_0) \delta \phi_c(y, \theta_0)} \cdot \frac{\partial \phi_c(y, \theta_0))}{\partial \theta_0} \\
&= \frac{\partial}{\partial \theta_0} \left(\frac{\partial S}{\delta \phi_c(x, \theta_0)} \right) \\
&= 0 \; .
\end{aligned} \tag{4.120}$$

The last line follows since $\frac{\partial S}{\partial \phi_c(x,\theta_0)} = 0$ for all θ_0.

These eigenmodes $\frac{\partial \phi_c(x,\theta_0)}{\partial \theta_0}$ with zero eigenvalues correspond to the Goldstone modes we have discussed previously. For example, in the $x-y$ model, θ_0 is the angle specifying the magnetization direction, and $\frac{\partial \phi_c}{\partial \theta_0}$ corresponds to a mode in which the magnetization moves around the circular minimum of the potential as sketched in Fig. 4.3. Since in field theory these modes correspond to zero mass particles, Goldstone modes are also referred to as massless modes.

The number of independent Goldstone modes is the order of the remaining symmetry group of the system in the ordered phase. For example, for the $O(n)$ spin model, the symmetry group of the ordered phase is $O(n-1)$, and there are $(n-1)$ Goldstone modes.

We will now calculate the quadratic corrections by evaluating the partition function

$$Z = e^{-S(\phi_c(x,\theta_0))} \int D\phi(x) e^{-\frac{1}{2}\int dx dy (\phi(x)-\phi_c(x,\theta_0))A(x,y)(\phi(y)-\phi_c(y,\theta_0))} \tag{4.121}$$

where $A(x,y)$ is defined in (4.119).

Since A has a zero eigenvalue for the mode $\frac{\partial \phi_c(x,\theta_0)}{\partial \theta_0}$, it is useful to separate the fluctuations of ϕ into a component proportional to $\frac{\partial \phi_c}{\partial \theta_0}$ and the remaining components orthogonal to $\frac{\partial \phi_c}{\partial \theta_0}$. Hence we define the function:

$$f(\theta_0) = \int dx \frac{\partial \phi_c(x,\theta_0)}{\partial \theta_0} \left(-\phi(x) + \phi_c(x,\theta_0)\right) \tag{4.122a}$$

so that fluctuations orthogonal to the zero mode $\frac{\partial \phi_c}{\partial \theta_0}(x,\theta_0)$ are defined by the condition

$$f(\theta_0) = 0 \ . \tag{4.122b}$$

This constraint may be included in the functional integral for the partition function by inserting the identity

$$\int d\theta_0 \ f'(\theta_0)\delta\left(f(\theta_0)\right) = 1 \tag{4.123}$$

in Eq. (4.121) to obtain

$$S = e^{-S_c} \int D\phi(x) d\theta_0 f'(\theta_0)\delta\left(f(\theta_0)\right) e^{-\frac{1}{2}\int dx dy (\phi(x)-\phi_c(x,\theta))A(x,y)(\phi(y)-\phi_c(y,\theta))} \ . \tag{4.124}$$

Using the Fourier representation of the δ-function, the identity (4.123) can be rewritten as

$$\begin{aligned} 1 = &\int \frac{d\theta_0 d\alpha}{2\pi} e^{i\alpha \int dx \frac{\partial \phi_c}{\partial \theta_0}(x,\theta_0)(\phi(x)-\phi_c(x,\theta_0))} \\ &\times \int dx \left(-\frac{\partial^2}{\partial \theta_0}\phi_c(x,\theta_0)(\phi(x)-\phi_c(x,\theta_0)) + \left(\frac{\partial \phi_c}{\partial \theta_0}(x,\theta_0)\right)^2 \right) \ . \end{aligned} \tag{4.125}$$

We make the change of variable $\psi(x) = \phi(x) - \phi_c(x, \theta_0)$ and substitute (4.125) in (4.124) to obtain

$$S = e^{-S_c} \int \frac{d\theta_0 d\alpha}{2\pi} \int D\phi(x) \int dx \left(-\frac{\partial^2 \phi_c(x,\theta_0)}{\partial\theta_0^2} \psi(x) + \left(\frac{\partial \phi_c}{\partial \theta_0}(x,\theta_0) \right)^2 \right) \times e^{-\frac{1}{2}\int dx dy \psi(x) A(x,y) \psi(y) + i\alpha \int dx \frac{\partial \phi_c}{\partial \theta_0}(x,\theta_0)\psi(x)} . \tag{4.126a}$$

The term linear in ψ in Eq. (4.126a) vanishes by parity since simultaneously changing ψ into $-\psi$ and α into $-\alpha$ leaves the exponent invariant while changing the sign of the linear term. Therefore,

$$Z = e^{-S_c} \int \frac{d\theta_0 d\alpha}{2\pi} \int D\psi(x) \int dx \left(\frac{\partial \phi_c}{\partial \theta_0}(x,\theta_0) \right)^2 \times e^{-\frac{1}{2} dx dy\, \psi(x) A(x,y) \psi(y) + i\alpha \int dx \frac{\partial \phi_c(x,\theta_0)}{\partial \theta_0} \psi(x)} . \tag{4.126b}$$

To evaluate the remaining Gaussian integral in which A has a zero eigenvalue, we define a new operator

$$A_\epsilon = \epsilon \frac{\left| \frac{\partial \phi_c}{\partial \phi_0} \right\rangle \left\langle \frac{\partial \phi_c}{\partial \phi_0} \right|}{\left\langle \frac{\partial \phi_c}{\partial \phi_0} \middle| \frac{\partial \phi_c}{\partial \phi_0} \right\rangle} + A_\perp \tag{4.127a}$$

where ϵ is a positive infinitesimal number which is eventually taken to zero, $|\frac{\partial \phi_c}{\partial \theta}\rangle$ denotes the zero mode, and $A_\perp$ is the projection of A on the subspace orthogonal to the zero mode. In block matrix form, A_ϵ and A may be written

$$A_\epsilon = \begin{bmatrix} \epsilon & 0 \\ 0 & A_\perp \end{bmatrix} \qquad A = \begin{bmatrix} 0 & 0 \\ 0 & A_\perp \end{bmatrix} . \tag{4.127b}$$

The operator A_ϵ has a finite inverse, and the Gaussian integral (4.126b) with A replaced by A_ϵ can be performed to obtain

$$Z = e^{-S_c} \int \frac{d\theta_0 d\alpha}{2\pi} (\det A_\epsilon)^{-\frac{1}{2}} \int dx \left(\frac{\partial \phi_c(x,\theta_0)}{\partial \theta_0} \right)^2 \times e^{-\frac{\alpha^2}{2} \int dx dy \frac{\partial \phi_c}{\theta_0}(x,\theta_0) A_\epsilon^{-1}(x,y) \frac{\partial \phi_c}{\partial \theta_0}(y,\theta_0)} . \tag{4.128a}$$

From Eq. (4.127), we see that

$$\det A_\epsilon = \epsilon \det A_\perp \tag{4.128b}$$

and that since $\frac{\partial \phi_c}{\partial \theta_0}$ is the zero mode,

$$\int dx dy \frac{\partial \phi_c(x,\theta_0)}{\partial \theta_0} A_\epsilon^{-1}(x,y) \frac{\partial \phi_c(y,\theta_0)}{\partial \theta_0} = \frac{1}{\epsilon} \int dx \left(\frac{\partial \phi_c(x,\theta_0)}{\partial \theta_0} \right)^2 . \tag{4.128c}$$

The integral over α may now be performed, yielding:

$$Z = \lim_{\epsilon\to 0} e^{-S_c} \int \frac{d\theta_0}{\sqrt{2\pi}} \frac{1}{\sqrt{\epsilon \det A_\perp}} \int dx \left(\frac{\partial\phi_c(x,\theta_0)}{\partial\theta_0}\right)^2 \frac{\sqrt{\epsilon}}{\sqrt{\int dx \left(\frac{\partial\phi_c(x,\theta_0)}{\partial\theta_0}\right)^2}} \tag{4.129a}$$

$$= e^{-S_c} \frac{1}{\sqrt{2\pi}} \cdot \frac{1}{\sqrt{\det A_\perp}} \int d\theta_0 \sqrt{\int dx \left(\frac{\partial\phi_c(x,\theta_0)}{\partial\theta_0}\right)^2} \ . \tag{4.129b}$$

Thus, the usual factor $\frac{1}{\sqrt{\det A}}$ for the quadratic fluctuations is replaced by the result

$$\frac{1}{\sqrt{\det A}} \to \frac{1}{\sqrt{2\pi}} \frac{1}{\sqrt{\det A_\perp}} \int d\theta_0 \sqrt{\int dx \left(\frac{\partial\phi_c(x,\theta_0)}{\partial\theta_0}\right)^2} \ . \tag{4.130}$$

The factor $\frac{1}{\sqrt{2\pi}}$ is the usual factor associated with each eigenvalue in a Gaussian integral, and the symbol $\int d\theta_0$ denotes the volume integral over the invariance group $\mathcal{G}$.

It is shown in Problem 4.6 that for the general case of a vector field $\vec{\phi}(x)$ in which the mean-field $\vec{\phi}_S(x,\vec{\theta})$ has an invariance group parameterized by the vector $(\theta_1,\ldots\theta_n)$ there are n Goldstone modes, $\frac{\partial\vec{\phi}}{\partial\theta_i}$, and the partition function is given by

$$Z = e^{-S_c}(\det A_\perp)^{-\frac{1}{2}} \int \prod_{i=1}^{n} \frac{d\theta_i}{\sqrt{2\pi}} \left[\det\left(\int dx \frac{\partial\vec{\phi}}{\partial\theta_i}\cdot\frac{\partial\vec{\phi}}{\partial\theta_j}\right)\right]^{\frac{1}{2}} \ . \tag{4.131}$$

These considerations also apply to the case of mean-field solutions which are localized in space and time, where the zero modes are associated with space and time translation. We will see in Chapter 7 how one sums over all possible combinations of such localized solutions to obtain physical results.

ONE-LOOP CORRECTIONS FOR THE $X-Y$ MODEL

To illustrate the physical effect of zero modes, we now evaluate the contributions of fluctuations for the two-component $X-Y$ model. Recall that the $X-Y$ model has two component spins, that is spins on the unit circle, at each lattice site. By the same argument used to arrive at Eq. (4.100) for Ising spins or, alternatively, by general symmetry considerations, the Landau-Ginzburg form of the partition function is

$$Z = \int D\vec{\phi}(x) e^{-\frac{1}{2}\int d^Dx\left((\vec{\nabla}\phi_1(x))^2 + (\vec{\nabla}\phi_2(x))^2\right) - \frac{r_0}{2}\int d^Dx \vec{\phi}^2(x) - \frac{u_0}{4}\int d^Dx \left(\vec{\phi}^2(x)\right)^2} \tag{4.132}$$

where ϕ_1 and ϕ_2 are the two components of the vector $\vec{\phi}$.

The mean field equation is

$$\nabla^2\vec{\phi} = (r_0 + u_0\vec{\phi}^2)\vec{\phi} \ . \tag{4.133a}$$

Below T_c, we know that r_0 is negative, and $\vec{\phi}^2$ is given by:

$$\vec{\phi}^2 = -\frac{r_0}{u_0} \quad . \tag{4.133b}$$

We will parameterize the degenerate mean field solutions as follows:

$$\vec{\phi}_{\mathrm{MF}}(\theta_0) = \begin{pmatrix} \phi_0 \cos\theta_0 \\ \phi_0 \sin\theta_0 \end{pmatrix} \tag{4.134a}$$

where θ_0 is a real parameter in the interval $[0, 2\pi]$ and $\phi_0 = \sqrt{\frac{-r_0}{u_0}}$. The invariance group is the group of two-dimensional rotations $SO(2)$, the volume of which is $\int_0^{2\pi} d\theta_0 = 2\pi$. Let us compute the quadratic fluctuations around the solution $\theta_0 = 0$ and denote the fluctuations relative to $\vec{\phi}_{\mathrm{MF}}(0)$ as $\vec{\psi}$

$$\vec{\phi}_{\mathrm{MF}}(0) = \begin{pmatrix} \phi_0 \\ 0 \end{pmatrix} \qquad \vec{\phi} = \vec{\phi}_{\mathrm{MF}}(x) + \vec{\psi} = \begin{pmatrix} \phi_0 + \psi_1 \\ \psi_2 \end{pmatrix} \quad . \tag{4.134b}$$

Retaining quadratic terms in $\vec{\psi}$, the partition function becomes:

$$Z = e^{\frac{\nu r_0^2}{4u_0}} \int \mathcal{D}\vec{\psi}(x)\, e^{-\frac{1}{2}\int d^Dx\left((\vec{\nabla}\psi_1)+(\vec{\nabla}\psi_2)^2-2r_0\psi_1^2\right)} \tag{4.135a}$$

or in Fourier representation,

$$Z = e^{\frac{\nu r_0^2}{4u_0}} \int \mathcal{D}\psi_{\vec{k}} e^{-\frac{1}{2}\frac{\nu}{(2\pi)^D}\int d^Dk\left(\vec{k}^2(\psi_1(\vec{k})\psi_1(-\vec{k})+\psi_2(\vec{k})\psi_2(-\vec{k}))-2r_0\psi_1(k)\psi_1(-k)\right)} \quad . \tag{4.135b}$$

In the language of field theory, given a functional integral of the form (4.135a), the square of the mass of the particle represented by the field $\vec{\psi}$ is the coefficient of $\frac{1}{2}\vec{\psi}^2$ in the Lagrangian. Therefore, in Eq. (4.135a) we see that the longitudinal fluctuations, represented by ψ_1 correspond to a mode of mass $\sqrt{-2r_0}$, whereas the transverse fluctuations represented by ϕ_2 have zero mass. This zero mass mode, resulting from the broken continuous symmetry is the previously mentioned Goldstone mode.

The quadratic fluctuation operator $A(x, y)$ may be written in block matrix form

$$A(x, y) = \begin{pmatrix} -\nabla^2\delta(x-y) - 2r_0\delta(x-y) & 0 \\ 0 & -\nabla^2\delta(x-y) \end{pmatrix} \tag{4.136a}$$

or in Fourier space

$$\tilde{A}(\vec{k}) = \begin{pmatrix} \vec{k}^2 - 2r_0 & 0 \\ 0 & \vec{k}^2 \end{pmatrix} \quad . \tag{4.136b}$$

The longitudinal modes, corresponding to fluctuations in the direction of the magnetization, and the transverse modes, corresponding to fluctuations perpendicular to

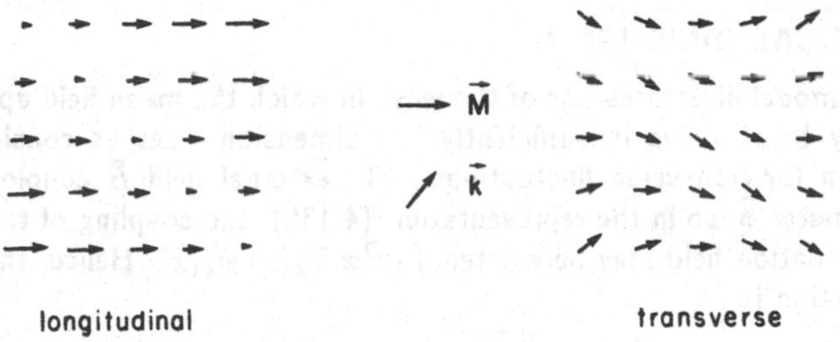

Fig. 4.7 Sketch of the spatial distribution of magnetization for longitudinal and transverse modes propagating in the direction $\vec{k}$.

the magnetization, are decoupled. The longitudinal mode ψ_L and its eigenvalue λ_L are given by

$$\psi_L(\vec{x}) = \begin{pmatrix} e^{i\vec{k}\cdot\vec{x}} \\ 0 \end{pmatrix} \qquad \lambda_L = \vec{k}^2 - 2r_0 \ . \tag{4.137a}$$

The transverse mode, ψ_T, known as the spin-wave mode, and its eigenvalues are given by

$$\psi_T(\vec{x}) = \begin{pmatrix} 0 \\ e^{i\vec{k}\cdot\vec{x}} \end{pmatrix} \qquad \lambda_T = \vec{k}^2 \ . \tag{4.137b}$$

The spin waves have an eigenvalue spectrum which goes to zero and correspond to the Goldstone mode associated with the broken continuous symmetry. In contrast, because $r_0 < 0$ below T_c, the longitudinal modes have strictly positive eigenvalues.

Since $\langle\vec{\phi}\rangle$ is proportional to the magnetization, one can associate these longitudinal and transverse modes with the fluctuations in the magnetization sketched in Fig. 4.7. Here, one observes that adding fluctuations ψ_1 in the direction of the magnetization $\vec{m}$ changes its magnitude and thus costs finite energy in the long wavelength limit whereas adding transverse fluctuations leaves the magnitude fixed and costs no energy in the long wavelength limit.

It is now straightforward to calculate the partition function, applying the general result for continuous symmetry, (4.130). Using the parameterization (4.135)

$$\frac{\partial}{\partial\theta_0}\vec{\phi}_{\rm MF}(\theta_0)\bigg|_{\theta_0=0} = \begin{bmatrix} 0 \\ \phi_0 \end{bmatrix} \tag{4.138a}$$

so that

$$\int d\theta_0 \left[\int dx \left(\frac{\partial\phi_{\rm MF}(\theta_0)}{\partial\theta_0}\right)^2\right]^{1/2} = 2\pi\sqrt{-\frac{\mathcal{V}r_0}{u_0}} \ . \tag{4.138b}$$

Finally, we write $[\det A_\perp]^{-\frac{1}{2}}$ as $-\frac{1}{2}\exp[tr \ln A_\perp]$ using the diagonal form (4.137b) and note that the phase space factor eliminates the $k = 0$ divergence for $D \geq 1$, so the partition function may be written

$$Z = \sqrt{\frac{-\mathcal{V}r_0 2\pi}{u_0}}\, e^{\mathcal{V}\left(\frac{r_0^2}{4u_0} - \frac{1}{2}\int\frac{d^Dk}{(2\pi)^D}\left(\ln\vec{k}^2 + \ln(\vec{k}^2 - 2r_0)\right)\right)} \ . \tag{4.139}$$

LOWER CRITICAL DIMENSION

The $x-y$ model illustrates one of the ways in which the mean field approximation may completely break down in sufficiently low dimension. Let us consider the correlation function for transverse fluctuations. The external field $\vec{H}$ couples linearly to the order parameter $\vec{\phi}$, so in the representation (4.135), the coupling of the transverse field to the fluctuation field may be written $\int d^D x\, H_2(x)\, \psi_2(x)$. Hence, the transverse correlation function is

$$\begin{aligned} G_T(x-y) &= -\frac{1}{\beta} \frac{\partial^2 F}{\partial H_2(x) \partial H_2(y)} \\ &= \langle \psi_2(\vec{x}) \psi_2(\vec{y}) \rangle \\ &= \int \frac{d^D k}{(2\pi)^D} \frac{e^{i\vec{k}\cdot(\vec{x}-\vec{y})}}{\vec{k}^2} \end{aligned} \tag{4.140}$$

where we have used the Fourier representations (4.135b) and (4.137b).

The convergence of the transverse correlation function depends on the dimension D. For more than two dimensions, $D > 2$, the integral (4.140) converges at $k = 0$. The Goldstone modes generate finite fluctuations which tend to disorder the system, but the phase space factor $d^D k$ renders their effect finite. In contrast, for $D \leq 2$, the integral diverges, and the Goldstone modes completely disorder the system. This result is the essence of the Mermin Wagner theorem (1966), which states that for an n-vector model, with $n \geq 2$, the lower critical dimension is 2. That is, for $D \leq 2$, spin waves destroy the long range order of the magnetization, and the mean field prediction of magnetic order is completely incorrect.

We can understand the lower critical dimension for magnetic systems somewhat more generally by physical arguments based on domain walls. This approach not only reproduces the spin wave instability just identified for vector spins, but also identifies the relevant mechanism for Ising spins. Whenever a system has degenerate states of broken symmetry, one must consider configurations in which the system breaks up into domains of different states separated by domain walls. Since in the vicinity of the walls, spins are not optimally aligned, domain walls increase the internal energy. However, because there are so many alternative positions for the walls, they may also significantly increase the entropy. Depending upon the competition between internal energy and entropy in the free energy, the system will either remain magnetically ordered or spontaneously break up into domains.

Let us heuristically estimate the entropy and internal energy of a plane domain in a D-dimensional system with linear dimension L. Since the plane may be located at any position along any of the D axes, the number of possible positions is L^D so the entropy behaves as

$$S \sim D \ln L \ . \tag{4.141}$$

The internal energy may be estimated by examining the interactions between successive layers of spins as sketched in Fig. 4.8. For Ising spins, a domain extends over only one lattice spacing, and the cost in energy relative to a uniformly magnetized state is proportional to the number of spins at the surface of the domain wall

$$E_{\text{Ising}} \sim L^{D-1} \ . \tag{4.142}$$

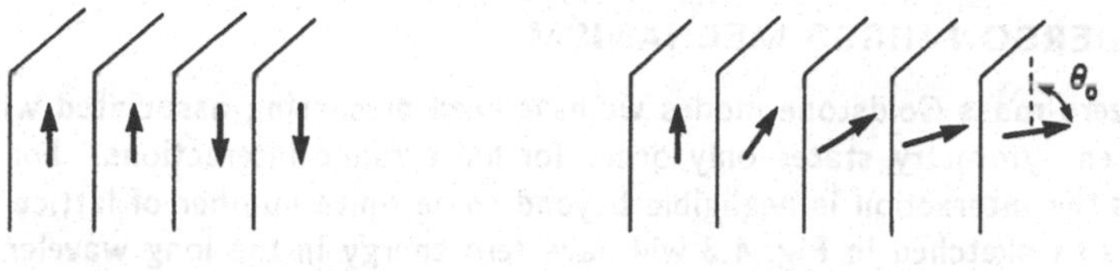

Fig. 4.8 Sketches of domain walls for Ising spins and for vector spins.

For vector spins, the orientation of the spin may change from $\theta = 0$, to $\theta = \theta_0$ over a domain A layers thick. Assuming the change in angle between two layers is $\frac{\theta_0}{A}$, the surface energy between the layers calculated by summing $-J_{ij}\left[\vec{S}_i \cdot \vec{S}_j - 1\right]$ for all sites in one layer and adjacent sites on the next layer is $L^{D-1}\left(1-\cos\frac{\theta_0}{A}\right)$. Therefore, summing over the A layers and expanding $\cos\frac{\theta}{A}$ for large A,

$$E_{\text{Vector}} \sim A\,L^{D-1}\left(1-\cos\frac{\theta_0}{A}\right) \sim A^{-1}L^{D-1}\theta_0^2 \ . \tag{4.143}$$

Using (4.141) and (4.142), the domain wall free energy in the case of Ising spins is

$$\Delta F_{\text{Ising}} \sim L^{D-1} - T\ \ln\ L \ . \tag{4.144}$$

If $D \leq 1$, the entropy dominates, the free energy is negative, and the system will spontaneously break up into domains. For $D > 1$, at sufficiently low T the internal energy dominates yielding a positive free energy and a state of uniform magnetic order is stable with respect to domain formation. Thus, for Ising spins, the lower critical dimension is 1.

For vector spins, we note that the domain wall energy (4.143) decreases with increasing wall thickness A. Since A is limited by the length L of the system, we have

$$E_{\text{Vector}} \simeq L^{D-2}\theta_0^2 \ . \tag{4.145a}$$

If we allow the wall thickness to approach the size of the system, its possible positions become restricted and the entropy estimate (4.141) becomes an overestimate

$$S_{\text{Vector}} \leq D\ \ln\ L \ . \tag{4.145b}$$

Hence, we have the following inequality for the free energy

$$\Delta F_{\text{Vector}} \geq L^{D-2}\theta_0^2 - T\ \ln\ L \ . \tag{4.145c}$$

If $D > 2$, we are assured that below some finite T, the free energy is positive and an ordered magnetized state is stable. Although the present inequality cannot distinguish whether the disordered case arises at $D = 2$ or lower, it is clear physically that the delocalized domain wall is essentially a spin wave. The domain wall argument is thus consistent with the spin wave calculation (4.140) which showed that the lower critical dimension is 2.

THE ANDERSON-HIGGS MECHANISM

The zero-mass Goldstone modes we have been discussing associated with degenerate broken symmetry states only occur for finite range interactions. For example, as long as the interaction is negligible beyond some finite number of lattice spacings, spin waves as sketched in Fig. 4.8 will have zero energy in the long wavelength limit. However, for infinite range interactions, spins with some finite relative rotation will produce a positive excitation energy at any distance, so the energy at zero wave number will be finite and the mode becomes massive. Thus, it is not surprising that systems with electromagnetic interactions and having broken symmetry do not have zero mass Goldstone modes.

What is remarkable, however, is the structure which emerges for a theory with spontaneously broken symmetry which also possesses a local gauge invariance. This phenomenon was originally discovered by Anderson (1958) in the context of an electromagnetic field in a superconductor, and was extended to more general gauge field theories by Anderson (1963) and Higgs (1964). We will illustrate the essential idea for a charged scalar field coupled to the electromagnetic field. From the discussion in Section 4.2, one may regard the scalar field as the order parameter for a superconductor and consider the present model as a Landau Ginzburg description of an electromagnetic field in a superconductor.

To facilitate the treatment of gauge invariance, it is convenient to extend our previous discussion of spatially dependent local fields to include space and time dependence on an equivalent basis, and to use the covariant notation of relativistic field theory. We will use the conventions $x^\mu = \{x^0, x^1, x^2, x^3\} = \{t, x, y, z\}$, $x_\mu = g_{\mu\nu}x^\nu$, $1 = g_{00} = -g_{11} = -g_{22} = -g_{33}$, and $\partial_\mu = \frac{\partial}{\partial x^\mu}$. As in the original presentation of path integrals, we will consider the path integral as an integral over the Lagrangian written in terms of the fields, and to study fluctuations, we will be concerned with expanding the Lagrangian to second order in the fluctuations around the stationary solution.

To appreciate the structure of the interacting theory, first consider separate, non-interacting scalar and electromagnetic fields. The charged scalar field is described by the real and imaginary components of a complex field, or equivalently by $\phi^*(x)$ and $\phi(x)$. The Lagrangian contains the kinetic term $\dot{\phi}^*\dot{\phi}$ and the potential terms include $-\nabla\phi^*\nabla\phi$ and the usual quadratic plus quartic polynomial needed for spontaneous symmetry breaking:

$$\mathcal{L}(\phi) = \partial^\mu\phi^*(x)\partial_\mu\phi(x) - r_0\phi^*(x)\phi(x) - \frac{u_0}{2}\left(\phi^*(x)\phi(x)\right)^2 \quad . \tag{4.146a}$$

The mean field equation from varying ϕ^* is

$$\left[\partial^\mu\partial_\mu + r_0 + u_0\phi^*(x)\phi(x)\right]\phi(x) = 0 \quad . \tag{4.146b}$$

Note that since $\mathcal{L}$ contains the negative of the potential, u_0 must be positive and the broken symmetry case corresponds to $r_0 < 0$, in which case the minimum energy configurations are given by

$$|\phi|^2 = -\frac{r_0}{u_0} \equiv \phi_0^2 \quad . \tag{4.146c}$$

To parameterize fluctuations parallel and perpendicular to the minimum, we write $\phi(x)$ and $\phi^*(x)$ in terms of two real fields $\xi(x)$ and $\eta(x)$

$$\begin{aligned}\phi(x) &= e^{i\xi(x)}\left(\phi_0 + \eta(x)\right)\\ \phi^*(x) &= e^{-i\xi(x)}\left(\phi_0 + \eta(x)\right) \quad .\end{aligned} \tag{4.147}$$

Expanding (4.146a) to second order in the fluctuations

$$\mathcal{L} = -\frac{r_0}{u_0}\partial^\mu \xi \partial_\mu \xi + \partial^\mu \eta \partial_\mu \eta + 2r_0\eta^2 \tag{4.148a}$$

so that the equations of motion are

$$\partial^\mu \partial_\mu \xi = 0 \tag{4.148b}$$

$$(\partial^\mu \partial_\mu - 2r_0)\,\eta = 0 \ . \tag{4.148c}$$

The Fourier transform of (4.148c) has the form $-p^\mu p_\mu + m^2 = -E^2 + \vec{p}^{\,2} + m^2 = 0$, so that $-2r_0$ plays the role of the mass squared. Thus, as expected from our previous treatment, of two component fields, we have one massive mode, η, perpendicular to the circular minimum of the potential and one massless mode, ξ, parallel to the minimum.

The Lagrangian for the electromagnetic potential A^μ has the familiar form

$$\mathcal{L}(A) = -\frac{1}{4}F^{\mu\nu}F_{\mu\nu} \tag{4.149a}$$

where

$$F^{\mu\nu} = \partial^\mu A^\nu - \partial^\nu A^\mu \tag{4.149b}$$

and variation of A^μ yields the free space Maxwell equations:

$$\partial^\nu F_{\nu\mu} = 0 \ . \tag{4.149c}$$

In particular, $\nabla \cdot \vec{E} = \nabla \cdot \vec{B} = 0$ or in momentum space $\vec{k}\cdot\vec{E} = \vec{k}\cdot\vec{B} = 0$ so there are two independent transverse massless modes of the electromagnetic field. For subsequent reference, these degrees of freedom and the associated masses are tabulated in Table 4.2.

The Lagrangian for interacting scalar and electromagnetic fields is

$$\begin{aligned}\mathcal{L} = &\left[(\partial^\mu + ieA^\mu)\,\phi(x)\right]^*\left[(\partial_\mu + ieA_\mu)\,\phi(x)\right] - r_0\phi^*(x)\phi(x)\\ &- \frac{u_0}{2}\left(\phi^*(x)\phi(x)\right)^2 - \frac{1}{4}F_{\mu\nu}F^{\mu\nu}\end{aligned} \tag{4.150}$$

where we have used Eqs. (4.146a) and (4.149a), and A^μ is coupled to ϕ via the minimal coupling $(\partial_\mu + ieA_\mu)$. The Lagrangian is invariant under local gauge transformations

$$\begin{aligned}\phi(x) &\to e^{-i\theta(x)}\phi(x)\\ \phi^*(x) &\to e^{i\theta(x)}\phi^*(x)\\ A^\mu(x) &\to A^\mu(x) + \frac{1}{e}\partial^\mu\theta(x)\end{aligned} \tag{4.151}$$

	Field	Modes	Number of Components	m^2
Non Interacting Fields	Scalar	η	1	$-2r_0$
		ξ	1	0
	em	A^ν	2 (transverse)	0
Interacting Fields Symmetric Phase ($r_0 > 0$)	Scalar	ϕ, ϕ^*	2	r_0
	em	A^μ	2 (transverse)	0
Interacting Fields Broken symmetry phase ($r_0 < 0$)	Scalar	η	1	$-2r_0$
	em	$\tilde{A}^n$	3	$-2e^2 r_0/u_0$

Table 4.2 Summary of the degrees of freedom for a charged scalar field and electromagnetic field.

as well as global gauge transformations in which θ is constant.

The mean field equations are obtained by variation with respect to ϕ^*, ϕ, and A^μ, with the results

$$(\partial^\mu + ieA^\mu)(\partial_\mu + ieA_\mu)\phi + r_0\phi + u_0\phi^*\phi\,\phi = 0 \tag{4.152a}$$

$$(\partial^\mu - ieA^\mu)(\partial_\mu - ieA_\mu)\phi^* + r_0\phi^* + u_0\phi^*\phi\,\phi^* = 0 \tag{4.152b}$$

$$-ie\phi^*(\partial_\mu + ieA_\mu)\phi + ie\phi(\partial^\mu - ieA^\mu)\phi^* + \partial^\nu F_{\nu\mu} = 0\ . \tag{4.152c}$$

For $r_0 > 0$, the stationary solution for ϕ with minimum potential is $\phi = \phi^* = 0$, and linearizing Eqs. (4.152) for infinitesimal fluctuations in ϕ, ϕ^*, and A_μ yields

$$(\partial^\mu\partial_\mu + r_0)\phi = 0 \tag{4.153a}$$

$$(\partial^\mu\partial_\mu + r_0)\phi^* = 0 \tag{4.153b}$$

$$\partial^\mu F^{\nu\mu} = 0\ . \tag{4.153c}$$

Thus, as tabulated in Table 4.1, the symmetric phase of the system has two massive modes for the scalar field and two massless transverse modes for the electromagnetic field.

For $r_0 < 0$, the stationary solutions with minimum potential correspond to the broken symmetry fields satisfying Eq. (4.146c) and $A_\mu = 0$. To parameterize fluctuations parallel and perpendicular to the minimum in the potential, we again use Eqs. (4.147) to write $\phi(x)$ and $\phi^*(x)$ in terms of two real fields $\xi(x)$ and $\eta(x)$. To evaluate the Lagrangian in terms of these new fields, it is convenient to make the gauge transformation (4.151) with $\theta(x) = \xi(x)$ to obtain

$$\begin{aligned}
\phi(x) &\to e^{-i\xi(x)}\phi(x) = e^{-i\xi(x)}\left[e^{i\xi(x)}\left(\phi_0 + \eta(x)\right)\right] = \phi_0 + \eta(x) \\
\phi^*(x) &\to e^{i\xi(x)}\phi^*(x) = e^{i\xi(x)}\left[e^{-i\xi(x)}\left(\phi_0 + \eta(x)\right)\right] = \phi_0 + \eta(x) \qquad (4.154) \\
A^\mu(x) &\to A^\mu(x) + \frac{1}{e}\partial^\mu\xi(x) = \tilde{A}^\mu(x) \ .
\end{aligned}$$

Note that the electromagnetic field tensor $F^{\mu\nu}$ is unchanged by the transformation from A^μ to $\tilde{A}^\mu$. In terms of the new fields η and $\tilde{A}^\mu$, the Lagrangian (4.150) becomes

$$\mathcal{L} = (\partial^\mu - ie\tilde{A}^\mu)(\phi_0+\eta)(\partial_\mu + ie\tilde{A}_\mu)(\phi_0+\eta) - r_0(\phi_0+\eta)^2 - \frac{u_0}{2}(\phi_0+\eta)^4 - \frac{1}{4}\tilde{F}_{\mu\nu}\tilde{F}^{\mu\nu} \qquad (4.155)$$

so that variation with respect to η and $\tilde{A}_\mu$ combined with Eq. (4.146c) for ϕ_0 yield the following linearized equations for infinitesimal fluctuations of the fields

$$(\partial^\mu\partial_\mu - 2r_0)\,\eta = 0 \qquad (4.156a)$$

$$-\frac{2e^2 r_0}{u_0}\tilde{A}^\mu + \partial_\nu F^{\nu\mu} = 0 \ . \qquad (4.156b)$$

Since $\partial_\mu\partial_\nu F^{\nu\mu} = 0$, Eq. (4.156b) implies the Lorentz gauge condition

$$\partial_\mu\tilde{A}^\mu = 0 \qquad (4.156c)$$

and (4.156b) may be rewritten

$$\left(\partial_\nu\partial^\nu - 2e^2\frac{r_0}{u_0}\right)A^\mu = 0 \ . \qquad (4.156d)$$

From the equations of motion (4.156), we observe two important results, which are also tabulated in Table 4.2.

- The mode $\xi(x)$, which in the non-interacting case was the massless component of the scalar field corresponding to fluctuations along the minimum of the potential, no longer appears explicitly in the action. Instead, it has been subsumed into the field $\tilde{A}(x)$ by the gauge transformation, Eq. (4.154). The massless Goldstone mode is said to have been "gauged away". Only the massive mode η of scalar field remains, and its mass is unaffected by the coupling to the electromagnetic field.
- The electromagnetic field $\tilde{A}^\mu(x)$ differs from the free-field solution to Maxwell's equation in two crucial respects. By Eq. (4.156d) it has a mass $m^2 = -2e^2\frac{r_0}{u_0}$. Furthermore, the four components $\tilde{A}^\mu$ have a single constraint, the Lorentz gauge condition (4.156c) so that there are three independent massive modes in contrast to the two transverse massless modes for Maxwell's equations. Thus, the massless Goldstone mode of the scalar field ξ has been effectively replaced by the massive longitudinal component of the electromagnetic field.

The Anderson-Higgs mechanism occurs in diverse physical systems. The example we have discussed applies directly to a superconductor in an electromagnetic field. The zero mass excitations of a neutral superconducting Fermi gas become longitudinal plasma modes of finite mass when the gas is charged. One physical manifestation of the finite mass is the Meissener effect, in which an externally applied magnetic field can only penetrate a depth equal to the inverse mass into a bulk superconductor. In the Weinberg-Salam theory of electromagnetic and weak interactions using non-Abelian gauge theory, all the masses of the gauge fields are generated from as yet unobserved Higgs fields by this mechanism.

Definition		Mean Field	$D=3$ Ising	$D=2$ Ising
$C_H = \frac{\partial U}{\partial T}\Big\|_H = -T\frac{\partial^2 F}{\partial T^2}\Big\|_H \underset{T\to T_c}{\sim} \|T-T_c\|^{-\alpha} + C$	α	0	0.11	0
$m \underset{T\to T_c}{\sim} (T_c - T)^\beta$	β	$\frac{1}{2}$	0.326	$\frac{1}{8}$
$\chi = \frac{\partial M}{\partial H} = -\frac{\partial^2 F}{\partial H^2} \underset{T\to T_c}{\sim} \|T-T_c\|^{-\gamma}$	γ	1	1.24	$\frac{7}{4}$
$m \underset{H\to H_c}{\sim} \|H-H_c\|^{1/\delta}$	δ	3	4.8	15
$G(\vec{r}) \underset{T-T_c}{\sim} \frac{e^{-r/\xi}}{\|r\|^{D-2+\eta}}$	η	0	0.037	$\frac{1}{4}$
$\xi \underset{T\to T_c}{\sim} \|T-T_c\|^{-\nu}$	ν	$\frac{1}{2}$	0.63	1

Table 4.3 Definitions of critical exponents and their values for spin systems. The mean field results, which are exact for $D \geq 4$ for any number of spin components, are compared with approximate numerical values for Ising spins in 3 dimensions (LeGuillou and Zinn-Justin, 1985) and the exact Ising values in 2 dimensions.

PROBLEMS FOR CHAPTER 4

The first two problems treat phenomenological Landau theory: Problem 1 explores the effect of a sixth order term in the Landau function and Problem 2 examines the effect of higher powers and gradients in the Landau-Ginzburg functional. Problems 3 and 5 fill in details of derivations omitted in the text for the mean field solution to the $O(n)$ spin model and the one-loop corrections to the Ising model, respectively. Problem 4 treats the liquid-gas phase transition in the canonical ensemble, and demonstrates how the same physical results emerge as in the grand canonical ensemble. The next two problems deal with corrections to the mean field approximation in the presence of zero modes. Problem 6 generalizes the one-loop corrections presented in the text to the case of N zero modes and Problem 7 shows how the general perturbation expansion is obtained. To complement the mean-field results developed throughout the chapter, Problem 8 presents an efficient and elegant exact solution of the two-dimensional Ising model using Grassmann variables.

PROBLEM 4.1 Landau Theory

Consider the Landau function including a sixth order contribution

$$\mathcal{L}(M, h, t) = -hM + tM^2 + b_4 M^4 + M^6 \quad .$$

Note, in comparing with Eq. (4.4), for simplicity we have set $c_1 = -1$, $d_2 = 1$ and $b_6 = 1$ by appropriately scaling h, t, and M and we assume M couples to h linearly.

a) Analyze the generic form of $\mathcal{L}$ in the $b_4 - T$ plane at $H = 0$ identifying regions of single, double, and various classes of triple minima.

b) Find the lines of first-order and second-order phase transitions at $H = 0$.

c) Locate the tricritical point and calculate the critical exponents.

PROBLEM 4.2 Landau-Ginzburg Theory

Consider a Landau-Ginzburg theory, with the functional:

$$\mathcal{L} = \int d^d x \left(\frac{1}{2} \left(\vec{\nabla} \phi \right) + \frac{r_0}{2} \phi^2(x) + \frac{g_r}{r!} \cdot \phi^r(x) \right) \quad .$$

a) By using dimensional analysis, show that there is a critical dimension $d_r = 2r/(r-2)$ such that for $d > d_r$, mean field theory is valid, whereas it is invalid near T_c for $d \leq d_r$.

b) Write the mean field equations for $r \geq 3$, and calculate the critical exponents.

c) Show, using dimensional analysis, that if $\mathcal{L}$ contains higher derivatives of ϕ or higher powers of $(\nabla \phi)$ these terms become negligible as $T \to T_c$.

PROBLEM 4.3 The Classical $\mathcal{O}(n)$ Spin Model

Consider the classical $\mathcal{O}(n)$ spin model defined in Eq. (4.50) for general numbers of components n. To study the mean field theory close to the critical point, expand $\ln(\int d\vec{S}\, \delta(\vec{S}^2 - 1) e^{\beta \vec{\phi}_i \cdot \vec{S}})$ through fourth order in ϕ and Legendre transform to obtain Eq. (4.75a). Explain how this result follows directly from symmetry arguments. Calculate all the critical exponents.

PROBLEM 4.4 The Liquid-Gas Phase Transition

Consider a classical gas of particles interacting via a two-body potential $v(\vec{r})$ subject to an external potential $U(r)$. The grand canonical partition function is given by Eq. (4.91a).

a) Write the mean field equation by applying the stationary-phase approximation to the functional integral (4.95). In the absence of an external potential $U(\vec{r})$, assuming periodic boundary conditions, one can seek a constant solution

$\phi(\vec{r}) = \phi_0$. Write the equation for ϕ_0 in terms of the volume integral of the interaction $V_0 = \int d^3r\, v(r)$.

b) Show that if $V_0 > 0$, the mean-field solution ϕ_0 is unique, whereas if $V_0 < 0$, there exists a critical temperature T_c below which the system can undergo a phase transition from a liquid phase to a gas phase. Compute the critical temperature, and the critical exponents.

c) Using methods similar for those used to derive Eq. (4.95), show that the canonical partition function for a system of $\mathcal{N}$ particles is:

$$Z = \frac{1}{N!} \int \mathcal{D}\phi\, e^{-\frac{\beta}{2}\int_\Omega d^3r d^3r' \phi(r) v^{-1}(r-r')\phi(r') + N \log\left(\int_\Omega d^3r dr\, e^{-\beta\phi(r)}\right)} .$$

d) Show that the constant mean field solution is given by $\phi_0 = \rho V_0$ where $\rho = \frac{N}{\Omega}$ is the particle density.

e) Calculate the pressure P and the chemical potential μ as a function of ρ. Show that when $V_0 < 0$, there exists a temperature T_c below which there are two values of ρ, denoted by ρ_G and ρ_L, for which the pressure is identical: $P(\rho_G) = P(\rho_L)$.

f) The condition for coexistence of two phases is the equality of their pressure and their chemical potential. Using ρ_G and ρ_L from part e), show that the condition $\mu(\rho_G) = \mu(\rho_L)$ is identical to the mean-field equation derived in the grand canonical case.

PROBLEM 4.5

Perform the Legendre transform for the Ising model including one-loop corrections to obtain Eq. (4.110). The algebra is greatly simplified by expanding consistently to first order in $\frac{1}{\ell}$ and avoiding explicit evaluation of the quantity $\frac{\partial A(\bar{\phi})}{\partial H_i}$ which cancels out of the final result.

PROBLEM 4.6 One-Loop Corrections in the Presence of Zero Modes

Consider the partition function

$$S = \int \mathcal{D}\phi(x) e^{-S(\phi(x))} \tag{1}$$

in the presence of a continuous symmetry for which the mean field equations

$$\left.\frac{\partial S(\phi(x))}{\partial\phi(x)}\right|_{\phi(x)=\phi_c} = 0 \tag{2}$$

have degenerate solutions $\phi_c(x, \vec{\theta})$ parameterized by an n-component vector $\vec{\theta} = \{\theta_1, \theta_2 \ldots \theta_n\}$. Expansion of the integral (1) around a specific mean field $\phi_c(x, \vec{\theta})$ as in Eq. (4.122) yields

$$Z = e^{-S(\phi(x,\vec{\theta}))} \int \mathcal{D}\phi(x)\, e^{-\frac{1}{2}\int dx dy \left(\phi(x) - \phi_c(x,\vec{\theta})\right) A(x,y) \left(\phi(y) - \phi_c(y,\vec{\theta})\right)} \tag{3}$$

where

$$A(x,y) = \frac{\delta^2 S}{\delta\phi(x,\vec{\theta})\delta\phi(y,\vec{\theta})} \; .$$

a) By taking derivatives of equation (2) with respect to θ_i, show that there are n Goldstone modes $\frac{\partial\phi_c(x,\vec{\theta})}{\partial\theta_i}$ which satisfy

$$\int dy \, A(x,y) \frac{\partial\phi_c(x,\vec{\theta})}{\partial\theta_i} = 0 \; .$$

b) As in Section 4.4, we will separate the transverse and longitudinal fluctuations.

If we define n functions $f_i(\vec{\theta})$:

$$f_i(\vec{\theta}) = \int dx \frac{\partial\phi_c}{\partial\theta_i}\left(\phi_c(x,\vec{\theta}) - \phi(x)\right)$$

then the fluctuations orthogonal to the Goldstone modes satisfy $f_i(\vec{\theta}) = 0, \quad i = 1, n$. Show that the constraints $f_i(\vec{\theta}) = 0$ may be included in the functional integral by including the following multidimensional generalization of the identity Eq. (4.123) in the partition function

$$\int \prod_{i=1}^{n} d\theta_i \prod_{i=1}^{n} \delta\left(f_i(\vec{\theta})\right) \, \det\left(\frac{\partial f_i}{\partial\theta_j}\right) = 1 \; .$$

c) Introducing an auxiliary field $h(x)$, show that the partition function (3) is given by:

$$Z = e^{-S\left(\phi_c(x,\vec{\theta})\right)} \det\left(\int dx \frac{\partial\phi_c}{\partial\phi_i}\cdot\frac{\partial\phi_c}{\partial\theta_j} - \int dx \frac{\partial^2\phi_c}{\partial\theta_i\partial\theta_j}\cdot\frac{\delta}{\delta h(x)}\right) \mathcal{L}\left(h(x)\right)\Big|_{h=0}$$

where

$$\mathcal{L}\left(h(x)\right) = \int \prod_{i=1}^{n} \frac{d\lambda_i d\theta_i}{2\pi} \int \mathcal{D}\phi \, e^{-\frac{1}{2}\int dx dy \phi(x) A(x,y)\phi(y) + \int dx \phi(x)\left(-i\sum_{i=1}^{n}\lambda_i \frac{\partial\phi_c}{\partial\theta_i} + h(x)\right)} \; . \tag{4}$$

d) Since the Goldstone modes $\frac{\partial\phi_c}{\partial\theta_i}$ are not necessarily orthonormal, let $\{\psi_i(x), \; i = 1, n\}$ denote an orthonormal basis of the zero eigenspace of A and define

$$A_\epsilon = \epsilon\sum_{i=1}^{n}|\psi_i\rangle\langle\psi_i| + A_\perp$$

analogous to Eq. (4.127a). Replacing A by A_ϵ in (4), and performing the ϕ integral, show that when ϵ goes to zero

$$\mathcal{L} \underset{\epsilon\to 0}{\sim} \frac{\epsilon^{-n/2}}{\sqrt{\det A_\perp}} \int \prod_{i=1}^{n} \frac{d\lambda_i d\theta_i}{2\pi} \, e^{\frac{1}{2\epsilon}\sum_{k=1}^{n}\left(\langle h|\psi_k\rangle - i\sum_{i=1}^{n}\lambda_i\left\langle\frac{\partial\phi_c}{\partial\theta_i}\middle|\psi_k\right\rangle\right)^2} \tag{5}$$

where $\langle f|g\rangle$ denotes $\int dx\, f(x)g(x)$.

e) Performing the $\{\lambda_i\}$ integrals in (5), show that

$$\mathcal{L} = \frac{1}{\sqrt{\det A_\perp}} \int \prod_{i=1}^{n} \frac{d\theta_i}{\sqrt{2\pi}} \cdot \frac{1}{\sqrt{\det\langle \frac{\partial\theta_c}{\partial\theta_i} | \frac{\partial\phi_c}{\partial\theta_j}\rangle}}$$

and thus

$$Z = \frac{e^{-S(\phi(x,\vec{\theta}))}}{\sqrt{\det A_\perp}} \int \prod_{i=1}^{n} \frac{d\theta_i}{\sqrt{2\pi}} \sqrt{\det(\langle \frac{\partial\phi_c}{\partial\theta_i} | \frac{\partial\phi_c}{\partial\theta_j}\rangle)} \ .$$

f) Show that in the case of a vector field $\vec{\phi}(x)$ this argument yields Eq. (4.131).

PROBLEM 4.7 Perturbation Theory in the Presence of Zero Modes

In order to perform a perturbation expansion, it is sufficient to be able to calculate Gaussian integrals, using the identity

$$\begin{aligned} Z &= \int \mathcal{D}\phi(x)\, e^{-\frac{1}{2}\int dxdy\, \phi(x)A(x,y)\phi(y) - \int dx\, V(\phi(x))} \\ &= e^{-\int dx\, V\left(\frac{\delta}{\delta j(x)}\right)} \int \mathcal{D}\phi(x) e^{-\frac{1}{2}\int dxdy\, \phi(x)A(x,y)\phi(y)} e^{\int dx\, j(x)\phi(x)} \Big|_{j=0} \ . \end{aligned}$$

Let us define

$$\mathcal{L}(j) = \int \mathcal{D}\phi(x) e^{-\frac{1}{2}\int dxdy (\phi(x)-\phi_c(x,\theta_0))A(x,y)(\phi(y)-\phi_c(y,\theta_0))} e^{\int dx\, j(x)\phi(x)} \qquad (1)$$

and $\partial\phi_c = \frac{\partial\phi_c}{\partial\theta_0}$ is a zero eigenvalue of $A(x,y)$. We regularize A as:

$$A = \epsilon \frac{|\partial\phi_c\rangle\langle\partial\phi_c|}{\langle\partial\phi_c|\partial\phi_c\rangle} + A_\perp \ .$$

and integrate over transverse fluctuations by introducing in Eq. (1) the identity

$$\int d\theta_0 f'(\theta_0)\delta\left(f(\theta_0)\right) = 1$$

where

$$f(\theta_0) = \int dx\, \frac{\partial\phi_c}{\partial\theta_0}(x,\theta_0)\left(-\phi(x) + \phi_c(x,\theta_0)\right) \ .$$

a) Show that

$$\mathcal{L}(j) = \int dx \left[-\frac{\partial^2\phi_c}{\partial\theta_0^2} \cdot \frac{\delta}{\delta j(x)} + \left(\frac{\partial\phi_c(x)}{\partial\theta_0}\right)^2 \right] \mathcal{L}_1(j)$$

where

$$\mathcal{L}_1(j) = \frac{1}{\sqrt{\det A_\perp}} \int \frac{d\theta_0}{\sqrt{2\pi}} \frac{1}{\sqrt{\int dx \left(\frac{\partial\phi_c}{\partial\theta_0}\right)^2}} \cdot e^{\frac{1}{2}\int dxdy j_\perp(x) A_\perp^{-1}(x,y) j_\perp(y)}$$

and

$$j_\perp(x) = j(x) - \frac{\langle\partial\phi_c|j\rangle}{\langle\partial\phi_c|\partial\phi_c\rangle} \frac{\partial\phi_c(x)}{\partial\theta_0} .$$

b) Show that:

$$\mathcal{L}(j) = \left[\int dx \left(\frac{\partial\phi_c(x,\theta_0)}{\partial\theta_0}\right)^2 - \int dxdy \frac{\partial^2\phi_c(x,\theta_0)}{\partial\theta_0^2} \cdot A_\perp^{-1}(x,y) j_\perp(y)\right] \mathcal{L}_1(j)$$

where we have used the fact that

$$A_\perp^{-1}|\partial\phi_c\rangle = 0 .$$

c) By expressing $\frac{\delta}{\delta j(x)}$ in terms of $\frac{\delta}{\delta j_\perp(x)}$, write the formal perturbation expansion of Z as the exponential of an operator $\left(\frac{\delta}{\delta j_\perp(x)}\right)$ acting on $\mathcal{L}(j)$, at $j = 0$. Evaluate the first order term of this perturbation expansion.

PROBLEM 4.8 The Two-Dimensional Ising Model

This problem outlines the solution of the two-dimensional Ising model using a method due to Samuel (1980).

Consider an Ising model on a square lattice of size $N = (2M_x + 1) \times (2M_y + 1)$, with nearest neighbor interaction J, and periodic boundary conditions. The partition function of the system is:

$$Z = \sum_{\{S_{\vec{r}}\}} e^{\beta J \sum_{\vec{r}} (S_{\vec{r}} S_{\vec{r}+\vec{a}_x} + S_{\vec{r}} S_{\vec{r}+\vec{a}_y})} \tag{1}$$

where $\sum_{\{S_{\vec{r}}\}}$ denotes a sum over all spin configurations $S_r = \pm 1$, the sum $\sum_{\vec{r}}$ runs over all lattice sites, and $\vec{a}_x$ and $\vec{a}_y$ are unit vectors along the horizontal and vertical bonds.

a) Using the fact that we are dealing with Ising spins ($S = \pm 1$), show that:

$$e^{\beta JSS'} = ch(\beta J)(1 + KSS')$$

where $K = th\beta J$. From this identity, show that

$$Z = (ch\beta J)^{2N} 2^N \varsigma$$

where

$$\varsigma = \frac{1}{2^N} \sum_{\{S_{\vec{r}}\}} \prod_{\vec{r}} (1 + KS_{\vec{r}} S_{\vec{r}+\vec{a}_x})(1 + KS_{\vec{r}} S_{\vec{r}+\vec{a}_y}) .$$

b) The high temperature expansion of Z is obtained by expanding ς in powers of K. Using the properties that:

$$\sum_{\{S_{\vec r}\}} S_{\vec r}^{2n+1} = 0 \qquad \sum_{\{S_{\vec r}\}} S_{\vec r}^{2n} = 2^N$$

show that ς is the sum of all closed polygons joining adjacent sites of the lattice, not connected and possibly self-intersecting. The weight associated with a graph is obtained in the following way: there is a factor K per bond, and there is an overall factor Γ, which is the number of ways in which the polygon can be drawn on the lattice.

Show that graphs 1, 2, 3 below contribute to ς, and graphs 4 and 5 do not.

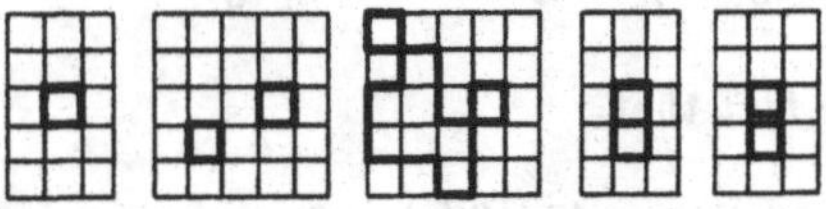

(1) (2) (3) (4) (5)

Show that the contributions of the first three graphs are $\varsigma_1 = NK^4$, $\varsigma_2 = \frac{1}{2}N(N-4)K^8$, and $\varsigma_3 = NK^{22}$

c) The essential idea of this method is to write a Grassmann functional integral which, when expanded, generates all the graphs of ς with the correct weights. Since each bond belongs to zero or one polygon, it is natural to try to associate Grassmann variables with bonds. Hence, we will let $\psi_r^{\dagger h}$ denote a creation operator for a horizontal bond between $\vec r$ and $\vec r + \vec a_x$, $\psi_r^{\dagger v}$ denote a creation operator for a vertical bond between $\vec r$ and $\vec r + \vec a_y$, and define the corresponding annihilation operators ψ_r^h and ψ_r^v. We associate with each of these operators the Grassmann variables η_r^{*h} η_r^{*v} η_r^h η_r^v . We will now prove the following formula:

$$\varsigma = \int d\mu \; e^{-\sum_r (\eta_r^{*h}\eta_r^h + \eta_r^{*v}\eta_r^v) + K\sum_r (\eta_r^{*h}\eta_{r+a_x}^h + \eta_r^{*v}\eta_{r+a_y}^v)} \; e^{\sum_r (\eta_r^{*h}\eta_r^v + \eta_r^{*v}\eta_r^h + \eta_r^{*v}\eta_r^{*h} + \eta_r^h\eta_r^v)} \tag{2}$$

where $d\mu = \prod_r d\eta_r^{*h} d\eta_r^h d\eta_r^{*h} d\eta_r^v$ by showing that it generates the same set of graphs as in (b) with the same weights. First, show that (2) can be rewritten as

$$\varsigma = \langle \prod_r \left(1 + \eta_r^{*h}\eta_r^v + \eta_r^{*v}\eta_r^h + \eta_r^{*v}\eta_r^{*h} + \eta_r^h\eta_r^v\right) \times \prod_r \left(1 + K\eta_r^{*h}\eta_{r+a_x}^h\right)\left(1 + K\eta_r^{*v}\eta_{r+a_y}^v\right)\rangle \tag{3}$$

where the brackets $\langle \ \rangle$ are defined by:

$$\langle A \rangle = \int d\mu e^{-\sum_r (\eta_r^{*h}\eta_r^h + \eta_r^{*v}\eta_r^v)} A \; . \tag{4}$$

In order to evaluate (3), we can use Wick's theorem, with the Gaussian weight given in (4). Show that the only non-vanishing contractions are $\langle \eta_r^{*h}\eta_{r'}^h\rangle = \delta_{rr'}$ $\langle \eta_r^{*v}\eta_{r'}^v\rangle = \delta_{rr'}$.

There are four terms in (3) that can be represented as corners:

 $= \eta_r^{*h}\eta_r^v$  $= \eta_r^{*v}\eta_r^h$  $= \eta_r^{*v}\eta_r^{*h}$ $= \eta_r^h\eta_r^v$

and two terms that can be represented as bonds:

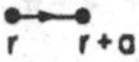

$= K\eta_r^{*h}\eta_{r+a_x}^h$ 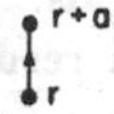

$= K\eta_r^{*v}\eta_{r+a_y}^v$.

Show by expanding (3) that the only non-vanishing graphs are the polygons discussed in (b). Finally, evaluate the factors associated with each graph. In particular, check that all graphs come with a plus sign and have the same weights as those of the Ising model.

d) In order to evaluate (2), it is convenient to diagonalize the quadratic form by going to Fourier representation. Using the following Fourier series:

$$a_{\vec{k}}^{*h,v} = \frac{1}{\sqrt{N}} \sum_{x=-M_x}^{M_x} \sum_{y=-M_y}^{M_y} e^{i(k_1x+k_2y)}\eta_{x,y}^{*h,v}$$

$$a_{\vec{k}}^{h,v} = \frac{1}{\sqrt{N}} \sum_{x=-M_x}^{M_x} \sum_{y=-M_y}^{M_y} e^{-i(k_1x+k_2y)}\eta_{x,y}^{h,v}$$

where $\vec{k} = (k_1, k_2)$, $k_1 = \frac{2i\pi}{2M_x+1}s$, $k_2 = \frac{2i\pi}{2M_y+1}t$ with $s = \{-M_x, \ldots, +M_x\}$ and $t = \{-M_y, \ldots, +M_y\}$, show that:

$$\begin{aligned} \varsigma = \int \prod_{s=-M_x}^{M_x} \prod_{t=-M_y}^{M_y} & da_{\vec{k}}^{*h} da_{\vec{k}}^{h} da_{\vec{k}}^{*v} da_{\vec{k}}^{v} e^{-\sum_{s,t}\left(a_{\vec{k}}^{*h}a_{\vec{k}}^{h}+a_{\vec{k}}^{*v}a_{\vec{k}}^{v}\right)} \\ & \times e^{K\sum_{s,t}\left(e^{ik_1}a_{\vec{k}}^{*h}a_{\vec{k}}^{h}+e^{ik_2}a_{\vec{k}}^{*v}a_{\vec{k}}^{v}\right)} e^{\sum_{s,t}\left(a_{\vec{k}}^{*h}a_{\vec{k}}^{v}+a_{\vec{k}}^{*v}a_{\vec{k}}^{h}+a_{\vec{k}}^{*v}a_{-\vec{k}}^{*h}+a_{\vec{k}}^{h}a_{-\vec{k}}^{v}\right)} . \end{aligned} \tag{5}$$

e) The quadratic form in (5) is not yet quite diagonalized due to the couplings of $\vec{k}$ to $-\vec{k}$. Show that it can be rewritten as:

$$\varsigma = \int \prod_{s=-M_x}^{M_x} \prod_{t=0}^{M_y} da_{\vec{k}}^{*h} da_{\vec{k}}^{h} da_{\vec{k}}^{*v} da_{\vec{k}}^{v} da_{-\vec{k}}^{*h} da_{-\vec{k}}^{h} da_{-\vec{k}}^{*v} da_{-\vec{k}}^{v}\, e^S$$

where

$$\begin{aligned} S = -\sum_{s=-M_x}^{M_x} \sum_{t=0}^{M_y} \left(1 - \frac{1}{2}\delta_{t,0}\right) & \Big[\left(a_{\vec{k}}^{*h}a_{\vec{k}}^{h} + a_{-\vec{k}}^{*h}a_{-\vec{k}}^{h} + a_{\vec{k}}^{*v}a_{\vec{k}}^{v} + a_{-\vec{k}}^{*v}a_{-\vec{k}}^{v}\right) \\ & + K\left(e^{ik_1}a_{\vec{k}}^{*h}a_{\vec{k}}^{h} + e^{-ik_1}a_{-\vec{k}}^{*h}a_{-\vec{k}}^{h} + e^{ik_2}a_{\vec{k}}^{*v}a_{\vec{k}}^{v} + e^{-ik_2}a_{-\vec{k}}^{*v}a_{-\vec{k}}^{v}\right) \\ & + \left(a_{\vec{k}}^{*h}a_{\vec{k}}^{v} + a_{\vec{k}}^{*v}a_{\vec{k}}^{h} + a_{\vec{k}}^{*v}a_{-\vec{k}}^{*h} + a_{\vec{k}}^{h}a_{-\vec{k}}^{v} + a_{-\vec{k}}^{*h}a_{-\vec{k}}^{v} + a_{-\vec{k}}^{*v}a_{-\vec{k}}^{h} + a_{-\vec{k}}^{*v}a_{\vec{k}}^{*h} + a_{-\vec{k}}^{h}a_{\vec{k}}^{v}\right)\Big] . \end{aligned}$$

Rewrite ς in terms of the new variables

$$\psi^*_{\vec{k}} = a^{*h}_{\vec{k}} \quad \psi_{\vec{k}} = a^h_{\vec{k}} \quad \phi^*_{\vec{k}} = a^{*v}_{\vec{k}} \quad \phi_{\vec{k}} = a^v_{\vec{k}}$$
$$f^*_{\vec{k}} = a^h_{-\vec{k}} \quad f_{\vec{k}} = a^{*h}_{-\vec{k}} \quad g^*_{\vec{k}} = a^v_{-\vec{k}} \quad g_{\vec{k}} = a^{*v}_{-\vec{k}}$$

and show that the quadratic form is diagonal in $\vec{k}$

d) Evaluate ς. In the thermodynamic limit, M_x, $M_y \to +\infty$ and the summations over $\vec{k}$ are replaced by integrals: $\sum_{\vec{k}} \to N \int_{-\pi}^{+\pi} \frac{dk_1}{2\pi} \int_{-\pi}^{+\pi} \frac{dk_2}{2\pi}$. Show that the free energy is given by the Onsager result:

$$\frac{\beta F}{N} = -\ln 2 - \frac{1}{2} \int_{-\pi}^{+\pi} \frac{dk_1}{2\pi} \int_{-\pi}^{+\pi} \frac{dk_2}{2\pi} \ln[ch^2(2\beta J) - sh(2\beta J)(\cos k_1 + \cos k_2)] \ . \quad (6)$$

g) The argument of the ln in (6) is minimal for $k_1 = k_2 = 0$. Show that for $k_1 = k_2$, the argument of the ln is positive, and that it vanishes when the temperature T reaches a critical value T_c. Calculate T_c. All the non-analyticities of F, if any, thus come from the vicinity of $\vec{k} = \vec{0}$ and $T = T_c$. By expanding the cosines in (6) around $\vec{k} = 0$, and rescaling appropriately, show that when T goes to T_c, the leading non-analyticity of F has the behavior:

$$F \simeq (T - T_c)^2 \ln |T - T_c| \ .$$

Show that the specific heat has a logarithmic divergence, and thus the critical exponent α vanishes.

CHAPTER 5

GREEN'S FUNCTIONS

5.1 INTRODUCTION

In Chapter 2, Green's functions emerged as natural quantities to characterize many-particle systems. On the one hand, they are defined as thermal or ground state averages of time-ordered products of operators which are directly calculable in perturbation theory, and on the other hand they may be related easily to experimental observables. Thus, although any particular observable could be calculated directly using the techniques of Chapters 2 and 3, it is worthwhile to develop the general properties of Green's functions which are applicable to a broad range of phenomena and are widely utilized in the literature.

DEFINITIONS

The n-body real-time Green's function in the Grand Canonical Ensemble was defined in Eq. (2.21) as the thermal trace of a time-ordered product of creation and annihilation operators in the Heisenberg representation:

$$\begin{aligned}\mathcal{G}^{(n)}&(\alpha_1 t_1, \cdots \alpha_n t_n | \alpha'_1 t'_1, \cdots \alpha'_n t'_n) \\ &= \frac{(-i)^n}{Z} \mathrm{Tr}\left(e^{-\beta(\hat{H}-\mu\hat{N})} T a^{(H)}_{\alpha_1}(t_1) \cdots a^{(H)}_{\alpha_n}(t_n) a^{(H)\dagger}_{\alpha'_n}(t'_n) \cdots a^{(H)\dagger}_{\alpha'_1}(t'_1)\right) \quad .\end{aligned} \tag{5.1a}$$

To simplify and unify the notation, it is convenient to replace operators $a^H_\alpha(t)$ in the α-representation by field operators defined in Eq. (1.89) $\psi(x,t) \equiv e^{i(\hat{H}-\mu\hat{N})t}\hat{\psi}(x)e^{-i(\hat{H}-\mu\hat{N})t}$ and to denote the thermal average by brackets:

$$\begin{aligned}\mathcal{G}^{(n)}&(x_1 t_1, \cdots x_n t_n | x'_1 t'_1 \cdots x'_n t'_n) \\ &= (-i)^n \langle T\hat{\psi}(x_1 t_1) \cdots \hat{\psi}(x_n t_n) \hat{\psi}^\dagger(x'_n t'_n) \cdots \hat{\psi}^\dagger(x'_1 t'_1)\rangle \quad .\end{aligned} \tag{5.1b}$$

Note, following the convention established in Section 1.1, any internal degrees of freedom such as spin or isospin, which may have been explicitly included in the label α, are now implicitly included in x. It will often be desirable to compress the notation still further, replacing the arguments $\{x_n t_n\}$ by n:

$$\mathcal{G}^{(n)}(1, \cdots n | 1', \cdots n') = (-i)^n \langle T\hat{\psi}(1) \cdots \hat{\psi}(n) \hat{\psi}^\dagger(n') \cdots \hat{\psi}^\dagger(1')\rangle \quad . \tag{5.1c}$$

Similarly, using field operators in the imaginary-time Heisenberg representation, Eq. (2.23), $a^H_\alpha(\tau) \to \hat{\psi}(x,\tau) = e^{(\hat{H}-\mu\hat{N})\tau}\hat{\psi}(x)e^{-(\hat{H}-\mu\hat{N})\tau}$ the thermal Green's function defined in Eq. (2.22) may be written

$$\mathcal{G}^{(n)}(x_1\tau_1, \cdots x_n\tau_n | x'_1\tau'_1, \cdots x'_n\tau'_n) = \langle T\hat{\psi}(x_1\tau_1) \cdots \hat{\psi}(x_n\tau_n) \hat{\psi}^\dagger(x'_n\tau'_n) \cdots \hat{\psi}^\dagger(x'_1\tau'_1)\rangle \tag{5.2}$$

where the brackets again denote the thermal average. Note that in this notation real-time and thermal Green's functions are distinguished by arguments t and τ, respectively.

Finally, the zero-temperature Green's function is defined as the ground state expectation value:

$$G^{(n)}(x_1t_1,\cdots x_nt_n|x_1't_1',\cdots x_n't_n')$$
$$= (-i)^n\langle\psi_0|T\hat{\psi}(x_1t_1)\cdots\hat{\psi}(x_nt_n)\hat{\psi}^\dagger(x_n't_n')\cdots\hat{\psi}^\dagger(x_1't_1')|\psi_0\rangle \tag{5.3a}$$

to emphasize its similarity in structure to the finite temperature Green's function, we will also write it

$$G^{(n)}(x_1t_1,\cdots x_nt_n|x_1't_1',\cdots x_n't_n') = (-i)^n\langle T\hat{\psi}(x_1t_1)\cdots\hat{\psi}(x_nt_n)\hat{\psi}^\dagger(x_n't_n')\cdots\hat{\psi}(x_1't_1')\rangle \tag{5.3b}$$

Note that finite and zero-temperature Green's functions are distinguished by the use of $\mathcal{G}$ and G, and that the brackets denote a thermal trace or ground state expectation value, respectively. Also, the inclusion of the factor $(-i)^n$ in Eqs. (5.1) and (5.3) is not a universal convention: whereas it is used by Abrikosov, Gorkov and Dzaloshinski (1963), Doniach and Sondheimer (1974), Fetter and Walecka (1971), and Kadanoff and Baym (1962), no factor of i is used by Thouless (1972), and Brown (1972) uses $(i)^n$.

Physically, the real-time Green's functions at finite and zero temperature have an obvious interpretation: they describe the propagation of disturbances created by injecting particles at the space-time points $\{x_1t_1\}\cdots\{x_nt_n\}$ and removing them at $\{x_1't_1'\}\cdots\{x_n't_n'\}$. Note also that the definitions are quite general, applying to finite or infinite systems, with no requirement of translational invariance.

Since one-particle Green's functions will be studied in detail in subsequent sections, it is useful to introduce several auxiliary definitions. For notational convenience, the superscript (n) will usually be omitted on one-particle Green's functions. The two time orders in real-time Green's functions at zero or finite temperature are distinguished by the definition

$$G(1|1') \equiv \theta(t_1 - t_1')G^>(1|1') + \theta(t_1' - t_1)G^<(1|1') \tag{5.4}$$

so that

$$iG^>(1|1') = \langle\hat{\psi}(1)\hat{\psi}^\dagger(1')\rangle \tag{5.5a}$$

and

$$iG^<(1|1') = \varsigma\langle\hat{\psi}^\dagger(1')\hat{\psi}(1)\rangle \tag{5.5b}$$

where ς is the usual factor denoting 1 for Bosons and -1 for Fermions. These quantities are combined to define retarded and advanced Green's functions as follows

$$\begin{aligned} iG^R(1|1') &= i\left(G^>(1|1') - G^<(1|1')\right)\theta(t_1 - t_1') \\ &= \langle[\hat{\psi}(1),\hat{\psi}^\dagger(1')]_{-\varsigma}\rangle\theta(t_1 - t_1') \end{aligned} \tag{5.6a}$$

$$\begin{aligned} iG^A(1|1') &= i\left(G^<(1|1') - G^>(1|1')\right)\theta(t_1' - t_1) \\ &= -\langle[\hat{\psi}(1),\hat{\psi}(1')]_{-\varsigma}\rangle\theta(t_1' - t_1) \quad . \end{aligned} \tag{5.6b}$$

Analogous quantities are defined at finite temperature by replacing G by $\mathcal{G}$ in Eqs. (5.4– 5.6)

EVALUATION OF OBSERVABLES

Although Green's functions were introduced because of their physical connection to response functions, they contain sufficient information to evaluate all ground state observables or thermal averages. Since the ground state expectation value of an n-body operator requires the matrix element of n creation operators to the left of n destruction operators, it may be obtained directly from the n-particle Green's function by evaluating all the times associated with creation operators at time t^+ infinitesimally later than the time t associated with the destruction operators:

$$
\begin{aligned}
G^{(n)}&\,(x_1t,\cdots x_nt|x_1't^+,\cdots x_n't^+) \\
&= (-i)^n\varsigma^n\langle\psi^\dagger(x_n't^+)\cdots\psi^\dagger(x_1't^+)\psi(x_1t)\cdots\psi(x_nt)\rangle \;.
\end{aligned}
\tag{5.7}
$$

A general one-body operator $\hat{\mathcal{O}} = \int dxdx'\mathcal{O}(x',x)\hat{\psi}^\dagger(x')\hat{\psi}(x)$ thus has the expectation value

$$
\langle\hat{\mathcal{O}}\rangle = i\varsigma\int dxdx'\mathcal{O}(x'x)G(xt|x't^+) \;. \tag{5.8}
$$

For example, using Eq. (1.96), the kinetic energy operator corresponds to $\mathcal{O}(x',x) = -\frac{1}{2m}\delta(x',x)\nabla_r^2$ so that

$$
\begin{aligned}
\langle\hat{T}\rangle &= i\varsigma\int dx\left[-\frac{1}{2m}\nabla_r^2G(xt|x't^+)\right]_{x'=x} \\
&= i\varsigma\int d^3r\sum_\sigma\left[-\frac{1}{2m}\nabla_r^2G(\vec{r}\sigma t|\vec{r}\sigma t^+)\right]_{\vec{r}'=\vec{r}}
\end{aligned}
\tag{5.9}
$$

where the explicit spin sum is written in the last line for the case of particles with spin. Similarly, for the spin density at position $\vec{r}''$, $\mathcal{O}(x',x) = \delta(\vec{r}'-\vec{r}'')\delta(\vec{r}-\vec{r}')\vec{\sigma}_{\sigma'\sigma}$ and

$$
\langle\vec{\sigma}(r'')\rangle = i\varsigma\sum_{\sigma\sigma'}\left[\vec{\sigma}_{\sigma'\sigma}G(\vec{r}''\sigma t|\vec{r}''\sigma t^+)\right] \;. \tag{5.10}
$$

For any system with a time-independent Hamiltonian, the Fourier frequency transform may be written

$$
G(xt|x't') = G(x,x';t-t') = \int\frac{d\omega}{2\pi}\tilde{G}(x,x';\omega)e^{-i\omega(t-t')} \tag{5.11}
$$

so that

$$
\langle\hat{\mathcal{O}}\rangle = i\varsigma\int dxdx'\mathcal{O}(x',x)\int\frac{d\omega}{2\pi}e^{i\omega\eta}\tilde{G}(x,x';\omega) \tag{5.12}
$$

where η is a positive infinitesimal. Finally, if the system is also translationally invariant, it may be useful to take the Fourier momentum transform, being careful to carry the spin dependence separately if necessary:

$$
G(xt|x't') = G_{\sigma\sigma'}(\vec{r}-\vec{r}';t-t') = \int\frac{d^3k}{(2\pi)^3}\frac{d\omega}{2\pi}\tilde{G}_{\sigma\sigma'}(\vec{k},\omega)e^{i\vec{k}(\vec{r}-\vec{r}')-i\omega(t-t')} \;. \tag{5.13}
$$

The kinetic energy may then be rewritten

$$\langle T\rangle = i\varsigma\frac{\mathcal{V}}{(2\pi)^4}\int d^3k d\omega\, e^{i\omega\eta}\frac{k^2}{2m}\sum_\sigma \tilde{G}_{\sigma\sigma}(\vec{k},\omega) \quad . \tag{5.14}$$

Finite-temperature results analogous to Eqs. (5.8–5.14) are obtained straightforwardly using thermal Green's function. The only changes are elimination of the factor i in the relation

$$\langle\hat{O}\rangle = \varsigma\int dx dx' O(x'x)\mathcal{G}(x\tau|x'\tau^+) \tag{5.15}$$

and replacement of the Fourier frequency integral in Eq. (5.11) by the Fourier series Eq. (2.129).

The ground state energy E_0, internal energy U, and grand potential Ω are of particular interest. Although for a Hamiltonian containing two-body forces, they can be calculated directly from the two-body Green's function, they may also be obtained from the one-body Green's function by using the Heisenberg equations of motion.

Using the field operator commutation relations, Eq. (1.88) and the commutator identity

$$[\hat{A},\hat{B}\hat{C}]_- = [\hat{A},\hat{B}]_\varsigma\hat{C} - \varsigma\hat{B}[\hat{A},\hat{C}]_\varsigma \tag{5.16}$$

the Heisenberg equation for $\hat{\psi}(x,t)$ yields

$$\begin{aligned} i\frac{\partial}{\partial t}\hat{\psi}(x,t) &= [\hat{\psi}(x,t),\hat{H}] \\ &= \left[\hat{\psi}(x,t), \int dz\hat{\psi}^\dagger(z)T(z)\hat{\psi}(z) + \frac{1}{2}\int dz dz'\hat{\psi}^\dagger(z)\hat{\psi}^\dagger(z')v(z,z')\hat{\psi}(z')\hat{\psi}(z)\right] \\ &= T_x\hat{\psi}(x,t) + \int dx''\hat{\psi}^\dagger(x'',t)v(x,x'')\hat{\psi}(x'',t)\hat{\psi}(x,t) \end{aligned} \tag{5.17}$$

where we have used the symmetry of $v(z,z')$ and T_x denotes the differential operator for the kinetic energy used above. Thus,

$$\begin{aligned} &\int dx\left[\left\langle\hat{\psi}^\dagger(x',t')\left(i\frac{\partial}{\partial t} - T_x\right)\hat{\psi}(x,t)\right\rangle\right]_{\substack{x'=x\\t'=t^+}} \\ &\quad = \int dx dx''\left[\left\langle\hat{\psi}^\dagger(x',t')\hat{\psi}^\dagger(x'',t)v(x,x'')\psi(x'',t)\psi(x,t)\right\rangle\right]_{\substack{x'=x\\t'=t^+}} \end{aligned} \tag{5.18}$$

so that

$$\langle V\rangle = \frac{1}{2}i\varsigma\int dx\left[\left(i\frac{\partial}{\partial t} - T_x\right)G(xt|x't^+)\right]_{x'=x} \quad . \tag{5.19}$$

The ground state energy may be written

$$E_0 = \langle T+V\rangle = \frac{1}{2}i\varsigma\int dx\left[\left(i\frac{\partial}{\partial t} - \frac{1}{2m}\nabla_r^2\right)G(xt|x't^+)\right]_{x'=x} \tag{5.20a}$$

or for translationally invariant systems

$$E_0 = \frac{1}{2} i\varsigma \frac{\mathcal{V}}{(2\pi)^4} \int d^3k d\omega \, e^{i\omega\eta} \left(\frac{k^2}{2m} + \omega\right) \sum_\sigma \tilde{G}_{\sigma\sigma}(\vec{k},\omega) \ . \tag{5.20b}$$

Analogous relations are obtained for the internal energy at finite temperature using thermal Green's functions and the Heisenberg equation $-\frac{\partial}{\partial t}\hat{\psi}(x,t) = [\hat{\psi}(x,\tau), \hat{H} - \mu\hat{N}]$ with the result

$$\langle H\rangle = \frac{1}{2}\varsigma \int dx \left[\left(-\frac{\partial}{\partial\tau} - \frac{1}{2m}\nabla_r^2 + \mu\right) \mathcal{G}(x\tau|x'\tau^+)\right]_{x=x'} \tag{5.21a}$$

$$= \frac{1}{2}\varsigma \frac{\mathcal{V}}{(2\pi)^3\beta} \int d^3k \sum_{\omega_n} e^{i\omega_n\eta} \left(\frac{k^2}{2m} + i\omega_n + \mu\right) \sum_\sigma \tilde{\mathcal{G}}_{\sigma\sigma}(\vec{k},\omega_n) \ . \tag{5.21b}$$

The connection between the expressions for E_0, Eq. (5.20), and the perturbation expansion of Chapter 3 is not evident because of the presence of the time derivatives or ω factors. To obtain a direct connection to diagrams, it is useful to recall the definition of the self-energy, Eq.(3.38)

$$G^{-1} = G_0^{-1} - \Sigma = i\frac{\partial}{\partial t} - T_x - \Sigma \ . \tag{5.22}$$

Thus, the combination $(i\frac{\partial}{\partial t} - T)\, G$ appearing in $\langle V\rangle$ as a result of the equation of motion can be rewritten $(G^{-1} + \Sigma)G = 1 + \Sigma G$. Since we have diagram expansions for both Σ and G they can easily be combined for the complete result. However, when $\langle T\rangle$ is added to obtain E_0, Eq. (5.20) no longer has the desired simple form.

A convenient technique, originally due to Pauli, to generate an expression for E_0 of the desired form is to consider a continuously parameterized set of Hamiltonians

$$H(\lambda) = T + \lambda V \tag{5.23}$$

having eigenvalues $E_0(\lambda)$, eigenfunctions $\Psi_0(\lambda)$, Green's functions $G^{(\lambda)}$, and self-energies $\Sigma^{(\lambda)}$. Using the familiar relation $\frac{d}{d\lambda}E_0(\lambda) = \langle\psi_0(\lambda)|V|\psi_0(\lambda)\rangle$ (which follows from evaluating $\frac{d}{d\lambda}E_0(\lambda) = \left(\frac{d}{d\lambda}\langle\psi_0|\right) H|\psi_0\rangle + \langle\psi_0|\left(\frac{d}{d\lambda}H\right)|\psi_0\rangle + \langle\psi_0|H\frac{d}{d\lambda}|\psi_0\rangle$ and noting the first and third terms yield $E_0(\lambda)\frac{d}{d\lambda}\langle\psi_0|\psi_0\rangle = 0$ since $|\psi_0\rangle$ is normalized), the difference between the energy of the fully interacting system $E_0 = E_0(1)$ and the energy of the noninteracting system $W_0 = E_0(0)$ may be written

$$E_0 - W_0 = \int_0^1 d\lambda \frac{d}{d\lambda} E_0(\lambda) = \int_0^1 \frac{d\lambda}{\lambda} \langle\psi_0(\lambda)|\lambda V|\psi_0(\lambda)\rangle$$

$$= \frac{i}{2}\varsigma \int_0^1 \frac{d\lambda}{\lambda} \int dx \left[\left(i\frac{\partial}{\partial t} - T_x\right) G^\lambda(xt|x't^+)\right]_{x=x'}$$

$$= \frac{i}{2}\varsigma \int_0^1 \frac{d\lambda}{\lambda} \int dx \left[\int dx''dt' \left([G^\lambda]^{-1}(xt|x''t') + \Sigma^\lambda(xt|x''t')\right) G^\lambda(x''t'|xt^+)\right]$$

$$= \frac{i}{2}\varsigma \int_0^1 \frac{d\lambda}{\lambda} \int dx dx' dt' \Sigma^\lambda(xt|x't') G^\lambda(x't'|xt^+) \tag{5.24a}$$

$$= \frac{i}{2}\varsigma \int_0^1 \frac{d\lambda}{\lambda} \int \frac{d\omega}{2\pi} \int dx dx' \Sigma^\lambda(x,x';\omega) G^\lambda(x',x;\omega) e^{i\omega\eta} \ .$$

For a translationally invariant system

$$E_0 - W_0 = \frac{i}{2}\varsigma \int_0^1 \frac{d\lambda}{\lambda} \frac{\mathcal{V}}{(2\pi)^4} \int d^3k d\omega \sum_{\sigma\sigma'} \Sigma^{\lambda}_{\sigma\sigma'}(k,\omega) G^{\lambda}_{\sigma'\sigma}(k,\omega) e^{i\omega\eta} \quad . \tag{5.24b}$$

Analogous expressions for the Grand potential are obtained in terms of thermal Green's functions using the relation $\frac{d}{d\lambda}\Omega(\lambda) = \langle V \rangle$ proved in Problem 2.11 and the definition of the self energy Eq. (2.178) (note the sign)

$$\begin{aligned} \Omega - \Omega_0 &= \int_0^1 \frac{d\lambda}{\lambda} \langle \lambda V \rangle \\ &= \frac{1}{2}\varsigma \int \frac{d\lambda}{\lambda} \int dx \left[\left(-\frac{\partial}{\partial \tau} - T_x + \mu \right) \mathcal{G}(x\tau|x'\tau^+) \right]_{x=x'} \\ &= \frac{1}{2}\varsigma \int \frac{d\lambda}{\lambda} \int dx dx' d\tau' \Sigma^{\lambda}(x\tau|x'\tau') \mathcal{G}^{\lambda}(x'\tau'|x\tau^+) \quad . \end{aligned} \tag{5.25}$$

Note that the price of obtaining direct contact with diagrams is the additional parametric integral over the coupling strength λ.

It is an instructive to exercise to combine the diagrams for Σ and $\mathcal{G}$ to obtain the familiar expansion for $\Omega - \Omega_0$ or $E_0 - W_0$, and the details are given in Problem 5.1. The trace $\Sigma^{\lambda}\mathcal{G}^{\lambda}$ is represented by a closed graph composed of $\mathcal{G}^{\lambda}$ connecting to the two external points of Σ^{λ}. An nth order contribution is obtained from an mth order term of Σ^{λ} combined with an $(n-m)^{\text{th}}$ order term of $\mathcal{G}$ and the parametric integral yields a factor $\int_0^1 \frac{d\lambda}{\lambda}\lambda^n = \frac{1}{n}$. It is seen in Problem 5.1 that a given graph in $\Omega - \Omega_0$ arises from combinations of many bits and pieces of Σ^{λ} and $\mathcal{G}^{\lambda}$. Thus, the Grand potential is an example of a quantity which may often be more easily calculated directly rather than by using Green's functions. In addition to being inconvenient, use of the one-particle Green's function may also be dangerous since a seemingly innocuous approximation having little effect on one-particle properties may have a large uncontrolled effect on Ω. Physically, this reflects the fact that the one-particle Green's function is directly related to single-particle propagation in the many body system, instead of the two-body correlations to which E_0 and Ω_0 may be strongly sensitive. The principal results of this section are summarized in Table 5.1.

5.2 ANALYTIC PROPERTIES

Single-particle Green's functions have important analytic properties which follow from general principles. We will establish the salient properties of zero-temperature, real-time finite temperature, and thermal Green's functions in turn.

ZERO TEMPERATURE GREEN'S FUNCTIONS

The spectral representation for the zero-temperature Green's functions (Lehmann, 1954) is obtained by inserting a complete set of eigenstates between the field operators in the definition of $G(xt|x't')$, Eq. (5.3a).

$$\begin{aligned} iG(xt|x't') &= \sum_n \theta(t-t') \langle \psi_0 | \hat{\psi}(x,t) | \psi_n^{N+1} \rangle \langle \psi_n^{N+1} | \hat{\psi}^{\dagger}(x',t') | \psi_0 \rangle \\ &+ \varsigma \sum_n \theta(t'-t) \langle \psi_0 | \hat{\psi}^{\dagger}(x't') | \psi_n^{N-1} \rangle \langle \psi_n^{N-1} | \hat{\psi}(x,t) | \psi_0 \rangle \quad . \end{aligned} \tag{5.26}$$

$G(xt\|x't')=-i\langle\psi_0\|T\hat{\psi}(xt)\hat{\psi}^\dagger(x't')\|\psi_0\rangle$	$\mathcal{G}(x\tau\|x'\tau')=\frac{1}{Z}\mathrm{Tr}\left(e^{-\beta(\hat{H}-\mu\hat{N})}T\hat{\psi}(xt)\hat{\psi}^\dagger(x't')\right)$
$\mathcal{G}_0^{-1}=i\frac{\partial}{\partial T}-H_0$	$\mathcal{G}_0^{-1}=\frac{\partial}{\partial\tau}+H_0-\mu$
$\mathcal{G}^{-1}=i\frac{\partial}{\partial t}-H_0-\Sigma$	$\mathcal{G}^{-1}=\frac{\partial}{\partial\tau}+H_0-\mu+\Sigma$
$G(t)=\frac{1}{2\pi}\int d\omega\, e^{-i\omega t}\tilde{G}(\omega)$	$\mathcal{G}(\tau)=\frac{1}{\beta}\sum_{\omega_n} e^{-i\omega_n\tau}\tilde{\mathcal{G}}(\omega_0)$
$\tilde{G}_0(\omega)=\frac{1}{\omega-\epsilon_k+i\eta\,\mathrm{sgn}(\epsilon_k-\epsilon_F)}$	$\tilde{\mathcal{G}}_0(\omega_n)=\frac{-1}{i\omega_n-(\epsilon_k-\mu)}$
$\langle O\rangle=i\varsigma\int dxdx'\,O(x'x)G(xt\|x't^+)$	$\langle O\rangle=\varsigma\int dxdx'\,O(x'x)\mathcal{G}(x\tau\|x'\tau^+)$
$\hat{\psi}(x,t)=e^{i\hat{H}t}\hat{\psi}(x)e^{-i\hat{H}t}$	$\hat{\psi}(x,\tau)=e^{(\hat{H}-\mu\hat{N})\tau}\tau\hat{\psi}(x)e^{-(\hat{H}-\mu\hat{N})\tau}$
$\hat{\psi}^\dagger(x,t)=e^{i\hat{H}t}\hat{\psi}^\dagger(x)e^{-i\hat{H}t}$	$\hat{\psi}^\dagger(x,\tau)=e^{(\hat{H}-\mu\hat{N})\tau}\hat{\psi}^\dagger(x)e^{-(\hat{H}-\mu\hat{N})\tau}$
$\langle H\rangle=\frac{1}{2}\varsigma\int dx\left[\left(i\frac{\partial}{\partial t}+T\right)G(xt\|x't^+)\right]_{x'=x}$	$\langle H\rangle=\frac{1}{2}\varsigma\int dx\left[\left(-\frac{\partial}{\partial\tau}+T+\mu\right)\mathcal{G}(x\tau\|x'\tau^+)\right]_{x=x'}$
$E-E_0=\frac{1}{2}\varsigma\int_0^1\frac{d\lambda}{\lambda}\int dx\left[\left(i\frac{\partial}{\partial t}-T\right)G(xt\|x't^+)\right]_{x'=x}$	$\Omega-\Omega_0=\frac{1}{2}\varsigma\int_0^1\frac{d\lambda}{\lambda}dx\left[\left(-\frac{\partial}{\partial\tau}-T+\mu\right)\mathcal{G}^\lambda(x\tau\|x'\tau^+)\right]_{x=x'}$
$=\frac{1}{2}\varsigma\int_0^1\frac{d\lambda}{\lambda}\int dxdx'dt'\,\Sigma^\lambda(xt\|x't')G^\lambda(x't'\|xt^+)$	$=\frac{1}{2}\varsigma\int_0^1\frac{d\lambda}{\lambda}\int dxdx'd\tau'\,\Sigma^\lambda(x\tau\|x'\tau')\mathcal{G}^\lambda(x'\tau'\|x\tau^+)$

Table 5.1 Summary of results for zero temperature and finite-temperature Green's Functions.

Note that the creation and annihilation operators only connect the N-particle ground state to $(N+1)$-particle states and $(N-1)$-particle states, respectively. The Heisenberg field operators acting on eigenstates $|\psi_n^M\rangle$ with energies E_n^M yield the explicit time-dependence $e^{-i(E_n^{N+1}-E_0)(t-t')}$ for the first term in Eq. (5.26) and $e^{-i(E_n^{N-1}-E_0)(t'-t)}$ for the second term. Using Eq. (3.30), the Fourier frequency transform of the Green's functions is

$$
\begin{aligned}
G(x,x';\omega) &= \sum_n \frac{\langle\psi_0|\hat{\psi}(x)|\psi_n^{N+1}\rangle\langle\psi_n^{N+1}|\hat{\psi}^\dagger(x')|\psi_0\rangle}{\omega-(E_n^{N+1}-E_0)+i\eta} \\
&\quad -\varsigma\sum_n \frac{\langle\psi_0|\hat{\psi}^\dagger(x)|\psi_n^{N-1}\rangle\langle\psi_n^{N-1}|\hat{\psi}(x)|\psi_0\rangle}{\omega+(E_n^{N-1}-E_0)-i\eta} \quad .
\end{aligned}
\tag{5.27}
$$

The analytic behavior of $G(\omega)$ has thus been clearly isolated and is sketched in Fig. 5.1. For each eigenstate $|\psi_n^{N+1}\rangle$ of the $(N+1)$-particle system, there is a pole in the lower half plane at $E^{N+1}-E_0$ with residue $\langle\psi_0|\hat{\psi}(x)|\psi_n^{N+1}\rangle\langle\psi_n^{N+1}|\hat{\psi}^\dagger(x)|\psi_0\rangle$ and for each eigenstate $|\psi_n^{N-1}\rangle$ there is a pole in the upper half plane at energy $-(E_n^{N-1}-E_0)$ with residue $\langle\psi_0|\hat{\psi}^\dagger(x)|\psi_n^{N-1}\rangle\langle\psi_n^{N-1}|\hat{\psi}(x)|\psi_0\rangle$. The poles in the lower half plane begin at $\mu^{N+1}=E_0^{N+1}-E_0$ and extend to $+\infty$ and those in the upper half plane extend from $-\infty$ up to $\mu^N=E_0-E_0^{N-1}$. Because it is often convenient to deal with functions which are analytic in the upper or lower half plane, it is frequently useful to use the retarded and advanced Green's functions, Eq. (5.6). The analytic structure for

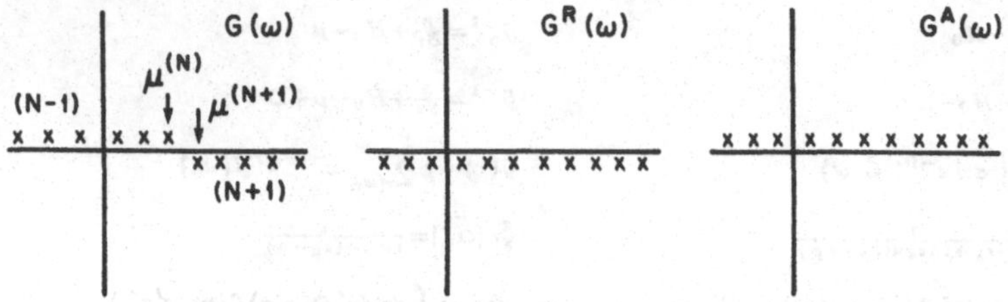

Fig. 5.1 Poles of $G(\omega)$, $G^R(\omega)$, and $G^A(\omega)$ in the complex ω plane.

$G^R_{(\omega)}$ and $G^A_{(\omega)}$ shown in Fig. 5.1 immediately follows from that of $G(\omega)$ by noting that $G^R(t-t')$ contains only $\theta(t-t')$ like the $(N+1)$-particle contribution to $G(t-t')$ and $G^A(t-t')$ contain $\theta(t'-t)$ like the $(N-1)$-particle contribution.

For notational simplicity, we will present the remaining results for a translationally invariant, infinite system. It is important to note, however, that all these results have obvious counterparts for finite systems. Since $\mu^{(N+1)} = \mu^{(N)} = \mu$ in the limit of large systems, it is convenient to measure all energies relative to the chemical potential and we define

$$\begin{aligned} E_n^{N+1} - E_0 &= E_n^{N+1} - E_0^{N+1} + E_0^{N+1} - E_0 \\ &\equiv \epsilon_n^{N+1} + \mu \end{aligned} \tag{5.28a}$$

and

$$\begin{aligned} E_n^{N-1} - E_0 &= E_n^{N-1} - E_0^{N-1} + E_0^{N-1} - E_0 \\ &\equiv \epsilon_n^{N-1} - \mu \,. \end{aligned} \tag{5.28b}$$

Writing the field operator in momentum representation $\hat{\psi}(x) = \sum_k \frac{1}{\sqrt{V}} e^{ik\cdot x}\hat{a}_k$ and writing eigenstates $\psi_n^{N\pm1}$ of total momentum $\pm k$, Fourier transformation of the Green's function Eq. (5.27) to momentum space yields

$$\left\{\begin{matrix} G(\vec{k},\omega) \\ G^R(\vec{k},\omega) \\ G^A(\vec{k},\omega) \end{matrix}\right\} = \sum_n \left[\frac{|\langle\psi_n^{N+1}|a_k^\dagger|\psi_0\rangle|^2}{\omega-\mu-\epsilon_n^{N+1}\left\{\begin{matrix} + \\ + \\ - \end{matrix}\right\} i\xi} - \varsigma \frac{|\langle\psi_n^{N-1}|a_k|\psi_0\rangle|^2}{\omega-\mu+\epsilon_n^{N+1}\left\{\begin{matrix} - \\ + \\ - \end{matrix}\right\} i\xi} \right] \tag{5.29}$$

where we have now indicated the advanced and retarded cases as well for completeness. The two simplifications for infinite systems are positive definite residues, representing the probability of finding $a_k^\dagger|\psi_0\rangle$ or $a_k|\psi_0\rangle$ in eigenstates of the $(N+1)$- or $(N-1)$-particle systems, respectively, and a common starting point μ for the $(N+1)$- and $(N-1)$-particle poles. For real ω, Eq. (5.29) shows that G, G^R, and G^A are related as follows

$$G^R(\omega)^* = G^A(\omega) \tag{5.30}$$

and

$$G(\omega)=\begin{cases} G^R(\omega) & \omega>\mu \\ G^A(\omega) & \omega<\mu \end{cases} \tag{5.31a}$$

or equivalently

$$G(\omega)=G^R(\omega)\theta(\omega-\mu)+G^A(\omega)\theta(\mu-\omega)\ . \tag{5.31b}$$

Since the poles of $G(k,\omega)$ become arbitrarily closely spaced in a large system, only averages can be measured and it is useful to define spectral weight functions

$$\begin{aligned} \rho^+(k,\omega) &= \sum_n |\langle\psi_n^{N+1}|a_k^\dagger|\psi_0\rangle|^2 2\pi\delta(\epsilon_n^{N+1}-\omega) \\ \rho^-(k,\omega) &= \sum_n |\langle\psi_n^{N-1}|a_k|\psi_0\rangle|^2 2\pi\delta(\epsilon_n^{N-1}-\omega) \end{aligned} \tag{5.32a}$$

and

$$\rho(k,\omega)=\theta(\omega)\rho^+(k,\omega)-\varsigma\theta(-\omega)\rho^-(k,-\omega)\ . \tag{5.32b}$$

In terms of these weight functions,

$$G(k,\omega)=\int_0^\infty \frac{d\omega'}{2\pi}\left[\frac{\rho^+(k,\omega')}{\omega-\mu-\omega'+i\eta}-\varsigma\frac{\rho^-(k,\omega')}{\omega-\mu+\omega'-i\eta}\right] \tag{5.33a}$$

and

$$\begin{aligned} \left\{\begin{matrix} G^R(k,\omega') \\ G^A(k,\omega') \end{matrix}\right\} &= \int_0^\infty \frac{d\omega'}{2\pi}\left[\frac{\rho^+(k,\omega')}{\omega-\mu-\omega'\pm i\eta}-\varsigma\frac{\rho^-(k,\omega')}{\omega-\mu+\omega'\pm i\eta}\right] \\ &= \int_{-\infty}^\infty \frac{d\omega'}{2\pi}\,\frac{\rho(k,\omega')}{\omega-\mu-\omega'\pm i\eta}\ . \end{aligned} \tag{5.33b}$$

From the relation $\frac{1}{\omega\pm i\xi}=P\frac{1}{\omega}\mp i\pi\delta(\omega)$, where P is the principal part, it follows that

$$\begin{aligned} \mathrm{Re}\left\{\begin{matrix} G(k,\omega) \\ G^R(k,\omega) \\ G^A(k,\omega) \end{matrix}\right\} &= P\int_{-\infty}^\infty \frac{d\omega'}{2\pi}\,\frac{\rho(k,\omega')}{\omega-\mu-\omega'} \\ \mathrm{Im}\left\{\begin{matrix} G(k,\omega) \\ G^R(k,\omega) \\ G^A(k,\omega) \end{matrix}\right\} &= \left\{\begin{matrix} -sgn(\omega-\mu) \\ - \\ + \end{matrix}\right\}\frac{1}{2}\rho(k,\omega-\mu) \end{aligned} \tag{5.34}$$

so that the Green's functions satisfy the dispersion relation

$$\mathrm{Re}\left\{\begin{matrix} G(k,\omega) \\ G^R(k,\omega) \\ G^A(k,\omega) \end{matrix}\right\} = \frac{P}{\pi}\int_{-\infty}^\infty \frac{d\omega'}{\omega-\omega'}\left\{\begin{matrix} -sgn(\omega'-\mu) \\ - \\ + \end{matrix}\right\}\mathrm{Im}\left\{\begin{matrix} G(k,\omega) \\ G^R(k,\omega) \\ G^A(k,\omega) \end{matrix}\right\}\ . \tag{5.35}$$

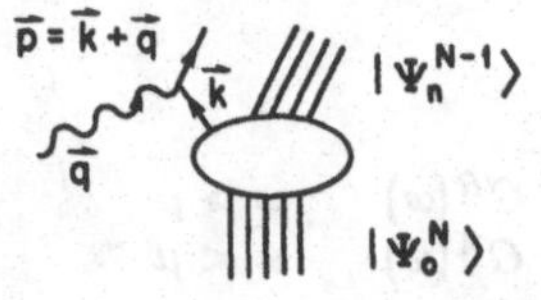

Fig. 5.2 Schematic representation of a (γ, p) or $(e, e'p)$ reaction.

The commutation relations of the creation and annihilation operators give rise to a sum rule for the spectral weight

$$\begin{aligned}1 = \langle\psi_0|\left[a_k, a_k^+\right]_{-\varsigma}|\psi_0\rangle &= \sum_n |\langle\psi_n^{N+1}|a_k^+|\psi_0\rangle|^2 - \varsigma\sum|\langle\psi_n^{N-1}|a_k|\psi_0\rangle|^2 \\ &= \int_0^\infty \frac{d\omega}{2\pi}\left[\rho^+(k,\omega) - \varsigma\rho^-(k,\omega)\right] \\ &= \int_{-\infty}^\infty \frac{d\omega}{2\pi}\rho(k,\omega) \ .\end{aligned} \tag{5.36}$$

Combined with Eqs. (5.33), this sum rule establishes the high frequency behavior of the Green's functions:

$$\left\{\begin{matrix} G(\omega) \\ G^R(\omega) \\ G^A(\omega)\end{matrix}\right\} \xrightarrow[\omega\to\infty]{} \int_0^\infty \frac{d\omega'}{2\pi}\,\frac{\rho^+(k,\omega') - \varsigma\rho^-(k,\omega')}{\omega} = \frac{1}{\omega} \ . \tag{5.37}$$

Experimentally, the spectral weight function is accessible through semi-inclusive experiments. Consider, for example, a (γ, p) or $(e, e'p)$ reaction on a nucleus, as sketched in Fig. 5.2 in which a real or virtual photon of momentum q is absorbed by a nucleus in its ground state $|\psi_0\rangle$ and a proton is ejected. In the final state, only the ejected proton having momentum $p = k + q$ is detected and the rest of the state is unresolved. In the impulse approximation, (in which the interactions of the ejected particles are neglected) the proton knocked out by the photon must have had an initial momentum of k, so a proton of momentum k has been removed and the cross section is

$$\begin{aligned}\sigma &= 2\pi\sum|\langle\psi_n^{N-1}|a_k|\psi_0\rangle|^2\delta(E_n^{N-1} + E_p - E_0 - E_\gamma) \\ &= \rho^-(k, E_\gamma + \mu^N - E_\rho) \ .\end{aligned} \tag{5.38}$$

By varying the kinematics and including corrections to the impulse approximation, much has been learned about the behavior of the spectral weights (Frullani and Mougey, 1984). An interesting energy weighted sum rule for $\rho^-(k,\omega)$ has been derived by Koltun (1972) and is derived in Problem 5.2.

FINITE TEMPERATURE GREEN'S FUNCTIONS

We now turn our attention to the analytic properties of finite-temperature Green's functions. Instead of regarding thermal and real-time Green's functions, Eq. (5.1– 5.2),

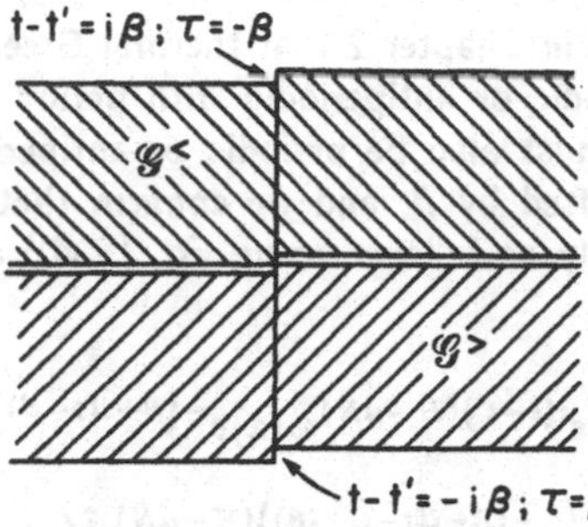

Fig. 5.3 Domains of complex $t - t'$ plane in which $\mathcal{G}^>(xt|x't')$ and $\mathcal{G}^<(xt|x't')$ may be continued. By the periodicity of $\mathcal{G}$, the domains are repeated indefinitely in the imaginary t direction.

as distinct entities, it is useful to consider $\mathcal{G}(xt|x't')$ as a function in the complex $t-t'$ plane. Under the change of variables $it \to \tau$,

$$\hat{\psi}(x,t) = e^{it(\hat{H}-\mu\hat{N})}\hat{\psi}(x)e^{-it(H-\mu\hat{N})} \to e^{\tau(\hat{H}-\mu\hat{N})}\hat{\psi}(x)e^{-\tau(\hat{H}-\mu\hat{N})} = \hat{\psi}(x,\tau)$$
$$\hat{\psi}^\dagger(x,t) \to \hat{\psi}^\dagger(x,\tau) \tag{5.39a}$$

so that

$$i\mathcal{G}(xt|x't') \to \mathcal{G}(x\tau|x'\tau') \tag{5.39b}$$

Thus, a single function of a complex time variable specifies the real time Green's function along the real-t-axis and the thermal Green's functions along the imaginary t-axis. The real-time Green's function, which describes the physical response of a system, may thereby be obtained from the thermal Green's function, which is calculable by the perturbation theory described in Chapter 2, by straightforward analytic continuation. To see the analytic structure of $\mathcal{G}$ in the complex t-plane, we write

$$\mathcal{G}(xt|x't') = \theta(t-t')\mathcal{G}^>(xt|x't'|) + \theta(t'-t)\mathcal{G}^<(xt|x't) \tag{5.40a}$$

where

$$\mathcal{G}^>(xt|x't') = \frac{-i}{Z}\mathrm{Tr}\left\{e^{-(\beta-i(t-t'))(\hat{H}-\mu\hat{N})}\hat{\psi}(x)e^{-i(t-t')(\hat{H}-\mu\hat{N})}\hat{\psi}^\dagger(x')\right\}$$
$$\mathcal{G}^<(xt|x't') = \frac{-\zeta i}{Z}\mathrm{Tr}\left\{e^{-(\beta+i(t-t'))(\hat{H}-\mu\hat{N})}\hat{\psi}^\dagger(x')e^{i(t-t')(\hat{H}-\mu\hat{N})}\hat{\psi}(x)\right\} \tag{5.40b}$$

In order for the thermodynamic traces to converge, $\mathcal{G}^>$ and $\mathcal{G}^<$ can only be continued into regions in which the factors multiplying $\hat{H}-\mu\hat{N}$ in the exponentials have negative real parts. (Note that $\hat{H}-\mu\hat{N}$ must be positive for the partition function to exist). Hence, $\mathcal{G}^>(t-t')$ can be continued if $\mathrm{Re}\{\beta - i(t-t')\} > 0$ and $\mathrm{Re}\{i(t-t')\} > 0$. Similarly, $\mathcal{G}^<(t-t')$ can be continued if $\mathrm{Re}\{\beta + i(t-t')\} > 0$ and $\mathrm{Re}\{i(t-t')\} < 0$. Thus, $\mathcal{G}^>$ and $\mathcal{G}^<$ may be continued from the real axis in the following domains, which are also sketched in Fig. 5.3.

$$\begin{aligned} -\beta < \mathrm{Im}(t-t') < 0 \quad & G^>(t-t') \\ 0 < \mathrm{Im}(t-t') < \beta \quad & G^<(t-t') \ . \end{aligned} \tag{5.41}$$

We have already shown in Chapter 2 that thermal Green's functions for the non-interacting system are periodic or antiperiodic with period β. Thus, the interacting $\mathcal{G}$ constructed from them must also be periodic or antiperiodic when the imaginary part of the argument is shifted by β, and we confirm that $G^<(t-t')$ is related to $G^>(t-t'-i\beta)$ appropriately as follows, using Eq. (5.40) and the cyclic property of the trace:

$$
\begin{aligned}
G^< (x,x';(t-t')) &= \frac{-i\varsigma}{Z}\mathrm{Tr}\left\{e^{i(t-t')(\hat{H}-\mu\hat{N})}\hat{\psi}(x)e^{-(\beta+i(t-t'))(\hat{H}-\mu\hat{N})}\hat{\psi}^\dagger(x')\right\} \\
&= \varsigma\left[\frac{-i}{Z}\mathrm{Tr}\left\{e^{-(\beta-i(t-t'-i\beta))(\hat{H}-\mu\hat{N})}\hat{\psi}(x)e^{-i(t-t'-i\beta)(\hat{H}-\mu\hat{N})}\psi^\dagger(x')\right\}\right] \\
&= \varsigma G^> (x,x';(t-t'-i\beta)) \quad . \qquad (5.42)
\end{aligned}
$$

This periodicity is consistent with the domain of continuation shown in Fig. 5.3, since the discontinuity between $G^>$ and $G^<$ adjacent to the real axis must be repeated again along the $t-t'=\pm i\beta$ axes.

To study the analytic properties in the ω plane, we insert a complete set of states to extract the explicit time dependence as in the zero-temperature case and Fourier transform. As before, the singularity structure is the same for finite and translationally invariant systems, and we only treat the translationally invariant case here for notational convenience. Writing the trace and completeness relation using a complete set of eigenstates $\{|\psi_m\rangle\}$ with all number of particles

$$
\begin{aligned}
i\mathcal{G}(xt|x't') = \theta(t-t')\frac{1}{Z}\sum_{m,n}\langle\psi_m|e^{-\beta(H-\mu N)}\hat{\psi}(x,t)|\psi_n\rangle\langle\psi_n|\hat{\psi}^\dagger(x't')|\psi_m\rangle \\
+\varsigma\theta(t'-t)\frac{1}{Z}\sum_{m,n}\langle\psi_n|e^{-\beta(H-\mu N)}\hat{\psi}^\dagger(x',t')|\psi_m\rangle\langle\psi_m|\hat{\psi}(x,t)|\psi_n\rangle \ .
\end{aligned}
\qquad (5.43)
$$

Extracting the time dependence as in Eq. (5.26), Fourier transforming to frequency and momentum space as in Eqs. (5.27) and (5.29) and treating the advanced and retarded cases in the same way, we obtain

$$
\left\{\begin{matrix} G(k,\omega) \\ G^R(k,\omega) \\ G^A(k,\omega)\end{matrix}\right\} = \frac{1}{Z}\sum_{m,n}|\langle\psi_n|a_k^\dagger|\psi_m\rangle|^2 \qquad (5.44)
$$

$$
\times\left\{\frac{e^{-\beta(E_m-\mu N_m)}}{\omega-(E_n-E_m-\mu)\left\{\begin{matrix}+\\+\\-\end{matrix}\right\}i\eta}-\varsigma\frac{e^{-\beta(E_n-\mu N_n)}}{\omega-(E_n-E_m-\mu)\left\{\begin{matrix}-\\+\\-\end{matrix}\right\}i\eta}\right\} .
$$

Note that we have used the fact that the only non-vanishing contributions arise for states in which the number of particles in state n, N_n, is one larger than the number in m, N_m. Since the two terms in Eq. (5.44) include all $N_n = N_m + 1$, they have a more symmetrical form than the corresponding terms in Eq. (5.2.4) which involve only $N+1$ and $N-1$. Thus, we may combine them to write the spectral weight function as

$$\rho(k,\omega) = \frac{1}{Z}\sum_{m,n}\Big\{\Big|\langle\psi_n|a_k^\dagger|\psi_m\rangle\Big|^2 \; e^{-\beta(E_m-\mu N_m)}(1-\varsigma e^{-\beta\omega}) \times 2\pi\delta[E_n - E_m - \mu - \omega]\Big\} \quad . \tag{5.45}$$

One observes that this expression effectively reduces to Eq. (5.32b) in the zero temperature limit as follows. For $\omega > 0$ the first term dominates and denoting by m_0 the state with minimum $E_m - \mu N_m$, $\lim_{\beta\to\infty}\frac{1}{Z}\sum_m e^{-\beta(E_m-\mu N_m)}|\langle\psi_n|a_k^\dagger|\psi_m\rangle|^2 = \langle\psi_n|a_k^\dagger|\psi_{m_0}\rangle|^2$. Analogously, for $\omega < 0$, the second term dominates and yields the factor $|\langle\psi_{m_0}|a_k^+|\psi_n\rangle|^2$.

Using Eq. (5.45), the spectral representation of $\mathcal{G}^R$ and $\mathcal{G}^A$ may be written in the same form as the zero-temperature case, Eq. (5.33b).

$$\left\{\begin{matrix}\mathcal{G}^R(k,\omega)\\ \mathcal{G}^A(k,\omega)\end{matrix}\right\} = \int_{-\infty}^{\infty}\frac{d\omega'}{2\pi}\,\frac{\rho(k,\omega')}{\omega-\omega'\pm i\eta} \; . \tag{5.46}$$

The real and imaginary parts obtained from Eq. (5.44) and (5.46) may be expressed in the following form using the identity $1+\eta e^{-\beta\omega} = \tanh\left(\frac{\beta\omega}{2}\right)^{-\eta}(1-\varsigma e^{-\beta\omega})$

$$\begin{aligned}
\mathrm{Re}\left\{\begin{matrix}\mathcal{G}(k,\omega)\\ \mathcal{G}^R(k,\omega)\\ \mathcal{G}^A(k,\omega)\end{matrix}\right\} &= P\int_{-\infty}^{\infty}\frac{d\omega'}{2\pi}\,\frac{\rho(k,\omega')}{\omega-\omega'}\\
\mathrm{Im}\left\{\begin{matrix}\mathcal{G}(k,\omega)\\ \mathcal{G}^R(k,\omega)\\ \mathcal{G}^A(k,\omega)\end{matrix}\right\} &= \left\{\begin{matrix}-\tanh\left(\frac{\beta\omega}{2}\right)^{-\varsigma}\\ -\\ +\end{matrix}\right\}\frac{1}{2}\rho(k,\omega) \; .
\end{aligned} \tag{5.47}$$

Hence, we obtain the dispersion relations

$$\mathrm{Re}\left\{\begin{matrix}\mathcal{G}(k,\omega)\\ \mathcal{G}^R(k,\omega)\\ \mathcal{G}^A(k,\omega)\end{matrix}\right\} = \frac{P}{\pi}\int_{-\infty}^{\infty}\frac{d\omega'}{\omega-\omega'}\left\{\begin{matrix}-\tanh\left(\frac{\beta\omega}{2}\right)^{-\varsigma}\\ -\\ +\end{matrix}\right\}\rho(k,\omega) \tag{5.48}$$

and the relation between $\mathcal{G}$, $\mathcal{G}^R$, and $\mathcal{G}^A$

$$\begin{aligned}
\mathcal{G}(k,\omega) &= \frac{\mathcal{G}^R(k,\omega)}{1-\varsigma e^{-\beta\omega}} + \frac{\mathcal{G}^A(k,\omega)}{1-\varsigma e^{\beta\omega}}\\
&= 1+\varsigma n(\omega+\mu)\mathcal{G}^R(k,\omega) - \varsigma n(\omega+\mu)\mathcal{G}^A(k,\omega)
\end{aligned} \tag{5.49}$$

where $n(\omega+\mu)$ is the familiar occupation probability, Eq. (2.75b). The last result reflects the fact that the poles in the upper and lower half plane overlap at finite temperature and only in the zero temperature limit does one recover the simple structure of non-overlapping poles reflected in Eq. (5.34b):

$$\mathcal{G}(k,\omega) \underset{\beta\to\infty}{\longrightarrow} \theta(\omega)\mathcal{G}^R(k,\omega) + \theta(-\omega)\mathcal{G}^A(k,\omega) \ . \tag{5.50}$$

Note in making this comparison that at finite temperature the combination $\omega + \mu$ occurred in Eq. (5.44) where only the frequency occurred at zero temperature in Eq. (5.27). Denoting the zero-temperature convention for the frequency as $\omega_0 = \omega + \mu$, we confirm that the factors $\theta(\pm\omega) = \theta\left(\pm(\omega_0 - \mu)\right)$ agree with Eq. (5.31b). The sum rule for $\rho(k,\omega)$ is obtained as in Eq. (5.36).

$$1 = \frac{1}{2}\mathrm{Tr}\{e^{-\beta(\hat{H}-\mu\hat{N})}[a_k, a_k^\dagger]_{-\varsigma}\} = \int_{-\infty}^{\infty}\frac{d\omega}{2\pi}\rho(k,\omega) \tag{5.51}$$

so that the finite temperature Green's functions have the high frequency behavior

$$\left\{\begin{matrix} \mathcal{G}(k,\omega) \\ \mathcal{G}^R(k,\omega) \\ \mathcal{G}^A(k,\omega) \end{matrix}\right\} \underset{\omega\to\infty}{\longrightarrow} \int \frac{d\omega'}{2\pi}\frac{\rho(k,\omega')}{\omega} = \frac{1}{\omega} \ . \tag{5.52}$$

Finally, we relate the thermal Green's function to the real-time Green's function through its spectral representation. Since $\mathcal{G}(x,x';\tau-\tau')$ is periodic or antiperiodic on the interval $(0,\beta)$, we expand in a Fourier series with Matsubara frequencies $\omega_n = \frac{2n\pi}{\beta}$ or $\frac{(2n+1)\pi}{\beta}$ defined in Eq. (3.129). Inserting a complete set of states in $\mathcal{G}(x\tau|x'0)$, Eq. (5.2), and evaluating the Fourier transform on the interval $0 < \tau < \beta$ as in Eq. (2.131) yields

$$\begin{aligned} \mathcal{G}(k,\omega_s) &= \frac{1}{Z}\int_0^\beta d\tau\, e^{i\omega_s\tau}\sum_{mn}|\langle\psi_n|a_k^\dagger|\psi_m\rangle|^2\, e^{\tau[E_m-E_n+\mu]}\, e^{-\beta(E_m-\mu N_m)} \\ &= \frac{1}{Z}\sum_{mn}|\langle\psi_n|a_k^\dagger|\psi_m\rangle|^2\, e^{-\beta(E_m-\mu N_m)}\left[\frac{-1+\varsigma\, e^{-\beta(E_n-E_m-\mu)}}{i\omega_s-(E_n-E_m-\mu)}\right] \\ &= -\int_{-\infty}^{\infty}\frac{d\omega'}{2\pi}\frac{\rho(k,\omega')}{i\omega_s-\omega'} \ . \end{aligned} \tag{5.53}$$

Comparison with Eq. (5.46) shows that $-\mathcal{G}(k,\omega_n)$, $\mathcal{G}^R(k,\omega_n)$ and $\mathcal{G}^A(k,\omega_n)$ are given by the same complex function specified by the weight $\rho(k,\omega)$ evaluated along the imaginary axis at the discrete Matsubara frequencies or infinitesimally above or below the real axis. Given the positions of the singularities of $\mathcal{G}^R$ and $\mathcal{G}^A$ sketched in Fig. 5.4, it is clear that $-\mathcal{G}(\omega_n)$ calculated in perturbation theory is to be continued in the upper half plane to determine $\mathcal{G}^R$ and in the lower half plane to determine $\mathcal{G}^A$. From $\mathcal{G}^R(\omega)$ and $\mathcal{G}^A$, $\mathcal{G}(\omega)$ is specified by Eq. (5.49). Although perturbation theory only specifies $\mathcal{G}(\omega_n)$ at a discrete set of points, the continuation is unique because of the requirement that $\mathcal{G}(\omega) \sim \frac{1}{|\omega|}$ at infinity.

Although we will subsequently present a detailed example of how this analytic continuation works in Section 5.5, we conclude this present section with the simple

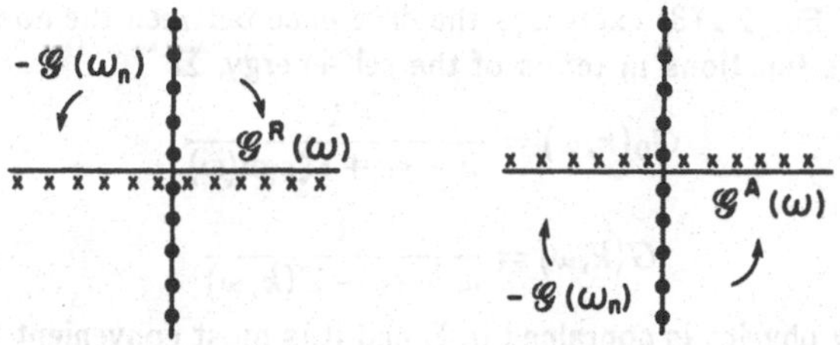

Fig. 5.4 Continuation of the function $\int \frac{d\omega'}{2\pi} \frac{\rho(\omega')}{\omega-\omega'}$ in the complex ω plane from points $\omega = i\omega_n$ along the imaginary axis where it equals $-\mathcal{G}(\omega_n)$ to points infinitesimally above the real axis where it yields $\mathcal{G}^R(\omega)$ and below the real axis where it gives $\mathcal{G}^A(\omega)$.

example of the Green's function for a non-interacting system. Equating the non-interacting thermal Green's functions, Eq. (2.131b) to the spectral representation Eq. (5.53), we obtain

$$\frac{-1}{i\omega_n - (\epsilon_k - \mu)} = -\int \frac{d\omega'}{2\pi} \frac{\rho(k,\omega')}{i\omega_n - \omega'} \tag{5.54a}$$

from which it follows that

$$\rho(k,\omega') = 2\pi\delta\left[\omega' - (\epsilon_k - \mu)\right] \tag{5.54b}$$

$$\left\{ \begin{matrix} \mathcal{G}^R(k,\omega) \\ \mathcal{G}^A(k,\omega) \end{matrix} \right\} = \int \frac{d\omega'}{2\pi} \frac{\rho(k,\omega')}{\omega - \omega' \pm i\eta} = \frac{1}{\omega - (\epsilon_k - \mu) \pm i\eta} . \tag{5.54c}$$

These manipulations are equivalent to simply replacing $i\omega_n$ in the thermal Green's function by $\omega \pm i\eta$ in the real-time Green's function (and including the overall minus sign from our conventions). Since there is only a single pole, there is no distinction between continuation from above or below the real axis. By Carlson's theorem, other functions, such as $\frac{e^{\beta\omega}}{\omega-(\epsilon_k-\mu)\pm i\eta}$ or $\frac{1}{\omega-(\epsilon_k-\mu)\pm+i\eta} + \sin\beta\omega$ which coincide with the thermal Green's function at the points $i\omega_n$ are ruled out because they do not converge as $\frac{1}{|\omega|}$ at large ω. Thus, shifting the frequency by μ to coincide with the zero-temperature convention and using Eq. (5.49) we obtain

$$\begin{aligned} \mathcal{G}(k,\omega-\mu) &= (1+\varsigma n(\omega))\, \mathcal{G}^R(\omega-\mu) - \varsigma n(\omega)\mathcal{G}^A(\omega-\mu) \\ &= \frac{1+\varsigma n(\epsilon_k)}{\omega - \epsilon_k + i\eta} - \frac{\varsigma n(\epsilon_k)}{\omega - \epsilon_k - i\eta} \end{aligned} \tag{5.55}$$

consistent with the zero temperature limit, Eq.(3.31)

5.3 PHYSICAL CONTENT OF THE SELF ENERGY

At this point, it is appropriate to complement the treatment of the formal properties of Green's functions by considering the physical content of a specific illustrative case: the one-particle Green's function for Fermions at zero temperature. Since

Dyson's equation, Eq.(2.178) expresses the difference between the non-interacting and interacting Green's functions in terms of the self-energy, Σ

$$G_0(k,\omega) = \frac{1}{\omega - \epsilon_k + i\xi\,\mathrm{sgn}(\omega)} \tag{5.56}$$

$$G(k,\omega) = \frac{1}{\omega - \epsilon_k - \Sigma(k,\omega)} \tag{5.57}$$

all the many-body physics is contained in Σ and it is most convenient to study $\Sigma(k,\omega)$ directly. Recall the convention from Chapter 3 that all energies and frequencies are defined relative to the Fermi energy ϵ_F. Also note once again that we assume translational invariance not of physical necessity but rather to simplify the notation by rendering equations diagonal in momentum space.

The essential features of Σ arise in the first two orders of perturbation theory, so we shall consider the approximate Green's function defined by the second order self-energy

$$G_2(k,\omega) = \frac{1}{\omega - \epsilon_k - \Sigma_1(k) - \Sigma_2(k,\omega)} \ . \tag{5.58}$$

The first-order self-energy is

$\Sigma_1(\alpha) =$  $= \sum_A \{\alpha A|v|\alpha A\}$ (5.59)

where, according to our standard conventions, the curly brackets denote an antisymmetrized matrix element, upper case letters denote occupied momentum states, lower case letters denote unoccupied momentum states, and the Greek letters (with the exception of ω which always denotes frequency) indicate an unrestricted momentum which may be above or below k_F. Because the eigenfunctions in a translationally invariant system are plane waves, $\Sigma_1(\alpha)$ coincides with the Hartree Fock potential, Eq. (2.180). Note that since the Hartree-Fock potential is instantaneous, it has no frequency dependence. The full generality of the structure in the self-energy arises in second order, for which we have previously evaluated the contributions of diagrams having two-particle one-hole intermediate states and two-hole one-particle intermediate states, Eqs. (3.58 – 3.59):

$\Sigma_2^{2p1h}(\alpha,\omega) =$ 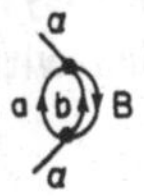 $= \frac{1}{2}\sum_{abB} \frac{|\{\alpha B|v|ab\}|^2}{\omega + \epsilon_B - \epsilon_a - \epsilon_b + i\eta}$

(5.60)

$\Sigma_2^{2h1p}(\alpha,\omega) =$  $= -\frac{1}{2}\sum_{ABb} \frac{|\{\alpha b|v|AB\}|^2}{\epsilon_A + \epsilon_B - \epsilon_a - \omega + i\eta}$.

Because these and all higher order diagrams have finite extent in time, they have explicit frequency dependence.

Both terms in Σ_2 have an infinite number of poles. The denominator of Σ^{2p1h} vanishes when $\omega = \epsilon_a + \epsilon_b - \epsilon_B$ which requires ω positive corresponding to an energy greater than ϵ_F. Similarly, Σ^{2h1p} has poles when $\omega = \epsilon_A + \epsilon_B - \epsilon_a$ corresponding to ω negative and energies less than ϵ_F. Since $\mathrm{Im}\,\Sigma^{2p1h} < 0$ and $\mathrm{Im}\,\Sigma^{2h1p} > 0$, the finite imaginary parts of Σ_2 replace the infinitesimal displacement $i\eta\,\mathrm{sgn}(\omega)$ required in the non-interacting Green's function.

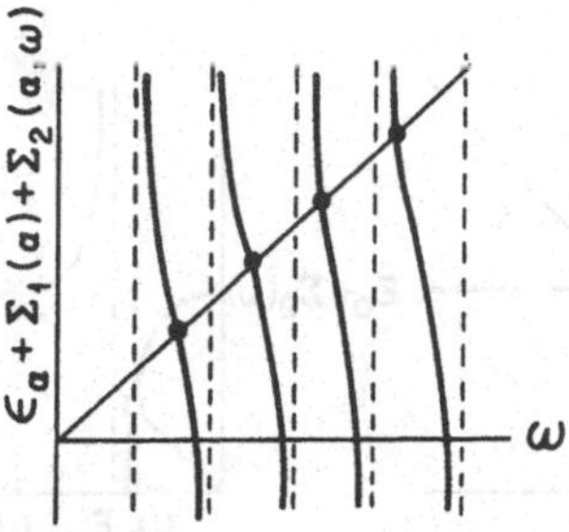

Fig. 5.5 Sketch of $\epsilon_\alpha + \Sigma_1(\alpha) + \Sigma_2(\alpha, \omega)$ as a function of ω. The dashed vertical asymptotes denote the poles of $\Sigma_2(\alpha, \omega)$ and the dots indicate the graphical solutions for the positions of the poles in $G_2(\alpha, \omega)$.

QUASIPARTICLE POLE

We now consider the poles in $G_2(\alpha, \omega)$. By our general arguments in Section 5.2, the poles represent the eigenstates of the interacting $N + 1$ particle system and the residues specify their overlap with $a_\alpha^\dagger|\psi_0\rangle$. If the system behaves as non-interacting particles, there will be a single pole with unit strength as in the case of $\mathcal{G}_0(\alpha, \omega)$. As the system becomes more and more strongly interacting, the strength will become fragmented between more and more complicated states, subject only to the sum rule that the integrated strength remains 1.

To analyze the pole structure of $G_2(\alpha, \omega)$, it is convenient to perform the graphical construction shown in Fig. 5.5, where $\epsilon_\alpha + \Sigma_1(\alpha) + \Sigma_2(\alpha, \omega)$ is sketched as a function of ω. We will first consider states above the Fermi sea. For every value $\omega = \epsilon_a + \epsilon_b - \epsilon_B$ at which $\Sigma_2^{2p1h}(\alpha, \omega)$ has a pole, a vertical asymptote is drawn in Fig. 5.5, and the function $\epsilon_\alpha + \Sigma_1(\alpha) + \Sigma_2^{2p1h}(\alpha, \omega)$ must smoothly decrease from $+\infty$ to $-\infty$ between even pair of asymptotes as shown. The condition for a pole in $G_2(\alpha, \omega)$ that $\omega = \epsilon_\alpha + \Sigma_1(\alpha) + \Sigma_2^{2p1h}(\alpha, \omega)$ is represented on the graph by the intersection of $\epsilon_\alpha + \Sigma_1(\alpha) + \Sigma_2^{2p1h}(\alpha, \omega)$ with the straight line at 45°.

Having appreciated the general structure, it is instructive to consider a schematic example. Instead of the infinite number of poles discussed above, we will assume $\Sigma_2(\omega)$ has only two poles

$$\Sigma_2(\omega) = \frac{A_1}{\omega - E_1} + \frac{A_2}{\omega - E_2} \tag{5.61}$$

and study the poles and residues of

$$G_2(\omega) = \frac{1}{\omega - E_0 - \Sigma_2(\omega)} \tag{5.62}$$

where for notational convenience $E_0 = \epsilon_\alpha + \Sigma_1$ and we have suppressed the $i\eta$. Further, we will assume that E_1 is above E_0. E_2 is below E_0 and that the residues in Σ_2 are very small, satisfying the conditions

$$\frac{A_i}{|E_i - E_0|\,|E_j - E_0|} < \delta << 1 \qquad \begin{matrix} i = 1, 2 \\ j = 1, 2 \end{matrix} \quad . \tag{5.63}$$

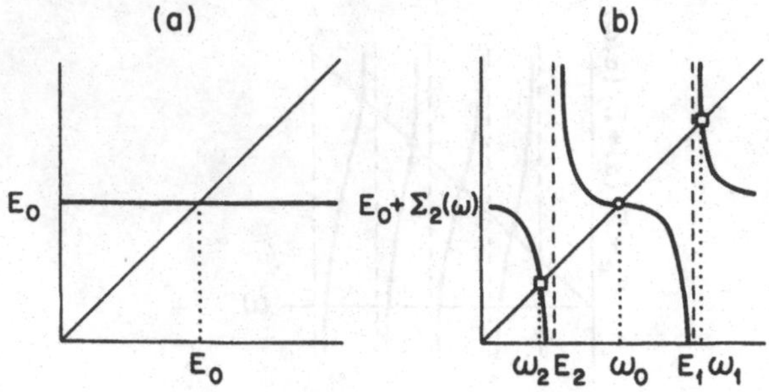

Fig. 5.6 Graphical solution for the poles of $G_1(\omega)$ and $G_2(\omega)$ in Eqs. (5.62) and (5.64).

For subsequent reference, since A_1 and A_2 are very weak so that $\Sigma_2(\omega)$ is in some sense small, we first neglect $\Sigma_2(\omega)$ entirely and perform the graphical construction for

$$G_1(\omega) = \frac{1}{\omega - E_0} \tag{5.64}$$

in part (a) of Fig. 5.6. Since E_0 has no ω-dependence, the graph is structureless and a single pole of unit residue occurs at E_0. In part (b) of Fig. 5.6, the graphical construction is repeated for $G_2(\omega)$. Far away from E_1 and E_2, $E_0+\Sigma_2(\omega)$ approaches the horizontal line sketched in part (a), and it is only very close to the singularities E_1 and E_2 that the curve diverges to $\pm\infty$. Instead of the single pole at E_0 in case (a), we now have three poles: ω_1 very close to E_1, ω_2 close to E_2, and ω_0 close to E_0.

We now expand $G_2(\omega)$ around each of the three poles ω_s which are solutions to the equation

$$\omega_s = E_0 + \Sigma_2(\omega_s) \qquad s = 0, 1, 2 \ . \tag{5.65}$$

Near the pole ω_s,

$$\begin{aligned} \omega - E_0 - \Sigma_2(\omega) &\approx \omega - E_0 - \Sigma_2(\omega_s) - (\omega - \omega_s)\Sigma_2'(\omega_s) \\ &= (\omega - \omega_s)(1 - \Sigma_2'(\omega_s)) \end{aligned} \tag{5.66a}$$

so that

$$G_2(\omega) \approx \frac{1}{(1 - \Sigma_2'(\omega_s))} \frac{1}{\omega - \omega_s} \ . \tag{5.66b}$$

Thus, in general the residue of each pole is $\frac{1}{1-\Sigma_2'(\omega_s)}$. For our schematic model, the assumption (5.63) makes it easy to evaluate the residue for each pole.

First, note using Eqs. (5.65) and (5.63) that the shifts $\omega_s - E_s$ are negligible relative to the energy spacing and

$$\begin{aligned} (\omega_i - E_0) &= (E_i - E_0)\,(1 + \mathcal{O}(\delta)) \\ &\qquad\qquad\qquad\qquad i = 1,2 \\ (E_i - \omega_0) &= (E_i - E_0)\,(1 + \mathcal{O}(\delta)) \\ (\omega_i - E_0) &= \left(\frac{A_i}{\omega_i - E_i}\right)(1 + \mathcal{O}(\delta)) \ . \end{aligned} \tag{5.67}$$

To leading order in δ, the residues are

$$\frac{1}{1-\Sigma_2'(\omega_0)} = \frac{1}{1+\sum_{i=1}^{2}\frac{A_i}{(\omega_0-E_i)^2}} \approx \frac{1}{1+\sum_{i=1}^{2}\frac{A_i}{(E_0-E_i)^2}} \approx 1-\sum_{i=1}^{2}\frac{A_i}{(E_0-E_i)^2} \tag{5.68a}$$

and

$$\frac{1}{1-\Sigma_2'(\omega_i)} = \frac{1}{1+\sum_{j=1}^{2}\frac{A_j}{(\omega_i-E_j)^2}} \approx \frac{1}{1+\frac{(\omega_i-E_0)^2}{A_i}} \approx \frac{A_i}{(E_0-E_i)^2} \qquad i=1,2 \ . \tag{5.68b}$$

Now, the full physical effect of switching on the matrix elements A_i is evident. Without this coupling, the system has a pole with unit residue corresponding to the propagation of a single particle, as sketched in Fig. 5.6a. After switching on the interaction, Fig. 5.6b, the strength is now fragmented between three poles. The system still possesses a fundamental excitation near the original energy E_0, but now the energy is shifted slightly to ω_0 and the residue $1-\sum_{i=1}^{2}\frac{A_i}{(E_0-E_i)^2}$ is less than one. This pole is called the quasiparticle pole: it still behaves very much like a single-particle excitation, but is no longer a true particle pole because of the medium modifications. Consistent with the sum rule, Eq. (5.36), the strength which has been removed from the quasiparticle pole has been distributed to the two new poles, with strength $\frac{A_i}{(E_0-E_i)^2}$ going to the pole at ω_i. These poles represent more complicated excitations of the many-particle medium, such as two-particle, one hole states.

With this schematic model as an introduction, it is now appropriate to return to the general expression for the second-order self-energy, Eq. (5.60), which has both two-particle one-hole and two-hole one-particle contributions. The residue of the quasiparticle pole corresponding to the state α, assuming a weak interaction v, is given by

$$\frac{1}{1-\frac{\partial}{\partial\omega}\Sigma_2(\alpha,\omega)} \approx 1-\frac{1}{2}\sum_{abB}\frac{|\{\alpha B|v|ab\}|^2}{(\omega+\epsilon_B-\epsilon_a-\epsilon_b)^2}-\frac{1}{2}\sum_{ABb}\frac{|\{\alpha b|v|AB\}|^2}{(\epsilon_A+\epsilon_B-\epsilon_a-\omega)^2} . \tag{5.69}$$

Thus, as shown in Fig. 5.7, the strength associated with a particle state is depleted by coupling both to two-particle one-hole states above the Fermi surface and to two-hole one-particle states below the Fermi surface. In the limit of a continuum, these two-particle one-hole and two-hole one-particle states yield a smooth background in addition to the simple quasiparticle excitations.

Having introduced quasiparticles, it is natural to ask if there is any regime in which they provide a useful and accurate description of a physical system or whether

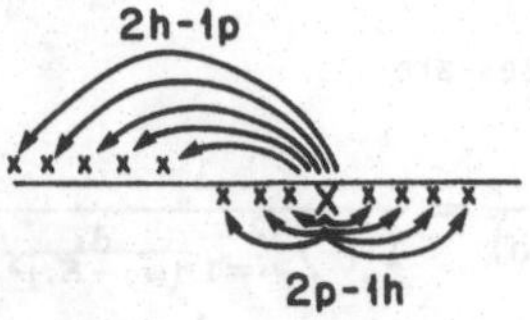

Fig. 5.7 Fragmentation of strength of the quasiparticle pole.

they always decay so quickly to more complicated states that they are of no practical significance. In fact, Landau's Fermi liquid theory is based upon the quasiparticle picture and as shown in Chapter 6 can provide an exact description of physical Fermion systems in an appropriate limit. The essential point can be seen simply by calculating the lifetime from the imaginary part of Σ_2 in Eq. (5.60). Since by the usual argument with outgoing wave boundary conditions, a state with complex energy $E = E_R - i\frac{\Gamma}{2}$ has lifetime $\tau = \frac{1}{\Gamma}$, the lifetime for a quasiparticle state α evaluated at the quasiparticle pole ϵ_α above the Fermi energy, *i.e.*, $\epsilon_\alpha > 0$, is given by

$$\frac{1}{\tau} = -2\,\mathrm{Im}\,\Sigma_2(\alpha,\epsilon_\alpha) = \frac{1}{2}\sum_{abB} |\{\alpha B|v|ab\}|^2\, 2\pi\delta(\epsilon_\alpha + \epsilon_B - \epsilon_a - \epsilon_b). \tag{5.70}$$

Note that if we had considered a hole state, the two-hole one-particle component of Σ_2 would have contributed instead of the two-particle one-hole component.

When the energy ϵ_α is close to the Fermi energy ϵ_F, phase space restrictions on the sum over a, b, and B severely limit the contributions to $\frac{1}{\tau}$. With the convention of measuring energies relative to ϵ_F, $\epsilon_B < 0$ and ϵ_α, ϵ_a, $\epsilon_b > 0$ so that the energy conservation condition in Eq. (5.70) is $\epsilon_\alpha = |\epsilon_a| + |\epsilon_b| + |\epsilon_B|$. Thus, neither ϵ_a nor ϵ_b may be greater than ϵ_α. Hence, letting $\rho(\epsilon)$ denote the density of states and defining the maximum values of $\rho(\epsilon)$ and $|\{\alpha B|\, v\,|ab\}|$ for $0 < \epsilon_a, \epsilon_b, |\epsilon_B| < \epsilon_\alpha$ as $\rho_{\max}$ and $V_{\max}$ we obtain the bound

$$\begin{aligned}\frac{1}{\tau} &\le \pi V_{\max}^2 \int_0^\infty d\epsilon_a \int_0^\infty d\epsilon_b \int_0^{-\infty} d\epsilon_B\, \rho(\epsilon_a)\rho(\epsilon_b)\rho(\epsilon_B)\delta\left(\epsilon_\alpha + \epsilon_B - \epsilon_a - \epsilon_b\right)\\ &= \pi V_{\max}^2 \int_0^{\epsilon_\alpha} d\epsilon_a \int_0^{\epsilon_\alpha} d\epsilon_b\, \rho(\epsilon_a)\rho(\epsilon_b)\rho\left(\epsilon_a + \epsilon_b - \epsilon_\alpha\right)\\ &\le \pi V_{\max}^2\, \rho_{\max}^3 \epsilon_\alpha^2.\end{aligned} \tag{5.71}$$

No matter how strong the two-body interaction, as long as its matrix elements and the density of states remain finite in the vicinity of the Fermi surface, $\frac{1}{\tau}$ is therefore bounded by a constant times ϵ_α^2. A completely analogous argument holds for a quasi-hole. So in general, as the energy ϵ of a quasiparticle or quasi-hole excitation approaches the Fermi Energy ϵ_F, the lifetime increases as

$$\tau(\epsilon) \propto |\epsilon - \epsilon_F|^{-2}. \tag{5.72}$$

Thus, under very general conditions, a strongly-interacting many-Fermion system will always have a domain sufficiently close to the Fermi surface in which quasiparticles have arbitrarily long lifetimes and are the appropriate degrees of freedom to describe the system. Note that nowhere in this argument have we invoked momentum conservation, as in most common derivations, so this result is clearly applicable to finite systems and non-translationally invariant systems.

EFFECTIVE MASSES

We now consider the effect of the energy and momentum dependence of the self-energy on quasiparticle propagation in an interacting Fermi system. Whereas the analysis pertains to a variety of interesting physical systems, such as a ^{3}He atom in liquid ^{3}He or a low energy nucleon propagating in a nucleus, we will illustrate the major points for the case of a nucleon in translationally invariant nuclear matter.

From the dispersion relation defining the quasiparticle pole

$$\epsilon = \frac{k^2}{2m} + \Sigma(k,\epsilon) \tag{5.73}$$

the density of states may be calculated as follows

$$\begin{aligned} \frac{d\epsilon}{dk} &= \frac{k}{m} + \frac{\partial\Sigma}{\partial k} + \frac{\partial\Sigma}{\partial\epsilon}\frac{d\epsilon}{dk} \\ &= \frac{k}{m}\left(1+\frac{m}{k}\frac{\partial\Sigma}{\partial K}\right)\left(1-\frac{\partial\Sigma}{\partial\epsilon}\right)^{-1} . \end{aligned} \tag{5.74a}$$

It is often convenient to subsume the complicated effect of the medium on a particular process into a suitably defined effective mass. The density of states may thus be expressed

$$\begin{aligned} \frac{d\epsilon}{dk} &= \frac{k}{m^*} \\ m^* &\equiv m\left(1+\frac{m}{k}\frac{\partial\Sigma}{\partial k}\right)^{-1}\left(1-\frac{\partial\Sigma}{\partial\epsilon}\right) . \end{aligned} \tag{5.74b}$$

Noting that m^* itself is the product of factors associated with the energy and momentum dependence of $\Sigma(k,\omega)$ it is useful to define the additional effective masses m_ϵ and m_k as follows (Jeukenne, Lejeune, and Mahaux 1976)

$$\begin{aligned} \frac{m_\epsilon}{m} &\equiv \left(1-\frac{\partial\Sigma}{\partial\epsilon}\right) \\ \frac{m_k}{m} &\equiv \left(1+\frac{m}{k}\frac{\partial\Sigma}{\partial k}\right)^{-1} \\ \frac{m^*}{m} &= \frac{m_\epsilon}{m}\times\frac{m_k}{m} \end{aligned} \tag{5.75}$$

and to study m_ϵ and m_k separately. Note that the factor $\frac{m}{m_\epsilon}$ is just the residue of the quasiparticle pole discussed in the last section.

The mass m_k reflects the spatial nonlocality of Σ, and may be understood qualitatively by considering the non-locality of the exchange term of the Hartree-Fock potential. As shown in Problem. 5.3, the general result for the exchange terms assuming a central potential $v(r)$ and spin degeneracy 2S+1 may be evaluated with plane wave states to obtain

$$\begin{aligned} \Sigma_1^{\text{exch}}(k) &= \frac{1}{2S+1}\sum_{|k'|<k_f}\langle kk'|v|k'k\rangle \\ &= -\frac{1}{2S+1}\int d^3r\, e^{ik\cdot r}v(r)\frac{3j_1(k_F r)\rho}{k_F r} . \end{aligned} \tag{5.76}$$

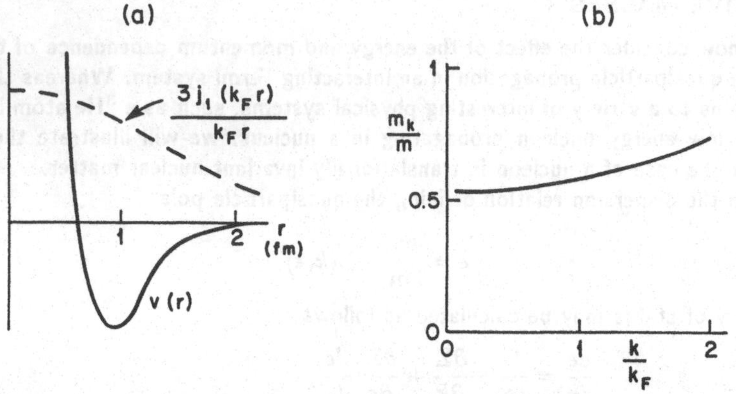

Fig. 5.8 Effective mass m_k in nuclear matter. Sketch (a) shows the two factors contributing to the Fourier transform of $\Sigma_1^{\rm exch}(k)$ in Eq. (5.77). The resulting effective mass $\frac{m_k}{m} = \left(1+\frac{m}{k}\frac{\partial\Sigma}{\partial k}\right)^{-1}$ is shown in (b).

Note that a central potential contributes to the exchange term with the opposite sign and a strength reduced by $\frac{1}{2S+1}$. If the potential is state-dependent, then different combinations of partial waves contribute to the direct and exchange terms. In particular, if we consider nucleons with two internal degrees of freedom, spin and isospin, and assume even partial waves interact with a potential $v_{\rm even}$ and odd partial waves interact with the potential $v_{\rm odd}$, then, from Problem 5.3, the exchange term is

$$\Sigma_1^{\rm exch}(k) = \int d^3r\, e^{ik\cdot r}\left[\frac{3}{8}v_{\rm even}(r) - \frac{5}{8}v_{\rm odd}(r)\right]\frac{3j_1(k_F r)}{k_F r} \quad . \tag{5.77}$$

Since the nucleon-nucleon interaction is strongly attractive in even partial waves and weakly repulsive in odd partial waves, the effective potential

$$v_{\rm exch}(r) \equiv \frac{3}{8}v_{\rm even}(r) - \frac{5}{8}v_{\rm odd}(r) \tag{5.78}$$

contributing to the exchange integral is strikingly different than in the state-independent case. Both contributions to $v_{\rm exch}(r)$ are attractive, so that the net attraction from the exchange term is larger than from the direct term, and the qualitative behavior of $v_{\rm exch}(r)$ and the Slater density $\frac{3j_1(k_F r)}{k_F r}$ are sketched in Fig. 5.8a. Since $\sum_1^{\rm exch}(k)$ is given by the Fourier transform of the product of these two factors, its momentum dependence is obvious. At low k, $\Sigma(k)$ is strongly attractive and when $k >> k_F$, the characteristic scale in the integrand, $\Sigma(k) \to 0$. Hence $\frac{\partial\Sigma(k)}{\partial k}$ is a positive decreasing function of k, so that $\frac{m_k}{m} = \left(1+\frac{m}{k}\frac{\partial\Sigma(k)}{k}\right)^{-1}$ has the behavior sketched in Fig. 5.8b. Note that at low momentum, the spatial nonlocality reduces m_k to roughly half of the bare mass m.

The mass m_ϵ reflects the nonlocality of Σ in time. Since the Hartree-Fock contribution to Σ is instantaneous, the leading contribution to m_ϵ arises from Σ_2, Eq.

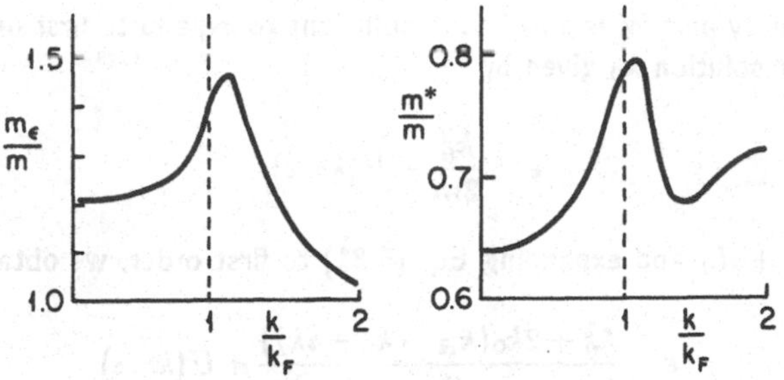

Fig. 5.9 Effective masses m_ϵ and m^* in nuclear matter.

(5.60). The qualitative behavior of this contribution may be understood by the following schematic argument (Bertsch and Kuo 1968). Represent the sum over all two-particle one-hole states by a single average state with excitation energy $\epsilon_a + \epsilon_b - \epsilon_B \sim E_x$ and an effective coupling matrix element V:

$$\Sigma_2^{2p1h}(k,\epsilon) = \frac{1}{2}\sum_{abB} \frac{|\{kB|v|ab\}|^2}{\epsilon + \epsilon_B - \epsilon_a - \epsilon_b} \approx \frac{1}{2}\,\frac{V^2}{\epsilon - E_x} \;. \tag{5.79a}$$

Similarly, represent the sum over two-hole one-particle states by a single average state, and further assume that near the Fermi surface particle and hole states are symmetric. Then, $\epsilon_A + \epsilon_B - \epsilon_a \sim -E_x$ and

$$\Sigma_2^{2h1p}(k,\epsilon) = -\frac{1}{2}\sum_{ABa} \frac{|\{ka|v|AB\}|^2}{\epsilon_A + \epsilon_B - \epsilon_a - \epsilon} \approx \frac{1}{2}\,\frac{V^2}{\epsilon + E_x} \;. \tag{5.79b}$$

Thus, both terms in Σ_2 become more negative as ϵ is increased from 0, yielding the following enhancement in m_ϵ at the Fermi surface.

$$\left.\frac{m_\epsilon}{m}\right|_{\epsilon_F} = 1 - \frac{\partial\Sigma}{\partial\epsilon} \approx 1 - \frac{d}{d\epsilon}\,\frac{1}{2}\left[\frac{V^2}{\epsilon - E_x} - \frac{V^2}{\epsilon + E_x}\right]_{\epsilon=0} = 1 + \frac{2V^2}{E_x{}^2} \;. \tag{5.80}$$

This schematic analysis is too crude to calculate m_ϵ away from the Fermi surface, but detailed calculations (Jeukenne, Lejeune, and Mahaux, 1976) yield the behavior graphed in Fig. 5.9a. The enhancement is very large at the Fermi surface, of the order 50%, and falls off significantly away from the Fermi surface. The combined effect of m_ϵ and m_k in the total effective mass m^* measured in the density of states has the structure shown in Fig. 5.9b. Note that because m_k is so small, on the average m^* is significantly less than m. However, near the Fermi surface, the peak in m_k brings m^* nearly up to m.

Although the product of m_ϵ and m_k appears in the density of states, other observables depend on m_ϵ and m_k separately. Consider, for example, the mean free path, which is calculated by specifying a real energy ϵ, and solving the dispersion relation (5.74) for a complex k. Denoting the real and imaginary parts of Σ by U and W, k is given by:

$$\epsilon = \frac{k^2}{2m} + U(k,\epsilon) + iW(k,\epsilon) \;. \tag{5.81}$$

Since the imaginary part W is small, it is sufficient to expand to first order in W about the zeroth-order solution k_0 given by

$$\epsilon = \frac{k_0^2}{2m} + U(k_0, \epsilon) \ . \tag{5.82}$$

Writing $k \equiv k_R + ik_I$ and expanding Eq. (5.81) to first order, we obtain

$$\begin{aligned} \epsilon = &\frac{k_0^2 + 2k_0(k_R - k_0 + ik_I)}{2m} + U(k_0, \epsilon) \\ &+ \left.\frac{\partial U}{\partial k}\right|_{k_0} (k_R - k_0 + ik_I) + iW(k_0, \epsilon) \end{aligned} \tag{5.83}$$

with the result

$$\begin{aligned} k_R &= k_0 \\ k_I &= -W(k_R, \epsilon)\left(\frac{k_R}{m} + \frac{\partial U}{\partial k}\right)^{-1} \\ &= -\frac{m_k}{k_R} W(k_R, \epsilon) \ . \end{aligned} \tag{5.84}$$

Since the attenuation factor for a complex wave vector is $\psi_k^2 \sim \left|e^{i(k_R + ik_I)\cdot r}\right| \sim e^{-2k_I r}$, the mean free path is

$$\lambda = \frac{1}{2k_I} = -\frac{k_R}{2m_k W(k_R, \epsilon)} \tag{5.85}$$

and is thus proportional to $\frac{1}{m_k}$.

Similarly, the lifetime of a quasiparticle excitation is obtained by specifying a real k and solving for the complex energy. The zeroth order equation is

$$\epsilon_0 = \frac{k^2}{2m} + U(k, \epsilon_0) \tag{5.86}$$

and writing $\epsilon = \epsilon_R - \frac{i\Gamma}{2}$ we obtain

$$\begin{aligned} \epsilon_R - \frac{i\Gamma}{2} = &\frac{k^2}{2m} + U(k, \epsilon_0) \\ &+ \left.\frac{\partial U}{\partial \epsilon}\right|_{\epsilon_0} \left(\epsilon_R - \epsilon_0 - i\frac{\Gamma}{2}\right) + iW(k, \epsilon_0) \end{aligned} \tag{5.87}$$

with the solution

$$\begin{aligned} \epsilon_R &= \epsilon_0 \\ \Gamma &= -2W(k, \epsilon_0)\left(1 - \frac{\partial U}{\partial \epsilon}\right)^{-1} = -2W(k, \epsilon_0)\frac{m}{m_\epsilon} \ . \end{aligned} \tag{5.88}$$

Thus, the lifetime $\tau = \frac{1}{\Gamma}$ is proportional to m_ϵ. The two results for λ and Γ in Eqs. (5.85) and (5.88) are consistent since λ and Γ are related through the group velocity v:

$$\lambda = \frac{v}{\Gamma}$$
$$v = \frac{dE}{dk} = \frac{k}{m^*} = \frac{k}{m}\frac{m}{m_k}\frac{m}{m_\epsilon}. \tag{5.89}$$

The effective masses play a quantitatively significant role in determining the mean free path of a nucleon in the nuclear medium and serve to resolve a long-standing discrepancy with experiment. The mean free path in a nucleus of neutrons in the energy range 50 – 150 MeV may be determined from the amplitude of shape resonances in total neutron scattering cross sections. In this analog of the atomic Ramsauer effect, interference between the incident and transmitted wave can only be observed if the mean free path is long enough for a neutron to pass through the nuclear medium, and in this way one measures $\lambda \sim 6\,\text{fm}$ for neutrons in this energy regime (Bohr and Mottelson, 1969).

The naive classical estimate of a mean free path $\lambda = \frac{1}{\bar{\sigma}\rho}$ which ignores the Pauli principle is far too low, with the average nucleon-nucleon cross section at 100 MeV $\bar{\sigma} = 5.5\,\text{fm}^2$ and nuclear density $\rho = 0.16\,\text{fm}^{-3}$yielding $\lambda = 1.1\,\text{fm}$. The simplest approximation to $W = \text{Im}\,\Sigma$ is obtained from Eq. (5.70) by replacing matrix elements of the potential $\{kB|v|ab\}$ by the free space T-matrix $\{kB|T|ab\}$ measured experimentally and replacing the energies ϵ_a in the medium by the free space energies $\frac{k_a^2}{2m}$:

$$W^T(k,\epsilon) = -\frac{\pi}{2}\sum_{abB}|\{kB|T|ab\}|^2\,\delta\left(\epsilon + \frac{k_B^2}{2m} - \frac{k_a^2}{2m} - \frac{k_b^2}{2m}\right). \tag{5.90}$$

Using this estimate for W and ignoring the effective mass in Eq. (5.85) yields $\lambda = -\frac{k}{2mW^T} \sim 3\,\text{fm}$, still far short of the experimental result of 6 fm. Thus, simply including the Pauli principle through the restriction $ab > k_F$, $B < k_F$ is insufficient. However, this formula omits two effective mass factors. The δ-function in (5.90) with free propagators simply includes the free Fermi gas density of states. Since the δ-function in Σ, Eq. (5.70), contain the energies in the medium, W^T should be multiplied by $\frac{m^*}{m}$ to include the density of states in the medium. Including the additional factor of m_k from Eq. (5.85), the correct T-matrix expression for the mean free path is

$$\lambda = -\left(\frac{m}{m_k}\right)\left(\frac{m}{m^*}\right)\frac{k}{2mW^T} \sim 6\,\text{fm}\ . \tag{5.91}$$

Thus, the medium dependence reflected in the two effective mass factors, each of the order of $\sim \frac{1}{0.7}$ is crucial to understanding the nucleon mean free path (Negele and Yazaki 1981, Fantoni, Friman and Pandharipande 1981).

OPTICAL POTENTIAL

Another important physical property of the self-energy is the fact that it specifies the optical potential for the scattering of a particle from a composite system made up of identical particles (Bell and Squires 1959).

Consider the elastic scattering of an electron from an atom or a nucleon from a nucleus. Since the composite system is required to remain in its ground state, the asymptotic scattering state may be described by a wave function $\phi(r)$ depending only on the relative coordinate between the composite system and the scattered particle. By definition, the optical potential is a one-particle potential producing phase shifts identically equal to those produced in $\phi(r)$ for the full many-body problem. Note that since any phase-shift equivalent potential is satisfactory, the optical potential is not unique.

The essential point in relating the optical potential to the self-energy is the observation that the wave function can be written in terms of the one-particle Green's function. Since $G(rt|r't')$ is the amplitude for adding a particle to a system at $r't'$ and detecting it at rt, it is clear that we should be able to express the scattering wave function in terms of G. To be precise, let $|\psi\rangle$ be the ground state of the N-body composite system and pick the zero of the energy scale such that its energy is zero. Then a scattering state may be generated by creating a particle at some point r' far away from the system at time t' and projecting onto a specific energy E by integrating over initial times

$$|S\rangle = \int dt'\, e^{-iEt'} \left[\theta(t'-t) + \theta(t-t')e^{-iH(t-t')}\psi^{\dagger}(r')\right] |\psi_0\rangle \ . \tag{5.92}$$

Since the optical model wave function is the amplitude for observing one particle at r and all other particles in the ground state $|\psi_0\rangle$, we obtain

$$\begin{aligned}
\phi(\vec{r},t) &= \langle 0|\hat{\psi}(r)|S\rangle \\
&= \int dt'\, e^{-iEt'} \langle\psi_0|\theta(t-t')e^{iHt}\hat{\psi}(r)e^{-iHt}e^{iHt'}\hat{\psi}(r')e^{-iHt'}|\psi_0\rangle \qquad (5.93) \\
&= \int dt' e^{-iEt'} \langle\psi_0|\theta(t-t')\hat{\psi}(r,t)\hat{\psi}^{\dagger}(r',t') - \theta(t'-t)\hat{\psi}^{\dagger}(r',t')\hat{\psi}(r,t)|\psi_0\rangle \\
&= e^{-iEt} iG(E;r,r')
\end{aligned}$$

In the second line, we have used the fact that the $\theta(t'-t)$ term in (5.92) does not contribute because $\langle 0|\psi(r)|0\rangle = 0$ and have inserted exponentials next to $|\psi_0\rangle$ because $H|\psi_0\rangle = 0$. The second time order in the third line required to obtain the Green's function could be inserted because $\langle 0|\hat{\psi}^{\dagger}(r',t') = (\psi(r',t')|0\rangle)^{\dagger} = 0$ since r' is far from the target and there are thus no particles for ψ to annihilate. For convenience, we will evaluate the wave function at time $t=0$ and drop the factor i so that $\phi(r) = G(E;r,r')$.

Consider, for reference, the problem in which the interacting Hamiltonian H is replaced by $H_0 = T + U$, where U is a one-body potential (such as the Hartree-Fock potential) producing a first approximation to the localized N-body target. Then, by the previous argument, scattering from the one-body potential U is described by the wave function $\phi_0(r) = G_0(E;r,r')$. Substituting $\phi(r)$ and $\phi_0(r)$ in the Dyson equation $G = G_0 + G_0\Sigma G$, we may write

$$\phi(r) = \phi_0(r) + \int dr''dr'''\, G_0(E;r,r'')\Sigma(E;r'',r''')\phi(r''') \ . \tag{5.94a}$$

Note from the spectral representation for G_0, Eq. (5.29), that for the positive energies relevant to the scattering problem the denominator in the $(N+1)$ particle term never vanishes so that $G_0(E)$ may be replaced by the retarded Green's function $G_0^R(E)$ having $E + i\eta$ in both terms. Thus, we may rewrite Eq. (5.94a) in the form

$$|\phi^+\rangle = |\phi_0^+\rangle + \frac{1}{E^+ - T - U}\Sigma|\phi^+\rangle \tag{5.94b}$$

where $|\phi^+\rangle$ and $|\phi_0^+\rangle$ denote the scattering wave function for the interacting and non-interacting problems $\phi(r)$ and $\phi_0(r)$ and $E^+ \equiv E + i\eta$. Because $|\phi_0^+\rangle$ is the scattering wave function for the potential U, it satisfies the Lippmann-Schwinger equation

$$|\phi_0^+\rangle = |\phi_0\rangle + \frac{1}{E^+ - T}U|\phi_0^+\rangle \tag{5.95}$$

where $|\phi_0\rangle$ denotes an incident plane wave. Substituting $|\phi_0^+\rangle$ from Eq. (5.94b) into (5.95) we obtain the desired Lippmann-Schwinger equation for $|\phi^+\rangle$

$$|\phi^+\rangle = |\phi_0\rangle + \frac{1}{E^+ - T}(U + \Sigma)|\phi^+\rangle \tag{5.96}$$

so that the optical potential is $U + \Sigma$.

When the self-energy is expanded in $H - H_0 = \frac{1}{2}\sum_{ij} v(r_i - r_j) - U$, the first order term containing $-U$ exactly cancels U and the first few time-ordered contributions to the optical potential are

$U + \Sigma =$

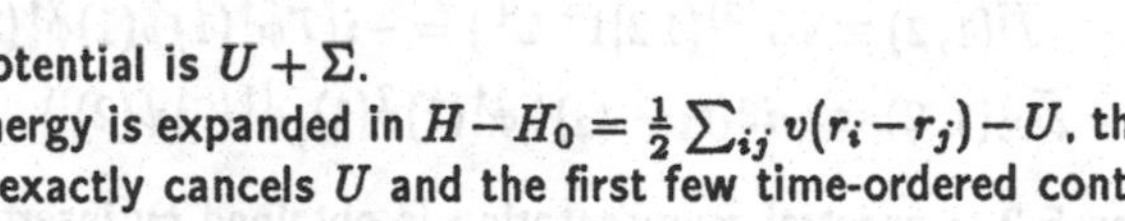

(5.97)

Diagram (a) describes propagation in the Hartree-Fock mean field and (b) represents the amplitude for coupling to a two-particle one-hole state and propagating in that state rather than in the single-particle state. Diagram (c) expresses the fact that in the interacting system, two normally occupied states A and B may be virtually excited to states a and k, thus blocking the addition of a particle in state k to the system. In contrast to other approaches to multiple scattering theory in which antisymmetry is either neglected or put in laboriously by hand, the self-energy systematically includes its effects through terms such as this. Diagrams (d) and (e) are representative of an infinite class of terms in which U and the Hartree-Fock potential enter with opposite signs. As usual, it is advantageous to cancel such terms identically by choosing U to be the Hartree-Fock potential.

Because one-particle irreducibility is defined in terms of Feynman diagrams rather than time-ordered diagrams, both of the following time-ordered diagrams are excluded from the optical potential

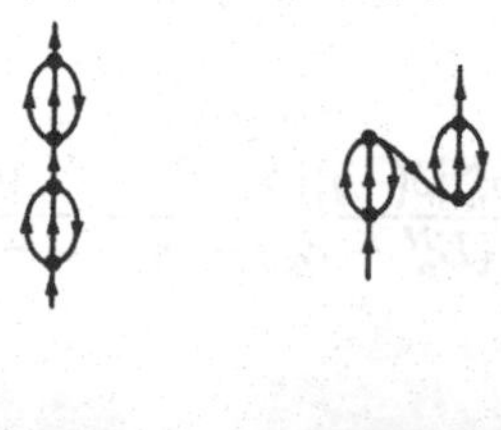

(5.98)

Whereas it is quite plausible that (A) represents a time history which will be generated when diagram (b) of (5.97) is iterated in the Lippmann-Schwinger equation, one might naively be tempted to regard (B) as a valid four-particle three-hole contribution to the optical potential. Clearly, the derivation shows it must not be included, which is associated with the fact that the Lippman Schwinger equation has no projector onto states above the Fermi surface and thus generates propagation in hole states as well as particle states.

5.4 LINEAR RESPONSE

THE RESPONSE FUNCTION

Section 2.1 showed how experimental observables could be expressed in terms of response functions, which we will now write in terms of two-particle Green's functions. To obtain the product of two one-body operators, we must consider a two-particle Green's function in which the creation operators are evaluated infinitesimally later than the annihilation operators and we define the density-density correlation function and its retarded counterpart as

$$\begin{aligned} \overline{D}(1,2) &\equiv iG^{(2)}(1\,2|1^+\,2^+) = -i\langle T\hat{\psi}^\dagger(1)\hat{\psi}(1)\hat{\psi}^\dagger(2)\hat{\psi}(2)\rangle \\ \overline{D}_R(1,2) &\equiv -i\theta(t_1-t_2)\langle[\hat{\psi}^\dagger(1)\hat{\psi}(1),\hat{\psi}^\dagger(2)\hat{\psi}(2)]\rangle \ . \end{aligned} \tag{5.99}$$

As in Section 5.2, a spectral representation is obtained by inserting a complete set of states and Fourier transforming. For brevity, we will only write the zero temperature results, and leave the analogous finite temperature expressions as a straightforward exercise:

$$\left\{ \begin{matrix} \overline{D}(x_1,x_2;\omega) \\ \overline{D}_R(x_1,x_2;\omega) \end{matrix} \right\} = \sum_n \left[\frac{\langle\psi_0|\hat{\rho}(x_1)|\psi_n^N\rangle\langle\psi_n^N|\hat{\rho}(x_2)|\psi_0\rangle}{\omega-(E_n^N-E_0)+i\eta} - \frac{\langle\psi_0|\hat{\rho}(x_2)|\psi_n^N\rangle\langle\psi_n^N|\hat{\rho}(x_1)|\psi_0\rangle}{\omega+(E_n^N-E_0)\mp i\eta} \right] \tag{5.100}$$

where the density operator is written $\psi^\dagger(x_1)\psi(x_1) \equiv \hat{\rho}(x_1)$. In momentum space, using

$$\begin{aligned} \hat{\rho}(q) &= \int e^{-iq\cdot x}\rho(x) \\ &= \int e^{-iq\cdot x}\left[\int \frac{d^3k}{(2\pi)^3}e^{-ik\cdot x}a_k^\dagger\right]\left[\int \frac{d^3k'}{(2\pi)^3}e^{+ik'\cdot x}a_{k'}\right] \\ &= \int \frac{d^3k}{(2\pi)^3}a_k^\dagger a_{k+q} \\ \hat{\rho}(q)^\dagger &= \hat{\rho}(-q) \end{aligned} \tag{5.101}$$

we obtain

$$\left\{ \begin{matrix} \overline{D}(k,\omega) \\ \overline{D}_R(k,\omega) \end{matrix} \right\} = \sum_n \left[\frac{|\langle\psi_n^N|\hat{\rho}(-q)|\psi_0\rangle|^2}{\omega-(E_n^N-E_0)+i\eta} - \frac{|\langle\psi_n^N|\hat{\rho}(q)|\psi_0\rangle|^2}{\omega+(E_n^N-E_0)\mp i\eta} \right] \ . \tag{5.102}$$

The essential difference relative to Eq. (5.27) is the presence of N-particle intermediate states rather than states with $N \pm 1$ particles, so that the density-density response function contains information about the eigenstates of the N-particle system itself.

From Eq. (5.100), it is evident how to calculate the observables discussed in Section 2.1. First $\overline{D}$ is calculated in an appropriate approximation using the perturbation theory we have derived for the Green's function G_2. A response function, Eq. (2.16), is then evaluated by calculating $\overline{D}_R$ from the relations

$$\begin{aligned} \operatorname{Re}\overline{D}(\omega) &= \operatorname{Re}\overline{D}_R(\omega) \\ \operatorname{Im}\overline{D}(\omega) &= \operatorname{sgn}(\omega - E_0)\operatorname{Im}\overline{D}_R \ . \end{aligned} \tag{5.103}$$

Analogous expressions can, of course, be written with an arbitrary one-body operator replacing the density operator. Similarly, the inclusive scattering cross section, Eq. (2.19) can be evaluated from the imaginary part of $\overline{D}$, since only the first term in Eq. (5.100) or (5.102) contributes for positive ω:

$$\begin{aligned} \sigma(q,\omega) &= 2\pi\sum_n \delta(E_n^N - E_0 - \omega)\tilde{v}(q)^2|\langle\psi_n^N|\hat{\rho}(-q)|\psi_0\rangle|^2 \\ &= -2\operatorname{Im}\overline{D}(q,\omega)\tilde{v}(q)^2 \ . \end{aligned} \tag{5.104a}$$

Because of its direct relation to experiment, the imaginary part of the response function is often referred to as the dynamic structure factor:

$$S(q,\omega) \equiv -\operatorname{Im}\overline{D}(q,\omega) = \frac{\sigma(q,\omega)}{2\tilde{v}(q)^2} \ . \tag{5.104b}$$

Finally, it is convenient and conventional to define the density fluctuation operator

$$\tilde{\rho} - \hat{\rho} - \langle\hat{\rho}\rangle \tag{5.105}$$

and its correlation functions

$$\begin{aligned} iD(1,2) &= \langle T\tilde{\rho}(1)\tilde{\rho}(2)\rangle \\ &= \langle T\hat{\psi}^\dagger(1)\hat{\psi}(1)\,\hat{\psi}^\dagger(2)\hat{\phi}(2)\rangle - \langle T\hat{\psi}^\dagger(1)\hat{\psi}(1)\rangle\langle T\hat{\psi}^\dagger(2)\psi(2)\rangle \\ &= i\overline{D}(1,2) - \langle\hat{\rho}(1)\rangle\langle\hat{\rho}(2)\rangle \end{aligned} \tag{5.106a}$$

$$iD_R(1,2) = \theta(t_1 - t_2)\langle[\tilde{\rho}(1),\tilde{\rho}(2)]\rangle = i\overline{D}_R(1,2) \ . \tag{5.106b}$$

Note that $D_R = \overline{D}_R$ since $\langle\hat{\rho}\rangle$ does not contribute to the commutator and that $\sigma(q,\omega) = -2\operatorname{Im}D(k,\omega)\tilde{v}^2(q)$ because $\langle\hat{\rho}\rangle$ does not connect $|\psi_0\rangle$ to any excited states in (5.104). The fluctuation-dissipation theorem is particularly clear in this language, since transport coefficients are specified by the expectation value of products of fluctuation operators. The advantage of Eq. (5.106a) is the fact that its diagram expansion contains all liked diagrams which connect the density operators at points 1 and 2. Taken by itself, $\overline{D}(1,2)$ contains linked diagrams of two generic types, those which connect the operators at 1 and 2 and those which do not:

$\overline{D}(1,2) =$ 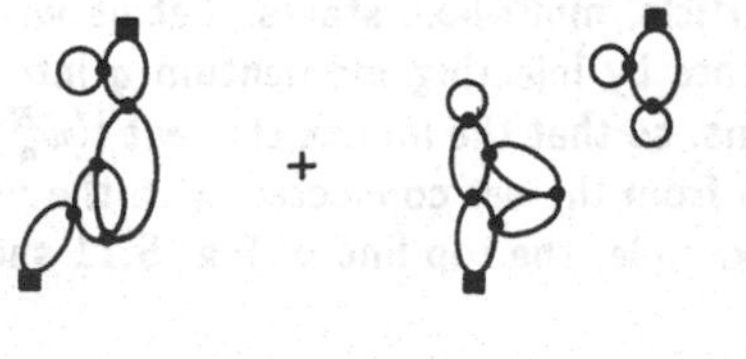 $+ \cdots$ (5.107)

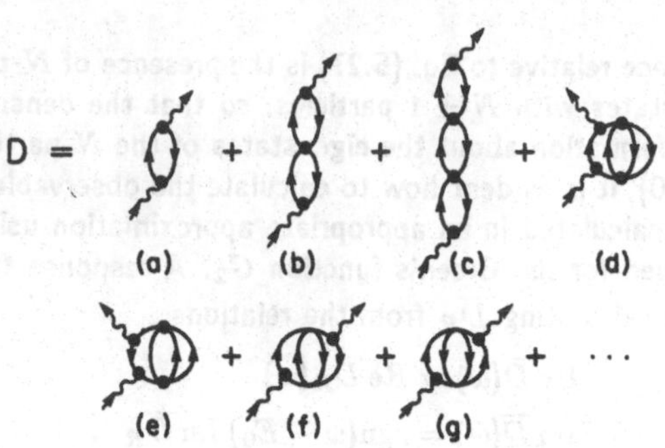

Fig. 5.10 Feynman diagrams for the response function D. Action of the density operators $\tilde{\rho}(1)$ and $\tilde{\rho}(2)$ are denoted by and which in momentum space may be regarded as injecting momentum $\vec{q}$ at (1) and removing it at (2). Note that in addition to the diagrams shown, the contributions through second order also include Hartree-Fock self-energy insertions on the propagators in diagrams (a) and (b).

The disconnected pieces in Eq. (5.107) simply correspond to all diagrams in the product $\langle\hat{\rho}(1)\rangle\langle\hat{\rho}(2)\rangle$ which are precisely subtracted off in $D(1, 2)$. Feynman diagrams corresponding to the first few orders of perturbation theory for $D(1, 2)$ are shown in Fig. 5.10.

The diagram rules for D follow immediately from the definition Eq. (5.99) and the zero-temperature Feynman rules of Section 4.1. Recall that i^2G_2 has propagators iG_0 for each line and the overall factor $(-i)^r(-1)^P(-1)^{n_L}$ where r is the number of interactions, n_L is the number of internal closed loops and $(-1)^P$ is the sign of the permutation of the external legs. For the identity permutation in which 1 connects to 1′ and 2 connects to 2′ the diagram for D has two additional closed loops relative to G_2. For all other permutations, connecting 1 to 2′ and 2 to 1′ creates one additional closed loop in D relative to G_2. Thus, associating a minus sign with every closed loop in D correctly accounts for both the factors $(-1)^P$ and $(-1)^{n_L}$. To include direct and exchange interactions on an equal footing as in Fig. 5.10, we will use Hugenholtz diagrams and include the factor $\frac{1}{2^{n_e}}$ for equivalent lines. Including the explicit i in the definition (5.99), the overall factor for a diagram contributing to D is $(-i)^{r+1}(-1)^{n'_L}\frac{1}{2^{n_e}}$ where r is the number of interactions, n'_L is the total number of closed loops in a diagram when it is drawn with direct matrix elements, and n_e is the number of equivalent lines. The propagators in the diagram are iG_0 and the interactions are antisymmetrized matrix elements $\{k_1\, k_2|v|k_3\, k_4\}$.

Additional insight into the physical content of the response function is obtained by considering the set of time-ordered diagrams corresponding to each Feynman diagram. For convenience, we will treat a translationally-invariant system and enumerate a complete set of multi-particle, multi-hole states. Let us write the amplitude for producing each intermediate state by injecting momentum q into the ground state as a sum of time-ordered diagrams, so that the matrix element $|\langle\psi_n^N|\hat{\rho}|\psi_0\rangle|^2$ is given by all possible pairs of one diagram from the set connected with the adjoint of another diagram from the same set. For example, the top line of Fig. 5.11 shows four time-histories leading

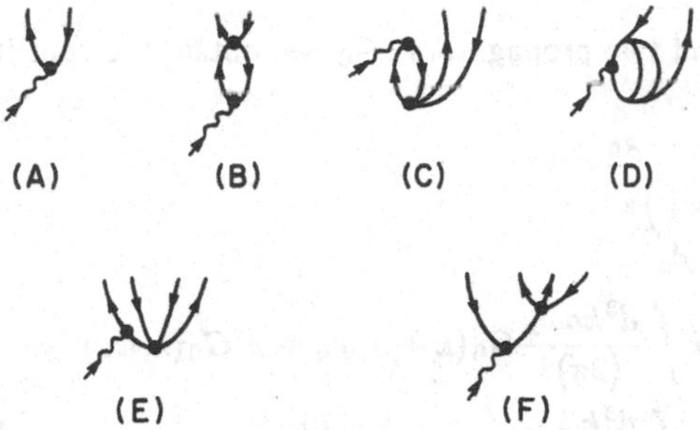

Fig. 5.11 Time-ordered diagrams for amplitudes contributing to the response function.

to a one-particle one-hole state. If diagram (A) is connected to ($A^\dagger$), which is the same diagram drawn upside down, we obtain one time order of diagram (a) of Fig. 5.10 in which a one-particle one-hole excitation is first created and then destroyed by the action of $\vec{q}$. In the language of scattering theory, diagrams (B) and (C) would be called final-state and initial state interaction corrections, respectively. Note how the cross terms between {A, B, C} and { $A^\dagger$ $B^\dagger$ $C^\dagger$} produce various time orders of the "chain" diagrams (a), (b), and (c), so that initial and final state interaction effects become totally entangled in the response function. Similarly, for the two-particle two-hole states in the second line, combining (E) with ($F^\dagger$) yields particular time-orders (d) and (g), but diagram (g) is also obtained by combining (D) and ($A^\dagger$). Thus, it is clear that the Feynman diagram expansion of the response function provides an economical description of a complicated process. The interplay of initial and final state interactions is treated consistently and the combinatoric of obtaining the same contribution from products of distinct amplitudes is automatically summarized by the rule that all symmetry factors are one for any Green's function. The flexibility to treat the response function in terms of time-ordered diagrams as well as Feynman diagrams may be exploited in specific applications, such as studying the scaling behavior of the response function at high momentum transfer (see Problem 4).

RANDOM PHASE APPROXIMATION

Any practical calculation of the response function must truncate the infinite diagram expansion. One alternative to simply stopping at some specific order of perturbation theory is to sum an appropriate infinite series of diagrams. Summing the set of all chain diagrams, the first three elements of which are shown in (a) – (c) of Fig. 5.10 yield the so-called random phase approximation or RPA. (Actually, there are two distinct approximations which are commonly referred to as the RPA: the present treatment of the response function and the summation of the ring diagram contributions, Eqs. (2.113) – (2.115) to the ground state correlation energy. The RPA correlation energy is treated in Problems 5.5 and 5.7.)

As a prelude to summing chains, let us first calculate a single link, diagram (a) of Fig. 10, which corresponds to the response function D_0 for a non-interacting system. For simplicity, we will treat a translationally invariant system in momentum space and assume spin-1/2 Fermions. Noting that diagram (a) has a closed loop, no interactions, no equivalent lines and two propagators iG_0, we obtain the contribution

$D_0(q,\omega) =$ 

$$= -2i\int \frac{d^3k\,d\omega_k}{(2\pi)^4} G_0(k+q,\omega_k+\omega)G_0(k,\omega_k) \tag{5.108}$$

$$= -2i\int \frac{d^3k\,d\omega_k}{(2\pi)^4}\left(\frac{1-n_{k+q}}{\omega_k+\omega-\epsilon_{k+q}+i\eta}+\frac{n_{k+q}}{\omega_k+\omega-\epsilon_{k+q}-i\eta}\right)$$
$$\times\left(\frac{1-n_k}{\omega_k-\epsilon_k+i\eta}+\frac{n_k}{\omega_k-\epsilon_k-i\eta}\right)$$
$$= 2\int\frac{d^3k}{(2\pi)^3}\left[\frac{(1-n_{k+q})n_k}{\omega+\epsilon_k-\epsilon_{k+q}+i\eta}-\frac{n_{k+q}(1-n_k)}{\omega+\epsilon_k-\epsilon_{k+q}-i\eta}\right].$$

Performing the momentum integrals in Eq. (5.108) with the zero-temperature occupation numbers $n_k=\theta(k_F-|k|)$ as outlined in Problem 5.6 yields the Lindhard function (Lindhard, 1954)

$$\mathrm{Re}\,D_0(\vec{q},\omega) = \frac{mk_F}{2\pi^2}\left\{-1+\frac{1}{2\tilde{q}}\left[1-\left(\frac{\tilde{\omega}}{\tilde{q}}-\frac{\tilde{q}}{2}\right)^2\right]\ln\left|\frac{1+\left(\frac{\tilde{\omega}}{\tilde{q}}-\frac{\tilde{q}}{2}\right)}{1-\left(\frac{\tilde{\omega}}{\tilde{q}}-\frac{\tilde{q}}{2}\right)}\right|\right.$$
$$\left.-\frac{1}{2\tilde{q}}\left[1-\left(\frac{\tilde{\omega}}{\tilde{q}}+\frac{\tilde{q}}{2}\right)^2\right]\ln\left|\frac{1+\left(\frac{\tilde{\omega}}{\tilde{q}}+\frac{\tilde{q}}{2}\right)}{1-\left(\frac{\tilde{\omega}}{\tilde{q}}+\frac{\tilde{q}}{2}\right)}\right|\right\} \tag{5.109a}$$

$$\mathrm{Im}\,D_0(\vec{q},\omega) = \left\{\begin{matrix} -\frac{mk_F}{4\pi\tilde{q}}\left[1-\left(\frac{\tilde{\omega}}{\tilde{q}}-\frac{\tilde{q}}{2}\right)^2\right] & \left|\frac{\tilde{q}}{2}-\tilde{q}\right|\le\tilde{\omega}\le\frac{\tilde{q}}{2}+\tilde{q} \\ -\frac{mk_F}{4\pi\tilde{q}}2\tilde{\omega} & 0\le\tilde{\omega}\le\tilde{q}-\frac{\tilde{q}}{2}^2 \end{matrix}\right\} \tag{5.109b}$$

where

$$\tilde{q}\equiv\frac{q}{k_F}\,, \qquad \tilde{\omega}\equiv\frac{\omega m}{k_F^2}\,. \tag{5.109c}$$

The imaginary part, which specifies the inclusive cross section, Eq. (5.104), for a non-interacting Fermi gas has a very simple physical interpretation: it simply counts the ways an occupied state $|\vec{k}|<k_F$ can be scattered to an unoccupied state $|\vec{k}+\vec{q}|>k_F$ by transferring momentum $\vec{q}$ and energy $\omega=\epsilon_{k+q}-\epsilon_k=\frac{\vec{k}\cdot\vec{q}}{m}+\frac{q^2}{2m}$. For $q>2k_F$, the Fermi sphere conditions are satisfied for all $k<k_F$, so as sketched in Fig. 5.12a, the extremal values $\omega=\frac{q^2}{2m}\pm\frac{qk_F}{m}$ are obtained for $|\vec{k}|=k_F$ and $\vec{k}$ aligned with or opposite to $\vec{q}$. The maximum occurs for $\vec{k}$ perpendicular to $\vec{q}$ corresponding to $\omega=\frac{q^2}{2m}$ since all transverse values of $|\vec{k}|<k_F$ can contribute in this case. For $q<2k_F$, the result is sketched in Fig. 5.12b. Whereas at large ω, one has the same parabola as in

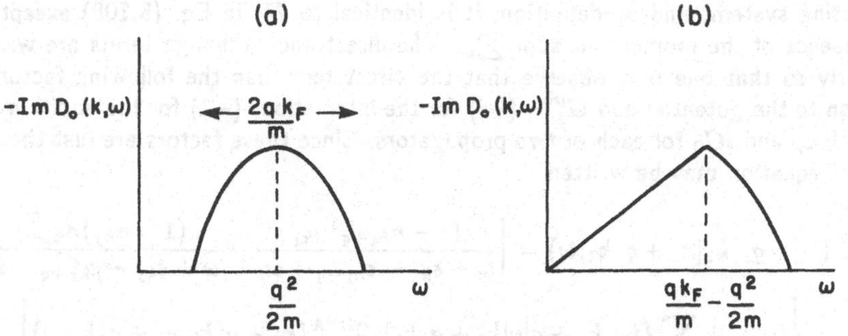

Fig. 5.12 Imaginary part of the response function $D_0(k,\omega)$ for $q > 2k_F$ (a) and $q < 2k_F$ (b).

case (a), at $\omega = \frac{qk_F}{m} - \frac{q^2}{2m}$ the simultaneous requirements $|\vec{k}| < k_F$ and $|\vec{k}+\vec{q}| > k_F$ further restrict the phase space, yielding the linear behavior shown for lower ω.

We may now sum the sequence of chain diagrams contributing to the response function by writing an integral equation which iterates the addition of a single link. Let $G^{\text{ph}}(k_1+q,\, k_1|k_2+q,\, k_2;\, \omega)$ denote the particle-hole Green's function obtained by separating the pairs of creation and annihilation operators which correspond to the external density operator in $D(q,\omega)$:

$$G^{\text{ph}}(k_1+q, k_1|k_2+q, k_2;\omega) \equiv$$

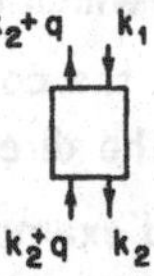

$$\equiv -i\int d(t_1-t_2)e^{i\omega(t_1-t_2)}\{\langle Ta^\dagger_{k_1}(t_1)a_{k_1+q}(t_1)a^\dagger_{k_2+q}a_{k_2}(t_2)\rangle - \langle a^\dagger_{k_1}a_{k_1+q}\rangle\langle a^\dagger_{k_2+q}a_{k_2}\rangle\} \quad (5.110)$$

with the property

$$D(q,\omega) = \sum_{k_1k_2} G^{\text{ph}}(k_1+q, k_1|k_2+q, k_2;\omega) \ . \quad (5.111)$$

Here and throughout this section where no ambiguity will arise, we abbreviate momentum integrals and spin sums by $\sum_k = \int \frac{d^3k}{(2\pi)^3}\sum_\sigma$ and neglect the normalization volume $\mathcal{V}$ which cancels out of all observables.

Chain diagrams are summed by the following integral equation which yields the particle-hole Green's functions in the random-phase approximation

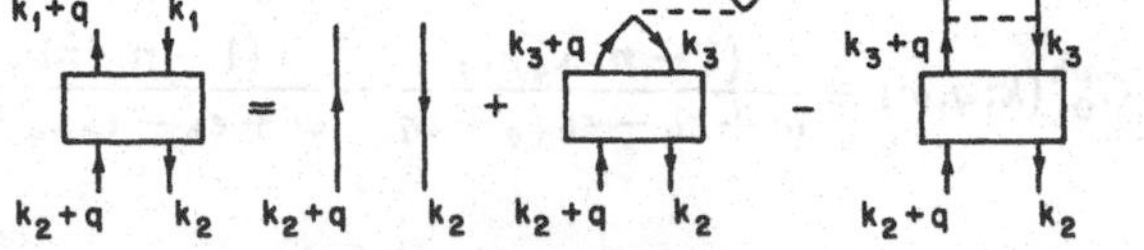

(5.112a)

The first term on the right-hand side is the particle-hole Green's function for the non-interacting system, and by definition, it is identical to D_0 in Eq. (5.108) except for the absence of the momentum sum $\sum_k$. The direct and exchange terms are written explicitly so that one may observe that the direct term has the following factors in addition to the potential and $G_\Lambda^{\rm ph}$: $(-i)$ for the interaction, (-1) for the addition of a closed loop and iG_0 for each of two propagators. Since these factors are just those in D_0, the equation may be written

$$G^{\rm RPA}(k_1+q,\ k_1|k_2+q,k_2;\omega) = \left[\frac{(1-n_{k_1+q})n_{k_1}}{\omega+\epsilon_{k_1}-\epsilon_{k_1+q}+i\eta} - \frac{(1-n_{k_1})n_{k_1+q}}{\omega+\epsilon_{k_1}-\epsilon_{k_1+q}-i\eta}\right]$$
$$\times\left[\delta_{k_1k_2} + \sum_{k_3}\{k_3,k_1+q|v|k_3+q,k_1\}\,G^{\rm RPA}(k_3+q,k_3|k_2+q,k_2;\omega)\right] \quad . \tag{5.112b}$$

This equation clearly has the general structure of the integral equation discussed in Section 2.4 and in fact corresponds to the Bethe Salpeter equation in the particle-hole channel with the vertex function approximated by the bare interaction. We will return to the more general case in Chapter 6 in connection with Landau Fermi liquid theory.

The direct matrix element in Eq. (5.112b) only depends on the momentum transfer q, $\langle k_3,k_1+q|v|k_3+q,k_1\rangle = \tilde{v}(-q)$ whereas the exchange matrix element $\tilde{v}(k_3-k_1)$ depends on the internal momenta. Hence, when it is physically justified to neglect the exchange term, the RPA equation becomes a simple algebraic equation. There are several circumstances under which the exchange term may be neglected. First, consider the Coulomb potential, for which the direct and exchange terms are $\frac{e\pi e^2}{q^2}$ and $\frac{4\pi e^2}{|k_1-k_3|^2}$, respectively. For small q, the conditions that $|k_i+q| > k_F$ and $|k_i| < k_F$ or vice versa force $|k_1| \sim k_F$ so the direct term generally dominates the exchange term by the factor $\left(\frac{k_F}{q}\right)^2$. A second example is use of the high-spin technique to treat Bosons, in which case the direct term dominates the exchange term by the factor N. A specific application to the correlation energy of the dilute Bose gas is given in Problem 5.7. Finally, in any system such that the momentum dependence of the potential is negligible between 0 and $2k_F$, the exchange term may be effectively included with the direct term. An example is the response function for the one-dimensional δ-function interaction in Problem 5.8.

The sum of all direct RPA chain diagrams is obtained by multiplying Eq. (5.112b) by $\tilde{v}(q)$, summing over k, and dropping all exchange terms, with the result

$$G^{\rm RPA}(k_1+q,k_1|k_2+q,k_2;\omega) = G_0^{\rm ph}(k_1;q,\omega)\left[\delta_{k_1k_2} + \frac{\tilde{v}(q)G_0^{\rm ph}(k_2;q,\omega)}{1-\tilde{v}(q)D_0(q,\omega)}\right] \tag{5.113}$$

where we have denoted the particle-hole Green's functions for the non-interacting system, which is the first term in Eq. (5.112b), by

$$G_0^{\rm ph}(k;q,\omega) \equiv \frac{(1-n_{k+q})n_k}{\omega+\epsilon_k-\epsilon_{k+q}+i\eta} - \frac{(1-n_k)n_{k+q}}{\omega+\epsilon_k-\epsilon_{k+q}-i\eta} \tag{5.114a}$$

and

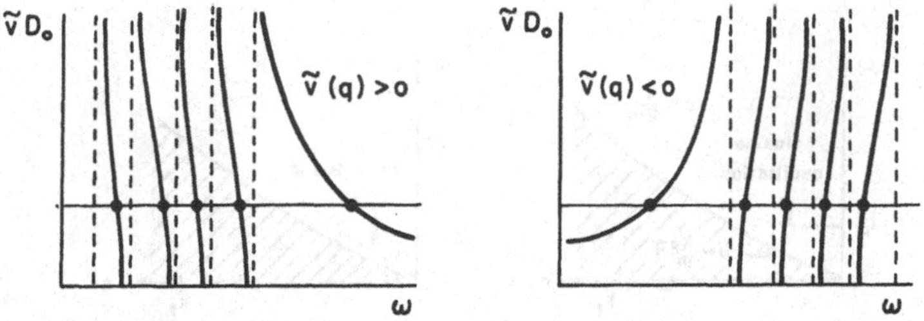

Fig. 5.13 Graphical solution for the RPA modes. The vertical asymptotes correspond to the particle-hole excitations $\omega = \epsilon_{k+q} - \epsilon_q$ and the curves show the qualitative behavior of $\tilde{v}(q)D_0(q,\omega)$ for the repulsive and attractive potentials, respectively.

$$D_0(q,\omega) = \sum_k G_0^{\mathrm{ph}}(k;q,\omega) \quad . \tag{5.114b}$$

Since the poles of G^{ph} (or D which is obtained by summing over k_1 and k_2) give the excited states of the system, in the random-phase approximation the excited states occur at ω such that

$$\tilde{v}(q)D_0(q,\omega) = 1 \quad . \tag{5.115}$$

To understand the structure of Eq. (5.115), it is useful to consider the graphical solutions sketched in Fig. 5.13. Since $D_0(q,\omega)$ is symmetric in ω (see Problem 5.6), it suffices to consider positive ω. For every value of $\vec{k}$ inside the Fermi sea such that $\vec{k}+\vec{q}$ is outside the Fermi sea, the first term in Eq. (5.114a) has a pole at $\omega = \epsilon_{k+q} - \epsilon_k$. For each $\vec{k}$ in the sum contributing to $D_0(q,\omega)$, asymptotes are plotted in Fig. 5.13, and $D_0(q,\omega)$ is a decreasing function between each pair of asymptotes. Thus, for a repulsive potential the solution to (5.115) is as sketched in the left part of Fig. 5.13. All except one of the solutions are trapped between states in the particle-hole continuum, and the remaining collective state is pushed above the highest state. Similarly, for an attractive potential, the collective state is brought down below the lowest particle-hole state.

In order to study the collective RPA modes in several physical systems at low q, it is useful to note the following behavior of $D_0(q,\omega)$ from Eq. (5.109)

$$\mathrm{Re}\, D_0(q,\omega) \xrightarrow[q\to 0]{} \begin{cases} \frac{k_F^3}{3\pi^2 m}\frac{q^2}{\omega^2} & \omega \text{ fixed} \\ \frac{mk_F}{\pi^2}\left(-1+\frac{s}{2}\ln\left|\frac{1+s}{1-s}\right|\right) & s=\frac{m\omega}{k_F q} \text{ fixed} \end{cases} \tag{5.116a}$$

$$\mathrm{Im}\, D_0(q,\omega) \xrightarrow[q\to 0]{} \begin{cases} 0 & \omega \text{ fixed} \\ 0 & s \text{ fixed} > 1 \\ -\frac{mk_F s}{2\pi} & s \text{ fixed} < 1 \end{cases} \quad . \tag{5.116b}$$

As the first example, consider the electron gas. Because of the $\frac{1}{q^2}$ behavior of the

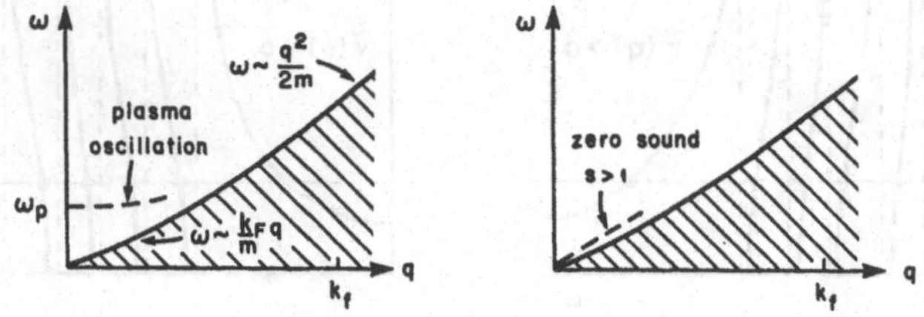

Fig. 5.14 Comparison of plasma oscillation and zero sound mode with the particle-hole continuum.

Coulomb force, we must take the limit in Eq. (5.116a) with finite ω, so that

$$\frac{4\pi e^2}{q^2}\,\frac{k_F^3}{3\pi^2 m}\,\frac{q^2}{\omega^2} = 1 \tag{5.117a}$$

yielding the familiar plasma frequency (see Problem 5.9)

$$\omega_p^2 = \frac{4\pi e^2}{m}\rho\ . \tag{5.117b}$$

The collective RPA mode, or plasmon, is a fundamental excitation of the electron gas which is readily observable experimentally. For example, if one measures the energy loss of electrons in solids, peaks are observed at multiples of ω_p. Since typical Fermi energies in metals are of the order $\epsilon_F \sim 5 - 20\,\text{eV}$, the plasma frequency is in the range $\omega_p \sim 7 - 27\,\text{eV}$. The spectrum of particle-hole excitations within the extremal values $\omega = \frac{q^2}{2m} \pm \frac{qk_F}{m}$ is compared with the plasma frequency in Fig. 5.14. Note that since ω_p is comparable to ϵ_F, the plasma mode is well-isolated from the particle-hole continuum and thus doesn't decay directly through one-particle one-hole states. The leading correction to ω_p, which we have not calculated here, is proportional to q^2

In contrast to the case of the repulsive Coulomb potential, many physical systems have scattering channels in which the interaction is attractive at the Fermi surface, in which case the collective mode is lowered in energy. Consider, for example the case of liquid ^{3}He. On the average, the interaction is repulsive at the Fermi surface since it is the short-range repulsive core of the two-body potential which keeps the liquid from collapsing to higher density. However, although there is negligible explicit spin dependence in the underlying potential, because two ^{3}He atoms in a spin triplet state must be spatially antisymmetric, they feel the hard core much less strongly than two atoms in a singlet state, and the interaction in the triplet channel is overall attractive. Thus, the spin-dependence tends to drive the system toward a ferromagnetic state and manifests itself in a low-energy collective RPA mode known as a paramagnon. In the limit in which the attraction between like spins were strong enough to drive the system ferromagnetic, this mode would occur at zero energy. Although it is not pushed down that far, it makes a significant contribution to the self-energy. In addition to the

Hartree-Fock and second order diagrams discussed in Section 5.3, RPA graphs such as

$$\Sigma_{\text{RPA}} =$$

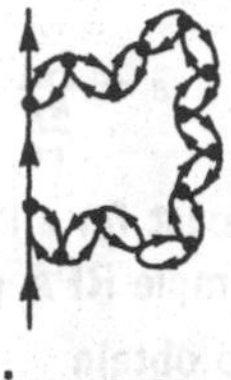

(5.118)

make a dramatic contribution to the effective mass. In the pressure range of 0.3 to 27 atm., $\frac{m^*}{m}$ ranges from 3.1 to 5.8, corresponding to the fact that in order to move a single ^{3}He atom through the medium, one must also drag along a large cluster of atoms with the same spin.

A second example is the case of pion condensation in neutron or nuclear matter at high density (occurring for example in neutron stars). Whereas the ^{3}He—^{3}He interaction is attractive in the $\sigma \cdot \sigma$ channel, by virtue of the spin-isospin coupling of pions to nucleons and the resulting strong tensor forces, the nucleon-nucleon interaction at the Fermi surface is strongly attractive in the $\sigma \cdot \sigma\, \tau \cdot \tau$ channel, giving rise to low energy collective excitations with the quantum numbers of the pion. Although uncertainties in the behavior of strong interactions at high density makes quantitative predictions uncertain, this mode is predicted to have zero energy and thus to lead to pion condensation at a density several times the density of nuclear matter.

ZERO SOUND

We now consider finite-range repulsive potentials, for which the low-q behavior of the collective mode is completely different than for the Coulomb potential. Since non-Coulombic potentials are less singular than $\frac{1}{q^2}$ at low q, the dispersion relation $\tilde{v}D = 1$ cannot be satisfied for a finite ω, so $\omega \to 0$ as $q \to 0$. Thus, modes which go to zero at large wavelength are the general rule, and the finite energy modes such as the plasmon are exceptional cases associated with the infinite range of the force. Note that we have already seen an example of the role of the force range in considering spin waves. Since orienting different spin domains in different directions costs no energy except at the domain boundaries, if the force range is finite, the effect of the boundaries becomes negligible in the long wavelength limit and the energy of such a Goldstone mode must go to zero. Only if the force range is infinite do the boundary contributions remain finite, yielding a finite excitation energy.

To simplify the discussion of the collective mode, which is called zero sound, let us consider the case in which $\tilde{v}(q)$ is independent of q at small q and define

$$f \equiv \tilde{v}(0) = \int d^3r\, v(r) \quad . \tag{5.119}$$

We now take the small q limit of the dispersion relation holding s fixed and greater than 1 to obtain real (undamped) solutions

$$\begin{aligned} \frac{1}{f} &= D_0(q,\omega) \\ &\xrightarrow[q\to 0]{} \frac{mk_F}{\pi^2}\left(-1 + \frac{s}{2}\ln\left|\frac{1+s}{1-s}\right|\right) \end{aligned} \tag{5.120}$$

where s defined in Eq. (5.116) is the ratio of the speed of the zero sound mode, $c_0 = \frac{\omega}{q}$, to the Fermi velocity, v_f,

$$s = \frac{\frac{\omega}{q}}{\frac{k_f}{m}} = \frac{c_0}{v_F} \quad . \tag{5.121}$$

Note that solutions with $s > 1$ exist for all positive f, that is, for all repulsive interactions. For large f, (where the simple RPA presented here is an inadequate approximation) we may expand $\ln\left|\frac{1+s}{1-s}\right|$ to obtain

$$\frac{1}{f} \approx \frac{mk_F}{\pi^2}\left[\frac{1}{3s^2} + \frac{1}{5s^4} + \cdots\right] \tag{5.122a}$$

so that

$$s \xrightarrow[f\to\infty]{} \left[\frac{mk_F}{3\pi^2} f\right]^{1/2} . \tag{5.122b}$$

In the limit of small f, s approaches 1 exponentially

$$s \xrightarrow[f\to 0]{} 1 + 2e^{-\frac{2\pi^2}{mk_F f}} \tag{5.122c}$$

and the velocity of zero sound is therefore close to, but slightly above, the Fermi velocity

$$c_0 \approx \frac{k_F}{m} . \tag{5.123}$$

The zero sound mode is sketched in part (b) of Fig. 5.14. Note that the fact $s > 1$ places the mode above the particle-hole continuum and prevents its direct decay through coupling to one-particle, one-hole states.

For comparison, compare this velocity with ordinary thermodynamic sound in a nearly non-interacting Fermi gas in d dimensions in which there are sufficient interactions to produce thermodynamic equilibrium, but these interactions contribute negligibly to the energy per particle. Noting that the Fermi gas energy per particle in d dimensions is $\frac{d}{d+2}\frac{k_F^2}{2m}$ and $\rho \propto k_F^d$, the usual thermodynamic arguments yield the velocity of thermodynamic, or first sound*

$$c_1 = \left[\frac{1}{m}\frac{dP}{d\rho}\right]^{1/2} = \frac{1}{\sqrt{d}}\frac{k_F}{m} . \tag{5.124}$$

Thus, in a weakly interacting system in three dimensions, c_1 is approximately $\frac{1}{\sqrt{3}}c_0$. Furthermore, at least the velocity suggests that the geometry of zero sound may be more one-dimensional than three dimensional, and we will now develop an appropriate language to examine the geometry of zero sound.

---

* For an infinitesimal change in density, the continuity equation requires $\frac{\partial}{\partial t}\delta\rho + \rho_0 \nabla \cdot v = 0$ and Newton's equation of motion requires $m\rho_0\frac{\partial v}{\partial t} = -\nabla P$. Hence, the wave equation is $\frac{\partial^2}{\partial t^2}\delta\rho = \frac{1}{m}\nabla^2 P(\rho_0 + \delta\rho) = \frac{1}{m}\frac{\partial P}{\partial \rho}\nabla^2\delta\rho$ and $c^2 = \frac{1}{m}\frac{dP}{\delta\rho}$.

Since zero sound is a coherent superposition of particle-hole excitations near the Fermi surface, we would like to simultaneously examine the behavior in momentum space and coordinate space. If we were dealing with a classical system, it would be natural to study its classical distribution function as a function of coordinates and momenta. For a quantal system, the best analog of the classical distribution function is the Wigner function and for our present purposes it is most useful to study the Wigner transform of the one-body density matrix:

$$f(\vec{p},\vec{R}) = \int d^3r\, e^{-i\vec{p}\cdot\vec{r}} \left\langle \psi^\dagger\left(\vec{R}-\tfrac{\vec{r}}{2}\right)\psi\left(\vec{R}-\tfrac{\vec{r}}{2}\right)\right\rangle \quad . \tag{5.125}$$

In addition to yielding the familiar distribution function in the classical limit, its moments have the following properties associated with the classical distribution function:

$$\begin{aligned}
\int \frac{d^3p}{(2\pi)^3} f(\vec{p},\vec{R}) &= \langle\psi^\dagger(\vec{R})\psi(\vec{R})\rangle = \langle\hat{\rho}(\vec{R})\rangle \\
\int d^3R\, f(\vec{p},\vec{R}) &= \int d^3r\, d^3R\, e^{-i\vec{p}(\vec{R}+\frac{\vec{r}}{2})}\, e^{i\vec{p}\cdot(\vec{R}-\frac{\vec{r}}{2})}\langle\psi^\dagger\left(\vec{R}-\tfrac{\vec{r}}{2}\right)\rho\left(\vec{R}+\tfrac{\vec{r}}{2}\right)\rangle \\
&= \langle a_p^\dagger a_p\rangle = \langle\hat{\rho}_p\rangle \qquad\qquad (5.126)\\
\int \frac{d^3p}{(2\pi)^3}\frac{\vec{p}}{m} f(\vec{p},\vec{R}) &= \int \frac{d^3p}{(2\pi)^3} d^3r\, e^{-i\vec{p}\cdot\vec{r}} \left\langle \frac{1}{im}\vec{\nabla}_R\left\{\psi^\dagger\left(\vec{R}-\tfrac{\vec{r}}{2}\right)\psi\left(\vec{R}+\tfrac{\vec{r}}{2}\right)\right\}\right\rangle \\
&= \frac{1}{2mi}\left\langle\psi^\dagger(\vec{R})\vec{\nabla}_R\psi(R) - \left(\nabla_R\psi^\dagger(\vec{R})\right)\psi(\vec{R})\right\rangle \\
&= \langle\hat{\jmath}(R)\rangle
\end{aligned}$$

and similarly for higher moments. Essentially the only property $f(\vec{r},\vec{R})$ does not share with the classical distribution function is positivity, and this deficiency will not affect the present argument.

Now, let us use the RPA response function to look at the response of the Wigner transform of the one-body density operator to a weak perturbation $\delta U(t_2)\rho_q(t_2)$ which couples to the zero sound mode. Using Eq. (2.15) and omitting the standard details in going from the retarded to time-ordered response function, which are irrelevant to the present point,

$$\begin{aligned}
\delta f(\vec{p},\vec{R};t) &= -i\int dt_2 \delta U(t_2)\int dr\, e^{-i\vec{p}\cdot\vec{r}}\langle T\psi^\dagger\left(\vec{R}-\frac{\vec{r}}{2},t_1\right)\psi\left(\vec{R}+\frac{\vec{r}}{2},t_1\right)\times \\
&\quad \times a^\dagger_{\vec{k}_2+\vec{q}}(t_2)a_{\vec{k}_2}(t_2)\rangle \\
&= -i\int dt_2\delta U(t_2)\int dr\, e^{-i\vec{p}\cdot\vec{r}}\sum_{k_1k'} e^{-i\vec{k}_1\cdot(\vec{R}-\frac{\vec{r}}{2})}\, e^{i\vec{k}'(\vec{R}+\frac{\vec{r}}{2})} \qquad (5.127)\\
&\quad \times \delta(\vec{k}_1+\vec{q}-\vec{k}')\left\langle Ta^\dagger_{\vec{k}_1}(t_1)a_{\vec{k}_1+\vec{q}}(t_1)a^\dagger_{\vec{k}_2+\vec{q}}(t_2)a_{\vec{k}_2}(t_2)\right\rangle \quad .
\end{aligned}$$

Noting that the zero sound pole arises from the second term in Eq. (5.113) and that the only dependence on k_1 enters through $G_0^{\mathrm{ph}}(k_1,q,\omega)$, substitution of the RPA response

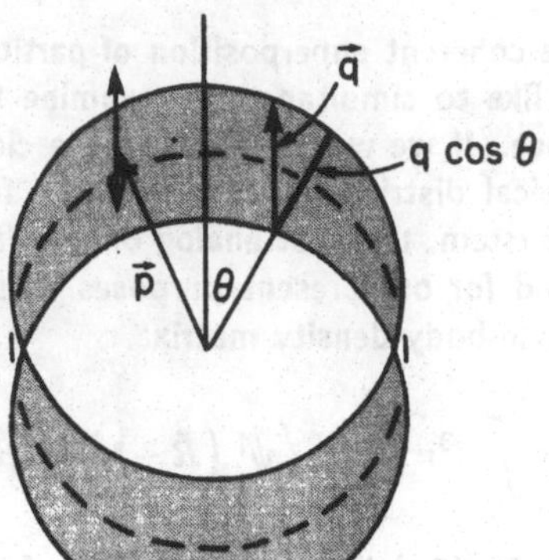

Fig. 5.15 Phase space contributing to $G_0^{\text{ph}}\left(\vec{p}-\frac{\vec{q}}{2}\right)$. The centers of the two solid spheres are displayed $\pm\frac{\vec{q}}{2}$ relative to the dashed Fermi sphere. Every vector $\vec{p}$ in the upper shaded region yields $\vec{p}+\frac{\vec{q}}{2}$ outside the dashed sphere and $\vec{p}-\frac{\vec{q}}{2}$ inside.

function in Eq. (5.127) yields

$$\delta f(\vec{p},\vec{R};\omega) \propto \int dr\, e^{-i\vec{p}\cdot\vec{r}} \sum_{k_1} e^{-i\vec{k}_1\cdot(\vec{R}-\frac{\vec{r}}{2})}\, e^{i(\vec{k}_1+\vec{q})\cdot(\vec{R}+\frac{\vec{r}}{2})} G_0^{\text{ph}}(\vec{k}_1)$$

$$= e^{i\vec{q}\cdot\vec{R}} G_0^{\text{ph}}\left(\vec{p}-\frac{\vec{q}}{2}\right) \tag{5.128}$$

$$= e^{i\vec{q}\cdot\vec{R}} \frac{(1-n_{\vec{p}+\frac{\vec{q}}{2}})n_{\vec{p}-\frac{\vec{q}}{2}} - (1-n_{\vec{p}-\frac{\vec{q}}{2}})n_{\vec{p}+\frac{\vec{q}}{2}}}{\omega+\epsilon_{\vec{p}-\frac{\vec{q}}{2}}-\epsilon_{\vec{p}+\frac{\vec{q}}{2}}} .$$

For small $\vec{q}$, the range of $\vec{p}$ which satisfy the conditions that $\vec{p}\pm\frac{\vec{q}}{2}$ be outside the Fermi sphere while $p\mp\frac{\vec{q}}{2}$ be inside can be calculated easily from the geometrical construction shown in Fig. 5.15. The shaded region defining the vectors $\vec{p}$ such that $\vec{p}\pm\frac{\vec{q}}{2}$ lies outside the dashed Fermi sphere while $\vec{p}\mp\frac{\vec{q}}{2}$ lies inside have radial thickness $q\cos\theta$ where θ is the angle between $\vec{p}$ and $\vec{q}$. Since the p's are all confined to be close to the Fermi surface, we may write the phase space factor as $\delta(k_F-|\vec{p}|)q\cos\theta$. Finally, writing the denominator as $\omega+\epsilon_{p-\frac{q}{2}}-\epsilon_{p+\frac{q}{2}}=\frac{sqk_F}{m}-\frac{qk_F\cos\theta}{m}$ and writing the explicit time-dependence, we obtain the desired result.

$$\delta f(\vec{p},\vec{R},t) \propto e^{i(\vec{q}\cdot\vec{R}-\omega t)} \frac{\delta(k_F-|\vec{p}|)\cos\theta}{s-\cos\theta} . \tag{5.129}$$

This result is also obtained using Landau Fermi Liquid Theory in Chapter 6. The factor $\frac{\cos\theta}{s-\cos\theta}$ specifies the change in the distribution function as a function of the angle between $\vec{p}$ and $\vec{q}$. Since s is close to 1, the magnitude of the peak at $\theta=0$

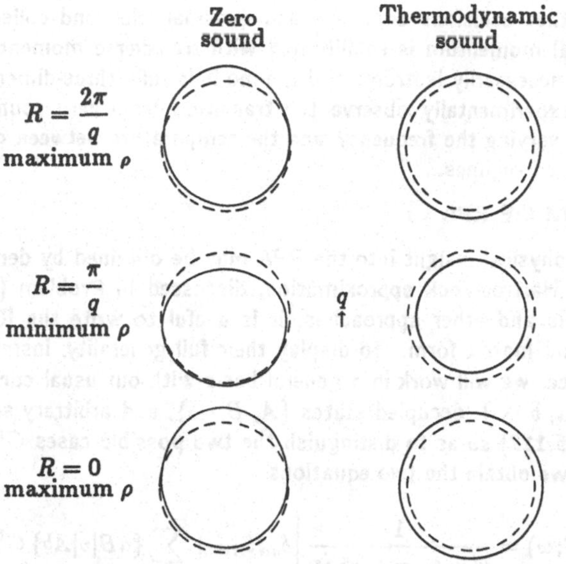

Fig. 5.16 The distribution function $f(p, R)$ for zero sound and thermodynamic sound. The equilibrium Fermi sphere is shown at each spatial position by the dashed line and the solid line denotes the momentum distribution of the propagating mode.

is much larger than at $\theta = \pi$, so the volume of the Fermi sphere and thus the local density changes.

Figure 5.16 compares the evolution of the distribution function for zero sound with that of thermodynamic sound. The factor $e^{i(\vec{q}\cdot\vec{R}-\omega t)}$ propagates the fluctuation of the zero sound distribution function in the $\vec{q}$ direction at speed $v_0 = s\frac{k_F}{m}$. Snapshots are shown of the momentum distribution at spatial points R separated by a half wavelength. The origin is selected such that the density is a maximum at $R = 0$, so the Fermi sphere distorts at successive displacements of $\frac{\pi}{q}$ by adding or subtracting $\frac{\cos\theta}{s-\cos\theta}$. The distribution is thus highly anisotropic and is concentrated in the forward $\vec{q}$ direction. In contrast, the Fermi sphere for thermodynamic sound just dilates and contracts spherically symmetrically. To the extend to which the zero sound distortion is concentrated in the longitudinal direction and the transverse radius of the Fermi surface is unaffected, the mode is more like one-dimensional thermodynamic sound than three-dimensional thermodynamic sound, consistent with Eq. (5.124).

Zero sound and thermodynamic sound thus represent two opposite extremes. Zero sound is built out of quasiparticles, so at finite temperatures the frequency must be high enough that the quasiparticles do not decay, that is, that collisions do not dominate. Quasiparticles travelling in the $\vec{q}$ direction are thus not scattered in the transverse directions and the mode can remain essentially one-dimensional. At zero temperature, for which we have performed our calculation, the lifetime becomes infinite as quasiparticles approach the Fermi surface, so zero sound propagates at all frequencies. In contrast, thermodynamic sound requires local equilibrium. Hence, the frequency must be low enough that collisions dominate and quasiparticles and collective modes decay. Longitudinal momentum is equilibrated with transverse momentum, so that the Fermi surface is necessarily isotropic and the mode is fully three-dimensional. In liquid ^{3}He, one may experimentally observe the transition from zero sound to thermodynamic sound by varying the frequency and the temperature between collision-free and collision-dominated regimes.

MATRIX FORM OF RPA

Additional physical insight into the RPA may be obtained by deriving it from the time-dependent Hartree-Fock approximation, discussed in Problem (5.10). To make contact with this and other approaches, it is useful to write the RPA equations in their conventional matrix form. To display their full generality, instead of working in momentum space, we will work in a general basis with our usual convention of unoccupied states $\{a, b \cdots\}$, occupied states $\{A, B \cdots\}$, and arbitrary states $\{\alpha, \beta \cdots\}$. Rewriting Eq. (5.112) so as to distinguish the two possible cases $G^{\mathrm{RPA}}(aA|\alpha\beta)$ and $G^{\mathrm{RPA}}(Aa|\alpha\beta)$ we obtain the two equations

$$G^{\mathrm{RPA}}(aA|\alpha\beta;\omega) = \frac{1}{\omega+\epsilon_A-\epsilon_a+i\eta}\left[\delta_{a\alpha}\delta_{A\beta}+\sum_{bB}\{aB|v|Ab\}\,G^{\mathrm{RPA}}(bB|\alpha\beta;\omega) + \sum_{bB}\{ab|v|AB\}\,G^{\mathrm{RPA}}(Bb|\alpha\beta;\omega)\right] \tag{5.130}$$

$$G^{\mathrm{RPA}}(Aa|\alpha\beta;\omega) = \frac{-1}{\omega+\epsilon_a-\epsilon_A-i\eta}\left[\delta_{A\alpha}\delta_{a\beta}+\sum_{bB}\{AB|v|ab\}\,G^{\mathrm{RPA}}(bB|\alpha\beta;\omega) + \sum\{Ab|v|aB\}\,G^{\mathrm{RPA}}(Bb|\alpha\beta;\omega)\right] .$$

With the definitions

$$\begin{aligned} A_{AaBb} &= (\epsilon_a-\epsilon_A)\delta_{AB}\delta_{ab}+\{aB|v|Ab\} \\ B_{AaBb} &\equiv \{ab|v|AB\} \end{aligned} \tag{5.131}$$

these equations may be written in the matrix form

$$\begin{pmatrix} -\omega+A & B \\ B^* & \omega+A^* \end{pmatrix}\begin{pmatrix} G(bB|\alpha\beta;\omega) \\ G(Bb|\alpha\beta;\omega)\end{pmatrix} = \begin{pmatrix} -\delta_{a\alpha}\delta_{A\beta} \\ -\delta_{A\alpha}\delta_{a\beta}\end{pmatrix} . \tag{5.132}$$

The poles of G are given by the eigenvalues of the generalized eigenvalue problem (note the minus sign on the right-hand side):

$$\begin{pmatrix} A & B \\ B^* & A^* \end{pmatrix}\begin{pmatrix} X^{(\nu)} \\ Y^{(\nu)}\end{pmatrix} = E^{(\nu)}\begin{pmatrix} X^{(\nu)} \\ -Y^{(\nu)}\end{pmatrix} \tag{5.133}$$

where $X^{(\nu)}$ is a column of elements $X^{(\nu)}_{bB}$ labeled by particle-hole labels $\{b, B\}$. Using the properties established in Problem (5.11), the particle-hole Green's function may be written as a sum over all positive energy modes

$$G^{\mathrm{RPA}} = \sum_{\nu} \frac{\begin{pmatrix} X^{(\nu)} \\ Y^{(\nu)} \end{pmatrix} (X^{(\nu)\dagger} Y^{(\nu)\dagger})}{\omega - E^{(\nu)} + i\delta} - \frac{\left[\begin{pmatrix} Y^{(\nu)} \\ X^{(\nu)} \end{pmatrix} (Y^{(\nu)\dagger} X^{(\nu)\dagger})\right]^*}{\omega + E^{(\nu)} - i\delta} \quad . \tag{5.134}$$

Since this is a spectral representation for the RPA response function of the same form as Eq. (5.100) for the exact response function, the $X^{(\nu)}$ and $Y^{(\nu)}$ should correspond to matrix elements of $a^\dagger_\alpha a_\beta$ between RPA approximates to the excited states, denoted $|\psi^{\mathrm{RPA}}_\nu\rangle$, and the RPA ground state, denoted $|\psi^{\mathrm{RPA}}_0\rangle$. Noting that the upper component $G^{\mathrm{RPA}}(bB|\alpha\beta)$ involves the matrix element $\langle a^\dagger_B a_b a^\dagger_\alpha a_\beta\rangle$ and $G^{\mathrm{RPA}}(Bb|\alpha\beta)$ corresponds to $\langle a^\dagger_b a_B a^\dagger_\alpha a_\beta\rangle$, it follows that

$$\begin{aligned} \left(X^{(\nu)}_{bB}\right)^* &= \langle\psi^{\mathrm{RPA}}_\nu|a^\dagger_b a_B|\psi^{\mathrm{RPA}}_0\rangle \\ Y^{(\nu)}_{bB} &= \langle\psi^{\mathrm{RPA}}_\nu|a^\dagger_B a_b|\psi^{\mathrm{RPA}}_0\rangle \quad . \end{aligned} \tag{5.135}$$

If $|\psi^{\mathrm{RPA}}_0\rangle$ were just the non-interacting ground state, $a^\dagger_B a_b$ would annihilate it yielding $Y^{(\nu)} = 0$. However, from Problems (5.5) and (5.10) it is clear that the RPA ground state contains multi particle-hole admixtures which produce a non-vanishing $Y^{(\nu)}$ amplitude. Equations (5.134– 5.135) thus relate the two distinct random phase approximations alluded to earlier for the response function and for individual eigenstates.

SUM RULES AND EXAMPLES

We conclude this section by stating several sum rules satisfied by the response function and presenting illustrative physical examples.

In addition to dispersion relations of the form of (5.35) which follow directly from the spectral representation of D, the imaginary part of the response function satisfies two important sum rules. The first follows from integrating out the energy conserving δ-function and using completeness and the field operator commutation relation

$$\begin{aligned} \int d\omega \,\mathrm{Im}\, D(x_1, x_2; \omega) &= -\pi \sum_n \int d\omega\, \delta(\omega - E^N_n + E_0) \\ &\quad \times \langle\psi_0|(\hat\rho(x_1) - \langle\hat\rho(x_1)\rangle)|\psi_n\rangle\langle\psi_n|(\hat\rho(x_2) - \langle\hat\rho(x_2)\rangle)|\psi_0\rangle \\ &= -\pi\big[\langle\rho(x_1)\rho(x_2)\rangle - \langle\rho(x_1)\rangle\langle\rho(x_2)\rangle\big] \\ &= -\pi[g(x_1, x_2) + \delta(x_1 - x_2)\rho(x_1)] \end{aligned} \tag{5.136}$$

where $g(x_1, x_2)$ is the two-body correlation function

$$g(x_1, x_2) \equiv \langle\psi_0\,|\psi^\dagger(x_1)\psi^\dagger(x_2)\psi(x_2)\psi(x_1)|\,\psi_0\rangle - \langle\psi_0\,|\hat\rho(x_1)|\,\psi_0\rangle\,\langle\psi_0\,|\hat\rho(x_2)|\,\psi_0\rangle \quad . \tag{5.137}$$

In momentum space, where $g(x_1, x_2) = g(x_1 - x_2)$

$$\begin{aligned} -\frac{1}{\pi N}\int d\omega\, \mathrm{Im}\, D(q, \omega) &\equiv S(q) \\ &= \rho \int dx\, e^{-iq\cdot x} g(x) + 1 \quad . \end{aligned} \tag{5.138}$$

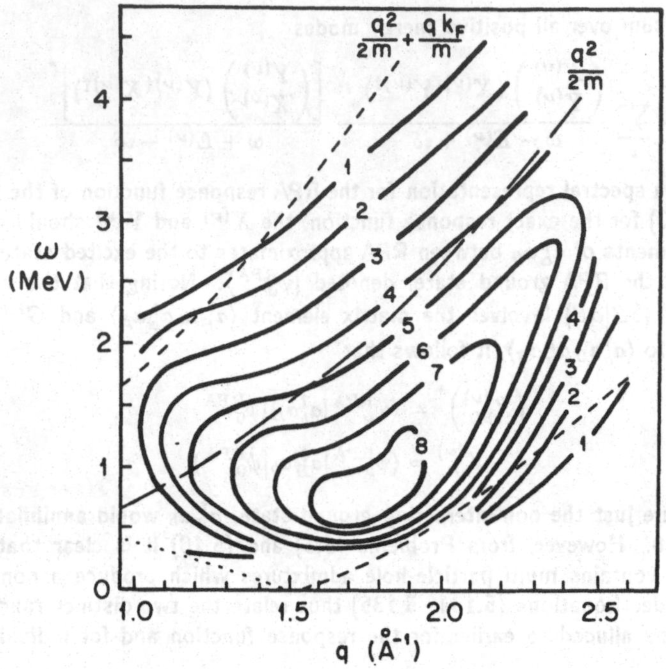

Fig. 5.17 **Sketch of cross section for inelastic neutron scattering from liquid ^{3}He based on the data of Stirling *et al* (1976). Solid contours denote the experimental cross sections, the long dashed curve indicates the Fermi gas maximum, $\frac{q^2}{2m}$, the short dashed curves show where the Fermi gas response goes to zero, $\frac{q^2}{2m} \pm \frac{qk_F}{m}$, and the dotted line denotes the point at which Pauli blocking begins.**

Since $\operatorname{Im} D(q, \omega)$ is directly measurable in inclusive scattering, Eq.. (5.138) shows that one may measure the Fourier transform of the two-body correlation function by simply integrating over ω.

An energy weighted sum rule is derived in Problem 5.13 by evaluating the expectation value of the double commutator $\langle [[H, \hat{\rho}_q], \rho_{-q}] \rangle$ with the result

$$-\frac{1}{\pi N} \int d\omega\, \omega \operatorname{Im} D(q, \omega) = \frac{q^2}{2m} \quad . \tag{5.139}$$

A variety of useful related sum rules may be obtained by choosing other operators in the double commutator instead of $\hat{\rho}_q$. For example, one may take the electric dipole transition operator and relate the integral of the dipole transition strength for a finite nucleus to the charge radius (plus corrections for velocity dependent and isospin dependent terms). This gives a total sum with which to compare the strength of an individual transition. The giant dipole resonance is "giant" because its strength exhausts most of the sum rule.

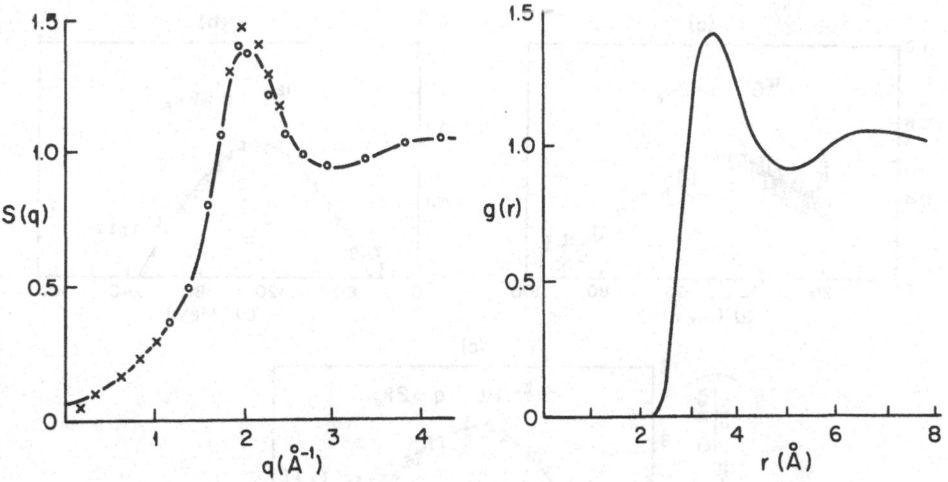

Fig. 5.18 Liquid structure function $S(q)$ and two-body correlation function $g(r)$ for liquid ^{4}He. The left-hand graph compares X-ray scattering data of Hallock (1972) and Rubkoff *et al.* (1979) (crosses) and neutron scattering data of Svensson *et al.* (1980) (circles) with the structure function obtained from Monte Carlo calculations (solid curves) of Kalos *et al.* (1981). The right-hand graph shows the two-body correlation function corresponding to this $S(q)$.

Experimental measurements of response functions for several physical systems are presented in Figs. 5.17 – 5.19. Fig. 5.17 shows a contour plot in the (q,ω) plane of the cross section for inelastic scattering of neutrons from liquid ^{3}He at 0.63° K (Stirling *et al.*, 1976). Since the neutron-^{3}He potential is effectively a δ-function on the scale of Angstroms relevant to the experiment, the cross section is directly proportional to $\mathrm{Im}\, D(q,\omega)$. Using the value of $k_F = 0.786$ Å^{-1} specified by the ground state density, the maximum $\frac{q^2}{2m}$ and zeros $\frac{q^2}{2m} \pm \frac{qk_F}{m}$ expected from $D_0(q,\omega)$ are indicated by dashed lines. Although the data has the qualitative structure of D_0, it differs quantitatively in detail. In particular, one sees clear evidence of a large enhancement in the effective mass. The experimental peak in the region of $q \sim 2k_F$ occurs at $\sim \frac{1}{3} \times \frac{q^2}{2m}$ so that $m^* \sim 3m$.

The structure factor for liquid ^{4}He, $S(q)$ defined in Eq. (5.138), has been measured both by X-ray scattering (Hallock, 1972; Robkoff *et al.* 1979) and neutron scattering (Svensson *et al.* 1980) Typical data are shown in Fig. 5.18, and are observed to agree quite well with each other and with the results obtained from Monte Carlo solution of the zero-temperature ground state of ^{4}He (Kalos *et al.* 1981) using the Aziz (1979) potential. As discussed in Chapter 8, the Monte Carlo solution provides an essentially exact solution for the ground state of N particles in a box with periodic boundary conditions, so the only error in the structure function is the departure from linearity at very low q associated with the finite box size. One thereby has an exceedingly pre-

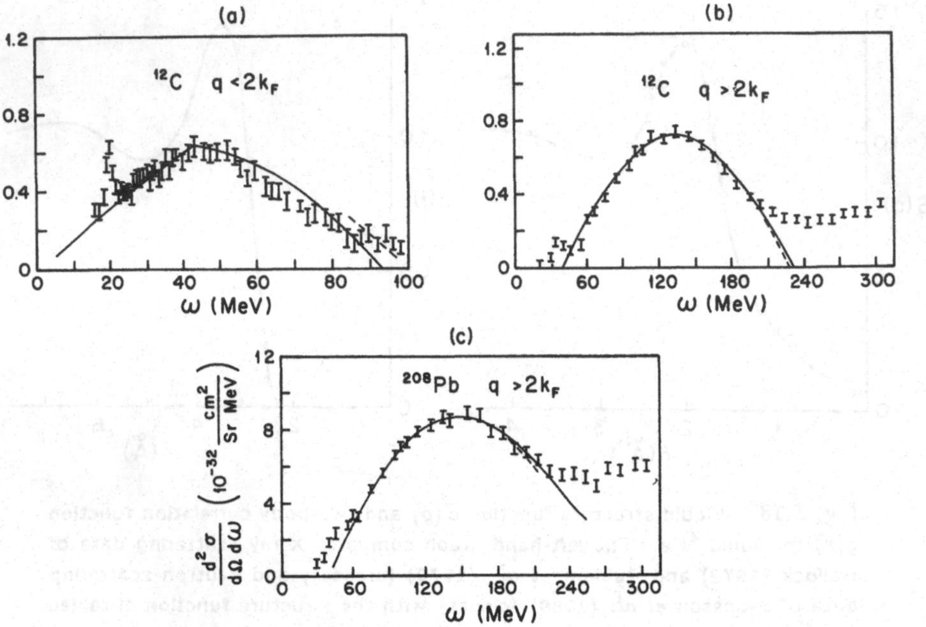

Fig. 5.19 Inclusive electron scattering from atomic nuclei. The inclusive cross sections as a function of energy loss ω for ^{12}C at incident energy $E = 198.5$ MeV and scattering angle $\theta = 135°$ in (a) correspond to momentum transfer q less than $2k_F$ whereas the $E = 500$ MeV data at $\theta = 60°$ shown for ^{12}C in (b) and ^{208}Pb in (c) correspond to momentum transfer greater than $2k_F$. The solid curves correspond to the Fermi gas response functions described in the text. The low energy data is from Leiss and Taylor (1963) and the high energy data from Moniz *et al.* (1971).

cise measurement of the two-body correlation function of a dense, strongly interacting quantum liquid, also shown in Fig. 5.18. One observes in particular the strong suppression of the wave function at short distance due to the strongly repulsive core of the He–He potential. Note that the strength of the interaction between the constituents in no way affects the applicability of the linear response analysis: all that matters is that the coupling of the external probe, in this case the X-ray or neutrons, is weak.

A final example is inelastic electron scattering from atomic nuclei, shown in Fig. 5.19. A good approximation to the response may be obtained using the response function for a non-interacting Fermi gas corrected for the medium dependence of the nucleon self-energy. Recall from Fig. 5.12 that when the momentum transfer $q > 2k_F$, there is no Pauli blocking and $\operatorname{Im} D_0(q,\omega)$ is an inverted parabola with center at $\omega = \frac{q^2}{2m}$ and width $\frac{2qk_F}{m}$. In contrast, when $q < 2k_F$, Pauli blocking yields linear dependence below $\omega = \frac{qk_F}{m} - \frac{q^2}{2m}$. This qualitative difference is observed in the low and high momentum transfer data for ^{13}C in (a) and (b) and the high momentum transfer data for ^{208}Pb in (c). Note that the excess cross section at high ω corresponds to the inelastic processes in which pions or Δ resonances are produced. In Section 5.3, we saw that the nuclear self-energy was strongly momentum dependent, with a nucleon within the Fermi sea having a strong attraction and nucleons well above the Fermi sea having much weaker attraction. This effect may be included approximately in the response function either by introducing an effective mass m^* or a binding correction $\bar{\epsilon}$ representing the average difference between the self-energy of a nucleon in and outside the Fermi sea. In either case, the data specify two parameters, k_F and either m^* or $\bar{\epsilon}$. The ^{12}C data of Leiss and Taylor (1960) in (a) are fit with $k_F = 1.19\,\mathrm{fm}^{-1}$ and $m^*/m = 0.7$ (Moniz 1969), consistent with the value m^*/m discussed in Section 5.3 and the fact that the average Fermi momentum in a small, surface dominated nucleus like ^{12}C is somewhat lower than the value $k_F = 1.31$ in fm^{-1} in bulk nuclear matter. The high momentum transfer data for ^{12}C in (b) is fit with $k_F = 1.14\,\mathrm{fm}^{-1}$ and $\bar{\epsilon} = 25\,\mathrm{MeV}$ (Moniz *et al.*, 1971) consistent with the previous value of k_F and the fact that the average of the observed single particle energies is of the order of 25 MeV. Finally, the same analysis of the ^{208}Pb data yields $k_F = 1.36\,\mathrm{fm}^{-1}$, in satisfactory agreement with bulk matter, and $\bar{\epsilon} = 44\,\mathrm{MeV}$ consistent with the average single-particle energy in ^{208}Pb. Thus, the Fermi gas response corrected for the known behavior of the self-energy gives a simple semi-quantitative physical description of the nuclear response function.

5.5 MAGNETIC SUSCEPTIBILITY OF A FERMI GAS

The magnetic susceptibility of a non-interacting Fermi gas is an instructive, illustrative application of many of the ideas presented in this chapter. Although the problem is in some aspects trivial, because it is an exactly solvable problem with one-body interactions, it clearly demonstrates the similarities and differences between zero and finite temperatures and between static and dynamic response. In addition, it will demonstrate some of the techniques involved in practical calculations.

The specific model to be solved is a system of non-interacting spin-1/2 Fermions coupled to a constant external magnetic field in the $\hat{z}$ direction

$$\begin{aligned} V_{\alpha\beta} &= -\mu_0\, H\, \sigma^{(z)}_{\alpha\beta} \\ V_{\alpha\beta}(q) &= -\mu_0 \sigma^{(z)}_{\alpha\beta} (2\pi)^3 \delta^3(q) \end{aligned} \tag{5.140}$$

where α, β denote spin indices, μ_0 is the magnetic moment and $\sigma^{(z)}$ is a Pauli spin matrix. We will now calculate the susceptibility four different ways and compare the results.

STATIC SUSCEPTIBILITY AT ZERO TEMPERATURE

The magnetic susceptibility is

$$\chi = \left.\frac{\partial \langle M \rangle}{\partial H}\right|_{H\to 0} \tag{5.141}$$

where, using Eq. (5.12), the magnetization may be written in terms of the one-particle Green's function

$$\langle M\rangle = -i\mu_0 \int \frac{d\omega d^3q}{(2\pi)^4}\, e^{i\omega\eta}\, \mathrm{Tr}\,\sigma G(q,\omega)\ . \tag{5.142}$$

In a one-body external potential, the Dyson equation is

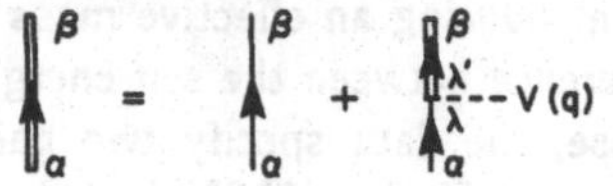

(5.143)

$$G(p',p;\omega)_{\beta\alpha} = \delta(p,p')G_0(p;\omega)_{\beta\alpha} + \int \frac{d^3q}{(2\pi)^3} G(p',p+q;\omega)_{\beta\lambda'} V_{\lambda'\lambda}(q) G_0(p;\omega)_{\lambda\alpha}$$

where we have explicitly written spin indices. Because V is constant in space and diagonal in spin, G is diagonal in spin and momentum. Using the familiar result $G_0^{-1} = G^{-1} + V$, Eq. (2.178), we obtain for $\alpha = \pm 1$

$$G_{\alpha\alpha} = \frac{\theta(k-k_F)}{\omega - \epsilon_k + \alpha\mu_0 H + i\epsilon} + \frac{\theta(k_F - k)}{\omega - \epsilon_k + \alpha\mu_0 H - i\epsilon}\ . \tag{5.144}$$

Hence

$$\begin{aligned}\langle M\rangle = -i\mu_0 \int \frac{d\omega d^3k}{(2\pi)^4} e^{i\omega\eta} \Big[&\frac{\theta(k-k_F)}{\omega - \epsilon_k + \mu_0 H + i\epsilon} + \frac{\theta(k_F-k)}{\omega - \epsilon_k + \mu_0 H - i\epsilon} \\ &- \frac{\theta(k-k_F)}{\omega - \epsilon_k - \mu_0 H + i\epsilon} - \frac{\theta(k_F-k)}{\omega - \epsilon_k - \mu_0 H - i\epsilon} \Big] \\ = 0 \qquad &\end{aligned} \tag{5.145}$$

since the contribution of the two poles in the upper half plane cancel, and the formalism has handed us the unphysical result that $\chi = \frac{\partial \langle M\rangle}{\partial H} = 0$.

STATIC SUSCEPTIBILITY AT FINITE TEMPERATURE

Let us see if the finite temperature theory is any different. Replacing G_0 by $\mathcal{G}_0$ and the factor i by (-1) in $\langle M\rangle$ (see Table 5.1) we obtain

$$\langle M\rangle = \frac{\mu_0}{\beta} \sum_{\omega_n} \int \frac{d^3k}{(2\pi)^3} e^{i\omega_n\eta} \left[\frac{1}{i\omega_n - (\epsilon_k - \mu - \mu_0 H)} - \frac{1}{i\omega_n - (\epsilon_k - \mu + \mu_0 H)} \right]\ . \tag{5.146}$$

The standard techniques for performing frequency sums such as that in Eq. (5.146) is to perform a contour integral with an integrand deliberately constructed to have poles at ω_n and residues equal to the summand. In the present case, we will calculate $\sum_{\omega_n} \frac{e^{i\omega_n\eta}}{i\omega_n - x}$ by evaluating

$$I \equiv \int_c \frac{e^{\omega\eta}}{\omega - x} \left(\frac{-\beta}{e^{\beta\omega} + 1} \right) d\omega \tag{5.147}$$

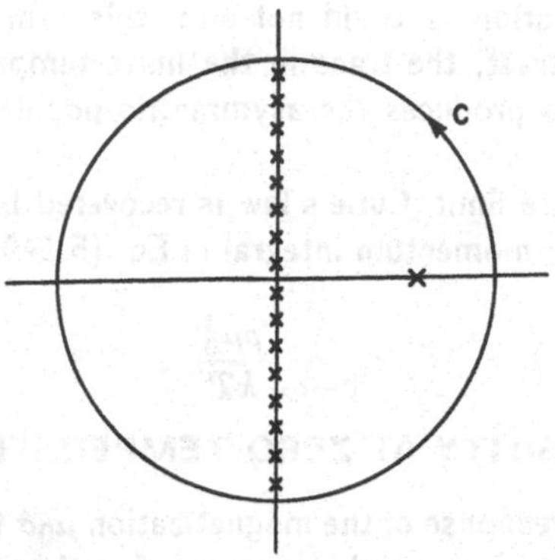

Fig. 5.20 Integration contour and poles of Eq. (5.147) for evaluating a frequency sum.

along the contour at infinity shown in Fig. 5.20. The integrand has poles at $\omega = x$ and at the Matsubara frequencies $i\omega_n = \frac{1}{\beta}(2n+1)\pi i$. As $\omega \to \infty$, the integrand decreases as $\frac{e^{\omega(\eta-\beta)}}{\omega}$ and as $\omega \to -\infty$ it decreases as $\frac{e^{-|\omega|\eta}}{\omega}$, so for a positive infinitesimal η, the contribution along the contour C vanishes. Equating the sum of residues to zero, we obtain the desired result:

$$\sum_{\omega_n} \frac{e^{i\omega_n \tau}}{i\omega_n - x} = \frac{\beta e^{x\eta}}{e^{\beta x}+1} \underset{\eta\to 0^+}{\longrightarrow} \beta n(x+\mu) \tag{5.148}$$

where, following the convention Eq. (2.75b), we have included the chemical potential μ in the Fermi occupation probability $n(\epsilon)$. Using this result to calculate the magnetization and taking the weak field limit

$$\begin{aligned}\langle M\rangle &= \mu_0 \int \frac{d^3k}{(2\pi)^3}\left[n(\epsilon_k - \mu_0 H) - n(\epsilon_k + \mu_0 H)\right] \\ &\underset{H\to 0}{\longrightarrow} -2\mu_0^2 H \int \frac{d^3k}{(2\pi)^3} \left.\frac{dn}{d\epsilon}\right|_{\epsilon_k} .\end{aligned} \tag{5.149}$$

The zero temperature limit is evaluated using $\frac{dn}{d\epsilon} \underset{T\to 0}{\longrightarrow} \frac{d}{d\epsilon}\theta(\epsilon_p - \epsilon) = -\delta(\epsilon - \epsilon_F)$ and changing the integration variable to energy $k = (2m\epsilon_k)^{1/2}$ with the result

$$\lim_{T\to 0} \chi = -2\mu_0^2 \int \frac{d^3k}{(2\pi)^3} \left.\frac{dn}{d\epsilon}\right|_{\epsilon_k} = \frac{3}{2}\frac{\rho}{\epsilon_F}\mu_0^2 \tag{5.150}$$

where ρ is the density.

On the basis of the general discussion in Section 3.3, it is simple to understand why the finite-temperature theory gives the correct result for the Pauli paramagnetism of a Fermi gas while the zero temperature calculation produced $\chi = 0$. The zero temperature theory finds the lowest eigenstate with the same symmetry as the non-interacting system. Since the non-interacting system had equal Fermi spheres for spin up and down, the perturbation H could not alter this symmetry and thus produced no magnetization. In contrast, the trace in the finite temperature theory has access to all symmetries and thus produces the asymmetric population of the Fermi spheres reflected in Eq. (5.149).

In the high-temperature limit, Curie's law is recovered by noting that $n(\epsilon) \propto e^{-\beta\epsilon}$ so that $\frac{dn}{d\epsilon} = -\beta n$ and the momentum integral in Eq. (5.149) simply yields $-\frac{\beta}{2}\rho$, with the result

$$\chi \xrightarrow[T\to\infty]{} \frac{\rho\mu_0^2}{kT} \; . \tag{5.151}$$

DYNAMIC SUSCEPTIBILITY AT ZERO TEMPERATURE

From Eq. (2.15), the response of the magnetization $\mu_0\hat{\sigma}$ to an infinitesimal external magnetic field $-\mu_0\delta H\hat{\sigma}$ is given by the response function $\langle[\sigma(1),\sigma(2)]\rangle$. Hence, the dynamic susceptibility, which is defined as the response of M to an infinitesimal field of frequency ω and wave vector q is given by

$$\chi(k,\omega) = -\mu_0^2 D(k,\omega) \tag{5.152}$$

where the spin-spin response function is given by

$D(k,\omega) =$

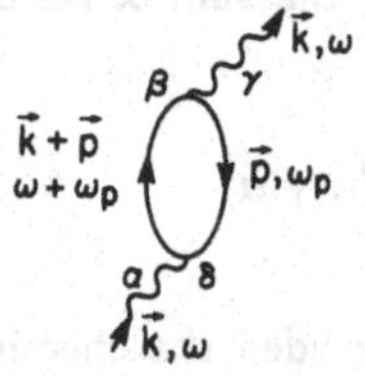

$$= -i\int \frac{d^3p\,d\omega_p}{(2\pi)^4} \sum_{\alpha\beta\gamma\delta} \sigma^*_{\alpha\delta} G_{\delta\gamma}(p,\omega_p)\sigma^*_{\gamma\beta} G_{\beta\alpha}(k+p,\omega+\omega^p)$$

$$= D_0(k,\omega) \; . \tag{5.153}$$

The last line follows because G is diagonal in spin so that the spin trace simply yields a factor of 2 equal to the spin factor already included in D_0. Hence, using Eq. (5.116a) the long wavelength limit of the zero frequency susceptibility is

$$\lim_{k\to 0} \chi(k,\omega=0) = -\mu_0^2\left(-\frac{mk_F}{\pi^2}\right) = \frac{3}{2}\frac{\rho}{\epsilon_F}\mu_0^2 \; . \tag{5.154}$$

Thus, although we are working at zero temperature, linear response theory allows the external field to break the symmetry of the non-interacting ground state, so the proper Pauli paramagnetism is again obtained.

DYNAMIC SUSCEPTIBILITY AT FINITE TEMPERATURE

Finally, although there is by now no question as to the outcome, we will show how the finite temperature response function produces the correct result. All of the results for linear response theory in Section 5.4 have finite temperature counterparts given by the correspondence in Table. 5.1. Since the spin trace is trivial as before, the finite temperature response function is given by

$$\mathcal{D}(k,\nu_m) =$$

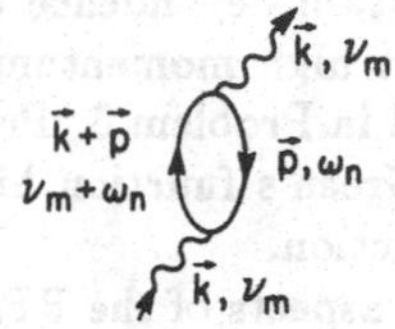

$$= 2\int \frac{d^3p}{(2\pi)^3}\frac{1}{\beta}\sum_{\omega_n}\frac{1}{i\omega_n-(\epsilon_p-n)}\,\frac{1}{i(\nu_m+\omega_n)-(\epsilon_{k+p}-\mu)}$$

$$= \frac{2}{\beta}\int\frac{d^3p}{(2\pi)^3}\sum_{\omega_n}\frac{e^{i\omega_n\eta}}{\epsilon_p-\epsilon_{k+p}+i\nu_m}\times \tag{5.155}$$

$$\left[\frac{1}{i\omega_n-(\epsilon_p-\mu)}-\frac{1}{i(\nu_m+\omega_m)-(\epsilon_{k+p}-\mu)}\right].$$

In the last line, the summand has been rewritten as the difference of two frequency sums which may be evaluated using Eq. (5.148). Whereas the original frequency sum converged because of the $\frac{1}{\omega_n^2}$ dependence, the sums in the last line are logarithmically divergent and a convergence factor $e^{i\omega_n\eta}$ has therefore been inserted. Performing the frequency sums,

$$\mathcal{D}(k,\nu_m) = 2\int\frac{d^2p}{(2\pi)^3}\,\frac{n(\epsilon_p)-n(\epsilon_{k+p})}{\epsilon_p-\epsilon_{k+p}+i\nu_m} \tag{5.156a}$$

and continuing to the real axis as in Fig. 5.4,

$$\mathcal{D}_R(k,\omega) = -2\int\frac{d^3p}{(2\pi)^3}\,\frac{n(\epsilon_{k+p})-n(\epsilon_p)}{\omega-(\epsilon_{k+p}-\epsilon_p)+i\eta}\;. \tag{5.156b}$$

Thus, in the limit $\omega = 0$, $k\to 0$

$$\begin{aligned}\lim_{k\to 0}\chi(k,\omega=0) &= \lim_{k\to 0} -\mu_0^2\mathcal{D}_R(k,0)\\ &= -2\mu_0^2\int\frac{d^3p}{(2\pi)^3}\,\frac{n(\epsilon_{k+p})-n(\epsilon_p)}{\epsilon_{k+p}-\epsilon_p}\\ &= -2\mu_0^2\int\frac{d^3p}{(2\pi)^3}\,\frac{dn}{d\epsilon}\\ &= \frac{3}{2}\frac{\rho}{\epsilon_F}\mu_0^2\end{aligned}$$

as before. The methods for treating frequency sums and analytic continuation illustrated in this very simple example are completely general and may be applied straightforwardly to any finite temperature calculation.

PROBLEMS FOR CHAPTER 5

The problems for this chapter elaborate and apply many of the basic ideas presented in the text. Problem 1 relates the Green's function expression for the

grand potential to the perturbation expansion in Chapter 2. Sum rules mentioned in the text are derived in Problems 2 and 11. Problem 3 explores the structure of the exchange term contributing to the momentum dependence of the self-energy. The relation between the structure function at high momentum transfer and the ground state momentum distribution is derived in Problem 5. Problem 12 explores several truncations of the Martin-Schwinger Green's function hierarchy to obtain approximations to the one particle Green's function.

The remaining problems deal with various aspects of the RPA. Problem 6 outlines the derivation of the Lindhard functions quoted in the text and its importance is emphasized by an *. The ground state RPA correlation energy is derived in Problem 5 and applied to the dilate Bose gas in Problem 7. The familiar one-dimensional δ-function system provides a simple pedagogical example of how the RPA reveals physical instabilities, and is also emphasized by an *. Problem 9 outlines the classical derivation of the plasma-oscillation which was obtained using the RPA in the text. Finally, Problem 10 outlines the derivation of the time-dependent Hartree Fock approximation and the equivalence to the RPA in the small-amplitude limit and Problem 11 establishes the properties of the RPA eigenstates used in the text.

PROBLEM 5.1 Evaluation of the Grand Potential from the One-Particle Green's Function

Show that Eq. (5.25) reproduces the Hugenholtz diagram expansion for $\Omega - \Omega_0$ developed in Chapter 2 through third order. First, note that $\int dx dx' d\tau' \Sigma^\lambda(x\tau|x'\tau')\mathcal{G}^\lambda(x'\tau'|x\tau')$ or its Fourier transform $\int d^3k\, d\omega \Sigma^\lambda(k,\omega)\mathcal{G}^\lambda(k,\omega)$ corresponds to the diagram

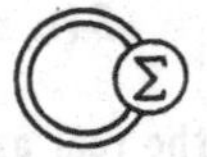

in which the two external points of Σ are connected by $\mathcal{G}$. Thus, the problem is to expand Σ^λ and $\mathcal{G}^\lambda$ consistently through third order and show that each linked diagram in Fig. 2.3 is obtained with the proper combinatorial factor.

Using Dyson's equation, show that to third order

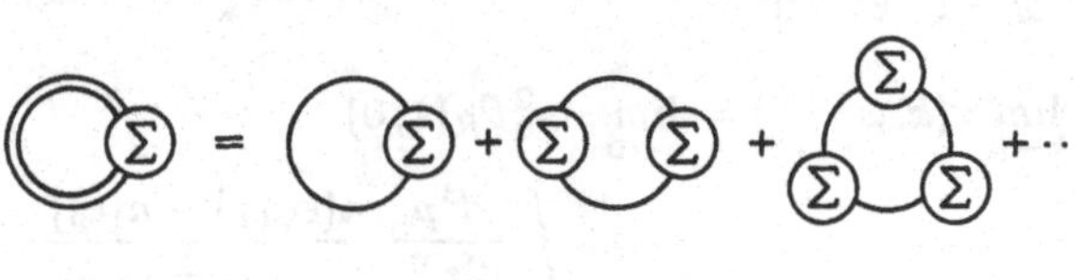

(1)

Use the fact that the self-energy is obtained from amputated one-particle irreducible Green's function diagrams and use the Hugenholtz rules that Green's functions have no symmetry factors and a factor of $\frac{1}{2}$ for each equivalent pair of lines to show that through third order

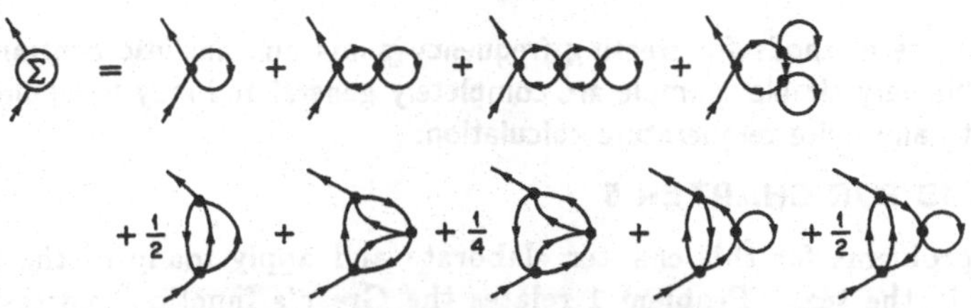

(2)

Finally, substitute Eq. (2) into Eq. (1), collect all equivalent diagrams through third order, for each diagram with n interactions include the factor $\frac{1}{n}$ from $\int \frac{d\lambda}{\lambda}$, and thus show that $\frac{1}{2}\int \frac{d\lambda}{\lambda}\Sigma^{\lambda}\mathcal{G}^{\lambda}$ reproduces the contributions of the linked diagrams in Fig. 2.3.

PROBLEM 5.2 Energy Sum Rule for the Spectral Function

For a system interacting via two-body forces, Eq. (5.20b) for the energy and Eq. (5.33a) yield

$$E_0 = \frac{i}{2}\eta\frac{\mathcal{V}}{(2\pi)^4}\int d^3k d\omega\, e^{i\omega\eta}\left(\frac{k^2}{2m}+\omega\right)\int_0^\infty \frac{d\omega'}{2\pi}$$
$$\times\left[\frac{\rho^+(k,\omega')}{\omega-\mu-\omega'+i\eta}-\varsigma\frac{\rho^-(k,\omega')}{\omega-\mu+\omega'-i\eta}\right].$$

By completing the ω contour in the upper half plane, show that the energy per unit volume is given by

$$\frac{E_0}{\mathcal{V}} = \frac{1}{2}\int\frac{d^3k}{(2\pi)^3}\int\frac{d\omega'}{(2\pi)}\left[\frac{k^2}{2m}+\mu-\omega'\right]\rho^-(k,\omega')$$
$$= \frac{1}{2}\int\frac{d^3k}{(2\pi)^3}\int\frac{d\epsilon}{2\pi}\left[\frac{k^2}{2m}+\epsilon\right]\rho^-(k,\mu-\epsilon)$$

where ϵ measures the energy up from the bottom of the fermi sea instead of downward from μ.

In schematic form, this sum rule may be written $E_0 = \left\langle\frac{k^2}{2m}+\epsilon\right\rangle$ where the brackets denote weighting with respect to the spectral function. This form, which is exact, is reminiscent of the result obtained in Problem 1.6 in the Hartree-Fock approximation $E = \sum_{k<k_F}\left(\frac{k^2}{2m}+\epsilon_k\right)$. Recover this old result by evaluating $\rho^-(k,\mu-\epsilon) = \sum_n |\langle\psi_n^{N-1}|a_k|\psi_0\rangle|^2 2\pi\delta\left(E_n^{N-1}-E_0+\epsilon\right)$ in the special case in which $|\Psi_0\rangle$ is an N-particle ground state Slater determinant and $|\psi_n^{N-1}\rangle$ is an $(N-1)$-particle Slater determinant obtained by removing one particle in orbit n from $|\psi_0\rangle$. Thus, the sum rule may be viewed as an exact generalization of the familiar Hartree-Fock energy expression.

PROBLEM 5.3 To understand the nonlocality of the Hartree-Fock potential and self-energy, it is instructive to evaluate the exchange term in a translationally invariant system.

a) Consider the sum $\sum_\beta \phi^*_\beta(x)\phi_\beta(y)$ appearing in the Hartree-Fock exchange term in Problem 1.6. (Recall $\phi(x)$ includes internal degrees of freedom such as spin and isospin). Perform the sum over all normalized plane wave spatial wave functions in a Fermi gas to obtain the Slater mixed density

$$\sum_{|k|<k_F}\frac{e^{-i\vec{k}\cdot\vec{x}}}{\sqrt{\mathcal{V}}}\frac{e^{i\vec{k}\vec{y}}}{\sqrt{\mathcal{V}}} = \frac{\rho}{2S+1}\rho_{sl}(k_F|x-y|)$$

where

$$\rho_{sl}(k_F r) \equiv \frac{3j_1(k_F r)}{k_F r}$$

and the total degeneracy associated with internal degrees of freedom $(2S+1)$ specifies the number of Fermi spheres which must be filled to obtain the total density ρ. Explain physically the value at $|x-y|=0$ and why k_F sets the scale at which the mixed density goes to zero.

b) Using the result in part (a), evaluate the two-body density $\rho_2(r_1,r_2) \equiv \langle\sum_{i\neq j}^{N}\delta(r_1-r_i)\delta(r_2-r_j)\rangle$ in a Fermi gas with degeneracy $(2S+1)$. Sketch $\rho_2(r_1-r_2)$ and note that the Slater density describes the "Fermi hole" in the vicinity of $r_1=r_2$ required by the Pauli-principle. Noting that the normalization condition on ρ_2 is $\int d^3r_2\rho_2(r_1,r_2)=(N-1)\rho(r_1)$, show that the "Fermi hole" excludes one particle.

c) Assume two different forces act in even and odd partial waves. Since even and odd partial waves are spatially symmetric and antisymmetric, a convenient representation is

$$v = v_{\text{even}}(r)\left(\frac{1+P_r}{2}\right) - v_{\text{odd}}(r)\left(\frac{1-P_r}{2}\right)$$

where the exchange operator P_r exchanges the spatial coördinates of two particles. Now consider nucleons with single-particle states characterized by spatial wave functions $U_\alpha(r_i)$, spin wave function $\chi_\sigma(i)$ and isospin wave function $\chi_\tau(i)$. Note

$$\begin{aligned}&P_rU_{\alpha_1}(r_1)\chi_{\sigma_1}(1)\chi_{\tau_1}(1)U_{\alpha_2}(r_2)\chi_{\sigma_2}(2)\chi_{\tau_2}(2)\\ =&U_{\alpha_1}(r_2)\chi_{\sigma_1}(1)\chi_{\tau_1}(1)U_{\alpha_2}(r_1)\chi_{\sigma_2}(2)\chi_{\tau_2}(2)\ .\end{aligned}$$

Show that in a Fermi gas of such wave functions

$$\frac{\langle V\rangle}{N} = \frac{\rho}{2}\int d^3r\left[\left(\frac{3}{8}v_{\text{even}}(r)+\frac{5}{8}v_{\text{odd}}(r)\right)+\left(\frac{3}{8}v_{\text{even}}(r)-\frac{5}{8}v_{\text{odd}}(r)\right)\rho_{\text{sl}}^2(k_Fr)\right]\ .$$

Note that the exchange potential, referred to in Eq. (5.78) reduces to the familiar result $-\frac{1}{4}v(r)$ for a central potential.

PROBLEM 5.4 Scaling Behavior of the Dynamic Structure Factor and Measurement of the Momentum Distribution

At high momentum transfer q, the liquid structure factor for a non-interacting system measures the longitudinal momentum distribution of the ground state. Physically, this is obvious because it just counts the ways a momentum k may be knocked out of the system by transferring a momentum q so large that it is assumed that the state $k+q$ must be unoccupied. Formally this may be written using Eq. (5.108)

$$\begin{aligned}S_0(q,\omega) &= -\text{Im}\,D_0(q,\omega)\\ &\underset{q\to\infty}{\longrightarrow} -\frac{m}{q}\int\frac{d^3k}{(2\pi)^3}\delta(\vec{k}\cdot\hat{q}-y)\left(2n(\vec{k})\right) \qquad (1)\\ &= -\frac{m}{q}\int\frac{dk_\parallel}{2\pi}\delta(k_\parallel-y)\int\frac{d^2k_\perp}{(2\pi)^2}2n(k_\parallel,k_\perp)\\ &= -\frac{m}{2\pi q}n_{0_\parallel}(y)\end{aligned}$$

where the longitudinal momentum distribution of the non-interacting system (including 2 spin states for each k) is

$$n_{0_{\parallel}}(k_{\parallel}) \equiv \int \frac{d^2k_{\perp}}{(2\pi)^2} 2n(k_{\parallel}, \vec{k}_{\perp}) \tag{2}$$

the scaling variable is defined

$$y \equiv \frac{m\omega}{q} - \frac{q}{2} \tag{3}$$

$k_{\parallel}$ is the component of k in the direction of $\hat{q}$, and $\vec{k}_{\perp}$ denotes the two perpendicular components. The question is whether the structure function for the interacting system is similarly related to the momentum distribution of the interacting system.

a) Analyze a general time-ordered diagram in which momentum q is injected at time 0 and removed at time T in the limit as $q \to \infty$ with y held fixed. Assume for the present that the two-body potential is smooth with a Fourier transform $v(q)$ which decreases exponentially in the $q \to \infty$ limit. Establish the following properties:

i) Unless $\vec{q}$ injected at time 0 can flow solely through particle propagators to be removed at time T, the diagram vanishes at least exponentially in q.

ii) Each global denominator between time 0 and time T has the form

$$E = \omega - \sum_{\text{particles}} \epsilon_P + \sum_{\text{holes}} \epsilon_h + i\varsigma$$
$$\to \frac{q}{m}\left[y - p\cdot\hat{q} + O\left(\frac{1}{q}\right)\right] + i\varsigma \;.$$

iii) Each energy denominator outside the range $(0, T)$ has no q dependence.

iv) The q-dependence of any time-ordered diagram with n interactions in the range $(0, T)$ and any number of interactions outside $(0, T)$ is $\frac{1}{q^{n+1}}$

v) Hence, show that the most general contributions of order $\frac{1}{q}$ to $S\left(q, \omega = \frac{qy}{m} + \frac{q^2}{2m}\right)$ are of the generic form

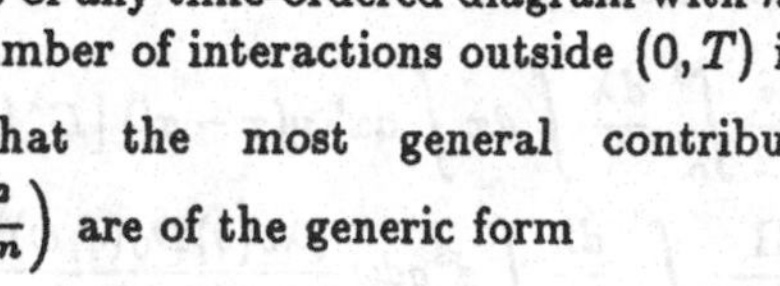

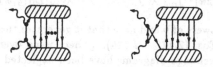

Relate these diagrams to the most general contributions to the momentum distribution and thus show

$$S\left(q, \omega = \frac{qy}{m} + \frac{q^2}{2m}\right) \underset{q\to\infty}{\longrightarrow} -\frac{m}{2\pi q} n_{\parallel}(y) \;.$$

b) Now consider the case of a potential with an infinitely hard core. Use the optical theorem to show that the imaginary part of the forward scattering amplitude grows linearly with q. Give two examples of diagrams which would be of order $\frac{1}{q^2}$ for smooth potentials but are of order $\frac{1}{q}$ for hard cores. Thus, although $qS\left(q, \omega = \frac{qy}{m} + \frac{q^2}{2m}\right)$ exhibits scaling by depending only on y, this scaling function is not the momentum distribution.

c) Finally, consider practical experiments on finite nuclei and liquid helium. Experimentally, the total cross section for nucleon-nucleon scattering is approximately constant for $0.4 \leq p_{\text{beam}} \leq 1.7\,\text{GeV}$. What does this say about measuring the momentum distribution for experiments with q in the range of 1 GeV? If the helium-helium interaction is described by the Lennard-Jones potential, is it practical to measure the momentum distribution for liquid helium? Some further details may be found in Weinstein and Negele (1982).

PROBLEM 5.5 RPA Correlation Energy

To see the relation of the ring diagram contributions to the ground state energy discussed in connection with Eqs. (2.113 – 2.115) to the RPA response function, it is useful to repeat the ground state energy derivation in Eqs. (5.23– 5.24) writing $\langle V\rangle$ in terms of D instead of G. Note, by the Fermion commutation relations, we may write

$$\langle\psi|V|\psi\rangle = \frac{1}{2}\int dx \int dx'\, v(x-x')\langle\psi|\hat{\psi}^\dagger(x')\hat{\psi}(x')\hat{\psi}^\dagger(x)\psi(x) - \delta(x-x')\hat{\psi}^\dagger(x)\psi(x)|\psi\rangle$$

and thus show in a translationally invariant system

$$\langle\psi_0(\lambda)|V|\psi_0(\lambda)\rangle - \langle\Phi_0|V|\Phi_0\rangle = \frac{1}{2}\int dx \int dx'\, v(x-x')\left[iD^\lambda(xt|x't) - iD_0(xt|x't)\right]$$

where $|\psi_0(\lambda)\rangle$ and D^λ denote the ground state and response function for the Hamiltonian Eq. (5.23) and $|\psi_0\rangle$ is the non-interacting ground state $|\psi_0(0)\rangle$. Hence, use Eq. (5.113) in the form $D^{\text{RPA}} = D_0 + \frac{D_0 v D_0}{1 - vD_0}$ and the relation $E_0 - \omega_0 = \int_0^1 \frac{d\lambda}{\lambda}\langle\psi_0(\lambda)\,|\lambda v|\,\psi_0(\lambda)\rangle$ to write

$$\begin{aligned} E_0 &= \langle\phi_0|T+V|\phi_0\rangle + \frac{i}{2}\int_0^1 \frac{d\lambda}{\lambda}\int dx \int dx'\, v(x-x')\left[D^\lambda(xt|x't) - D_0(xt|x't)\right] \\ &= \langle\phi_0|\hat{H}|\phi_0\rangle + \frac{i}{2}\frac{\Omega}{(2\pi)^4}\int_0^1 \frac{d\lambda}{\lambda}\int d^3q\,d\omega \frac{(\lambda v(q)D_0(q,\omega))^2}{1-\lambda v(q)D_0(q,\omega)} \ . \end{aligned} \tag{1}$$

Expand this result in powers of v and note that it generates the direct ring diagrams with the factor $\frac{1}{2n}$ in Eqs. (2.112 – 2.115). (There is no overcounting problem in second order because exchange diagrams have been neglected throughout.)

PROBLEM 5.6* Lindhard Function

Evaluate the momentum integrals in the response function for a non-interacting Fermi gas, Eq. (5.108), to obtain Eq. (5.109). To calculate the real part, it is convenient to change variables in the second term of Eq. (5.108) by defining $k' = -k - q$ and write

$$\int d^3k \frac{n_{k+q}(1-n_k)}{\omega + \epsilon_k - \epsilon_{k+q} - i\eta} = \int d^3k' \frac{n_{-k'}\left(1 - n_{(-k'-q)}\right)}{\omega + \epsilon_{(-k'-q)} - \epsilon_{(-k')} - i\varsigma}$$

so that identical θ-functions appear in both terms. The imaginary part is simply proportional the phase space such that $\vec{k}+\vec{q}$ is outside a Fermi sphere while $\vec{k}$ is inside a Fermi sphere and energy is conserved, $\frac{\vec{k}\cdot\vec{q}}{m}+\frac{q^2}{2m}=\omega$. Thus, Im D_0 may be calculated geometrically by computing the area of the circle in the plane defined by $\vec{k}\cdot\hat{q}+\frac{q}{2}=m\omega$ such that $|q+k|>k_F$ and $k<k_F$.

PROBLEM 5.7 RPA Correlation Energy for Dilute Bose Gas

Using the RPA correlation energy expression from Problem 5.5, Eq. (1), sum all the ring diagrams for the dilute Bose gas treated in Problem 3.3. The integrals can be performed analytically and should reproduce the exact leading order result.

PROBLEM 5.8* RPA Instability of Uniform Gas

In Problem 1.9 the uniform gas solution for Fermions interacting with repulsive δ-function forces in one dimension was shown to have a binding energy per particle about 10% less than a collection of clusters of $(2S+1)$ Fermions in the ground state. This suggests that the gas phase is unstable against fluctuations. Note, as seen for example in Problem 3.4, that perturbation theory for the ground state energy does not reveal instabilities, so instead it is useful to examine the response function. If all the RPA modes have real positive frequencies, the system is stable with respect to all infinitesimal fluctuations. If at some q, there is a mode with $\omega=0$, the mode is self-sustaining: it costs no energy to create it. In regions of q for which ω is complex, the mode will grow exponentially and the system is unstable.

a) Calculate the RPA response function to search for instabilities at all q and ρ. Show that as $q\to 0$, the long wavelength stability criterion is identical to that derived in Problem 1.7 and applied already in Problem 1.9d.

b) Determine the range of q for which the equilibrium density gas is unstable. Compare the wavelength of the instabilities with the size of a ground state cluster of $(2S+1)$ particles and explain this result physically.

PROBLEM 5.9 Classical Plasma Oscillations

Show that the RPA mode (5.118) in an electron gas corresponds to the plasma frequency derived classically. Linearize the continuity equation

$$\frac{\partial\rho}{\partial t}+\nabla\cdot\rho\vec{v}=0$$

and equation of motion

$$m\frac{\partial}{\partial t}(\rho\vec{v})+m\vec{v}\cdot\nabla(\rho v)=-e\rho E$$

to leading order in the density fluctuation $\delta\rho=\rho-\rho_0$ and velocity. Combine these results with the Maxwell equation

$$\nabla\cdot E=4\pi\delta\rho$$

to obtain the result

$$\frac{\partial^2}{\partial t^2}\delta n=-\frac{4\pi\rho_0}{m}\delta n$$

with frequency $\omega=\frac{4\pi\rho_0}{m}$.

PROBLEM 5.10 Time-Dependent Hartree-Fock Approximation

a) Using the equation of motion for $\hat{\psi}(x,t)$, Eq. (5.17), derive the following exact equation of motion for $G^{(1)}$

$$\left(i\frac{\partial}{\partial t_1} - T_{x_1}\right) G^{(1)}(1,1') = \delta(1,1') - i\int dx_2\, v(x_1 - x_2) G^{(2)}(1\,2|1'2^+)\Big|_{t_1=t_2} \quad . \quad (1)$$

This is the first of an infinite heirarchy of equations (Martin and Schwinger, 1959) relating the evolution of $G^{(n)}$ to $G^{(n+1)}$ and $G^{(n-1)}$

Now, make the time-dependent Hartree-Fock (TDHF) approximation by assuming that the two particles propagate independently in $G^{(2)}$, that is

$$G^{(2)}(1\,2|1'2') \approx G^{(1)}(1|1')G^{(1)}(2|2') - G^{(1)}(1|2')G^{(1)}(2|1')$$

to obtain

$$i\frac{\partial}{\partial t_1} G(x_1 t_1|x_2 t_2) = \int d^3x_3\, h(x_1, x_3; t_1) G(x_3 t_1|x_2 t_2) + \delta(1,\,2)$$

where the TDHF single particle Hamiltonian is

$$h(x_1, x_3; t) \equiv \langle x_1|T|x_3\rangle + \int\int dx_2 dx_4 \{x_1 x_2|v|x_3 x_4\}\, \rho(x_4, x_2; t) \qquad (2)$$

and the one-body density matrix is defined

$$\rho(x, x'; t) \equiv \langle \psi^\dagger(x' t)\psi(x,t)\rangle \quad .$$

Combining the equation for G with its adjoint, the TDHF equation may equivalently be written purely in terms of the one-body density matrix

$$i\frac{\partial}{\partial t}\rho(x_1, x_2; t) = \int d^3x_3\, (h(x_1, x_3; t)\rho(x_3, x_2; t) - \rho(x_1, x_3; t)h(x_3, x_2; t))$$

or in matrix form

$$i\dot{\rho} = [h, p] \quad . \qquad (3)$$

b) For a Slater determinant, $\rho(x, x') = \sum_A \phi_A(x)\phi_A^*(x')$. Thus, show that Eq. (1) is equivalent to the single-particle equation of motion

$$i\dot{\phi}_A = h\phi_A \quad .$$

This equation will be obtained by using the time-dependent variational principle in Problem 7.5.

c) Recover the RPA by considering infinitesimal TDHF fluctuations around the Hartree-Fock ground state (Goldstone and Gottfried, 1959). Let ρ_0 denote the static density matrix and ρ_1 the infinitesimal fluctuations in the density matrix.

The dependence of the TDHF Hamiltonian on ρ is indicated explicitly by writing Eq. (2) in the form

$$h = T + W[\rho]$$

and the static HF basis is defined

$$(T + W[\rho_0])\,|\alpha\rangle = E_\alpha|\alpha\rangle \ .$$

Linearize Eq. (3) in ρ_1 to obtain

$$i\dot{\rho}_1 = [W[\rho_1], \rho_0] + [T + W[\rho_0], \rho_1]$$

so that, taking matrix elements in the HF basis

$$\left(i\frac{\partial}{\partial t} - (E_\alpha - E_\beta)\right)\langle\alpha|\rho_1|\beta\rangle = \sum_{\gamma\delta\epsilon}\{\alpha\gamma|v|\epsilon\delta\}\langle\delta|\rho_1|\gamma\rangle\langle\epsilon|\rho_0|\beta\rangle$$
$$+ \{\epsilon\gamma|v|\beta\delta\}\,\langle\delta|\rho_1|\gamma\rangle\langle\alpha|\rho_0|\epsilon\rangle \ .$$

Finally, note that the most general infinitesimal change in the ground state wave function may be written as a superposition of particle-hole excitations so that an excitation of frequency ω is written in the Hermitian form $\rho_1 e^{-i\omega t} + \rho_1^\dagger e^{i\omega t}$ with ρ_1 only having particle-hole matrix elements, and thus obtain Eq. (5.133) with $X_{aA} = \langle A|\rho_1^\dagger|a\rangle$ and $Y_{aA} = \langle a|\rho^\dagger|A\rangle$. Show that the TDHF single particle wave functions for this mode have the form

$$\phi_A(x) = e^{-iE_A t}\left[\phi_A^{\rm HF}(x) + \sum_a \left(X_{aA}e^{-i\omega t} + Y_{aA}^* e^{i\omega t}\right)\phi_a^{\rm HF}(x)\right] \ . \tag{4}$$

PROBLEM 5.11 Properties of RPA Eigenfunctions Derive the following properties of the RPA eigenfunctions defined by Eq. (5.133).

a) Orthogonality. Left-multiply Eq. (5.133) by $[X^{(\mu)\dagger}Y^{(\mu)\dagger}]$ and compare with the adjoint equations to show

$$\left(X^{(\mu)+}Y^{(\mu)+}\right)\begin{pmatrix} X^{(\nu)} \\ -Y^{(\nu)} \end{pmatrix} = 0$$

if $E^{(\nu)} \neq E^{(\mu)*}$.

b) Show that if $\left\{\begin{matrix} X^{(\nu)} \\ Y^{(\nu)} \end{matrix}\right)$ has energy $E^{(\nu)}$ then $\begin{pmatrix} Y^{(\nu)*} \\ X^{(\nu)*} \end{pmatrix}$ has energy $-E^{(\nu)*}$. It is therefore convenient to normalize eigenvectors by the condition

$$\left(X^{(\nu)\dagger}Y^{(\nu)\dagger}\right)\begin{pmatrix} X^{(\nu)} \\ -Y^{(\nu)} \end{pmatrix} = \mathrm{sgn}\, E^{(\nu)}$$

so that $\begin{pmatrix} X^{(\nu)} \\ Y^{(\nu)} \end{pmatrix}$ and $\begin{pmatrix} Y^{(\nu)*} \\ X^{(\nu)*} \end{pmatrix}$ have the same norm.

c) Completeness. Let an arbitrary vector F be expanded in eigenstates $F = \sum_\nu a_\nu \begin{pmatrix} X^{(\nu)} \\ -Y^{(\nu)} \end{pmatrix}$. Verify the completeness relation

$$\sum_\mu \mathrm{sgn}(E^{(\mu)}) \begin{pmatrix} X^{(\mu)} \\ -Y^{(\mu)} \end{pmatrix} \left(X^{(\mu)\dagger} Y^{(\mu)\dagger} \right) = 1$$

by applying it to F.

d) Green's Function. Show that the Green's function

$$G = \sum_\nu \mathrm{sgn}(E^{(\nu)} \frac{\begin{pmatrix} X^{(\nu)} \\ Y^{(\nu)} \end{pmatrix} \left(X^{(\nu)\dagger} Y^{(\nu)\dagger} \right)}{\omega - E^{(\nu)} \pm i\delta}$$

satisfies the equation (5.132). Thus, derive Eq. (5.134) (with the appropriate $\pm i\varsigma$).

PROBLEM 5.12 Green's Function Summations of Ladder Diagrams.

Multiply the first equation in the Martin-Schwinger Green's function hierarchy, Eq. (1) of Problem 5.10 by G_0 to obtain the exact equation

$$G^{(1)}(1|1') = G_0^{(1)}(1|1') + iG_0^{(1)}(1|4)\,\{45|v|23\}\,G^{(2)}(2\,3|5^+1')$$

where 5^+ denotes a positive infinitesimal time added to t_5, v contains a δ-function in time, and repeated coordinates are summed.

In diagrams, this may be written

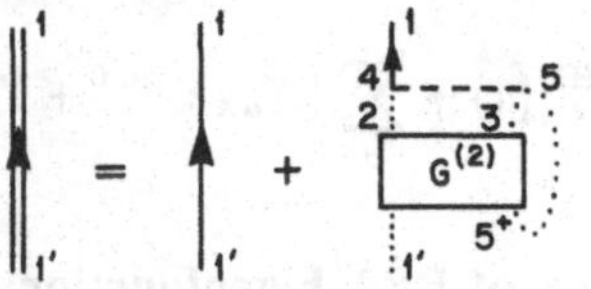

Instead of making the Hartree-Fock approximation as in Problem 5.10, consider more general summations defined by the equation

$$G^{(2)}(1\,2, 1'2') = G^{(1)}(1\,1')G^{(1)}(2\,2') - G^{(1)}(1\,2')G^{(1)}(2\,1') + \Lambda(1\,2, 3\,4)\,\{3\,4|v|5\,6\}\,G^{(2)}(5\,6, 1'2')$$

which may be represented diagrammatically as follows:

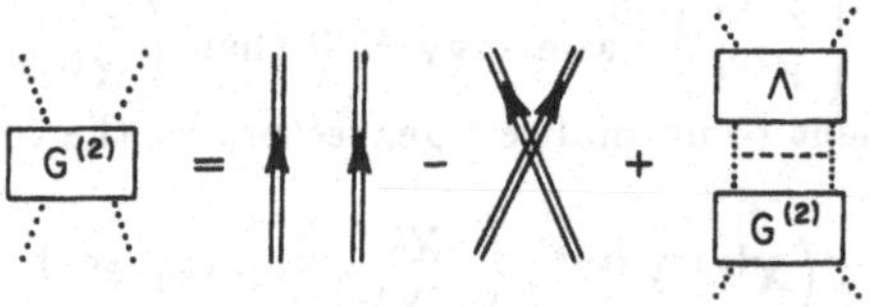

Note that in these diagrams the dotted lines are not propagators but just indicate how the various parts of the diagram are connected up, whereas the solid lines are really G_0 propagators.

We will deal with two approximations for Λ:

$$\Lambda_{00}(12,1'2') = iG_0^{(1)}(11')G_0^{(1)}(22')$$

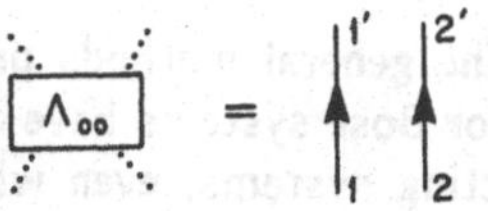

and

$$\Lambda_{11}(12,1'2') = iG^{(1)}(11')G^{(1)}(22')$$

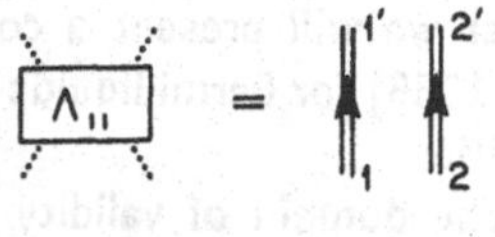

a) Treat the two equations as a system of closed equations and solve for the self energy Σ in the Λ_{00} approximations. Retain terms sufficient to obtain contributions to the ground state energy through third oder in part b, below. Recall Σ satisfies the equation:

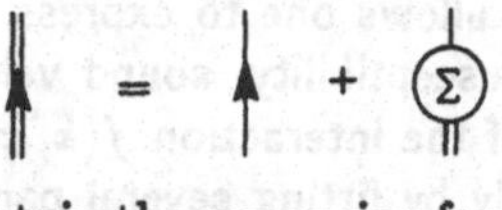

b) From $G^{(1)}$ and Σ, obtain the expansion for the ground state energy through third order. Compare with the Goldstone expansion. (Here you will need to remember that Green's functions include all relative time orderings and both particle and hole propagators. You will need to be very careful with factors.) State explicitly which graphs are counted correctly, incorrectly, and omitted.

c) Now, make contact with Brueckner theory (see Eq. (3.64)) by determining which ladder graphs are included in the Λ_{00} and Λ_{11} approximations to all orders. Comparison is easiest in the low density limit where graphs with two independent hole lines dominate those with any higher number of hole lines. For two-hole line graphs, do Λ_{11} and Λ_{00} correspond exactly to any prescription for the potential in Brueckner theory? What is the essential difference at higher density (more than two hole lines)?

PROBLEM 5.13 Energy Weighted Sum Rules

Derive the sum rule Eq. (5.139) by evaluating the double commutator $[[H,\hat{\rho}_q],\rho_{-q}]$. First, show $[\hat{V},\hat{\rho}_q] = 0$. Then show that $[\hat{T},\hat{\rho}_q] = \sum_p \left(\frac{p^2}{2m} - \frac{(p+q)^2}{2m}\right) a_p^\dagger a_{p+q}$ from which it follows that $\left[[\hat{T},\hat{\rho}_q],\hat{\rho}_q\right] = -\sum_p \frac{q^2}{m} a_p^\dagger a_p$. Finally, insert a complete set of states in the matrix element $\langle\psi_0|\,[[H,\hat{\rho}_q],\hat{\rho}_{-q}]\rangle$ to obtain the sum rule.

CHAPTER 6

THE LANDAU THEORY OF FERMI LIQUIDS

The general methods presented in previous chapters to calculate observables in Fermi or Bose systems have been based on perturbation theory. In the case of strongly interacting systems, even when physically motivated resummations have been made, it is difficult to assess the reliability of the resulting approximations. Hence, in this chapter we will present a completely different approach developed by Landau (1956, 1957, 1958) for Fermi liquids which provides exact relations between certain observable quantities.

The domain of validity of the Landau theory is restricted to phenomena which involve excitations very close to the Fermi surface. In this domain, the fundamental degrees of freedom of the system are quasiparticles which interact with a quasiparticle interaction $f(k, k')$. Although the theory does not specify any properties of the strongly interacting ground state, it describes small departures from the ground state and response functions. It thus allows one to express a number of observables such as the specific heat, magnetic susceptibility, sound velocity, zero sound velocity, and transport coefficients in terms of the interaction $f(k, k')$. This function can either be parameterized phenomenologically by fitting several parameters to experiment, or else it can be calculated microscopically, in which case the theory becomes exact.

The chapter is organized as follows. In the first section we present the basic assumptions of the theory in a heuristic and phenomenological manner and in the next section we show how to calculate thermodynamical observables with this set of assumptions. We emphasize that the arguments in these sections will not be rigorously deduced from general principles; rather, at key points assumptions or postulates will simply be introduced following Landau's original development. For those who wish to pursue this approach to Landau theory in further detail, the review of Baym and Pethick (1976) is particularly clear and pedagogical. The last section presents the microscopic justification of the theory. This microscopic understanding both provides a clear foundation for the postulates of the earlier sections and establishes the relation of the phenomenological interaction $f(k, k')$ to the two-body vertex function.

6.1 QUASIPARTICLES AND THEIR INTERACTIONS

Let us consider a uniform gas of spin-$\frac{1}{2}$ Fermions, containing N particles in a volume $\mathcal{V}$. If the Fermions are not interacting, the ground state of the system consists of a Fermi sea of plane waves of momenta $|\vec{k}| < k_F$. The Fermi momentum k_F is related to the density by:

$$\rho = \frac{k_F^3}{3\pi^2} \quad . \tag{6.1}$$

The total energy of the system is given by

$$E = \sum_{\vec{k}} \frac{\vec{k}^2}{2m} \cdot n(\vec{k}) \tag{6.2}$$

where $n(\vec{k})$ is the occupation number of the plane wave state $|\vec{k}\rangle$,

$$n(\vec{k}) = 2\theta\left(k_F - |\vec{k}|\right) \quad . \tag{6.3}$$

If an arbitrary weak external field is coupled to the system, the net effect will be a variation of the occupation numbers, and the corresponding variation of the total energy can be written:

$$\delta E = \sum_{\vec{k}} \frac{\vec{k}^2}{2m} \cdot \delta n(\vec{k}) \quad . \tag{6.4}$$

Further, is the field is very weak, it can only excite states close to the Fermi surface, so that $\delta n(\vec{k})$ will be sharply peaked around k_F.

Let us now adiabatically turn on the interaction between the particles. A normal Fermi liquid is defined as a system in which the non-interacting ground state evolves into the interacting ground state and there is a one-to-one correspondence between the bare particle states of the original system and the dressed or quasiparticle states of the interacting system. A quasiparticle state with $|k_p| > k_F$ is defined as the state obtained from a non-interacting Fermi sea plus a plane wave $|k_p\rangle$ by switching on the interaction, and similarly, a quasihole state is obtained by starting with a Fermi sea with a hole in state $|\vec{k}_h\rangle$.

Note that in order for a system to be a normal Fermi liquid, switching on the interaction must not produce bound states. For example, in a superconductor, the non-interacting ground state does not evolve into the BCS ground state and the formation of Cooper pairs destroys the one-to-one correspondence between non-interacting states and quasiparticle states. Throughout this chapter, we will tacitly assume a normal state, either by virtue of a totally repulsive interaction or by treating temperatures above the superfluid or superconducting transition temperatures. Normal Fermi liquids to which the theory is commonly applied include liquid ^{4}He, the electron gas in metals, and nuclear matter.

We have seen in Eq. (5.72) that the quasiparticle lifetime is proportional to $(\epsilon - \epsilon_F)^2$, so that even in a normal Fermi system, quasiparticles far away from the Fermi surface are ill-defined, unstable states. If one tried to prepare them by turning on the interaction slowly enough to be adiabatic, they would decay before the process was complete. Hence, the Landau theory is only applicable to low-lying excited states of the system, which are made of superpositions of quasiparticle excitations close to the Fermi surface. In particular, it can describe neither the ground state energy, which would require summation over all occupied states in the sea, nor highly excited states.

Let us temporarily suppress spin and denote the quasiparticle energy by $\epsilon^0_{\vec{k}}$ and the interaction energy of quasiparticles of momentum $\vec{k}$ and $\vec{k}'$ by $f(\vec{k},\vec{k}')$. If we apply a weak perturbation to the system which takes it away from its ground state, it induces a change $\delta n(\vec{k})$ in the occupation number of quasiparticle $\vec{k}$, and Landau postulated that the change in the total energy of the system is given by

$$\delta E = \sum_{\vec{k}} \epsilon^0_{\vec{k}} \delta n(\vec{k}) + \frac{1}{2\mathcal{V}} \sum_{\vec{k},\vec{k}'} f(\vec{k},\vec{k}') \delta n(\vec{k}) \delta n(\vec{k}') \quad . \tag{6.5}$$

The energy of quasiparticle $\vec{k}$, when it is surrounded by other quasiparticles, is given by:

$$\epsilon_{\vec{k}} = \frac{\delta E}{\delta n(\vec{k})} = \epsilon^0_{\vec{k}} + \frac{1}{\mathcal{V}} \sum_{\vec{k}'} f(\vec{k}, \vec{k}') \delta n(\vec{k}') \ . \tag{6.6}$$

The interaction $f(\vec{k}, \vec{k}')$ of quasiparticles $\vec{k}$ and $\vec{k}'$ is given by

$$\frac{1}{\mathcal{V}} f(\vec{k}, \vec{k}') = \frac{\delta^2 E}{\delta n(\vec{k}) \delta n(\vec{k}')} \tag{6.7a}$$

and it is thus symmetric in $\vec{k}$ and $\vec{k}'$

$$f(\vec{k}, \vec{k}') = f(\vec{k}', \vec{k}) \ . \tag{6.7b}$$

Although the Landau theory is much more general than the Hartree-Fock approximation, it is instructive to note that Eq. (6.5) is of the same form as the total Hartree Fock energy where $\epsilon^0_{\vec{k}}$ is the kinetic energy, $\epsilon_{\vec{k}}$ is the self-consistent single-particle energy and $\frac{1}{\mathcal{V}} f(\vec{k}, \vec{k}')$ is the antisymmetrized matrix element of the bare two-body interaction.

Since quasiparticles are adiabatically evolved from Fermions, they obey Fermi-Dirac statistics and the distribution functions at finite temperature is given by the usual Fermi distribution

$$n(\vec{k}) = \frac{1}{e^{\beta(\epsilon(\vec{k}) - \mu)} + 1} \tag{6.8}$$

where μ is the chemical potential.

At zero temperature, we recover a step function distribution, and at non-zero temperature, Eq. (6.8) becomes a set of self-consistent equations since $\epsilon(\vec{k})$ depends on $n(\vec{k})$ through (6.6).

We now extend the notation to include spin. Formulas (6.5) to (6.8) are unchanged, except that $\vec{k}$ should be replaced by $(\vec{k}, \sigma)$ where σ is the spin projection on a quantization axis and the sums over $\vec{k}$ are replaced by $\sum_{\sigma = \pm 1/2} \sum_{\vec{k}}$. In the absence of a magnetic field, by symmetry $\epsilon(\vec{k}, \sigma)$ is independent of σ, so that $\epsilon(\vec{k}, \sigma) = \epsilon(\vec{k})$. Similarly, $f(\vec{k}\sigma, \vec{k}'\sigma')$ can depend only on the product $\sigma \cdot \sigma'$, and may be parametrized as

$$f(\vec{k}\sigma, \vec{k}'\sigma') = f(\vec{k}, \vec{k}') + 4\vec{\sigma} \cdot \vec{\sigma}' \varphi(\vec{k}, \vec{k}') \tag{6.9a}$$

or

$$f(\vec{k}\sigma, \vec{k}'\sigma') = f_o(\vec{k}, \vec{k}') + \delta_{\sigma\sigma'} f_e(\vec{k}, \vec{k}') \tag{6.9b}$$

with the relation

$$\begin{aligned} f_o &= f - \varphi \\ f_e &= 2\varphi \ . \end{aligned} \tag{6.9c}$$

Note that here σ denotes a spin $\frac{1}{2}$ matrix so that a Pauli matrix would be 2σ.

A crucial simplification occurs for the translationally invariant systems. Since Landau theory is restricted to phenomena involving only quasiparticles close to the Fermi surface, $f(k, k')$ only enters observables with the two vectors k and k' on the

Fermi surface. Thus, f will depend only on the angle θ between $\vec{k}$ and $\vec{k}'$, and it is convenient to expand it on the basis of Legendre polynomials as:

$$\begin{aligned} f(\vec{k}\sigma, \vec{k}'\sigma')\big|_{|\vec{k}|=|\vec{k}'|=k_f} &= f(\theta, \sigma, \sigma') \\ &= f(\theta) + 4\vec{\sigma}\cdot\vec{\sigma}'\varphi(\theta) \\ &= \sum_{L=0}^{\infty} \left(f_L + 4\vec{\sigma}\cdot\vec{\sigma}'\cdot\varphi_L\right) P_L(\cos\theta) \ . \end{aligned} \tag{6.10}$$

The orthogonality relation for the Legendre polynomials:

$$\frac{2L+1}{2}\int_{-1}^{1} P_L(\cos\theta)P_{L'}(\cos\theta)d(\cos\theta) = \delta_{LL'} \tag{6.11}$$

implies that

$$\begin{Bmatrix} f_L \\ \varphi_L \end{Bmatrix} = \frac{2L+1}{4\pi}\int d\Omega\, P_L(\cos\theta) \begin{Bmatrix} f(\vec{k},\vec{k}') \\ \phi(\vec{k},\vec{k}') \end{Bmatrix}\Bigg|_{|\vec{k}|=|\vec{k}'|=k_f} . \tag{6.12}$$

In practical applications, the Legendre coefficients f_L and ϕ_L decrease sufficiently rapidly with L that truncation after several terms yields an adequate phenomenology. In this case, a small number of empirical coefficients describes a much larger number of experimental observables and the theory has non-trivial physical content and predictive power. For metals with non-spherical Fermi surfaces or finite systems, $f(k, k')$ depends on many more parameters and the theory is correspondingly less powerful.

6.2 OBSERVABLE PROPERTIES OF A NORMAL FERMI LIQUID

We shall now calculate a number of observable properties of a normal Fermi liquid in terms of the effective interaction $f(k, k')$, assuming the validity of Eqs. (6.5– 6.6). The microscopic justification of the theory and the relation of $f(k, k')$ to the bare interaction will be deferred to the following section.

EQUILIBRIUM PROPERTIES

A fundamental parameter of the theory is the effective mass. As in Eq. (5.74), m^* is defined in terms of the density of states or group velocity at the Fermi surface

$$\frac{k_F}{m^*} \equiv \frac{d\epsilon_k^0}{d\vec{k}}\bigg|_{k_F} = v_F \ . \tag{6.13a}$$

By the definition of a quasiparticle as the eigenstate obtained from a non-interacting Fermi sea with a particle is state $|k\rangle$, $\epsilon_{k_F}^0$ must be the chemical potential μ and expanding in the neighborhood of the Fermi surface

$$\begin{aligned} \epsilon_k^0 &= \mu + (k - k_F)\frac{\partial\epsilon^0}{\partial k}\bigg|_{k_F} \\ &= \mu + (k - k_F)v_F \\ &= \mu + (k - k_F)\frac{k_F}{m^*} . \end{aligned} \tag{6.13b}$$

Specific Heat

The specific heat at constant volume is defined

$$c_V = \frac{1}{\mathcal{V}} \frac{\partial E}{\partial T}\bigg|_V \quad . \tag{6.14}$$

A change in temperature induces a change in occupation numbers, so that c_V is given by

$$\begin{aligned} c_V &= \frac{1}{\mathcal{V}} \sum_{\vec{k},\sigma} \frac{\delta E}{\delta n(\vec{k},\sigma)} \frac{\delta n(\vec{k},\sigma)}{\delta T} \\ &= \frac{1}{\mathcal{V}} \sum_{\vec{k},\sigma} \epsilon_{\vec{k}} \frac{\delta n(\vec{k},\sigma)}{\delta \epsilon_{\vec{k}}} \left(-\frac{\epsilon_k - \mu}{T} + \frac{\partial}{\partial T}(\epsilon_k - \mu) \right) \quad . \end{aligned} \tag{6.15}$$

At low temperature, the sum $\sum_{\vec{k}'} f(\vec{k},\vec{k}')\delta n(\vec{k}')$ in Eq. (6.6) is of order T^2 so that the leading contribution to c_V may be obtained by replacing $\epsilon_{\vec{k}}$ by $\epsilon^0_{\vec{k}}$ and converting the integral over k to an integral over $\epsilon^0_{\vec{k}}$ using Eq. (6.13). The standard method for calculating the resulting integral using the low temperature expansion for the derivative of the Fermi function

$$\frac{\partial}{\partial \epsilon}\left[\frac{1}{e^{\beta(\epsilon-\mu)}+1}\right] \underset{\beta\to\infty}{\sim} -\delta(\epsilon-\mu) - \frac{\pi^2}{6\beta^2}\frac{\partial^2}{\partial \epsilon^2}\delta(\epsilon-\mu) + O(\beta^{-4}) \tag{6.16}$$

is reviewed in Problem 6.1. The result to lowest order in T is

$$c_V = \frac{1}{3} m^* k_F k_B^2 T \quad . \tag{6.17}$$

Physically, since the effective mass is proportional to the density of state at the Fermi surface, we expect the specific heat to depend on m^*. The result, Eq. (6.17), in fact shows that it depends on no other parameter of the theory. That is, to leading order in T the specific heat is equal to that of non-interacting Fermions of mass m^* and thus provides a direct experimental measurement of m^*. For liquid ^{3}He, the effective mass ranges from roughly 3m at zero pressure to over 5m at 27 atm, where the primary experimental uncertainty in measuring the specific heat arises from uncertainties in the temperature scale. These results as well as subsequent experimental Landau parameters for liquid ^{3}He are tabulated in Table 6.1 at the end of the chapter. So-called "heavy Fermion systems" such as $CeCu_2Si_2$, UPt_3, U_2Zn_{17}, and $CeAl_3$ have effective masses m^*/m of the order of $10^2 - 10^3$ (see Stewart, 1984).

Effective Mass

Having seen that m^* may be measured experimentally, we now relate it to the quasiparticle interaction $f(k,k')$. It is clear that if one postulates that Eqs. (6.5–6.6) hold in any Galilean frame, the different relative contributions of $\epsilon^0_{\vec{k}}$ and $f(k,k')$ in different frames will give constraints relating $\epsilon^0_{\vec{k}}$ and $f(k,k')$. Problem 6.2 explicitly uses Galilean invariance to relate m^* to f. Here, we present Landau's original argument which equates the momentum of a unit volume of the liquid to the mass flow.

The momentum per unit volume is given by

$$\vec{P} = \frac{1}{\mathcal{V}} \sum_{\vec{k},\sigma} \vec{k}\, n(\vec{k},\sigma) \quad . \tag{6.18a}$$

On the other hand, since the velocity of a quasiparticle is $\frac{\partial \epsilon_{\vec{k}}}{\partial \vec{k}}$, and the number of quasiparticles equals the number of particles, the total momentum can also be expressed as:

$$\vec{P} = \frac{1}{\mathcal{V}} \sum_{\vec{k},\sigma} m \frac{\partial \epsilon_{\vec{k}}}{\partial \vec{k}} n(\vec{k},\sigma) \tag{6.18b}$$

where m is the bare mass of a particle.

Equating (6.18a) and (6.18b), we have:

$$\sum_{\vec{k},\sigma} \vec{k}\, n(\vec{k}) = \sum_{\vec{k},\sigma} m \frac{\partial \epsilon_k}{\partial \vec{k}} n(\vec{k},\sigma) \quad . \tag{6.19}$$

Since (6.19) is an identity, we may take a functional derivative with respect to $n(\vec{k},\sigma)$

$$\vec{k} = m \frac{\partial \epsilon_{\vec{k}}}{\partial \vec{k}} + m \sum_{\sigma'} \int \frac{d^3k'}{(2\pi)^3} \frac{\partial}{\partial \vec{k}'} \left(\frac{\delta \epsilon_{\vec{k}'}}{\delta n(\vec{k},\sigma)} \right) n(\vec{k}',\sigma') \tag{6.20a}$$

which we rewrite:

$$\begin{aligned} \frac{\vec{k}}{m} &= \frac{\partial \epsilon_{\vec{k}}}{\partial \vec{k}} + \sum_{\sigma'} \int \frac{d^3k'}{(2\pi)^3} \frac{\partial f(\vec{k}\sigma, \vec{k}'\sigma')}{\partial \vec{k}'} n(\vec{k}',\sigma') \\ &= \frac{\partial \epsilon_{\vec{k}}}{\partial \vec{k}} - \sum_{\sigma'} \int \frac{d^3k'}{(2\pi)^3} f(\vec{k}\sigma, \vec{k}'\sigma') \frac{\partial n(\vec{k}',\sigma')}{\partial \vec{k}'} \quad . \end{aligned} \tag{6.20b}$$

In (6.20), we have used the usual replacement of the sum $\sum_{\vec{k}}$ by an integral $\mathcal{V} \int \frac{d^3k}{(2\pi)^3}$. At zero temperature

$$\frac{\partial n(\vec{k},\sigma)}{\partial \vec{k}} = -\hat{k}\, \delta(k_F - k) \tag{6.21}$$

where $\hat{k}$ is the unit vector along $\vec{k}$. Thus, using (6.13) and (6.20) and taking $\vec{k}$ on the Fermi surface, we obtain:

$$\frac{\vec{k}}{m} = \frac{\vec{k}}{m^*} + \sum_{\sigma'} \int \frac{d^3k'}{(2\pi)^3} f(\vec{k}\sigma, \vec{k}'\sigma') \hat{k}'\, \delta(k_F - k') \quad . \tag{6.22a}$$

Finally, denoting by θ the angle between $\vec{k}$ and $\vec{k}'$, and using Eqs. (6.10) – (6.12) we obtain

$$\begin{aligned} \frac{1}{m} &= \frac{1}{m^*} + \frac{k_F}{(2\pi)^3} \sum_{\sigma'} 4\pi \int d(\cos\theta) \sum_L (f_L + 4\sigma \cdot \sigma' \phi_L) P_L(\cos\theta) \cos\theta \\ &= \frac{1}{m^*} + \frac{k_F}{3\pi^2} f_1 \quad . \end{aligned} \tag{6.22b}$$

Because this and all subsequent integrals over $f(k, k')$ are evaluated at the Fermi surface and thus are ultimately multiplied by the density of states at the Fermi surface, it is conventional to define new Legendre expansion coefficients which include the density of states. Using Eq. (6.13b), the density of states at the Fermi surface is

$$\begin{aligned} N(0) &= \frac{1}{\mathcal{V}}\sum_{k,\sigma}\delta\left(\epsilon^0(k)-\mu\right) \\ &= \frac{2}{(2\pi)^3}4\pi\int k^2 dk\,\delta\left(\epsilon^0(k)-\mu\right) \\ &= \frac{m^* k_F}{\pi^2} \end{aligned} \tag{6.23}$$

and we define the normalized expansion coefficients*

$$F_L \equiv \frac{k_F m^*}{\pi^2} f_L \tag{6.24a}$$

$$Z_L \equiv \frac{k_F m^*}{\pi^2}\phi_L \ . \tag{6.24b}$$

With these definitions, we see from Eq. (6.22) that the effective mass directly specifies the $L = 1$ component of the spin independent quasiparticle interaction

$$\frac{m^*}{m} = \left(1 + \frac{F_1}{3}\right) \ . \tag{6.25}$$

As is also evident in Problem 6.2, we observe from Eq. (6.22) that the current associated with a quasiparticle is not simply $\frac{k}{m^*}$. Because of the interaction with all the other particles in the medium, there is another piece coming from the interaction f which represents the other particles dragged along with the quasiparticle, or the backflow of other particles around the quasiparticle.

Compressibility and Sound Velocity

The compressibility χ of a liquid characterizes the change of pressure with volume according to the standard definition†

$$\frac{1}{\chi} = -\mathcal{V}\frac{\partial P}{\partial \mathcal{V}} = \rho\frac{\partial P}{\partial \rho} \tag{6.26}$$

where $\mathcal{V}$ is the volume of the system and the density is $\rho = \frac{N}{\mathcal{V}}$. The velocity of thermodynamic sound, c_1, provides a convenient experimental means of measuring χ, since by Eq. (5.124)

$$c_1^2 = \frac{1}{m}\frac{\partial P}{\partial \rho} = \frac{1}{m\rho\chi} \ . \tag{6.27}$$

---

* A more modern notation for F_L is F_L^s and for Z_L is F_L^a, where the s and a denote symmetric and antisymmetric, spin combinations. Note also that our definition of Z differs by a factor of 4 from some of the early literature.

† Frequently, the compressibility of nuclear matter is expressed in terms of the compression modulus, $K \equiv k_F^2 \frac{\partial^2}{\partial k_F^2}\left(\frac{E}{A}\right)$. At equilibrium, $\frac{1}{\chi} = \frac{\rho K}{9}$.

From the observation in Fig. 5.16 that thermodynamic sound corresponds to a spherically symmetric change in the occupation of states at the Fermi surface, it is clear physically that χ and c_1 will depend on a spherically symmetric average of $f(k,k')$ and thus on the Landau parameter F_0. We now derive the precise relation.

Because the free energy is extensive, we may write it in terms of a free energy per unit volume as follows:

$$F(T,\mathcal{V},N) \equiv \mathcal{V}\, f\left(T, \frac{N}{\mathcal{V}}\right) . \tag{6.28}$$

Hence,

$$P = -\frac{\partial F}{\partial \mathcal{V}} = f - \rho\frac{\partial f}{\partial \rho} \tag{6.29}$$

and thus

$$\frac{1}{\chi} = \rho^2 \frac{\partial^2}{\partial \rho^2} f(T,\rho) . \tag{6.30}$$

Since it is convenient to express the final result in terms of the chemical potential, we use Eq. (6.28) to obtain

$$\mu = \frac{\partial F}{\partial N} = \frac{\partial}{\partial \rho} f(T,\rho) \tag{6.31}$$

so that

$$\frac{1}{\chi} = \rho^2 \frac{\partial \mu}{\partial \rho} . \tag{6.32}$$

In order to calculate $\frac{\partial \mu}{\partial \rho}$, we note that

$$\mu = \epsilon(k_F, n(\vec{k},\sigma)) \tag{6.33}$$

and thus, using Eq. (6.6),

$$\frac{\partial \mu}{\partial \rho} = \frac{\partial \epsilon^0}{\partial k_F}\frac{\partial k_F}{\partial \rho} + \sum_{\sigma'} \int \frac{d^3k'}{(2\pi)^3} f(k_F, \vec{k}') \frac{\delta n(k')}{\delta k_F}\frac{\partial k_F}{\partial \rho} . \tag{6.34a}$$

Using the relations $\rho = \frac{k_F^3}{3\pi^2}$ from Eq. (6.1), $\frac{\partial \epsilon^0}{\partial k_F} = \frac{k_F}{m^*}$ from Eq. (6.13), $\frac{\partial n(k')}{\partial k_F} = \delta(k' - k_F)$ from Eq. (6.3) and $\int \frac{d\Omega}{4\pi} f(\theta,\sigma,\sigma') = f_0 = \frac{\pi^2}{m^* k_F} F_0$ from Eqs. (6.12) and (6.24), we obtain

$$\begin{aligned}\rho\frac{\partial \mu}{\partial \rho} &= \rho\frac{\partial k_F}{\partial \rho}\left[\frac{k_F}{m^*} + \frac{k_F^2}{(2\pi)^3}\sum_{\sigma'}\int d\Omega f(\theta,\sigma,\sigma')\right] \\ &= \frac{k_F^2}{3m^*}[1+F_0]\end{aligned} \tag{6.34b}$$

from which it follows from Eqs. (6.32) and (6.27) that

$$\frac{1}{\chi} = \frac{\rho k_F^2}{3m^*}(1+F_0) = \frac{\rho k_F^2}{m}\,\frac{(1+F_0)}{(3+F_1)}$$

and

$$c_1^2 = \frac{k_F{}^2}{3mm^*}(1+F_0) = \frac{k_F{}^2}{m^2}\frac{(1+F_0)}{(3+F_1)} \ . \tag{6.35}$$

As anticipated, the compressibility depends upon F_0, coming from the spherical average of $f(\theta)$, as well as F_1, arising from the effective mass. Note that the compressibility becomes infinite, that is, the system becomes unstable against density oscillations, when $F_0 = -1$. This is a specific example of the general stability criteria $F_L > -(2L+1)$, and $Z_L > -(2L+1)$ derived in Problem 6.3.

Magnetic Susceptibility

In contrast to the previous properties which were independent of the Fermion spin, we now consider the magnetization which is induced by an external magnetic field. The magnetic interaction energy of a Fermion with an external field H is $-\sigma\gamma H$, where for a point particle with no anomalous moment, the gyromagnetic ratio is $\gamma = \frac{e\hbar}{mc}$ and $\sigma = \pm\frac{1}{2}$ is the spin projection along H. Hence, in the presence of an external field H, the energies of the spin $+\frac{1}{2}$ particles are lowered and the energies of the spin $-\frac{1}{2}$ particles are raised, causing the equilibrium occupation of the spin $+\frac{1}{2}$ states to be larger than the spin $-\frac{1}{2}$ states and thus producing net magnetization.

To calculate the induced magnetization, we need to calculate how $n(\sigma)$, the density of particles with spin projection σ, changes in the presence of a weak external field H. Since the total number of particles is fixed, we may write

$$\rho(\sigma) = \rho_0 + \sigma\Delta\rho \ . \tag{6.36}$$

Since the chemical potential of spin state $\sigma, \mu(\sigma)$, depends on both $n(\sigma)$ and $n(-\sigma)$, the equilibrium condition $\mu(\sigma) = \mu(-\sigma)$ poses a constraint which we will use to specify Δn. The chemical potential for each spin is equal to the quasiparticle energy at the corresponding Fermi surface

$$\mu(\sigma) = \epsilon\left(k_F(\sigma), \sigma\right) \tag{6.37}$$

where, in the presence of an external field, the quasiparticle energy is

$$\epsilon(\vec{k},\sigma) = \epsilon^0(\vec{k},\sigma) - \sigma\gamma H + \frac{1}{\mathcal{V}}\sum_{\vec{k}'\sigma'} f(\vec{k}\sigma, \vec{k}'\sigma')\delta n(\vec{k}'\sigma') \ . \tag{6.38}$$

Since both $k_F(\sigma)$ and the distribution function $n(k,\sigma)$ are specified by the density $\rho(\sigma)$, $\mu(\sigma)$ is a function of $\rho(\sigma)$ and $\rho(-\sigma)$. For a weak external field, we may expand the chemical potential to first order around the result μ_0 at $H = 0$.

$$\mu(\sigma) \approx \mu_0 - \sigma\gamma H + \sum_{\sigma'} \frac{\partial\mu(\sigma)}{\partial\rho(\sigma')}\sigma'\Delta\rho \ . \tag{6.39}$$

The derivatives of the chemical potential $\mu(\sigma)$ with respect to the densities for each spin state are evaluated as in Eq. (6.34), where we now distinguish the individual contributions of each spin population

$$\begin{aligned}\frac{\partial\mu(\sigma)}{\partial\rho(\sigma')} &= \frac{\partial k_F(\sigma')}{\partial\rho(\sigma')}\left[\frac{\partial\epsilon^0(\sigma)}{\partial k_F(\sigma)}\delta(\sigma,\sigma') + \int\frac{d^3k'}{(2\pi)^3} f\left(k_F\sigma, k'\sigma'\right)\frac{\partial n(k',\sigma')}{\partial k_F(\sigma')}\right] \\ &= \frac{2\pi^2}{k_F^2(\sigma')}\left[\frac{k_F}{m^*}\delta(\sigma,\sigma') + \frac{k_F}{2m^*}\left(F_0 + 4\sigma\sigma' Z_0\right)\right] \ .\end{aligned} \tag{6.40}$$

Using Eqs. (6.39– 6.40), the equilibrium condition $\mu(\sigma) = \mu(-\sigma)$ provides the desired constraint on $\Delta\rho$

$$\gamma H = \frac{2\pi^2}{k_F m^*}(1 + Z_0)\Delta\rho \ . \tag{6.41}$$

Using Eqs. (6.36) and (6.41) the magnetization density is

$$\begin{aligned} m &= \sum_{\sigma=-\frac{1}{2}}^{\frac{1}{2}} (\rho_0 + \sigma\Delta\rho)\,\sigma\gamma = \frac{\gamma}{2}\Delta\rho \\ &= \frac{\gamma^2 H k_F m^*}{4\pi^2(1+Z_0)} \ . \end{aligned} \tag{6.42}$$

From which it follows that the magnetic susceptibility is

$$\chi_M = \frac{\partial m}{\partial H} = \frac{\gamma^2 k_F m^*}{4\pi^2(1+Z_0)} \ . \tag{6.43}$$

As in the case of the compressibility, the $L = 0$ Legendre coefficient of the quasiparticle interaction enters because of the spherically symmetric deformation of the Fermi sphere. However, since the spin $+\frac{1}{2}$ sphere expands while the spin $-\frac{1}{2}$ sphere decreases, the difference, Z_0, between the interactions of like and unlike spins contributes rather than the sum F_0. Again, we see a specific example of the general stability criterion $Z_L > (2L+1)$ of Problem 6.3. In liquid ^{3}He, as tabulated in Table 6.1, Z_0 ranges between -0.67 at 0 atm and -0.76 at 27 atm indicating that the strong spin dependence of the effective interaction makes the system nearly unstable with respect to ferromagnetic ordering. The physical origin of this dependence is simple. By antisymmetry, the spatial wave function of a spin singlet is S-wave and appreciably samples the short-range repulsive core of the He-He potential, whereas the spatial wave function of a spin triplet is P-wave and experiences much less repulsion.

NONEQUILIBRIUM PROPERTIES AND COLLECTIVE MODES

The dynamics of a Fermi liquid close to equilibrium is governed by a Boltzmann equation for the quasiparticle distribution. Two fundamental assumptions are required. The first is that we consider only long wavelength, low-energy excitations. Then, instead of considering the quantum Wigner distribution function introduced in Section 5.4, we may treat the non-equilibrium distribution function at position $\vec{r}$ as a classical distribution function $n(\vec{k},\vec{r},t)$. The second major assumption is that the local quasiparticle energy at the position $\vec{r}$, $\epsilon(\vec{k},\vec{r})$, plays the role of the quasiparticle Hamiltonian, so that

$$\begin{aligned} \dot{r} &= \frac{\partial\epsilon(k,r)}{\partial k} \\ \dot{k} &= -\frac{\partial\epsilon(k,r)}{\partial r} \ . \end{aligned} \tag{6.44}$$

Then, the Boltzmann equation is obtained by equating the total time derivative of $n(\vec{k},\vec{r},t)$ to a quasiparticle collision integral $I(n)$

$$\frac{\partial n}{\partial t} + \frac{\partial n}{\partial \vec{r}}\cdot\frac{\partial\epsilon}{\partial\vec{k}} - \frac{\partial n}{\partial\vec{k}}\cdot\frac{\partial\epsilon}{\partial\vec{r}} = I(n) \ . \tag{6.45}$$

Since we will be interested in small deviations from equilibrium, we expand n and ϵ around their equilibrium values as follows

$$n(\vec{k},\vec{r},t) = n_0(\vec{k}) + \delta n(\vec{k},\vec{r},t)$$
$$\epsilon(\vec{k},\vec{r},t) = \epsilon^0_{\vec{k}} + \frac{1}{\mathcal{V}}\sum_{k} f(\vec{k},\vec{k}')\delta n(\vec{k}',\vec{r},t) \tag{6.46}$$
$$\frac{\partial \epsilon}{\partial \vec{r}} = \frac{1}{\mathcal{V}}\sum_{k'} f(\vec{k},\vec{k}')\frac{\partial}{\partial \vec{r}}\delta n(\vec{k}',\vec{r},t) \; .$$

Expanding Eq. (6.45) to first order in δn, we obtain

$$\frac{\partial}{\partial t}\delta n(\vec{k},\vec{r},t) + \frac{\vec{k}}{m^*}\frac{\partial \delta n}{\partial r} - \frac{\partial n_0}{\partial k}\frac{1}{\mathcal{V}}\sum_{k'} f(k,k')\frac{\partial}{\partial r}\delta n(k',r,t) = I(n) \tag{6.47a}$$

where throughout this section where no ambiguity arises, spin sums will be understood to be included in sums over k. Using

$$\frac{\partial n_0}{\partial \vec{k}} = -\delta(\mu - \epsilon_k)\frac{\partial \epsilon_k}{\partial \vec{k}} = -\frac{\vec{k}}{m^*}\delta(\mu - \epsilon_k) \tag{6.47b}$$

the linearized Boltzmann equation is written

$$\frac{\partial}{\partial t}\delta n + \frac{\vec{k}}{m^*}\cdot\left(\frac{\partial}{\partial \vec{r}}\delta n + \delta(\mu - \epsilon_k)\frac{1}{\mathcal{V}}\sum_{\vec{k}'} f(\vec{k},\vec{k}')\frac{\partial}{\partial \vec{r}}\delta n(\vec{k}',\vec{r},t)\right) = I(n) \; . \tag{6.48}$$

Conservation Laws

Using the relation

$$\frac{\partial n}{\partial r_j}\frac{\partial \epsilon}{\partial k_j} - \frac{\partial n}{\partial k_j}\frac{\partial \epsilon}{\partial r_j} = \frac{\partial}{\partial r_j}\left(n\frac{\partial \epsilon}{\partial k_j}\right) - \frac{\partial}{\partial k_j}\left(n\frac{\partial \epsilon}{\partial r_j}\right) \; , \tag{6.49}$$

summing (6.45) over momentum and spin, and noting the sum over the collision integral vanishes yields

$$\begin{aligned}\frac{\partial}{\partial t}\sum_{k} n &= -\frac{\partial}{\partial r_j}\sum_{k} n\frac{\partial \epsilon}{\partial k_j} + \sum_{k}\frac{\partial}{\partial k_j}\left(n\frac{\partial \epsilon}{\partial r_j}\right) \\ &= -\nabla\cdot\vec{J}(r,t)\end{aligned} \tag{6.50a}$$

where the current is defined

$$J_i = \sum_{\vec{k}} n(\vec{k},\vec{r},t)\frac{\partial \epsilon}{\partial k_i} \; . \tag{6.50b}$$

Linearization using Eqs. (6.46) and integration by parts yields

$$\delta J_i = \frac{1}{\mathcal{V}}\sum_{\vec{k}}\left(\delta n(\vec{k})\frac{\partial \epsilon^0(\vec{k})}{\partial k_i} - \frac{\partial n_0(k)}{\partial k_i}\frac{1}{\mathcal{V}}\sum_{\vec{k}'} f(\vec{k},\vec{k}')\delta n(\vec{k}')\right) \; . \tag{6.50c}$$

In particular, if δn corresponds to the distribution function for a single quasiparticle of momentum $\vec{k}$, we recover the result, Eq. (6.22a) for the quasiparticle current including the drag term

$$\vec{J}(\vec{k})_{\text{quasiparticle}} = \frac{\vec{k}}{m^*} + \frac{1}{V}\sum_{\vec{k}'} f(\vec{k},\vec{k}')\hat{k}'\,\delta(k_F - k') \ . \tag{6.50d}$$

The equation of momentum conservation is obtained by multiplying the Boltzmann equation (6.45) by k_i, integrating over momentum and spin using (6.49), and noting that the collision integral vanishes by momentum conservation

$$\frac{\partial}{\partial t}\left(\int \frac{d^3k}{(2\pi)^3} k_i n\right) + \int \frac{d^3k}{(2\pi)^3} k_i \left[\frac{\partial}{\partial r_j}\left(n\frac{\partial \epsilon}{\partial k_j}\right) - \frac{\partial}{\partial k_j}\left(n\frac{\partial \epsilon}{\partial r_j}\right)\right] = 0 \ . \tag{6.51}$$

Integrating the last term by parts and using the fundamental relation Eq. (6.6) that $\epsilon_k = \frac{\delta E}{\partial n(k)}$, we may write

$$\begin{aligned} -\int \frac{d^3k}{2\pi^3} k_i \frac{\partial}{\partial k_j}\left(n\frac{\partial \epsilon}{\partial r_j}\right) &= \int \frac{d^3k}{(2\pi)^3} n \frac{\partial \epsilon}{\partial r_i} \\ &= \frac{\partial}{\partial r_i}\left(\int \frac{d^3k}{(2\pi)^3} n\epsilon\right) - \int \frac{d^3k}{(2\pi)^3}\frac{\partial n}{\partial r_i}\frac{\partial E}{\partial n} \\ &= \frac{\partial}{\partial r_i}\left(\int \frac{d^3k}{(2\pi)^3} n\epsilon - E\right) \ . \end{aligned} \tag{6.52}$$

Hence, the momentum conservation equation may be written

$$\frac{\partial}{\partial t}\int \frac{d^3k}{(2\pi)^3} k_i n + \frac{\partial}{\partial r_j}\Pi_{ij} = 0 \tag{6.53a}$$

where the momentum flux tensor is given by

$$\Pi_{ij} = \int \frac{d^3k}{(2\pi)^3} k_i n \frac{\partial \epsilon}{\partial k_j} + \delta_{ij}\left[\int \frac{d^3k}{(2\pi)^3} n\epsilon - E\right] \ . \tag{6.53b}$$

The law of conservation of energy is derived analogously by multiplying (6.45) by $\epsilon(k,r)$, using Eq. (6.49), and noting that the integral of the collision term vanishes by energy conservation

$$\int \frac{d^3k}{(2\pi)^3}\epsilon\frac{\partial n}{\partial t} + \int \frac{d^3k}{(2\pi)^3}\epsilon\left[\frac{\partial}{\partial r_j}\left(n\frac{\partial \epsilon}{\partial k_j}\right) - \frac{\partial}{\partial k_j}\left(n\frac{\partial \epsilon}{\partial r_j}\right)\right] = 0 \ . \tag{6.54}$$

Integrating the last term by parts and using Eq. (6.6) yields

$$\frac{\partial}{\partial t}E + \frac{\partial}{\partial r_j}Q_j = 0 \tag{6.55a}$$

where

$$Q_j = \int \frac{d^3k}{(2\pi)^3} n \epsilon \frac{\partial \epsilon}{\partial k_j} \; . \tag{6.55b}$$

Linearizing Eq. (6.55b) using Eqs. (6.46) and integrating by parts yields

$$\delta Q_j = \frac{1}{V} \sum_k \left(\delta_n(k) \epsilon^0(k) \frac{k_j}{m^*} - \frac{\partial n_0}{\partial k_i} \epsilon^0 \frac{1}{V} \sum_{k'} f(k,k') \delta n(k') \right) \; . \tag{6.55c}$$

Like the current, the energy current has a drag term accounting for the energy flow of the particle interacting with a quasiparticle.

Zero Sound

The zero sound mode discussed in Section 5.4 in connection with the RPA linear response theory was originally discovered by Landau (1957) as a collective oscillation of the Fermi liquid. In this section we will show how this mode emerges naturally as a solution to the linearized Boltzmann equation.

Recall that Eq. (6.27) for thermodynamic sound relies on the hypothesis of local equilibrium and thus requires that the sound frequency ν be small compared to the inverse collision time $\nu << \frac{1}{\tau}$. By the general phase space argument used in Eq. (5.72) for the quasiparticle lifetime, we know that the collision time is of order $\tau \sim (\epsilon - \epsilon_F)^{-2}$. Finally, since the quasiparticles excited at temperature T have energies $(\epsilon_k - \epsilon_F) \sim kT$, it follows that ordinary sound propagates for

$$\nu << T^2 \tag{6.56}$$

and thus cannot propagate for sufficiently low temperatures.

At temperatures sufficiently low that ordinary sound ceases to propagate because collisions are negligible, we may neglect the collision term $I(n)$ in the Boltzmann equation and solve for the relevant collective modes in the collisionless regime. We will seek a solution of the form

$$\delta n(\vec{k}, \vec{r}, t) = e^{i(\vec{q}\cdot\vec{r} - \omega t)} \phi_k \tag{6.57a}$$

in which case the linearized Boltzmann equation Eq. (6.48) with $I(n) = 0$ becomes

$$(\vec{q} \cdot \vec{v}_k - \omega)\, \phi_{\vec{k}} + \vec{q} \cdot \vec{v}_k \delta(\mu - \epsilon_k) \frac{1}{V} \sum_{k'} f(k,k') \phi_{k'} = 0 \tag{6.57b}$$

where $\vec{v}_k = \frac{\vec{k}}{m^*}$. Since ϕ_k is proportional to $\delta(\mu - \epsilon_k)$, it is convenient to define

$$\phi_{\vec{k}} \equiv \delta(\epsilon_k - \mu) v_F\, u(\vec{k}) \tag{6.57c}$$

and using the fact that $|v_k| = v_F = \frac{k_F}{m^*}$ for momenta restricted to the Fermi surface, we obtain

$$(\vec{q} \cdot \vec{v}_k - \omega)\, u(\vec{k}) + \vec{q} \cdot \vec{v}_k \frac{1}{V} \sum_{k'} f(k,k') \delta(\epsilon_{k'} - \mu) u(k') = 0 \; . \tag{6.58}$$

Note that physically, u_k is the displacement of the Fermi surface at momentum k since, by Eqs. (6.57)

$$\begin{aligned} n(\vec{k},\vec{r},t) &= n_0(\vec{k}) + \delta n(\vec{k},\vec{r},t) \\ &= n_0(\vec{k}) + e^{i(\vec{q}\cdot\vec{r}-\omega t)}\delta(\epsilon_k - \mu)v_F u(\vec{k}) \\ &= n_0(\vec{k}) - e^{i(\vec{q}\cdot\vec{r}-\omega t)}\frac{\partial n_0(\vec{k})}{\partial|k|}u(\vec{k}) \ . \end{aligned} \tag{6.59}$$

Since momenta are restricted to the Fermi surface, $u(\vec{k})$ only depends on the spin σ and the direction of $\vec{k}$, which we denote by polar coordinates $\Omega = (\theta, \phi)$ with the polar axis in the direction of $\vec{q}$. As in Eq. (5.121), we also define the ratio of the velocity of the mode $\frac{\omega}{q}$ to the quasiparticle velocity at the Fermi surface.

$$s = \frac{\omega}{qv_F} \tag{6.60}$$

so that Eq. (6.58) becomes

$$\begin{aligned} (s-\cos\theta)u(\Omega,\sigma) &= \cos\theta \sum_{\sigma'}\int\frac{d^3\vec{k}'}{(2\pi)^3}\delta\left(\frac{k_F}{m^*}(k-k_F)\right) f(\vec{k}\sigma,\vec{k}'\sigma')u(\Omega',\sigma') \\ &= \cos\theta\int\frac{d\Omega'}{4\pi}\frac{1}{2}\sum_{\sigma}F(\vec{k}\sigma,\vec{k}'\sigma')u(\Omega',\sigma') \end{aligned} \tag{6.61a}$$

where, as in Eq. (6.24), F includes the density of states at the Fermi surface

$$\begin{aligned} F(\vec{k}\sigma,\vec{k}'\sigma') &= \frac{m^* k_F}{\pi^2} f(\vec{k}\sigma,\vec{k}'\sigma') \\ &= \sum_L (F_L + 4\sigma\cdot\sigma' Z_L)P_L(\cos\theta_{\Omega\Omega'}) \ , \end{aligned} \tag{6.61b}$$

$\theta_{\Omega\Omega'}$ denotes the angle between $\vec{k}$ and $\vec{k}'$, and we have explicitly displayed the spin dependence for future reference.

Let us first consider the case in which the Fermi surfaces of both spin states oscillate in phase, in which case Eq. (6.61a) simplifies to

$$(s-\cos\theta)u(\Omega) = \cos\theta\int\frac{d\Omega'}{4\pi}\sum_L F_L P_L(\cos\theta_{\Omega\Omega'})u(\Omega') \ . \tag{6.62a}$$

Making the further assumption that the quasiparticle interaction is independent of angle simplifies the equation still further

$$(s-\cos\theta)u(\Omega) = \cos\theta F_o\int\frac{d\Omega'}{4\pi}u(\Omega') \ . \tag{6.62b}$$

The solution has the form

$$u(\Omega) = \frac{\cos\theta}{s-\cos\theta}C \tag{6.63a}$$

where

$$C = F_0 \int \frac{d\Omega'}{4\pi} u(\Omega') \ . \tag{6.63b}$$

Substituting the solution (6.63a) in (6.63b) and allowing for the possibility that s may have an infinitesimal positive imaginary part, we obtain the condition

$$\frac{1}{F_0} = \frac{s}{2} \ln\left(\frac{s+1}{s-1}\right) - 1 - \frac{i\pi}{2} s\theta(1-|s|) \tag{6.64}$$

which is identical to Eq. (5.120). As we argued in Chapter 5, in order to obtain a real solution to (6.64), corresponding to an undamped mode, we must have $s > 1$ which requires $F > 0$. As F runs from 0 to ∞, s acquires all values greater than 1, with the limiting behavior

$$\begin{aligned} s &\xrightarrow[F\to 0]{} 1 + 2e^{-\frac{2}{F_0}} \\ &\xrightarrow[F\to\infty]{} \sqrt{\frac{F_0}{3}} \ . \end{aligned} \tag{6.65}$$

Relation (6.63a) shows that the Fermi surface for the longitudinal zero sound mode is deformed into the surface of revolution sketched in Fig. 5.16, with the elongated end pointing in the $\vec{q}$ direction.

Let us now consider a more general case in which all multipoles F_l are included, but we still assume both spin states oscillate in phase and that $u(\Omega)$ is independent of the azimuthal angle ϕ. Then, Eq. (6.62a) becomes

$$(s - \cos\theta) u(\theta) = \cos\theta \int \frac{d(\cos\theta')}{2} \sum_L F_L P_L \cos\theta_{\Omega\Omega'} u(\theta') \ . \tag{6.66a}$$

As shown in Problem 6.4, expanding $u(\theta)$ in Legendre polynomials

$$u(\theta) = \sum_L u_L P_L(\cos\theta) \tag{6.66b}$$

and using the addition theorem in Eq. (6.66a) yields the coupled equation

$$\frac{u_L}{2L+1} = -\sum_{L'} D_{LL'} \frac{F_{L'} u_{L'}}{2L'+1} \tag{6.66c}$$

where

$$D_{LL'} = \int_{-1}^{1} \frac{d(\cos\theta)}{2} P_L(\cos\theta) P_{L'}(\cos\theta) \frac{\cos\theta}{\cos\theta - s} \ . \tag{6.66d}$$

Using the explicit results of Problem 6.4

$$\begin{aligned} D_{00} &= 1 - \frac{s}{2} \ln\left(\frac{s+1}{s-1}\right) \equiv D \\ D_{01} &= D_{10} = sD \\ D_{11} &= S^2 D + \frac{1}{3} \end{aligned} \tag{6.67a}$$

we find for the case in which only F_0 and F_1 are non-zero the simultaneous equation

$$\begin{aligned} -u_0 &= DF_0 u_0 + sDF_1 \frac{u_1}{3} \\ -\frac{u_1}{3} &= sDF_0 u_0 + \left(s^2 D + \frac{1}{3}\right) F_1 \frac{u_1}{3} \end{aligned} \tag{6.67b}$$

which have a solution when s satisfies

$$\frac{1}{2} s \ln\left(\frac{s+1}{s-1}\right) - 1 = \frac{1 + \frac{1}{3}F_1}{F_0 + s^2 F_1 + \frac{1}{3} F_0 F_1} \quad . \tag{6.67c}$$

Yet another solution arises if we allow for azimuthal dependence. Let us keep only the first two terms in the Legendre expansion

$$\begin{aligned} F &= F_0 + F_1 \cos\theta_{\Omega\Omega'} \\ &= F_0 + F_1 \left[\cos\theta \cos\theta' + \frac{1}{2} \sin\theta \sin\theta' \left(e^{i(\phi-\phi')} + e^{i(\phi'-\phi)}\right)\right] \end{aligned} \tag{6.68a}$$

and assume a circularly polarized solution

$$u(\theta, \phi) = w(\theta) e^{i\phi} \quad . \tag{6.68b}$$

Using this form of F and u in Eq. (6.62a) yields the solution

$$w(\theta) = \frac{\cos\theta \sin\theta}{s - \cos\theta} C' \tag{6.69a}$$

where

$$C' = \frac{F_1}{4} \int_{-1}^{1} d(\cos\theta') \sin\theta' \, w(\theta') \quad . \tag{6.69b}$$

Note that in contrast to the forward peaked zero sound mode in Eq. (6.63a), this mode has the largest deformation at $\theta \sim \frac{\pi}{4}$. Substituting (6.69a) in (6.69b) yields the equation for the propagation velocity of the mode

$$\frac{4}{F_1} = -\frac{4}{3} + 2s^2 - s\left(s^2 - 1\right) \ln\left(\frac{s+1}{s-1}\right) \quad . \tag{6.70}$$

In order for this mode to propagate, we must have $s > 1$ which requires $F_1 > 6$, and the limiting behavior is given by

$$\begin{aligned} s &\underset{F_1 \to \infty}{\longrightarrow} \sqrt{\frac{F_1}{15}} \\ (s-1) \ln\left(\frac{s-1}{2}\right) &\underset{F_1 \to 6+}{\longrightarrow} \frac{6 - F}{18} \quad . \end{aligned} \tag{6.71}$$

Zero sound has been observed in liquid ^{3}He at 0.32 atm with a velocity relative to thermodynamic sound of $\left.\frac{c_0 - c_1}{c_1}\right|_{0.32\ \mathrm{atm}} = 0.035 \pm .003$ (Abel, Anderson and Wheatley,

1966). Assuming this ratio of velocities changes slowly with pressure, we will compare it with the prediction of Fermi liquid theory at 0.28 atm, where the Landau parameters from measurement of the specific heat and thermodynamic sound (Wheatley, 1966) are $F_0 = 10.77$ and $F_1 = 6.25$. From Eqs. (6.35) and (6.60), the measured value of $\frac{c_0 - c_1}{c_1}$ corresponds to $s = 3.60 \pm 0.01$. Using only the single multipole F_0 in Eq. (6.64) yields a $s = 2.05$, in poor agreement with experiment. Including both F_0 and F_1 in Eq. (6.67c) yields $s = 3.60$, in perhaps fortuitously good agreement considering the omission of still higher multipoles. Although Table 6.1 shows that F_1 increases significantly above six at high pressure, the circularly polarized mode has not yet been observed.

Spin Waves

Having explored in detail zero sound modes in which both spin states oscillate in phase, it is straightforward to extend the analysis to spin waves in which the spins oscillate out of phase. Spin waves are obtained from solutions of the general equations (6.61a) of the form

$$u_s(\Omega, \sigma) = u_s(\Omega) 2\sigma \quad . \tag{6.72}$$

Substituting this *ansatz* in Eq. (6.61a), using Eq. (6.61b) and performing the spin sums yields the equation for $u_s(\Omega)$

$$(s - \cos\theta) u_s(\Omega) = \cos\theta \int \frac{d\Omega'}{4\pi} \sum_L Z_L P_L(\cos\theta_{\Omega\Omega'}) u(\Omega') \quad . \tag{6.73}$$

This result is of the same form as Eq. (6.62a) with F_L replaced by Z_L and $u(\Omega)$ replaced by $u_s(\Omega)$, so that all the solutions of the previous section apply to spin modes as well. In the case of liquid ^{3}He, however, the relevant Landau parameters shown in Table 6.1 are negative so that there are no real solutions to the equations (6.64) or (6.67c). Since s is not real, the spin wave modes are damped and do not propagate. A general theorem by Mermin (1967) discussed in Problem 6.5 shows that at least one of these two possibilities, spin waves or density waves, must always propagate in a Fermi liquid.

In the case of nuclear systems, in addition to spin degrees of freedom, there is also isospin. Thus, the preceding analysis is generalized to include terms of the form $G(\cos\theta_{\Omega\Omega'})\tau \cdot \vec{\tau}$ and $G'(\cos\theta_{\Omega\Omega'})\sigma \cdot \vec{\sigma}\tau \cdot \vec{\tau}$ in the quasiparticle interaction and one obtains a rich variety of spin, isospin, and spin-isospin modes. Electromagnetic probes couple to the proton much more strongly than to the neutrons in nuclei, and thus strongly excite isospin modes, of which the giant dipole resonance is an outstanding example. At energies of several hundred MeV, (p, n) reactions, in which a proton is scattered from a nucleus and transfers one unit of charge to produce an outgoing neutron, produce spin-isospin excitations. Of particular interest in dense nuclear and neutron star matter is the spin-isospin mode with the quantum numbers of the pion. As the density increases, the energy of the mode decreases until at a critical density the mode becomes self-sustaining and gives rise to pion condensation (see, for example, Midgal, 1978).

To conclude this section, let us emphasize again that Landau theory describes the properties of the lowest excited states of a normal Fermi liquid in terms of two functions $f(\theta)$ and $\varphi(\theta)$ which parameterize quasiparticle interactions on the Fermi surface. The coefficients of the expansion of these functions in Legendre polynomials can be related to simple thermodynamic observables, and Landau theory can be used as a phenomenological theory where the coefficients $F_1, F_1, \cdots$ and $Z_0, Z_1, \cdots$, are measured by simple experiments.

In the next section, we show that the quasiparticle interaction is related to the two-particle irreducible vertex function taken on the Fermi surface, that is, to the forward scattering amplitude of two quasi-particles on the Fermi surface, and we show how it can be computed from the microscopic interaction.

6.3 MICROSCOPIC FOUNDATION

The microscopic foundation of Fermi liquid theory at zero temperature and the definition of the quasiparticle interaction may be seen by studying the exact integral equation for the two-body vertex function. The basic idea is simple. We know from the study of Green's functions that the poles of $G^{(2)}$ and thus of the vertex function $\Gamma^{(2)}$ describe the excited states of the N-body and $N \pm 2$-body systems. By restricting our attention to the relevant poles of the N-body system and analyzing the structure of the exact integral equation for the vertex function in the vicinity of these poles, we will obtain an integral equation for the fundamental excitations of precisely the same form as Eq. (6.56a) in the last section describing zero sound and collective excitations. From this correspondence, we will be able to relate the quasiparticle interaction to the vertex function at the Fermi surface in the particle-hole channel, normalized by the residue of the quasiparticle pole. In addition, we will also be able to identify the quasiparticle forward scattering amplitude, and use the restriction posed by antisymmetry to obtain a sum rule on the Landau parameters. Since all the essential ideas are clearly presented in the original paper by Landau (1958), we will follow his development and notation.

RELATION OF THE QUASIPARTICLE INTERACTION AND VERTEX FUNCTION

Using the notation of earlier chapters, we will write the one-particle Green's function as

$$G(1,2) = \frac{1}{i}\langle\psi_0|T(\hat{\psi}(1)\hat{\psi}^\dagger(2))|\psi_0\rangle \tag{6.74}$$

where $|\psi_0\rangle$ is the ground state and 1 and 2 denote space, time, and spin coordinates. For notational simplicity, we will suppress spin sums where convenient, and reinstitute them at the very end. In a translationally invariant system in the absence of a spin-dependent external field, the momentum and frequency transform may be written

$$G(\vec{P})\delta_{\sigma_1,\sigma_2} = \int d1\, G(1,2)\; e^{i(\vec{p}\cdot(\vec{r}_1-\vec{r}_2)-\omega(t_1-t_2))} \tag{6.75}$$

where $\vec{P}$ denotes $(\vec{p},\omega)$. Using the results of Chapter 6, $G(\vec{P})$ may be written

$$G(\vec{p}) = \frac{1}{\omega - \frac{p^2}{2m} - \Sigma(\vec{p},\omega)}\ . \tag{6.76}$$

Expanding $\Sigma(\vec{p},\omega)$ around the Fermi momentum k_F and Fermi energy μ, using Eq. (5.74), and using the fact that the imaginary part of Σ goes to $-\eta\,\mathrm{sgn}\,(\omega-\mu)$ as one approaches the Fermi surface from above or below, where η is a positive infinitesimal, we obtain

$$G(\vec{P}) = \frac{\left(1-\frac{\partial\Sigma}{\partial\epsilon}\right)^{-1}}{\omega-\mu-\frac{k_F}{m}\left(1-\frac{\partial\Sigma}{\partial\epsilon}\right)^{-1}\left(1+\frac{m}{k_F}\right)(p-k_F)+i\eta\,\mathrm{sgn}(\omega-\mu)} \equiv \frac{a}{\omega-\mu-v_F(p-k_F)+i\eta\,\mathrm{sgn}(\omega-\mu)} \tag{6.77}$$

where $v_F = \frac{k_F}{m^*}$, and we use Landau's notation for the residue at the quasiparticle pole, $a = \left(1-\frac{\partial\Sigma}{\partial\epsilon}\right)^{-1}$.

As in Section 2.4 for thermal Green's functions, the real time two-body Green's function

$$G^{(2)}(1,2|3,4) = \langle\psi_0|T(\hat{\psi}(1)\hat{\psi}(2)\hat{\psi}^\dagger(3)\hat{\psi}^\dagger(4))|\psi_0\rangle \tag{6.78}$$

is related to the vertex function $\Gamma^{(2)}(1,2|3,4)$ by the relation

$$\begin{aligned} G^{(2)}(1,2|3,4) &= G(1,3)G(2,4) - G(1,4)G(2,3) \\ &+ i\int d1'd2'd3'd4'\,G(1,1')G(2,2')\Gamma^{(2)}(1',2'|3',4')G(3',3)G(4',4) \end{aligned} \tag{6.79a}$$

which may be written diagrammatically

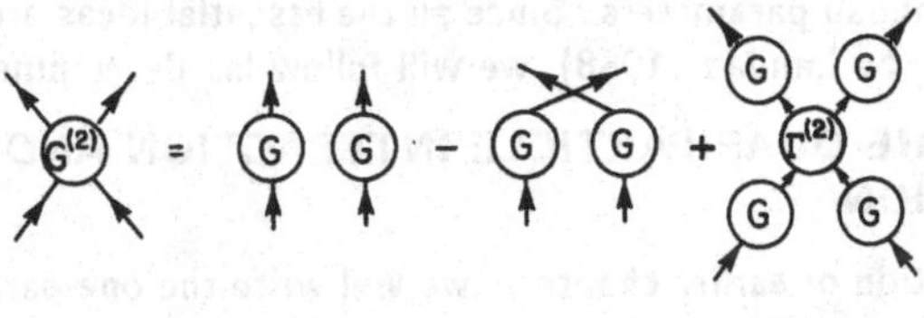

(6.79b)

Recall from Section 2.4 that $\Gamma^{(2)}$ is the sum of all one-particle irreducible amputated diagrams with two incoming and two outgoing legs. The factor of i in Eq. (6.79a) for real time follows from the diagram rules of Table 3.1. Since the first order contribution to $i^2G^{(2)}$ is $i\int GG\,v\,GG$, in order for $\Gamma^{(2)}$ to reduce to v in lowest order, it must enter Eq. (6.79a) with a factor of i.

The Fourier transform of Eq. (6.79a) is

$$\begin{aligned} G^{(2)}(\vec{P}_1,\vec{P}_2|\vec{P}_3,\vec{P}_4) &= (2\pi)^8 G(\vec{P}_1)G(\vec{P}_2)(\delta(\vec{P}_1-\vec{P}_3)\delta(\vec{P}_2-\vec{P}_4)-\delta(\vec{P}_1-\vec{P}_4)\delta(\vec{P}_2-\vec{P}_3)) \\ &+ iG(\vec{P}_1)G(\vec{P}_2)G(\vec{P}_3)G(\vec{P}_4)\Gamma^{(2)}(\vec{P}_1,\vec{P}_2|\vec{P}_3,\vec{P}_4)\ . \end{aligned} \tag{6.79c}$$

From the conservation of energy and momentum at each vertex, we know that both $G^{(2)}$ and $\Gamma^{(2)}$ are proportional to $(2\pi)^4\delta\left(\vec{P}_1+\vec{P}_2-\vec{P}_3-\vec{P}_4\right)$, and with the definition

$$\Gamma^{(2)}\left(\vec{P}_1,\vec{P}_2|\vec{P}_3,\vec{P}_4\right) = (2\pi)^4\delta\left(\vec{P}_1+\vec{P}_2-\vec{P}_3-\vec{P}_4\right)\Gamma(\vec{P}_1,\vec{P}_2|\vec{P}_3,\vec{P}_4) \tag{6.80a}$$

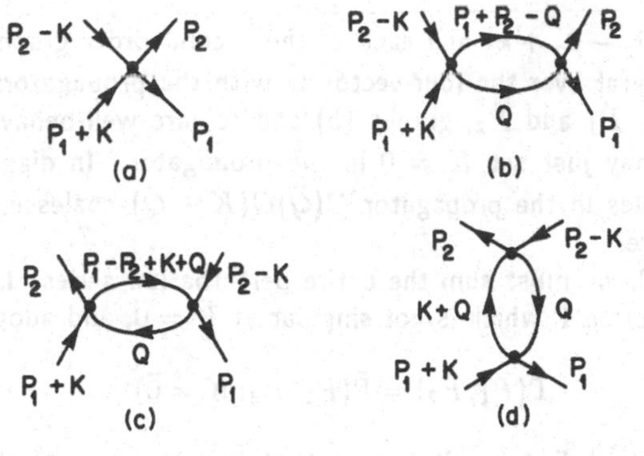

Fig. 6.1 First and second order graphs contributing to Γ.

Eq. (6.79a) becomes

$$\begin{aligned}
&G^{(2)}\left(\vec{P}_1, \vec{P}_2|\vec{P}_3, \vec{P}_4\right) \\
&\quad = (2\pi)^8 G(\vec{P}_1)G(\vec{P}_2)\big(\delta(\vec{P}_1 - \vec{P}_3)\delta(\vec{P}_2 - \vec{P}_4) - \delta(\vec{P}_1 - \vec{P}_4)\delta(\vec{P}_2 - \vec{P}_3)\big) \\
&\quad + (2\pi)^4 \delta(\vec{P}_1 + \vec{P}_2 - \vec{P}_3 - \vec{P}_4) iG(\vec{P}_1)G(\vec{P}_2)G(\vec{P}_3)G(\vec{P}_4)\Gamma(\vec{P}_1, \vec{P}_2|\vec{P}_3, \vec{P}_4) \ .
\end{aligned} \tag{6.80b}$$

As is the case for the Green's function, the vertex part Γ is antisymmetric with respect to the exchange of two particles. By the Lehman representation arguments of Chapter 5, the vertex function $\Gamma(\vec{P}_1, \vec{P}_2|\vec{P}_3, \vec{P}_4)$, like $G^{(2)}$, has poles that correspond to states of systems with different particle number depending on the time ordering of the field operators. For example, the time ordering $\psi_1\psi_2\psi_3^\dagger\psi_4^\dagger$ defines the particle-particle channel and has holes in the energy variable $\omega_1 + \omega_2$ corresponding to $(N+2)$-particle states. In contrast, the particle-hole channel is defined by the time ordering $\psi_1\psi_3^\dagger\psi_2\psi_4^\dagger$ and has poles in the energy variable $\omega_3 - \omega_1$ corresponding to N-particle states. Just as we restricted our attention to the particle-hole channel in Chapter 5 to study N-particle excited states in linear response theory, to make contact with Landau's theory of low-lying collective excitations in the N-particle system, we will examine the low lying excitations in the particle-hole channel by studying the poles of Γ in the variable $\omega_3 - \omega_1$. Because of the restriction to low excitation energy and long wavelength, we need to consider the vertex function for small values of $\omega_3 - \omega_1$ and $\vec{p}_3 - \vec{p}_1$. Thus, we shall consider the vertex function for

$$\begin{aligned}
\vec{P}_3 &= \vec{P}_1 + \vec{K} \\
\vec{P}_4 &= \vec{P}_2 - \vec{K}
\end{aligned} \tag{6.81a}$$

and define

$$\Gamma\left(\vec{P}_1, \vec{P}_2; K\right) \equiv \Gamma\left(\vec{P}_1\vec{P}_2|\vec{P}_3\vec{P}_4\right) \tag{6.81b}$$

where $\vec{K} = (\vec{k}, \omega)$ is a small four-vector. This corresponds to nearly forward scattering.

The first and second-order graphs contributing to Γ are drawn in Fig. 6.1 in the Hugenholtz representation. The first order graph (a) is just the antisymmetrized matrix element $v(\vec{k}) - v(\vec{p}_1 - \vec{p}_2 + \vec{k})$ and each of the second order graphs (b), (c) and (d) involve a loop integral over the four-vector $\vec{Q}$ with the propagators indicated in Fig. 6.1. With arbitrary $\vec{P}_1$ and $\vec{P}_2$, graphs (b) and (c) are well-behaved at $\vec{K} = 0$, and for small $\vec{K}$, we may just set $\vec{K} = 0$ in the propagator. In diagram (d), however, as $\vec{K} \to 0$, the poles in the propagator $G(\vec{Q})G(\vec{K} + \vec{Q})$ coalesce, and this diagram requires special care.

To calculate Γ, we must sum the entire perturbation series. Let us denote by $\tilde{\Gamma}$ the part of the function Γ which is not singular at $\vec{K} = \vec{0}$, and adopt the notation

$$\tilde{\Gamma}(\vec{P}_1, \vec{P}_2) = \tilde{\Gamma}(\vec{P}_1, \vec{P}_2; \vec{K} = \vec{0}) \quad . \tag{6.82}$$

By analyzing a general diagram, it is seen that $\tilde{\Gamma}$ is the sum of all the graphs of Γ, which cannot be decomposed into two components connected only by two propagators differing in four-momenta by K. Graphs (a), (b) and (c) of Fig. 6.1 are the lowest order contributions to $\tilde{\Gamma}$.

By the construction of $\tilde{\Gamma}$, it follows that the entire series for Γ is generated by the integral equation

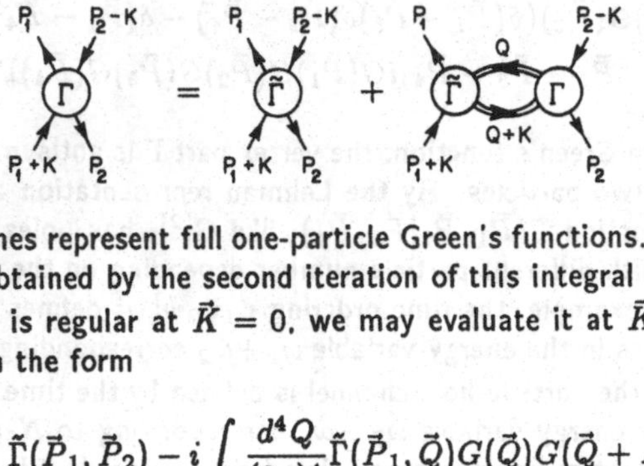

(6.83a)

where the double lines represent full one-particle Green's functions. For example, graph (d) of Fig. 6.1 is obtained by the second iteration of this integral equation with graph (a) for $\tilde{\Gamma}$. Since $\tilde{\Gamma}$ is regular at $\vec{K} = 0$, we may evaluate it at $\vec{K} = 0$ and write the integral equation in the form

$$\Gamma(\vec{P}_1, \vec{P}_2; K) = \tilde{\Gamma}(\vec{P}_1, \vec{P}_2) - i \int \frac{d^4Q}{(2\pi)^4} \tilde{\Gamma}(\vec{P}_1, \vec{Q}) G(\vec{Q}) G(\vec{Q} + \vec{K}) \Gamma(\vec{Q}, \vec{P}_2; \vec{K}) \quad . \tag{6.83b}$$

Using Eq. (6.77) and writing $\vec{Q} = (\vec{q}, \epsilon)$, the product of propagators in (6.83b) is

$$\begin{aligned} G(\vec{Q})G(\vec{K} + \vec{Q}) = {} & \frac{a}{(\epsilon - \mu - v_F(q - k_F) + i\eta \,\mathrm{sgn}(\epsilon - \mu))} \\ & \times \frac{a}{\left(\omega + \epsilon - \mu - v_F(|\vec{k} + \vec{q}| - k_F) + i\eta \,\mathrm{sgn}(\omega + \epsilon - \mu)\right)} \quad . \end{aligned} \tag{6.84a}$$

When $\vec{k}$ and ω go to zero, the singularities of this product approach $\epsilon = \mu$ and $q = k_F$, and we may therefore approximate it by the form

$$G(\vec{Q})G(\vec{K} + \vec{Q}) \underset{\vec{K} \to 0}{\sim} A(\theta)\delta(\epsilon - \mu)\delta(q - k_F) + \phi(\vec{Q}) \tag{6.84b}$$

where $\phi(\vec{Q})$ corresponds to the non-singular principal value contribution. The coefficient $A(\theta)$, which depends on the angle θ between $\vec{k}$ and $\vec{q}$, may be evaluated by integrating $G(\vec{Q})G(\vec{K}+\vec{Q})$ over $\vec{Q}$ leading to

$$A(\theta) = \int d\epsilon\, dq \frac{a}{(\epsilon - \mu - v_F(q - k_F) + i\eta\,\text{sgn}(\epsilon - \mu))} \times \frac{a}{\left(\epsilon + \omega - \mu - v_F(|\vec{k} + \vec{q}| - k_F) - i\eta\,\text{sgn}(\epsilon + \omega - \mu)\right)} . \tag{6.84c}$$

Let us first perform the integral over ϵ. We note that at the poles $\text{sgn}(\epsilon - \mu) = \text{sgn}(q - k_F)$ and $\text{sgn}(\epsilon + \omega - \mu) = \text{sgn}(|\vec{k} + \vec{q}| - k_F)$. Thus, if the quantities $(p - k_F)$ and $|k+q| - k_F$ have the same sign, the two poles of (6.84c) lie in the same half plane so that the integration contour may be closed in the other half plane yielding zero. Therefore, the only contribution arises when $q - k_F$ and $|\vec{q} + \vec{k}| - k_F$ have opposite signs.

We first treat the case that $\vec{q} \cdot \vec{k} = qk\cos\theta$ is positive. Noting that

$$\begin{aligned} |\vec{k} + \vec{q}| &= \left(\vec{q}^{\,2} + \vec{k}^{\,2} + 2kq\cos\theta\right)^{1/2} \\ &\underset{|\vec{k}| \to 0}{\sim} q + k\cos\theta \end{aligned} \tag{6.85a}$$

we observe that $q - k_F$ and $|\vec{k} + \vec{q}| - k_F$ have opposite signs if

$$q - k_F < 0 < q - k_F + k\cos\theta \tag{6.85b}$$

so that

$$k_F - k\cos\theta < q < k_F\ . \tag{6.85c}$$

Closing the ϵ integral in one half plane and evaluating the residue yields

$$\begin{aligned} A(\theta) &\underset{k \to 0}{\sim} \int_{k_F - k\cos\theta}^{k_F} dq \frac{2i\pi a^2}{\omega - v_F(|\vec{k} + \vec{q}| - q)} \\ &= \frac{2i\pi a^2 k\cos\theta}{\omega - v_F k\cos\theta}\ . \end{aligned} \tag{6.86}$$

Similarly, for the case $\cos\theta < 0$, the conditions for $(q - k_F)$ and $|\vec{k} + \vec{q}| - k_F$ to have opposite signs is

$$q + k\cos\theta - k_F < 0 < q - k_F \tag{6.87}$$

which again yields the result Eq. (6.86). Thus, the product of propagators may finally be written

$$G(\vec{Q}G(\vec{K} + \vec{Q}) = \frac{2i\pi a^2 \hat{q}\cdot\vec{k}}{\omega - v_F \hat{q}\cdot\vec{k}}\delta(\epsilon - \mu)\delta(q - k_F) + \phi(\vec{Q}) \tag{6.88}$$

where $\hat{q}$ denotes a unit vector in the direction of $\vec{q}$ and the integral equation (6.83b) becomes

$$\begin{aligned} \Gamma(\vec{P}_1, \vec{P}_2; \vec{K}) &= \tilde{\Gamma}(\vec{P}_1, \vec{P}_2) - i\int \frac{d^4Q}{(2\pi)^4}\tilde{\Gamma}(\vec{P}_1, \vec{Q})\phi(\vec{Q})\Gamma(\vec{Q}, \vec{P}_2; \vec{K}) \\ &+ \frac{a^2 k_F^2}{(2\pi)^3}\int d\Omega\, \tilde{\Gamma}(P_1, \vec{Q})\frac{\hat{q}\cdot\vec{k}}{\omega - v_F\,\hat{q}\cdot\vec{k}}\Gamma(\vec{Q}, \vec{P}_2; \vec{K}) \end{aligned} \tag{6.89a}$$

where

$$d^4Q = d\epsilon q^2 dq d\Omega \tag{6.89b}$$

and $d\Omega$ is the element of solid angle in the $\hat{q}$ direction.

We are now in a position to study the singularity of Γ at $\vec{K} = 0$. In order to take this limit, because of the kernel $\frac{\hat{q}\cdot\vec{k}}{\omega - v_F\hat{q}\cdot\vec{k}}$ of (6.89a), we must specify how the $\omega \to 0$ and $\vec{k} \to 0$ limits are taken, since they do not commute. Let us first consider the case $\frac{k}{\omega} \to 0$ and define

$$\Gamma^\omega(\vec{P}_1, \vec{P}_2) = \lim_{\omega\to 0}\lim_{k\to 0}\Gamma(\vec{P}_1, \vec{P}_2; \vec{K}) \ . \tag{6.90}$$

This limit requires that momentum be strictly conserved while allowing for small energy transfer, and we shall see that this limit corresponds to the quasiparticle excitations composing a collective excitation. The kernel goes to zero in this limit

$$\lim_{\omega\to 0}\left(\lim_{k\to 0}\frac{\hat{q}\cdot k}{\omega - v_F\hat{q}\cdot k}\right) = 0 \tag{6.91}$$

so the integral equation becomes

$$\Gamma^{(\omega)}(\vec{P}_1, \vec{P}_2) = \tilde{\Gamma}(\vec{P}_1, \vec{P}_2) - i\int\frac{d^4Q}{(2\pi)^4}\tilde{\Gamma}(\vec{P}_1, \vec{Q})\phi(\vec{Q})\Gamma^\omega(\vec{Q}, \vec{P}_2) \tag{6.92a}$$

which we rewrite in the obvious symbolic notation

$$\Gamma^\omega = \tilde{\Gamma} - i\tilde{\Gamma}\phi\Gamma^\omega \ . \tag{6.92b}$$

We may solve for $\tilde{\Gamma}$ in terms of Γ^ω as follows

$$\begin{aligned}\tilde{\Gamma} &= \Gamma^\omega\frac{1}{1 - i\phi\Gamma^\omega} \\ &= \frac{1}{1 - i\Gamma^\omega\phi}\Gamma^\omega \ .\end{aligned} \tag{6.93}$$

Similarly, we may rewrite Eq. (6.89a) in the symbolic form

$$\Gamma = \tilde{\Gamma} - i\tilde{\Gamma}\phi\Gamma + \tilde{\Gamma}\frac{\hat{q}\cdot\vec{k}}{\omega - v_F\hat{q}\cdot\vec{k}}\Gamma \tag{6.94a}$$

and substitution of Eq. (6.93) yields

$$(1 - i\Gamma^\omega\phi)\Gamma = \Gamma^\omega - i\Gamma^\omega\phi\Gamma + \Gamma^\omega\frac{\hat{q}\cdot\vec{k}}{\omega - v_F\hat{q}\cdot\vec{k}}\Gamma \tag{6.94b}$$

which in explicit integral form is written

$$\Gamma(\vec{P}_1, \vec{P}_2; K) = \Gamma^{(\omega)}(\vec{P}_1, \vec{P}_2) + \frac{a^2k_F^2}{(2\pi)^3}\int d\Omega\,\Gamma^\omega(\vec{P}, Q)\frac{\hat{q}\cdot\vec{k}}{\omega - v_F\hat{q}\cdot\vec{k}}\Gamma(\vec{Q}, \vec{P}_2; \vec{K}) \ . \tag{6.94c}$$

A second limit of interest is the case $\frac{\omega}{k} \to 0$, and we define

$$\Gamma^k(\vec{P}_1, \vec{P}_2) = \lim_{k\to 0}\lim_{\omega\to 0}\Gamma(\vec{P}_1, \vec{P}_2; K) \ . \tag{6.95}$$

This limit requires that energy be strictly conserved, and thus pertains to the physical scattering of two quasiparticles. In this limit, the kernel of (6.89a) approaches a constant

$$\lim_{k\to 0}\left(\lim_{\omega\to 0}\frac{\hat{q}\cdot\vec{k}}{\omega - v_F\hat{q}\cdot\vec{k}}\right) = -\frac{1}{v_F} \tag{6.96}$$

so that

$$\begin{aligned}\Gamma^k(\vec{P}_1, \vec{P}_2) = \tilde{\Gamma}(\vec{P}_1, \vec{P}_2) - i\int\frac{d^4Q}{(2\pi)^4}\tilde{\Gamma}(\vec{P}, \vec{Q})\phi(\vec{Q})\Gamma^k(\vec{Q}, \vec{P}_2) \\ - \frac{a^2}{(2\pi)^3}\frac{k_F^2}{v_F}\int d\Omega\,\tilde{\Gamma}(\vec{P}_1, \vec{Q})\Gamma^k(\vec{Q}, \vec{P}_2) \ .\end{aligned} \tag{6.97a}$$

Using Eq. (6.93) as before, we may eliminate $\tilde{\Gamma}$ from this result and obtain the following relation between the two limits:

$$\Gamma^k(\vec{P}_1, \vec{P}_2) = \Gamma^\omega(\vec{P}_1, \vec{P}_2) - \frac{a^2}{(2\pi)^3}\frac{k_F^2}{v_F}\int d\Omega\,\Gamma^\omega(\vec{P}_1, \vec{Q})\Gamma^k(\vec{Q}, \vec{P}_2) \ . \tag{6.97b}$$

With these results, we are now in a position to study the poles of $\Gamma(\vec{P}_1, \vec{P}_2; K)$ corresponding to the low lying, long wavelength excited states of the system. Since we are interested in the lowest lying excited states of the system, in addition to taking the limit as $\vec{K}$ goes to zero, we shall restrict $\vec{P}_1$ and $\vec{P}_2$ to the Fermi surface and use the notation

$$\begin{aligned}\vec{P}_1 &= (\mu,\ k_F\hat{p}_1) \\ \vec{P}_2 &= (\mu,\ k_F\hat{p}_2)\end{aligned} \tag{6.98}$$

where $\hat{p}_1$ and $\hat{p}_2$ are unit vectors. From Eq. (6.92), we observe that Γ^ω is a regular function, and therefore when $\vec{K}$ goes to zero, it is negligible compared to Γ which becomes infinite. Thus, Γ^ω may be neglected in Eq. (6.94c), and we obtain

$$\Gamma(\vec{P}_1, \vec{P}_2; \vec{K}) \underset{K\to 0}{\approx} \frac{a^2k_F^2}{(2\pi)^3)}\int d\Omega\,\Gamma^\omega(\vec{P}_1, \vec{Q})\frac{\vec{k}\cdot\hat{q}}{\omega - v_F\vec{k}\cdot\hat{q}}\Gamma(\vec{Q}, \vec{P}_2; \vec{K}) \ . \tag{6.99a}$$

Note that $\vec{P}_2$ and $\vec{K}$ only enter Eq. (6.99) as parameters and by Eq. (6.98), the only variable in $\vec{P}_1$ is the unit vector $\hat{p}_1$, so we may rewrite Eq. (6.99) as the eigenvalue equation

$$\chi(\hat{p}) = \frac{a^2k_F^2}{(2\pi)^3}\int d\Omega\,\Gamma^\omega(\hat{p}, \hat{q})\frac{\vec{k}\cdot\hat{q}}{\omega - v_F\vec{k}\cdot\hat{q}}\chi(\hat{q}) \ . \tag{6.99b}$$

Finally, introducing the function

$$\mu(\hat{p}) \equiv \frac{\hat{p}\cdot\vec{k}}{\omega - v_F\hat{p}\cdot\vec{k}}\chi(\hat{p})$$

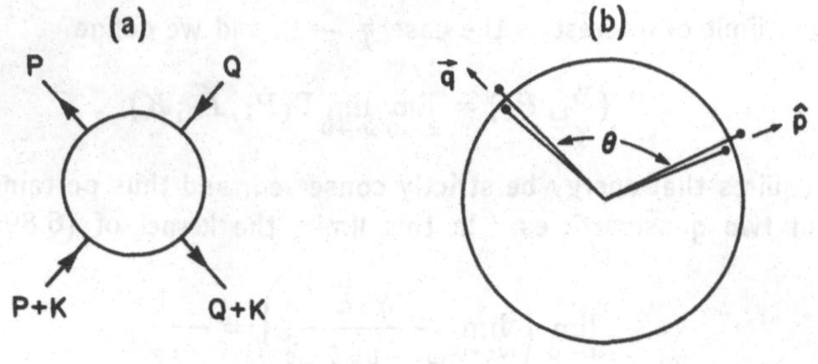

Fig. 6.2 The geometry of the momenta in the vertex function (a) defining the quasiparticle interaction is shown in (b).

and dividing through by kv_F yields

$$\left(\frac{\omega}{kv_F} - \cos\theta\right) u(\hat{p}) = \cos\theta \frac{m^* k_F}{\pi^2} \frac{1}{2} \int \frac{d\Omega}{4\pi} a^2 \Gamma^\omega(\hat{p},\hat{q}) u(\hat{q}) \tag{6.99c}$$

which, when the suppressed spin sums are restored, is identical with Eq. (6.61a) for the collective oscillations of the system in Fermi liquid theory. In addition to justifying the fundamental equation for the zero sound modes, the derivation provides a microscopic definition for the quasiparticle interaction

$$f(\hat{p},\hat{q}) = a^2 \Gamma^\omega(\hat{p},\hat{q}) \quad . \tag{6.100}$$

The geometry of the momenta defining $\Gamma^\omega(\hat{p},\hat{q})$ is sketched in Fig. 6.2. The vectors $\vec{p}$ and $\vec{p}+\vec{k}$ are on the Fermi surface in the direction $\hat{p}$, $\vec{q}$ and $\vec{q}+\vec{k}$ are on the Fermi surface in the direction $\hat{q}$. We know further from our restriction to the particle hole channel, as reflected in propagator poles on opposite sides of the real axis, that one of the momenta in each direction is a particle state and the other is a hole state. Thus, the vertex function simply describes the scattering of a particle hole pair at one point on the Fermi surface to any other part. The two factors of a, the residue of the quasiparticle pole, in Eq. (6.100) represent the amplitude for the intermediate state in the integral equation to be in the simple two-quasiparticle state which has a pole, rather than in some complicated non-singular background configuration.

Finally, let us consider the physical scattering amplitude for forward scattering of quasiparticles on the Fermi surface. Whereas Γ^ω represents virtual excitations with small energy transfer, Γ^k describes the physical forward scattering amplitude $(\vec{p}_1, \vec{p}_2 \to \vec{p}_1, \vec{p}_2)$ in which collisions involve changes of momentum with strictly conserved energy. Using Eqs. (6.97b) and (6.100) and displaying the spin dependence explicitly we may express Γ^k in terms of f as follows

$$a^2\Gamma^k(\hat{p}\sigma,\hat{p}'\sigma') = f(\hat{p}\sigma,\hat{p}'\sigma') - \frac{k_F m^*}{\pi^2}\frac{1}{2}\sum_{\sigma''}\int \frac{d\Omega}{2\pi} f(\hat{p}\sigma,\hat{p}''\sigma'')a^2\Gamma^k(\hat{p}''\sigma'',\hat{p}'\sigma') \tag{6.101}$$

where all momenta are on the Fermi surface with direction denoted by unit vector $\hat{p}$. Just as we have parametrized the spin dependence, included the density of states at the Fermi surface, and Legendre expanded the quasiparticle interaction f

$$\begin{aligned}\frac{m^* k_F}{\pi^2} f(\hat{p}\sigma, \hat{p}'\sigma') &= F(\hat{p}, \hat{p}') + 4\sigma\cdot\sigma'\Phi(\hat{p}, \hat{p}') \\ &= \sum_\ell F_\ell P_\ell(\cos\theta) + 4\sigma\cdot\sigma'\sum_\ell Z_\ell P_\ell(\cos\theta)\end{aligned} \tag{6.102a}$$

where $\cos\theta = \hat{p}\cdot\hat{p}'$, we write the physical scattering amplitude Γ^k as

$$\begin{aligned}\frac{m^* k_F}{\pi^2} a^2\Gamma^k(\hat{p}\sigma, \hat{p}'\sigma') &= B(\hat{p}, \hat{p}') + 4\sigma\cdot\sigma' C(\hat{p}, \hat{p}') \\ &= \sum_\ell B_\ell P_\ell(\cos\theta) + 4\sigma\cdot\sigma'\sum_\ell C_\ell P_\ell(\cos\theta) \ .\end{aligned} \tag{6.102b}$$

Equation (6.101) then becomes

$$\begin{aligned}B(\hat{p}, \hat{p}') &= F(\hat{p}, \hat{p}') - \int \frac{d\Omega''}{4\pi} F(\hat{p}, \hat{p}'')B(\hat{p}'', \hat{p}') \\ C(\hat{p}, \hat{p}') &= \Phi(\hat{p}, \hat{p}') - \int \frac{d\Omega''}{4\pi} \Phi(\hat{p}, \hat{p}'')C(\hat{p}'', \hat{p}')\end{aligned} \tag{6.103a}$$

from which we may solve for the Legendre coefficients

$$\begin{aligned}B_\ell &= \frac{F_\ell}{1+\frac{F_L}{2L+1}} \\ C_\ell &= \frac{Z_\ell}{1+\frac{Z_L}{2L+1}} \ .\end{aligned} \tag{6.103b}$$

One useful application of these results for the scattering amplitude is to obtain the physical constraint on the Landau parameters implied by the Pauli principle. By antisymmetry, one can show that the following limit of the scattering amplitude for particles with parallel spin must vanish

$$\lim_{\vec{P}'\to\vec{P}} \Gamma^k(\vec{P}'\sigma, \vec{P}\sigma) = 0 \tag{6.104a}$$

which, from Eqs. (6.102b) and (6.103b) yields the Landau sum rule†

$$\begin{aligned}0 &= \sum_L (B_L + C_L) \\ &= \sum_L \left(\frac{F_L}{1+\frac{F_L}{2L+1}} + \frac{Z_L}{1+\frac{Z_L}{ZL+1}}\right) \ .\end{aligned} \tag{6.104b}$$

† The limit in Eq. (6.104a) is crucial. In addition to $\Gamma^k(p\sigma, p, \sigma')$, $\Gamma^\omega(p\sigma, p\sigma')$ also vanishes by antisymmetry, from which one might erroneously conclude an additional sum rule $\Sigma_L(F_L + Z_L) = 0$ which is inconsistent with Eq. (6.104b). Because of the singularities in the propagators in (6.83a) one must carefully calculate the limit of

$$\Gamma(\vec{P}_1\vec{P}_2|\vec{P}_1+\vec{K}, \vec{P}_2+\vec{K}) - \Gamma(\vec{P}_2\vec{P}_1|\vec{P}_1+\vec{K}, \vec{P}_2+\vec{K})$$

as $P_1 \to P_2$ which as shown by Mermin (1967) vanishes when $\frac{\omega}{k} = 0$ but not when $\frac{k}{\omega} = 0$.

CALCULATION OF THE QUASIPARTICLE INTERACTION

Having seen that the quasiparticle interaction corresponds to the particle-hole vertex function $\Gamma(\vec{P}_1, \vec{P}_2; \vec{K})$ in the limit $\frac{k}{\omega} \to 0$ with $|\vec{P}_1|$ and $|\vec{P}_2| = k_F$, it is instructive to see how the same set of diagrams arises from taking the second functional derivative of the energy with respect to the occupation numbers.

To see how the essential features arise, let us consider the following low order contributions to the ground state energy

$E = E_0 +$ 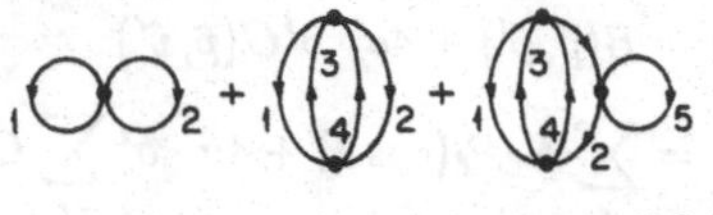 . (6.105a)

The first and second order contributions may be written explicitly as

$$
\begin{aligned}
E = \sum_{1} \frac{k_1^2}{2m} n_1 + \frac{1}{2} \sum_{12} \langle 1\,2|v|1\,2\rangle\, n_1 n_2 \\
+ \frac{1}{4} \sum_{1234} \delta_{\vec{k}_1+\vec{k}_2, \vec{k}_3+\vec{k}_4} \frac{|\langle 1\,2|v|3\,4\rangle|^2}{\frac{1}{2m}\left(k_1^2+k_2^2-k_3^2-k_4^2\right)} n_1 n_2 (1-n_3)(1-n_4)
\end{aligned}
\qquad (6.105b)
$$

where we have used the antisymmetrized matrix elements and factors appropriate for Hugenholtz diagrams, (1) denotes the momentum and spin variables $(\vec{k}_1, \sigma_1)$, and n_1 is the occupation number $\theta(k_F - k_1)$ for $\sigma = \pm 1$. We thus obtain

$$
\begin{aligned}
(k_{\bar1}, \sigma_{\bar1}, k_{\bar2}, \sigma_{\bar2}) &= \frac{\delta^2 E}{\delta n_{\bar1} \delta n_{\bar2}} \\
&= \langle \bar1\bar2|v|\bar1\bar2\rangle + \frac{1}{2}\sum_{34} \delta_{\vec{k}_{\bar1}+\vec{k}_{\bar2}, \vec{k}_3+\vec{k}_4} \frac{|\langle \bar1\bar2|v|3\,4\rangle|^2}{\frac{1}{2m}\left(k_{\bar1}^2+k_{\bar2}^2-k_3^2-k_4^2\right)}(1-n_3)(1-n_4) \\
&\quad + \frac{1}{2}\sum_{12} \delta_{\vec{k}_1+\vec{k}_2, \vec{k}_{\bar1}+\vec{k}_{\bar2}} \frac{|\langle 1\,2|v|\bar1\bar2\rangle|^2}{\frac{1}{2}\left(k_1^2+k_2^2-k_{\bar1}^2-k_{\bar1}^2\right)} n_1 n_2 \\
&\quad - \sum_{24} \delta_{\vec{k}_{\bar1}+\vec{k}_2, \vec{k}_{\bar2}+\vec{k}_4} \frac{|\langle \bar1 2|v|\bar2 4\rangle|^2}{\frac{1}{2}\left(k_{\bar1}^2+k_2^2-k_{\bar2}^2-k_4^2\right)} n_2(1-n_4)
\end{aligned}
\qquad (6.106)
$$

which may be represented diagrammatically as

$f(\bar1\bar2) =$ 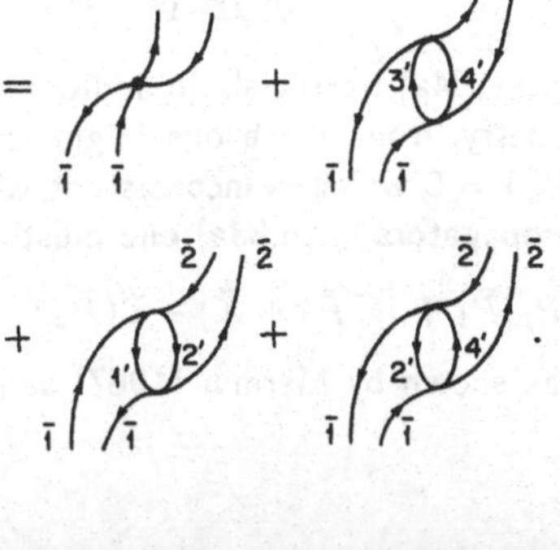

Pressure	0 atm		27 atm	
m^*/m	3.01	2.76	5.63	5.17
$F_0\,(F_0^s)$	10.07	9.15	74.38	68.22
$F_1\,(F_1^s)$	6.04	5.27	13.90	12.50
$Z_0\,(F_0^a)$	-0.67	-0.70	-0.74	-0.76
$Z_1\,(F_1^a)$	-0.67	-0.55	-0.53	-0.99

Table 6.1 Fermi liquid parameters for liquid ^{3}He. The left column of values for each pressure is taken from Wheatley's compilation (1975). The right column in each case is taken from Greywall (1983) who obtains different values of the specific heat and thus m^* from using a different temperature scale, and the other parameters are scaled accordingly. The estimates of Z_1, are obtained from the forward scattering sum rule (6.104b) assuming all parameters with $L > 0$ vanish.

Clearly, for a general diagram, the effect of differentiation with respect to $n_{\bar{1}}$ is to break open any propagator which we draw as a particle-hole pair entering from the bottom of the diagram and differentiation with respect to $n_{\bar{2}}$ breaks a second propagator which we draw as a particle-hole pair entering from the top. The usual time-ordered diagram rules produce the correct energy denominators, since between the initial and final times of the original propagator, there is a single propagator of the same momentum in the same direction, and outside this time interval the particle and hole energies cancel.

To see how the factor a corresponding to the renormalization of the quasiparticle pole arises, consider the contribution from the third order diagram in Eq. (6.105a) in which differentiation breaks open propagators 2 and 5:

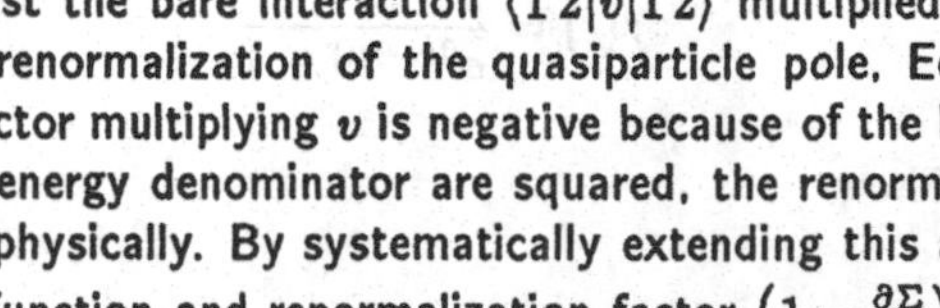

The result is just the bare interaction $\langle \bar{1}\bar{2}|v|\bar{1}\bar{2}\rangle$ multiplied by the lowest order contribution to the renormalization of the quasiparticle pole, Eq. (5.66). Since the overall sign of the factor multiplying v is negative because of the Fermion loop and the matrix element and energy denominator are squared, the renormalization reduces the strength as it must physically. By systematically extending this analysis, one can see how the full vertex function and renormalization factor $\left(1-\frac{\partial\Sigma}{\partial\epsilon}\right)^{-1}$ are generated by differentiation of the energy.

In practical microscopic calculations of the Landau parameters, one typically performs infinite summations of diagrams. For example, one may use the G-matrix to sum all two-body ladder diagrams, and make a self-consistent definition of the single-particle energies in terms of this G-matrix. In the low density limit, as shown in Problem 6.6 it is possible to evaluate the second order contribution in Eq. (6.106) analytically to obtain a closed form expression (Abrikosov and Khalatnikov, 1957) in terms of the experimental scattering length.

This chapter has only presented a few of the basic ideas and elementary results of Fermi liquid theory, and there is a substantial literature on additional aspects of the theory. A clear introduction to the treatment of transport coefficients is given by Baym and Pethick (1976, Chapter 1). Attempts have been made to calculate Landau parameters both in liquid helium and in nuclear matter (Feenberg, 1969; Dickhoff, 1983). The theory we have described for normal Fermi systems has also been extended to a variety of other systems, including solutions of ^{3}He in ^{4}He (Baym and Pethick, 1976, Chapter 2), superfluids (Leggett, 1966 and 1968), finite Fermi systems (Migdal, 1962), and relativistic systems (Baym and Chin, 1976).

PROBLEMS FOR CHAPTER 6

The first four problems elaborate results discussed in the text: the derivation of the specific heat, an alternative evaluation of the effective mass using Galilean invariance, the stability criteria for the Landau parameters, and derivation of the general formula for the velocity of zero sound. Problems 2 and 3 are particularly important and are designated with an *. Problem 5 proves that there must always be a zero sound mode, either in the form of a density oscillation or a spin oscillation. Problem 6 evaluates the quasiparticle interaction in second order perturbation theory and from it recovers the well-known result for the binding energy of a dilute Fermi gas. The last problem specializes the general results of Fermi liquid theory to one spatial dimension and shows that the sound velocity is uniquely specified by the specific heat.

PROBLEM 6.1 Specific Heat of a Fermi Liquid

a) Consider the following integral

$$I = \int_{-\infty}^{\infty} g(\epsilon) f(\epsilon) d\epsilon$$

where $f(\epsilon)$ is Fermi function

$$f(\epsilon) = \frac{1}{e^{\beta(\epsilon-\mu)} + 1}$$

and $g(\epsilon) \underset{\epsilon\to-\infty}{\longrightarrow} 0$. Show

$$I = \int_{-\infty}^{\mu} g(\epsilon) d\epsilon + \sum_n \frac{1}{\beta^{2n}} g^{(2n-1)}(\mu) \left(2 - \frac{1}{2^{2n-2}}\right) \varsigma(2n)$$

$$= \int_{-\infty}^{\mu} g(\epsilon) d\epsilon + \frac{\pi^2}{6\beta^2} g^{(1)}(u) + \frac{7\pi^4}{360\beta^4} g^{(3)}(\mu) + O(\beta^{-6})$$

where $\varsigma(m)$ is the Riemann ς-function

$$\varsigma(m) = \sum_{k=0}^{\infty} \frac{1}{k^m} \ .$$

To obtain this result, define $G(\epsilon) = \int_{-\infty}^{\epsilon} g(\epsilon) d\epsilon$ and integrate by parts

$$I = \int_{-\infty}^{\infty} G(\epsilon) \left(-\frac{df}{d\epsilon}\right) d\epsilon \ .$$

Then expand $G(\epsilon)$ as a Taylor series around μ, expand the Fermi function as a geometric series

$$\frac{1}{e^x + 1} = \frac{e^{-x}}{1 + e^{-x}} = \sum_{m=1}^{\infty} (-1)^{m+1} e^{-xm}$$

and integrate term by term. From the result for I, obtain Eq. (6.16).

b) Use the result in (a) to calculate the specific heat of a non-interacting Fermi gas. For a fixed chemical potential, μ, calculate the density and internal energy at temperature T.

$$n = \frac{2}{(2\pi)^3} \int d^3k \, f\left(\frac{k^2}{2m}\right) = \frac{\sqrt{2} m^{3/2}}{\pi^2} \int_0^{\infty} d\epsilon \, f(\epsilon) \epsilon^{1/2}$$

$$u = \frac{2}{(2\pi)^3} \int d^3k \frac{k^2}{2m} f\left(\frac{k^2}{2m}\right) = \frac{\sqrt{2} m^{3/2}}{\pi^2} \int_0^{\infty} d\epsilon \, f(\epsilon) \epsilon^{3/2} \ .$$

Show that to keep the density constant, the chemical potential at temperature T must be

$$\mu(T) = \mu_0 - (k_B T)^2 \frac{\pi^2}{12\mu_0} + O(T^4) \ .$$

Using this temperature dependence of μ, show

$$u(T) = u(T=0) + (k_B T)^2 \frac{m k_F}{6}$$

from which

$$c_V = \frac{du}{dT} = \frac{1}{3} m k_F k_B^2 T \ .$$

c) Now evaluate the integral Eq. (6.15) for a Fermi liquid two ways. Using (6.13b), show as above that to maintain constant density $\frac{\partial \mu}{\partial T} = \frac{-(k_B \pi)^2 T 2m^*}{3k_F^2}$ and evaluate

$$c_V = \frac{m^*}{\pi^2 k_F} \int d\epsilon \left(\frac{m^*}{k_F}(\epsilon - \mu) + k_F\right)^2 \epsilon$$

$$\times \left[\delta(\epsilon - \mu) + \frac{(\pi k_B T)^2}{6} \frac{\partial^2}{\partial \epsilon^2} \delta(\epsilon - \mu)\right] \left[\frac{1}{T}(\epsilon - \mu) - \frac{(k_B \pi)^2 T 2m^*}{3k_F^2}\right]$$

to obtain Eq. (6.17).

An easier way is to note that since $\frac{\delta n}{\delta T}$ must integrate to zero, we may shift $\frac{\delta E}{\delta n}$ by an arbitrary constant so Eq. (6.15) may be rewritten

$$c_V = \frac{1}{\mathcal{V}} \sum_{k\sigma} (\epsilon_k - \mu) \frac{\delta n(\vec{k}, r)}{\delta \epsilon_{\vec{k}}} \left(-\frac{\epsilon_k - \mu}{T} + \frac{\partial}{\partial T}(\epsilon_k - \mu) \right) .$$

Note that the change in chemical potential no longer contributes to leading order and Eq. (6.17) is obtained directly.

PROBLEM 6.2* Galilean Invariance and the Effective Mass

Obtain the relation between m^* and F_1, Eq. (6.25) by using Galilean invariance. Consider the energy in a new frame moving at a velocity $\frac{q}{m}$ relative to the original frame.

a) Since the fluid has been given a velocity $\frac{q}{m}$, explain why the energy should increase by $\rho\frac{q^2}{2m}$ relative to the original energy.

b) Now, calculate the energy change using

$$\delta E = \sum_k \epsilon_k^0 \delta n(k) + \frac{1}{2} \sum_{\vec{k}\vec{k}'} f(k, k') \delta n(\vec{k}) \delta n(\vec{k}')$$

and Eq. (6.28), where $\delta n(k)$ is the difference between the occupation numbers of the original Fermi sphere and one whose origin is shifted by $\vec{q}$. Show that the first term yields $\rho\frac{q^2}{2m^*}$ and the second term yields $\rho\frac{q^2}{2m^*}\frac{F_1}{3}$. Equating parts a) and b) yields the desired result. Yet another instructive derivation is given by Baym and Pethick (1976).

PROBLEM 6.3* Stability Criteria

Show that in order for the ground state energy to be a minimum rather than simply stationary, the Landau parameter must satisfy the conditions $F_L > -(2L+1)$ and $Z_L > -(2L+1)$.

First, parameterize a general distortion of the Fermi surface by a polar angle and spin-dependent Fermi momentum $k_F(\theta, \sigma)$, so that the distribution function may be written $n(k, \sigma) = \theta(k_F(\theta, \sigma) - k)$. Then, to second order,

$$\Delta(E - \mu n) = \frac{1}{\mathcal{V}} \sum_{k,\sigma} (\epsilon_k^0 - \mu)\, \Delta n(k, \sigma) + \frac{1}{2\mathcal{V}^2} \sum_{\substack{k\sigma \\ k'\sigma'}} f(k\sigma, k'\sigma') \Delta n(k\sigma) \Delta n(k'\sigma')$$

where

$$\Delta n(k, \sigma) = \delta(k_F - k) \delta k_F(\theta, \sigma) - \frac{1}{2} \frac{\partial}{\partial k} \delta(k_F - k) \left(\delta k_F(\theta, \sigma)\right)^2 .$$

The first order change in $E - \mu n$ vanishes by stationarity and the second order change must be positive for stability. Show that the second order change is

$$\Delta^{(2)}(E - \mu n) = \frac{1}{4} N(0) v_F^2 \frac{1}{2} \sum_\sigma \int d(\cos\theta)$$

$$\times \left[\left(\delta k_F(\theta, \sigma)\right)^2 + \frac{1}{2} \sum_{\sigma'} \int d(\cos\theta')\, f(\theta\sigma, \theta'\sigma') \delta k_F(\theta, \sigma) \delta k_F(\theta'\sigma') \right] .$$

Expand $\delta k_F(\theta,\sigma) = \Sigma k_L(\sigma)P_L(\cos\theta)$ and use the addition theorem to obtain

$$\Delta^{(2)} = \frac{1}{8}N(0)v_F^2 \sum_L \left\{ \left(k_L\left(\tfrac{1}{2}\right) + k_L\left(-\tfrac{1}{2}\right)\right)^2 \left(1 + \frac{F_L}{2L+1}\right) + \left(k_L\left(\tfrac{1}{2}\right) - k_L\left(-\tfrac{1}{2}\right)\right)^2 \left(1 + \frac{Z_L}{2L+1}\right) \right\}$$

which is positive only if F_L and Z_L are greater than $-(2L+1)$.

PROBLEM 6.4 Zero Sound Mode for General $f(k, k')$

Legendre expand the zero sound solution $U(\theta) = \sum_L U_L P_L(\cos\theta)$ and use the addition theorem to derive Eq. (6.66c)

$$\frac{U_L}{2L+1} = -\sum_{L'} D_{LL'} \frac{F_{L'}U_{L'}}{2L'+1}$$

where

$$D_{LL'} = \int \frac{dx}{2} \frac{x}{x-s} P_L(x)P_{L'}(x) \ .$$

Use the properties of Legendre functions to show

$$\int \frac{dx}{2} \frac{x}{x-s} P_L(x)P_{L'}(x) = \frac{\delta_{LL'}}{2L+1} - sP_{L'}(s)Q_L(s)$$

where the Legendre function of the second kind $Q_L(s)$ may be written

$$Q_L(s) = \frac{1}{2}P_L(s) \ln\left(\frac{1+s}{1-s}\right) - W_{L-1}(s)$$

and

$$W_{L-1}(s) = \begin{cases} \sum_{M=1}^{L} P_{M-1}(s)P_{L-M} & L \geq 1 \\ 0 & L = 0 \end{cases} \ .$$

Thus, obtain Eq. (6.67a) for $D_{LL'}$ and Eq. (6.67c) for zero sound assuming non-zero F_0 and F_1.

PROBLEM 6.5 Existence of Zero Sound Modes

The object of this problem is to prove that in the collisionless regime, a Fermi liquid must have at least one zero sound mode, either in the form of a density oscillation or a spin oscillation. The ingredients for the proof are the following four relations, which we rewrite in a compact operator notation:

The kinetic equation, (6.61a)

$$(s - \cos\theta)u(\hat{k}) = \cos\theta \int \frac{d\hat{k}}{4\pi} F(\hat{k}\cdot\hat{k}')u(\hat{k}')$$

or

$$(s - \cos)u = \cos\theta\, Fu \ . \tag{1}$$

The relation of the forward scattering amplitude to the quasiparticle interaction, Eq. (6.103a)

$$B(\hat{k}\cdot\hat{k}') = F(\hat{k}\cdot\hat{k}') - \int \frac{d\hat{k}''}{4\pi} F(\hat{k}\cdot\hat{k}'')B(\hat{k}''\cdot\hat{k}')$$

or

$$B = F - FB \ . \tag{2}$$

The stability condition from Problem 6.3

$$\int \frac{d\hat{k}}{4\pi}\frac{d\hat{k}'}{4\pi}\,\delta(\hat{k})\left[1 + F(\hat{k}\cdot\hat{k}')\right]\delta(\hat{k}') > 0$$

or

$$1 + F > 0 \ . \tag{3}$$

The vanishing of the forward scattering amplitude, Eq. (6.104b)

$$B(1) + C(1) = 0 \ . \tag{4}$$

Note that each relation $(1-3)$ has its counterpart for spin modes with $F \to \phi$ and $B \to C$.

a) Begin by observing that (4) requires that either $B(1)$ or $C(1)$ be positive (We exclude the case of an accidental degeneracy $B(1) = C(1) = 0$ which could be eliminated by a slight change in pressure.) Hence, we need to show that if $B(1) > 0$, then relations $1-3$ imply an undamped zero sound mode. Identical reasoning would then require an undamped spin mode if $C(1) > 0$.

b) Define

$$v \equiv (1 + F)u$$

and show that relations (1) and (2) imply that

$$H_s v = v$$

where

$$H_s \equiv \frac{\cos\theta}{s} + B \ .$$

Therefore, there will be an undamped zero sound mode if H_s has the eigenvalue 1 for some real s between 1 and ∞.

c) Use relations (2) and (3) to show that the maximum eigenvalue of B, and thus of H_∞, is less than 1. Therefore, if we can show that at some finite $s > 1$ the maximum eigenvalue of H_s is greater than 1, by continuity there must be a value of s for which the eigenvalue equals 1 and there exists a zero sound mode.

d) Note that the following variational expression provides a bound on the largest eigenvalue λ_s

$$\lambda_s > (\chi, H\chi) = \frac{1}{s}\int \frac{d\hat{k}}{4\pi}|\chi(\hat{k})|^2\cos\theta + \int \frac{d\hat{k}}{4\pi}\frac{d\hat{k}'}{4\pi}\chi^*(\hat{k})B(\hat{k}\cdot\hat{k}')\chi(\hat{k}') \tag{5}$$

where

$$\int \frac{d\hat{k}}{4\pi} |\chi(\hat{k})|^2 = 1 \ . \tag{6}$$

Since $B(1) > 0$, one can find a positive B_0 and a θ_0 sufficiently small that $B(x) > B_0 > 0$ for $\cos\theta_0 < x < 1$. Therefore, use the following trial function

$$\chi(\theta, \phi) = \begin{cases} \frac{A}{2-\cos\theta} & 0 < \theta < \theta_0 \\ 0 & \text{otherwise} \end{cases}$$

with A fixed by the normalization condition (6) to show

$$\lambda_s > 1 + \frac{1}{s} \left[\frac{\int_{\cos\theta_0}^{1} \frac{dx}{s-x}}{\int_{\cos\theta_0}^{1} \frac{dx}{(s-x)^2}} \right] \left[B_0 \frac{s}{2} \int_{\cos\theta_0}^{1} \frac{dx}{s-x} - 1 \right] \ .$$

Thus, show that there exists an s such that $\lambda_s > 1$.

e) In the case of both spin and isospin degrees of freedom, what can you say about the existence of zero sound?

PROBLEM 6.6 Second Order Perturbation Theory for the Dilute Fermi Gas

The object of this problem is to calculate properties of a dilute Fermi gas using the second order expansion Eq. (6.106). As in Problem 3.3 for the dilute Bose gas, we will assume that the density is sufficiently low that the range of the interaction is short compared with the wavelength. The potential may, therefore, be approximated by a δ-function in coordinate space or a constant in momentum space

$$V(r) = U\delta^3(r)$$

and in lowest order, U is related to the s-wave scattering length a by

$$U = \frac{4\pi a}{m} \ .$$

Calculate the scattering length to second order in perturbation theory, remembering throughout the problem that with zero range forces only opposite spins interact. Although your expression is infinite, the infinite terms will cancel out when second order observables are expressed in terms of a.

Evaluate the quasiparticle interaction using Eq. (6.106) and show

$$f(\theta) = \frac{2\pi a}{m} \left[1 + 2 \left(\frac{3}{\pi} \right)^{1/3} a N^{1/3} \left(2 + \frac{\cos\theta}{2\sin\frac{\theta}{2}} \ln \left(\frac{1+\sin\frac{\theta}{2}}{1-\sin\frac{\theta}{2}} \right) \right) \right]$$
$$- \frac{8\pi a}{m} \vec{\sigma}_1 \cdot \vec{\sigma}_3 \left[1 + 2 \left(\frac{3}{\pi} \right)^{1/3} a N^{1/3} \left(1 - \frac{\sin\frac{\theta}{2}}{2} \ln \left(\frac{1+\sin\frac{\theta}{2}}{1-\sin\frac{\theta}{2}} \right) \right) \right] \ .$$

Note that although this expression has a logarithmic singularity at $\theta = \pi$, it is not important for integrals of $f(\theta)$ with regular functions.

Thus, obtain the results

$$\frac{m}{m^*} = 1 - \frac{8}{15} a^2 \rho^{2/3} \left(\frac{3}{\pi}\right)^{2/3} (7 \ln 2 - 1)$$

and

$$c_1^2 = \frac{\pi^{4/3}}{3^{1/3}} \frac{\rho 2/3}{m^2} + \frac{2\pi a \rho}{m^2} \left[1 - \frac{4}{15}\left(\frac{3}{4}\right)^{1/3} a\rho^{1/3}(11 - 2 \ln 2)\right] .$$

Finally, use the relation

$$c^2 = \frac{\rho}{m} \frac{\partial \mu}{\partial \rho}$$

which follows from Eqs. (6.27) and (6.32) to find the total energy

$$E = \int \mu \, f \rho = E^0 + \frac{\pi a \rho^2}{m} \left[1 + \frac{6}{35}\left(\frac{3}{\pi}\right)^{1/3} a\rho^{1/3}(11 - 2 \ln 2)\right] .$$

This may also be obtained by evaluating Eq. (6.105) and agrees with the result of Lee and Yang (1957).

PROBLEM 6.7 Landau Theory in One Spatial Dimension

Consider a Fermi liquid in one spatial dimension, in which case the Fermi surface is comprised of two points, $k = \pm k_F$. The quasiparticle interaction in any spin channel is characterized by two values, $f(k_F\sigma, k_F\sigma')$ and $f(k_F\sigma, -k_F\sigma')$ and it is convenient to establish contact with our previous three-dimensional notation by defining

$$f(k\sigma, k'\sigma') = f(k, k') + 4\vec{\sigma} \cdot \vec{\sigma}' \phi(k, k')$$

$$F_{\left\{ {0 \atop 1} \right\}} = N(0) \frac{1}{2} \left(f(k_F, k_F) + f(k_F, -k_F)\right)$$

$$Z_{\left\{ {0 \atop 1} \right\}} = N(0) \frac{1}{2} \left(\phi(k_F, k_F) + \phi(k_F, -k_F)\right)$$

where the one-dimensional density of states is $\frac{g}{\pi} \frac{m^*}{k_F}$ and σ represents some internal symmetry such as spin or isospin with degeneracy g. Note that the symmetric and antisymmetric combinations F_0 and F_1 correspond to the first two terms in the Legendre expansion (6.10) since $P_{\left\{ {0 \atop 1} \right\}}(1) = 1$ and $P_{\left\{ {0 \atop 1} \right\}}(-1) = \pm 1$.

a) Derive the following results for the effective mass

$$\frac{m^*}{m} = 1 + F_0$$

specific heat

$$c_V = \frac{C_V}{L} = \frac{\pi k_B^2 T m^*}{3 k_F}$$

sound velocity

$$c^2 = \frac{k_F^2 (1 + F_0)}{m m^*}$$

and forward scattering sum rule

$$\sum_{i=1}^{2}\left(\frac{F_i}{1+F_i}+\frac{Z_i}{1+Z_i}\right)=0 \ .$$

b) Now, consider the special case of no internal symmetry. The only two parameters of the theory F_0 and F_1 then are related by the forward scattering sum rules, so that the specific heat is uniquely specified by the sound velocity. Show

$$\left(\frac{c}{c^{(0)}}\right)^2=\frac{1}{2\left(\frac{C_V}{C_V^{(0)}}\right)-1}$$

where $c^{(0)}$ and $C_V^{(0)}$ denote the sound velocity and specific heat of the non-interacting system.

c) Express c and C_V in terms of appropriate derivatives of the grand potential Ω. Using the first-order perturbation theory result for Ω from Chapter 2, verify that in one dimension the result in part (b) is satisfied to first order in the potential. Is there an analogous relation between c and C_V in first order for higher dimensions?

CHAPTER 7

FURTHER DEVELOPMENT OF FUNCTIONAL INTEGRALS

Thus far, functional integral methods have been used to derive more efficiently, elegantly, or physically, results which could also be obtained using other techniques. This chapter and the next address a range of problems for which functional integrals provide unique insights and approximations.

In the first section of this chapter, we will show how to generate a variety of alternative functional integrals for the evolution operator which give rise to different physical approximations in the stationary-phase approximation. In Section 2, we show how to perform the general saddle point approximation around a static mean field solution and derive an expansion in terms of the fundamental vibrational excitations of the system. In subsequent sections, we show how time-dependent mean field solutions may be used to calculate transition amplitudes between specific states, approximate quantum eigenstates of large amplitude collective motion, study barrier penetration and spontaneous fission, and evaluate the asymptotic behavior of large orders of perturbation theory.

7.1 REPRESENTATIONS OF THE EVOLUTION OPERATOR

Different functional integral representations for the evolution operator produce different physical approximations in the stationary phase approximation. For example, the Feynman path integral obtained by inserting complete sets of coordinate states yields the classical equations of motion whereas the functional integral derived from inserting Boson coherent states produce the Hartree equations of motion for the Bose condensate wave function. Thus, in this section we will explore the freedom available in formulating functional integrals for the evolution operator and how to exploit this freedom in making physical approximations. To focus the discussion, we will consider Fermion systems. The modifications to treat Bosons are straightforward.

THE AUXILIARY FIELD

Consider a system of Fermions with Hamiltonian:

$$H = H_0 + V \tag{7.1a}$$

where H_0 is a one-body operator

$$H_0 = \sum_{\alpha\beta} T_{\alpha\beta} a^\dagger_\alpha a_\beta \tag{7.1b}$$

and V is a two-body interaction:

$$V = \frac{1}{2} \sum_{\alpha\beta\gamma\delta} (\alpha\beta|v|\gamma\delta)\, a^\dagger_\alpha a^\dagger_\beta a_\delta a_\gamma \quad . \tag{7.1c}$$

We wish to evaluate a matrix element of the evolution operator between two states:

$$W(t_f, t_i) = \langle \psi_f | e^{iH(t_f - t_i)} | \psi_i \rangle \quad . \tag{7.2}$$

It is useful to rearrange the Hamiltonian as follows:

$$H = \sum_{\alpha\beta} K_{\alpha\beta}\hat{\rho}_{\alpha\beta} + \frac{1}{2}\sum_{\alpha\beta\gamma\delta} v_{\alpha\beta\gamma\delta}\hat{\rho}_{\alpha\gamma}\hat{\rho}_{\beta\delta} \tag{7.3a}$$

where

$$\hat{\rho}_{\alpha\gamma} = a^\dagger_\alpha a_\gamma \ , \tag{7.3b}$$

$$K_{\alpha\delta} = T_{\alpha\delta} - \frac{1}{2}\sum_\beta v_{\alpha\beta\beta\delta} \tag{7.3c}$$

and

$$v_{\alpha\beta\gamma\delta} = (\alpha\beta|v|\gamma\delta) \ . \tag{7.3d}$$

The unphysical self-interaction term grouped with the kinetic energy in K arises from anticommuting the creation and annihilation operators in the normal-ordered two-body interaction, V, and will ultimately cancel out of physical observables.

We now express the evolution operator containing the two-body Hamiltonian as an integral over one-body evolution operators using a transformation introduced by Stratonovich (1957) and Hubbard (1958). This transformation is simply an operator form of the familiar Gaussian identity. The non-commutitivity of the two terms in H is handled in the usual way by dividing the integral $(t_f - t_i)$ in N slices of length $\epsilon = \frac{t_f - t_i}{N}$. To order ϵ^2, we obtain

$$\begin{aligned} e^{-i\epsilon H} &= e^{-i\epsilon\left(\sum_{\alpha\beta} K_{\alpha\beta}\hat{\rho}_{\alpha\beta} + \frac{1}{2}\sum_{\alpha\beta\gamma\delta} v_{\alpha\beta\gamma\delta}\hat{\rho}_{\alpha\gamma}\hat{\rho}_{\beta\delta}\right)} \\ &\underset{\epsilon\to 0}{\sim} e^{-i\epsilon\sum_{\alpha\beta} K_{\alpha\beta}\hat{\rho}_{\alpha\beta}}\, e^{-\frac{i\epsilon}{2}\sum_{\alpha\beta\gamma\delta} v_{\alpha\beta\gamma\delta}\hat{\rho}_{\alpha\gamma}\hat{\rho}_{\beta\delta}} \ . \end{aligned} \tag{7.4}$$

To simplify notation, we denote $m = (\alpha\gamma)$, $n = (\beta\delta)$, and $v_{\alpha\beta\gamma\delta} = V_{mn}$. Then, using the Gaussian integral (1.179), we obtain:

$$e^{-\frac{i\epsilon}{2}\sum_{m,n}\hat{\rho}_m V_{mn}\hat{\rho}_n} = (\det \epsilon V^{-1}_{mn})^{1/2}\int\prod_m \frac{d\sigma_m}{\sqrt{2\pi i}}\, e^{\frac{i}{2}\epsilon\sum_{m,n}\sigma_m V^{-1}_{mn}\sigma_n - i\epsilon\sum_m \sigma_m\hat{\rho}_m} \tag{7.5a}$$

or alternatively, changing the integration variable by defining $\tilde{\sigma}_n = V^{-1}_{nm}\sigma_m$,

$$e^{-i\frac{\epsilon}{2}\sum\limits_{m,n}\hat{\rho}_m V_{mn}\hat{\rho}_n} = (\det \epsilon V_{mn})^{1/2}\int\prod_m \frac{d\tilde{\sigma}_m}{\sqrt{2\pi i}}\, e^{\frac{i}{2}\epsilon\sum\limits_{mn}\tilde{\sigma}_m V_{mn}\tilde{\sigma}_n - i\epsilon\sum\limits_{mn}\tilde{\sigma}_m V_{mn}\hat{\rho}_n} \ . \tag{7.5b}$$

It is often convenient to go back and forth between forms (7.5a) and (7.5b), and we will do so freely throughout this chapter.

Since this transformation has to be performed at each time step, we introduce a time index k in σ, and write

$$\begin{aligned} e^{-iH(t_f - t_i)} = \frac{1}{\mathcal{N}}\int\prod_{k=1}^{N}\prod_{\alpha\beta} d\sigma_{\alpha\beta}(k) e^{i\frac{\epsilon}{2}\sum_{k=1}^{N}\sum_{\alpha\beta\gamma\delta}\sigma_{\alpha\beta}(k)v_{\alpha\gamma\beta\delta}\sigma_{\gamma\delta}(k)} \\ \times\prod_{k=1}^{N} e^{-i\epsilon\sum_{\alpha\beta}\left(K_{\alpha\beta} + \sum_{\gamma\delta}\sigma_{\gamma\delta}(k)v_{\gamma\alpha\delta\beta}\right)\hat{\rho}_{\alpha\beta}} \end{aligned} \tag{7.6a}$$

where $\mathcal{N}$ is a normalization constant, given by

$$\frac{1}{\mathcal{N}} = \prod_{k=1}^{N} \frac{(\det \epsilon v_{\alpha\gamma\beta\gamma})^{1/2}}{\prod_{\alpha\beta}(2i\pi)^{1/2}} . \tag{7.6b}$$

Note that since Eq. (7.4) is valid to order ϵ^2, we expect Eq. (7.6) to be valid to order $N\epsilon^2 = \epsilon$.

Since we take the limit $\epsilon \to 0$, it is natural to replace the time index k by a continuous variable t, which runs between t_i and t_f, and the sums over time by integrals. With obvious notation we write the evolution operator

$$\begin{aligned} e^{-H(t_f-t_i)} = \int \prod_{\alpha\beta} \mathcal{D}\sigma_{\alpha\beta}(t)\, e^{\frac{1}{2}\int_{t_i}^{t_f} dt\, \sigma_{\alpha\beta}(t) v_{\alpha\gamma\beta\delta}\sigma_{\gamma\delta}(t)} \\ T\left[e^{-i\int_{t_i}^{t_f} dt \left(K_{\alpha\beta}+\sigma_{\gamma\delta}(t)v_{\gamma\alpha\delta\beta}\right)\hat{\rho}_{\alpha\beta}} \right] \end{aligned} \tag{7.7a}$$

where there is an implicit summation over repeated indices. The time-ordered exponential is the continuous generalization of the product over k in (7.6a) and the measure is defined by:

$$\prod_{\alpha\beta} \mathcal{D}\sigma_{\alpha\beta}(t) = \lim_{\epsilon\to 0} \prod_{k=1}^{N} (\det \epsilon v_{\alpha\beta\gamma\delta})^{1/2} \prod_{\alpha\beta} \frac{d\sigma_{\alpha\beta}(k)}{(2i\pi)^{1/2}} . \tag{7.7b}$$

We will omit the single particle indices α whenever it will cause no ambiguity and reinstate them for specific equations. For example, we will use the abbreviated notation

$$W(t_f,t_i) = \int \mathcal{D}\sigma\, e^{\frac{1}{2}\int_{t_i}^{t_f} dt\, \sigma v \sigma}\, W_\sigma^H(t_f,t_i) \tag{7.8a}$$

with

$$W_\sigma^H(t_f,t_i) = \langle \psi_f | T[e^{-i\int_{t_i}^{t_f}(K+\sigma(t)v)\hat{\rho}}] | \psi_i \rangle . \tag{7.8b}$$

Thus, the evolution operator is the functional integral over an auxiliary field σ of the evolution operator for a one-body time-dependent Hamiltonian $\hat{h}_\sigma^H(t)$, defined by

$$\hat{h}_\sigma^H(t) = \sum_{\alpha\beta}\left(K_{\alpha\beta} + \sum_{\gamma\delta} \sigma_{\gamma\delta}(t) v_{\gamma\alpha\delta\beta}\right) a_\alpha^\dagger a_\beta \tag{7.8c}$$

and weighted by a Gaussian factor. As we will subsequently verify, the superscript H in Eqs. (7.8) indicates that in the stationary phase approximation, the single particle Hamiltonian represents evolution in the Hartree mean field.

From (7.6), we see that the $\sigma_{\alpha\beta}(t)$ are independent integration variables, and thus the trajectory $\sigma_{\alpha\beta}(t)$ is not a continuous function. This differs from the case of the Feynman path integral, in which the kinetic energy term in the action, $\int dt\dot{q}^2$, forces all relevant trajectories to be continuous, but non-differentiable. Let us also note that the form (7.7a) of the Hamiltonian implies that $\sigma(t)$ is a classical field conjugate to $v\hat{\rho}$, and similarly in Eq. (7.5a), σ is a field conjugate to $\hat{\rho}$.

There is a large amount of arbitrariness in the choice of the rearrangement of the Hamiltonian. For example, had we written:

$$H = \sum_{\alpha\beta} \tilde{K}_{\alpha\beta}\hat{\rho}_{\alpha\beta} - \frac{1}{2}\sum_{\alpha\beta\gamma\delta} v_{\alpha\beta\delta\gamma}\hat{\rho}_{\alpha\gamma}\hat{\rho}_{\beta\delta} \tag{7.9a}$$

with

$$\tilde{K}_{\alpha\delta} = T_{\alpha\delta} + \frac{1}{2}\sum v_{\alpha\beta\delta\beta} \tag{7.9b}$$

we would have obtained an alternative functional integral:

$$W(t_f - t_i) = \int \mathcal{D}\sigma\, e^{-i\int_{t_i}^{t_f} dt\, \sigma_{\alpha\beta}(t) v_{\alpha\gamma\delta\beta}\sigma_{\gamma\delta}(t)} W_\sigma^F(t_f, t_i) \tag{7.10a}$$

with

$$W_\sigma^F(t_f, t_i) = \langle\psi_f|T[e^{-i\int_{t_i}^{t_f} dt\left(\tilde{K}_{\alpha\beta} - \sigma_{\gamma\delta}(t) v_{\gamma\alpha\beta\delta}\right)\hat{\rho}_{\alpha\beta}}]|\psi_i\rangle \tag{7.10b}$$

where the superscript F denotes evolution in the exchange or Fock term of the mean field. Similarly, writing

$$H = \sum_{\alpha\beta} T_{\alpha\beta}\hat{\rho}_{\alpha\beta} + \frac{1}{2}\sum_{\alpha\beta\gamma\delta} \hat{\Delta}^\dagger_{\alpha\beta} v_{\alpha\beta\gamma\delta}\hat{\Delta}_{\gamma\delta} \tag{7.11a}$$

where

$$\Delta^\dagger_{\alpha\beta} = a^\dagger_\alpha a^\dagger_\beta \quad \text{and} \quad \Delta_{\delta\gamma} = a_\gamma a_\delta \tag{7.11b}$$

one obtains:

$$W(t_f, t_i) = \int \mathcal{D}\left(\chi^*_{\alpha\beta'}\chi_{\alpha\beta}\right) e^{\frac{i}{2}\int_{t_i}^{t_f} dt\, \chi^*_{\alpha\beta}(t) v_{\alpha\beta\gamma\delta}\chi_{\gamma\delta}(t)} W^P(t_f, t_i) \tag{7.12a}$$

where

$$W^P(t_f, t_i) = \langle\psi_f|T[e^{-i\int_{t_i}^{t_f} dt\left[T_{\alpha\beta}\hat{\rho}_{\alpha\beta} + v_{\alpha\beta\gamma\delta}\left(\chi^*_{\alpha\beta}\hat{\Delta}_{\gamma\delta} + \hat{\Delta}^\dagger_{\alpha\beta}\chi_{\gamma\delta}\right)\right]}]|\psi_i\rangle \tag{7.12b}$$

and the integration is over pairs of complex fields $\chi^*_{\alpha\beta}(t)$ and $\chi_{\alpha\beta}(t)$. In this case, as denoted by the superscript P, $\chi^*_{\alpha\beta}(t)$ and $\chi_{\alpha\beta}(t)$ play the role of classical pairing fields.

Although each of the functional integrals (7.8), (7.10), and (7.12) represents the exact evolution operator and reproduces the correct perturbation expansion for all observables, in the stationary phase approximation, they produce the Hartree, Fock, or pairing mean field, respectively. As shown in Problem 7.1, there is additional flexibility in representing the evolution operator by taking combinations of these rearrangements of the Hamiltonian, which may be exploited to obtain Hartree-Fock or Hartree-Fock-Bogoliubov mean fields. In physical applications, the choice of the auxiliary field representation should be guided by the physics of the problem. For simplicity, in derivations in subsequent sections, we will use the Hartree form, (7.8).

In addition to this arbitrariness in breaking up the Hamiltonian, there is further freedom in formulating the auxiliary field functional integral (Kerman, Levit and Troudet, 1982). In the limit when $\epsilon \to 0$, Eq. (7.5) may be written

$$\begin{aligned} e^{-i\frac{\epsilon}{2}\sum_{m,n}\hat{\rho}_m V_{m,n}\hat{\rho}_n} &\underset{\epsilon\to 0}{\sim} (\det \epsilon\, V_{mn})^{1/2} \int \prod_n \frac{d\sigma_n}{\sqrt{2\pi i}}\, e^{i\frac{\epsilon}{2}\sum_{m,n}\sigma_m V_{mn}\sigma_n} \\ &\times \left(1 - i\epsilon \sum_{m,n}\sigma_m V_{mn}\hat{\rho}_n - \frac{\epsilon^2}{2}\left(\sum_{m,n}\sigma_m V_{mn}\hat{\rho}_n\right)^2\right) . \end{aligned} \tag{7.13}$$

The term linear in σ on the right hand side integrates to zero, since it is odd and the quadratic term integrates to the contribution of order ϵ of the left hand side. The quadratic term is actually of order ϵ because the dominant values of σ with the Gaussian weighting factor $e^{i\frac{\epsilon}{2}\sigma V\sigma}$ are of order $\epsilon^{-1/2}$. The fact that the linear term in σ integrates to zero allows one to modify its coefficient arbitrarily. Therefore, let us consider the equivalent expression

$$\begin{aligned} e^{-i\frac{\epsilon}{2}\sum_{mn}\hat{\rho}_m V_{mn}\hat{\rho}_n} &\underset{\epsilon\to 0}{\sim} (\det \epsilon V_{mn})^{1/2} \int \prod_m \frac{d\sigma_n}{\sqrt{2\pi i}}\, e^{i\frac{\epsilon}{2}\sum_{mn}\sigma_m V_{mn}\sigma_n} \\ &\times \left(1 - i\epsilon \sum_{mn}\sigma_m W_{mn}\hat{\rho}_n - \frac{\epsilon^2}{2}\left(\sum_{mn}\sigma_m V_{mn}\hat{\rho}_n\right)^2\right) \end{aligned} \tag{7.14a}$$

where W is arbitrary. Although Eqs. (7.13) and (7.14a) are identical when the integral over σ is performed exactly, they differ completely in the stationary phase approximation. As shown in Problem 2, the one-body evolution operator for a continuous stationary auxiliary field $\sigma^{(s)}$ is

$$U_{\sigma^s} = \lim_{\epsilon\to 0} \prod_{t_i} \left(1 - i\epsilon \sum_{mn} \sigma_m^{(s)} W_{mn}\hat{\rho}_n\right) = e^{-i\int d\tau \sum_{mn}\sigma_m^{(s)} W_{mn}\hat{\rho}_n} \tag{7.14b}$$

so that one may specify the mean field at will by the choice of W. In particular, the choice $W_{\alpha\beta\gamma\delta} = v_{\alpha\beta\gamma\delta} - v_{\alpha\beta\delta\gamma}$ yields the Hartree-Fock mean field. Alternatively, one may choose W to be a G-matrix, corresponding to the sum of ladder diagrams, or some other physically motivated effective interaction. In contrast to (7.14a), where $\sigma \sim \epsilon^{-1/2}$, the stationary solution $\sigma^{(s)}$ is independent of ϵ so the quadratic term does not contribute to (7.14b) in the continuum limit. Beyond the stationary phase level, the theory does not have a well-defined continuum limit and as shown in Problem 7.2, one must use discrete time slices with finite ϵ.

OVERCOMPLETE SETS OF STATES

An alternative and powerful method for generating functional integrals is to use overcomplete sets of states. A set of states is overcomplete if it redundantly generates the Hilbert space $\mathcal{H}$. We shall assume that the set of states is continuous, and denote it by $\{|\varphi\rangle\}$. Thus, from its definition, there exists a measure $d\mu(\varphi)$ on the set $\{|\varphi\rangle\}$ such that any state $|\psi\rangle$ of the Hilbert space can be decomposed as

$$|\psi\rangle = \int d\mu(\varphi)\ \langle\varphi|\psi\rangle|\varphi\rangle\ . \tag{7.15}$$

The overcompleteness implies that the decomposition (7.15) is not unique. Examples of overcomplete sets of states are the Boson and Fermion coherent states introduced in Chapter 1.

From (7.15), it follows that there exists a decomposition of unity:

$$1 = \int d\mu(\varphi)|\varphi\rangle\langle\varphi| \tag{7.16a}$$

and this is the basic formula to construct functional integrals. As in the case of Boson or Fermion coherent states, we will assume that the set $\{|\varphi\rangle\}$ is parametrized by a complex variable φ, and the adjoint set $\{\langle\varphi|\}$ by the conjugate variable φ^*. The measure must be a function of φ^* and φ and we shall write*

$$\begin{aligned} 1 &= \int d\mu(\varphi^*, \varphi)|\varphi\rangle\langle\varphi| \\ &= \int \frac{d\varphi^* d\varphi}{2i\pi}\gamma(\varphi^*\varphi)|\varphi\rangle\langle\varphi| \ . \end{aligned} \tag{7.16b}$$

By repeated insertion of Eq. (7.16b) in the Trotter formula

$$e^{-iH(t_f-t_i)} = \lim_{\epsilon\to 0}(1-i\epsilon H)^N \tag{7.17}$$

where $\epsilon = \frac{t_f-t_i}{N}$, the matrix element of the evolution operator between two states $|\varphi_i\rangle$ and $|\varphi_f\rangle$ which belong to the overcomplete set is given by:

$$\begin{aligned} W_{fi} &= W(t_f, t_i) \\ &= \langle\varphi_f|e^{-H(t_f-t_i)}|\varphi_i\rangle \\ &= \lim_{\epsilon\to 0}\int\prod_{k=1}^{N-1}\frac{d\varphi_k^* d\varphi_k}{2i\pi}\gamma(\varphi_k^*\varphi_k)\langle\varphi_f|1-i\epsilon H|\varphi_{N-1}\rangle\langle\varphi_{N-1}|(1-i\epsilon H)|\varphi_{N-2}\rangle\dots \\ &\quad \dots\langle\varphi_1|(1-i\epsilon H)|\varphi_i\rangle \ . \end{aligned} \tag{7.18}$$

If the measure $\gamma(\varphi^*\varphi)$ decreases fast enough when $|\varphi^*\varphi|$ goes to infinity, it is legitimate to reexponentiate each matrix element in (7.18) using:

$$\langle\varphi|1-i\epsilon H|\varphi'\rangle \sim \langle\varphi|\varphi'\rangle e^{-i\epsilon H(\varphi^*,\varphi')} \tag{7.19a}$$

where we use the notation:

$$H(\varphi^*, \varphi') = \frac{\langle\varphi|H|\varphi'\rangle}{\langle\varphi|\varphi'\rangle} \ . \tag{7.19b}$$

* By making a gauge transformation $\varphi^* \to e^{-i\alpha}\varphi^*$, $\varphi \to e^{i\alpha}\varphi$, one sees that the measure may depend only on the combination $\varphi^*\varphi$.

Using the definition $|\delta\phi_k\rangle = |\phi_k\rangle - |\phi_{k-1}\rangle$ and using continuum notation, we may rewrite (7.18) as:

$$W_{fi} = \lim_{\epsilon\to 0} \int \prod_{k=1}^{N-1} \frac{d\varphi_k^* d\varphi_k}{2i\pi} \gamma(\varphi_k^* \varphi_k) \prod_{k=1}^{N} \langle\varphi_k|\varphi_k\rangle$$
$$e^{\sum_{k=1}^{N}\left[\ln\left(1-\frac{\langle\varphi_k|\delta\varphi_k\rangle}{\langle\varphi_k|\varphi_k\rangle}\right)-i\epsilon H(\varphi_k^*,\varphi_{k-1})\right]} \tag{7.20a}$$
$$= \int_{\substack{\varphi_{t_i}=\varphi_i \\ \varphi_{t_f}^*=\varphi_f^*}} \mathcal{D}\mu(\varphi_t^*, \varphi_t)\langle\varphi_f|\varphi(t_f)\rangle e^{iS}$$

where

$$\mathcal{D}\mu(\varphi_t^*, \varphi_t) = \prod_{t_i<t<t_f} \left\{\gamma(\varphi_t^* \varphi_t)\langle\varphi_t|\varphi_t\rangle \frac{d\varphi_t^* d\varphi_t}{2i\pi}\right\} \tag{7.20b}$$

is the functional integral measure, and

$$S = \frac{1}{i}\int_{t_i}^{t_f} dt \left\{\frac{1}{dt}\cdot \ln\left(1 - dt\frac{\langle\varphi_t|\frac{\partial\varphi}{\partial t}\rangle}{\langle\varphi_t|\varphi_t\rangle}\right) - i\frac{\langle\varphi_t|H|\varphi_{t-dt}\rangle}{\langle\varphi_t|\varphi_{t-dt}\rangle}\right\} \tag{7.20c}$$

is the action.

Note that as in the coherent state functional integrals discussed in Chapter 2, the continuum notation is only a convenient shorthand for the original discrete expressions. The trajectories have no reason to be differentiable or even continuous, and thus, the time derivative only represents a finite difference, $\frac{\partial\varphi}{\partial t} = \frac{\varphi_{t+dt}-\varphi_t}{dt}$. Also, although φ_{t_i} and $\varphi_{t_f}^*$ are held fixed, the variables $\varphi_{t_i}^*$ and φ_{t_f} are integrated over.

Let us note that when expanding (7.20a) using the stationary-phase method, we shall seek stationary solutions of the action (7.20c). We shall use only solutions which are either constant or differentiable with respect to time. In that case, Eq. (7.20c) becomes:

$$S = \int_{t_i}^{t_f} dt \, \frac{\langle\varphi_t|i\frac{\partial}{\partial t} - H|\varphi_t\rangle}{\langle\varphi_t|\varphi_t\rangle} \tag{7.21}$$

and the stationary-phase approximation thus reduces to minimizing this action with respect to the wave functions $|\varphi_t\rangle$ and $\langle\varphi_t|$.

To illustrate this method and provide the background for the use of Slater determinants for the many-Fermion problem, it is useful to examine several simple examples. The simplest application of this general formalism is in the case of Boson or Fermion coherent states. As shown in Problem 7.3, it is easy to see that Eq. (7.20) reduces to the coherent state results obtained in Section 2.2.

We next write the one-body problem as a functional integral over an overcomplete set of wave functions. The set of all square integrable wave-functions is identical to the total Hilbert space, and it is thus an overcomplete set. The closure relation in the Hilbert space is given by

$$\int \mathcal{D}(\psi^*, \psi)\, e^{-\int d\mathbf{x}|\psi(\mathbf{x})|^2} |\psi\rangle\langle\psi| = 1 \tag{7.22a}$$

where the measure $\mathcal{D}(\psi^*,\psi)$ is equal to $\frac{1}{\mathcal{N}}\prod_{\vec{x}} d\psi^*(\vec{x})d\psi(\vec{x})$ and $\mathcal{N}$ is a normalization constant chosen so that

$$\int \mathcal{D}(\psi^*,\psi)\, e^{-\int d\vec{x}|\psi(\vec{x})|^2} = 1 \,. \tag{7.22b}$$

The closure relation is easily proven by taking a matrix element of Eq. (7.22a) between $\langle\vec{x})$ and $|\vec{y}\rangle$, and noting that with the normalization (7.22b), the contraction $\overbrace{\psi(\vec{x})\psi^*}(\vec{y})$ is equal to $\delta(\vec{x}-\vec{y})$.

Therefore, equation (7.20) yields

$$W_{fi} = \int\limits_{\substack{\psi(\vec{x},t_i)=\psi_i(\vec{x})\\ \psi^*(\vec{x},t_f)=\psi_f^*(\vec{x})}} \mathcal{D}\mu\left(\psi^*(\vec{x},t),\psi(\vec{x},t)\right)\langle\psi_f|\psi(t_f)\rangle\, e^{iS} \tag{7.23a}$$

where

$$\mathcal{D}\mu\left(\psi^*(\vec{x},t),\psi(\vec{x},t)\right) = \prod_{t_i<t<t_f}\left\{e^{-\langle\psi(t)|\psi(t)\rangle}\langle\psi(t)|\psi(t)\rangle\mathcal{D}(\psi^*,\psi)\right\} \tag{7.23b}$$

and S is given by (7.20c). Note that according to (7.21), the stationary $|\psi(t)\rangle$ is just the solution of the usual Schödinger equation.

We now treat the many-Fermion problem using an overcomplete set of Slater determinants. The set of all Slater determinants with N particles, denoted by $|\varphi_1\ldots\varphi_N\rangle$, spans the whole Hilbert space of antisymmetrized wave-functions. The closure relation is

$$\frac{1}{N!}\int\prod_{\alpha=1}^{N}\mathcal{D}(\varphi_\alpha^*,\varphi_\alpha)\, e^{-\sum_{\alpha=1}^{N}\int dx|\varphi_\alpha(x)|^2}|\varphi_1\ldots\varphi_N\rangle\langle\varphi_1\ldots\varphi_N| = 1 \,. \tag{7.24a}$$

As in the previous case, this relation is proved by calculating the matrix element of (7.24) between the antisymmetrized states $\langle x_1\ldots x_N|$ and $|y_1\ldots y_N\rangle$, and showing that it is equal to $\det\left(\langle x_i|y_j\rangle\right)$. Since it is often useful to consider Slater determinants composed of orthonormal wave functions, one may use the alternative closure relation

$$\frac{1}{\mathcal{N}}\int\prod_{\alpha=1}^{N}\mathcal{D}(\phi_\alpha^*,\phi_\alpha)\prod_{\alpha\beta}\delta\left(\langle\phi_\alpha|\phi_\beta\rangle-\delta_{\alpha\beta}\right)|\phi_1\ldots\phi_N\rangle\langle\phi_1\ldots\phi_N| = 1 \tag{7.24b}$$

where $\mathcal{N}$ is a normalized constant. The details of the derivation of Eqs. (7.24a,b) are outlined in Problem 7.4.

Using (7.24), the functional integral becomes

$$W_{fi} = \int\limits_{\substack{\varphi_\alpha(x,t_i)=\varphi_{\alpha,i}(x)\\ \varphi_\alpha^*(x,t_f)=\varphi_{\alpha,f}^*(x)}} \mathcal{D}\mu\left(\varphi^*,\varphi\right)\det\left(\langle\varphi_{\alpha,f}|\varphi_\beta(t_f)\rangle\right)\, e^{iS} \tag{7.25a}$$

where

$$\mathcal{D}\mu(\varphi^*,\varphi)=\prod_{\alpha=1}^{N}\prod_{t_i<t<t_f}e^{-\int dx|\varphi_\alpha(x,t)|^2}\mathcal{D}(\varphi^*_\alpha(t),\varphi_\alpha(t))\det\langle\varphi_\alpha(t)|\varphi_\beta(t)\rangle \quad (7.25b)$$

or

$$\mathcal{D}\mu(\phi^*,\phi)=\prod_{\alpha=1}^{N}\prod_{t_i<t<t_f}\mathcal{D}(\phi^*_\alpha(t),\phi_\alpha(t))\prod_{\alpha\beta}(\langle\phi_\alpha|\phi_\beta\rangle-\delta_{\alpha\beta}) \quad (7.25c)$$

and S is given in (7.20c).

In the case of continuous trajectories, the action (7.21) becomes:

$$S=\int_{t_i}^{t_f}dt\frac{\langle\varphi_1\ldots\varphi_N|i\frac{\partial}{\partial t}-H|\varphi_1\ldots\varphi_N\rangle}{\langle\varphi_1\ldots\varphi_N|\varphi_1\ldots\varphi_N\rangle}\ . \quad (7.26)$$

It is shown in Problem 7.5 that the stationarity of S with respect to the Slater determinant $|\varphi_1\ldots\varphi_N\rangle$ yields the time-dependent Hartree-Fock equations of motion. Consequently, stationary phase evaluation of (7.25a) leads to the time-dependent Hartree-Fock approximation, and provides a simple, straightforward way to consistently include direct and exchange terms in a mean field approximation.

7.2 GROUND STATE PROPERTIES FOR FINITE SYSTEMS

We will now perform the complete stationary phase expansion around localized time-independent solutions. For simplicity, we will use the functional integral (7.7) in terms of the auxiliary σ-field, which will yield the Hartree approximation and a loop expansion in terms of Hartree phonons. It is possible to do the same calculations using functional integrals over Slater determinants, Eq. (7.25), which yield an expansion around the Hartree-Fock solution. This is cumbersome, due to the lack of a continuum limit, and is discussed in Problem 7.6. We will also defer for the moment the important physical question of the validity of the stationary phase expansion and the nature of the small parameter in which one is expanding. After seeing the structure of the resulting theory, we will return to this question in Section 7.4.

THE RESOLVENT OPERATOR

All the information about the eigenvalues of a Hamiltonian H and the matrix elements of an operator $\mathcal{O}$ in eigenstates of H is contained in the resolvent operator:

$$G_{\mathcal{O}}(E)=\mathrm{tr}\left(\frac{\mathcal{O}}{E-H+i\eta}\right) \quad (7.27a)$$

where η is positive infinitesimal. Denoting by $|\Phi_n\rangle$ the eigenvectors of H with energies E_n, we see that:

$$G_{\mathcal{O}}(E)=\sum_n\frac{\langle\Phi_n|\mathcal{O}|\Phi_n\rangle}{E-E_n+i\eta}\ . \quad (7.27b)$$

Thus, the eigenstates of H are poles of $G_O(E)$, and the matrix elements $\langle\Phi_n|O|\Phi_n\rangle$ are the corresponding residues. To evaluate the ground state energy of a finite Fermion system, we may choose $O = 1$ and calculate

$$\begin{aligned} G(E) &= \operatorname{tr}\left(\frac{1}{E - H + i\eta}\right) \\ &= -i\int_0^\infty dT e^{iET} \operatorname{tr} U(\tfrac{T}{2}, \tfrac{-T}{2}) \end{aligned} \tag{7.28a}$$

where we have used the Fourier transform (3.31b).

We now express the evolution operator in terms of a functional integral over an auxiliary field. For notational simplicity, it is convenient to use the form (7.5a) in which v^{-1} occurs in the Gaussian factor and no v occurs in the one-body operator. Thus, we write

$$G(E) = -i\int_0^\infty dT\, e^{iET} \int \frac{\mathcal{D}\sigma}{\mathcal{N}}\, e^{\frac{i}{2}\int_{-T/2}^{T/2} dt\, \sigma v^{-1}\sigma} \operatorname{tr} U_\sigma(\tfrac{T}{2}, \tfrac{-T}{2}) \tag{7.28b}$$

where

$$U_\sigma(\tfrac{T}{2}, \tfrac{-T}{2}) = T\big[e^{-i\int_{-T/2}^{T/2} dt\, h_\sigma(t)}\big] \tag{7.28c}$$

$$h_\sigma(t) = (K + \sigma(t))\,\hat{\rho} \tag{7.28d}$$

and

$$\mathcal{N} = \int \mathcal{D}\sigma\, e^{\frac{i}{2}\int_{-T/2}^{T/2} dt\, \sigma v^{-1}\sigma} \quad . \tag{7.28e}$$

It is convenient to evaluate $\operatorname{tr} U_\sigma$ in the basis of eigenfunctions of U_σ. Let us denote this basis by $|\Psi_N\rangle$, keeping in mind that it depends on σ. Since U_σ is the evolution operator of a time-dependent one-body Hamiltonian, its eigenstates are Slater determinants, composed of single-particle eigenstates of U_σ. We will denote by $\{|\varphi_k\rangle\}$ the set of orthonormal single-particle eigenstates of U_σ:

$$T\big[e^{-i\int_{-T/2}^{T/2} dt\, h_\sigma(t)}\big]|\varphi_k\rangle = e^{-i\alpha_k(\sigma,T)}|\varphi_k\rangle \tag{7.29a}$$

and an N-body eigenstate of U_σ is given by:

$$|\Psi_{k_1\ldots k_N}\rangle = |\varphi_{k_1}\cdots\varphi_{k_N}\rangle \tag{7.29b}$$

with eigenvalue $e^{-i\sum_{i=1}^N \alpha_{k_i}(\sigma,T)}$. In the N-particle space, the unit operator 1 and U_σ may be represented as:

$$1 = \sum_{\{k_1\ldots k_N\}} |\Psi_{k_1\ldots k_N}\rangle\langle\Psi_{k_1\ldots k_N}| \tag{7.30a}$$

and

$$U_\sigma = \sum_{\{k_1\ldots k_N\}} |\Psi_{k_1\ldots k_N}\rangle e^{-i\sum_{i=1}^N \alpha_{k_i}(T,\sigma)} \langle\Psi_{k_1\ldots k_N}| \quad . \tag{7.30b}$$

By its definition (7.28) U_σ is a unitary operator:

$$U_\sigma^\dagger(\tfrac{T}{2},\tfrac{-T}{2})U_\sigma(\tfrac{T}{2},\tfrac{-T}{2}) = 1 \tag{7.31}$$

so its eigenvalues have unit modulus, and the $\alpha_k(\sigma, T)$ are real. Using this basis, the resolvent in the space of N particles is

$$G(E) = -i \sum_{\{k_1 \ldots k_N\}} \int_0^\infty dT\, e^{iET} \int \frac{\mathcal{D}\sigma}{\mathcal{N}}\, e^{\frac{i}{2}\int_{-T/2}^{T/2} dt\, \sigma v^{-1}\sigma - i\sum_{i=1}^N \alpha_{k_i}(\sigma,T)} \tag{7.32}$$

and the resolvent in the full Fock space is obtained by summing over all N. This is the basic formula that we shall use in the following sections.

STATIC HARTREE APPROXIMATION

In applying the stationary phase approximation to the $\{\sigma\}$ integral in (7.32), one should consider general time-dependent stationary solutions $\sigma_s(t)$. In this section, we will consider the simplest case of time-independent solutions, denoted σ_0.

The stationarity condition is

$$v^{-1}\sigma_0 = \frac{1}{T}\sum_{i=1}^N \frac{\delta\alpha_{k_i}(\sigma_0, T)}{\delta\sigma_0} \quad . \tag{7.33}$$

For a time-independent σ_0, Eq. (7.29a) becomes:

$$e^{-iT h_{\sigma_0}}|\varphi_k\rangle = e^{-i\alpha_k(\sigma_0,T)}|\varphi_k\rangle \tag{7.34a}$$

so that $|\varphi_k\rangle$ is an eigenstate of h_{σ_0} with eigenvalues $\frac{\alpha_k}{T}$:

$$h_{\sigma_0}|\varphi_k\rangle = \frac{\alpha_k}{T}|\varphi_k\rangle \quad . \tag{7.34b}$$

Taking a derivative of this equation with respect to σ_0, and projecting on the left with $\langle\varphi_k|$, we obtain

$$\langle\varphi_k|h_{\sigma_0}\frac{\delta}{\delta\sigma_0}|\varphi_k\rangle + \langle\varphi_k|\frac{\delta h_{\sigma_0}}{\delta\sigma_0}|\varphi_k\rangle = \frac{1}{T}\frac{\delta\alpha_k}{\delta\sigma_0}\langle\varphi_k|\varphi_k\rangle + \frac{\alpha_k}{T}\langle\varphi_k|\frac{\delta}{\delta\sigma_0}|\varphi_k\rangle \tag{7.34c}$$

which, combined with the adjoint of (7.34b), yields

$$\frac{\delta\alpha_k}{\delta\sigma_0} = T\langle\varphi_k|\frac{\delta h_{\sigma_0}}{\delta\sigma_0}|\varphi_k\rangle \quad . \tag{7.35}$$

Using the stationarity equation (7.33) and the definition of h_σ, Eq. (7.28d) we obtain

$$\sigma_{0\alpha\beta} = \sum_{i=1}^N \sum_{\gamma,\delta} (\alpha\gamma|v|\beta\delta)\,\langle\gamma|\varphi_{k_i}\rangle\langle\varphi_{k_i}|\delta\rangle \tag{7.36a}$$

or in the $\{\vec{r}\}$ representation

$$\sigma_0(\vec{r}) = \int dr' v(\vec{r}-\vec{r}') \sum_{i=1}^{N} |\varphi_{k_i}(\vec{r}')|^2 . \tag{7.36b}$$

This result shows that σ_0 is the Hartree mean field generated by the Slater determinant $|\varphi_{k_1} \dots \varphi_{k_N}\rangle$. The alternative quantity $\tilde{\sigma} = v^{-1}\sigma$ introduced in Eq. (7.5b) corresponds to the density. Note that for each set of occupied states $\{k_1 \dots k_N\}$ appearing in the sum (7.32), we obtain a different stationary σ. From Eq. (7.34b), we see that the single-particle wave functions $|\psi_k\rangle$ satisfy the self-consistent Hartree equations

$$\left(-\frac{1}{2m}\nabla_r^2 - \frac{1}{2}v(0) + \sum_{i=1}^{N}\int dr' v(r-r')|\varphi_{k_i}(r')|^2\right)\varphi_k(r) = \epsilon_k \varphi_k(r) \tag{7.37a}$$

where the quantity

$$\epsilon_k = \frac{\alpha_k(\sigma_0, T)}{T} \tag{7.37b}$$

is time-independent and plays the role of the single-particle energy. Note the presence of the $\frac{1}{2}v(0)$, which shifts all single-particle energies, but does not modify the wave-functions.

Using these results, the stationary phase approximation to the resolvent, Eq. (7.32), is

$$\begin{aligned} G(E) &\approx -i \sum_{\{k_1 \dots k_N\}} \int_0^\infty dT\, e^{iET}\, e^{i\frac{T}{2}\int dr dr' \sigma_0(r) v^{-1}(r-r')\sigma_0(r')}\, e^{-iT\sum_{i=1}^{N}\epsilon_{k_i}} \\ &= \sum_{\{k_1 \dots k_N\}} \frac{1}{E - \left(\sum_{i=1}^{N}\epsilon_{k_i} - \frac{1}{2}\int dr dr' \sigma_0(r) v^{-1}(r-r')\sigma_0(r)\right) + i\eta} . \end{aligned} \tag{7.38}$$

In this static approximation, the total energy is thus:

$$E_H = \sum_{i=1}^{N}\epsilon_{k_i} - \frac{1}{2}\int drdr'\, \sigma_0(r)v^{-1}(r-r')\sigma_0(r') . \tag{7.39a}$$

Using the single-particle equation (7.12a), with the notation $|k\rangle \equiv |\varphi_k\rangle$, we see that

$$\epsilon_k = \langle k|K|k\rangle + \sum_{i=1}^{N}(k\, k_i|v|k\, k_i) \tag{7.39b}$$

so that the total energy is:

$$E_H = \sum_{i=1}^{N}\langle k_i|K|k_i\rangle + \frac{1}{2}\sum_{i,j}(k_i k_j|v|k_i k_j) . \tag{7.39c}$$

Equation (7.39c) is the familiar Hartree approximation to the total energy, and arises from the fact that the factor $\frac{1}{2}\int \sigma_0 v^{-1}\sigma_0$ properly corrected the overcounting of the potential energy which occurred in the sum $\sum_i \epsilon_{k_i}$. Since one can obtain static self-consistent solutions for any set of occupied states $\{k_1, \ldots k_N\}$, we see that $G(E)$ has poles at the Hartree ground state and all multiparticle-multihole excitations relative to it. Let us note at this stage, that if we had used a functional integral representation on the set of Slater determinants, as in Problem 7.5, the stationary-phase approximation would have yielded the Hartree-Fock, rather than Hartree approximation. Using the Hartree form of the auxiliary field, the exchange term in the energy is one of the corrections included in the quadratic fluctuations which we will now evaluate.

RPA CORRECTIONS

In order to calculate the corrections to the stationary Hartree contribution, it is convenient to evaluate the trace of U_σ in a fixed basis of Slater determinants $|\psi_{\{k\}}\rangle$ where $\{k\}$ denotes the set of quantum numbers of the occupied single-particle states. Since the basis is fixed and independent of σ, we can no longer use Eq. (7.32), but rather must use (7.28b) which becomes

$$G(E) = \sum_{\{k\}} G_{\{k\}}(E) \tag{7.40a}$$

where

$$G_{\{k\}}(E) = -i\int_0^\infty dT\, e^{iET} \int \frac{\mathcal{D}\sigma}{\mathcal{N}}\, e^{\frac{i}{2}\int_{-T/2}^{T/2} dt\,\sigma v^{-1}\sigma + \ln\langle\Psi_{\{k\}}|U_\sigma|\Psi_{\{k\}}\rangle}\,. \tag{7.40b}$$

The expansion must be performed for each set of quantum numbers $\{k\}$ and we will show how it works around one particular state $|\Psi_{\{k\}}\rangle$. To optimize the expansion for this specific state we define the fixed basis to be Slater determinants composed of eigenstates of the Hartree Hamiltonian in which σ_0 is defined self-consistently for the specific occupation numbers $\{k\}$. Thus, $|\Psi_{\{k\}}\rangle$ is a self-consistent determinant of Hartree eigenfunctions which we shall denote as $|\Psi_H\rangle$. Note that for all the other occupation numbers $\{k'\}$ which we are not trying to calculate, the Slater determinant $|\Psi_{\{k'\}}\rangle$ is not a self-consistent Hartree wave function and the expansion is not optimal.

We make the change of variable

$$\sigma(r,t) = \sigma_0(r) + \eta(r,t) \tag{7.41a}$$

where $\sigma_0(r)$ is the Hartree mean field corresponding to the term $\{k\}$ that we are expanding. The contribution independent of η generates E_H, the terms linear in η vanish, and we obtain

$$G_{\{k\}} = \int_0^\infty dT\, e^{i(E-E_H)T} \int \frac{\mathcal{D}\eta(r,t)}{\mathcal{N}}\, e^{\frac{i}{2}\int_{-T/2}^{+T/2} dt\,\eta v^{-1}\eta + \sum_{n=2}^{\infty}\frac{1}{n!}\int d1\ldots dn \left.\frac{\delta^n \ln\langle\Psi_H|U_\sigma|\Psi_H\rangle}{\delta\sigma(1)\ldots\delta\sigma(n)}\right|_{\sigma=\sigma_0}\eta(1)\ldots\eta(n)} \tag{7.41b}$$

where the notation 1 stands for (r_1, t_1). The η-integral is rewritten as:

$$Z = \int \frac{D\eta(r,t)}{\mathcal{N}} e^{-\frac{1}{2}\int_{-T/2}^{+T/2} dt dt' dr dr' \eta(rt)\Gamma^{-1}(rt,r't')\eta(r't') + V(\eta)} \tag{7.42a}$$

where

$$\Gamma^{-1}(r,t,r't') = -i\delta(t-t')v^{-1}(r-r') - \left.\frac{\delta^2 \ln\langle\Psi_H|U_\sigma|\Psi_H\rangle}{\delta\sigma(rt)\delta\sigma(r't')}\right|_{\sigma=\sigma_0} \tag{7.42b}$$

and

$$V(\eta) = \sum_{n=3}^{\infty} \frac{1}{n!} \int d1 \ldots dn \left.\frac{\delta^n \ln\langle\Psi_H|U_\sigma|\Psi_H\rangle}{\delta\sigma(1)\ldots\delta\sigma(n)}\right|_{\sigma=\sigma_0} \eta(1)\ldots\eta(n) \ . \tag{7.42c}$$

Using the definition Eq. (7.28c,d) of U_σ and the result in Chapter 2 that the derivative of the logarithm of the evolution operator generates connected Green's functions, we obtain

$$\begin{aligned}\left.\frac{\delta^n \ln\langle\Psi_H|U_\sigma|\Psi_H\rangle}{\delta\sigma(1)\ldots\delta\sigma(n)}\right|_{\sigma=\sigma_0} &= \left.\frac{\delta^n \ln\langle\Psi_H|T\, e^{-i\int_{-T/2}^{T/2} dt (K+\sigma(t))\hat{\rho}}|\Psi_H\rangle}{\delta\sigma(1)\ldots\delta\sigma(n)}\right|_{\sigma=\sigma_0} \\ &= (-i)^n \langle\Psi_H|TU_{\sigma_0}(\tfrac{T}{2},\tfrac{-T}{2})\hat{\rho}(1)\ldots\hat{\rho}(n)|\Psi_H\rangle_c\end{aligned} \tag{7.43a}$$

where the Heisenberg representation density operator is defined

$$\hat{\rho}(r,t) \equiv e^{ith_{\sigma_0}}\hat{\rho}e^{-ith_{\sigma_0}} \tag{7.43b}$$

$$U_{\sigma_0}(\tfrac{T}{2},\tfrac{-T}{2}) = e^{-iTh_{\sigma_0}} \tag{7.43c}$$

h_{σ_0} is the Hartree Hamiltonian, and c denotes a connected expectation value.

We now evaluate the quadratic fluctuations given by

$$\begin{aligned} Z &= \frac{(\det\Gamma^{-1})^{-1/2}}{\mathcal{N}} \\ &= \left[\frac{\det\Gamma^{-1}}{\det(-i\delta v^{-1})}\right]^{-1/2} \\ &= \left[\det\left(i(\delta v)\Gamma^{-1}\right)\right]^{-1/2} \ .\end{aligned} \tag{7.44}$$

The derivative appearing in Γ may be evaluated as follows, using Wick's theorem in the Slater determinant $|\Psi_H\rangle$

$$\begin{aligned}\left.\frac{\delta^2 \ln\langle\Psi_H|U_\sigma|\Psi_H\rangle}{\delta\sigma(1)\delta\sigma(1')}\right|_{\sigma_0} &= -\frac{\langle\Psi_H|T\left[U_{\sigma_0}\hat{\rho}(1)\hat{\rho}(1')\right]|\Psi_H\rangle}{\langle\Psi_H|T\left[U_{\sigma_0}\right]|\Psi_H\rangle} \\ &\quad + \frac{\langle\Psi_H|T\left[U_{\sigma_0}\hat{\rho}(1)\right]|\Psi_H\rangle}{\langle\Psi_H|T\left[U_{\sigma_0}\right]|\Psi_H\rangle}\frac{\langle\Psi_H|T\left[U_{\sigma_0}\hat{\rho}(1')\right]|\Psi_H\rangle}{\langle\Psi_H|T\left[U_{\sigma_0}\right]|\Psi_H\rangle} \\ &= -\langle\Psi_H|T\left[\hat{\psi}^\dagger(1)\hat{\psi}(1)\hat{\psi}^\dagger(1')\hat{\psi}(1')\right]|\Psi_H\rangle \\ &\quad + \langle\Psi_H|\hat{\psi}^\dagger(1)\hat{\psi}(1)|\Psi_H\rangle\langle\Psi_H|\hat{\psi}^\dagger(1')\hat{\psi}(1')|\Psi_H\rangle \\ &= -\langle\Psi_H|T\left[\hat{\psi}^\dagger(1)\hat{\psi}(1')\right]|\Psi_H\rangle\langle\Psi_H|T\left[\hat{\psi}(1)\hat{\psi}^\dagger(1')\right]\Psi_H\rangle\end{aligned} \tag{7.45a}$$

where $\hat{\psi}^\dagger(1)$ is a creation operator at (r_1, t_1) in the Heisenberg representation of the Hartree Hamiltonian

$$\hat{\psi}^\dagger(1) = e^{ih_{\sigma_0} t_1} \hat{\psi}^\dagger(r_1) e^{-ih_{\sigma_0} t_1} \ . \qquad (7.45b)$$

Note that in the limiting case of equal time, $t_1' = t_1 - \epsilon$, Eq. (7.45a) becomes equal to

$$\begin{aligned} -\langle \Psi_H | \hat{\psi}^\dagger(r_1) \hat{\psi}(r_1') | \Psi_H \rangle \langle \Psi_H | \hat{\psi}(r_1) \hat{\psi}^\dagger(r_1') | \Psi_H \rangle \\ = -\delta(r_1 - r_1') \langle \Psi_H | \hat{\psi}^\dagger(r_1) \hat{\psi}(r_1') | \Psi_H \rangle \\ + \langle \Psi_H | \hat{\psi}^\dagger(r_1) \hat{\psi}(r_1') | \Psi_H \rangle \langle \Psi_H | \hat{\psi}^\dagger(r_1') \hat{\psi}(r_1) | \Psi_H \rangle \ . \end{aligned} \qquad (7.45c)$$

Denoting by $G(1,1')$ the Hartree Green's function

$$G(1,1') = \frac{1}{i} \langle \Psi_H | T \left[\psi(1) \psi^\dagger(1') \right] | \Psi_H \rangle \qquad (7.46a)$$

and by D_0 the Hartree particle-hole propagator:

$$iD_0(1,1') = G(1,1')G(1',1) = \qquad (7.46b)$$

we see that if $t_1 \neq t_1'$:

$$\begin{aligned} \Gamma^{-1}(1,1') &= -i\delta(t_1 - t_1') v^{-1}(r_1 - r_1') + iD_0(1,1') \\ &= -i\delta v^{-1} [1 - \delta v D_0] \end{aligned} \qquad (7.46c)$$

and the quadratic fluctuations, Eq. (7.44) may be written

$$Z = [\det(1 - (\delta v) D_0)]^{-1/2} \ . \qquad (7.46d)$$

Using the identity $\det A = \exp(\operatorname{tr} \ln A)$ we write

$$\begin{aligned} Z &= e^{-\frac{1}{2} \operatorname{tr} \ln(1 - (\delta v) D_0)} \\ &= e^{\frac{1}{2} \sum_{n=1}^{\infty} \frac{1}{n} \operatorname{tr}((\delta v D_0)^n)} \\ &= e^{\sum_{n=1}^{\infty} F^{(n)}} \ . \end{aligned} \qquad (7.47a)$$

The nth term in the exponent has the explicit form

$$\begin{aligned} F^{(n)} = \frac{1}{2n} \int_{-T/2}^{T/2} & dt_1 dr_1 dr_1' \dots dt_n dr_n dr_n' \ v(r_1 - r_1') D_0(r_1' t_1, r_2 t_2) \\ & \times v(r_2 - r_2') D_0(r_2' t_2, r_3 t_3) \dots v(r_n' - r_n) D_0(r_n' t_n, r_1 t_1) \end{aligned} \qquad (7.47b)$$

and can be represented by the diagram:

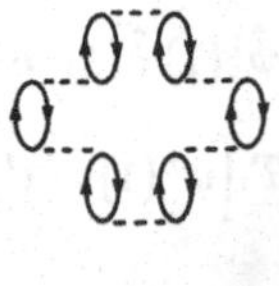

(7.47c)

This contribution is the nth order ring diagram that enters the RPA approximation. Note that the factor $\frac{1}{2n}$ is just the symmetry factor of this unlabelled ring diagram as discussed in Section 2.3.

The $n = 1$ term, which involves equal time propagators, is calculated as follows using Eq. (7.45c)

$$\begin{aligned}
F^{(1)} &= i\frac{T}{2}\int drdr'v(r-r')\big[\langle\Psi_H|\hat{\psi}^\dagger(r)\hat{\psi}(r')|\Psi_H\rangle\langle\Psi_H|\hat{\psi}^\dagger(r')\hat{\psi}(r)|\Psi_H\rangle \\
&\quad -\delta(r-r')\langle\Psi_H|\hat{\psi}^\dagger(r)\psi(r')|\Psi_H\rangle\big] \\
&= \frac{iT}{2}\sum_{i,j}^{N}\int drdr'v(r-r')\psi^*_{k_i}(r)\psi^*_{k_j}(r')\psi_{k_i}(r')\psi_{k_j}(r) - i\frac{T}{2}\sum_{i=1}^{N}\int drv(0)|\psi_{k_i}(r)|^2 \\
&= \frac{iT}{2}\sum_{i,j}^{N}(\psi_{k_i}\psi_{k_j}|v|\psi_{k_j}\psi_{k_i}) - i\frac{T}{2}Nv(0) \ .
\end{aligned} \tag{7.48}$$

The $n = 1$ term thus makes two important corrections. It removes the unphysical self-energy $\frac{1}{2}v(0)$ from the operator K, and it provides the exchange energy, which was absent at the Hartree level.

The sum of all the ring diagrams for $n \geq 2$ is the usual expression for the RPA energy with direct ring diagrams.

$$E_{\mathrm{RPA}} = \frac{1}{2T}\sum_{n=2}^{\infty}\frac{1}{n}\operatorname{tr}(D_0V)^n \tag{7.49}$$

where V denotes $\delta(t-t')v(r-r')$. Note that the overall contribution to the energy of an nth order ring diagram is $\frac{1}{T(2n)}i^{n+1}(iG_0iG_0)^n$, in agreement with the factor $\frac{1}{TS}(-i)^n(-1)^{n_\ell}$ specified in Table 3.1 since the number of loops is $n_\ell = n$ and the symmetry factor is $S = 2n$. Collecting together Eqs. (7.47), (7.48) and (7.49), we see that when quadratic corrections are included, the new poles have energy:

$$E^{(2)} = \sum_{i=1}^{N}\langle k_i|T|k_i\rangle + \frac{1}{2}\sum_{i,j}\langle k_ik_j|v|k_ik_j\rangle + E_{\mathrm{RPA}} \tag{7.50a}$$

or in terms of diagrams

$E^{(2)} - E^{(1)} =$ 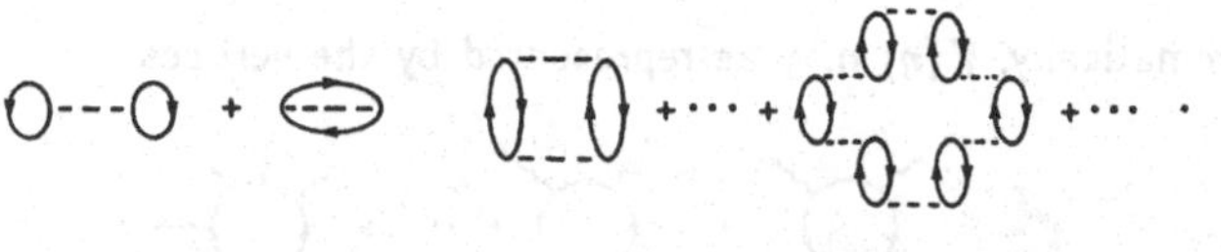 (7.50b)

This result begins to reveal the physical content of the expansion. Physically, we expect the fluctuations around the mean field to describe the fundamental collective excitations of a system, such as giant dipole, quadrupole, or octupole vibrations in a finite nucleus. Furthermore, from the study of response functions in Chapter 5, we know that the RPA phonon is the leading approximation to collective excitations. Thus, it is physically reasonable that the principal effect of the quadratic corrections is to include the contribution to the ground state energy of a single loop composed of an RPA phonon. In higher order, we shall see that the effects of multiple collective excitations are included. Since the expansion is systematic and equivalent to perturbation theory, along the way all of the other physics besides collective effects is also built in. We have seen two concrete examples in second order: the $\frac{1}{2}v(0)$ term and the exchange term.

Note that in the present formulation, since the stationary solution is the Hartree approximation, even though the direct and exchange terms appear in the energy, the propagators are Hartree propagators and RPA phonons have only direct ring diagrams. For some systems, this is physically inappropriate. For example, in the case of nuclear forces, the exchange term in the potential is much more attractive than the direct term and the Hartree approximation with a realistic effective interaction does not even produce binding. In this case, it is necessary to use a functional integral representation which yields the Hartree-Fock approximation at the stationary phase level and sums direct and exchange ring diagrams. One way to do this using a functional integral over Slater determinants is discussed in Problem 7.6.

THE LOOP EXPANSION

We now evaluate the corrections beyond quadratic order in (7.42) by perturbation theory. Using (7.43), $V(\eta)$ may be written

$$V(\eta) = \sum_{n=3}^{\infty} \frac{(-i)^n}{n!} \int d1 \ldots dn \langle T[\hat{\psi}^\dagger(1)\hat{\psi}(1)\ldots\hat{\psi}^\dagger(n)\hat{\psi}(n)]\rangle_c \eta(1)\ldots\eta(n) \quad (7.51a)$$

where $\langle\ \rangle_c$ again denotes the connected expectation value in the Hartree state. Expanding the expectation value by means of Wick's theorem, we note that the completely linked contributions correspond to closed loops in which $\psi^\dagger(n_1)$ at one point n_1 is contracted with $\psi(n_2)$ at one of the $(n-1)$ remaining points n_2. $\psi^\dagger(n_2)$ is contracted with $\psi(n_3)$ at one of the $(n-2)$ remaining points n_3 and so on until $\psi^\dagger(n_n)$ at the last point n_n is contracted with $\psi(n_1)$. These $(n-1)!$ contractions contribute equally to Eq. (7.51), and using the Hartree propagator $G(m,n) = -i\langle T\hat{\psi}(m)\hat{\psi}^\dagger(n)\rangle$ as before, $V(\eta)$ may be rewritten

$$V(\eta) = \sum_{n=3}^{\infty} \frac{-1}{n} \int d_1 \ldots d_n \eta(1)G(1,2)\eta(2)G(2,3)\ldots G(n-1,n)\eta(n)G(n,1) \ . \quad (7.51b)$$

Diagrammatically, $V(\eta)$ may be represented by the vertices

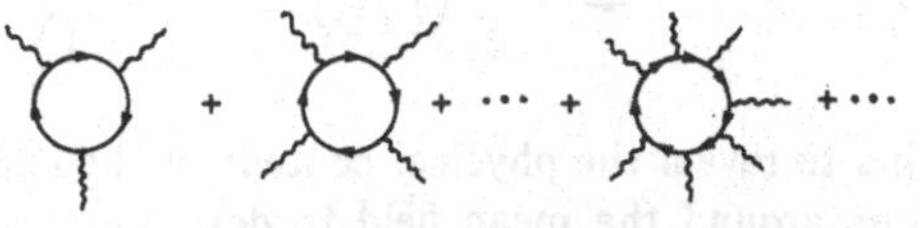
(7.51c)

where a solid line denotes a Hartree propagator and a wavy line denotes an η variable.

Finally, to perform the perturbation expansion, we need to compute the η propagators, that is, the contraction of two η variables. Taking the quadratic term of Eq.

(7.42a), the contraction is defined

$$\overbrace{\eta(1)\eta(1')} = \frac{\int D\eta\, \eta(1)\eta(1')e^{-\frac{1}{2}d\tau\eta\Gamma^{-1}\eta}}{\int D\eta\, e^{-\frac{1}{2}\int dt\eta\Gamma^{-1}\eta}} = \Gamma(1,1') \;. \tag{7.52}$$

The relation $\Gamma^{-1} = -iV^{-1}[1 - V\,D_0]$, Eq. (7.46c), may be inverted and expanded as follows

$$\begin{aligned} \Gamma &= i(1 - V\,D_0)^{-1}V \\ &= i[V + V\,D_0V + V\,D_0V\,D_0\,V + \ldots] \\ &\equiv i\mathcal{V}_{\mathrm{RPA}} \;. \end{aligned} \tag{7.53a}$$

We recognize $\mathcal{V}_{\mathrm{RPA}}$ as the propagator of the RPA phonon introduced in Section 5.4 and its expansion (7.53a) may be represented diagramatically as follows

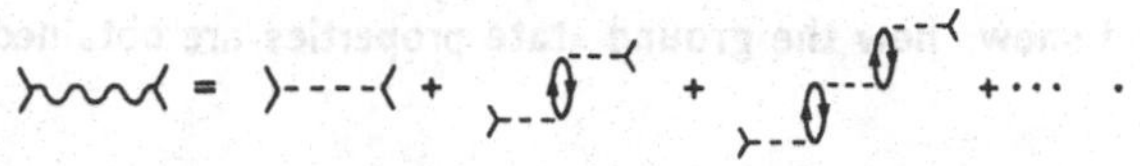
(7.53b)

The diagram expansion for Z is now straightforward to generate. Expanding $e^{V(\eta)}$ yields any number of closed Fermion loops of the form (7.51c) with three or more wavy lines emanating from each loop. The Gaussian integral $\int D\eta\, e^{-\frac{1}{2}\int d1d2\eta(1)\Gamma^{-1}(12)\eta(2)}$ connects these wavy lines with propagators $i\mathcal{V}_{\mathrm{RPA}}$ in all possible ways. By the linked cluster theorem, only the completely linked diagrams contribute to the total energy. Thus

$$E = E^{(2)} + \lim_{T\to\infty} \frac{-1}{iT} \sum \text{linked diagrams} \;. \tag{7.54}$$

Each Fermion loop has $n \geq 3$ wavy lines and contributes a factor $\frac{-1}{n}G(12)G(23)\ldots G((n-1)n)$. The total number of wavy lines must be even and these lines are joined with propagators $i\mathcal{V}_{\mathrm{RPA}}(nm)$ in all possible ways such that the entire diagram is linked. For a diagram with m Fermion loops, there is a factor $\frac{1}{m!}$. The expansion is organized according to the number of RPA phonons. That is, the "loops" of the loop expansion are the wavy phonon propagators, not the closed Fermion propagators. For example, at the two phonon level, there are two distinct graphs,

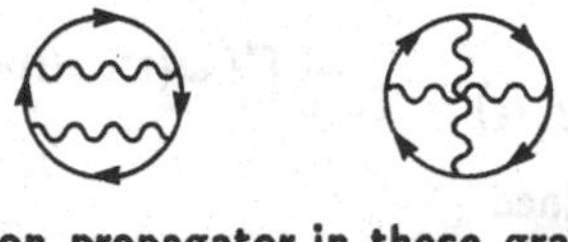
(7.55a)

Expansion of the RPA phonon propagator in these graphs yields an infinite sequence of diagrams in the bare interaction of the following generic form

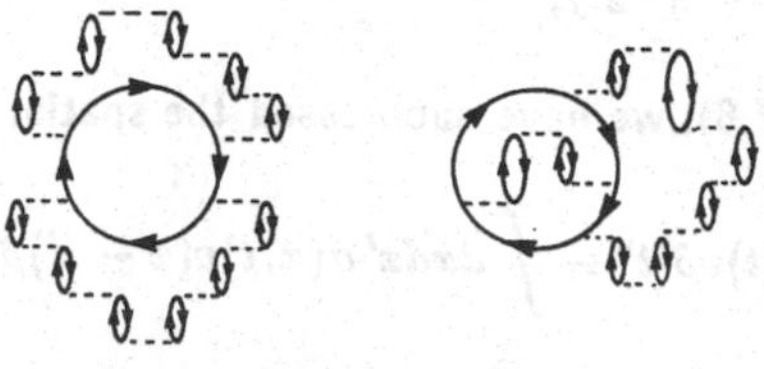
(7.55b)

Similarly, the three distinct three-phonon graphs are the following

(7.56)

The final result is an exact rearrangement of perturbation theory in which the basic elements are dressed Hartree particles and RPA phonons. Such an effective field theory in particle and vibrational degrees of freedom is useful in understanding the low-lying states of nuclei, for example, where the fundamental excitations are single particle and vibrational modes (Reinhardt, 1978). To understand the loop expansion in detail, it is instructive to derive the precise correspondence between the diagrams contributing to Eq. (7.54) and the expansion of the ground state energy in Feynman diagrams as outlined in Problem 7.7.

To conclude this section, it is important to note that the stationary phase expansion we have presented here for the ground state may also be carried out at finite temperature. Problem 7.8 outlines the evaluation of the partition function at finite temperature and shows how the ground state properties are obtained in the zero temperature limit.

7.3 TRANSITION AMPLITUDES

In contrast to the treatment of the ground state in the previous section, the stationary phase approximation to an arbitrary transition amplitude is in general time-dependent. We will now show how a time-dependent mean field approximation is obtained for general transition amplitudes and S-matrix elements. For brevity, we treat only the auxiliary field path integral here following closely the work of Levit (1980) and present a detailed example using coherent states in Problem 7.9.

Using the form of the Hubbard Stratonovich transformation in (7.5b), the transition matrix element between an initial state $|\Psi_i\rangle$ and final state $\langle\Psi_f|$ may be written

$$\begin{aligned}\langle\Psi_f|U(t_f,t_i)|\Psi_i\rangle &= \int \mathcal{D}\sigma\, e^{\frac{i}{2}\int_{t_i}^{t_f} dt\sigma v\sigma}\langle\Psi_f|\hat{U}_\sigma(t_f,t_i)|\Psi_i\rangle \\ &\equiv \int \mathcal{D}\sigma\, e^{iS_{\text{eff}}(\sigma)}\end{aligned} \tag{7.57a}$$

where

$$\hat{U}_\sigma(t_f,t_i) = \mathrm{T}e^{-i\int_{t_i}^{t_f} dt(K+\sigma(t)v)\hat{\rho}(t)} \tag{7.57b}$$

and the effective action is defined

$$S_{\text{eff}}(\sigma) = \frac{1}{2}\int_{t_i}^{t_f} dt\sigma v\sigma - i\ln\langle\Psi_f|U_\sigma(t_f,t_i)|\Psi_i\rangle \ . \tag{7.57c}$$

Note that as in Eq. (7.8), we have suppressed the spatial labels and for example

$$\sigma(t)v\hat{\rho}(t) \equiv \int dxdx'\,\sigma(x,t)v(x-x')\hat{\rho}(x',t') \ . \tag{7.57d}$$

The stationary values of the auxiliary field $\sigma_c(x,t)$ are obtained from the condition $\delta S_{\rm eff}(\sigma) = 0$, with the result

$$\sigma_c(x,t) = \frac{\langle \Psi_f | U_{\sigma_c}(t_f,t)\hat{\rho}(x,t)U_{\sigma_c}(t,t_i)|\Psi_i\rangle}{\langle \Psi_f | U_{\sigma_c}(t_f,t_i)|\Psi_i\rangle} \quad . \tag{7.58}$$

This result is a self-consistent equation for σ_c.

The solution σ_c is a matrix element of $\hat{\rho}$ taken between the states $U_{\sigma_c}(t,t_i)|\Psi_i\rangle$ and $\langle\Psi_f|U_{\sigma_c}(t_f,t)$, where the evolution operator U_{σ_c} defining these states evolves $|\Psi_i\rangle$ and $\langle\Psi_f|$ in the mean field specified by σ_c. We thus have a form of mean field approximation in which the one-body mean field depends upon the initial and final states defining the transition.

If the initial and final states are momentum eigenstates

$$\hat{P}|\Psi_i\rangle = P_i|\Psi_i\rangle \qquad \hat{P}|\Psi_f\rangle = P_f|\Psi_f\rangle \tag{7.59a}$$

then if $\sigma_c(x,t)$ is a solution, $\sigma_c(x-a,t)$ is also a solution:

$$\begin{aligned}\sigma_c(x-a,t) &= \frac{\langle\Psi_f|e^{-i\hat{P}a}\,e^{i\hat{P}a}U_{\sigma_c}(x)e^{-i\hat{P}a}e^{i\hat{P}a}\hat{\rho}(x,t)e^{-i\hat{P}a}\,e^{i\hat{P}a}U_{\sigma_c}(x)e^{-i\hat{P}a}e^{i\hat{P}a}|\Psi_i\rangle}{\langle\Psi_f|e^{-i\hat{P}a}\,e^{i\hat{P}a}U_{\sigma_c}(x)e^{-i\hat{P}a}\,e^{i\hat{P}a}|\Psi_i\rangle} \\ &= \frac{\langle\Psi_f|U_{\sigma_c}(x-a)\hat{\rho}(x-a,t)U_{\sigma_c}(x-a)|\Psi_i\rangle}{\langle\Psi_f|U_\sigma(x-a)|\Psi_i\rangle} \quad .\end{aligned} \tag{7.59b}$$

Hence, summation over all stationary solutions yields momentum conservation

$$\begin{aligned}\langle\Psi_f|U|\Psi_i\rangle &\approx \int da\, e^{\frac{i}{2}\int \sigma_c(x-a)v\sigma_c(y-a)}\langle\Psi_f|U_{\sigma(x-a)}|\Psi_i\rangle \\ &= \int da\, e^{-i(P_f-P_i)a}\, e^{\frac{i}{2}\int\sigma_c v\sigma_c}\langle\Psi_f|U_{\sigma_c}|\Psi_i\rangle \\ &= \delta(P_f-P_i)\, e^{\frac{i}{2}\int\sigma_c v\sigma_c}\langle\Psi_f|U_{\sigma_c}|\Psi_i\rangle \quad .\end{aligned} \tag{7.59c}$$

Energy conservation follows analogously from time translation invariance.

Since the self-consistent solution $\sigma_c(x,t)$ is not necessarily real, the evolution operator U_{σ_c} need not be unitary. Hence, it is often convenient to make an alternative stationary phase approximation by varying only the phase of Eq. (7.57a), that is, varying $\mathrm{Re}\, S_{\rm eff}(\sigma)$, with the result

$$\sigma_c(x,t) = \mathrm{Re}\frac{\langle\Psi_f|U_{\sigma_c}(t_f,t)\hat{\rho}(x,t)U_{\sigma_c}(t,t_i)|\Psi_i\rangle}{\langle\varphi_f|U_{\sigma_c}(t_f,t_i)|\varphi_i\rangle} \quad . \tag{7.60}$$

Now consider the important special case in which the initial and final states are Slater determinants, which we write in terms of single-particle wave functions as

$$|\Psi_i\rangle = |\{\psi_k^i\}\rangle \quad \text{and} \quad |\Psi_f\rangle = |\{\tilde{\psi}_k^f\}\rangle \quad .$$

If we define solutions to the time-dependent Schrödinger equation in the mean field generated by the real σ_c of (7.60)

$$i\frac{\partial\psi_k(x,t)}{\partial t} = (K+\sigma_c(t)v)\,\psi_k(x,t) \tag{7.61a}$$

$$-i\frac{\partial\tilde{\psi}_k(x,t)}{\partial t} = (K+\sigma_c(t)v)\,\tilde{\psi}_k(x,t) \tag{7.61b}$$

with the boundary conditions

$$\psi_k(x,t_i) = \psi_k^i(x) \qquad \tilde{\psi}_k(x,t_f) = \tilde{\psi}_k^f(x) \tag{7.61c}$$

then

$$\begin{aligned} U_{\sigma_c}(t,t_i)|\Psi_i\rangle &= |\{\psi_k(x,t)\}\rangle \\ \langle\Psi_f|U_\sigma(t_f,t) &= \langle\{\tilde{\psi}_k(x,t)\}| \ . \end{aligned} \tag{7.62}$$

Hence, the stationarity condition gives $\sigma_c(x,t)$ as a simple determinantal matrix element in the mixed basis $\{\psi\}\{\tilde{\psi}\}$

$$\sigma_c(x,t) = \mathrm{Re}\frac{\langle\{\tilde{\psi}_k(x,t)\}|\hat{\rho}(x)|\{\psi_k(x,t)\}\rangle}{\langle\{\tilde{\psi}_k(x,t)\}|\{\psi_k(x,t)\}\rangle} \tag{7.63}$$

where the denominator is easily seen to be time-independent.

It is instructive to see the relation of this general stationary-phase approximation result to the time-dependent Hartree approximation (which is the analog of time-dependent Hartree Fock with the Hartree form of the auxiliary field). Observe that with σ_c real, the stationary-phase approximation

$$\langle\{\tilde{\phi}_k^f\}|U(t_f,t_i)|\{\phi_k^i\}\rangle \approx e^{\frac{i}{2}\int dt\sigma_c v\sigma_c}\langle\{\tilde{\phi}_k^f\}|\{\phi_k(x,t_f)\}\rangle \tag{7.64}$$

has modulus ≤ 1 and the equality only obtains if the final determinant equals $|\{\phi_k(x,t_f)\}\rangle$ multiplied by an irrelevant phase factor. If the final determinant is equal to that obtained by evolving the initial determinant with $\sigma_c(t)$, then the two sets of wave functions $\{\phi_k\}$ and $\{\tilde{\phi}_k\}$ defined in (7.61) coincide. By (7.63), $\sigma_c(x,t) = \sum_k|\phi_k(x,t)|^2$ so that the determinant evolves with the time-dependent Hartree equation. Thus, if we consider the evolution of any determinant in the stationary-phase approximation, the most probable final state is found by solving the time-dependent Hartree equation. The transition to any other less likely final state may be found by the general result (7.63), and it is physically clear that the mean field $\sigma_c(x,t)$ which best approximates the transition must depend upon the final state.

S-MATRIX ELEMENTS

The basic idea of how this approach can be applied to the calculation of S-matrix elements is illustrated by the calculation of the response of a many-Fermion system to an external potential (Alhassid and Koonin, 1981). The S-matrix may be written as the limit of the following evolution operator

$$S = \lim_{T\to\infty} U_0(0,T)\,U(T,-T)U_0(-T,0) \tag{7.65}$$

where U_0 denotes evolution under the many-body Hamiltonian (7.1) and U denotes evolution under the many-body Hamiltonian plus an external field $\hat{W} = \sum_{\alpha\beta} W_{\alpha\beta}(t)\hat{\rho}_{\alpha\beta}$ where $W(t) \to 0$ as $t \to \pm\infty$. Each evolution operator may be written in terms of a functional integral over an auxiliary field in the usual way with the result

$$\langle\Psi_f|S|\Psi_i\rangle = \lim_{T\to\infty} \int\int\int D[\sigma_f]D[\sigma]D[\sigma_i]\, e^{\frac{i}{2}\left[\int_T^0 dt\,\sigma_f v\sigma_f + + \int_{-T}^{T} dt\,\sigma v\sigma + \int_0^{-T} dt\,\sigma_i v\sigma_i\right]}$$
$$\times \langle\Psi_f|U_{\sigma_f}(0,T)U_\sigma(T,-T)U_{\sigma_i}(-T,0)|\Psi_i\rangle \tag{7.66a}$$

where

$$U_{\sigma_{f,i}}(t',t) = \mathrm{T}\left[e^{-i\int_t^{t'} dt(K+\sigma_{f,i}v)\hat{\rho}}\right] \tag{7.66b}$$

and

$$U_\sigma(t',t) = \mathrm{T}\left[e^{-i\int_t^{t'} dt(K+W+\sigma_{f,i}v)\hat{\rho}}\right] . \tag{7.66c}$$

One may view (7.66) as an auxiliary field evolution operator of the usual form with a time variable which runs from 0 to $-T$ without W, from $-T$ to $+T$ with W, and from T to 0 without W. Requiring stationarity of the phase of (7.66a) yields the stationary solution, analogous to (7.60),

$$\sigma_\ell^0(t) = \mathrm{Re}\frac{\langle\Psi_f|\tilde{\mathrm{T}}[\hat{\rho}U_{\sigma_f^0}U_{\sigma^0}U_{\sigma_i^0}]|\Psi_i\rangle}{\langle\Psi_f|\mathrm{T}[U_{\sigma_f^0}U_{\sigma^0}U_{\sigma_i^0}]|\Psi_i\rangle} \tag{7.67}$$

where σ_ℓ denotes any one of σ_f, σ, or σ_i and $\tilde{T}$ inserts $\hat{\rho}$ into the corresponding interval $(0,T)$, $(T,-T)$, or $(-T,0)$, respectively at time t.

Since the physical S-matrix is independent of T as $T \to \infty$, time reversal invariant if $W(x,t)$ has time reversal symmetry, and unitary, one may ask which of these properties is also retained by the stationary phase approximation. To see that (7.67) is independent of T, note for times $-t$ prior to the interaction period, $\sigma_i^0(-t) = \sigma^0(-t)$ is a solution and evolution by U_{σ_i} prior to the interaction time is precisely compensated by U_σ. Hence, once $-T$ is earlier than the interaction period, it does not enter the stationary approximation to the S-matrix. Similarly, the mean-field solution may be verified to be time reversal invariant if W has time reversal symmetry. However, because the mean field solution satisfies a different non-linear self-consistent equation for every final state $\langle\Psi_f|$, the approximation for each final state is obtained by a different evolution operator $U_{\sigma_f^0}U_{\sigma^0}U_{\sigma_i^0}$ and the result in general is not unitary.

Formulation of a complete scattering theory involves additional technical complications which do not arise in considering the response to an external potential. One must define asymptotic channel states for the target, projectile and all the reaction products, and project then onto states of specified center of mass momentum. These and other aspects of scattering theory are addressed by Reinhardt (1982).

7.4 COLLECTIVE EXCITATIONS AND TUNNELING

We now consider general eigenstates of a many-body system in which application of the stationary-phase approximation to the resolvent operator leads to time-dependent periodic mean field equations. In Section 7.2, we have already seen the special case of static Hartree solutions which are trivially periodic and describe the ground state and multiple particle-hole excitations relative to it. Having introduced the idea of self-consistent mean field equations specifying time-dependent solutions for specific transition matrix elements in the preceding section, we will now develop the periodic mean field equations describing quantum eigenstates of large amplitude collective motion and spontaneous decay of a metastable state following the approach of Levit, Negele and Paltiel (1980a,b). To illustrate the basic ideas, it is useful to begin with the simple problem of quantum mechanics in a one-dimensional potential.

EXAMPLE WITH ONE DEGREE OF FREEDOM

We will calculate the eigenstate in a general potential $V(q)$ using the resolvent operator introduced in Eq. (7.28). The resolvent operator may be written as a Feynman path integral as follows

$$\begin{aligned} G(E) &= -i\int_0^\infty dT\, e^{iET}\int dq\, \langle q|e^{-iHT}|q\rangle \\ &= -i\int_0^\infty dT\, e^{iET}\int dq \int_{\substack{q(0)=q\\ q(T)=q}} \mathcal{D}[q(t)]\, e^{iS[q(t)]} \end{aligned} \tag{7.68a}$$

where

$$S[q(t)] = \int_0^T dt\left[\frac{m}{2}\dot{q}(t)^2 - V\left(q(t)\right)\right] \tag{7.68b}$$

is the classical action and the trajectory satisfies the boundary conditions

$$q(0) = q(T) = q\ . \tag{7.68c}$$

The exact eigenstates are specified by the poles of $G(E)$ when the three integrals over T, q, and $q(t)$ are performed exactly. Here, we consider the approximation obtained by evaluating each of these three integrals in the stationary-phase approximation.

Application of the stationary-phase approximation to the functional integral over $q(t)$ yields

$$\langle q|e^{iS[q(t)]}|q\rangle \approx A\, e^{iS[q_c(t)]} \tag{7.69a}$$

where $q_c(T)$ is the solution to the classical Euler-Lagrange equation

$$m\frac{d^2}{dt^2}q_c(t) = -\nabla V\left(q_c(t)\right) \tag{7.69b}$$

subject to the boundary condition (7.68c). The factor A denotes the result of integrating the quadratic fluctuations around $q_c(t)$ and its explicit form is not required for the present argument.

To apply the stationary-phase approximation to the integral over q, note that $S[q_c]$ depends upon q through the end points of $q_c(t)$. Since the derivative of the action with respect to the end point yields the momentum, stationarity of $S[q_c]$ requires

$$0 = \frac{\partial S[q_c]}{\partial q} = \frac{\partial S[q_c]}{\partial q(T)} - \frac{\partial S[q_c]}{\partial q(0)} = p(T) = p(0)\ . \tag{7.70}$$

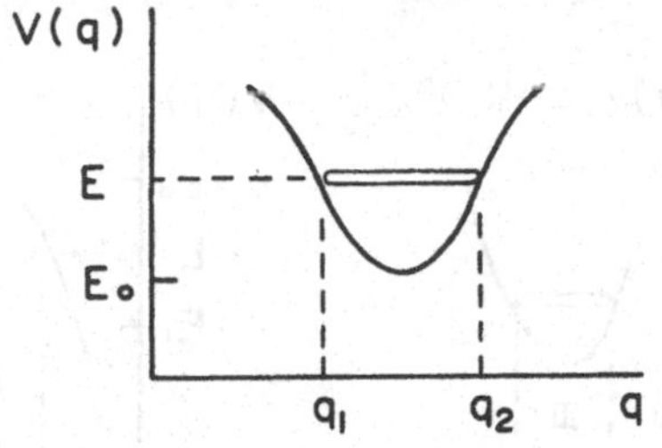

Fig. 7.1 Sketch of potential $V(q)$ with a single minimum and the periodic trajectory at energy E between classical turning points q_1 and q_2 which contributes in the stationary-phase approximation.

Thus, both q and $\dot{q}$ are equal at the end points and since the classical equation of motion is second order, the trajectory is periodic. Hence, the resolvent may be rewritten in the form

$$G(E) \approx -i\int_0^\infty dT\, A' e^{iET+iS(T)} \equiv -i\int_0^\infty dT\, A' e^{iW(T)} \tag{7.71}$$

where $S(T)$ denotes the classical action for a periodic trajectory of period T and an additional quadratic fluctuation factor has been included in A'. The quantity $W(T) \equiv ET + S(T)$ appearing in this and subsequent results is called the reduced action.

The derivative of the classical action $S(T)$ with respect to the final time T yields the negative of the classical energy. Thus, application of the stationary phase approximation to the remaining integral over T leads to the condition

$$E = -\frac{\partial S(T)}{\partial T} = E_c(T) \tag{7.72}$$

where $E_c(T)$ is the energy of a classical periodic orbit of period T. The final stationary phase result obtained by summing over all periods T_m which yield classical periodic solutions with energy E is

$$G(E) \approx -i\sum_m A'_m\, e^{iW(T_m)} \tag{7.73}$$

where A'_m denotes the factor obtained by integrating the quadratic fluctuations around the stationary point T_m.

The structure of the final result, Eq. (7.73), is particularly simple for the case of a potential $V(q)$ with a single minimum as sketched in Fig. 7.1. The classical periodic trajectory may be visualized in terms of the motion of a particle on a curve $V(q)$ in a uniform gravitational field. When it is released from position q_1 it executes periodic classical motion between the turning points q_1 and q_2 as shown. Using energy conservation to express $\frac{dq}{dt}$ in terms of $E-V(q)$, the fundamental period for a trajectory with energy E is

$$T(E) = 2\int_{q_1}^{q_2}\left[\frac{m}{2[E-V(q)]}\right]^{1/2} dq\ . \tag{7.74}$$

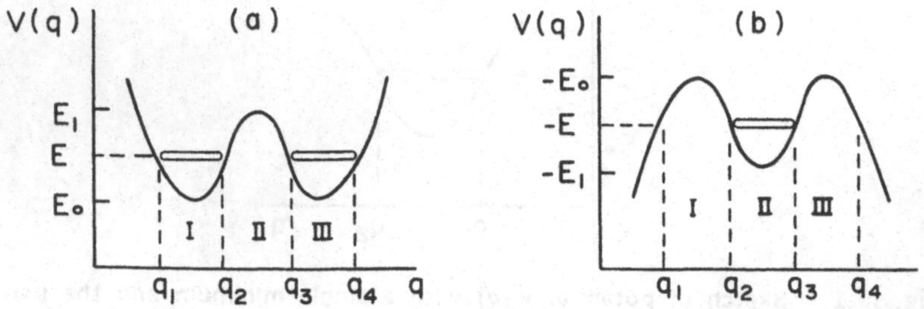

Fig. 7.2 Sketches of a double well potential (a), and the inverted potential obtained by continuation to imaginary time (b). The trajectories contributing in the stationary-phase approximation in the classically allowed and forbidden regions are shown in (a) and (b), respectively.

Any integer multiple of $T(E)$ also satisfies the stationarity condition, so that $T_m = mT(E)$. It is convenient to write the reduced action in (7.73) in the following form

$$W(T_m) = ET_m + \int_0^{T_m} (p\dot{q} - H) \equiv mW[T(E)] , \tag{7.75a}$$

where

$$\begin{aligned} W[T(E)] &= \int_0^{T(E)} p\dot{q}\, dt = \oint p\, dq \\ &= 2\int_{q_1}^{q_2} \{2m[E - V(q)]\}^{1/2}\, dq \ . \end{aligned} \tag{7.75b}$$

If we temporarily ignore the dependence of the quadratic fluctuation factors A'_m on the multiple m, Eq. (7.73) yields the following series

$$\begin{aligned} G(E) &= -iA' \sum_m \left(e^{i\oint W[T(E)]}\right)^m \\ &= \frac{-A'\, e^{iW[T(E)]}}{1 - e^{iW[T(E)]}} \ , \end{aligned} \tag{7.76a}$$

which has poles for all energies satisfying the quantization condition

$$n2\pi = W[T(E)] = \oint p\, dq \ . \tag{7.76b}$$

As shown by Gutzwiller (1978), the factors A' turn out to contribute the same magnitude for each m and a phase of $\frac{\pi}{2}$ for each classical turning point, so their inclusion simply changes the minus sign in the demoninator of Eq (7.76a) to a plus and thereby yields the correct Bohr-Sommerfeld quantization rule

$$(2n+1)\pi = W(T(E)) = \oint p\, dq \ . \tag{7.76c}$$

The lowest state in the well thus acquires its proper zero-point energy in this approximation.

An important new feature arises in the analogous treatment of the double-well potential sketched in Fig. 7.2. General application of the saddle-point approximation to the time integral in Eq. (7.71) requires summation over isolated stationary points in the complex T plane. Whereas trajectories in classically allowed regions yield real stationary points T_m as discussed above, we will see that classically forbidden regions introduce complex stationary points. To be specific, we will treat an energy E below the barrier height E_1 and consider the three regions I, II, and III indicated in Fig. 7.2a.

The analysis of periodic trajectories in the classically allowed regions I and III is the same as for the single well considered above. We will denote the fundamental periods given by Eq. (7.74) with the appropriate end points by T_{I} and T_{III} and the corresponding reduced actions given by Eq. (7.75) by W_{I} and W_{III}.

In the classically forbidden region II, the existence of a classical periodic solution with purely imaginary period may be understood by continuation of the classical equation of motion to imaginary time. Replacing (it) by τ, Eq. (7.69b) may be rewritten

$$m\frac{d^2}{d\tau^2}q_c = -\nabla[-V(q_c)] \,. \tag{7.77}$$

Since continuation of this second-order equation introduces an overall minus sign, a picturesque way to visualize the stationary solutions is to consider the classical solutions in the inverted potential sketched in Fig. 7.2b. Calculation of the stationary solution in region II corresponding to one classical oscillation from q_2 to q_3 and back and repetition of the steps performed above for real solutions give rise to the period

$$T_{\text{II}} = 2i\int_{q_2}^{q_3}\left[\frac{m}{2[V(q)-E]}\right]^{1/2} dq \tag{7.78a}$$

and a contribution to the resolvent of $e^{-W_{\text{II}}}$, where

$$W_{\text{II}} = 2\int_{q_2}^{q_3}\{2m[V(q)-E]\}^{1/2}dq \,. \tag{7.78b}$$

A general periodic trajectory in the double well of Fig. 7.2a is thus composed of any number of closed orbits in each of the three regions connected in any order, so that the stationary phase approximation to the resolvent is

$$G(E) = -i\sum_{k\ell m} A'_{k\ell m}\, e^{ikW_{\text{I}}(E)-\ell W_{\text{II}}(E)+imW_{\text{III}}(E)} \,. \tag{7.79}$$

The technical question of phases in the quadratic correction factor $A'_{k\ell m}$ is addressed by Bender *et al.* (1978), with the result that $A'_{k\ell m}$ is a constant times $(-1)^{k+\ell+m}$. Calculating the sum over all trajectories beginning in any of the three regions and containing all combination of cycles in each region yields

$$G(E) \propto \sum_{n=1}^{\infty}\left[e^{iW_{\text{I}}}\sum_{m=0}^{\infty}\left[-e^{-W_{\text{II}}}\sum_{\ell=0}^{\infty}(-e^{iW_{\text{III}}})^{\ell}\right]^{m}\right]^{n}$$

$$+\sum_{m=1}^{\infty}\left[\sum_{n=0}^{\infty}(-e^{iW_{\rm I}})^n(-e^{-W_{\rm I}})\sum_{\ell=0}^{\infty}(-e^{iW_{\rm III}})^\ell\right]^m \tag{7.80}$$
$$+\sum_{n=1}^{\infty}\left[-e^{iW_{\rm III}}\sum_{m=0}^{\infty}\left[-e^{W_{\rm II}}\sum_{\ell=0}(-e^{iW_{\rm III}})^\ell\right]^m\right]^n$$
$$=\frac{-2e^{i(W_{\rm I}+W_{\rm III})}-e^{iW_{\rm I}}-e^{-W_{\rm II}}-e^{iW_{\rm III}}}{(1+e^{iW_{\rm I}})(1+e^{iW_{\rm III}})+e^{-W_{\rm II}}}\ .$$

To understand the physical content of this result for general double well, it is instructive to examine two specific special cases. We first consider the symmetric double well, for which $W_{\rm II}=W_{\rm III}$, and will recover the familiar WKB expression for the splitting of nearly degenerate even- and odd-parity states. In lowest approximation, if the central barrier were very high, the problem would reduce to a degenerate pair of single-well problems, yielding degenerate solutions satisfying Eq. (7.76c):

$$W_{\rm I}(E_n^{(0)})=(2n+1)\pi\ , \tag{7.81}$$

where $E_n^{(0)}$ denotes the zeroth approximation to the $n^{\rm th}$ eigenstate. This result follows immediately from Eq. (7.80), since in the case of a high barrier, $W_{\rm II}$ is very large, rendering $e^{-W_{\rm II}}$ exponentially small, and the demoninator reduces to double poles at energy $E_n^{(0)}$. In next approximation, we may write $E_n=E_n^0+\Delta E_n$ and expand the condition for a pole in Eq. (7.80)

$$\left(1+e^{iW_{\rm I}(E_n^0+\Delta E_n)}\right)^2=-e^{-W_{\rm II}(E_n^0+\Delta E_n)}\ , \tag{7.82}$$

to first order in E_n. Using Eq. (7.81), noting from Eqs. (7.74) and (7.75) that

$$\frac{\partial W(E)}{\partial E}=T=\frac{2\pi}{\omega}\ , \tag{7.83}$$

and observing that $e^{-W_{\rm II}(E_n^0)}\Delta E_n$ is second-order small, we obtain

$$\left[\frac{\partial W_1}{\partial E}\Delta E_n\right]^2\approx e^{-W_2(E_n^0)}\ , \tag{7.84}$$

which yields the familiar WKB result for the energy splitting

$$E_n\approx E_n^0\pm\frac{\omega}{2\pi}e^{-W_2(E_n^0)/2}\ . \tag{7.85}$$

Thus, in this application, the physics of tunneling in the classically forbidden region, has been described by the periodic imaginary time solutions corresponding to classical motion in the inverted well.

A second example is the decay of a metastable state. In this case, the right well is distorted to extend to the edge of an arbitrarily large normalization box as sketched in

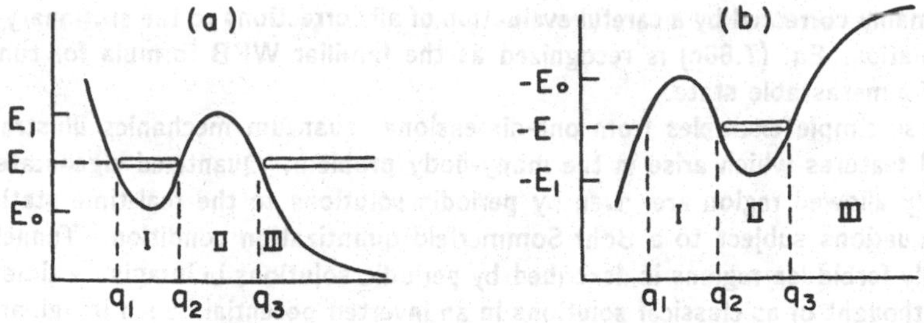

Fig. 7.3 Sketches of a potential for a metastable state (a) and the corresponding inverted potential (b).

Fig. 7.3a. Instead of calculating the resolvent, it is convenient to evaluate the smoothed level density (Balian and Bloch, 1974)

$$\rho_\gamma(E) = \frac{1}{\pi}\operatorname{Im}\operatorname{tr}(H - E - i\gamma)^{-1} = \frac{1}{\pi}\operatorname{Im} G(E + i\gamma) \tag{7.86}$$

where the finite width γ is smaller than any physical width in the problem but larger than the level spacing in the normalization box. In this case, we obtain periodic stationary solutions in region I as before, and in lowest approximation these yield the result Eq. (7.76c) for the energies of the quasistable states. Also, as in the case of the symmetric potential, periodic imaginary-time trajectories are obtained in region II, corresponding to solution of the classical equations of motion in the inverted potential sketched in Fig. 7.3b. The role of periodic solutions in region III is quite different in the present problem. Since

$$\operatorname{Im} W_{\mathrm{III}} = \operatorname{Im}\int_{q_3}^{L} dq\,[2m(E + i\gamma - V(q))]^{\frac{1}{2}} \underset{L\to\infty}{\sim} \left(\frac{m}{2E}\right)^{1/2}\gamma L \tag{7.87}$$

and γ is finite, $e^{-iW_{\mathrm{III}}}$ does not contribute in the limit as the length of the normalization box, L, goes to infinity.

Thus, the smoothed density of states has poles at complex energies $E_n^0 + \Delta E_N$ satisfying

$$1 + e^{iW_{\mathrm{I}}(E_n^0 + \Delta E_n)} = e^{-W_{\mathrm{II}}(E_n^0 + \Delta E_n)} \quad , \tag{7.88a}$$

and expansion to first order in ΔE_n as before yields

$$\Delta E_n = -\frac{i\Gamma_n}{2} \quad , \tag{7.88b}$$

where

$$\Gamma_n = 2\frac{\omega(E_n)}{2\pi}e^{-W_2(E_n)} \quad . \tag{7.88c}$$

Near E_n, the level density is therefore proportional to

$$\left[(E - E_n)^2 + \left[\frac{\Gamma_n}{2}\right]^2\right]^{-1} \quad , \tag{7.88d}$$

so that Γ_n is the inverse lifetime of the metastable state. To within the factor 2, which is presumably corrected by a careful evaluation of all corrections to the stationary phase approximation, Eq. (7.88c) is recognized as the familiar WKB formula for tunneling decay of a metastable state.

These simple examples from one-dimensional quantum mechanics illustrate the essential features which arise in the many-body problem. Quantized eigenstates in a classically allowed region are given by periodic solutions to the real-time stationary-phase equations subject to a Bohr-Sommerfeld quantization condition. Tunneling in classically forbidden regions is described by periodic solutions in imaginary time which may be thought of as classical solutions in an inverted potential. Such imaginary time solutions for many degrees of freedom were first introduced by Langer (1969) in the treatment of bubble formation and have been used extensively in field theory (see for example Polyakov, 1977 and Coleman, 1977) where solutions connecting degenerate vacua are called "instantons" and the solution describing the decay of a metastable state is called a "bounce".

EIGENSTATES OF LARGE AMPLITUDE COLLECTIVE MOTION

To generalize the calculation of eigenstates in a one-dimensional potential to the case of the many-Fermion problem, we will use the form of the resolvent and notation given in Eqs. (7.28) and (7.32):

$$G(E) = -i \sum_{\{k_1 \ldots k_N\}} \int_0^\infty dT\, e^{iET} \int D\sigma\, e^{\frac{i}{2}\int_{-T/2}^{T/2} dt\, \sigma v^{-1} \sigma} \times \langle \Psi_{k_1 \ldots k_N} | \mathrm{T} e^{-i\int_{-T/2}^{T/2} dt\, h_\sigma(t)} | \Psi_{k_1 \ldots k_n} \rangle$$
$$= -i \sum_{\{k_1 \ldots k_n\}} \int_0^\infty dT\, e^{iET} \int D\sigma\, e^{\frac{i}{2}\int_{-T/2}^{T/2} dt\, \sigma v^{-1} \sigma - i \sum_{i=1}^N \alpha_{k_i}(\sigma, T)} \tag{7.89a}$$

where

$$h_\sigma(t) = (K + \sigma(t))\, \hat{\rho} \tag{7.89b}$$

and $|\Psi_{k_1 \ldots k_N}\rangle$ is the Slater determinant of eigenstates $\{|\psi_k\rangle\}$ of U_0 satisfying

$$\mathrm{T} e^{-i\int_{-T/2}^{T/2} dt\, h_\sigma(t)} |\psi_k\rangle = e^{-i\alpha_k(\sigma, T)} |\psi_k\rangle \ . \tag{7.89c}$$

Equation (7.89c) is equivalent to the boundary-value problem

$$i \frac{\partial}{\partial t} \psi_k(r, t) = (K + \sigma(r, t))\, \psi_k(r, t) \tag{7.90a}$$

with the boundary conditions

$$\psi_k\left(r, \frac{T}{2}\right) = e^{-i\alpha_k(\sigma, t)}\, \psi_k\left(r, -\frac{T}{2}\right) \ . \tag{7.90c}$$

Defining new single-particle functions with a phase factor removed

$$\phi_k(r,t) \equiv e^{i\frac{t}{T}\alpha_k(\sigma,T)}\psi_k(r,t) \tag{7.91a}$$

we obtain an equivalent boundary value problem

$$\left[i\frac{\partial}{\partial t} - K - \sigma(r,t)\right]\phi_k(r,t) = -\frac{\alpha_k(\sigma,T)}{T}\phi_k(r,t) \tag{7.91b}$$

with the periodic boundary condition

$$\phi_k\left(r,\frac{T}{2}\right) = \phi_k\left(r,-\frac{T}{2}\right) . \tag{7.91c}$$

The stationary solution for the auxiliary field $\sigma_0(x,t)$ is obtained by requiring that the exponent in the resolvent, Eq. (7.89a) be stationary with respect to variation of $\sigma(x,t)$

$$\frac{\delta}{\delta\sigma(r,t)}\left[\frac{1}{2}\int_{-T/2}^{T/2} dt\,\sigma v^{-1}\sigma - \sum_{i=1}^{n}\alpha_{k_i}(\sigma,T)\right] = 0 . \tag{7.92}$$

In order to evaluate the functional derivative $\frac{\delta\alpha_k}{\delta\sigma(x,t)}$ in a form sufficiently general to apply to both real and imaginary time, we introduce a biorthogonal basis $\{\overline{\phi}_k\phi_k\}$ where the ϕ_k are solutions to (7.91b) and the $\overline{\phi}_k$ are determined by the adjoint operator

$$\left[i\frac{\partial}{\partial t} - K - \sigma(r,t)\right]^{\dagger}\overline{\phi}_k(r,t) = -\frac{\overline{\alpha}_k}{T}\overline{\phi}_k(r,t) \tag{7.93a}$$

with the normalization convention

$$\int_{-T/2}^{T/2} dt \int dr\,\overline{\phi}_k^*(r,t)\phi_k(r,t) = T . \tag{7.93b}$$

Note that from Eqs. (7.91b) and (7.93a) that $\overline{\alpha}_k^* = \alpha_k$ and $\int dr\,\overline{\phi}_k(r,t)\phi_k(r,t)$ is time independent and thus equal to unity. Taking the derivative of Eq. (7.91b) with respect to $\sigma(r',t)$ and projecting onto $\overline{\phi}_k$, we obtain

$$\begin{aligned}\int dr\,dt\,\overline{\phi}^*(r,t)\frac{\delta}{\delta\sigma(r',t)}&\left[\left(i\frac{\partial}{\partial t} - K - \sigma(r,t)\right)\phi_k(r,t)\right]\\ = \int dr\,dt\overline{\phi}^*(r,t)\frac{\delta}{\delta\sigma(r',t')}&\left[-\frac{\alpha_k}{T}\phi_k(r,t)\right]\end{aligned} \tag{7.94a}$$

which yields the desired functional derivative

$$\frac{\delta\alpha_k(\sigma,T)}{\delta\sigma(r',t')} = \sum_{k=1}^{N}\overline{\phi}_k^*(r',t')\phi_k(r',t') . \tag{7.94b}$$

Hence, from Eq. (7.92), we obtain

$$\sigma^0(r,t) = \int dr'\, v(r-r') \sum_{i=1}^{N} \overline{\phi}_k^*(r',t)\phi_k(r',t) \quad . \tag{7.95a}$$

This result for the stationary solution $\sigma^0(r,t)$ is the many-Fermion counterpart of the periodic classical solution obtained in the stationary phase approximation to the one-dimensional path integral. Since ϕ and $\overline{\phi}$ are periodic, σ^0 is periodic. The form of solution $\psi = e^{-i\alpha t/T}\phi$ with ϕ periodic then follows from Floquet's theorem. As in the analogous case of Bloch's theorem for crystals, in which periodic functions are multiplied by exponentials containing a quasimomentum, the quantity α/T has the physical interpretation of a quasienergy.

For real time, t, it follows from (7.91b), (7.93a) and (7.95a) that $\overline{\phi}_k(r,t) = \phi_k(r,t)$ and $\overline{\alpha}_k = \alpha_k$. Hence, $\sigma_0(r,t)$ has the familiar form of the Hartree mean field

$$\sigma^0(r,t) = \int dr'\, v(r-r') \sum_{i=1}^{N} \phi_k^*(r',t)\phi_k(r',t) \tag{7.95b}$$

and the $\{\phi_k(r,t)\}$ satisfy the self-consistent coupled equation

$$\begin{aligned} \left(-i\frac{\partial}{\partial t} - \frac{1}{2m}\nabla_r^2 - \frac{1}{2}v(0) + \int dr'\, v(r-r') \sum_k \phi_k^*(r',t)\phi_k(r',t)\right)\phi_k(r',t) \\ = \frac{\alpha_k}{T}\phi_k(r,t) \end{aligned} \tag{7.96}$$

with the boundary conditions that ϕ vanish on the spatial boundaries and be periodic in time with period T. One particular solution to these equations is, of course, the static Hartree solution discussed in Section 7.2.

We now consider the approximate eigenstates given by the poles of the resolvent when it is evaluated in the stationary phase approximation. Using the stationary solution $\sigma_0(t)$ in Eq. (7.89a), and ignoring quadratic corrections for simplicity, we may write

$$\begin{aligned} G(E) &\sim \sum_{\{k_1\ldots k_n\}} \int_0^\infty dT\, e^{i(ET+S[\sigma_0,T])} \\ &\equiv \sum_{\{k_1\ldots k_n\}} \int_0^\infty dT\, e^{iW(T)} \end{aligned} \tag{7.97a}$$

where

$$S[\sigma_0,T] = \frac{1}{2}\int_{-T/2}^{T/2} dt\, \sigma^0(t) v^{-1} \sigma^0(t) - \sum_k \alpha_k(\sigma^0,T) \quad . \tag{7.97b}$$

For each set of occupation numbers $\{k_1 \ldots k_n\}$ the integral over period T has the same structure as Eq. (7.71) for the one-dimensional example, and we obtain poles by applying the stationary-phase approximation to the T integral as before. Stationarity with respect to T yields

$$E = -\frac{\partial S[\sigma,T]}{\partial T} \quad . \tag{7.98}$$

The derivative of $\alpha_k(\sigma^0, T)$ with respect to T is conveniently evaluated by rescaling the time variable according to $t = \eta T$ and writing the eigenvalue equations as

$$\left[i\frac{\partial}{\partial \eta} - T\left(K + \sigma(r,\eta)\right)\right]\phi_k(r,\eta) = -\alpha_k(\sigma,T)\phi_k(r,\eta) \quad . \tag{7.99}$$

Then, taking the derivative of both sides with respect to T, and projecting with $\phi_k^*(r,\eta)$ as before yields

$$\frac{\partial \alpha_k}{\partial T} = \int dr d\eta\, \phi_k^*(r,\eta)\left(K + \sigma(r,\eta)\right)\phi_k(r,\eta) \quad . \tag{7.100}$$

The term $\frac{\partial}{\partial T}\left[\frac{1}{2}T\int_{-1/2}^{1/2} d\eta\, \sigma^0(\eta) v^{-1} \sigma^0(\eta)\right]$ just subtracts off the overcounting of the potential energy as in Eq. (7.39c) for the static case, so that after rescaling back from η to t, stationarity with respect to T yields the condition

$$E = \frac{1}{T}\int_{-T/2}^{T/2} \mathcal{H}(\phi_k^*, \phi_k) = E_H \tag{7.101a}$$

where $\mathcal{H}(\phi_k^*, \phi_k)$ is the Hartree Hamiltonian density

$$\begin{aligned} \mathcal{H}(\phi_k^* \phi_k) \equiv & \int_{-T/2}^{T/2} dt \int dr\, \phi_k^*(r,t) \times \\ & \times \left[K + \frac{1}{2}\int dr' \sum_{k'} \phi_{k'}^*(r',t)\phi_{k'}(r',t) v(r-r')\right]\phi_k(r,t) \quad . \end{aligned} \tag{7.101b}$$

Using the equations of motion for ϕ_k and ϕ_k^*, it is easy to show that the Hartree energy, $E_H = \mathcal{H}(\phi_k^*, \phi_k)$ is conserved (see Problem 7.5). Equation (7.101a) setting E equal to the Hartree energy of the periodic mean field solution in the many-Fermion case is analogous to Eq. (7.72) for the Feynman path integral requiring that E equal the energy of the classical periodic solution.

Finally, using the expression for E in Eq. (7.101) we find that

$$\begin{aligned} W(T) &\equiv ET + S[\sigma_0, T] \\ &= i\int_{-T/2}^{T/2} dt \int dr \sum_k \phi_k^*(r,t)\frac{\partial}{\partial t}\phi_k(r,t) \end{aligned} \tag{7.102}$$

since ET cancels all the terms in $S(T)$ except those involving the time derivatives of single particle wave functions. From this point on, derivation of the quantization condition follows precisely as in the one-dimensional example, Eqs. (7.73) – (7.76). Denoting $T(E)$ as the fundamental period which gives rise to periodic time-dependent Hartree solutions with energy E, stationary points in Eq. (7.97) are obtained for all integral multiples, $mT(E)$, with reduced action $W[mT(E)] = mW[T(E)]$. The same geometric series, Eq. (7.76a), arises, yielding the quantization condition

$$\begin{aligned} n2\pi &= W[T(E)] \\ &= i\int_{T(E)/2}^{T(E)/2} dt \int dr \sum_k \phi_k^*(r,t)\frac{\partial}{\partial t}\phi_k(r,t) \ . \end{aligned} \tag{7.103}$$

The similarity of the quantization conditions in the example with one degree of freedom, Eq. (7.76b) and the many-Fermion problem, Eq. (7.103) is particularly clear when one observes, as shown in Problem 7.5, that the time-dependent Hartree Fock equations are of Hamiltonian form with $\pi_k \equiv i\phi_k^*$ playing the role of the momentum conjugate to ϕ_k so that $\sum_k \oint dt \pi_k \dot{\phi}_k$ is a natural generalization of $\oint dt\, p\dot{q}$.

At the SPA level, quantum states of large-amplitude collective motion are fully specified by the self-consistent equations for periodic time-dependent Hartree solutions, Eqs. (7.96), and the quantization condition, Eq. (7.103). In general, Eqs. (7.96) specify time-dependent oscillations having amplitudes which depend continuously upon T, and the quantization condition singles out a discrete set of amplitudes and energies corresponding to quantum eigenstates. The theory represents the simplest available approximation that has all the physical elements of a fundamental theory of quantum collective motion. All the degrees of freedom of the one-body density are accessible, and the dynamical equations and quantization condition arise with no further prescriptions concerning collective variables, inertial parameters, or quantization procedure. Furthermore, corrections to the SPA may in principle be systematically evaluated.

Two special limits of these time-dependent equations should be noted. Time-independent solutions to Eqs. (7.96) correspond to $n = 0$ in Eq. (7.103), so the static Hartree states discussed previously are automatically included. As shown in Problem 7.10, for infinitesimal periodic fluctuations about the static Hartree solutions, the time-dependent quantized theory reduces to the familiar random-phase approximation (RPA).

BARRIER PENETRATION AND SPONTANEOUS FISSION

In the case of a single particle in a potential, the stationary-phase solution in a classically forbidden region was given by the classical equation of motion in imaginary time and corresponded to classical motion in the inverted well. This section presents the analogous theory for the many-Fermion problem and applies it to the example of spontaneous fission.

First, consider the stationary solution for the auxiliary field in the case of purely imaginary time, which we write in terms of a real variable $\tau \equiv it$. Equations (7.91b) and (7.93a) for the biorthogonal basis yield

$$\left(-\frac{\partial}{\partial\tau} - K - \sigma\left(r,\frac{\tau}{i}\right)\right)\phi_k\left(r,\frac{\tau}{i}\right) = -\frac{\alpha_k}{T}\phi_k\left(r,\frac{\tau}{i}\right) \tag{7.104a}$$

$$\left[\left(\frac{\partial}{\partial\tau} - K - \sigma\left(r,\frac{\tau}{i}\right)\right)\overline{\phi}_k\left(r,\frac{\tau}{i}\right)\right]^* = \left[-\frac{\bar{\alpha}_k}{T}\overline{\phi}_k\left(r,\frac{\tau}{i}\right)\right]^* \tag{7.104b}$$

from which it follows that $\overline{\phi}_k\left(r, \frac{\tau}{i}\right)^* = \phi_k\left(r, -\frac{\tau}{i}\right)$. Denoting the imaginary time solution $\tilde{\phi}_k(r,\tau) \equiv \phi_k\left(r, \frac{\tau}{i}\right)$ and using Eq. (7.95a) for σ^0 we may write the self-consistent periodic equations in imaginary time as follows

$$\left(\frac{\partial}{\partial\tau} - \frac{1}{2m}\nabla_r^2 - \frac{1}{2}v(0) + \int dr' v(r-r') \sum_{k'} \tilde{\phi}_{k'}(r', -\tau)\tilde{\phi}_{k'}(r', \tau)\right)\tilde{\phi}_k(r,\tau)$$
$$= \frac{\tilde{\alpha}_k}{T}\tilde{\phi}_k(r,\tau) \tag{7.105a}$$

where $\tilde{\phi}_k(r,\tau)$ satisfy the periodic condition

$$\tilde{\phi}_k\left(r, \frac{T}{2}\right) = \tilde{\phi}_k\left(r, -\frac{T}{2}\right) \tag{7.105b}$$

and the orthonormality relation

$$\int dr\, \tilde{\phi}_j(r, -\tau)\tilde{\phi}_k(r,\tau) = \delta_{jk} \quad . \tag{7.105c}$$

Several observations should be made concerning these equations. First, although this self-consistent eigenvalue problem is not Hermitian, it has a set of real solutions with real eigenvalues, and general solutions differ from these real solutions by the trivial phase factors $e^{i(2\pi n/T)\tau}$ and eigenvalue shifts $i(2\pi n/T)$ where n is any integer. The real solutions will yield the "bounce" solution governing spontaneous fission. Second, note that if $\tilde{\phi}(r,\tau)$ is real, the combination $\tilde{\phi}_k(r,-\tau)\tilde{\phi}_k(r,\tau)$ entering $\sigma_0(r,\tau)$ may be written in terms of the original function $\phi_k\left(r, t = \frac{\tau}{i}\right)$ as $\phi_k(r,t^*)^*\phi_k(r,t)$, so that it indeed represents analytic continuation of the real time expression (7.95b) to complex time. In addition, observe that although $\tilde{\phi}_k(r,\tau)$ and $\tilde{\phi}_k(r,-\tau)$ grow and decay exponentially in time, the combination $\tilde{\phi}_k(r,-\tau)\tilde{\phi}_k(r,\tau)$ remains normalized so that σ_0 retains the physical interpretation of a mean field. Finally, note that the static Hartree solutions are valid τ-independent solutions of these equations and in fact, the bounce solution will approach the static solution asymptotically at large time.

To establish contact with the example of a single particle in an inverted potential, we must express the action for both a single particle and for many Fermions in analogous Hamiltonian form. First, consider the Lagrangian for a particle in one dimension with a position-dependent mass:

$$\mathcal{L} = \frac{1}{2}m(q)\dot{q}^2 - V(q) \quad . \tag{7.106}$$

Note that although our previous one-dimensional example had a constant mass term, the more general form in (7.106) is useful in identifying the structure of the many-body problem. In the usual way, the momentum is $p = \frac{\partial\mathcal{L}}{\partial\dot{q}} = m(q)\dot{q}$ and the Hamiltonian is $H = p\dot{q} - \mathcal{L} = \frac{1}{2m(q)}p^2 + V(q)$. Thus, the action $S = \int dt\,\mathcal{L}$ may be written in the Hamiltonian form

$$S(p,q) = \int dt\left(p\dot{q} - \frac{1}{2m(q)}p^2 - V(q)\right) \quad . \tag{7.107}$$

The action $S(\psi^*,\psi) = \int dt \langle \psi_1 \cdots \psi_N | i\frac{\partial}{\partial t} - H | \psi_1 \cdots \psi_N \rangle$, which under variation yields the time-dependent Hartree-Fock equation, is derived in Eq. (7.21) and discussed in Problem 7.5. The analogous action which yields the time-dependent Hartree equation in real time is

$$S_H(\psi^*,\psi) = \int dt \int dr \sum_k \psi_k^*(r,t) \times \tag{7.108}$$

$$\times \left(i\frac{\partial}{\partial t} + \frac{1}{2m}\nabla^2 - \int dr'\, v(r-r') \sum_{k'} \psi_{k'}^*(r',t)\psi_{k'}(r,t) \right) \psi_k(r,t) \ .$$

To obtain a form analogous to Eq. (7.107), we must perform a canonical transformation from the variables $i\psi_k^*$ and ψ_k to new canonical variables and momenta which are time-even and time-odd, respectively. Let us choose the time origin at a classical turning point and define

$$\psi_k(r,t) \equiv \sqrt{\rho_k(r,t)}\, e^{i\chi_k(r,t)} \tag{7.109a}$$

where ρ_k, analogous to a coordinate, is real and time-even

$$\rho_k(r,t) = \rho_k(r,-t) \equiv g(r,t^2) \tag{7.109b}$$

and χ_k, corresponding to a momentum, is real and time-odd

$$\chi_k(r,t) = -\chi_k(r,-t) \equiv t f(r,t^2) \ . \tag{7.109c}$$

Transforming the Hartree action (7.108) to these new variables yields

$$S_H(\chi,\rho) = \int dr\, dt \sum_k \left(\chi_k \dot{\rho}_k - \chi_k \overleftarrow{\nabla} \frac{\rho_k}{2m} \overrightarrow{\nabla} \chi_k - V(\rho) \right) \tag{7.110a}$$

where

$$V(\rho) = \mathcal{H}(\sqrt{\rho}\sqrt{\rho})$$
$$= \int dr \sum_k \sqrt{\rho_k(r)} \left(-\frac{1}{2m}\nabla^2 + \int dr' \sum_{k'} \rho_{k'}(r) v(r-r') \right) \sqrt{\rho_k(r)} \tag{7.110b}$$

is the Hartree energy, Eq. (7.101b), for the time-even single-particle wave function $\{\sqrt{\rho_k}\}$. The action $S_H(\chi,\rho)$, Eq. (7.110a), is precisely of the form of $S(p,q)$ in Eq. (7.107), where $\sqrt{\rho_k}$ corresponds to the coordinate q, χ_k replaces the momentum p, $\overleftarrow{\nabla} \frac{\rho_k}{2m} \overrightarrow{\nabla}$ is the coordinate-dependent mass $\frac{1}{2m(q)}$, and the Hartree energy $\mathcal{H}(\sqrt{\rho},\sqrt{\rho}) = V(\rho)$ plays the role of the potential.

The potential $V(\rho)$ provides a useful way to visualize the time-dependent Hartree problem. Think of the Hartree energy surface in the multidimensional space of all determinants composed of time-even wave functions. This surface has local minima at all the stable Hartree solutions, and in general, for an energy above that of a local minimum, the configuration will evolve in some classically allowed domain around the minimum. A convenient way to characterize the gross features of this surface is to

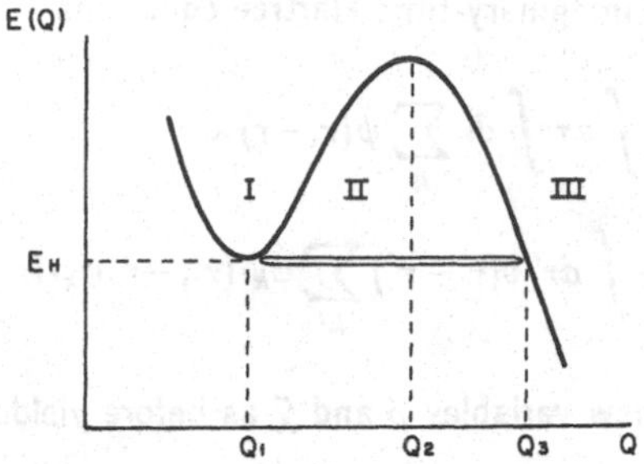

Fig. 7.4 Sketch of the constrained Hartree energy $E(Q)$ as a function of quadrupole moment Q for a fissile nucleus. At any value of Q, $E(Q)$ is defined as a minimum of the Hartree energy, Eq. (7.101b), in the space of time-even determinants having quadrupole moment Q.

evaluate the constrained energy of deformation surface defined by minimizing $V(\rho)$ with respect to all determinants satisfying one or more constraints. A familiar example to which we will return in discussing nuclear fission is the constrained energy curve as a function of deformation sketched in Fig. 7.4. The ground state of a deformed, fissile A-particle nucleus has a ground state quadrupole moment Q_1. If an external field is applied to increase the quadrupole moment, the energy increases since the cost in surface energy is greater than the savings in Coulomb energy. For any $Q > Q_1$, there exists some minimum value obtained by searching over all determinants of time-even wave functions having total quadrupole moment Q, and solid curve $E(Q)$ is the envelope of all such minima. Beyond the saddle point Q_2, which is a stationary point of the action, further increase in the deformation can decrease the Coulomb energy more than it increases the surface energy. Eventually, at very large Q, the original nucleus breaks into two separated fragments with $A/2$ particles, with an energy lower than the original energy E_H. The regions I, II and III of this figure correspond to those in Fig. 7.3b. Introducing more constraints would generalize Fig. 7.4 to a multidimensional surface in which one could visualize the projected motion.

Consider now the changes which occur when Eqs. (7.100– 7.101) are continued to imaginary time. We define

$$\tilde{\rho}_k(r,\tau) \equiv \rho_k\left(r, t = \frac{\tau}{i}\right) = g(r, -\tau^2) = \tilde{\rho}_k(r, -\tau) \tag{7.111a}$$

and

$$\tilde{\chi}_k(r,\tau) \equiv -i\chi\left(r, t = \frac{\tau}{i}\right) = -\tau f(r, -\tau^2) = -\tilde{\chi}(r, -\tau) \tag{7.111b}$$

in terms of which

$$\tilde{\psi}_k(r,\tau) = \sqrt{\tilde{\rho}_k(r,\tau)}\, e^{-i\tilde{\chi}_k(r,t)} \ . \tag{7.111c}$$

Writing the action for the imaginary-time Hartree equations

$$S_H\left(\tilde{\psi}(r,-\tau)\tilde{\psi}(r,\tau)\right) = \int d\tau \int dr \sum_k \tilde{\psi}(r,-\tau)\times$$
$$\times\left(-\frac{\partial}{\partial\tau}+\frac{1}{2m}\nabla^2-\int dr'\, v(r-r')\sum_{k'}\tilde{\psi}_{k'}(r',-\tau)\psi_{k'}(r',\tau)\right)\psi_k(r,t) \tag{7.112}$$

and transforming to the new variables $\tilde{\rho}$ and $\tilde{\chi}$ as before yields

$$S_H(\tilde{\chi},\tilde{\rho}) = \int dt\, dr\left(-\tilde{\chi}_k\dot{\tilde{\rho}}_k + \tilde{\chi}_k\vec{\nabla}\frac{\tilde{\rho}}{2m}\vec{\nabla}\tilde{\chi}_k - V(\tilde{\rho})\right) \tag{7.113a}$$

where

$$V(p) = \mathcal{H}\left(\sqrt{\tilde{\rho}},\sqrt{\tilde{\rho}}\right) \quad . \tag{7.113b}$$

Since the overall sign of S does not affect the Euler-Lagrange equation, we note that the only difference between the real-time equations of motion for $\{\chi_k,\rho_k\}$ from Eq. (7.110) and the imaginary-time equation of motion for $\{\tilde{\chi}_k,\tilde{\rho}_k\}$ from (7.113) is that $\mathcal{V}(\rho)$ and $\mathcal{V}(\tilde{\rho})$ enter with opposite signs. Thus, the multidimensional energy surface $\mathcal{V}(\rho)$ and the constrained projection $E(Q)$ sketched in Fig. 7.2 are simply inverted as in the one-dimensional example, and even in the many-Fermion problem, we may continue to think of the imaginary-time solutions as periodic trajectories in an inverted potential.

One last observation facilitated by the transformation to the $\{\rho,\chi\}$ representation is the meaning of the quasiperiodic boundary condition $\psi_k\left(r,\frac{T}{2}\right) = e^{-i\alpha_k}\psi_k\left(r,-\frac{T}{2}\right)$, Eq. (7.90b), and the corresponding condition in imaginary time. Using the symmetry properties, Eq. (7.109), this condition implies $2\chi_k\left(r,\frac{T}{2}\right) = -\alpha_k$ so that $\nabla\chi_k\left(r,\pm\frac{T}{2}\right) = 0$. Hence, the kinetic term in the Hamiltonian, $\frac{\rho_k}{2m}\left(\nabla\chi_k\right)^2$, vanishes and $\frac{T}{2}$ is a classical turning point.

With this understanding of the structure of the real and imaginary time periodic solutions, the stationary-phase approximation to the spontaneous fission of a nucleus is nearly analogous to the treatment of the lifetime of a metastable state in the one-dimensional potential, Fig. 7.1d. Recall that in calculating the smoothed level density, Eq. (7.86), trajectories in region III did not contribute, and summation of periodic trajectories in regions I and II yielded the WKB result, Eq. (7.88c), in which the penetrability was given by the reduced action of the trajectory in region II having an energy equal to that of an eigenstate in region I. Thus, in principle, we should sum all trajectories in regions I and II of the multidimensional analog of Fig. 7.4.

The essential elements in describing fission are the so-called "bounce" solutions, the periodic self-consistent solutions to the imaginary-time equations Eq. (7.105a). Like the real-time solutions discussed in the preceding section, the periodic solutions in region II of the inverted well of Fig. 7.4 depend continuously on the period T. As T increases, the amplitude grows and the energy decreases until as $T\to\infty$, the energy approaches the static Hartree energy E_H and the solution at the classical turning point approaches the static Hartree solution. In contrast to the simple one-dimensional

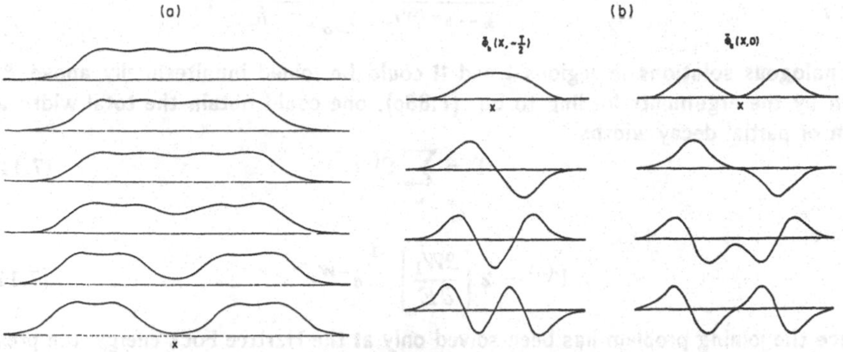

Fig. 7.5 Bounce solution for a fissioning one-dimensional nuclear model. Part (a) shows density profiles at evenly spaced time intervals from $\tau = -\frac{T}{2}$ to 0. Part (b) shows the four distinct spatial wave functions at time $\tau = -\frac{T}{2}$ and at $\tau = 0$.

example, in general there are several distinct, well-separated bounce solutions corresponding to symmetric fission, asymmetric fission, and more complicated many-body breakup. Each such solution evolves from the Hartree ground state through a saddle point to some distinct configuration at the boundary of the classically allowed regime, and all solutions involving any combination of these trajectories should be summed in the stationary-phase approximation. To help visualize the bounce, the solution for symmetric fission of a model nuclear system in one spatial dimension is shown in Fig. 7.5. The sequence of density profiles $\rho(x,\tau) = \sum_k \phi_k(x,-\tau)\tilde{\phi}_k(x,\tau)$ shows how the system evolves from the Hartree solution for the original nucleus at time $\tau = -\frac{T}{2}$ to two nearly separated symmetric fragments at $\tau = 0$. Note that although the continuous variable τ has nothing to do with the physical time (it is just a formal variable in the path integral to deal with the non-commutivity of T and V) the sequence of densities through which the system evolves corresponds to the most probable fission path for a given channel and is physically significant.

Let us now sum all the trajectories at the Hartree energy, E_H, for which all the bounce solutions in region II join into the static Hartree solution in region I. Denoting the reduced action in region I, Eq. (7.102), by W_I and defining the imaginary-time counterpart in a specific decay channel α by

$$W_{\rm II}^{\alpha} = \int_{-\frac{T}{2}}^{\frac{T}{2}} d\tau\, dr \sum_k \tilde{\phi}_k^{\alpha}(r,-\tau)\frac{\partial}{\partial\tau}\tilde{\phi}_k^{\alpha}(r,\tau) \tag{7.114}$$

we obtain the following contribution of all stationary solutions to the resolvent

$$\begin{aligned} G(E_H + i\gamma) = &\sum_{k=1}^{\infty}\left[e^{iW_I}\sum_{n=0}^{\infty}\left[\sum_{\alpha} e^{-W_{\rm II}^{\alpha}}\right]^n\right]^k \\ &+\sum_{k=1}^{\infty}\left[\sum_{\alpha} e^{-W_{\rm II}^{\alpha}}\sum_{n=0}^{\infty}\left[e^{iW_I}\right]^n\right]^k \end{aligned} \tag{7.115}$$

$$= \frac{1}{1 - e^{-iW_{\rm I}} - \sum_\alpha e^{-W^\alpha_{\rm II}}} - 1 \ .$$

If analogous solutions in regions I and II could be joined infinitesimally above $E_{\rm HF}$, then by the arguments leading to Eq. (7.88c), one could obtain the total width as a sum of partial decay widths

$$\Gamma = \sum_a \Gamma^{(a)} \tag{7.116a}$$

where

$$\Gamma^{(a)} = 2\left[\frac{\partial W_{\rm I}}{\partial E}\right]^{-1} e^{-W^{(a)}_{\rm II}} \ . \tag{7.116b}$$

Since the joining problem has been solved only at the Hartree Fock energy, the present treatment yields only the penetrability $e_{\rm II}^{-W^{(a)}}$, and the premultiplying factor must be derived by other means. A derivation of the premultiplying factor using the dilute instanton gas approximation is outlined in Problem 7.11.

It is instructive to note how the essential physical aspects of the fission problem enter the stationary-phase approximation to the lifetime. Both the gross competition between volume, surface, and Coulomb energies and quantitatively significant single-particle shell effects are included through the evolution of a determinantal Hartree-Fock wave function. The bounce solution determines both the relevant collective degrees of freedom for the most probable fission path and the corresponding conjugate momenta. Finally, the competition between alternative decay channels, such as symmetric or asymmetric fission, is manifested in the distinct solutions governing each partial width. Clearly, this functional integral approach to tunneling problems is quite general and has been applied to a variety of physical problems, ranging from bubble formation in a phase transition (Langer 1977) to the structure and decay of the vacuum (Polyakov 1977, Coleman 1977).

CONCEPTUAL QUESTIONS

Whereas the applications of functional integrals in this chapter are physically and intuitively appealing, they are subject to limitations which require comment.

The most salient problem is the absence, except in specifically constructed models, of an explicit small parameter in which to generate an asymptotic expansion. Recall that the Feynman path integral has an explicit factor of $\frac{1}{\hbar}$ multiplying the action and thus generates an asymptotic expansion in powers of $\hbar$. Unfortunately, the Hartree or Hartree-Fock action obtained using an auxiliary field, coherent states, or an overcomplete set of Slater determinants contains $\hbar$ in the kinetic energy as well as an overall multiplicative factor. Hence, to take advantage of the physics of the mean field, one must relinquish a strict semiclassical expansion in powers of $\hbar$.

A second alternative is to generate a $1/N$ expansion. One may imagine a class of theories which differ from one another by their interaction strength and the number N of internal degrees of freedom. In certain special cases, requiring that the class of theories has a sensible limit for large N specifies the N dependence of the interaction strength and allows one to rescale the integration variables such that an explicit factor of N will multiply the action. The resulting formal expansion in powers of $1/N$ may then be useful if the physical Hamiltonian embodies a sufficiently large number of internal degrees of freedom. A detailed example is discussed in Problem 7.12. In the case of nuclear physics, where there are four spin-isospin degrees of freedom, one might naively invoke a $1/N$ expansion with $N = 4$. Unfortunately, the spin and isospin dependence of the forces is so large that the argument is inapplicable. For example, the leading contribution to the potential energy, the direct Hartree potential, is considerably weaker than the Fock exchange term, which formally should be a factor of $1/N$ smaller. Similarly, in $SU(N)$ non-Abelian gauge field theories, where N is the number of colors, there is a significant question as to the relevance of the large N limit to the physical case $N = 3$.

The lack of an explicit expansion parameter does not necessarily preclude application of the stationary-phase approximation. Indeed, the functional integrals we evaluate may well possess saddle points with very large second derivatives in the directions of steepest descent, giving rise to useful and accurate low-order approximations. For example, in the case of very collective states, it is physically plausible that many particles participate in motion characterized by the appropriate collective variable, and that the action for this variable is thus multiplied by a suitably large constant reflecting this collectivity. However, in this case, there is as yet no quantitative measure of the accuracy of the expansion.

A second aspect associated with the lack of expansion parameter is the freedom to write a variety of different exact expressions, each of which yields a different lowest-order approximation. In the absence of a formal expansion parameter, one must be guided by the physics of the problem, and for example choose a formulation which includes the physically relevant combinations of direct, exchange, and pairing mean field contributions. Similarly, when the presence of strong short-range correlations requires that mean-field theory be formulated in terms of some effective interaction such as a G-matrix, it is essential to reformulate the functional integral in terms of the effective interaction instead of the bare interaction. Since the choice of the functional integral representation determines the starting point for a systematic expansion, it is analogous to the choice of the decomposition of H into a non-interacting Hamiltonian H_0 defining the unperturbed basis and the residual interaction V in which one expands perturbatively.

A final question concerns the correct counting of quantum states when one considers expansions around all stationary solutions. The problem is illustrated by considering the RPA limit. Some RPA states correspond to nearly pure single particle-hole excitations which are also generated by a static Hartree-Fock solution with an appropriate set of occupation numbers. Thus, in this case, the same physical state is approximated both by a static self-consistent solution and a time-dependent periodic solution. In general, there thus exists the problem of overcounting the physical quantum states of a system by considering all stationary-phase solutions. Since substantial mathematical difficulties arise in attempting to add the contributions of distinct stationary solutions, one must again be guided by physical considerations in selecting the form of stationary solutions with which to approximate a particular physical state.

7.5 LARGE ORDERS OF PERTURBATION THEORY

The final application of functional integral methods we shall present is their use as a tool to study the behavior of large orders of perturbation theory. The motivation for studying the behavior of large orders is the conviction that deeper understanding of the nature of the asymptotic behavior will lead to the development of new methods to extract physics from the divergent series. We will briefly discuss two such possibilities, Borel summation and Padé approximants.

Historically, Bender and Wu (1969, 1971, 1972) first calculated the asymptotic expansion of the energy of the anharmonic harmonic oscillator as a function of the coupling constant using the WKB method, which has no natural extension to problems with large numbers of degrees of freedom. Subsequently, Lipatov (1976, 1977) developed a functional integral method which is directly applicable to many-body theory and field theory and which has been applied to several examples by Brézin *et al.* (1977). We will illustrate the basic ideas on the simple integral introduced in Section 2.1 where we first discussed asymptotic expansions. Then, we will discuss Borel summation and finally use these methods to evaluate the asymptotic behavior of the anharmonic oscillator. Because the developments in this section are particularly formal, we present only the essential ideas in the text and relegate many of the mathematical details to the problems.

STUDY OF A SIMPLE INTEGRAL

Consider again the integral of Section 2.1, Eq. (2.24)

$$Z(g) = \frac{1}{\sqrt{2\pi}} \int_{-\infty}^{\infty} dx\, e^{-\frac{x^2}{2} - \frac{g}{4}x^4} \tag{7.117}$$

which corresponds to the classical partition function of a particle in a quartic potential. Physically, we have already seen that $Z(g)$ has an essential singularity at $g = 0$. Mathematically, $Z(g)$ is analytic in the complex g plane with a cut along the negative real axis. The asymptotic expansion of $Z(g)$ in powers of g obtained in Eq. (2.26) is

$$Z(g) = \sum_{k=0}^{\infty} Z_k g^k \tag{7.118a}$$

where

$$Z_k = \frac{(-1)^k}{2^{4k}} \frac{(4k)!}{k!\,(2k)!} \; . \tag{7.118b}$$

Using Stirling's formula $k! \sim \sqrt{2\pi} k^{k+\frac{1}{2}} e^{-k}$, it is useful to write the asymptotic behavior of Z_k in the following equivalent form for future reference

$$Z_k \underset{k\to\infty}{\sim} \frac{(-1)^k}{\sqrt{k\pi}} \left(\frac{4k}{e}\right)^k \tag{7.118c}$$

$$\underset{k\to\infty}{\sim} \frac{(-1)^k \, 4^k \, k!}{\pi\sqrt{2}\, k} \; . \tag{7.118d}$$

As noted in Chapter 2, the physical origin of the divergence of this alternating series is the fact that the number of contractions or diagrams at each order grows like $k!$.

In preparation for the treatment of functional integrals, we now show how to obtain this asymptotic behavior of Z_k by applying the stationary-phase approximation to the integral for $I(g)$. Expanding the exponential $e^{-\frac{g}{4}x^4}$ in Eq. (7.117) and rescaling the integration variable $x = \sqrt{k}y$ yields

$$\begin{aligned} Z_k &= \frac{(-1)^k}{k!4^k} \int_{-\infty}^{\infty} \frac{dx}{\sqrt{2\pi}} e^{-\frac{x^2}{2}} x^{4k} \\ &= \frac{(-1)^k}{k!4^k} e^{2k \ln k} \sqrt{k} \int_{-\infty}^{\infty} \frac{dy}{\sqrt{2\pi}} e^{-k\left(\frac{y^2}{2} - 2\ln y^2\right)} \\ &\propto \int dy\, e^{-kA(y)} \end{aligned} \tag{7.119a}$$

where

$$A(y) = \frac{y^2}{2} - 2\ln y^2 \ . \tag{7.119b}$$

Hence, as $k \to \infty$, the asymptotic expression for Z_k is obtained by the stationary-phase approximation. The stationarity condition

$$0 = \frac{\partial A}{\partial y} = y - \frac{4}{y} \tag{7.120a}$$

has two stable saddle points

$$y_s = \pm 2 \tag{7.120b}$$

with curvature

$$\left.\frac{\partial^2 A}{\partial y^2}\right|_{y_s} = 1 + \frac{4}{y_s^2} = 2 \ . \tag{7.120c}$$

Adding the stationary contribution and quadratic fluctuations of these two saddle points yields

$$\begin{aligned} Z_k \underset{k\to\infty}{\sim} & 2\frac{(-1)^k}{k!4^k} e^{2k\ln k}\sqrt{k}\, e^{-k(2-2\ln 4)} \frac{1}{\sqrt{2\pi}} \sqrt{\frac{\pi}{k}} \\ \sim & \frac{(-1)^k}{\sqrt{\pi k}} e^{k \ln 4k - k} \end{aligned} \tag{7.121}$$

in agreement with Eq. (7.118c). Higher order terms in powers of $\frac{1}{k}$ are evaluated in Problem 7.13.

BOREL SUMMATION

As in the preceding example of a simple integral, one frequently encounters divergent series of the form

$$Z(x) = \sum_{k=0}^{\infty} Z_k x^k \tag{7.122}$$

which diverge because some combinatorial factor makes Z_k grow like $k!$. Hence, it is useful to consider the Borel transform in which the kth term in the original series is divided by $k!$

$$B(x) = \sum_{k=0}^{\infty} \frac{Z_k}{k!} x^k \ . \tag{7.123}$$

If $B(x)$ converges for all values of x, one may then extract a finite result from the divergent series (7.122) by the inverse Borel transform

$$Z_B(x) = \int_0^\infty dt\, e^{-t} B(xt) \ . \tag{7.124}$$

Obviously, if the original series converges, Z_B yields the correct result since

$$Z_B(x) = \sum_{k=0}^{\infty} \frac{\int_0^\infty dt\, e^{-t} t^k}{k!} Z_k x^k = Z(x) \ . \tag{7.125}$$

A simple example of how Borel summation works for a divergent series is provided by the power series for $\frac{1}{1+x}$ for which $Z_k = (-1)^k$. Then

$$B(x) = \sum_{k=0}^{\infty} \frac{(-1)^k}{k!} x^k = e^{-x} \tag{7.126a}$$

and

$$Z_B(x) = \int_0^\infty dt\, e^{-t} B(x,t) = \int_0^\infty dt\, e^{-t(1+x)} = \frac{1}{1+x} \tag{7.126b}$$

which is the correct result. Further mathematical aspects of Borel summation are given by Hardy (1948). Physically, the obvious question is whether the finite result extracted from the divergent perturbation series by Borel summation is the correct physical solution, and one must argue on physical grounds that pathological terms such as $e^{-\frac{1}{x}}$ with vanishing derivatives are excluded.

Let us now use Borel summation to sum the leading contributions of high orders of perturbation theory. Denoting the asymptotic coefficient in Eq. (7.118d) by $\tilde{Z}_k$, we wish to sum the divergent series

$$\begin{aligned} \tilde{Z}(g) &= \sum_{k=1}^{\infty} \tilde{Z}_k g^k \\ &= \sum_{k=1}^{\infty} \frac{(-1)^k\, 4^k\, k!}{\pi\sqrt{2}\, k} g^k \ . \end{aligned} \tag{7.127}$$

Hence, we consider the Borel transform

$$B(g) = \sum_{k=1}^{\infty} \frac{(-1)^k\, (4g)^k}{\pi\sqrt{2}\, k} \tag{7.128a}$$

$$= -\frac{1}{\pi\sqrt{2}} \ln(1+4g) \tag{7.128b}$$

for which the inverse transform is

$$\tilde{Z}_B(g) = -\int_0^\infty \frac{dt}{\pi\sqrt{2}} e^{-t} \ln(1+4gt) \ . \tag{7.129}$$

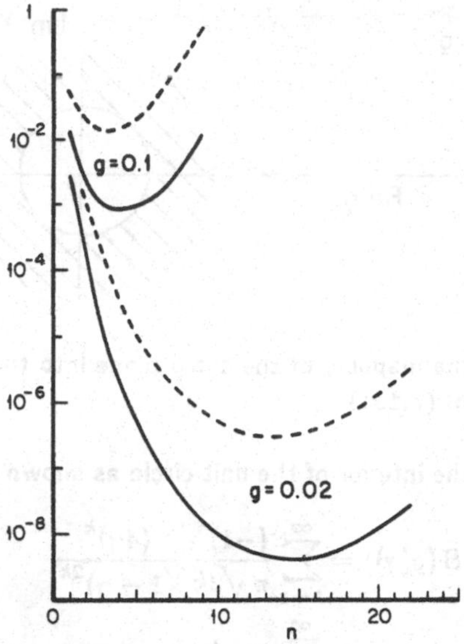

Fig. 7.6 Asymptotic expansion of $Z(g)$ for $g = 0.1$ and 0.02. The dashed lines indicate the magnitude of the residual error $R_n = \left|Z(g) - \sum_{k=1}^{n} Z_k g^k\right|$ which was plotted in Fig. 2.2 and the solid lines denote the residual $R_n = \left|Z(g) - Z^{(n)}(g)\right|$ using the Borel sum in Eq. (7.81).

One practical way to use this result is to approximate $Z(g)$ by evaluating a finite number of orders of the asymptotic expansion exactly and then use $\tilde{Z}(g)$ to approximate the sum of all the remaining orders. Thus, we may define the approximant

$$Z^{(n)}(g) = \sum_{k=1}^{n} Z_k g^k + \sum_{k=n+1}^{\infty} \tilde{Z}_k g^k$$
$$= \sum_{k=1}^{n} \left(Z_k - \tilde{Z}_k\right) g^k + \tilde{Z}_B(g) \quad . \tag{7.130}$$

As seen in Fig. 7.6, the approximant $Z^{(n)}(g)$ yields several orders of magnitude improvement relative to the first n terms of the asymptotic expansion.

Note that the series (7.128a) has a singularity at $g = -\frac{1}{4}$ and thus has a finite radius of convergence about the origin. Hence, the Borel transform, which involves an integral on the interval $[0, +\infty[$, required the analytic continuation in Eq. (7.128b). For cases in which the analytic continuation cannot be done by inspection, one may alternatively use a conformal transformation. In the present example, Eq. (7.128b) shows there is a cut along $]-\infty, -1/4]$. Hence, we may use the transformation

$$\gamma = \frac{(1+4g)^{1/2} - 1}{(1+4g)^{1/2} + 1} \tag{7.131a}$$

$$g = \frac{\gamma}{(1-\gamma)^2} \tag{7.131b}$$

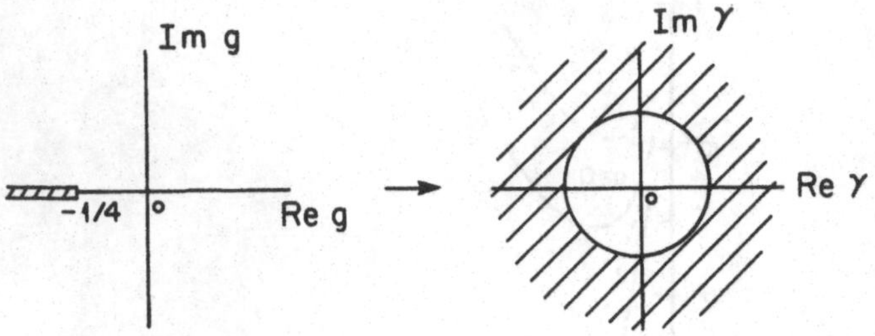

Fig. 7.7 Sketch of the mapping of the cut g plane into the unit circle using the transformation, Eq. (7.131).

to map the cut plane into the interior of the unit circle as shown in Fig. 7.7. The Borel transform

$$\begin{aligned} B\left(g(\gamma)\right) &= \sum_{k=1}^{\infty} \frac{(-1)^k}{\pi\sqrt{2k}} \frac{(4\gamma)^k}{(1-\gamma)^{2k}} \\ &\equiv \sum_{k=1}^{\infty} B_k \gamma^k \end{aligned} \tag{7.132}$$

is thus analytic in the unit circle of γ. By performing the inverse transformation, we obtain

$$\tilde{Z}(g) = \int_0^\infty e^{-t} \sum_{k=1}^{\infty} B_k \left(\frac{(1+4gt)^{1/2}-1}{(1+4gt)^{1/2}+1} \right)^k \tag{7.133}$$

where, by construction, the integrand has now been continued to the entire cut plane. The general procedure for Borel summation of a divergent series thus consists of three steps: evaluation of the Borel transform of the series, analytic continuation to the interval $[0, \infty[$, and inverse transformation.

THE ANHARMONIC OSCILLATOR

Now consider the quantum anharmonic oscillator with the Hamiltonian

$$H(g) = \frac{p^2}{2} + \frac{x^2}{2} + \frac{g}{4} x^4 \tag{7.134}$$

and the expansion of the ground state energy in powers of g

$$E_0(g) = \sum_{k=0}^{\infty} E_k g^k \quad . \tag{7.135}$$

We will evaluate the ground state energy by calculating the normalized partition function

$$Z(g) = \frac{\operatorname{tr} e^{-\beta(H(g))}}{\operatorname{tr} e^{-\beta(H(0))}} \tag{7.136a}$$

and taking the zero temperature limit of the free energy

$$E_0(g) = E_0(0) - \lim_{\beta\to\infty} \frac{1}{\beta} \ln Z(g) \quad . \tag{7.136b}$$

The Feynman path integral for $Z(g)$ is

$$Z(g) = \int_{x(\frac{\beta}{2})=x(-\frac{\beta}{2})} \mathcal{D}[x(t)]\, e^{-\int_{-\beta/2}^{\beta/2}\left(\frac{1}{2}\dot{x}^2+\frac{1}{2}x^2+\frac{g}{4}x^4\right)} \tag{7.137}$$

where the measure is normalized such that $Z(0) = 1$. Expanding the exponential in powers of g and rescaling the integration variable $x \to \sqrt{k}x$ yields the expansion for the partition function

$$Z(g) = \sum_{k=0}^{\infty} Z_k g^k \tag{7.138a}$$

where

$$\begin{aligned} Z_k &= \frac{(-1)^k}{k!\,4^k} \int_{x(\frac{\beta}{2})=x(-\frac{\beta}{2})} e^{-\int_{-\beta/2}^{\beta/2}\frac{dt}{2}(\dot{x}^2+x^2)} \left[\int_{-\beta/2}^{\beta/2} dt\, x^4\right]^k \\ &= \frac{(-1)^k\, k^{2k}}{k!4^k} \int_{x(\frac{\beta}{2})=x(-\frac{\beta}{2})} \mathcal{D}[x(t)]\, e^{-k\left[\int_{-\beta/2}^{\beta/2}\frac{dt}{2}(\dot{x}^2+x^2)-\ln\left(\int_{-\beta/2}^{\beta/2}dt x^4\right)\right]} \end{aligned} \tag{7.138b}$$

and the measure is normalized such that the integral of the quadratic term in the exponent is 1.

Saddle-Point Contribution

As in the example of the one-dimensional integral, the stationary-phase approximation yields the asymptotic expansion of Z_k for large k. The stationarity condition for the exponent is

$$-\ddot{x}_s + x_s - \frac{4x_s^3}{\int_{-\beta/2}^{\beta/2} dt x_s^4(t)} = 0 \tag{7.139a}$$

with the periodic boundary conditions

$$x\left(\tfrac{\beta}{2}\right) = x\left(-\tfrac{\beta}{2}\right) \quad ; \quad \dot{x}\left(\tfrac{\beta}{2}\right) = \dot{x}\left(-\tfrac{\beta}{2}\right) \ . \tag{7.139b}$$

Recall from Eq. (7.70) that variation of the end point yields periodicity of the momentum and thus $\dot{x}$. This equation may be simplified by the change of variable:

$$q(t) = \frac{x_s(t)}{\lambda} \tag{7.140a}$$

and

$$\lambda = \frac{1}{2}\left[\int_{-\beta/2}^{\beta/2} dt\, x_s^4(t)\right]^{1/2} = 2\left[\int_{-\beta/2}^{\beta/2} dt\, q^4(t)\right]^{1/2} \tag{7.140b}$$

with the result

$$-\ddot{q} + q - q^3 = 0 \tag{7.141a}$$

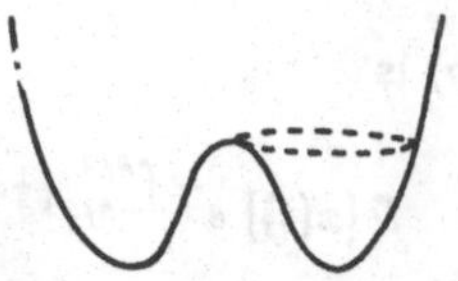

Fig. 7.8 Sketch of the effective potential for which stationary solutions for the action correspond to classical periodic trajectories. The bounce trajectory is indicated by the dotted line.

with

$$q\left(\tfrac{\beta}{2}\right) = q\left(-\tfrac{\beta}{2}\right) \; ; \qquad \dot{q}\left(\tfrac{\beta}{2}\right) = \dot{q}\left(-\tfrac{\beta}{2}\right) \; . \tag{7.141b}$$

As in Section 7.4, it is useful to visualize the periodic stationary solutions by regarding the stationarity equation as the classical equation of motion of a particle in an effective potential. Hence, we write

$$\begin{aligned} \ddot{q} &= q - q^3 \\ &\equiv -\nabla\left(V_{\text{eff}}(q)\right) \end{aligned} \tag{7.142a}$$

where

$$V_{\text{eff}}(q) = -\frac{1}{2}q^2 + \frac{1}{4}q^4 \; . \tag{7.142b}$$

The effective potential is sketched in Fig. 7.8. Note that its relation to the physical potential in the path integral in Eq. (7.137) is slightly different from the simple inversion encountered in Fig. 7.2. In the present case, the quadratic term has the usual sign change associated with a Euclidean path integral. The quartic term, however, comes from the logarithm in (7.138b) and is positive irrespective of the sign of g. Hence, whereas the physical potential has a single minimum, the effective potential is a double well. Using the fact that the energy is a constant of the motion (as may be verified by calculating dE/dt and using Eq. (7.141a))

$$E = \frac{1}{2}\dot{q}^2 - \frac{1}{2}q^2 + \frac{1}{4}q^4 \tag{7.143}$$

the period of a periodic solution may be written as in Eq. (7.74)

$$\frac{\beta}{2} = \int_{q_-(E)}^{q_+(E)} dq \frac{1}{\sqrt{2E + q^2 - \frac{1}{2}q^4}} \tag{7.144}$$

where $q_\pm(E)$ are the classical turning points at energy E.

We now seek the stationary solutions which provide the dominant asymptotic contributions to Z_k as $\beta \to \infty$. As shown in Problem 7.14, the trivial constant solution $q = \pm 1$ is unstable and thus does not correspond to a minimum of the action. The relevant solution for our present purposes is the bounce solution already considered in Fig. 7.3 for the decay of a metastable state. As shown in Problem 7.15, in the low temperature limit, $\beta \to \infty$, the periodic trajectory in Eq. (7.144) approaches the zero energy solution exponentially

$$E \underset{\beta\to\infty}{\sim} e^{-\beta} \; . \tag{7.145}$$

The zero-energy bounce solution has the analytic form

$$q(t) = \frac{\sqrt{2}}{\cosh(t - t_0)} \tag{7.146}$$

where t_0 is an arbitrary parameter specifying the time at which the trajectory reaches the classical turning point A. This degeneracy of the bounce with respect to translation in time will produce a zero mode in the quadratic fluctuation matrix which we will subsequently treat using the techniques of Chapter 4.

Note that in addition to a single bounce, there are additional stationary solutions corresponding to n well-separated bounces in which the trajectory runs from 0 to A and 0 to B n times. Since the action S_n of an n-bounce trajectory is n times the action of a single bounce S_1, the contributions to Z_k of n-bounce trajectories are subdominant for large k

$$Z_k \sim \sum_n A_n\, e^{-knS_n} \underset{k\to\infty}{\sim} A_1\, e^{-kS_1} \tag{7.147}$$

so we only need to consider the single bounce trajectory, Eq. (7.146). Note, however, that when the dilute instanton gas approximation is made in Problem 7.11 to calculate the splitting in a double well or lifetime of a metastable state, the coefficients have n-dependence $A_n \sim \frac{1}{n!}$ and the complete series will be summed to obtain an exponential.

The action for the single bounce (7.146) can be calculated using Eq. (7.138b), the change of variables (7.140), and the identities

$$\begin{aligned} \int_{-\infty}^{\infty} \frac{1}{\cosh^2 t} &= 2 \\ \int_{-\infty}^{\infty} \frac{1}{\cosh^4 t} &= \frac{4}{3} \end{aligned} \tag{7.148}$$

with the result that the stationary contribution to Z_k is

$$Z_k \underset{k\to\infty}{\sim} (-1)^k \left(\frac{3}{4}\right)^k k^k e^{-k} . \tag{7.149}$$

As in the case of the simple illustrative integral, the asymptotic behavior is given by an alternating series with coefficients which grow as $k!$, and is therefore Borel summable.

Fluctuation Contributions

In order to evaluate the quadratic fluctuations, we write $x(1) = x_S(t) + \eta(t)$, expand the action in (7.138b) to second order around the stationary bounce trajectory

$$x_S(t) = \sqrt{\frac{3}{2}}\, \frac{1}{\cosh t} , \tag{7.150}$$

and evaluate the fluctuation integral

$$F = \int D(\eta(t))\, e^{-\frac{k}{2}\int_{-\beta/2}^{\beta/2} dt\, dt'\, \eta(t) A(t,t') \eta(t')} \tag{7.151a}$$

where

$$A(t,t') = \left(-\frac{\partial^2}{\partial t^2} + 1 - \frac{6}{\cosh^2 t}\right)\delta(t-t') + \frac{6}{\cosh^3 t\,\cosh^3 t'} \quad . \tag{7.151b}$$

As previously noted, because of the degeneracy with respect to translation of the bounce in time, A has a zero eigenvalue associated with the mode

$$\frac{\partial x_S(t-t_0)}{\partial t_0} = -\sqrt{\frac{3}{2}}\,\frac{\sinh t}{\cosh^2 t} \tag{7.152}$$

which may easily be verified by evaluating $\int dt'\, A(t-t')\frac{\partial x_S(t')}{\partial t_0}$. Hence, we project out the zero mode using Eq. (4.130)and obtain

$$\begin{aligned} F &= \frac{1}{\sqrt{2\pi}}\,\frac{\sqrt{\int_{-\infty}^{\infty} dt\left(\frac{\partial x_S}{\partial t_0}\right)^2}}{\sqrt{\det A_\perp}}\int_{-\beta/2}^{\beta/2} dt_0 \\ &= \frac{\beta}{\sqrt{2\pi}\,\sqrt{\det A_\perp}} \end{aligned} \tag{7.153}$$

where $A_\perp$ denotes the determinant of A in the subspace orthogonal to its zero eigenfunction. There are various techniques for evaluating determinants, one of which is explained in detail in Problem 7.16. The result of combining the stationary contribution of Eq. (7.149) with the quadratic fluctuations is

$$\begin{aligned} Z_k &= (-1)^k \beta \left(\frac{3}{4}\right)^{k+1/2} \frac{\Gamma(k+1/2)}{\pi\Gamma(1/2)} 2^{3/2} \\ &= (-1)^k \beta \sqrt{\frac{6}{\pi^3}} \left(\frac{3}{4}\right)^k \frac{1}{\sqrt{k}}\left(k^k e^{-k}\sqrt{2\pi k}\right) \quad . \end{aligned} \tag{7.154}$$

Finally, the expansion for the ground state energy is obtained by expanding the logarithm in Eq. (7.136b) as follows

$$\begin{aligned} E_0(g) - E_0(0) &= -\lim_{\beta\to\infty} \ln\left(1 + \sum_{k=1}^{\infty} Z_k g^k\right) \\ &= -\lim_{\beta\to\infty}\left[\sum_{k=1}^{\infty} Z_k g^k - \frac{1}{2}\left(\sum_{k=1}^{\infty} Z_k g^k\right)^2 + \frac{1}{3}\left(\sum_{k=1}^{\infty} Z_k g^k\right)^3 + \cdots\right] \\ &\equiv \sum_{k=1}^{\infty} E_k g^k \end{aligned} \tag{7.155a}$$

where the coefficient of g^k is given by

$$E_k = \lim_{\beta\to\infty}\left[Z_k - Z_1 Z_{k-1} + \left(-Z_2 + Z_1^2\right) Z_{k-2} + \left(-Z_3 + 2Z_1 Z_2 - Z_1^3\right) Z_{k-3} + \cdots\right] . \tag{7.155b}$$

Since Z_k grows as $k!$, the dominant term in E_k is Z_k and thus, using Eq. (7.154), the asymptotic behavior of E_k is given by

$$E_k \approx (-1)^{k+1}\left(\frac{6}{\pi^3}\right)^{1/2}\left(\frac{3}{4}\right)^k \frac{k!}{\sqrt{k}}\left(1+O\left(\frac{1}{k}\right)\right) . \tag{7.156}$$

This result is identical to the WKB result of Bender and Wu. Further details of the instanton or bounce calculations for this and related problems are given by Zinn-Justin (1984).

Resummation of the Series

We conclude this section by briefly considering the alternatives available to approximate the energy when one knows exactly a finite number of coefficients E_n of the perturbation expansion. We have already considered two options. One is to calculate the finite sum $\sum_{k=1}^{n} E_n g^n$ as in the case of the illustrative one-dimensional integral and terminate this sum when the contributions begin to increase. The other is to approximate all coefficients beyond $k = N$ by the leading contribution given in Eq. (7.154) and use Borel summation to sum these contributions to all orders as in Fig. 7.6.

Another alternative is to use Padé approximants to represent $E_0(g)$. The $[N, M]$ Padé approximant to a function $f(x) = \sum_{n=0}^{\infty} a_n x^n$ is defined by the ratio of polynomials

$$f^{[N,M]}(x) = \frac{P^{[N,M]}(x)}{Q^{[N,M]}(x)} \tag{7.157}$$

where $P^{[N,M]}$ and $Q^{[N,M]}$ are polynomials of degrees N and M, respectively, such that the first $M + N$ coefficients in the expansion of $f^{[N,M]}(x)$ and $f(x)$ in powers of x agree; that is,

$$f^{[N,M]}(x) - \sum_{n=0}^{N+M} a_n x^n = O\left(x^{N+M+1}\right) . \tag{7.158}$$

For example, the [1, 1] Padé approximant to e^x is given by

$$(e^x)^{[1,1]} = \frac{1+\frac{x}{2}}{1-\frac{x}{2}} . \tag{7.159}$$

A general discussion of Padé approximants is given by Baker (1965).

To illustrate the effectiveness of Padé approximants in extracting the energy from a finite number of terms of a divergent perturbation series, consider the energy of the anharmonic oscillator for $g = 2$. Using the exact coefficients given by Bender and Wu (1969), the first few terms in the expansion of the energy, Eq. (7.155a), are given by

$$E_0(g) = 0.5 + 0.1875\,g - 0.164025\,g^2 + 0.325195313\,g^3 + \cdots \tag{7.160}$$

and the energy for $g = 2$ is $E_0(2) = 0.696175\cdots$. For g as large as 2, the magnitude of the correction at each order increases, so the optimal asymptotic approximation is obtained at first order. The relative error at each order is shown in the first column of

N	$\sum_{k=0}^{N} E_k g^k$	$E^{[N,N]}(g)$	$E_B^{[N,N]}(g)$
1	0.26	8.6×10^{-2}	4.9×10^{-2}
2	0.69	3.2×10^{-2}	4.9×10^{-3}
3	3.1	1.3×10^{-2}	
4	18.6	6.1×10^{-3}	3.1×10^{-4}
6		1.6×10^{-3}	1.7×10^{-5}

Table 7.1 Relative error in three approximations to the ground state energy of the anharmonic oscillator with $g=2$. The first column shows the sum of the first N orders of the perturbation series, the second presents the $[N,N]$ Padé approximant, and the third shows the Padé approximant of the Borel transform defined in Eq. (7.161b). In each case, the magnitude of the relative error $\left|\frac{E-E_{\text{exact}}}{E_{\text{exact}}}\right|$ is tabulated.

Table 7.1, from which we observe that the minimum error is 26% and conclude that simple summation of the asymptotic s eries is not quantitatively useful. In contrast, Loeffel *et al.* (1969) have proved that the $[N,N]$ diagonal Padé approximants converge to the ground state energy and have shown that even low values of N give accurate approximations. The relative errors for their results with $g=2$ are shown in the second column of Table 7.1. An even better resummation is obtained by evaluating the Borel transform

$$B(g)=\sum_k \frac{E_k}{k!}g^k \ , \tag{7.161a}$$

approximating it by the $[N,N]$ Padé approximant, $B^{[N,N]}(g)$, and then performing the inverse Borel transform

$$E_B^{[N,N]}(g)\equiv\int_0^\infty dt\, e^{-t}B^{[N,N]}(gt) \ . \tag{7.161b}$$

The results obtained by Graffi *et al.* (1970) are shown in the third column of Table 7.1. Note that $E_B^{[2,2]}$ requires only four coefficients of the original perturbation series and yields a relative error less than one half percent and that the error may be reduced to the order of 10^{-5}. Although realistic systems have yet to be analyzed to the same extent as the anharmonic oscillator, these resummation techniques may be of more general utility in extracting the physical content of divergent perturbation expansions.

PROBLEMS FOR CHAPTER 7

The first two problems show how the freedom in introducing an auxiliary field may be exploited to obtain a physically relevant mean field approximation. Problems 3 and 4 derive properties of overcomplete sets of states cited in the text. Problem 5 elucidates the structure of the time-dependent Hartree Fock approximation as a classical field theory. Since time dependent mean field theory is the physical foundation underlying the stationary-phase expansions in this chapter, this problem is particularly important and is therefore accentuated with an *. Corrections to the stationary solution at zero temperature are treated in Problems 6 and 7 and Problem 8 outlines the stationary-phase approximation to the partition function at finite temperature. As an alternative to the treatment of transition amplitudes in Section 7.3 using an auxiliary field, Problem 9 shows how to use coherent states. The next two problems apply the general theory of collective excitations and tunneling in Section 7.4 to analytically solvable cases: the RPA and tunneling in a quartic potential. The $\frac{1}{N}$ expansion is discussed in Problem 12 and is of sufficient importance to merit an *. The next three problems treat aspects of the asymptotic behavior of large orders of perturbation theory which were not derived in Section 7.5. Finally, Problem 16 shows how to evaluate determinants in one dimension analytically. This problem is essential for calculating fluctuations such as in Eq. (7.154) and in Problem 11 and therefore is also emphasized with an *.

PROBLEM 7.1 Auxiliary Field Representations. This problem explores the freedom to take combinations of the decomposition of the Hamiltonian (7.2), (7.9), and (7.11) to obtain physical combinations of the Hartree, Fock and pairing fields in the stationary phase approximation.

a) Following the arguments in the text for the Hartree case, show that stationarity of

$$S^F(\sigma) = \frac{1}{2}\int dt\sigma_{\alpha\beta}(-v_{\alpha\gamma\delta\beta})\sigma_{\gamma\delta} - \ln\langle\psi|T\,e^{-i\int dt\left(\tilde{K}_{\alpha\beta}-\sigma_{\gamma\delta}v_{\gamma\alpha\beta\delta}\right)\hat{\rho}_{\alpha\beta}}|\psi\rangle$$

leads to a mean field which contains the Fock exchange term and no direct term. Write the result in coordinate representation for a spin-dependent potential

$$V_{\alpha\beta\gamma\delta} = -\delta(r_\alpha - r_\delta)\delta(r_\beta - r_\gamma)\delta_{\sigma_\alpha\sigma_\delta}\delta_{\sigma_\beta\sigma_\gamma}V(r_\alpha - r_\beta)$$

where σ_α denotes the spin component for state α. Similarly, vary an analogous expression to show that (7.12) yields a pairing mean field.

b) Try to obtain the Hartree-Fock mean field by replacing $v_{\alpha\beta\gamma\delta}$ in (7.2) by the antisymmetrized combination $\tilde{v}_{\alpha\beta\gamma\delta} = \frac{1}{2}v_{\alpha\beta\gamma\delta} - \frac{1}{2}v_{\alpha\beta\delta\gamma}$. Show that whereas this gives an exact representation of the evolution operator, the mean field is one half the normal Hartree-Fock mean field.

c) Suppose we decompose the two body interaction into the sum of direct and exchange terms

$$\hat{v} \equiv \hat{v}^D + \hat{v}^E \tag{1}$$

and require that the Hartree potential generated by $\hat{v}^E$ vanish and the exchange potential generated by $\hat{v}^D$ vanish

$$\sum_k \langle k'k|\hat{v}^E|k''k\rangle = \sum_k \langle k'k|\hat{v}^D|kk''\rangle = 0 \ . \tag{2}$$

Show that the mean field is the Hartree-Fock potential, the stationary phase energy (7.39) is the Hartree-Fock energy, the self-energy term in K vanishes, and the $n = 1$ RPA correction, Eq. (7.48), vanishes.

A simple way to implement Eqs. (1) and (2) for spin saturated systems, that is, systems in which corresponding single particle states are occupied with spin up and spin down Fermions, is to use spin algebra. For a central potential $v(r_1 - r_2)$ show that the following decomposition has the desired properties

$$v^D == \frac{1}{2}(2 - P_S)v(r_1 - r_2) = (1 - \frac{1}{3}\hat{\sigma}_1 \cdot \hat{\sigma}_2)v(r_1 - r_2)$$
$$v^E = \frac{1}{3}(2P_S - 1)v(r_1 - r_2) = \frac{1}{3}\hat{\sigma} \cdot \hat{\sigma}_2\, v(r_1 - r_2)$$

where P_S is the spin exchange operator and $\hat{\sigma}$ denotes a Pauli matrix. When the potential is written in this form, it is clear that the auxiliary field may be expressed in terms of four independent spatial functions

$$\sigma(r_1, \sigma_1; r_2\sigma_2) = \sigma_0(r_1 r_2) + \sum_{i=1}^{3} (\sigma_2|\hat{\sigma}^{(i)}|\sigma_1)\sigma_i(r_1, r_2) \quad .$$

Show that the equations for the spatial components σ_i are identical so there are only two independent functions which generate the direct and exchange components of the mean field.

Since this introduction of spin dependence may appear artificial, it is instructive to consider the relation between an auxiliary field functional integral for nucleons interacting via a static two-body potential and a field theory of mesons coupled to nucleons. Note that the integral $\int \mathcal{D}r\, e^{-i\sigma v \hat{\rho} + \frac{i}{2}\sigma v \sigma}$ looks like a path integral for a scalar field coupled to the Fermion field $\hat{\rho} = \psi^\dagger \psi$ with the free scalar field action $\sigma v \sigma$. Thus, it is natural to associate various components of the σ field with various meson fields. Since the $\sigma \cdot \nabla$ coupling of pions to nucleons generates a $\sigma_1 \cdot \sigma_2$ contribution to the nuclear potential, the pion field is naturally associated with the exchange potential v^E. Similarly, the direct potential v^D is associated with scalar meson exchange.

d) Finally, show that the functional integral

$$W = \int \mathcal{D}\sigma_{\alpha\beta} \mathcal{D}\chi^*_{\alpha\beta} \mathcal{D}\chi_{\alpha\beta}\, e^{\frac{i}{2}\int dT \left(\sigma_{\alpha\beta}\left[v^D_{\alpha\gamma\beta\delta} - v^E_{\alpha\gamma\delta\beta}\right]\sigma_{\gamma\delta} + \chi^*_{\alpha\beta} v^P_{\alpha\beta\gamma\delta}\chi_{\gamma\delta}\right)}$$
$$\times \langle\psi| T\, e^{-i\int dt \left[\left(T_{\alpha\beta} + \sigma_{\gamma\delta}\left(v^D_{\gamma\alpha\delta\beta} - v^E_{\gamma\alpha\beta\delta}\right)\right]\hat{\rho}_{\alpha\beta} + v^P_{\alpha\beta\gamma\delta}\left(\chi^*_{\alpha\beta}\hat{\Delta}_{\gamma\delta} + \hat{\Delta}^\dagger_{\alpha\beta}\chi_{\gamma\delta}\right)} |\psi\rangle$$

produces the Hartree-Fock-Bogoliubov mean field for a spin-saturated system interacting with a central potential in the stationary phase approximation, where

$$\hat{v}^D = (1 - \hat{\sigma}_1^y \hat{\sigma}_2^y)v(r_1 - r_2)$$
$$\hat{v}^E = \frac{1}{2}(\hat{\sigma}_1^y \hat{\sigma}_2^y + \hat{\sigma}_1^z \hat{\sigma}_2^z)v(r_1 - r_2)$$
$$\hat{v}^P = \frac{1}{2}(\hat{\sigma}_1^y \hat{\sigma}_2^y - \hat{\sigma}_1^z \hat{\sigma}_1^z)v(r_1 - r_2) \quad .$$

PROBLEM 7.2 Alternative Auxiliary Field. This problem examines the alternative auxiliary field formulation obtained by writing the evolution operator as a product of terms of the form (7.14a):

$$e^{-iHT} = \frac{1}{\mathcal{N}} \int \prod_{k=1}^{N} \prod_{n} d\sigma_n(k)\, e^{i\frac{\epsilon}{2} \sum_{m,n,k} \sigma_m(k) V_{mn} \sigma_n(k)} \times \prod_k \left[1 - i\epsilon \sum_n \left(K_n + \sum_m \sigma_m W_{mn} \right) \hat{\rho}_n - \frac{\epsilon^2}{2} \left(\sum_{mn} \sigma_m V_{mn} \hat{\rho}_n \right)^2 \right] \tag{1}$$

where $m \equiv (\alpha\gamma)$ and we have included the one-body operator K.

a) Write the stationary-phase equations for a matrix element $\langle \psi | e^{-HT} | \psi \rangle$ with finite ϵ. Take the limit $\epsilon \to 0$ carefully and show that the stationary solution $\sigma^{s}_m(t)$ satisfies

$$\sigma^{s}_m(t) = \frac{\langle \psi | T e^{-i \int dt \sum_n [K_n + \sum_m \sigma^{(s)}_m(t) W_{mn}] \hat{\rho}_n} \hat{\rho}_m(t) | \psi \rangle}{\langle \phi | T e^{-i \int dt [K + \sigma W] \hat{\rho}} | \psi \rangle} \;.$$

Calculate the ground state energy in the stationary-phase approximation as in Section 7.2 and obtain the Hartree-Fock approximation for the choice $W_{\alpha\beta\gamma\delta} = v_{\alpha\beta\gamma\delta} - v_{\alpha\beta\delta\gamma}$.

b) Calculate the quadratic corrections to the static solution for finite ϵ. Show that the limit of this expression as $\epsilon \to 0$ gives direct and exchange ring diagrams. Remember the overcounting problems associated with antisymmetrized matrix elements in ring diagrams discussed in connection with (2.124) and note how the present formalism deals with it.

PROBLEM 7.3 Overcomplete Sets of Coherent States. In the case of Boson or Fermion coherent states, show how Eq. (7.20e) reduces to the functional integral representation of Chapter 2.

PROBLEM 7.4 Closure Relations for Slater Determinants.

a) Derive Eq. (7.24a) by taking its matrix element between $\langle x_1 \dots x_N |$ and $| y_1 \dots y_N \rangle$, and using Wick's theorem in the form

$$\overbrace{\varphi^*_i(x)\varphi_j(y)} = \delta_{ij}\delta(x-y) \;.$$

b) Prove Eq. (7.24b) as follows. Write the δ-functions as

$$\delta(\langle \phi_i | \phi_j \rangle - \delta_{ij}) = \int d\, A_{ij} e^{iA_{ij}\left(\delta_{ij} - \int dx\, \phi^*_i(x)\phi_j(x)\right)}$$

and show that the matrix element of the left-hand side between $\langle x_1 \dots x_N |$ and $| y_1 \dots y_N \rangle$ is equal to

$$\frac{1}{\mathcal{N}} \int \prod_{i,j} \frac{dA_{ij}}{2\pi} e^{i \sum_i A_{ii}} (\det i A_{ij})^{-1} \sum_{P,P',Q} \epsilon_P \epsilon_{P'} \prod_{i=1}^{N} A^{-1}_{iQi} \prod_{i=1}^{N} \delta(x_{Pi} - y_{P'Qi})$$

where P, P', Q are permutations of N elements. Show that this expression is proportional to $\det\delta(x_i x_j)$ and write an expression for the proportionality constant. An alternative derivation is given by Blaizot and Orland (1981).

PROBLEM 7.5* Variational Derivation of the Time-Dependent Hartree-Fock Equations. Consider the action

$$S[\phi^*,\phi] = \int dt \langle \phi_1 \ldots \phi_N | i\frac{\partial}{\partial t} - H | \phi_1 \ldots \phi_N \rangle$$

where $|\phi_1 \ldots \phi_N\rangle$ is a Slater determinant of orthonormal wave functions.

a) Show that the action may be rewritten

$$S = \int dt \left[\sum_\alpha \int dx \phi_\alpha^*(x,t) i\frac{\partial}{\partial t}\phi_\alpha(x,t) - \mathcal{H}(\phi^*(t),\phi(t)) \right]$$

where the Hartree-Fock Hamiltonian density is

$$\begin{aligned}
&\mathcal{H}(\phi^*(t),\phi(t)) \\
&= \sum_\alpha \int dx \phi_\alpha^*(x,t) \frac{-\hbar^2}{2m} \nabla^2 \phi_\alpha(x,t) \\
&+ \sum_{\alpha\beta} \int dx dy \, \phi_\alpha^*(x,t)\phi_\beta^*(y,t) v(x-y) \left(\phi_\alpha(x,t)\phi_\beta(y,t) - \phi_\beta(x,t)\phi_\alpha(y,t) \right) \quad .
\end{aligned}$$

By variation with respect to ϕ_α^* and ϕ_α, obtain the Euler-Lagrange equations

$$i\frac{\partial \phi_\alpha}{\partial t} = \frac{\partial \mathcal{H}}{\partial \phi_\alpha^*} \qquad i\frac{\partial \phi_\alpha^*}{\partial t} = -\frac{\partial \mathcal{H}}{\partial \phi_\alpha} \quad .$$

Note that these equations are of Hamiltonian form with ϕ_α and $i\phi_\alpha^*$ playing the role of conjugate coordinates and momenta. Write out $\frac{\partial \mathcal{H}}{\partial \phi_\alpha^*}$ explicitly to obtain the time-dependent Hartree-Fock equation of motion for $\phi_\alpha(x,t)$

b) Show that the overlap matrix of the single-particle wave functions is a constant of the motion

$$\frac{d}{dt}\int dx \, \phi_\alpha^*(x,t)\phi_\beta(x,t) = 0 \quad .$$

Show that the time dependence of any operator O having no intrinsic time dependence may be written

$$i\frac{d}{dt}\langle \phi_1 \ldots \phi_n | O | \phi_1 \ldots \phi_N \rangle = \sum_\alpha \int dr \left[\frac{\delta\langle O\rangle}{\delta\phi_\alpha(r)} \frac{\delta \mathcal{H}}{\delta \phi_\alpha^*(r)} - \frac{\delta\langle O\rangle}{\delta\phi_\alpha^*(r)} \frac{\delta \mathcal{H}}{\delta \phi_\alpha(r)} \right] \quad .$$

Hence, conclude that the time-dependent Hartree-Fock energy $\langle H\rangle = \mathcal{H}$ is conserved. Also, show that any one-body operator which commutes with H is conserved

c) In applying the stationary-phase approximation to the functional integral (7.25a) with the orthonormal measure (7.25b), we may enforce the orthonormality constraint by introducing Lagrange multipliers and varying the modified action

$$S' = S - \sum_{\alpha\beta} \int dt \int dx \phi_\alpha^*(x,t)\phi_\beta(x,t) A_{\alpha\beta}(t) \quad .$$

Write the resulting equations of motion and the initial conditions satisfied by ϕ and ϕ^*. Now consider the special case in which $\{\phi_{\alpha i}\}$ are a set of orthonormal wave functions and $\{\phi_{\alpha f}\}$ are the wave functions obtained by evolving $\{\phi_{\alpha i}\}$ by the time-dependent Hartree-Fock equations of (a) for time T. Show what if we set $A_{ij}(t) = 0$, then

$$\frac{\partial}{\partial t}\langle\phi_\alpha(t)|\phi_\beta(t)\rangle = 0$$

and the wave functions remain orthonormal in time. Thus, conclude that for these boundary conditions, the stationary phase approximation yields the time-dependent Hartree-Fock equations.

d) Write the resolvent (7.28a) using the functional integral in (b) and look for time-independent orthonormal stationary solutions. Show that these solutions satisfy the static Hartree-Fock equations and yield poles in the resolvent at the Hartree-Fock energies.

PROBLEM 7.6 Quadratic Corrections for Slater Determinant Path Integrals. In Problem 7.5, it was shown that calculating the resolvent using the Slater determinant path integral (7.25a) yields the Hartree Fock energy in the stationary-phase approximation. Now, calculate the quadratic fluctuations around the stationary solution and obtain the direct and exchange ring diagrams with Hartree-Fock propagators. Note that it is simplest to work with orthonormal determinantal wave functions.

PROBLEM 7.7 The Loop Expansion and Feynman Diagrams. To understand the loop expansion in detail, it is useful to show explicitly how the diagrams contributing to Eq. (7.54) reproduce all the signs and factors occurring in the standard Feynman diagram expansion for the ground state energy.

a) First, enumerate the linked Feynman diagrams contributing to the ground state energy through third order. The zero temperature diagram rules are summarized in Table 3.1 and the diagrams and symmetry factors are given in Figures 2.4 and 2.5.

b) Collect the diagrams from $E^{(2)}$ of the loop expansion, Fig. 7.1, and expand out the RPA propagators in (7.55a) and (7.56) to identify all the terms through third order. Carefully evaluate the factors associated with each term and show that the results in (a) are exactly reproduced.

PROBLEM 7.8 Stationary Phase Approximation at Finite Temperature. Consider a grand canonical ensemble of Fermions, with chemical potential μ, at temperature T. The grand partition function of the system is given by:

$$Z = \int_{\varphi(r,\beta)=-\varphi(r,0)} \mathcal{D}\left(\varphi^*(r,t),\varphi(r,t)\right) e^{-\int_0^\beta dr dr\, \varphi^*(r,t)\left(\frac{\partial}{\partial t}-\frac{\hbar^2}{2m}\nabla^2-\mu\right)\varphi(r,t)}$$

$$\times\, e^{-\frac{1}{2}\int_0^\beta dr dx dy |\varphi(x,t)|^2 v(x-y)|\varphi(y,t)|^2} .$$

a) Using an auxiliary field to eliminate the quartic term, show that:

$$Z = \int \frac{\mathcal{D}U(r,t)}{\mathcal{N}} e^{\frac{1}{2}\int_0^\beta dt dr dr'\, U(r,t)v^{-1}(r-r')U(r',t)+\ln\det\left(\frac{\partial}{\partial t}-\frac{\hbar^2}{2m}\nabla^2-\mu+U(r,t)\right)}$$

where the determinant is evaluated on a set of antiperiodic functions, of antiperiod β, and

$$\mathcal{N} = \int \mathcal{D}U(r,t) e^{\frac{1}{2}\int_0^\beta drdrdt'\, U(r,t)v^{-1}(r-r')U(r',t)} \; .$$

b) We will perform a stationary phase expansion, considering only time-independent solutions U. Show that for a time-independent U_o, the action is given by

$$S_0 = \frac{\beta}{2}\int drdr'\, U_0(r)v^{-1}(r-r')U_0(r') + \sum_\alpha \ln\left(1 + e^{-\beta(\epsilon_\alpha - \mu)}\right)$$

where ϵ_α are the eigenvalues of the equation:

$$\left(-\frac{\hbar^2}{2m}\nabla^2 + U_0(r)\right)\varphi_\alpha(r) = \epsilon_\alpha\varphi_\alpha(r)$$

and the φ_α are normalized to one.

c) Show that the stationarity condition on S_0 reduces to the finite temperature Hartree equations:

$$\left(-\frac{\hbar^2}{2m}\nabla^2 + \sum_{\alpha'}\int dr'\, v(r-r')\left|\varphi_{\alpha'}(r')\right|^2 f_{\alpha'}\right)\varphi_\alpha(r) = \epsilon_\alpha\varphi_\alpha(r)$$

where f_α is the occupation number of state α:

$$f_\alpha = \frac{e^{-\beta(\epsilon_\alpha - \mu)}}{1 + e^{-\beta(\epsilon_\alpha - \mu)}} \; .$$

Note the additional self-consistency requirements which occur at finite temperature.

d) Assuming that the system is translationally invariant, we make the *ansatz* that plane waves are Hartree solutions. Verify this assumption, by showing that

$$\varphi_{\vec{k}}(\vec{r}) = \frac{1}{\sqrt{\Omega}}e^{i\vec{k}\vec{r}}$$

$$\epsilon_{\vec{k}} = \frac{\hbar^2}{2m}\vec{k}^2 + \rho V_0$$

is a solution, where $V_0 = \int d^3r\, v(\vec{r})$, and ρ is the average density of particles. Calculate the stationary action S_0.

e) We want to evaluate the RPA corrections to the Hartree approximation. Show that the contribution of the quadratic fluctuations is given by

$$e^{-\frac{1}{2}\ln\det\left(\delta(r-r')\delta(t-t') - \int d\bar{r}\, v(r-\bar{r})G_0(\bar{r}t,r't')G_0(r't',\bar{r}t)\right)}$$

where G_0 is the Hartree propagator, given by

$$\left(\frac{\partial}{\partial t} - \frac{\hbar^2}{2m}\nabla^2 - \mu + \rho V_0\right)G_0(rt|r't') = \delta(r-r')\delta(t-t')$$

with antiperiodic boundary conditions.

f) In order to evaluate the quadratic fluctuations, one has to diagonalize the operator:

$$A(rt|r't') = \delta(r-r')\delta(t-t') - \int d\bar{r} \cdot v(r-r')G_0(\bar{r}-r'|t-t')G_0(t'-\bar{r}|t'-t) \ .$$

Show that A is periodic in time, with period β, and that it can be expanded in a Fourier series as:

$$A(r-r'|t-t') = \int \frac{d^3k}{(2\pi)^3} \sum_n e^{i\vec{k}(\vec{r}-\vec{r}')-i\omega_n(t-t')} A(\vec{k},\omega_n)$$

where

$$A(\vec{k},\omega_n) = 1 + v(\vec{k}) \int \frac{d^3k'}{(2\pi)^3} \cdot \frac{f(\vec{k}') - f(\vec{k}-\vec{k}')}{i\omega_n - \Delta_{k\,k'}}$$

with $\omega_n = \frac{2n\pi}{\beta}$ and $\Delta_{k\,k'} = \epsilon_{\vec{k}'} - \epsilon_{\vec{k}-\vec{k}'}$. Symbolically, this equation may be written

$$A = 1 - \beta V F G$$

where

$$F(\vec{k},\vec{k}') = f(\vec{k}') - f(\vec{k}-\vec{k}')$$

and

$$\beta G(\vec{k},\vec{k}'; \omega_n) = -\frac{1}{i\omega_n - \Delta_{kk'}} \ .$$

Using the factorization:

$$A = \left((\beta G)^{-1} - VF\right)\beta G$$

show that

$$A = \left(\frac{\partial}{\partial t} - \Delta - VF\right)\left(\frac{\partial}{\partial t} - \Delta\right)^{-1}$$

the operator A being defined with periodic boundary conditions.

g) We now evaluate the determinant

$$\det A = \frac{\det\left(\frac{\partial}{\partial t} - \Delta - VF\right)}{\det\left(\frac{\partial}{\partial t} - \Delta\right)}$$

by diagonalizing the operators in the numerator and demoninator. Let $|\xi_\nu\rangle$ be the eigenfunctions of the operator $\Delta + VF$ with eigenvalue E_ν. Write down explicitly the eigenvalue equation for $|\xi_\nu\rangle$. These are the temperature dependent RPA equations. Show that $e^{i2\pi n\frac{t}{\beta}}|\xi_\nu\rangle$ is an eigenvector of the operator $\frac{\partial}{\partial t} - (\Delta + VF)$, with the proper boundary condition, and with eigenvalue $E_\nu + \frac{2ni\pi}{\beta}$. Show that:

$$\det A = \frac{\prod_\nu \prod_{n=-\infty}^{+\infty}\left(E_\nu + \frac{2ni\pi}{\beta}\right)}{\prod_{(k,k')} \prod_{n=-\infty}^{+\infty}\left(\epsilon_k - \epsilon_{k'} + \frac{2ni\pi}{\beta}\right)} \ .$$

Using Euler's formula:

$$\frac{\sinh \varsigma}{\varsigma} = \prod_{n=1}^{\infty} \left(1 + \frac{\varsigma^2}{n^2\pi^2}\right)$$

show that

$$\det A = \frac{\prod_\nu \sinh\left(\frac{\beta E_\nu}{2}\right)}{\prod_{(k,k')} \sinh \frac{\beta}{2}(\epsilon_k - \epsilon_{k'})} \ .$$

h) Give a physical interpretation of this expression in terms of harmonic oscillators. Show that these expressions reduce to the usual RPA in the limit when β goes to infinity.

PROBLEM 7.9 Transition Amplitude for the Forced Harmonic Oscillator. An alternative to the treatment of transition amplitudes in Section 7.3 is to use coherent states. The basic idea is illustrated by the case of the forced harmonic oscillator described by the Hamiltonian

$$H = \omega a^\dagger a - \gamma(t) a^\dagger - \gamma^*(t) a \ .$$

Using the coherent states of Chapter 1 and coherent state path integral of Chapter 2, show that the transition matrix element between an initial harmonic oscillator eigenstate $|n_i\rangle$ and final state $|n_f\rangle$ may be written

$$\begin{aligned}\langle n_f|e^{-iH(t_f-t_i)}|n_i\rangle &= \int \frac{d\phi_f^* d\phi_f}{2\pi i} \int \frac{d\phi_i^* d\phi_i}{2\pi i} \langle n_f|\phi_f\rangle e^{-\phi_f^*\phi_f} \\ &\quad \times \langle \phi_f|e^{-iH(t_f-t_i)}|\phi_i\rangle e^{\phi_i^*\phi_i} \langle \phi_i|n_i\rangle \\ &= \int \frac{d\phi_f^* d\phi_f}{2\pi i} \int \frac{d\phi_i^* d\phi_i}{2\pi i} \frac{1}{\sqrt{n_f n_i}} e^{n_f \ln \phi_f + n_i \ln \phi_i^* - \phi_f^*\phi_f - \phi_i^*\phi_i + S(\phi_f^*\phi_i)}\end{aligned}$$

where

$$e^{S(\phi_f^*\phi_i)} = \int_{\phi(t_i)=\phi_i}^{\phi^*(t_f)=\phi_f^*} \mathcal{D}\phi^*\phi \, e^{\phi^*(t_f)\phi(t_f) + i\int_{t_i}^{t_f} dt \left[i\phi^*(t)\frac{\partial}{\partial t}\phi(t) - H(\phi^*(t)\phi(t))\right]} \ .$$

Now, evaluate the integrals over ϕ^* and ϕ in the stationary-phase approximation. For t between t_i and t_f, derive the equations of motion

$$i\dot{\phi}(t) = \frac{\partial H}{\partial \phi^*} = \omega\phi(t) - \gamma(t)$$

and

$$-i\dot{\phi}^*(t) = \frac{\partial H}{\partial \phi} = \omega\phi^*(t) - \gamma^*(t)$$

and analytic solutions

$$\phi(t) = \phi_i \, e^{i\omega(t_i - t)} + ie^{-i\omega t} \int_{t_i}^{t} e^{i\omega t'} \gamma(t') dt'$$

$$\phi^*(t) = \phi_f^* \, e^{i\omega(t - t_f)} + ie^{i\omega t} \int_{t}^{t_f} e^{-i\omega t'} \gamma^*(t') dt' \ .$$

Hence, show

$$\phi_f = \phi_i e^{i\omega(t_i - t_f)} + ie^{-i\omega t_f}\Gamma \tag{1}$$

and

$$\phi_i^* = \phi_f^* e^{i\omega(t_i - t_f)} + ie^{i\omega t_i}\Gamma^* \tag{2}$$

where

$$\Gamma = \int_{t_i}^{t_f} e^{i\omega t'}\gamma(t')dt' \ .$$

Show that stationarity of the action with respect to ϕ_f and ϕ_i^* yield the condition

$$n_f = \phi_f^*\phi_f \tag{3}$$
$$n_i = \phi_i^*\phi_i \ . \tag{4}$$

The four conditions (1–4) fully specify the stationary solutions. Note that because we have applied the stationary-phase approximation, ϕ^* need not be the complex conjugate of ϕ. Show that for $\left(\sqrt{n_i} - |\Gamma|\right)^2 < n_f < \left(\sqrt{n_i} + |\Gamma|\right)^2$, there are two conjugate stationary solutions which interfere whereas outside that domain there are only exponentially decaying solutions.

The structure of this harmonic oscillator example is representative of the general structure of the analogous coherent state path integral for the many-body problem. Time dependent mean field equations of the form $i\dot{\phi}_k(t) = \frac{\partial H}{\partial \phi_k^*}$ evolve the $\{\phi_k\}$ from t_i to t_f, as in (1), the quantization condition for the final state relates $\{\phi_k(t_f)\}$ to $\{\phi_k^*(t_f)\}$ as in (3), the time-dependent mean field equations $-i\dot{\phi}_k^* = \frac{\partial H}{\partial \phi_k}$ evolve the $\{\phi_k^*\}$ from t_f to t_i as in (2) and finally the quantization on the initial state closes the loop by relating $\{\phi_k(t_i)\}$ to $\{\phi_k^*(t_i)\}$ as in (4).

PROBLEM 7.10 RPA Limit for Quantized Eigenstates. Show that the small-amplitude limit of this theory for eigenstates of collective motion reproduces the Random Phase Approximation of Section 5.4 as follows.

Write a solution to the self-consistent periodic mean field equations (7.96) which differs infinitesimally from the static Hartree wave function as follows.

$$\phi_K(x,t) = e^{-i\epsilon_K t}\left[\psi_K(x) + \sum_m C_{mK}\psi_M(x)\right] + \sum_M Q_{MK}\psi_M(x) \tag{1}$$

where ϵ_K and $\psi_K(x)$ denote Hartree eigenvalues and eigenfunctions, upper case and lower case state labels denote occupied and unoccupied states, respectively, and the expansion coefficients C_{mK} are assumed small. Show that orthornomality of the $\{\phi_K(x,t)\}$ requires

$$Q_{MK} = -\frac{1}{2}\, e^{-i\epsilon_M t}\sum_m C_{mM}^* C_{mK} + O(C^3) \ . \tag{2}$$

Write C_{mK} in the conventional form

$$C_{mK} \equiv X_{mK}\, e^{-i\omega t} + Y_{mK}^*\, e^{i\omega t}$$

and show that linearization of the equation of motion yields the RPA matrix equation

$$\begin{pmatrix} A & B \\ B^* & A^* \end{pmatrix} \begin{pmatrix} X \\ Y \end{pmatrix} = \omega \begin{pmatrix} X \\ -Y \end{pmatrix}$$

where

$$\begin{aligned} A_{mM,\,nN} &= (\epsilon_m - \epsilon_M)\delta_{mn}\delta_{MN} + \langle mN|V|Mn\rangle \\ B_{mM,\,nN} &= \langle mn|V|MN\rangle \ . \end{aligned}$$

Denoting the RPA eigenvalues and eigenvectors by ω_ν and $(X^{(\nu)}, Y^{(\nu)})$, show that the period $\omega_\nu T = 2\pi$ and that $\alpha_K = \epsilon_K T$.

Show that the quantization condition (7.103) yields the RPA normalization condition

$$\sum_{ab} \left(\left| X^{(\nu)}_{bB} \right|^2 - \left| Y^{(\nu)}_{bB} \right|^2 \right) = m$$

and that the corresponding energy of the state is

$$E^{(\nu)}_{\mathrm{RPA}} = E_H + m\omega_\nu \ .$$

Note that since both the quantization integral and energy involve terms of $O(C^2)$ one must use the condition (2) in Eq. (1) and consistently carry all terms through $O(C^2)$.

PROBLEM 7.11 Dilute Instanton Gas Approximation. This problem develops an alternative to the treatment of tunneling in Section 7.4.

a) Consider a symmetric double well $V(x)$ of the form shown in Fig. 7.2 and let the positions of the minima be denoted $\pm x_0$. The Feynman path integral for the propagator is expanded in the stationary-phase approximation as

$$\langle x_f|e^{-\frac{1}{\hbar}TH}|x_i\rangle = Ne^{-\frac{1}{\hbar}S[\overline{x}(\tau)]}\left[\det\left(-\partial_\tau^2 + V''(\overline{x})\right)\right]^{-1/2}\left(1 + O\left(\frac{1}{\hbar}\right)\right)$$

where $\overline{x}(\tau)$ denotes a stationary solution. As in Section 7.5, in addition to the exact single-instanton solution for large β, approximate stationary solutions are obtained by joining together sequences of N instantons and antiinstantons and assuming that they are well-separated (*i.e.* dilute). Show that the one-instanton contribution is

$$\langle -x_0|e^{-\frac{1}{\hbar}TH}|x_0\rangle\Big|_{\text{one-instanton}} = \left(\frac{\omega}{\pi\hbar}\right)^{1/2} e^{-\frac{1}{2}\omega T} e^{-S_0/\hbar} KT$$

where

$$K = \left(\frac{S_0}{2\pi\hbar}\right)^{1/2} \left[\frac{\det\left(-\frac{\partial^2}{\partial\tau^2} + \omega^2\right)}{\det'\left(-\frac{\partial^2}{\partial\tau^2} + V''(\overline{x})\right)}\right]$$

$\omega^2 = V''(\pm x_0)$, S_0 is the action for the instanton, and det′ is the determinant after projection of the zero-energy translational mode.

Sum over all numbers of instantons to obtain

$$\langle \pm x_0 | e^{-\frac{1}{\hbar}TH} | x_0 \rangle = \left(\frac{\omega}{\pi\hbar} \right)^{1/2} e^{-\frac{1}{2}\omega T} \frac{1}{2} \left[e^{TKe^{-\frac{1}{\hbar}S_0}} \pm e^{-TKe^{-\frac{1}{\hbar}S_0}} \right] .$$

By comparing with the exact result

$$\langle \pm x_0 | e^{-\frac{1}{\hbar}TH} | x_0 \rangle = \sum_n \langle \pm x_0 | n \rangle \langle n | x_0 \rangle e^{-\frac{1}{\hbar}E_n T}$$

show that the two lowest eigenvalues in the dilute instanton gas approximation are

$$E\pm = \frac{1}{2}\hbar\omega \pm \hbar K e^{-S_0/\hbar} .$$

Compare this result with Eq. (7.85). Note that the treatment of the zero-point energy is different in the two expressions. The classical action S_0 is defined for a trajectory from the bottom of one well to the bottom of the other well, whereas $\frac{1}{2}W_2(E_n^0)$ is the action for a trajectory joining the classical turning points at finite energy. Also note the difference in premultiplying factors.

b) Now, consider the specific case of a particle of mass m in the potential $V(x) = (x^2 - 1)^2$. Show that the exact instanton solutions are given by $\overline{x}(t) = \tanh \gamma t$ where $\gamma = \sqrt{\frac{2}{m}}$ and that $S_0 = \frac{4}{3}\sqrt{2m}$ and $K = \frac{8}{\sqrt{\pi}} \left(\frac{2}{m}\right)^{1/4}$. One method for evaluating the ratio of determinants in K is presented in Problem 7.16.

c) Now use the same method to treat the case of fission using an auxiliary field path integral. Show that the contribution of a single bounce to the evolution operator is

$$\langle \phi_H | e^{-HT} | \phi_H \rangle \Big|_{\text{one-bounce}} = e^{-E_0 T} K \int d\tau \, e^{-\int_{-T/2}^{T/2} d\tau \int dr \sum_k \tilde{\phi}_k^{B}(r,-\tau) \frac{\partial}{\partial \tau} \tilde{\phi}_k^{B}(r,\tau)}$$

where

$$K = \frac{1}{2} a \left[\frac{\det \left(\frac{\delta^2 S(\sigma_H)}{\delta \sigma^2} \right)}{\det' \left(\frac{\delta^2 S(\sigma_B)}{\delta \sigma^2} \right)} \right]^{1/2} .$$

det′ denotes the determinant omitting the zero eigenvalue, σ_H and σ_B denote the static Hartree and time dependent bounce solutions, respectively, a is a factor obtained from projecting out the zero mode, the factor 1/2 arises from integrating over only one-half of the Gaussian peak in applying the steepest descent approximation to the variable corresponding to the negative eigenvalue of $\frac{\delta^2 S(\sigma_B)}{\delta \sigma^2}$, and K is pure imaginary due to the single negative eigenvalue. Finally, sum over all numbers of instantons to obtain

$$\operatorname{Im} E = \frac{\Gamma}{2} = |K| e^{-\int dt \int dr \sum_k \tilde{\phi}_k^{B}(r,-\tau) \tilde{\phi}_k^{B}(r,\tau)} .$$

PROBLEM 7.12* 1/N Expansion for the One-Dimensional δ-Function Problem. To explore the nature of the expansion parameter when the stationary-phase approximation is applied to a functional integral over an auxiliary field, it is instructive to reexamine the $1/N$ expansion discussed in Problem 3.5.

a) Consider the trace of the evolution operator

$$\operatorname{tr} e^{-i\beta H} = \int \mathcal{D}(\xi^*_\alpha \xi_\alpha)\, e^{i\int dt\,dx \sum_\alpha \xi^*_\alpha(x,t)\left[i\frac{\partial}{\partial t}+\frac{1}{2}\frac{d^2}{dx^2}+\mu-\frac{g}{2}\sum_\beta \xi^*_\beta(x,t)\xi_\beta(x,t)\right]\xi_\alpha(x,t)} \tag{1}$$

where α denotes one of the $(2S+1)$ spin projections and, for the moment, ξ_α denotes a complex field for a Boson coherent state. Show that by introducing an auxiliary field, performing the Gaussian integral over $\xi^*_\alpha \xi_\alpha$, and rescaling the coupling constant and auxiliary field according to $g' \equiv gN$ and $\sigma'(x,t) = \sigma(x,t)/N$ that

$$\begin{aligned}\operatorname{tr} e^{-i\beta H} &= \int \mathcal{D}(\sigma)\mathcal{D}(\xi^*_\alpha \xi_\alpha)\, e^{i\int dt\,dx\, g\sigma^2(x,t)} \\ &\quad\times e^{i\int dt\,dx \sum_\alpha \xi^*_\alpha(x,t)\left[i\frac{\partial}{\partial t}+\frac{1}{2}\frac{d^2}{dx^2}+\mu-g\sigma(x,t)\right]\xi_\alpha(x,t)} \\ &= \int \mathcal{D}(\sigma')e^{N\left[\ln\det\left(\frac{\partial}{\partial t}-i\left(\frac{1}{2}\frac{d^2}{dx^2}+\mu-g'\sigma'\right)\right)+\frac{i}{2}\int dt\,dx g'\left(\sigma'(x,t)\right)^2\right]}\ .\end{aligned} \tag{2}$$

Note that since N appears as an explicit multiplicative factor in the exponent, the stationary-phase approximation yields an expansion in $1/N$.

b) Use the relation $\frac{\delta}{\delta A_{ij}} \det A = A^{-1}_{ji} \det A$ to show that stationarity of the exponent in (2) yields

$$\begin{aligned}\sigma'(x,t) &= -i\left[i\frac{\partial}{\partial t}+\frac{1}{2}\frac{d^2}{dx^2}+\mu-g'\sigma'\right]^{-1} \\ &= iG(xt,xt^+) = \rho(x)\ .\end{aligned} \tag{3}$$

Identify the stationary contribution and leading $\frac{1}{N}$ corrections with diagrams in (7.50b). Note that classification of diagrams according to N^{C-I}, where C is the number of closed particle propagators and I is the number of interactions, follows immediately from the fact that the coupling constant scales as N^{-1} and the Green's functions scales as N.

c) Repeat the preceding analysis for Fermions, in which case $\{\xi^*_\alpha \xi_\alpha\}$ are Grassmann variables. Note that minus signs arise both from the Gaussian integral and from the definition of the Green's functions.

d) Finally, consider the case of a general finite-range potential. Formally define $v'_{\alpha\beta\gamma\delta} = v_{\alpha\beta\gamma\delta}N$ and repeat the analysis in Eqs. (2) and (3). Whereas the argument was exact for the δ-function because it had no length scale, to make the $1/N$ expansion for a finite-range interaction one must consider a hypothetical class of theories in which the interaction strength varies with N and have a physical reason to believe that the limit as $N \to \infty$ is meaningful.

PROBLEM 7.13 Corrections to the Leading Behavior of High Orders of Perturbation Theory. Consider the integral

$$Z_k = \frac{(-1)^k}{k!4^k}e^{2k\ln k}\int_{-\infty}^{\infty}\frac{dx}{\sqrt{2\pi}}e^{-k\left(\frac{x^2}{2}-2\ln x^2\right)}$$

which appears in Eq. (7.119a) for the expansion of the simple illustrative integral. Expand the exponent around the saddle point $x_s = 2$ and show that

$$Z_k = \frac{(-1)^k e^{k \ln 4k - k}}{\sqrt{\pi k}} e^{\{\text{sum of all connected graphs}\}}$$

where the connected graphs have n-point vertices for all $n \geq 3$ with value $\frac{4(-1)^{n+1}}{n 2^n k^{\frac{n}{2}-1}}$ connected by propagators with value $\frac{1}{2}$. Thus, evaluate the corrections to Z_k of order $\frac{1}{k}$ and $\frac{1}{k^2}$.

PROBLEM 7.14 Consider the functional integral (7.138b) for the coefficient Z_k in the expansion of the partition function for the anharmonic oscillator. Show that $x_s = \pm 2/\sqrt{\beta}$ is a stationary solution. By expanding around either solution, show that these solutions are not minima of the action and are thus unstable.

PROBLEM 7.15 Use Eq. (7.144) to show that the energy of the bounce solution for the anharmonic oscillator approaches zero as $e^{-\beta}$ when $\beta \to \infty$. First, evaluate the turning points $q_\pm(E)$ as $E \to 0$. Then, rescale the integration variable in Eq. (7.114) according to $q = \sqrt{-2E}\, x$ and take the limit $E \to 0^-$ to show that $\frac{\beta}{2} \underset{E\to 0^-}{\longrightarrow} \ln \frac{1}{\sqrt{-E}}$.

PROBLEM 7.16* Evaluation of Determinants. The determinants arising from fluctuations around a stationary solution in one dimension may be evaluated analytically. The method presented in this problem follows that of Coleman (1977).

The Fredholm determinant of an operator $A(t, t')$ is the product of the eigenvalues of A normalized so as to render it finite. We will evaluate det A by considering the quantity

$$D(\lambda) \equiv \frac{\det(A - \lambda)}{\det(A_0 - \lambda)} \tag{1}$$

where A_0 is a reference operator.

a) First, consider the case of a first-order differential operator

$$A = \frac{\partial}{\partial t} + h(t)$$

$$A_0 = \frac{\partial}{\partial t} + h_0(t)$$

with periodic or antiperiodic boundary conditions. Let $\psi_n(t)$ and λ_n be the eigenfunctions and eigenvalues of A

$$\left(\frac{\partial}{\partial t} + h(t)\right) \psi_n(t) = \lambda_n \psi_n(t)$$

with boundary condition

$$\psi_n\left(\frac{\beta}{2}\right) = \varsigma \psi_n\left(-\frac{\beta}{2}\right)$$

and denote the corresponding eigenfunctions for A_0 by $\psi_n^{(0)}$ and $\lambda_n^{(0)}$. As usual, ς is $+1$ for the periodic case and -1 for antiperiodic boundary conditions.

i) Show that $D(\lambda)$ is a meromorphic function with simple zeros at $\lambda = \lambda_n$ and simple poles at $\lambda = \lambda_n^{(0)}$ and that $D(\lambda) \underset{|\lambda|\to\infty}{\longrightarrow} 1$ except along the positive real axis.

ii) Now consider the ratio

$$\Delta(\lambda) = \frac{\psi_\lambda\left(\frac{\beta}{2}\right) - \varsigma}{\psi_\lambda^{(0)}\left(\frac{\beta}{2}\right) - \varsigma}$$

where $\psi_\lambda(t)$ is the solution of the equation

$$\left(\frac{\partial}{\partial t} + h(t)\right)\psi_\lambda(t) = \lambda\psi_\lambda(t)$$

with the boundary condition

$$\psi_\lambda\left(-\frac{\beta}{2}\right) = 1$$

and $\psi_\lambda^{(0)}(t)$ is the corresponding solution for A_0. Note that $\psi_\lambda(t)$ is an eigenfunction with eigenvalue λ when $\psi_\lambda\left(\frac{\beta}{2}\right) = \varsigma$. Thus, show that $\Delta(\lambda)$ is a meromorphic function with the same poles and zeros as $D(\lambda)$ and that $\Delta(\lambda) \underset{|\lambda|\to\infty}{\longrightarrow} 1$ except along the positive real axis. Thus, conclude that $\frac{D(\lambda)}{\Delta(\lambda)}$ is an analytic function that goes to 1 as $\lambda \to \infty$ in any direction except along the positive real axis and therefore is equal to 1 so that

$$\frac{\det\left(\frac{\partial}{\partial t} + h(t)\right)}{\det\left(\frac{\partial}{\partial t} + h_0(t)\right)} = \frac{\psi_{\lambda=0}\left(\frac{\beta}{2}\right) - \varsigma}{\psi_{\lambda=0}^{(0)}\left(\frac{\beta}{2}\right) - \varsigma} .$$

iii) Observe that the first order equation for $\psi_\lambda(t)$ may be integrated directly to obtain $\psi_\lambda(t) = e^{-\int_{-\frac{\beta}{2}}^{t} d\tau (h(\tau)-\lambda)}$ so that

$$\frac{\det\left(\frac{\partial}{\partial t} + h(t)\right)}{\det\left(\frac{\partial}{\partial t} + h_0(t)\right)} = \frac{e^{-\int_{-\frac{\beta}{2}}^{\frac{\beta}{2}} d\tau\, h(\tau)} - \varsigma}{e^{-\int_{-\frac{\beta}{2}}^{\frac{\beta}{2}} d\tau\, h_0(\tau)} - \varsigma} . \tag{2}$$

Also, calculate the ratio of determinants by explicitly evaluating the products of eigenvalues. Show that the eigenvalues may be written

$$\lambda_n = \frac{2\pi i}{\beta}\begin{Bmatrix} n \\ n+\frac{1}{2} \end{Bmatrix} + \frac{1}{\beta}\int_{-\frac{\beta}{2}}^{\frac{\beta}{2}} d\tau\, h(\tau)$$

where the integer and half-integer values correspond to $\varsigma = 1$ and -1, respectively. Evaluate the products using the Euler formulas

$$\sinh x = x \prod_{k=1}^{\infty} \left(1 + \frac{x^2}{k^2\pi^2}\right)$$

$$\cosh x = \prod_{k=0}^{\infty} \left(1 + \frac{4x^2}{(2k+1)^2\pi^2}\right) .$$

The fact that the result obtained in this way differs from Eq. (2) by a multiplicative constant demonstrates the need to define the regularization of the Fredholm determinant carefully. One option is to replace the differential equation by a difference equation with N mesh points and take the limit as $N \to \infty$.

b) Now, treat the case of a second order differential operator with zero boundary conditions at $\pm\frac{\beta}{2}$.

$$A = -\frac{\partial^2}{\partial r^2} + W(t) .$$

The eigenfunctions of A satisfy

$$\left(-\frac{\partial^2}{\partial t^2} + W(t)\right) \psi_n(t) = \lambda_n \psi_n(t)$$

with boundary conditions

$$\psi_n\left(\frac{\beta}{2}\right) = \psi_n\left(-\frac{\beta}{2}\right) = 0$$

and we define the function $\psi_\lambda(t)$ satisfying

$$\left(-\frac{\partial^2}{\partial t^2} + W(t)\right) \psi_n(t) = \lambda \psi_n(t)$$

with boundary conditions

$$\psi_\lambda\left(-\frac{\beta}{2}\right) = 0 \qquad \frac{\partial}{\partial t}\psi_\lambda\left(-\frac{\beta}{2}\right) = 1 .$$

With analogous definitions for $A^{(0)}$, we now compare $D(\lambda)$ defined in Eq. (1) with the ratio

$$\Delta(\lambda) = \frac{\psi_\lambda\left(\frac{\beta}{2}\right)}{\psi_\lambda^0\left(\frac{\beta}{2}\right)} .$$

By arguments analogous to those in parts i) and ii) of a), show that $D(\lambda) = \Delta(\lambda)$ and thus

$$\frac{\det\left(-\frac{\partial^2}{\partial t^2} + W(t)\right)}{\det\left(-\frac{\partial^2}{\partial t^2} + W_0(t)\right)} = \frac{\psi_{\lambda=0}\left(\frac{\beta}{2}\right)}{\psi_{\lambda=0}^{(0)}\left(\frac{\beta}{2}\right)} .$$

c) Next, treat the case in which the second-order operator A has an eigenfunction ψ_0 with zero eigenvalue

$$\left(-\frac{\partial^2}{\partial t^2} + W(t)\right)\psi_0(t) = 0$$

with

$$\psi_0\left(-\frac{\beta}{2}\right) = \psi_0\left(\frac{\beta}{2}\right) = 0$$

and in which A_0 has no such zero eigenvalue. The quantity of interest is

$$D = \frac{\det A_\perp}{\det A_0}$$

where $A_\perp$ is the projection of the operator A in the subspace orthogonal to the zero eigenmode.

i) Show that

$$\begin{aligned} D &= \lim_{\lambda\to 0} -\frac{1}{\lambda}\frac{\det(A-\lambda)}{\det(A_0-\lambda)} \\ &= \lim_{\lambda\to 0} -\frac{\frac{1}{\lambda}\psi_\lambda\left(\frac{\beta}{2}\right)}{\psi_0^{(0)}\left(\frac{\beta}{2}\right)} . \end{aligned}$$

ii) To evaluate the limit in which λ goes to zero, we expand $\psi_\lambda(t)$ in powers of λ as follows:

$$\psi_\lambda(t) = \psi_0(t) + \lambda\varphi(t) .$$

Show that:

$$D = \frac{\varphi\left(\frac{\beta}{2}\right)}{\psi_0^{(0)}\left(\frac{\beta}{2}\right)}$$

where $\varphi(t)$ satisfies the equation:

$$\left(-\frac{\partial^2}{\partial t^2} + W(t)\right)\varphi(t) = \psi_0(t)$$

with boundary conditions

$$\varphi\left(-\frac{\beta}{2}\right) = 0$$

$$\frac{\partial}{\partial t}\varphi\left(-\frac{\beta}{2}\right) = 0 .$$

d) Finally, consider the case in which $\beta \to +\infty$. In this case, the eigenstates of A are defined as solutions of the equation:

$$\left(-\frac{\partial^2}{\partial t^2} + W(t)\right)\psi_n(t) = \lambda_n\psi_n(t)$$

with modified boundary conditions

$$\psi_n(t) \underset{|t|\to\infty}{\sim} e^{-\sqrt{w-\lambda_n}|t|}$$

for bound states and

$$\psi_n(t) \underset{t\to+\infty}{\sim} e^{i\sqrt{\lambda_n - w}\, t}$$

for outgoing wave scattering states. The quantity w is defined by

$$w = \lim_{|t|\to+\infty} W(t) \ .$$

Consider the wavefunction $\psi_\lambda(r)$ defined by

$$\left(-\frac{\partial^2}{\partial t^2} + W(t)\right)\psi_\lambda(t) = \lambda\psi_\lambda(t)$$

with boundary conditions

$$\psi_\lambda(t) \underset{t\to+\infty}{\sim} e^{i\sqrt{\lambda - w}\, t}$$

where we use the convention that $\sqrt{-1} = i$.

i) Show that

$$D(\lambda) = \lim_{t\to-\infty} \frac{\psi_\lambda^{(0)}(t)}{\psi_\lambda(t)}$$

where we also assume that $W_0(t)$ goes to w when $|t|$ goes to infinity. Note that $\psi^{(0)}$ now appears in the numerator and ψ in the denominator since the boundary conditions were fixed at $t = +\infty$.

ii) Show that in the case of a zero eigenvalue, the result for $\frac{\det A_\perp}{\det A_0}$ becomes

$$D = \frac{\det A_\perp}{\det A_0} = -\frac{\psi_0^{(0)}(-\infty)}{\varphi(-\infty)}$$

where $\varphi(t)$ satisfies the equation:

$$\left(-\frac{\partial^2}{\partial t^2} + W(t)\right)\varphi(t) = \psi_0(t)$$

with boundary condition

$$\varphi(t) \underset{t\to+\infty}{\sim} \frac{1}{2\sqrt{-w}}\, t e^{i\sqrt{-w}\, t} \ .$$

CHAPTER 8

STOCHASTIC METHODS

Functional integrals reduce the quantum many-body problem to quadrature. All that remains to calculate the observable properties of physical systems is evaluation of a very complicated multidimensional integral. In situations in which analytical methods such as the stationary-phase approximation or perturbation theory are uncontrolled or impractical, Monte Carlo evaluation of path integrals is a powerful and appealing alternative. It provides precise calculations of physical observables which are exact within controllable systematic and stochastic sampling errors. This chapter describes the range of techniques available for the stochastic evaluation of path integrals and how this flexibility may be effectively exploited to incorporate one's understanding of the underlying physics in the method.

8.1 MONTE CARLO EVALUATION OF INTEGRALS

We will eventually need to evaluate path integrals of the form $\int \Pi_{i\tau} d\vec{r}_i^{(\tau)}\; e^{-S(\vec{r})}\mathcal{O}(\vec{r})$ where $\vec{r}_i^{(\tau)}$ denotes the coordinate $\vec{r}$ of the ith particle on the τth time slice or integrals over fields of the form $\Pi_{ijk\tau} d\sigma_{ijk}^{(\tau)}\; e^{-S(\sigma)}\mathcal{O}(\sigma)$ where $\sigma_{ijk}^{(\tau)}$ is a field defined on a space-time lattice. It is easy to see that conventional quadrature methods appropriate for integrals in few dimensions are completely useless for such problems.

First, consider the number of points at which the integrand must be sampled on a single time slice for a system like the nucleus ^{208}Pb (or equivalently, a drop of liquid Helium containing several hundred particles). Since the rms radius is 7 fm and the two-body potential varies significantly on the scale of 0.2 fm, one requires at least 100 mesh points in each cartesian direction for each coordinate, so that the integrand must be evaluated at $(100)^{3\times 208} = 10^{1248}$ points. Even at the fastest computer speeds of the order of 10^7 evaluations per second or 10^{14} evaluations per year, this corresponds to 10^{1234} years in a universe whose age is only 2×10^{10} years. Similarly, simply tabulating the integrand at 1000 values per page in a 1000 page book weighing 1Kg would require a mass of 10^{1242} Kg whereas the mass of our galaxy is only of the order 4×10^{41} Kg. Clearly, even ignoring the need to treat multiple time slices, the situation would be hopeless if physics required sampling of the integrand in this degree of detail. (Note that this same counting indicates that solution of the many-body Schrödinger equation as a finite difference equation requires diagonalization of a $10^{1248}\times 10^{1248}$ matrix.

The essence of the problem is the unfavorable way in which the accuracy of conventional quadrature formulae depends on the number of points at which the integrand is sampled in high dimensions. Consider, for example, Simpson's rule in one dimension. Since the integrand is locally approximated quadratically and odd terms vanish, the leading error in a single interval $(-h,h)$ is $\int_{-h}^{h} dx\, x^4 = \mathcal{O}(h^5)$. If the total interval is divided into N mesh points, $h\sim\frac{1}{N}$ and the total error is of order $Nh^5\sim\frac{1}{N^4}$. Similarly, in two dimensions, the error in a single square cell is $\int_{-h}^{h} dx \int_{-h}^{h} dy\, x^4 = \mathcal{O}(h^6)$

and in d-dimensions, the error per cell is $\mathcal{O}(h^{4+d})$. If the integrand in d-dimensions is sampled at N points, then $h \sim \frac{1}{N^{1/d}}$ and the total error is of order $Nh^{4+d} = \mathcal{O}\left(\frac{1}{N^{4/d}}\right)$. Thus, as the dimension d becomes large the error falls off increasingly slowly with N. Indeed, it is the fact that N must vary as $(\mathrm{const})^d$ to keep the error fixed which led to the impractical estimate for ^{208}Pb with $d = 3 \times 208$.

The basic idea of the Monte Carlo method is to sample the integral statistically, so that independent of the dimension of the integral, the sampling errors decrease as $\frac{1}{\sqrt{N}}$ where N is the number of points at which the integrand is sampled. Note that there is no reason in principle why the evaluation of physical observables should require the incredible level of detail cited in the quadrature example for ^{208}Pb. For example, in evaluating the potential energy, one only needs to know the two-body correlation function accurately within the range of the two-body potential. The detailed behavior of distant particles is totally irrelevant and even if calculated to high precision, identically cancels out of the numerator and denominator in calculating $\langle\psi|v|\psi\rangle/\langle\psi|\psi\rangle$. Hence, the fact that we will use values of N immensely smaller than the value $(\mathrm{const})^d$ required to fully specify the many-particle wave function poses no problem in principle.

CENTRAL LIMIT THEOREM

The Monte Carlo method for evaluating an integral is based on the familiar central limit theorem. Consider an integral of the following general form

$$I = \int d^d x \, f(\vec{x})\, P(\vec{x}) \tag{8.1}$$

where $\vec{x}$ is a vector in d-dimensions and $P(\vec{x})$ is a probability distribution satisfying the condition

$$\begin{gathered} P(\vec{x}) \geq 0 \\ \int d^d x \, P(\vec{x}) = 1 \ . \end{gathered} \tag{8.2}$$

Note that there is infinite freedom, which we will subsequently exploit, to decompose the integrand of any multidimensional integral in the form of Eq. (8.1).

We will now try to approximate I by forming the average of N independent samples of the probability distribution $P(\vec{x})$

$$X = \frac{1}{N} \sum_{\substack{i=1 \\ \vec{x}_i \epsilon P(\vec{x})}}^{N} f(\vec{x}_i) \tag{8.3}$$

where throughout this chapter, the notation $\vec{x}_i \epsilon P(\vec{x})$ indicates that the variable $\vec{x}_i$ is sampled according to the function $P(\vec{x})$. In order to use such an average as a controlled approximation for I, we need to know the probability distribution for the variable X, and especially how this distribution behaves for large N. To simplify notation, it is convenient to define the mean value of a function $g(\vec{x})$ with respect to the distribution $P(\vec{x})$ as follows

$$\langle g \rangle_P \equiv \int d^d x \, g(\vec{x})\, P(\vec{x}) \ . \tag{8.4}$$

By definition, $\langle f \rangle_P = I$.

The probability for obtaining a particular value X when each $\vec{x}$ is distributed according to $P(\vec{x})$ may be written as follows:

$$\mathcal{P}(X) = \int \prod_{i=1}^{N} d^d x_i \; P(\vec{x}_i) \; \delta\left(\frac{1}{N}\sum_{j=1}^{N} f(\vec{x}_j) - X\right) \quad . \tag{8.5}$$

Using the integral representation for the δ-function, the probability distribution for X may be written

$$\begin{aligned} \mathcal{P}(X) &= \int \prod_{i=1}^{N} d^d x_i \; P(\vec{x}_i) \int \frac{N d\lambda}{2\pi} \; e^{iN\lambda X - i\lambda \sum_{j=1}^{N} f(\vec{x}_j)} \\ &= \frac{N}{2\pi} \int d\lambda \; e^{iN\lambda X + N \ln\left(\int d^d y \; e^{-i\lambda f(\vec{y})} P(\vec{y})\right)} \\ &\equiv \frac{N}{2\pi} \int d\lambda \; e^{N F(\lambda, X)} \end{aligned} \tag{8.6a}$$

where

$$F(\lambda, X) \equiv i\lambda X + g(\lambda) \tag{8.6b}$$

$$g(\lambda) \equiv \ln\left(\int d^d y \; e^{-i\lambda f(\vec{y})} P(\vec{y})\right) \quad . \tag{8.6c}$$

For a specified value of X, we will perform the λ integral using the stationary phase approximation and denote the stationary value of λ for a given X as $\tilde{\lambda}(X)$. The stationarity condition $\frac{\partial F}{\partial \lambda}\Big|_{\tilde{\lambda}(X)} = 0$ yields

$$X = i g'(\tilde{\lambda}(X)) = \frac{\int d^d y \; f(\vec{y}) \; e^{-i\tilde{\lambda}(X) f(\vec{y})} P(\vec{y})}{\int d^d y \; e^{-i\tilde{\lambda}(X) f(\vec{y})} P(\vec{y})} \tag{8.7}$$

which implicitly defines $\tilde{\lambda}(X)$. Including the quadratic correction $\frac{d^2 F}{d\lambda^2}\Big|_{\tilde{\lambda}(X)} = g''(\tilde{\lambda}(X))$, the probability distribution for large N is

$$\mathcal{P}(X) = \left[\frac{N}{-g''(\tilde{\lambda}(X)) 2\pi}\right]^{\frac{1}{2}} e^{N F(\tilde{\lambda}(X), X)} \left(1 + \mathcal{O}\left(\frac{1}{N}\right)\right) \quad . \tag{8.8}$$

To display the X-dependence of the exponent explicitly, we first find the extrema

$$0 = \frac{dF(\tilde{\lambda}(X), X)}{dX} = \frac{\partial F}{\partial X} + \frac{\partial F}{\partial \lambda}\frac{d\tilde{\lambda}}{dX} = i\tilde{\lambda}(X) \tag{8.9}$$

where we have used the stationarity condition $\frac{\partial F}{\partial \lambda} = 0$. Note that there is only one solution $\tilde{\lambda}(X) = 0$ for which by Eq. (8.7) $X = \langle f \rangle_P$. Expanding to second order around this point

$$\begin{aligned}\left.\frac{d^2F(\tilde{\lambda}(X),X)}{dX}\right|_{\tilde{\lambda}=0} &= i\left.\frac{d\tilde{\lambda}(X)}{dX}\right|_{\tilde{\lambda}=0}\\ &= \frac{1}{g''(\tilde{\lambda}=0)} \\ &= \frac{-1}{\langle f^2\rangle_P - \langle f\rangle_P^2}\end{aligned} \tag{8.10}$$

where we have differentiated Eq. (8.7) with respect to X to evaluate $\frac{d\tilde{\lambda}}{dX}$. $\mathcal{P}(X)$ thus has a single maximum at $X = \langle f\rangle_P$ and is monotone decreasing away from this point with curvature specified by the variance $\langle f^2\rangle_P - \langle f\rangle_P^2$. In the limit as $N \to \infty$, higher terms in the expansion of the exponent contribute negligibly to the tails of the distribution. In the sense of distributions, $\mathcal{P}(X)$ thus approaches a normal distribution

$$\mathcal{P}(X) \underset{N\to\infty}{\longrightarrow} \sqrt{\frac{N}{2\pi(\langle f^2\rangle_P - \langle f\rangle_P^2)}}\, e^{-\frac{(X-\langle f\rangle_P)^2}{\frac{2}{N}\left(\langle f^2\rangle_P-\langle f\rangle_P^2\right)}} \,. \tag{8.11}$$

Equation (8.11) is the desired result which provides the foundation for the Monte Carlo method. No matter what the distribution $P(\vec{x})$, the function $f(\vec{x})$, or the dimension d, for large N the average Eq. (8.3) becomes normally distributed about I with standard deviation $\sigma = \left[\frac{\langle f^2\rangle_P - \langle f\rangle_P^2}{N}\right]^{\frac{1}{2}}$. Thus, a general integral may be approximated

$$\int d^dx\, f(\vec{x})P(\vec{x}) = \frac{1}{N}\sum_{\substack{i=1\\ \vec{x}_i\in P(\vec{x})}}^{N} f(\vec{x}_i) \pm \frac{1}{\sqrt{N}}\left[\langle f^2\rangle_P - \langle f\rangle_P^2\right]^{1/2} \tag{8.12}$$

and the variance may be evaluated by the usual unbiased estimate*

$$\langle f^2\rangle_P - \langle f\rangle_P^2 \approx \frac{N}{N-1}\left[\frac{1}{N}\sum_i f(\vec{x}_i)^2 - \left(\frac{1}{N}\sum_i f(\vec{x}_i)\right)^2\right] \,. \tag{8.13}$$

Note that the factor $\frac{N}{N-1}$ to remove the bias is irrelevant for large N so that one effectively uses Eq. (8.12) to evaluate both $\langle f^2\rangle_P$ and $\langle f\rangle_P^2$.

IMPORTANCE SAMPLING

Although in principle, Eq. (8.12) is applicable to any multi-dimensional integral which could arise in many-particle systems, its practical utility is governed by the signal to noise ratio. The ratio of the variance to the mean must be sufficiently small that the fractional error may be controlled with a feasible sample size. For some problems in

---

* The origin of the factor $\frac{N}{N-1}$ is clear from the extreme case of two random variables x and y having mean zero and standard deviation σ. The naive estimate of the variance $\frac{1}{2}(x^2+y^2) - \left(\frac{x+y}{2}\right)^2 = \frac{x^2}{4} + \frac{y^2}{4} - \frac{1}{2}xy$ has mean value $\frac{1}{2}\sigma^2$ instead of σ^2.

which the integrand has nearly cancelling positive and negative contributions which are separately very large in magnitude, the Monte Carlo method becomes very difficult to apply. Such difficulties will be encountered subsequently for Fermions in more than one dimension. For problems with dominantly positive integrands, however, the freedom to decompose the integrand into factors $f(\vec{x})$ and $P(x)$ may be effectively exploited to reduce the variance. Selection of $P(x)$ to reduce the variance is known as importance sampling.

The basic idea is most simply demonstrated for a one-dimensional integral with a non-negative integrand. Integrals with both positive and negative contributions can be treated in the same way as the difference of two non-negative integrals as long as one integral clearly dominates, and the generalization to higher dimension is obvious.

Consider the integral

$$I = \int_0^1 f(x)dx \equiv \int_0^1 \left(\frac{f(x)}{P(x)}\right) P(x)dx \tag{8.14}$$

which has been written in the form of Eq. (8.1) for an arbitrary probability distribution $P(x)$. By Eq. (8.12), it may be approximated

$$\int_0^1 f(x)dx = \frac{1}{N} \sum_{\substack{i=1 \\ x_i \epsilon P(x)}}^{N} \frac{f(x_i)}{P(x_i)} \pm \frac{1}{\sqrt{N}} \left[\left\langle\left(\frac{f}{P}\right)^2\right\rangle_P - \left\langle\frac{f}{P}\right\rangle_P^2\right]^{1/2} . \tag{8.15}$$

The question, then, is what we can do to reduce the variance

$$\begin{aligned} \sigma_P^2 &\equiv \left\langle\left(\frac{f}{P}\right)^2\right\rangle_P - \left\langle\frac{f}{P}\right\rangle_P^2 \\ &= \left\langle\left(\frac{f}{P} - I\right)^2\right\rangle_P . \end{aligned} \tag{8.16}$$

Obviously, the best choice is to use $P(x) = \frac{f(x)}{I}$, for which case the variance is zero and each term in the average $\frac{f(x_i)}{P(x_i)}$ is simply I. This optimal choice is only of academic interest, since if we knew I and were able to sample $f(x)$, we wouldn't be in the business of calculating the integral in the first place. However, it suggests the extremely useful and practical strategy of choosing $P(x)$ to be a simple function which is easy to sample and which is as similar as possible to $f(x)$.

The quantitative utility of such a choice is illustrated by a simple numerical example. Consider the function sketched in Fig. 8.1

$$f(x) = \frac{e^x - 1}{e - 1} \tag{8.17}$$

for which

$$I = \int_0^1 f(x)dx = \frac{e-2}{e-1} = 0.418 . \tag{8.18}$$

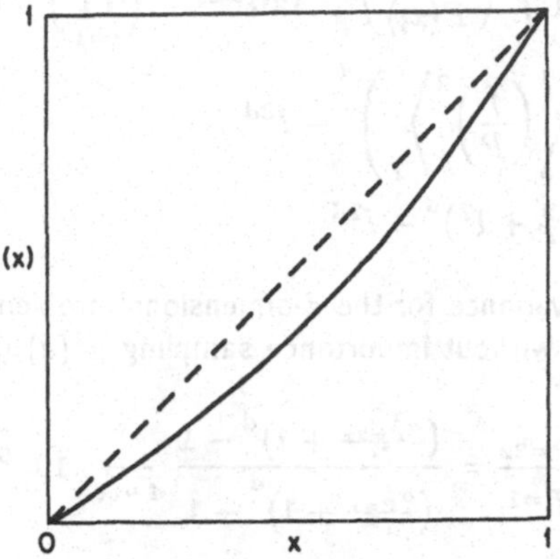

Fig. 8.1 Sketch of the example function $f(x) = \frac{e^x - 1}{e-1}$ (solid curve) and a linear importance sampling function (dashed line.)

If $P(x)$ is taken to be a uniform distribution on the interval $(0, 1)$, then the variance $\sigma^2_{P=1} = \int (f(x) - I)^2 dx$ leads to the statistical error

$$0.418 = \sum_{\substack{i=1 \\ x_i \epsilon P}}^{N} f(x_i) \pm \frac{0.286}{\sqrt{N}} \quad . \tag{8.19}$$

Thus, the sample size must be of the order $N \sim 4.7 \times 10^3$ to obtain 1% accuracy. Now, suppose we use a simple linear probability distribution $P(x) = 2x$, which we will show in the next section is easy to sample. As seen in Fig. 8.1, although a linear distribution does not reproduce $f(x)$ in detail, it does effectively concentrate the points in the region of the most "important" contributions, thus fulfilling the intent of importance sampling. With this choice, the variance $\sigma^2_{P=2x} = \int_0^1 \left(\frac{f(x)}{2x} - I\right)^2 2x dx$ leads to a sampling error

$$0.418 = \sum_{\substack{i=1 \\ x_i \epsilon P}}^{N} \frac{f(x_i)}{2x_i} \pm \frac{0.0523}{\sqrt{N}} \tag{8.20}$$

which requires only of the order $N \sim 155$ samples to obtain the same 1% accuracy. In this one-dimensional example, introduction of even the simplest importance sampling reduced the sample size for a given accuracy by a factor of $\frac{1}{30}$. In higher dimensions, where a progressively smaller fraction of the phase space contributes significantly, the advantages grow dramatically. For instance, if the previous example is generalized to a d-dimensional separable problem

$$f_d(\vec{f}) = \prod_{i=1}^{d} f(x_i) \qquad P_d(\vec{x}) = \prod_{i=1}^{d} P(x_i) \tag{8.21}$$

then the variance is

$$\sigma^2(d) = \prod_{i=1}^{d} \int \left(\frac{f(x_i)}{P(x_i)}\right)^2 P(x_i)dx_i - \left(\prod_{i=1}^{d} \int f(x_i)dx_i\right)^2$$
$$= \left(\left\langle\left(\frac{f}{P}\right)^2\right\rangle_P\right)^d - I^{2d} \tag{8.22}$$
$$= \left(\sigma_P^2 + I^2\right)^d - I^{2d}$$

so that the ratio of the variance for the d-dimensional problem with importance sampling $\sigma^2(d)_{P=2x}$ to that without importance sampling $\sigma^2(d)_{P=1}$ is

$$\frac{\sigma^2(d)_{P=2x}}{\sigma^2(d)_{P=1}} = \frac{\left(\frac{\sigma_{P=2x}^2}{I^2} + 1\right)^d - 1}{\left(\frac{\sigma_{P=1}^2}{I^2} + 1\right)^d - 1} \underset{d\to\infty}{\longrightarrow} 10^{-0.16d} \quad . \tag{8.23}$$

Even for a modest dimension $d = 50$, the reduction of the variance and the necessary sample size by 8 orders of magnitude becomes quite impressive.

8.2 SAMPLING TECHNIQUES

Application of the Monte Carlo Method requires the generation of statistically independent variables $\{\vec{x}\}$ distributed according to a specified probability distribution $P(\vec{x})$. Because of the diversity of distributions which may arise in many-body problems, it is useful to survey a variety of sampling techniques.

SAMPLING SIMPLE FUNCTIONS

The fundamental building block for all sampling methods is a random number generator for variables uniformly distributed on the interval (0, 1). Computers can, at best, generate a sequence of pseudorandom numbers which are determined by a sufficiently complicated rule that they appear to be random for all practical purposes.

One common algorithm implemented on most computers is the linear congruential method which generates a sequence of pseudorandom numbers according to the rule

$$x_{i+1} = (ax_i + c)_{\text{mod } m} \quad . \tag{8.24}$$

When a and c are appropriately chosen, this generates a non-recurring sequence of length m where m is the largest integer the computer can represent (*i.e.*, $m = 2^{31} - 1$ for a 32-bit word). Although this algorithm may be adequate for many purposes if optimal values of a and c are used, it is important to realize that sequential values of x_i have definite correlations. To see these correlations geometrically, let k sequential points define a point in a k-dimensional space: that is, $\vec{r}_1 = \{x_1, x_2, \ldots, x_k\}$, $\vec{r}_2 = \{x_{k+1}, x_{k+2}, \ldots x_{2k}\}$ and so on. It may be shown that these points lie on a sequence of at most $m^{1/k}$ $k-1$ dimensional planes(Press *et. al.*,1985). In three dimensions, at best, there are $2^{31/3} = 1290$ planes for a 32-bit word and only $2^{15/3} = 32$ planes for a 16-bit word. In practice, algorithms provided with commercial machines are often

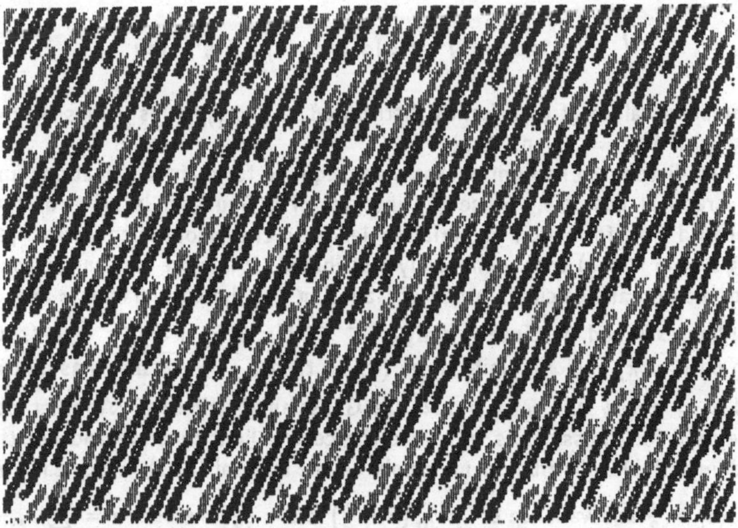

Fig. 8.2 Correlation between the triplets of points $\vec{r}_n = \{x_{3n}, x_{3n+1}, x_{3n+2}\}$ generated by a pseudorandom number generator. The value of the third coordinate is represented by the intensity of the point.

less than optimal. Figure 8.2 shows the 15 planes generated by the random number generator on the IMB PC* and one notorious random number generator on an earlier main-frame computer produced only 11 planes. It should be obvious that sampling the integrand of a three-dimensional integral which is highly peaked between these planes would give totally erroneous results. One effective cure for such correlations is to store pseudorandom numbers in a look-up table and a particularly efficient procedure is given by Knuth(1981). For the subsequent discussion, it will be assumed that a satisfactory random number generator is available to provide uniformly distributed random numbers.

Given a uniform distribution of random numbers, it is easy to sample any distribution specified by an integrable function by simple change of variables. If we define

$$y(x) = \int^{x} dx' \; P(x') \tag{8.25}$$

then $\frac{dy}{dx} = P(x)$ and we may write

$$\int dx P(x) \; f(x) = \int dy \; f(x(y)) \quad . \tag{8.26}$$

Thus, when the variable y is sampled uniformly, the inverse function $x(y)$ is distributed according to the desired probability distribution $P(x)$. For instance, the example used

* This figure is generated by repeating the following BASIC instructions under DOS 2.0: X=320*RND: Y=200*RND: C=4*RND: PSET(X,Y),C.

to illustrate importance sampling required

$$\int_0^1 f(x)2x dx = \int_0^1 f(x)d(x^2) = \int_0^1 f(y^{1/2})dy$$

so that sampling values of y uniformly distributed on (0, 1) yields values of $x = y^{1/2}$ distributed with $P(x) = 2x$ on (0, 1)

Gaussian distributed random numbers are required for many applications and thus provide a practical example. ' Since the integral of a Gaussian is an error function, straightforward transformation of variables leads to the inconvenient prescription of calculating $x = erf^{-1}(y)$ with y uniformly distributed. It is more convenient to use the Box-Muller method and calculate a pair of Gaussian distributed variables x and y using polar coordinates θ and ρ. Since $\int dx dy\, e^{-\frac{x^2}{2}-\frac{y^2}{2}} f(x,y) = \int d\theta d\left(e^{-\frac{\rho^2}{2}}\right) f(\rho,\theta)$, if θ is uniform on $(0, 2\pi)$ and ω is uniform on (0, 1), then $\rho = \sqrt{-2 \ln \omega}$ has distribution $e^{-\frac{\rho^2}{2}}$ and $x = \rho\cos\theta$ and $y = \rho\sin\theta$ are Gaussian distributed. Another common alternative used for Gaussians is to invoke the central limit theorem and use the sum of 12 random variables distributed uniformly on the interval (0, 1). Since each variable has mean $\langle x\rangle = \frac{1}{2}$ and variance $\sigma^2 = \int_0^1 (x-\frac{1}{2})^2 dx = \frac{1}{12}$, the sum $s = \sum_{i=1}^{12} x_i - 6$ has mean 0, variance 1, and is approximately Gaussian.

A final elementary method for sampling simple functions is von Neumann rejection illustrated in Fig. 8.3. Let $F(x)$ be a positive function everywhere greater than or equal to a probability distribution $P(x)$ to be sampled. If points are generated uniformly in the plane below $F(x)$ and only those which are also below $P(x)$ are accepted, then the accepted points will be distributed according to $P(x)$. Operationally, a random variable x_i is selected with probability proportional to $F(x)$ (*i.e.*, $P(x) = F(x)/\int_0^1 F(x)$) and a random number ξ uniformly distributed on the interval $(0,1)$ is generated. The value x_i is accepted if $P(x_i)/F(x_i) > \xi$; otherwise it is rejected. Clearly, this procedure is only efficient if one can find a function $F(x)$ which can be efficiently sampled and is close enough to $P(x)$ that a large fraction of the points are accepted.

MARKOV PROCESSES

A Markov chain of variables $\{x^{(1)}, x^{(2)}, x^{(3)}\ldots\}$ is generated by a rule which specifies the probability distribution for the $(n+1)^{\text{th}}$ element $x^{(n+1)}$ solely on the basis of the nth element $x^{(n)}$. We will use the notation that $P(x \to y)$ denotes the probability to reach y starting from x. In this section, we will be interested in the probability distribution of elements of a Markov chain and, in particular, will seek algorithms such that the distribution of the x's converges to a specified distribution $e^{-S(x)}$ (The notation $e^{-S(x)}$ is used to emphasize our eventual interest in sampling the action and the distribution is assumed to be normalized.) Although in general $x^{(n)}$ may be a vector with an arbitrary number of components, vector notation will be suppressed whenever possible.

For a Markov process to sample $e^{-S(x)}$, it is sufficient to require that the rule $P(x \to y)$ have eventual access to every configuration in the system and satisfy the microreversibility condition

$$e^{-S(x)} P(x \to y) = e^{-S(y)} P(y \to x) \quad . \tag{8.27}$$

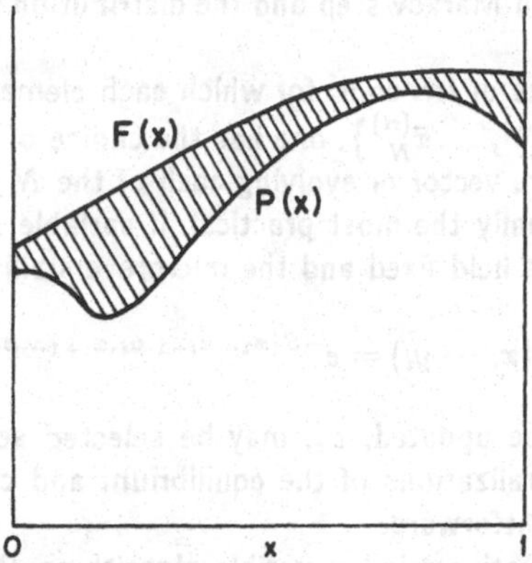

Fig. 8.3 Geometrical illustration of von Neumann rejection. Points are rejected in the shaded area and accepted in the unshaded area.

The proof proceeds in two steps. The first step is to show that $e^{-S(x)}$ is an equilibrium solution: that is, if x is distributed according to $e^{-S(x)}$ at one step of a Markov process then the distribution for the next step $\mathcal{P}(y) = \int dx\, e^{-S(x)} P(x \to y)$ is $e^{-S(y)}$. This property follows immediately from microreversibility and the normalization condition $\int dx\, P(y \to x) = 1$:

$$\begin{aligned}\mathcal{P}(y) &= \int dx\, e^{-S(x)} P(x \to y) \\ &= e^{-S(y)} \int dx\, P(y \to x) \\ &= e^{-S(y)} \,.\end{aligned} \tag{8.28}$$

The second step is to show that the distribution converges to $e^{-S(x)}$. A useful measure of the deviation between two distributions $P_A(x)$ and $P_B(x)$ is given by $\int dx |P_A(x) - P_B(x)|$. If the distribution at one step of the Markov process is $M(x)$, its deviation from equilibrium is $D_{\text{old}} = \int dx |M(x) - e^{-S(x)}|$ and the distribution at the next step $\mathcal{P}(y) = \int dx\, M(x)\, P(x \to y)$ has deviation $D_{\text{new}} = \int dy | \int dx\, M(x)\, P(x \to y) - e^{-S(y)}|$. It follows that the deviation is non increasing:

$$\begin{aligned}D_{\text{new}} &= \int dy \left| \int dx\, M(x)\, P(x \to y) - e^{-S(y)} \right| \\ &= \int dy \left| \int dx\, \left(M(x) - e^{-S(x)}\right) P(x \to y) \right| \\ &\leq \int dy \int dx \left| M(x) - e^{-S(x)} \right| P(x \to y) \\ &= \int dx \left| M(x) - e^{-S(x)} \right| = D_{\text{old}} \,.\end{aligned} \tag{8.29}$$

Since the strict equality only occurs in equilibrium $\left(M(x) = e^{-S(x)}\right)$ or if some configurations are not accessible to $P(x \to y)$ (which is excluded by hypothesis), the deviation decreases with each Markov step and the distribution approaches the desired equilibrium solution $e^{-S(x)}$.

In the case of N degrees of freedom, for which each element of the Markov chain is a vector $\vec{x}^{(n)} \equiv \{\vec{x}_1^{(n)}, \vec{x}_2^{(n)}, \ldots \vec{x}_N^{(n)}\}$, one has the choice of using a microreversible algorithm to evolve the entire vector or evolving each of the N variables separately. In the latter case, which is usually the most practical, if variable x_i is being updated, all other variables x_j, $j \neq i$ are held fixed and the microreversibility condition is

$$e^{-S(x_1 \ldots x_{i-1}, x_i, x_{i+1} \ldots x_n)} P(x_i \to y_i) = e^{-S(x_1 \ldots x_{i-1}, y_i, x_{i+1} \ldots x_n)} P(y_i \to x_i) \quad . \quad (8.30)$$

The degree of freedom to be updated, x_i, may be selected sequentially or randomly and in either case the generalizations of the equilibrium and convergence proofs, Eq. (8.28) and (8.29) are straightforward.

There are several alternative microreversible algorithms $P(x \to y)$ for generating Markov chains. One of the simplest is the so-called heat bath method $P(x \to y) = e^{-S(y)}$ for which the result is independent of the previous variable. For a single variable, the method is of no practical significance since it assumes we can already sample $S(y)$. However, for a system with many degrees of freedom and nearly local action, such as spins on lattice with nearest-neighbor interactions, it may well be possible to sample values of y distributed according to $e^{-S(x_1 \ldots x_{i-1}, y_i, x_{i+1} \ldots x_N)}$ with all x's fixed. If we regard $S(\vec{x})$ as $\beta H(\vec{x})$, this corresponds to fixing all the neighbors of y_i at the specified values x_j and letting y_i come to thermal equilibrium with a "heat bath" at inverse temperature β. By the general proof, sequentially solving the one-body problem for a single variable y_i in a heat bath with its neighbors fixed generates the equilibrium solution for the full many-body problem.

The method of Metropolis, Rosenbluth, Rosenbluth, Teller, and Teller (1953), commonly called the Metropolis method, is similar in spirit to von Neumann rejection, and like the latter, applies to an arbitrary distribution. We will first state the rule for $P(\vec{x} \to \vec{y})$ and then demonstrate that is is microreversible. A more general algorithm is discussed in Problem 8.1. Given $\vec{x}$, a tentative value $\vec{x}^T$ is generated by some convenient symmetrical probability distribution $F(\vec{x} \to \vec{x}^T) = F(\vec{x}^T \to \vec{x})$. This tentative value is accepted, that is, $\vec{y}$ is set equal to $\vec{x}^T$ with probability $\min\left\{1, e^{-S(\vec{x}^T)}/e^{-S(\vec{x})}\right\}$. Otherwise, it is rejected, that is, $\vec{y}$ is set equal to $\vec{x}$. In other words, if the change reduces the action, it is always accepted; if it increases the action, it is accepted with probability $e^{-\Delta S}$. One simple implementation of this rule is to reject $\vec{x}^T$ if $e^{-S(\vec{x}^T)}/e^{-S(\vec{x})} < \xi$ for ξ uniformly distributed on (0, 1). The most common implementation is sequential updating of the variables $\{x_i\}$ by defining $x_i^T = x_i + \xi \Delta x$ with ξ uniformly distributed on the interval $\left(-\frac{1}{2}, \frac{1}{2}\right)$. The characteristic size of the step, Δx, is chosen as a compromise between two inefficient extremes. If Δx is very small, ΔS is small, the move is almost always accepted, but x takes a very long time to explore the full space. If ΔS is very large, it is very likely that the step will move out of the vicinity of the minimum of the action and thus be rejected. A reasonable compromise between ineffectively small steps that never explore the space and moves that are never accepted is to select Δx such that roughly half the moves are accepted.

Microreversibility of the algorithm is readily verified. First, consider the case $e^{-S(\vec{y})} > e^{-S(\vec{x})}$ for which the move $\vec{x} \to \vec{y}$ is always accepted and the move $\vec{y} \to$

$\vec{x}$ is accepted with probability $e^{-S(\vec{x})}/e^{-S(\vec{y})}$. Since F is symmetric, the ratio of probabilities satisfies detailed balance as follows:

$$\frac{P(\vec{x}\to\vec{y})}{P(\vec{y}\to\vec{x})}=\frac{F(\vec{x}\to\vec{y})}{F(\vec{y}\to\vec{x})}\,\frac{1}{e^{-S(\vec{x})}/e^{-S(\vec{y})}}=\frac{e^{-S(\vec{y})}}{e^{-S(\vec{x})}}\ . \tag{8.31}$$

Verification for the case $e^{-S(\vec{y})}\le e^{-S(\vec{x})}$ is analogous.

It is useful to compare the heat bath algorithm with M iterations of the Metropolis method for updating one variable x_i of a many-body problem with all other variables x_j, $j\ne i$ fixed. For a single step of the Metropolis algorithm, $M=1$, a new variable y cannot differ from x_i by more than Δx. In the limit, $M\to\infty$, the distribution of y approaches $e^{-S(x_1\ldots x_{i-1},y,x_{i+1}\ldots x_n)}$ and the limit of a large number of Metropolis updates of a single site becomes equivalent to a single application of the heat bath algorithm. Hence, if the effort required for one heat bath update is less than for several Metropolis steps, the heat bath is preferable.

A final microreversible alternative is integration of the Langevin equation. Consider the variables in the action to be a function of a formal continuous variable τ, not to be confused with physical time or inverse temperature, and let them satisfy the following partial differential equation

$$\frac{\partial x_i}{\partial\tau}=-\Gamma\frac{\partial S}{\partial x_i}+\xi_i(\tau) \tag{8.32}$$

where Γ is an arbitrary scale parameter and $\xi_i(\tau)$ denotes a set of Gaussian distributed random variables which are independent at each i and τ and have variance

$$\langle\xi_i(\tau)\xi_j(\tau')\rangle=2\Gamma\delta_{ij}\delta(\tau-\tau')\ . \tag{8.33}$$

In practice, the continuum Langevin equation is replaced by a discrete finite difference equation $x_i(\tau_{n+1})=x_i(\tau_n)+\Delta\tau\left[-\Gamma\frac{\partial S}{\partial x_i(\tau_n)}+\xi_i(\tau_n)\right]$, so that the variable $x_i^{(n)}=x_i(\tau_n)$ corresponds to the nth element in a Markov chain and the probability distribution for $\xi_i(\tau_n)$ corresponding to Eq. (8.33) is

$$\mathcal{P}\{\xi_i(\tau_n)\}=\prod_i\sqrt{\frac{\Delta\tau}{4\pi\Gamma}}e^{-\sum_i\frac{\Delta\tau}{4\Gamma}\xi_i(\tau_n)^2}\ . \tag{8.34}$$

The rule for the Markov chain is that the probability to go from $\vec{x}$ to $\vec{y}$ is equal to the probability for the noise ξ_i to equal $\frac{y_i-x_i}{\Delta\tau}+\Gamma\frac{\partial S}{\partial x_i}$. In the limit $\Delta\tau\to 0$, it is straightforward to see that the evolution is microreversible:

$$\begin{aligned}\frac{P(\vec{x}\to\vec{y})}{P(\vec{y}\to\vec{x})}&\xrightarrow[\Delta\tau\to0]{}\frac{\prod_i\mathcal{P}\left(\xi_i=\frac{y_i-x_i}{\Delta\tau}+\Gamma\frac{\partial S}{\partial x_i}\right)}{\prod_i\mathcal{P}\left(\xi_i=\frac{x_i-y_i}{\Delta\tau}+\Gamma\frac{\partial S}{\partial x_i}\right)}\\&=\frac{e^{-\frac{\Delta\tau}{4\Gamma}\sum_i\left(\frac{y_i-x_i}{\Delta\tau}+\Gamma\frac{\partial S}{\partial x_i}\right)^2}}{e^{-\frac{\Delta\tau}{4\Gamma}\sum_i\left(\frac{x_i-y_i}{\Delta\tau}+\Gamma\frac{\partial S}{\partial x_i}\right)^2}}\\&=e^{-\sum_i(y_i-x_i)\frac{\partial S}{\partial x_i}}\\&\xrightarrow[\Delta\tau\to0]{}\frac{e^{-S(\vec{y})}}{e^{-S(\vec{x})}}\ .\end{aligned} \tag{8.35}$$

An obvious advantage of this method is its simplicity and generality. A potential disadvantage is the fact that any finite difference approximation to Eq. (8.32) will lead to errors of some order in $\Delta\tau$ in the distribution of $\vec{x}$. In special cases, such as field theory and critical phenomena, these errors may correspond to renormalization of bare coupling constants and introduction of irrelevant variables and thus be of no consequence(see, for example, Batrouni *et. al.*,1985).

NEUMANN-ULAM MATRIX INVERSION

A special problem frequently arising in stochastic solutions of many-body problems is sampling of the inverse of a matrix. For example, the Green's Function Monte Carlo method for determining the ground state wave functions requires repeated sampling of $\mathcal{P}(r) = \langle r|\frac{1}{H}|r'\rangle$. Similarly, when Grassmann variables have been integrated out of a many-Fermion problem, evaluation of the change of the resulting effective action when Boson or auxiliary field variables are updated requires calculation of $\frac{\delta}{\delta M_{ij}}$ ln Det $M = M_{ji}^{-1}$. Hence, it is useful to define a random walk which will evaluate the inverse of a matrix.

As a prelude to matrix inversion, consider evaluation of the Nth power of a matrix, T^N, as arises in repeated application of the transfer matrix or calculation of $\left(e^{-\epsilon H}\right)^N$. Let T_{ij} be decomposed into the product of a transition probability and a residual weight as follows:

$$T_{ij} \equiv P_{ij}W_{ij} \tag{8.36a}$$

with

$$\sum_j P_{ij} = 1 \qquad P_{ij} \geq 0 \ . \tag{8.36b}$$

Define an N-step random walk on the domain of integers labelling the basis $i \rightarrow i_2 \rightarrow i_3 \rightarrow \ldots \rightarrow i_N$ such that the probability of going from state $k \rightarrow \ell$ is $P_{k\ell}$ and for each walk define the score

$$S_{ij} = \begin{cases} W_{ii_2}W_{i_2i_3}\ldots W_{i_{N-1}i_N} & i_N = j \\ 0 & i_N \neq j \ . \end{cases} \tag{8.37}$$

Since the probability of the score is $P_{ii_2}P_{i_2i_3}\ldots P_{i_{N-1}i_N}$, it follows that the mean value of the score is the desired matrix element of $(T^N)_{ij}$

$$\langle S_{ij}\rangle = (T^N)_{ij} \ . \tag{8.38}$$

The Neumann-Ulam method of evaluating the inverse of a matrix is a simple variant of this procedure (Forsythe and Leibler, 1950). Let the matrix M be scaled and shifted if necessary such that $M = 1 - T$ with the magnitudes of all the eigenvalues of $T < 1$. The inverse is then given by the von Neumann series $M^{-1} = \sum\limits_{n=0}^{\infty} T^n$. To sum the series, T_{ik} is decomposed slightly differently:

$$T_{ij} = P_{ij}W_{ij} \tag{8.39}$$

with transition probabilities P_{ij} between states i and j defined such that

$$\sum_j P_{ij} < 1 \qquad P_{ij} > 0 \tag{8.39b}$$

and with a probability for stopping at site i defined by

$$P_i = 1 - \sum_j P_{ij} \; . \tag{8.39c}$$

A random walk of indefinite length is now defined such that the probability of going from state $k \to \ell$ is $P_{k\ell}$ and the probability of terminating the walk at point k is P_k. For a walk of k steps beginning at i, the score is defined

$$S_{ij} = \begin{cases} W_{i i_2} W_{i_2 i_3} \dots W_{i_{k-1} i_k} \frac{1}{P_{i_k}} & i_k = j \\ 0 & i_k \neq j \; . \end{cases} \tag{8.40}$$

The probability of attaining this score is $P_{i i_2} P_{i_2 i_3} \dots P_{i_{k-1} i_k} P_{i_k}$ so that the mean value of all scores of length k is $(T^k)_{ij}$ and the mean value of all scores of all lengths is

$$\langle S_{ij} \rangle = \sum_{k=0}^{\infty} (T^k)_{ij} = (M^{-1})_{ij} \; . \tag{8.41}$$

Variations of this basic method are discussed by Kuti and Polonyi (1982) and in Problem 8.2.

MICROCANONICAL METHODS

The microcanonical Monte Carlo method and the method of molecular dynamics utilize the ideas of classical statistical mechanics in the canonical ensemble to sample an arbitrary distribution.

One can understand the microcanonical Monte Carlo method (Creutz,1983) for sampling an arbitrary distribution $e^{-\beta S(\vec{x})}$, by regarding $S(\vec{x})$ as the energy of a classical system for the configuration $\vec{x} = \{x_i\}$. Imagine that this system is in equilibrium with a thermal reservoir and consider the microcanonical partition function for the combination of the system plus reservoir

$$Z = \sum_{\{x_i\}} \sum_{E_R} \delta\left[S(\{x_i\}) + E_R - E\right] \tag{8.42}$$

where E_R is the energy of the reservoir, E is the fixed total energy of the system plus reservoir, and the zero of the energy scale has been defined such that $E_R \geq 0$. Since the only role of the reservoir is to accept or provide energy to conserve the total energy E, it is very easy to sample Z. As in the case of the Metropolis algorithm, let all components of $\vec{x}$ be held fixed except x_i, and define the trial moves $x_i^T = x_i + \xi$ where ξ is uniformly distributed on $\left(-\frac{\Delta x}{2}, \frac{\Delta x}{2}\right)$ and $E_R^T = E_R + S(x_1 \dots x_j \dots x_N) - S(x_1 \dots x_j^T \dots x_N)$. This trial move is accepted if $E_R^T > 0$, since it corresponds to an accessible state of the system plus reservoir, and is rejected if $E_R^T < 0$, since the move would not correspond to an accessible state. By the usual arguments of statistical mechanics, if the average value of E_R is much larger than the average value of the energy of the system, the accepted moves sample $e^{-\beta S(\vec{x})}$ for the system and the distribution of energy in the reservoir is $\mathcal{P}(E_R) \propto e^{-\beta E_R}$. Since E rather than β is specified in the algorithm, β must be evaluated explicitly. For a continuous spectrum, this is most easily accomplished by calculating $\beta = \frac{1}{\langle E_R \rangle}$; otherwise, the exponential $e^{-\beta E_R}$ may be evaluated.

Molecular dynamics uses similar ideas from statistical mechanics to sample an arbitrary distribution, but differs conceptually by invoking ergodicity rather than a random number generator to sample representative accessible states (Callaway and Rahmann, 1982). To understand the use of molecular dynamics to sample the distribution $e^{-\beta S(\vec{x})}$, it is useful to think of $S(\vec{x})$ as the classical potential energy for canonical coordinates $\{x_i\}$, to introduce a canonically conjugate momenta $\{P_i\}$ and to define a classical Hamiltonian $H = \frac{1}{2}\sum_i P_i^2 + S(\vec{x})$. Then, the mean value in the canonical ensemble of an observable $\mathcal{O}\{\vec{x}\}$ depending only on the coordinates $\{x_i\}$ yields precisely the desired average of $\mathcal{O}\{\vec{x}\}$ with respect to the distribution of $e^{-\beta S(\vec{x})}$

$$
\begin{aligned}
\langle \mathcal{O} \rangle_{\text{canonical}} &\equiv \frac{\int d\vec{p}d\vec{x}\; e^{-\beta\left(\frac{1}{2}\sum_i p_i^2 + S(\vec{x})\right)} \mathcal{O}(\vec{x})}{\int d\vec{p}d\vec{x}\; e^{-\beta\left(\frac{1}{2}\sum_i p_i^2 + S(\vec{x})\right)}} \\
&= \frac{\int d\vec{x}\; e^{-\beta S(\vec{x})} \mathcal{O}(\vec{x})}{\int d\vec{x}\; e^{-\beta S(\vec{x})}} \\
&\equiv \langle \mathcal{O} \rangle_{e^{-\beta S}} \;.
\end{aligned}
\tag{8.43}
$$

Next, one uses the fact that for an extensive system in the thermodynamics limit, the mean values of an observable in the canonical and microcanonical ensemble become equal. That is, the integral $\int d\vec{p}d\vec{x}$ may be carried out on the hypersurface of states with constant energy $E = H(\vec{p}, \vec{x})$. Finally, one considers the classical Hamiltonian equations of motion for $\vec{x}$ and $\vec{p}$ in a formal time variable τ (not to be confused with physical time) and invokes ergodicity to let a representative sample of variables $\vec{p}$ and $\vec{q}$ with constant energy be generated by the classical trajectories $\vec{p}(\tau)$, $\vec{q}(\tau)$. Thus, this full chain of reasoning yields

$$
\begin{aligned}
\langle \mathcal{O} \rangle_{e^{-\beta S}} &= \langle \mathcal{O} \rangle_{\text{canonical}} \\
&\underset{\mathcal{V}\to\infty}{\longrightarrow} \langle \mathcal{O} \rangle_{\text{microcanonical}} \\
&= \lim_{T\to\infty} \frac{1}{T} \int_0^T \mathcal{O}(\vec{x}(\tau))\, d\tau
\end{aligned}
\tag{8.44}
$$

where the trajectory $\vec{x}(\tau)$ is determined by integrating the equation of motion

$$
\frac{d^2}{d\tau^2} x_i = \frac{d}{d\tau} p_i = -\frac{\partial S(\vec{x})}{\partial x_i}
\tag{8.45}
$$

with an initial condition corresponding to energy E. The name molecular dynamics is motivated by the direct analogy of this procedure to the method which has been extensively used to sample the classical partition function for molecular system by evolving classical equations of motion. In practice, as in the case of the Langevin equation, a discrete finite difference approximation to the differential equation, Eq. (8.45), is evolved yielding a discrete set of coordinates $\vec{x}^{(n)} \equiv \vec{x}(\tau_n)$. The similarity between the molecular dynamics and Langevin methods and advantages of combining them are discussed in Problem 8.3.

A simple variant of this method (Polonyi and Wyld, 1983) is especially useful for sampling an effective action of the form $\det M(\vec{x})\, e^{-\beta S(\vec{x})}$ in which a Fermion determinant arises from integrating out Grassmann variables $\int d\xi^* d\xi\, e^{-\xi^* M(x)\xi}$ and the remaining functional integral is over Boson coordinates or an auxiliary field denoted by $\vec{x}$. In addition to the formal momenta p_i conjugate to x_i introduced above, additional canonical variables P_i and X_i are introduced to define the Hamiltonian

$$H = \frac{1}{2}\sum_{ij} P_i M(\vec{x})^{-2}_{ij} P_j + \frac{1}{2}\omega \sum_i X_i^2 + \frac{1}{2}\sum_i p_i^2 + S(\vec{x}) \quad . \tag{8.46}$$

The corresponding Lagrangian yields the classical equation of motion in the formal variable τ

$$\frac{d}{d\tau}\left(M(\vec{x})^2_{ij}\frac{dX_j}{d\tau}\right) = -\omega X_i \tag{8.47a}$$

$$\frac{d^2 x_i}{d\tau^2} = -\frac{\partial S}{\partial x_i} + \frac{1}{2}\sum_{m,n}\frac{dX_m}{d\tau}\left[\frac{\partial M^2(\vec{x})}{\partial x_i}\right]_{mn}\frac{dX_n}{d\tau} \quad . \tag{8.47b}$$

Following the previous reasoning, the average values of an observable $O(\vec{x})$ may be sampled using the solution to Eq. (8.47), $\vec{x}(\tau)$,

$$\begin{aligned}\langle O\rangle_{\det M e^{-\beta S}} &= \frac{\int d\vec{p}\,d\vec{x}\,d\vec{P}\,d\vec{X}\; e^{-\beta H(\vec{p},\vec{x},\vec{P},\vec{X})}\, O(\vec{x})}{\int d\vec{p}\,d\vec{x}\,d\vec{P}\,d\vec{X}\; e^{-\beta H(\vec{p},\vec{X},\vec{P},\vec{X})}} \\ &\underset{\substack{\mathcal{V}\to\infty\\ T\to\infty}}{\longrightarrow} \frac{1}{T}\int_0^T O(\vec{x}(\tau))\,d\tau \,.\end{aligned} \tag{8.48}$$

Using the equipartition theorem, the temperature β may be calculated conveniently by evaluating the kinetic energy

$$\lim_{T\to\infty}\frac{1}{T}\int_0^T d\tau \frac{1}{2}\sum_i\left(\frac{dx_i}{dt}\right)^2 = \frac{N_{\vec{x}}}{2\beta} \tag{8.49}$$

where $N_{\vec{x}}$ is the number of degrees of freedom associated with the coordinate $\vec{x}$. Note that this treatment may be generalized straightforwardly to the case of a complex matrix M with real determinant by using $P_i^*\left(M^\dagger(\vec{x})M(\vec{x})\right)^{-1}_{ij} P_j + \omega^2 \sum_i x_i^* x_i$ in the Hamiltonian and that in the case of lattice gauge theory, local gauge symmetry of the variable $\vec{x}$ reduces the number of degrees of freedom $N_{\vec{x}}$ to be used in evaluating β.

8.3 EVALUATION OF ONE-PARTICLE PATH INTEGRAL

The stochastic techniques described above may be applied to many-body problems in a variety of ways. The principal intellectual challenge is to exploit the freedom and to utilize physical insight to incorporate as much of the essential physics of the problem as possible into the stochastic algorithm.

There are three main categories of choices to be made: the observables to evaluate, whether to sample the complete action or solve an initial value problem, and the form of functional integral to calculate. The first two sets of choices may be illustrated most simply by the Feynman path integral for a single degree of freedom discussed in this section. In addition, even this simple example illustrates two of the primary limitations of stochastic methods: the need to calculate a dominantly positive integral and the need to independently sample the entire range of configurations

OBSERVABLES

In order to sample dominantly positive functions, we must evaluate Euclidean rather than real time path integrals. One obvious quantity to sample is the trace of the evolution operator $\sum_n \langle n|e^{-\beta H}|n\rangle$ which provides all thermodynamic information at inverse temperature β. The ground state expectation value of an operator $\mathcal{O}$ may be calculated by taking the limit in which $\frac{1}{\beta}$ is small compared to the excitation energy of the first excited stated:

$$\begin{aligned}\langle \mathcal{O}\rangle_\beta &= \frac{\sum_n \langle n|\mathcal{O}e^{-\beta H}|n\rangle}{\sum_n \langle n|e^{-\beta H}|n\rangle} \\ &\xrightarrow[\beta(E_1-E_0)>>1]{} \langle 0|\mathcal{O}|0\rangle \ .\end{aligned} \tag{8.49}$$

The imaginary-time response function:

$$\begin{aligned}C(\tau) &= \frac{\mathrm{Tr}\{T\mathcal{O}(\tau)\mathcal{O}(0)e^{-\int_0^\beta d\tau H}\}}{\mathrm{Tr}\{e^{-\beta H}\}} \\ &= \frac{\sum_{m,n} e^{-E_m(\beta-\tau)-E_n\tau}\,|\langle m|\mathcal{O}|n\rangle|^2}{\sum_m e^{-E_m\beta}} \\ &\xrightarrow[\beta\to\infty]{} \sum_n |\langle 0|\mathcal{O}|n\rangle|^2\, e^{-(E_n-E_0)\tau}\end{aligned} \tag{8.50}$$

may be used to extract energies and transition matrix elements for low lying excited states. However, it is not practical in general to analytically continue the decaying exponentials in imaginary time into oscillatory functions in real time to extract much useful information about the physical real time response function.

Instead of calculating the trace, it is often preferable to evaluate a specific matrix element of $e^{-\beta H}$ which is optimal for calculation of a specific observable. Given an approximate wave function ϕ_a, the ground state energy may be written

$$E = \lim_{\beta\to\infty} \frac{\langle\phi_a|He^{-\beta H}|\phi_a\rangle}{\langle\phi_a|e^{-\beta H}|\phi_a\rangle} \ . \tag{8.51}$$

Clearly, the more physics that is incorporated in ϕ_a, the less work $e^{-\beta H}$ must do to filter out excited state components. Note that the ability to calculate the ground state energy may be used to evaluate the scattering phase shifts for a finite-range potential by placing an infinite wall at an arbitrary position x_w in the exterior region. Since E specifies the phase shift δ, $E(x_w)$ determines $\delta(k)$. Similarly, the expectation value of a general operator O may be evaluated using ϕ_a either exactly

$$\langle O\rangle = \lim_{\beta\to\infty} \frac{\langle\phi_a|e^{-\frac{\beta H}{2}}Oe^{-\frac{\beta H}{2}}|\phi_a\rangle}{\langle\phi_a|e^{-\beta H}|\phi_a\rangle} \tag{8.52a}$$

or perturbatively assuming $|\phi_0\rangle - |\phi_a\rangle$ is small

$$\langle O\rangle \approx 2\frac{\langle\phi_a|Oe^{-\beta H}|\phi_a\rangle}{\langle\phi_a|e^{-\beta H}|\phi_a\rangle} - \frac{\langle\phi_a|O|\phi_a\rangle}{\langle\phi_a|\phi_a\rangle} \ . \tag{8.52b}$$

Finally, the splitting ΔE between the lowest two nearly degenerate states in a double well separated by a high barrier may be determined by evaluating the ratio

$$R = \frac{\langle\phi_L|e^{-\beta H}|\phi_R\rangle}{\left[\langle\phi_L|e^{-\beta H}|\phi_L\rangle\langle\phi_R|e^{-\beta H}|\phi_R\rangle\right]^{\frac{1}{2}}} \tag{8.53}$$

where ϕ_L and ϕ_R are wave functions localized in left and right wells, respectively. For the simple case of a symmetric well with the definition $\alpha \equiv \left(\langle 1|\phi_L\rangle/\langle 0|\phi_L\rangle\right)^2 = \left(\langle 1|\phi_R\rangle/\langle 0|\phi_R\rangle\right)^2$

$$R \underset{\beta(E_2-E_0)>>1}{\longrightarrow} \frac{1-\alpha e^{-\beta\Delta E}}{1+\alpha e^{-\beta\Delta E}} = \tanh\left(\frac{\beta\Delta E - \ln\alpha}{2}\right) \tag{8.54}$$

from which ΔE may be extracted easily.

SAMPLING THE ACTION

One straightforward way to evaluate the observables is to sample $P_S(\vec{x}) = e^{-S(x_1\ldots x_N)}/\sum_{x_1\ldots x_N} e^{-S(x_1\ldots x_N)}$ as a probability distribution to be used to calculate the appropriate mean value of operators. For the trace of a local operator $O(x)$, Eq. (8.49), $S(\vec{x}) = \sum_{n=1}^{N} \frac{m}{2\epsilon}(x_n - x_{n-1})^2 + \epsilon V(x_n)$ with $x_0 \equiv x_N$ so that all time slices are equivalent by cyclic symmetry and O may be averaged over the entire interval

$$\langle O\rangle_P = \frac{1}{M} \sum_{\substack{m=1 \\ x^{(m)}\epsilon P_s(x)}}^{M} \frac{1}{N}\sum_{n=1}^{N} O(x_n^{(m)}) \ . \tag{8.55}$$

For a matrix element of the form Eq. (8.52a), $S(\vec{x})$ contains $\ln\phi_a(x_N) + \ln\phi_a(x_1)$ and the operator must be evaluated only at the midpoint, $O(x_{N/2}^{(m)})$.

Because this approach requires sampling the global trajectory at all times and we must consider the limit $\epsilon \to 0$, it is useful to insure that the leading error in the path integral is of order ϵ^2 rather than $O(\epsilon)$ or even $O(\sqrt{\epsilon})$. We have already seen in Problem 2.4 that for finite potentials the symmetric decomposition $e^{-\epsilon H} = e^{-\frac{\epsilon}{2}V} e^{-\epsilon T} e^{-\frac{\epsilon}{2}V}$ is accurate to $O(\epsilon^2)$. By regrouping terms in products of $e^{-\epsilon H}$ and a local operator $O(x)$, it follows that Eq. (8.55) is also accurate to $O(\epsilon^2)$. In the case of a hard wall boundary condition, simply putting an infinite potential into the action introduces an error of $O(\sqrt{\epsilon})$ as shown in Problem 8.4. Hence, to implement a hard wall boundary condition at position $x = a$, the path integral should be written using states which are manifestly odd about $x = a$, with the result

$$\begin{aligned}\left[\langle x_n| - \langle 2a - x_n|\right] e^{-\epsilon H} |x_{n-1}\rangle = \sqrt{\frac{m}{2\pi\epsilon}} e^{-\frac{m}{2\epsilon}(x_n - x_{n-1})^2} e^{-\frac{\epsilon}{2}(V(x_n)+V(x_{n-1}))} \\ \times \left(1 - e^{-\frac{2m}{\epsilon}(a-x_n)(a-x_{n-1})}\right) \ .\end{aligned} \tag{8.56}$$

This antisymmetrization just subtracts off the Green's function corresponding to the first image across the boundary and one can explicitly verify for the first odd state of the harmonic oscillator that the leading error is $O(\epsilon^2)$ when the first image is included (Problem 8.4).

Non-local operators, such as the kinetic energy operator are less convenient to evaluate than local operators. As shown in Problem 8.5, two straightforward ways to evaluate the kinetic energy are to evaluate $\langle x_{n+1}|\hat{p}^2 e^{-\epsilon H}|x_n\rangle$ or $\langle x_{n+1}|e^{-\epsilon H} p|x_n\rangle\langle x_n|p e^{-\epsilon H}|x_{n-1}\rangle$ with the results

$$\langle T\rangle = \langle \frac{1}{2\epsilon} - \frac{m}{2\epsilon^2}(x_n - x_{n-1})^2\rangle \tag{8.57a}$$

and

$$\langle T\rangle = \langle -\frac{m}{2\epsilon^2}(x_n - x_{n-1})(x_{n-1} - x_{n-2})\rangle \ . \tag{8.57b}$$

The factor $e^{-\frac{m}{2\epsilon}(x_n - x_{n-1})^2}$ in the action makes the naive second difference diverge in the limit $\epsilon \to 0$, $\frac{m}{2}\left(\frac{x_n - x_{n-1}}{\epsilon}\right)^2 \to \frac{1}{2\epsilon} + O(1)$, reflecting the fact that the dominant trajectories in the path integral are non-differentiable. Both the remaining finite term in Eq. (8.57a) and the split-difference in (8.57b) are subject to large variance, so it is preferable to use the virial theorem to replace $\langle T\rangle$ by a local expressing. If $|\psi_\alpha\rangle$ is an eigenstate of H defined on the domain $[a, b]$ then for any x_0

$$\begin{aligned}\langle\psi_\alpha|[\hat{H}, \hat{p}(\hat{x} - x_0)]|\psi_\alpha\rangle &= \langle\psi_\alpha|2T - (x - x_0)V'(x)|\psi_\alpha\rangle \\ &= \frac{\hbar^2}{2m}(x - x_0)|\psi'_\alpha(x)|^2\Big|_a^b \ .\end{aligned} \tag{8.58}$$

As long as ψ'_α vanishes on at least one endpoint, x_0 may be selected to cancel the boundary term at the other endpoint and $\langle T\rangle = \frac{1}{2}\langle(x - x_0)V'(x)\rangle$.

As mentioned previously, one standard way of sampling the action $P_s(x_1, x_2 \ldots x_N)$ is to sequentially make a tentative change in each of the x_i's in turn by setting $x_i^T = x_i + \xi\Delta$ and to accept or reject the change using the Metropolis algorithm. However, other alternatives are possible and sometimes desirable. For example, one may introduce a correlated additive trial change for the entire trajectory

$$F(\vec{x} \to \vec{x}^T) = \frac{1}{2} \prod_i \left(\delta(x_i^T - x_i - f_i) + \delta(x_i^T - x_i + f_i) \right) \quad . \tag{8.59}$$

That is, one adds to or subtracts from each coordinate a (physically motivated) global change f_i defined on each of the time slices. Another alternative is a multiplicative move of the form

$$F(\vec{x} \to \vec{x}^T) = \frac{1}{1 + \prod_i \alpha_i} \left[\prod_i \delta\left(x_i^T - \frac{1}{\alpha_i} x_i \right) + \prod_i \delta\left(\frac{1}{\alpha_i} x_i^T - x_i \right) \right] \tag{8.60}$$

which can scale the overall trajectory to a larger or smaller spatial extent. Note that any sequence of reversible moves may be offered in any order to facilitate equilibrium and complete sampling of the entire space.

INITIAL VALUE RANDOM WALK

Instead of sampling the global action, it is often preferable to calculate matrix elements of the form $\langle \phi_a | e^{-\beta H} | \phi_b \rangle$ appearing in the observables (8.51) – (8.53) by solving an initial-value problem. Starting with an initial ensemble of points distributed according to $\phi_b(x)$ (which must be positive), the ensemble is evolved by successive applications of $e^{-\epsilon H}$ to obtain an ensemble distributed according to $\psi(x) = e^{-\beta H} \phi_b(x)$ which is then used to calculate the overlap $\langle \phi_a | \psi \rangle$. This alternative is particularly useful for evaluating ground state properties since instead of having to retain an entire time history over many time slices, an ensemble of configurations is simply refined by $e^{-\epsilon H}$ the required number of times. The method has the additional advantage that it can conveniently incorporate physical understanding of the nature of the solution.

The simplest method for evaluation of a matrix element of the Euclidean evolution operator as an initial value problem for an ensemble of coordinates is analogous to Eqs. (8.36- 8.37) for evaluation of T^N. Let us decompose the infinitesimal evolution operator into the product of a Gaussian probability and a residual weight as follows:

$$\langle x_n | e^{-\epsilon(\hat{H} - E)} | x_{n-1} \rangle \equiv P(x_n, x_{n-1}) W(x_{n-1}) \tag{8.61a}$$

where

$$P(x_n, x_{n-1}) = \sqrt{\frac{m}{2\pi\epsilon}} \, e^{-\frac{m}{2\epsilon}(x_n - x_{n-1})^2} \tag{8.61b}$$

so that

$$W(x_{n-1}) = e^{-\epsilon(V(x_{n-1}) - E)} \tag{8.61c}$$

so that

$$\begin{aligned} \langle \phi_a | e^{-\beta(H-E)} | \phi_b \rangle = \int dx_n \dots dx_1 \phi_a(x_n) P(x_n, x_{n-1}) \\ \times W(x_{n-1}) \dots P(x_3, x_2) W(x_2) P(x_2, x_1) W(x_1) \phi_b(x_1) \quad . \end{aligned} \tag{8.62}$$

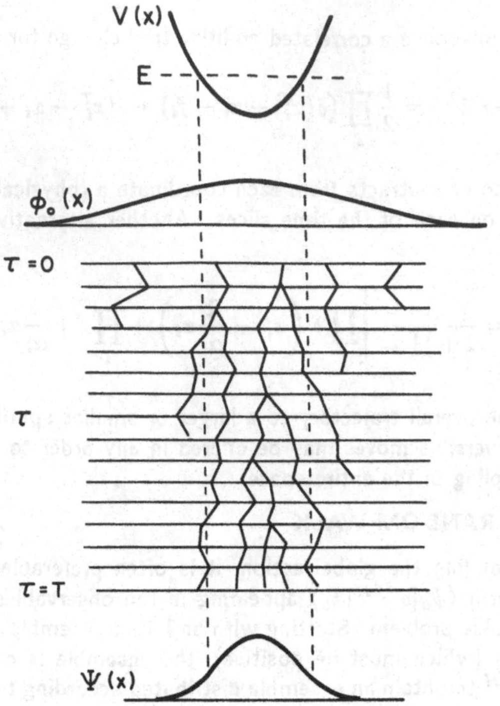

Fig. 8.4 Initial-value random walk for the ground state wave function $\psi(x)$ in the potential $V(x)$. In this example, an initial ensemble of five points is sampled from $\phi_a(x)$ which is more delocalized than $\psi(x)$. At each τ step, the points undergo both diffusion and replication, with points being removed outside the classical turning points and created in the classically allowed region. By the final time, the combination of diffusion and replication yields the distribution representative of $\psi(x)$ as shown.

This product of probabilities and weights is evaluated as follows. First, x_1 is randomly selected according to the distribution function $\phi_b(x_1)$, which may be chosen to be positive, and the temporary value of the score is defined to be $W(x_1)$. Given x_1, x_2 is chosen to be Gaussian distributed about x_1 according to the probability $P(x_2, x_1)$ and the score is multiplied by $W(x_2)$. This procedure is repeated for all n and finally x_n is chosen to be Gaussian distributed about x_{n-1} and the score is multiplied by $\phi_a(x_n)$. For an ensemble of such calculations, each score $\phi_a(x_n)\prod_{i=1}^{n-1} W(x_i)$ is obtained with probability $\prod_{i=1}^{n-1} P(x_{i+1}, x_i)\phi_b(x_1)$ so that the average value of the score for a large ensemble of random samples approaches $\langle\phi_a|e^{-\beta(H-E)}|\phi_b\rangle$.

The statistical accuracy of this basic method may be improved greatly by replicating points at each step with probability proportional to $W(x_n)$ instead of accumulating the weight $W(x_n)$ in the score. The calculation may then be viewed as a diffusion process with a source-sink term $W(x)$ and diffusion term $P(x_m, x_{m-1})$, as sketched in Fig. 8.4 An initial ensemble of points $\{x^i\}$ distributed according to $\phi_b(x)$ is first diffused by the Gaussian $P(x_2, x_1)$. In regions where $V(x) > E$, $W(x^i) < 1$ and the point x^i is deleted from the ensemble with probability $1 - W(x^i)$. When $W(x^i) > 1$ the point x^i is always replicated $[W(x^i)]$ times (where $[W]$ denotes the greatest integer in W) and with probability $W(x^i) - [W(x^i)]$ it is replicated one additional time. In each successive step, points diffuse according to $P(x^i_m, x^i_{m-1})$ and are replicated according to $W(x^i_m)$. Thus, elements of the ensemble are created in regions of attractive $V - E$ and deleted in regions of repulsive $V - E$ such that the final ensemble of points $\{x_n\}$ is distributed according to $\prod_{m=1}^{n-1} P(x_{m+1}, x_m)W(x_m)\phi(x_1)$ and $\langle\phi_a|e^{-\beta(H-E)}|\phi_b\rangle$ is given by the average value of $\phi_a(x)$ evaluated with the ensemble $\{x_n\}$. Whereas the first method retains ensembles having products of weights $\prod_{i=1}^{n} W(x_i)$ which may vary over many orders of magnitude, with corresponding loss of statistical accuracy, in the replication method each member of the final ensemble $\{x_n\}$ contributes with the same weight. In practice, the value of E is selected to maintain a constant average ensemble size and yields an independent evaluation of the ground state energy. To show how such an initial value Monte Carlo calculation is actually carried out in practice, a schematic outline of the principal steps corresponding to the process depicted in Fig. 8.4 is shown in Fig. 8.5.

The method may be improved still further by guiding the random walk using an approximate trial wave function $\phi(x,\tau)$ which contains as much of the essential physics as may be understood at the outset. Instead of evolving the solution to the imaginary-time Schrödinger equation

$$-\dot{\psi}(x,\tau) = \left[-\frac{1}{2m}\frac{d^2}{dx^2} + V(x)\right]\psi(x,\tau) \tag{8.63}$$

consider instead evolution of the product

$$f(x,\tau) = \phi(x,\tau)\psi(x,\tau) \quad . \tag{8.64a}$$

As derived in Problem 8.6, the equation of motion for $f(x,\tau)$ is

$$-\dot{f}(x,\tau) = -\frac{1}{2}\frac{\partial^2}{\partial x^2}f + \frac{1}{m}\frac{\partial}{\partial x}\left(\frac{\phi'}{\phi}f\right) + \frac{-\frac{1}{2m}\phi'' + V\phi - \dot{\phi}}{\phi}f \tag{8.64b}$$

which includes a drift term $\frac{1}{m}\frac{\partial}{\partial x}\left(\frac{\phi'}{\phi}f\right)$ and a source-sink term $\frac{\left(-\frac{1}{2m}\frac{\partial^2}{\partial x^2}+V-\frac{\partial}{\partial\tau}\right)\phi}{\phi}$. The infinitesimal evolution operator for $\phi(x,\tau)\psi(x,\tau)$ is $\phi(\hat{x},\tau+\epsilon)e^{-\epsilon\hat{H}}\frac{1}{\phi(\hat{x},\tau)}$ which, as shown in Problem 8.6 has the matrix element

$$\langle x_n|\phi(\hat{x},\tau+\epsilon)e^{-\epsilon\hat{H}}\frac{1}{\phi(\hat{x},\tau)}|x_{n-1}\rangle = \sqrt{\tfrac{m}{2\pi\hat{\epsilon}}}\, e^{-\frac{m}{2\hat{\epsilon}}\left(x_n - x_{n-1} - \frac{\epsilon}{m}\,\frac{\phi'}{\phi}\right)^2 - \epsilon\frac{(H-\frac{\partial}{\partial\tau})\phi}{\phi}} \tag{8.65a}$$

where

$$\hat{\epsilon} = \hat{\epsilon}(x_{n-1}) \equiv \epsilon\left[1 + \frac{\epsilon}{m}\frac{d^2}{dx^2}\ln\phi\right]_{x_{n-1}} \tag{8.65b}$$

and ϕ, ϕ', and $(H - \frac{\partial}{\partial\tau})\phi$ are evaluated at x_{n-1}.

Initialization

Sample $\phi_a(x)$ to obtain N_0 Coordinates $x_i^{(0)}$
(Metropolis method)

Evolution from m^{th} to $(m+1)^{\text{th}}$ generation

Evolve N_m coordinates y_i from $x_i^{(m)}$
according to $P(x_i^{(m)} \to y_i)$
(Gaussian random number generator)

Replicate or delete according to weight
$W(y_i)$ to obtain N_{m+1} coordinates $x_i^{(m+1)}$
(rejection with uniform random number)

Adjust E if necessary to maintain
average population $N_{m+1} \sim N$

Evaluate $\sum_i H\phi_a(x_i)$ and $\sum_i \phi_a(x_i)$ every M_I^{th} generation
after M_E equilibration steps.

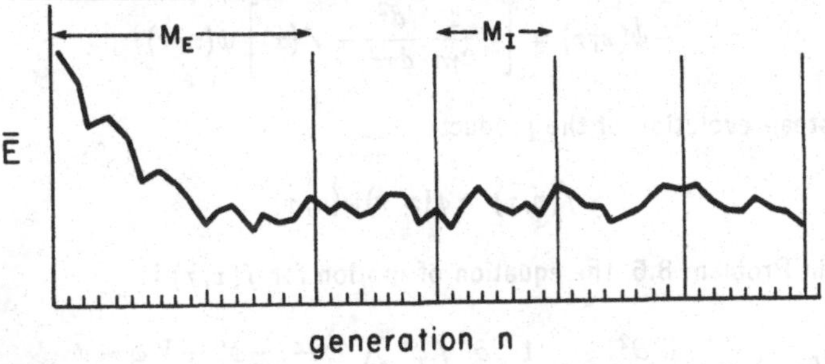

Fig. 8.5 Schematic representation of an initial value Monte Carlo calculation of the ground state energy. The principal steps in generating x_i distributed according to $\psi(x)$ and calculating $\overline{E} = \frac{\langle H\hat{\phi}_a|\psi\rangle}{\langle\phi_a|\psi\rangle}$ are shown in the upper diagram. The lower graph shows the schematic form of $\overline{E}$ as a function of generation. Initial equilibration to the ground state requires M_E initial steps. Thereafter, essentially independent samples are obtained by calculating $\overline{E}$ every M_I generations.

To understand the role of the drift and source terms, it is instructive to compare sampling the ground state wave functions using the unguided walk, Eq. (8.62), and using the guided walk, Eq. (8.65), in the special case that the trial function is constructed from the exact ground state wave function ψ_0 and energy E_0, $\phi(x,\tau) = e^{E_0\tau}\psi_0(x)$. Whereas even in equilibrium, points are continually added and deleted in the unguided walk by the source-sink term $(V(x) - E)$ leading to a large variance for rapidly varying potentials, the source-sink term in the guided walk $\frac{(H-\frac{\partial}{\partial\tau})\phi}{\phi}$ vanishes, yielding zero variance. The Gaussian diffusion term is shifted by the drift term $\frac{\epsilon}{m}\frac{\phi'}{\phi}$ which guides members of the ensemble away from regions where the wave function is small toward where it is largest. In this special case, $f = \phi(x,t)\psi(x,\tau)$ is just the ground state density $(\psi_0(x))^2$.

Even when the wave function is not known exactly, to the extent to which an approximate ϕ incorporates much of the essential physics, the evolution is guided by that portion of the physics through the drift term and the stochastic treatment of the source term is only required to treat the remnant of the physics which is left out of the trial function. In the extreme case in which nothing is known about the wave function, a constant trial function reproduces the unguided walk. Ratios of matrix elements such as those in Eqs. (8.51) and (8.52b) have a particularly convenient form for guided walks, since

$$\lim_{\beta\to\infty}\frac{\langle\phi|Oe^{-\beta H}|\phi\rangle}{\langle\phi|e^{-\beta H}|\phi\rangle} = \frac{\langle\phi|O|\psi_0\rangle}{\langle\phi|\psi_0\rangle} = \frac{\int dx \frac{O\phi(x)}{\phi(x)} f(x)}{\int dx \frac{\phi(x)}{\phi(x)} f(x)}$$
$$= \langle \frac{O\phi(x)}{\phi(x)} \rangle_f \quad . \tag{8.66}$$

By the previous arguments, in the case of a general matrix element $\langle\phi_a|e^{-\beta H}|\phi_b\rangle$, the optimal guiding function is the solution to the time-reversed equation of motion with the initial condition $\phi_a(x)$ at β (see for example Pollack and Ceperley 1984).

$$\phi(x,\tau) = e^{(\tau-\beta)H}\phi_a \quad . \tag{8.67}$$

The initial value random walk is closely related to the Green's Function Monte Carlo Method which is described in detail by Ceperley and Kalos (1979). Whereas path integral Monte Carlo projects out the ground state by successive applications of the filter $e^{-\epsilon(H-E)}$, the Green's function method iterates the filter $\frac{1}{H-E}$ (which may be thought of as an integral over exponentials $\int_0^\infty e^{-\tau(H-E)}d\tau$). Importance sampling is included in the same way as above by filtering $f = \phi\psi$ using Eq. (8.64). In practical calculations, the two methods are roughly comparable. Sampling the inverse $\frac{1}{H-E}$ by some interactive technique such as the Neumann Ulam method entails more computation than a single application of $e^{-\epsilon(H-E)}$, but the entire calculation does not need to be repeated several times to obtain the $\epsilon \to 0$ limit.

TUNNELING

Tunneling problems illustrate some potential pitfalls of the stochastic methods we have discussed and physically motivated remedies. Consider the splitting between the two nearly degenerate lowest states in the symmetric double well

$$\left[-\frac{1}{2m}\frac{d^2}{dx^2} + \left(x^2 - 1\right)^2\right]\phi = E\phi \tag{8.68}$$

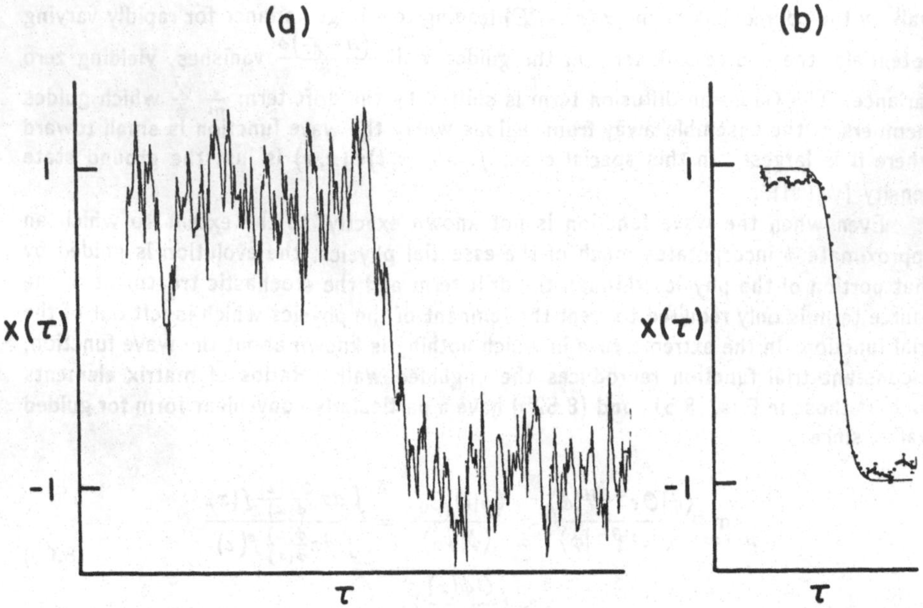

Fig. 8.6 Tunneling trajectories in a double well. A portion of a trajectory containing two instantons connecting the left and right wells is shown in (a), and (b) compares a classical instanton (solid curve) with the average over many Monte Carlo histories (error bars).

where the spatial coordinate has been scaled to yield minima at $x = \pm 1$, the energy has been scaled such that the barrier height is 1, and the penetrability is controlled by the one remaining parameter m. As discussed in Chapter 7, in the stationary phase approximation the leading contribution to tunneling properties is given by the instanton solution which satisfies the classical equations of motion in the inverted well. Using the results of Problem 7.11 for the symmetric well in Eq. (8.68), the instanton solution is

$$x(\tau) = \tanh \sqrt{\frac{2}{m}}\tau$$

with action

$$S_0 = \frac{4}{3}\sqrt{2m} \ .$$

Thus, we expect a general trajectory to be composed of segments fluctuating around each minimum, connected by trajectories across the classically forbidden region which fluctuate around the classical instanton. Figure 8.6 shows a segment of a typical Monte-Carlo trajectory connecting the two minima and demonstrates that it has the expected instanton behavior.

Given a periodic trajectory localized in the left well, consider the problem of sequentially updating the configuration with the Metropolis algorithm to randomly sample the action. When the time step is so large that it is comparable to the width of an

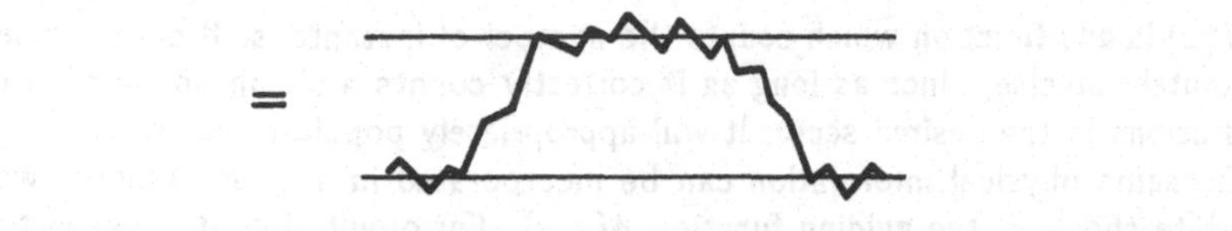

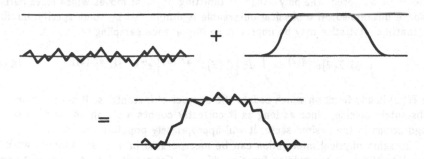

Fig. 8.7 Sketch of two instanton trial move showing the result of adding a classical two-instanton configuration to a trajectory originally localized in one well.

instanton, $\epsilon \sim \sqrt{m}$, an instanton-anti-instanton pair can be created by moving a single coordinate x_m from one minimum to the other. The change in the kinetic contribution to the action $\Delta S \sim \frac{4m}{\epsilon} = 4\sqrt{m}$, is then comparable to the action for the instanton pair $\frac{8}{3}\sqrt{2m} \simeq 3.8\sqrt{m}$. Thus, a pair will be created directly with roughly the correct probability, and once x_m has crossed, there is no additional penalty for subsequent coordinates to cross. When $\epsilon << \sqrt{m}$, as required for the continuum limit, the formation of instantons in a single step is exponentially suppressed and many variables must cross the barrier in a highly correlated configuration with a shape close to that of the classical instanton to avoid an exponentially prohibitive action. As $\epsilon \to 0$, the probability of generating an instanton solution by a specific sequence of random updates becomes arbitrarily small and the number of Metropolis iterations required to change the number of instantons goes to infinity. Within a sector of fixed instanton number, the sequential updating efficiently sums the local fluctuations around the classical solution, even when higher than quadratic terms are important. For example, in Fig. 8.6b, one observes that the average behavior of trajectories in the frame of the instanton closely follows the classical result in the forbidden region and is displaced away from the potential minima in the allowed region, reflecting the significant contributions of anharmonic terms. Thus, we observe a general potential pitfall of the Metropolis method: it may accurately sum all local fluctuations in some subspace, but becomes effectively trapped in that space for any practical number of iterations.

Once the essential problem with sequential updates has been recognized and the underlying physics has been understood, it is simple to generalize the algorithm appropriately. One alternative, sketched in Fig. 8.7, is to use the correlated additive trial move, Eq. (8.59), in which the entire trajectory is shifted by a two-instanton configuration $f_i = \tanh\sqrt{\frac{2}{m}}(\tau_i - \tau_\ell) + \tanh\sqrt{\frac{2}{m}}(\tau_a - \tau_i)$ and the positions of the instanton τ_ℓ, and antiinstanton τ_a, are picked randomly. The cost in action for such a global change is distributed around the classical result, thus assuring that the correct equilibrium distribution of instantons is built up. Alternately offering sequential and global updates thus effectively sums local fluctuations and samples all instanton sectors. Note that offering an unphysical global move will not bias the result because it will never be accepted: the only danger is omitting physical moves which leave part of the space unsampled. If a physical observable is dominated by some specific number of instantons, statistics may be improved by importance sampling

$$\int d\vec{x}\,O(\vec{x})e^{-S(\vec{x})} = \int d\vec{x}\left[O(\vec{x})e^{-\alpha N(\vec{x})}\right]e^{-(S(\vec{x})-\alpha N(\vec{x}))} \quad . \tag{8.69}$$

Here $N(\vec{x})$ is any function which counts the number of instantons. It does not need to be absolutely precise, since as long as it correctly counts a significant portion of the configurations in the desired sector it will appropriately populate that sector.

The same physical information can be incorporated in a guided random walk by appropriate choice of the guiding function $\phi(x,\tau)$. For orientation, it is useful to note that if one wishes to sample the ground state wave function beginning with an ensemble of points in the right well, one should definitely not use the ground state wave function since the drift term $\frac{\phi'}{\phi}$ will inhibit rather than encourage tunneling to the other well. Heuristically, it is clear that something like $e^{-\alpha x}$ is required in the forbidden region to force the ensemble across. The optimal choice $\phi(x,\tau) = e^{(\tau-\beta)H}\phi_a$ always depends on the specific final state ϕ_a appearing in the matrix element $\langle\phi_a|e^{-\beta H}|\phi_b\rangle$ being evaluated. Problem 8.7 presents a detailed example of how points are migrated across the barrier and collected in the region of maximal statistical importance for the matrix element. Note that in contrast to sampling the action, where an unphysical collective move will not be accepted, an inappropriate trial function will migrate the ensemble in the wrong direction. If the trial function becomes too unphysical, the guided random walk may become a biased random walk.

The basic ideas described here for the symmetric double well are applicable to a variety of physical tunneling problems, ranging from the lifetime of many-particle metastable state described in Problem 8.8 to the contribution of the ground state energy of solid ^3He of cyclic interchanges of rings of atoms.

8.4 MANY PARTICLE SYSTEMS

The stochastic methods developed in the preceding section may be applied to a variety of many particle systems. In addition to all the freedom encountered in treating one degree of freedom, we also have the possibility of utilizing alternative functional integrals for the evolution operator.

PATH INTEGRAL IN COORDINATE REPRESENTATION

One obvious approach to the many-particle problem is to try to generalize the methods of Section 8.3 to a Feynman path integral for many particles. We begin by considering the initial value random walk to calculate the ground state of a many-Boson system.

Recall that for a single particle in a potential, the filter $e^{-\beta H(p,x)}$ acts in the space of all one particle states to select out the ground state. Furthermore, the ground state wave function may be chosen positive so that $\psi(x)$ may be sampled as a probability distribution Eq. (8.67) and the diffusion equation for ψ, Eq. (8.63) yields a positive solution. In the many-particle case, $e^{-\beta H(p_1\ldots p_N,\, x_1\ldots x_N)}$ acts in the full space of N-particle states, including all totally symmetric states, totally antisymmetric states, and states of mixed symmetry to select the lowest energy state. For ordinary systems interaction with Coulomb, atomic, or nuclear forces, the low kinetic energy associated with spatial symmetry causes the lowest eigenstate of H to be totally symmetric with respect to particle interchange, so that $e^{-\beta H}$ will automatically project out the Boson ground state. Thus, for a system like liquid ${}^4He$, the ground state may be sampled by repeated application of the infinitesimal evolution operator

$$\left(x_1 \dots x_N \middle| e^{-\epsilon H} \middle| y_1 \dots y_N\right) = \left(\frac{m}{2\pi\epsilon}\right)^{N/2} e^{-\frac{m}{2\epsilon}\sum_i (x_i - y_i)^2 - \frac{\epsilon}{2}\sum_{ij} v(y_i - y_j)} \tag{8.70}$$

to a positive trial function $\phi_T(x_1 \dots x_N)$. Because of the strong short-range repulsion and long-range attraction of the Helium-Helium atomic potential, allowing $v(y_i - y_j)$ to serve as a source-sink term introduces prohibitively large variance. Since, as emphasized in the discussion in Section 8.1, importance sampling becomes especially important in high dimensions, it is essential to use a guided random with a trial function $\phi(x_1 \dots x_n)$ which has physically reasonable short-range correlations. One convenient possibility is the Jastrow form $\phi(\vec{x}) = \prod_{i \neq j} f(x_i - x_j)$ where f is determined variationally.

As derived in Problem 8.6, the infinitesimal evolution operator for a guided random walk in the many-particle case is

$$\left(x_1 \dots x_N \middle| \phi e^{-\epsilon \hat{H}} \frac{1}{\phi} \middle| y_1 \dots y_N\right) = \prod_i \left[\frac{m}{2\pi\epsilon_i}\right]^{1/2} e^{-\sum_i \frac{m}{2\epsilon_i}(x_i - y_i - D_i(\vec{y}))^2 - S} \tag{8.71a}$$

where the source term is

$$S = \epsilon \frac{\left(H - \frac{\partial}{\partial\tau}\right)\phi}{\phi}\bigg|_{\vec{y}} - \frac{1}{2}\sum_{ij}(x_i - y_i)(x_j - y_j)\frac{\partial}{\partial y_i}\frac{\partial}{\partial y_j}\ln\phi(\vec{y}) \tag{8.71b}$$

the drift term is

$$D_i(\vec{y}) = \frac{\epsilon}{m}\frac{\partial}{\partial y_i}\ln\phi(\vec{y}) \tag{8.71c}$$

and the effective time step is

$$\epsilon_i = \epsilon\left[1 + \frac{\epsilon}{m}\frac{\partial^2}{\partial y_i^2}\ln\phi\right] \quad . \tag{8.71d}$$

In contrast to the case of Bosons, formidable problems are encountered in formulating the initial value random walk for many-Fermion systems. The essential difficulty is the fact that $e^{-\beta H}\phi_T$ filters out the lowest state of any symmetry, so that at large times the component of the lowest antisymmetric state $e^{-\beta\epsilon_A}\langle\psi_A|\phi_T\rangle\,|\psi_A\rangle$ is exponentially suppressed relative to the component of the lowest symmetric state $e^{-\beta E_S}\langle\psi_S|\phi_T\rangle|\psi_S\rangle$. At some point in the calculation, it is necessary to project onto antisymmetric states. If this projection could be done exactly, there would be no problem. If it is only done stochastically, there is competition between the exponential growth of the symmetric noise $\sim e^{\beta(E_A - E_0)}$ and the stochastic projection error which only decreases as $\frac{1}{\sqrt{N}}$.

To project onto the antisymmetric space, we write the infinitesimal evolution operator between antisymmetric coordinate states

$$\begin{aligned}\{x_1 \dots x_n|e^{-\epsilon H}|y_1 \dots y_N\} &= \left(\frac{m}{2\pi\epsilon}\right)^{N/2} \sum_P (-1)^P e^{-\frac{m}{2\epsilon}\sum_i (x_{P_i}-y_i)^2 - \frac{\epsilon}{2}\sum_{ij} v(y_i-y_j)} \\ &= \left(\frac{m}{2\pi\epsilon}\right)^{N/2} \operatorname{Det} M \; e^{-\frac{m}{2\epsilon}\sum_i (x_i-y_i)^2 - \frac{\epsilon}{2}\sum_{ij} v(y_i-y_i)}\end{aligned} \tag{8.72}$$

where

$$M_{ij} = e^{-\frac{m}{2\epsilon}\left[(x_j-y_i)^2-(x_i-y_i)^2\right]} \quad .$$

The problem now arises from the minus signs in the determinant $\operatorname{Det} M$. To see the effect of these signs most clearly, consider the special case of a two-Fermion system, which when written in relative coordinates corresponds to solving for the lowest odd state in a symmetric potential. The antisymmetrized evolution operator in this case, with relative coordinate $\vec{x} = \vec{x}_1 - \vec{x}_2$ and reduced mass μ, is

$$\begin{aligned}(\langle\vec{x}| - \langle-\vec{x}|)\, e^{-\epsilon H}|\vec{y}\rangle &= \sqrt{\frac{\mu}{2\pi\epsilon}}\left[e^{-\frac{\mu}{2\epsilon}(\vec{x}-\vec{y})^2} - e^{-\frac{\mu}{2\epsilon}(\vec{x}+\vec{y})^2}\right] e^{-\epsilon v(|\vec{y}|)} \\ &= \sqrt{\frac{\mu}{2\pi\epsilon}}\left(1 - e^{-\frac{2\mu}{\epsilon}\vec{x}\cdot\vec{y}}\right) e^{-\frac{\mu}{2}(\vec{x}-\vec{y})^2 - \epsilon v(|\vec{y}|)} \quad .\end{aligned} \tag{8.73}$$

As sketched in Fig. 8.8, one may now view the evolution as the diffusion of both positive and negative points. The kinetic contribution to the propagator for a point at y to x is composed of two parts as shown in Fig. 8.8a. Gaussian diffusion about the point y with no change in sign and Gaussian diffusion about the point $-y$ with the sign of the point changed. This sum is the exact odd state propagator shown by the dashed line. The sequence of configurations obtained by starting with an arbitrary odd wave function and diffusing both positive and negative populations is shown in Fig. 8.8b. Both the positive and negative solutions separately approach the symmetric ground state and the sum, which corresponds to first odd state, becomes exponentially suppressed as shown by the dotted line in the bottom graph.

In the special case of one spatial dimension, the sign cancellation problem may be circumvented by working in the ordered subspace $x_1 < x_2 < \dots < x_N$ with the boundary condition that the wave function vanish whenever any two adjacent coordinates coincide. The wave function in this sector is positive and the complete wave function may be obtained from it by antisymmetry. In the two particle example, this corresponds to evolution in the domain of positive relative coordinate where the wave function may be chosen positive and the kinetic propagator is given by the dashed line in Fig. 8.8a. Note that one may group the factor $\left(1 - e^{-\frac{2\mu}{\epsilon}xy}\right) = e^{-V_p}$ with the potential term and view it as a repulsive "potential" V_p arising from the Pauli principle.

Spin cancellations for Fermions in more than one dimension, however, cannot be avoided. In two or more dimensions, the configuration obtained in a single step with a sign change by interchanging particles i and j may also be reached in a series of steps with no sign change in which i and j simply circle around each other. This difference between one and higher dimensions reflects the fundamental difference in

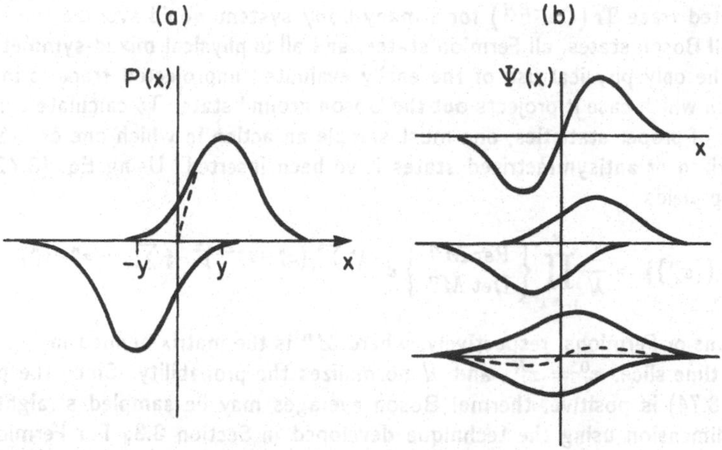

Fig. 8.8 Lowest odd state in a symmetric potential. The odd-state kinetic propagator is shown in (a) and a sequence of configurations obtained by diffusion by Eq. (8.73) is shown in (b).

how nodal surfaces and the surfaces defined by antisymmetry depend on dimension. In d dimension, the condition $\psi(\vec{r}_1, \vec{r}_2 \ldots \vec{r}_n) = 0$ defines a $dn-1$ dimensional nodal surface, whereas antisymmetry specifies $\frac{n(n-1)}{2}$ constraints of the form $\vec{r}_i = \vec{r}_j$ corresponding to $(dn-d)$ dimensional surfaces. Thus, for $d=1$, the dimensions of these surfaces coincide, and antisymmetry alone can completely specify the nodal surface of the wave function. For higher dimension, however, the nodal surfaces have higher dimension $(dn-1)$ than the surfaces specified by antisymmetry $(dn-n)$, so that antisymmetry alone cannot specify the nodal surfaces. For example, for two particles in two dimensions with relative coordinate $\vec{r}$, antisymmetry of the relative wave function $\psi(\vec{r}) = -\psi(-\vec{r})$ only specifies a single point $\psi(0) = 0$, whereas the nodal surface is a line in the $x-y$ plane passing through the origin. Hence, for $d>1$ antisymmetry alone cannot uniquely specify a sector in which the wave function is positive.

Whereas there is no complete solution to the sign problem for Fermions in more than one dimension, there are two useful strategies. One possibility is to fix the nodes by selecting the best possible trial wave function and to perform a guided random walk in which no particles are permitted to cross the nodal surfaces defined by the trial function. Within each region bounded by nodal surfaces, the problem is effectively Bosonic, and the result is a variational calculation and upper bound on the energy corresponding to the best possible wave function with the specified nodes. A second possibility is to perform a transient estimate starting with the best fixed-node distribution and releasing the fixed-node restrictions. As points cross nodes, cancellations will give rise to statistical errors which increase exponentially with evolution time while the system relaxes to its ground state. The early stages of the relaxation may then be calculated by using sufficiently large sample size.

The same symmetry considerations encountered in the initial value diffusion problem for many particle systems also arise when the global action is sampled. The unrestricted trace $\mathrm{Tr}\left(\mathcal{O}e^{-\beta H}\right)$ for a many-body system would average the observable $\mathcal{O}$ over all Boson states, all Fermion states, and all unphysical mixed-symmetry states. Hence, the only physical use of the easily evaluated unprojected trace is in the limit $\beta \to \infty$ in which case it projects out the Boson ground state. To calculate true thermal averages of proper statistics, one must sample an action in which one or more sets of symmetrized or antisymmetrized states have been inserted. Using Eq. (8.72) at each time step yields

$$P_S\left(\{x_i^n\}\right) = \frac{1}{\mathcal{N}} \prod_{n=1}^{N} \left\{ \begin{matrix} \mathrm{Per}\, M^n \\ \mathrm{Det}\, M^n \end{matrix} \right\} e^{-\frac{m}{2\epsilon}\sum_i (x_i^n - x_i^{n-1})^2 - \frac{\epsilon}{2}\sum_{ij} v(x_i^n - y_i^n)} \tag{8.74}$$

for Bosons or Fermions, respectively, where M^n is the matrix defined in Eq. (8.72) for the nth time slice, $x_i^0 = x_i^N$, and $\mathcal{N}$ normalizes the probability. Since the permanent in Eq. (8.74) is positive, thermal Boson averages may be sampled straightforwardly in any dimension using the technique developed in Section 8.3. For Fermions, negative contributions of the determinant may also be avoided in the special case of one dimension by working in the ordered subspace as before.

Finally, before considering alternative functional integrals for the many-body evolution operator, it is useful to note that the path integral Monte Carlo approach illustrated above for the case of N particles in coordinate representation may be applied to a variety of other physical systems with many degrees of freedom. For example, although the most common way of solving lattice gauge theories is to sample the global Lagrangian action using the Metropolis algorithm, one may also treat the Hamiltonian form of lattice gauge theory as a many-body problem. The degrees of freedom are the elements of a group defined in each of the links of a lattice in d spatial dimensions (see, for example, Creutz, 1983). In the simple case of $U(1)$, corresponding to quantum electrodynamics, a group element at a link originating at site n in direction k may be written $U = e^{iA_k(n)}$ and the resulting Hamiltonian is

$$H = \sum_{k,n} \frac{1}{2} E_k^2(n) + \lambda \sum_{kn} [1 - \cos B_k(n)] \tag{8.75a}$$

where B is the sum of the link variables around an elementary plaquette corresponding to the lattice curl

$$B_i(n) = \epsilon_{ijk}\left[A_k(n + e_j) - A_k(n)\right] \tag{8.75b}$$

and the commutation relations $[A_k(n),\ E_{k'}(n')] = i\delta_{kk'}\delta_{nn'}$ imply $E_k(n) = -i\frac{\partial}{\partial A_\mu(n)}$. Note that conventional electrodynamics is recovered when the lattice spacing $\to 0$ and the $\frac{1}{2}B^2$ term from the cosine dominates. Discrete $U(1)$ lattice gauge theory thus looks just like a many-body problem with coordinates x_i replaced by compact variables $0 < A_k(n) < 2\pi$ and a four-body potential connecting links on common plaquettes. The only difference relative to the evolution operator (8.70) arises from the kinetic term for the compact variable A, which follows by using a complete set of periodic states $\langle A|m\rangle = e^{imA}$ and the Poisson sum formula

$$\langle A'|e^{-\frac{\epsilon}{2}E^2}|A\rangle = \sum_m \frac{1}{2\pi}\langle A'|e^{-\frac{\epsilon}{2}E^2}|m\rangle\langle m|A\rangle$$

$$= \sum_m \frac{1}{2\pi} e^{im(A'-A)} e^{-\frac{\epsilon}{2}m^2} \qquad (8.76)$$

$$= \sum_n \frac{1}{\sqrt{2\pi\epsilon}} e^{-\frac{1}{2\epsilon}(A'-A+2\pi n)^2} .$$

In the limit $\epsilon \to 0$, the contributions of the periodic images outside the fundamental interval $[0, 2\pi]$ are exponentially suppressed, and the evolution for A is precisely the familiar Gaussian diffusion encountered in the many-body problem. The infinitesimal evolution operator for non-Abelian groups has analogous structure. For example, in the case of $SU(2)$, group elements may be written in the form $U = a_0 I + i \sum_{i=1}^{3} a_i \tau_i$ where the τ_i denote Pauli matrices and the real coefficients a_j satisfy $\sum_{j=0}^{4} a_j^2 = 1$ and are thus constrained to the surface of a three-sphere. In the limit $\epsilon \to 0$, the kinetic term becomes $\frac{1}{(2\pi\epsilon)^{3/2}} e^{-\frac{\Delta S^2}{2\epsilon}}$, where ΔS is the distance between $U(\vec{a})$ and $U(\vec{a}')$ and produces Gaussian diffusion on the three-sphere.

The generalization of the potential term in (8.75) is the sum over all plaquettes of $\mathrm{Re}\,\mathrm{Tr}(UUUU)$ where the product of the U's is taken around each plaquette. Note that this term still has the form of a four-body potential in the compact variables and that $\mathrm{Re}(e^{iA}e^{iA}e^{iA}e^{iA})$ reproduces the $\cos B$ potential, (Eq. (8.75)) for $U(1)$. Further details are provided by Chin *et al.* (1984, 1985). Yet another example of this path integral Monte Carlo approach is the case of spin systems discussed subsequently, in which a random walk in the space of continuous coordinates x_i is replaced by a random walk in the space of discrete spin projections m_i.

FUNCTIONAL INTEGRALS OVER FIELDS

It is natural to ask whether any of the functional integral representations of the many-body evolution operator using fields are preferable to the particle coordinate path integral for stochastic calculations. *A priori*, it appears counterproductive to resort to field variables instead of particle coordinates, since continuum fields must be approximated on a discrete mesh. The field representation is thus no longer exact in principle, and in order to obtain reasonable accuracy, the number of degrees of freedom required is immensely larger than the number of particle coordinates. Recall, for example, the case of ^{208}Pb at the beginning of Section 8.1, which for even moderate accuracy would require a $(100)^3$ lattice with 10^6 degrees of freedom to describe the physics of 624 physical degrees of freedom. Hence, a functional integral over fields will only be useful when it circumvents an otherwise insurmountable problem, such as the Fermion sign problem in more than one dimension.

Consider first the possibility of sampling the action for the coherent state partition function Eq. (2.60)

$$Z = \int \mathcal{D}(\phi^* \phi)\, e^{-S(\phi^*, \phi)} \qquad (8.77)$$

where ϕ_α is a complex variable or Grassmann number and α denotes the point of a space-time lattice and any internal quantum numbers. For Bosons, the integrand is unavoidably complex, since even the one-body piece $\phi_\alpha^* \left(\frac{\partial}{\partial \tau} - \mu + T\right)_{\alpha\beta} \phi_\beta$ is non-Hermitian, and one is thus confronted with the usual sign problem in sampling the real part. Note that although the coherent state path integral for Bosons is formally similar to the field theory of a scalar field, only the latter is suitable for Monte Carlo sampling since the action for a scalar field may be written in terms of a real field ϕ and contains the Hermitian contribution $\phi_\alpha\left(\frac{\partial^2}{\partial\tau^2} - \nabla^2\right)_{\alpha\beta}\phi_{\alpha\beta}$. For Fermions, practical methods have not been developed for Monte Carlo integration over Grassmann variables. Recall, for example, that a matrix representation of Grassmann numbers with n generators requires $2^n \times 2^n$ matrices.

Thus, in order to obtain a suitable functional integral over field variables for either Bosons or Fermions, it is necessary to introduce an auxiliary field and integrate out the complex or Grassmann variables ϕ_α as in Chapter (7).

$$
\begin{aligned}
Z &= \int \mathcal{D}(\phi^*\phi) e^{-\Sigma \int_0^\beta d\tau \phi^*_\alpha \left(\frac{\partial}{\partial\tau} - \mu + T\right)_{\alpha\beta}\phi_\gamma - \Sigma \frac{1}{2}\int_0^\beta d\tau \phi^*_\alpha \phi_\gamma v_{\alpha\beta\gamma\delta}\phi^*_\beta\phi_\delta} \\
&= \int \mathcal{D}(\phi^*\phi)\mathcal{D}(\sigma) e^{\frac{1}{2}\Sigma\int_0^\beta d\tau \sigma_{\alpha\gamma}v_{\alpha\beta\gamma\delta}\sigma_{\beta\delta} - \Sigma\int_0^\beta d\tau\phi^*_\alpha\left(\frac{\partial}{\partial\tau} - \mu + T + \frac{1}{2}v_{\alpha\beta\gamma\delta}\sigma_{\beta\delta}\right)\phi_\gamma} \qquad (8.78) \\
&= \int \mathcal{D}(\sigma)\left[\mathrm{Det}\left(\frac{\partial}{\partial\tau} - \mu + T + v_{\alpha\beta\gamma\delta}\sigma_{\beta\delta}\right)\right]^{-\varsigma} e^{\frac{1}{2}\Sigma\int_0^\beta d\tau\sigma_{\alpha\gamma}v_{\alpha\beta\gamma\delta}\sigma_{\beta\delta}} \ .
\end{aligned}
$$

Note that in order for the Gaussian integral over σ to converge, all the eigenvalues of v must be negative so it may be necessary to introduce a constant or diagonal shift in the potential (see Sugiyama and Koonin, 1986).

The essential question is when the resulting determinant is positive so that the action may be sampled without sign cancellations. Although positivity is not assured in general, the integrand is positive for one important class of problems: those in which the potential is independent of an internal symmetry having an even number of states. Consider the case of Fermions with a spin independent interaction and write the space, time, and spin labels included in α explicitly $\phi_\alpha = \phi_m(x,\tau)$. Then the interaction is $\int dx\, dx' \sum_m \phi^*_m(x)\phi_m(x)\ v(x-x') \sum_{m'} \phi^*_{m'}(x')\phi_{m'}(x')$, only a single spin-independent auxiliary field is required and

$$
\begin{aligned}
Z &= \int \mathcal{D}(\phi^*\phi)\mathcal{D}(\sigma) e^{\frac{1}{2}\int dx\, dx' d\tau\, \sigma(x,\tau)v(x-x')\sigma(x',\tau)} \\
&\quad \times e^{-\sum_m \int dx\, dx' d\tau\, d\tau' \phi^*_m(x,\tau)\left[\frac{\partial}{\partial\tau} - \mu + T + \frac{1}{2}\int dy\, v(x-y)\sigma(y,\tau)\right]_{x\tau,\, x'\tau'}\phi_m(x',\tau')} \\
&= \int \mathcal{D}(\sigma)\left[\mathrm{Det}\, M(\sigma)\right]^{2S+1} e^{-S(\sigma)} \qquad (8.79a)
\end{aligned}
$$

where

$$
M_{x\tau;\,x'\tau'} \equiv \left[\frac{\partial}{\partial\tau} - \mu + T + \frac{1}{2}\int dy\, v(x-y)\sigma(y,\tau)\right]_{x\tau;\,x'\tau'} \qquad (8.79b)
$$

and

$$
S(\sigma) = -\frac{1}{2}\int dx\, dx'\, d\tau\ \sigma(x,\tau)\, v(x-x')\, \sigma(x',\tau) \ . \qquad (8.79c)
$$

Since Fermions have half-integer spins, only even powers $2S+1$ of the determinant arise and the integrand is positive. The same argument applies for any other internal symmetries such as isospin, color, or flavor and ground state properties for even numbers of Bosons may be obtained by treating them as Fermions with spin degeneracy N as in Section 3.1.

Once the partition function has been expressed as an integral over σ, Eq. (8.79), observables are evaluated as the sum of contractions using Wick's theorem in the usual way (see Eq. (2.84)):

$$\begin{aligned}
&\langle\hat{\psi}(x_1\tau_1)\ldots\hat{\psi}(x_n\tau_n)\hat{\psi}^\dagger(x'_n\tau'_n)\ldots\hat{\psi}^\dagger(x'_1\tau'_1)\rangle \\
&\quad=\frac{1}{Z}\int \mathcal{D}(\phi^*\phi)\mathcal{D}(\sigma)e^{-S(\sigma)-\sum_m\int dxd\tau dx'd\tau'\ \phi^*_m(x\tau)M_{x\tau;\,x'\tau'}\phi^*_m(x\tau)} \\
&\qquad\times\phi_{m_1}(x_1\tau_1)\ldots\phi_{m_n}(x_n\tau_n)\phi^*_{m'_n}(x'_n\tau'_n)\ldots\phi^*_{m'_1}(x'_1\tau'_1) \qquad (8.80)\\
&\quad=\frac{1}{Z}\int\mathcal{D}(\sigma)[\mathrm{Det}\,M(\sigma)]^{2S+1}e^{-S(\sigma)} \\
&\qquad\times\sum_P\varsigma^P M^{-1}_{x_{P1}\tau_{P1};x'_1\tau'_1}(\sigma)\delta_{m_{P1}m'_1}\ldots M^{-1}_{x_{Pn}\tau_{Pn};x'_n\tau_n}(\sigma)\delta_{m_{Pn}m'_n}\,.
\end{aligned}$$

Note that even though $[\mathrm{Det}\,M(\sigma)]^{2S+1}\,e^{-S(\sigma)}$ is positive, the signs ς^P arising in Wick's theorem may cause uncontrollable cancellation for some operators. The biggest practical problems in applying Eq. (8.80) are encountered in evaluating $\mathrm{Det}\,M$ and M^{-1}. The quantity $\mathrm{Det}\,M(\sigma)e^{-S(\sigma)}$ may be sampled directly using the microcanonical method, Eq. (8.48), or the exponent $\ln\mathrm{Det}\,M(\sigma)-S(\sigma)$ may be updated using the relation $\frac{\delta}{\delta\sigma}\ln\mathrm{Det}\,M(\sigma)=\sum\frac{\delta M_{ij}}{\delta\sigma}M^{-1}_{ji}$. The inverses arising in matrix elements, Eq. (8.80) or in updating the action may be evaluated by conventional sparse matrix techniques such as the conjugate gradient method, or by stochastic methods such as the Neumann-Ulam method (8.41) or sampling a Gaussian integral $M^{-1}_{qr}=\int d\{\phi\}e^{-\frac{1}{2}\phi_i M_{ij}\phi_j}\phi_q\phi_r/\int d\{\phi\}e^{-\frac{1}{2}\phi_i M_{ij}\phi_j}$.

Another closely related method is to use the operator form of the Hubbard Stratonovich transformation and replace the trace by a matrix element of Slater determinants. (Sugiyama and Koonin, 1985). Thus, for example, the ground state energy is calculated by evaluating

$$E_0=\lim_{\beta\to\infty}\frac{\int\mathcal{D}(\sigma)e^{-S(\sigma)}\langle\Phi|U_\sigma|\Phi\rangle\frac{\langle\Phi|H\,U_\sigma|\Phi\rangle}{\langle\Phi|U_\sigma|\Phi\rangle}}{\int\mathcal{D}(\sigma)e^{-S(\sigma)}\langle\Phi|U_\sigma|\Phi\rangle} \qquad (8.81)$$

where

$$U_\sigma=T\,e^{-\sum_m\int dxdx'd\tau d\tau'\ \hat{\psi}^+_m(x,\tau)\left[T+\frac{1}{2}\int dy\,v(x-y)\,\sigma(y,\tau)\right]_{x\tau,\,x'\tau}\hat{\psi}_m(x',\tau)}$$

and $|\Phi\rangle$ denotes an N-particle Slater determinant. Since U_σ corresponds to evolution in a time-dependent one-body field specified by $\sigma(x,\tau)$, $U_\sigma|\phi\rangle$ is a Slater determinant and $\langle\phi|U_\sigma|\phi\rangle$ is a determinant of overlaps between the single-particle wave functions in $\langle\phi|$ and $U_\sigma|\phi\rangle$. Again for spin-independent interactions, one obtains the overlap determinant to the $(2S+1)^{\text{th}}$ power.

8.5 SPIN SYSTEMS AND LATTICE FERMIONS

Additional interesting possibilities arise in the stochastic solution of quantum spin systems on a lattice and of closely related lattice Fermion models.

The magnetic properties of solids may be described by a quantum spin Hamiltonian since the exchange interaction between electrons gives rise to an effective spin-spin interaction (see, for example Ashcroft and Mermin, 1976). For illustrative purposes, it suffices to consider the simple case of the Heisenberg Hamiltonian for spin one-half with nearest neighbor interactions

$$H = -J \sum_{\langle ij \rangle} \vec{S}_i \cdot \vec{S}_j \tag{8.82}$$

where positive J corresponds to ferromagnetic coupling and the spin one-half operator is written in terms of the Pauli matrices $\vec{S}_i = \frac{1}{2}\{\sigma_i^x, \sigma_i^y, \sigma_i^z\}$. Extension of the basic ideas to anisotropic couplings, external fields, and higher spin is generally straightforward, and further details may be found in the review by DeRaedt and Lagendijk (1985).

The connection in one dimension between lattice spin Hamiltonians and lattice Fermion models, such as the Hubbard model (1963, 1964) is seen by noticing that the raising and lowering operators

$$\begin{aligned} a_j^\dagger &= \frac{1}{2}\sigma_j^+ = \frac{1}{2}\left(\sigma_j^x + i\sigma^y{}_j\right) \\ a_j &= \frac{1}{2}\sigma_j^- = \frac{1}{2}\left(\sigma_j^x - i\sigma^y{}_j\right) \end{aligned} \tag{8.83a}$$

behave nearly like Fermions. They have Fermion commutation relations on sites

$$\{a_i, a_i^\dagger\} = 1 \qquad a_i^2 = a_i^{\dagger 2} = 0 \tag{8.83b}$$

but have Bosonic commutations relations between sites

$$[a_i, a_j^\dagger] = [a_i, a_j] = [a_i^\dagger, a_j^\dagger] = 0 \ . \tag{8.83c}$$

True Fermion operators may be constructed from the a's via the Jordan Wigner transformation

$$c_i = e^{i\pi \sum_{j=1}^{i-1} a_j^\dagger a_j} a_i \qquad c_i^\dagger = e^{-i\pi \sum_{j=1}^{i-1} a_j^\dagger a_j} a_i^\dagger \tag{8.84}$$

which inserts the requisite minus signs by counting the number of occupied sites to the left of the ith site (see Problem 11). Thus, for example, we may rewrite a lattice Fermion model as a spin $\frac{1}{2}$ model as follows

$$\begin{aligned} &\sum_i \left[-T\left(c_i^\dagger c_{i+1} + c_{i+1}^\dagger c_i\right) + U c_i^\dagger c_i c_{i+1}^\dagger c_{i+1}\right] \\ &= \sum_i \left[-2T\left(S_i^x S_{i+1}^x + S_i^y S_{i+1}^y\right) + U\left(S_i^z S_{i+1}^z + S_i^z + \frac{1}{4}\right)\right] \ . \end{aligned} \tag{8.85}$$

In general, except for technical details such as treating the Jordan Wigner phases with periodic boundary conditions one may pass freely between the spin and Fermion language, and it is often useful to think of the spin-projection representation as an occupation number representation.

CHECKERBOARD DECOMPOSITION

One alternative to evaluate the evolution operator for spin systems is to use the Trotter formula as we have for other path integrals (Suzuki, 1976). To break up the evolution operator into a sequence of infinitesimal steps which are particularly easy to evaluate, it is useful to decompose the Hamiltonian into sums of non-overlapping cell Hamiltonians. To illustrate the decomposition, which may be performed for any lattice in any dimension, we consider a one-dimensional chain and break up H into nearest neighbor interactions beginning on odd and even sites

$$H \equiv H_{\text{odd}} + H_{\text{even}}$$

$$H_{\left\{\substack{\text{odd}\\\text{even}}\right\}} = -J \sum_{i\left\{\substack{\text{odd}\\\text{even}}\right\}} \vec{S}_i \cdot \vec{S}_{i+1} \; . \tag{8.86}$$

The infinitesimal operator is written in the usual way $e^{-\epsilon H} \simeq e^{-\frac{\epsilon}{2}H_{\text{odd}}} e^{-\epsilon H_{\text{even}}} e^{-\frac{\epsilon}{2}H_{\text{odd}}}$, so that when complete sets of intermediate states are inserted, the evolution operator has the form

$$\sum_{M^1 \ldots M^N} \langle M' | e^{-\frac{\epsilon}{2}H_{\text{odd}}} | M^N \rangle \ldots \langle M^3 | e^{-\epsilon H_{\text{odd}}} | M^2 \rangle \\ \times \langle M^2 | e^{-\epsilon H_{\text{even}}} | M^1 \rangle \langle M^1 | e^{-\frac{\epsilon}{2}H_{\text{odd}}} | M \rangle \; . \tag{8.87a}$$

The essential advantage of this result is that since H_{odd} and H_{even} are sums of commuting operators each of the matrix elements of $e^{-\epsilon H_{\text{odd}}}$ and $e^{-\epsilon H_{\text{even}}}$ breaks up into products of independent two-spin matrix elements. For example

$$\langle M^2 | e^{-\epsilon H_{\text{even}}} | M^1 \rangle = \prod_{i\,\text{even}} \langle m_i^2 m_{i+1}^2 | e^{\epsilon J S_i \cdot S_{i+1}} | m_i^1 m_{i+1}^1 \rangle \qquad 8.87b$$

where $\{M\}$ denotes the set of spin projections on the sites $\{m_1 m_2 \ldots m_N\}$. The sequential action of H_{even} and H_{odd} is sketched in Fig. 8.9. Since the total spin projection is conserved, as emphasized by Hirsch *et al.* (1981), it is convenient to view the evolution in terms of world lines representing the trajectories of the up spins or equivalently, lattice Fermions occupation numbers. Using the familiar identity relating $\vec{\sigma}_i \cdot \vec{\sigma}_j$ to the spin exchange operator P_{ij}

$$P_{ij} = \frac{1}{2}(\vec{\sigma}_i \cdot \vec{\sigma}_j + 1)$$

$$= \begin{pmatrix} 1 & 0 & 0 & 0 \\ 0 & 0 & 1 & 0 \\ 0 & 1 & 0 & 0 \\ 0 & 0 & 0 & 1 \end{pmatrix} \quad \begin{array}{|rr} 1 & 1 \;\rangle \\ 1 & -1 \;\rangle \\ -1 & 1 \;\rangle \\ -1 & -1 \;\rangle \end{array} \tag{8.88}$$

and the fact $P^2 = 1$, the infinitesimal evolution operator Eq. (8.87) may be written

$$e^{\frac{\epsilon J}{4}\sigma_i \sigma_j} = e^{-\frac{\epsilon J}{4}} e^{\frac{\epsilon J}{2} P_{ij}}$$

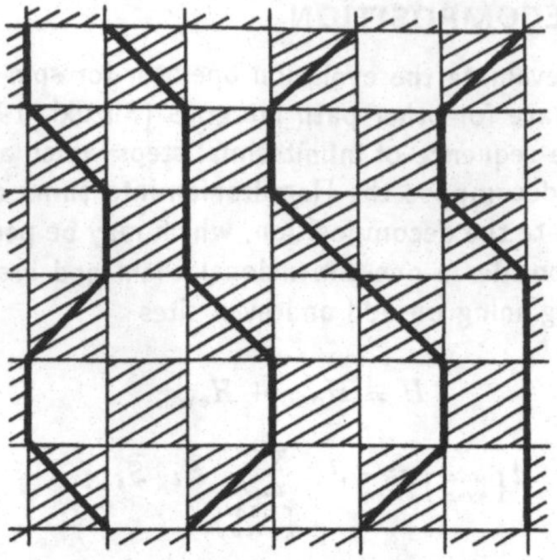

Fig. 8.9 Checkerboard decomposition of $e^{-\beta H}$ in one space dimension. The shaded regions indicate sites connected by H_{even} and H_{odd}. The solid world lines denote one possible trajectory for the up spins under the sequential action of H_{even} and H_{odd} in a small portion of the space-time lattice. Because at each time step the system evolves either under H_{even} or H_{odd}, the world lines are constrained to the shaded regions.

$$
\begin{aligned}
&= e^{-\frac{\epsilon J}{4}}\left[\cosh\frac{\epsilon J}{2} + \sinh\frac{\epsilon J}{2} P_{ij}\right] \\
&= e^{-\frac{\epsilon J}{4}}\begin{bmatrix} e^{\frac{\epsilon J}{2}} & 0 & 0 & 0 \\ 0 & \cosh\frac{\epsilon J}{2} & \sinh\frac{\epsilon J}{2} & 0 \\ 0 & \sinh\frac{\epsilon J}{2} & \cosh\frac{\epsilon J}{2} & 0 \\ 0 & 0 & 0 & e^{\frac{\epsilon J}{2}} \end{bmatrix} \\
&\xrightarrow[\epsilon\to 0]{} \begin{bmatrix} 1+\frac{\epsilon J}{4} & 0 & 0 & 0 \\ 0 & 1-\frac{\epsilon J}{4} & \frac{\epsilon J}{2} & 0 \\ 0 & \epsilon\frac{J}{2} & 1-\frac{\epsilon J}{4} & 0 \\ 0 & 0 & 0 & 1+\frac{\epsilon J}{4} \end{bmatrix} .
\end{aligned}
\tag{8.89}
$$

For the Heisenberg ferromagnet, $J > 0$, all the matrix elements are positive and the initial-value random walk for any dimension lattice proceeds like the coordinate space path integral for the many-Boson problem with the world lines for the up spins corresponding to the trajectories for particles.

The antiferromagnet naively appears to have sign problems analogous to the many-Fermion problem because of the negative off-diagonal matrix elements appearing in Eq. (8.89) for $J < 0$. However, a cure is suggested by the fact that classical magnets and ferromagnets on bipartite lattices are equivalent. By definition, a bipartite lattice is composed of two sublattices such that all the nearest neighbors of a site on one sublattice belong to the other sublattice and *vice versa*. Common examples include the linear chain, square, cubic, and body-centered cubic lattices. Classically, if all the spins on one sublattice are mapped into new variables with a sign change $S_A \rightarrow -\tilde{S}_A$ while the signs are unchanged on the other lattice $S_B \to \tilde{S}_B$, then $J\sum_{ij}\vec{S}_i\cdot\vec{S}_j \to -J\sum_{ij}\tilde{S}_i\tilde{S}_j$. Quantum mechanically, the ferromagnet and antiferromagnet are fundamentally different and therefore cannot be mapped exactly into each other. However, we are still free to choose different bases on the two sublattices for our own convenience. The appropriate transformation for our present purpose is to rotate all states on one sublattice by 180° around the z-axis $|\tilde{m}\rangle = e^{i\frac{\pi}{2}\sigma_z}|m\rangle$. The transformation of the spin operators under this operation is $\sigma^x \to -\sigma^x$, $\sigma^y \to -\sigma^y$ and $\sigma^z \to \sigma^z$ so that $\vec{\sigma}_i\cdot\vec{\sigma}_j \to -\vec{\sigma}_i\cdot\vec{\sigma}_j + 2\sigma_i^z\sigma_j^z$ and the troublesome off-diagonal minus signs are removed. The ferromagnetic and antiferromagnetic cases are conveniently combined by defining the matrix

$$P(\eta) \equiv \begin{bmatrix} 1 & 0 & 0 & 0 \\ 0 & 0 & \eta & 0 \\ 0 & \eta & 0 & 0 \\ 0 & 0 & 0 & 1 \end{bmatrix} \qquad \eta = \pm 1 \tag{8.90}$$

which corresponds to the matrix elements of P_{ij} in the original basis for $\eta = 1$ and the matrix elements in the transformed basis for $\eta = -1$. Since $P(\eta)^2 = 1$, we may write

$$\begin{aligned} e^{\frac{\epsilon\eta|J|}{4}\sigma_i\sigma_j} &= e^{-\frac{\epsilon\eta|J|}{2}P(\eta)} \\ &= e^{-\frac{\epsilon\eta|J|}{4}} \begin{bmatrix} e^{\frac{\epsilon\eta|J|}{2}} & 0 & 0 & 0 \\ 0 & \cosh\frac{\epsilon|J|}{2} & \sinh\frac{\epsilon|J|}{2} & 0 \\ 0 & \sinh\frac{\epsilon|J|}{2} & \cosh\frac{\epsilon|J|}{2} & 0 \\ 0 & 0 & 0 & e^{\frac{\epsilon\eta|J|}{2}} \end{bmatrix} \\ &\xrightarrow[\epsilon\to 0]{} \begin{bmatrix} 1+\frac{\eta\epsilon|J|}{4} & 0 & 0 & 0 \\ & 1-\frac{\eta\epsilon|J|}{4} & \frac{\epsilon|J|}{2} & 0 \\ 0 & \frac{\epsilon|J|}{2} & 1-\frac{\eta\epsilon|J|}{4} & 0 \\ 0 & 0 & 0 & 1+\frac{\eta\epsilon|J|}{4} \end{bmatrix} \end{aligned} \tag{8.91}$$

where $\eta = 1$ for a ferromagnet and $\eta = -1$ for an antiferromagnet. The explicit η dependence of the diagonal matrix elements contains the physical difference between quantum ferromagnets and antiferromagnets. Note that in contrast to the Fermion sign problem which was cured only in one spatial dimension, the sublattice transformation has eliminated minus signs for bipartite lattices of any spatial dimension. For other antiferromagnetic problems, however, such as the triangular lattice which is important because it is the simplest example of frustrated spins and the problem of cyclic exchanges in solid ^{3}He, the minus signs cannot be removed completely.

When the trace of the evolution operator is evaluated instead of a specific matrix element as in Eq. (8.87), one must again address the problem of trapping in a restricted subset of all spin states. Sequential updating of individual spins only samples states of fixed total spin projection (or total Fermion number) since H commutes with S_z. In order to sample the full space, microreversible global moves must also be included which offer the possibility of adding or removing an entire world line. For Fermions with periodic boundary conditions, which may be represented by world lines on a torus, one must similarly allow for transitions between configurations of different winding number n, where n is the number of times a point on a world line must circle the torus before returning to its original position.

SPECIAL METHODS FOR SPINS

A method by Hanscomb (1962) exploits the special properties of spin $\frac{1}{2}$ algebra to avoid the finite ϵ errors inherent in path integrals and represents the thermal trace as an exact average. Using Eq. (8.88), the Hamiltonian may be rewritten as $H = -J\sum_{\langle ij\rangle} \vec{S}_i \cdot \vec{S}_j = -\frac{1}{2}\sum_{\langle ij\rangle} P_{ij} + \frac{1}{4}N_b J$ where P_{ij} permeates the spin states at i and j and N_b is the total number of bonds. To calculate the partition function, one may omit the constant $\frac{1}{4}N_b J$ and expand $e^{-\beta H}$ as a power series with the result

$$Z = \sum_{n=0}^{\infty} \frac{1}{n!}\left(\beta\frac{J}{2}\right)^n \sum_{Cn} \mathrm{Tr}(P_{m_1}P_{m_2}\ldots P_{m_n}) \tag{8.92}$$

where $m_j = 1,\ldots N_b$ denotes each nearest-neighbor pair and $C_n \equiv (P_{m_1}, P_{m_2}\ldots P_{m_n})$ denotes a particular sequence of n permutation operators. The Monte Carlo problem then corresponds to a random walk in the sample space of permutations $\{C_n;\ n = 1, \infty\}$ and thermal averages are given by

$$\langle O\rangle = \frac{\sum_{n,C_n} O(C_n)\Pi(C_n)}{\sum_{n,C_n}\Pi(C_n)} \tag{8.93a}$$

where

$$\Pi(C_n) \equiv \frac{1}{n!}\left(\beta\frac{J}{2}\right)^n \mathrm{Tr}\,(P_{m_1}P_{m_2}\ldots P_{m_n}) \tag{8.93b}$$

and

$$O(C_n) \equiv \frac{\mathrm{Tr}\,(O\,P_{m_1}P_{m_2}\ldots P_{m_n})}{\mathrm{Tr}\,(P_{m_1}P_{m_2}\ldots P_{m_n})}\ . \tag{8.93c}$$

The spin traces of permutation operators are straightforward to evaluate, and for ferromagnets, $\Pi(C_n)$ is a positive probability. A random walk to sample $\Pi(C_n)$ may be defined by a sequence of steps in which a permutation is microreversibly added to or deleted from the sequence C_n. For example, following Lyklema (1982) one may define a trial move for which the probability for adding a permutation is

$$F\,(C_n \rightarrow C_{n+1}) = \frac{1}{N_b}\,\frac{1}{n+1}f(n) \qquad 8.94a$$

where $\frac{1}{N_b}$ is the probability of choosing a particular permutation $P_{m'}$, $\frac{1}{n+1}$ is the probability of choosing a particular position in the sequence C_n, and $f(n)$ is a distribution chosen to specify the probability of adding a permutation to a sequence of length n. The corresponding probability for removing a permutation is

$$F\,(C_{n+1} \rightarrow C_N) = \frac{1}{n+1}(1 - f(n+1)) \tag{8.94b}$$

where $\frac{1}{n+1}$ is the probability of choosing the specific permutation to remove. Using the generalized Metropolis algorithm of Problem 8.1, the process is microreversible if the trial move to configuration C_{n+1} is accepted with probability

$$P(C_{n+1}) = \min\left(1,\ \frac{1-f(n+1)}{\frac{1}{N_B}f(n)}\,\frac{\Pi(C_{n+1})}{\Pi(C_n)}\right)\ . \tag{8.94c}$$

For an antiferromagnet, Eq. (8.93) is unsuitable as it stands, since each odd term in n is negative giving rise to sign cancellations which cannot be handled stochastically except at high temperatures. Thus, following Lee *et al.* (1984), it is useful to rearrange the series using the identity

$$P_{ij} - 1 = h_{ij} - h_{ij}^2 \tag{8.95a}$$

$$h_{ij} = \frac{1}{4}\left(\sigma_i^+\sigma_j^- + \sigma_i^-\sigma_j^+\right) \quad . \tag{8.95b}$$

In the representation of Eq. (8.90)

$$(P_{ij} - 1) = \begin{bmatrix} 0 & 0 & 0 & 0 \\ 0 & -1 & \eta & 0 \\ 0 & \eta & -1 & 0 \\ 0 & 0 & 0 & 0 \end{bmatrix} \qquad h = \begin{bmatrix} 0 & 0 & 0 & 0 \\ 0 & 0 & \eta & 0 \\ 0 & \eta & 0 & 0 \\ 0 & 0 & 0 & 0 \end{bmatrix} \tag{8.95c}$$

from which we note that h_{ij} interchanges spins if i and j have opposite spin states and yields zero for identical spin states and that the minus sign of h may be removed for a biparticle lattice. Ignoring irrelevant constants, the partition function for $J < 0$ may be expanded as follows

$$\begin{aligned} Z &= \operatorname{tr} e^{-\beta\frac{J}{2}(h_{ij}^2 - h_{ij})} \\ &= \sum_{n=0}^{\infty} \frac{1}{n!}\left(\frac{-J\beta}{2}\right)^n \sum_{C_n} \operatorname{Tr}\left(O_{m_1} O_{m_2} \ldots O_{m_n}\right) \end{aligned} \tag{8.96}$$

where $O = h_{ij}^2$ or $-h_{ij}$. Since only closed loops with an even number of h_s's contribute or equivalently by virtue of the sublattice transformation, all C_n's yield positive contributions and the evaluation of thermal expectation values and the random walk in the space of operator sequences C_n proceeds as before.

Another novel feature of spin and lattice Fermion systems is the possibility of using a discrete rather than continuous auxiliary field. If some operator A can only assume a finite set of discrete values, it is always possible to replace the integral over a continuous auxiliary variable in the Hubbard Stratanovich transformation $e^{\frac{1}{2}bA^2} = \sqrt{\frac{b}{2\pi}} \int d\sigma\, e^{-\frac{1}{2}b\sigma^2 + \sigma bA}$ by a discrete sum. For example, consider an operator A which can only assume the three values $\{-1, 0, 1\}$, in which case a single sum over an Ising variable suffices

$$e^{bA^2} = \frac{1}{2}\sum_{\sigma=\pm 1} e^{\sigma a A} \qquad A = \{-1, 0, 1\} \tag{8.97}$$

where

$$\tanh^2 \frac{a}{2} = \tanh \frac{b}{2} \ .$$

This discrete Ising auxiliary field has been used by Hirsch (1983, 1984) to replace the two-body interaction $U n_\uparrow n_\downarrow$ in the Hubbard model by a sum over one-body fields. Using the fact that this occupation number for spin up and spin down particles, $n_\uparrow$

and $n_\downarrow$, are 1 or 0, the operator A may be chosen to be $(n_\uparrow - n_\downarrow)$ or $(n_\uparrow + n_\downarrow - 1)$, yielding the following real identities for positive and negative coupling, respectively.

$$
\begin{aligned}
e^{-\epsilon U n_\uparrow n_\downarrow} &= \frac{1}{2}\sum_{\sigma=\pm 1} e^{2a\sigma(n_\uparrow - n_\downarrow) - \frac{1}{2}U\epsilon(n_\uparrow + n_\downarrow)} \\
\tanh^2 a &= \tanh\left(\frac{\epsilon U}{4}\right) \qquad U > 0
\end{aligned}
\tag{8.98}
$$

and

$$
\begin{aligned}
e^{-\epsilon U n_\uparrow n_\downarrow} &= \frac{1}{2}\sum_{\sigma=\pm 1} e^{2b\sigma(n_\uparrow + n_\downarrow - 1) - \frac{1}{2}U\epsilon(n_\uparrow + n_\downarrow - 1)} \\
-\tanh^2 b &= \tanh\left(\frac{\epsilon U}{4}\right) \qquad U < 0 \ .
\end{aligned}
\tag{8.99}
$$

When the Fermion trace is taken over the one-body operator, the resulting Ising action has much less phase space than the continuous Gaussian action and is more convenient to sample.

The basic stochastic methods presented in this chapter are currently being applied to a broad range of quantum systems involving many degrees of freedom, including both many-particle systems and field theories. Although it is impractical to provide an exhaustive list of references, the following list provides a starting point for exploring applications to specific systems.

Because of the absence of sign problems for Bosons, Monte Carlo studies of liquid and solid ^{4}He have been the most definitive, and representative results are given by Kalos *et al.* (1974), Whitlock *et al.* (1979) and Kalos *et al.* (1981). Extensive results are becoming available for a variety of many-Fermion systems including molecules (Anderson, 1981; Moskowitz *et al.*, 1982; and Reynolds *et al.*, 1982), the electron gas (Ceperley and Alder, 1980), liquid and solid Hydrogen (Ceperley and Alder, 1981), liquid ^{3}He (Lee *et al.*, 1981) and light nuclei (Zabolitzky and Kalos, 1981; Zabolitzky *et al.* 1982). Applications to spin systems and models such as the Hubbard model are cited in the review by DeRaedt and Lagendijk (1985). Because of the simplifications which arise for Fermions in one spatial dimension, particular attention has been devoted to one-dimensional models such as a potential model for nuclei (Negele, 1979), meson-nucleon field theory (Serot *et al.* 1983), a quark model of hadronic matter (Horowitz *et al.* 1985), and the Gross-Neveu model (Hirsch *et al.*, 1982). Lattice gauge theory has been studied both in the Hamiltonian form described in this chapter (Heys and Stump, 1983, 1985; Chin *et al.*, 1984, 1985) and in Lagrangian form as reviewed, for example, by Creutz (1983) and Kogut (1985). Recent references to these and other topics may be found in the conference proceedings edited by Gubernatis *et al.* (1985).

PROBLEMS FOR CHAPTER 8

The most crucial problem for learning to apply the ideas developed in this chapter is Problem 9, which leads one through the salient practical problems encountered in an actual calculation and then requires a real numerical Monte Carlo calculation. The remaining problems derive, apply, or generalize results presented in the text.

PROBLEM 8.1 Generalized Metropolis Algorithm The standard algorithm of Metropolis *et al* described in Section 8.2 may be generalized in several ways.

a) Let $F(\vec{x} \rightarrow \vec{x}_T)$ be any distribution, not necessarily symmetrical, and define

$$q(\vec{x} \rightarrow \vec{x}_T) \equiv \frac{F(\vec{x}_T \rightarrow \vec{x})e^{-S(\vec{x}_T)}}{F(\vec{x} \rightarrow \vec{x}_T)e^{-S(\vec{x})}} \quad .$$

Show that if a tentative step $\vec{x}_T$ is generated by $F(\vec{x} \rightarrow \vec{x}_T)$ and accepted with probability $\min\{1, q(\vec{x} \rightarrow \vec{x}_T)\}$, then the evolution satisfies the microreversibility condition (8.27). Note that this algorithm reduces to the standard method if F is symmetric.

b) Show that acceptance with probability $\frac{q(\vec{x}\rightarrow\vec{x}_T)}{1+q(\vec{x}\rightarrow\vec{x}_T)}$ is also microreversible.

PROBLEM 8.2 Variant of the Neumann-Ulam Method Show that when M^{-1} is calculated by evaluating the mean value $\langle S_{ij}\rangle$, Eq. (8.40), the variance is $\langle(S_{ij} - M_{ij}^{-1})^2\rangle = \left[(1-K)^{-1}\right]_{ij} P_j^{-1} - (M^{-1})_{ij}^2$ where $K_{ij} = T_{ij}R_{ij}$. Thus, the method has two problems: each walk only contributes to a single matrix element of the inverse and the variance increases inversely with the stopping probability.

A useful alternative is to define a separate score $\hat{S}_{ij}$ for each j as follows. At each step k when the walk has progressed to state i_k, the product of all the residue $R_{ii_2}R_{i_2i_3}\ldots R_{i_{k-1}i_k}$ is added to $\hat{S}_{i\,i_k}$. Thus, the stop probability is removed from the score although it still affects the length of the walk and each walk contributes to many matrix elements. Show that $\langle\hat{S}_{ij}\rangle = M_{ij}^{-1}$.

PROBLEM 8.3 Comparison of Langevin and Molecular Dynamics Methods Write the lowest order finite difference approximations to the Langevin equation, Eq. (8.32) with $\Gamma = \frac{\Delta\tau}{2}$ so that (8.34) has unit standard deviation and the molecular dynamics equation of motion, Eq. (8.45), to obtain

$$x_i^{n+1} = x_i^n - \frac{1}{2}(\Delta\tau)^2\frac{\partial S}{\partial x_i} + \Delta\tau\ \xi_i \qquad \text{(Langevin)}$$

$$x_i^{n+1} = x_i^n - \frac{1}{2}(\Delta\tau)^2\frac{\partial S}{\partial x_i} + \frac{1}{2}\left(x_i^{n+1} - x_i^{n-1}\right) \quad \text{(Molecular dynamics)} \quad .$$

Thus observe that the stochastic noise enters in the same way as the velocity $\xi_i \sim \frac{x_i^{n+1}-x_i^{n-1}}{2\Delta\tau}$.

Comment on the efficiency of the two methods in sampling the local space and in sampling the global space. Thus, explain the potential advantages of alternating between the two evolution algorithms. A specific example is given by Kogut (1985).

PROBLEM 8.4 Discrete Time Errors

a) To see that the symmetric decomposition $e^{-\frac{\epsilon}{2}T}e^{-\epsilon V}e^{-\frac{\epsilon}{2}T}$ really yields leading errors of $\mathcal{O}(\epsilon^2)$ and asymmetric choices yield $\mathcal{O}(\epsilon)$, find the eigenfunctions of $U_\eta(x,y) \equiv e^{-\frac{1}{2\epsilon}(x-y)^2 - \frac{\omega^2\epsilon}{2}\left[\eta y^2 + (1-\eta)x^2\right]}$. Since it is clear that a quadratic form will satisfy $\int dy\, U_\eta(x,y)\psi(y) = e^{-\epsilon\lambda}\psi(x)$, one straightforward method is to try a general quadratic form and to solve for the parameters. Verify that for the symmetric case $\eta = \frac{1}{2}$ you obtain a harmonic oscillator wave function $e^{-\frac{1}{2}\tilde{\omega}x^2}$ with energy $\frac{\tilde{\omega}}{2}$

where $\tilde{\omega} = \omega\left[1 + \left(\frac{\omega\epsilon}{2}\right)^2\right]^{1/2}$. Note that if $\left(e^{-\epsilon T} e^{-\epsilon V}\right)^n$ is used as a filter to obtain the ground state wave function, an extra order of ϵ accuracy is obtained if it is multiplied by $e^{-\frac{\epsilon}{2}V}$ whenever it is sampled for an observable.

b) Now consider an oscillator potential with an infinite wall at the origin, *i.e.* $V(x) = \frac{\omega}{2}x^2 \quad x > 0$, and $V(x) = +\infty \quad x \leq 0$. Show that simply adding an infinite potential for $x < 0$ in the infinitesimal evolution operator yields errors of $\mathcal{O}(\sqrt{\epsilon})$.

c) Use the propagator with images, Eq. (8.56) to solve for the first odd eigenstate of the harmonic oscillator as in part a) with $\eta = \frac{1}{2}$ and show this hard wall boundary condition yields $\mathcal{O}(\epsilon^2)$ accuracy.

PROBLEM 8.5 Evaluation of Kinetic Energy

a) Derive eqs. (8.49a,b) for $\langle T \rangle$ by inserting p^2 into the path integral two ways: $\langle x_n | e^{-\frac{\epsilon}{2}V} e^{-\epsilon T} \hat{p}^2 e^{-\frac{\epsilon}{2}V} | x_{n-1} \rangle$ and $\langle x_n | e^{-\frac{\epsilon}{2}V} e^{-\epsilon T} e^{-\frac{\epsilon}{2}V} \hat{p} | x_{n-1} \rangle \langle x_{n-1} | \hat{p} e^{-\frac{\epsilon}{2}V} e^{-\epsilon T} e^{-\frac{\epsilon}{2}V} | x_{n-2} \rangle$. What is the order of the leading error in ϵ in each case?

b) Derive the continuum form of the virial theorem, Eq. (8.58), being careful to allow for non-Hermitian boundary conditions. In the special case that ψ' vanishes at the upper and lower limits, note that $\langle T \rangle = \frac{1}{2}\langle (x - x_0) V'(x) \rangle$ for any choice of x_0. Show explicitly that $\langle V' \rangle = 0$ in this case. How should x_0 be chosen to optimize statistics for the case of the ground state of the harmonic oscillator?

c) Derive the virial theorem for the discrete evolution operator as follows. Let ϕ_n denote any eigenfunction of the discrete evolution operator $e^{-\epsilon\frac{V}{2}} e^{-\epsilon T} e^{-\epsilon\frac{V}{2}}$. Hence, $\langle \phi_n | [p\hat{x}, e^{-\epsilon\frac{V}{2}} e^{-\epsilon T} e^{-\epsilon\frac{V}{2}}] | \phi_n \rangle = 0$ so that the commutator is zero in the trace or ground state expectation value. Evaluate the commutator to obtain

$$\langle x | [p\hat{x}, e^{-\epsilon\frac{V}{2}} e^{-\epsilon T} e^{-\epsilon\frac{V}{2}}] | y \rangle = \langle x | e^{-\epsilon\frac{V}{2}} \frac{p^2}{2m} e^{-\epsilon T} e^{-\epsilon\frac{V}{2}} | y \rangle + \frac{1}{2}\left(xV'(x) + yV'(y)\right) e^{-\frac{m}{2\epsilon}(x-y)^2 - \frac{\epsilon}{2}(V(x)+V(y))} .$$

Since to $\mathcal{O}(\epsilon^2)$ the evolution operator may also be written $e^{-\frac{\epsilon}{2}T} e^{-\epsilon V} e^{-\frac{\epsilon}{2}T}$, show that $\langle T \rangle = \langle \frac{1}{N} \sum_i x_i V'(x_i) \rangle_{P_S}$ to $\mathcal{O}(\epsilon^2)$, where S is the action for the trace.

PROBLEM 8.6 Guided Random Walks

a) Using the Schrödinger equation in imaginary time for $\psi(x, \tau)$, derive Eq. (8.64) for the evolution of the product $f(x, \tau) \equiv \phi(x, \tau)\, \psi(x, \tau)$.

b) Derive the infinitesimal evolution operator for f, Eq. (8.65). Note that this equation has the form naively expected from the diffusion equation Eq. (8.64), except that ϵ is replaced by $\hat{\epsilon}$ in the Gaussian. One method is to show

$$\langle x | \phi(\hat{x}) e^{-\epsilon \hat{T}} e^{-\epsilon \hat{V}} \frac{1}{\phi(\hat{x})} | y \rangle = \sqrt{\frac{m}{2\pi\epsilon}}\, e^{\ln\phi(x) - \frac{m(x-y)^2}{2\epsilon} - \epsilon V(y) - \ln\phi(y)} .$$

in the usual way and to expand $\ln\phi(x)$ around y being careful to retain all terms of order ϵ and to normalize the Gaussian appropriately. Another alternative is to use the Baker-Hausdorf identity $e^{A+B} = e^A e^B e^{-\frac{1}{2}[A,B] + \ldots}$ on $e^{-\epsilon\left[\frac{1}{2m}\hat{p}^2 + \frac{1}{m}\hat{p}\frac{\phi'}{\phi} + \frac{(H - \partial_\tau)\phi}{\phi}\right]}$ and note $p^2 \sim \frac{1}{\epsilon}$ so that terms like $\epsilon^2 p^2$ must be retained.

Observe that in the case of the harmonic oscillator, $\hat{\epsilon}$ is a constant corresponding to a trivial rescaling of the time step. The role of $\hat{\epsilon}$ in practical calculations is discussed by Moskowitz *et al.* (1982).

c) Derive the infinitesimal evolution operator for a many-Boson system with wave function $\psi(x_1 \dots x_n, \tau)$ and guiding function $\phi(x_1 \dots x_n, \tau)$. Do any new complications arise in the many-particle case?

PROBLEM 8.7 Time-Dependent Guiding Function To understand how a time-dependent guiding function, Eq. (8.64), guides a random walk, consider a symmetric double well with a high barrier at the origin and minima at $\pm D/2$. Let $\phi_a(x)$ be a narrow Gaussian of width Δx centered at $-D/2$, let $\phi_b(x)$ be a positive wave function localized in the right well, and consider the matrix element $\langle \phi_a | e^{-\beta H} | \phi_b \rangle$.

a) Explain why the guiding function $\phi(x, \tau) = e^{(\tau-\beta)H}\phi_a$ is positive at all times. Determine its behavior at times very close to β and explain how the drift term $\frac{\epsilon\phi'}{m\phi}$ forces the ensemble of points into the region of maximum statistical importance.

b) Consider ϕ at early times such that $\beta - \tau > \frac{1}{E_2}$ for which $|\phi(\beta - \tau)\rangle \approx |0\rangle\langle 0|\phi_b\rangle + |1\rangle\langle 1|\phi_b\rangle e^{-(\beta-\tau)\Delta}$ where Δ and E_2 denote the excitation energies of the first and second excited states and $|0\rangle$ and $|1\rangle$ denote the ground state and first excited state. Sketch ϕ for $(\beta - \tau) \ll \frac{1}{\Delta}$ and for $(\beta - \tau) \gg \frac{1}{\Delta}$. Note that $|0\rangle$ and $|1\rangle$ correspond to the lowest eigenfunctions ψ_e and ϕ_o in the right-hand well satisfying even and odd boundary conditions at the origin and thus show that $\psi_e(0)\psi_o'(0) = 2m\Delta \int_0^\infty \psi_o(x)\phi_e(x)$. Hence, evaluate the drift term at the origin and show

$$\frac{\epsilon}{m}\frac{\phi'(0)}{\phi(0)} \approx \frac{2\epsilon\Delta}{\psi_e^2(0)} e^{-\beta\Delta} \; .$$

Note that since the density at the origin $\psi_e^2(0)$ is much less than the average density, $D\psi_e^2(0) \ll 1$ and the drift term will migrate the propulation from the right well to the left well in characteristic time much less than $\frac{1}{\Delta}$

$$\beta_D = \frac{D\psi_e^2(0)}{2}\frac{1}{\Delta} \ll \frac{1}{\Delta} \; .$$

In summary, explain qualitatively the role of the guiding function in each of the three time domains: $\frac{1}{\Delta} \leq (\beta - \tau)$, $\frac{1}{E_2} < (\beta - \tau) \ll \frac{1}{\Delta}$, and $(\beta - \tau) < \frac{1}{E_2}$.

c) Define a localized eigenstate ϕ_L as the solution in a single well defined equal to $V(x)$ for $x < 0$ and equal to $V(0)$ for $x > 0$. Let ϕ_R be an analogous eigenstate in the right well. How does this analysis change if one calculates the overlap $\langle \phi_L | e^{-\beta H} | \phi_R \rangle$ used to determine the gap in Eq. (8.53)?

PROBLEM 8.8 Lifetime of a Metastable State Although in general it may be difficult to express a physical observable in real time in terms of quantities which can be evaluated in imaginary time, it is possible to do so for the lifetime of a metastable state.

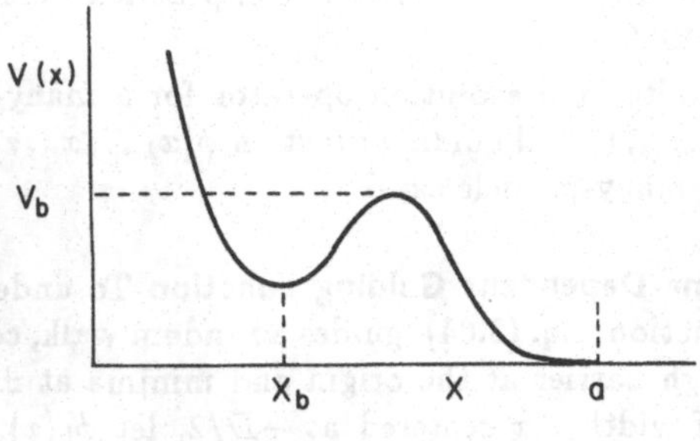

a) Use the asymptotic form of the wave function in the potential sketched in this problem to obtain the familiar Breit-Wigner phase shift

$$\delta(k) = \delta_0 + kb + \tan^{-1}\frac{\gamma}{k_0 - k}$$

where $k = \sqrt{2mE}$, $k_0 = \sqrt{2mE_0}$, E_0 is the energy of the metastable state, $\gamma = \frac{m}{2k_0\Gamma}$, and b is the origin with respect to which δ is defined. Using the argument in Section 8.3, explain how a hard wall boundary condition may be imposed at an arbitrary position a in the exterior region and used to determine γ. What are the limitations of this method?

b) Since for high barriers, $\frac{\gamma}{k_0} \ll 1$, it is impractical to measure γ from the phase shift, show how to extract it from an energy splitting as follows. For a hard wall fixed at positive a, consider the two lowest eigenstates specified by $\delta(k_n) = n\pi - k_n a$ in the limit $\gamma \ll |k_0 - k_n|$. Vary a to minimize the splitting between k_1 and k_2 and show that at the minimum value a_m,

$$\gamma = \frac{a_m + b}{4}(k_2 - k_1)^2\Big|_{a_m}\left(1 + \mathcal{O}\left(\frac{\gamma}{k_0}\right)\right)$$

Now a left well $V_L(x) = \theta(x_b - x)V(x) + \theta(x - x_b)V_b$ and a right well $V_R(x) = \theta(x_b - x)V_b + \theta(x - x_b)V(x)$ with ground state energies E_L and E_R. Show that to order $\frac{\gamma}{k_0}$, a_m corresponds to the value of a for which $E_L = E_R$ and that $b + a = \frac{-2E_R}{dE_R/da}\Big|_{a_m}$ Hence show

$$\gamma = -\left[\frac{m}{4\frac{dE_R}{da}}(\Delta E)^2\right]_{E_L = E_R}\left(1 + \mathcal{O}\frac{\gamma}{k_0}\right)$$

that is, the lifetime is given by the energy splitting in a suitably defined double well. Note that since $\Delta E \sim \langle L|v|R\rangle$ this result has the familiar Fermi golden rule form of a transition matrix element squared times a density of states.

PROBLEM 8.9* Monte Carlo Calculation To fully understand and appreciate the stochastic methods described in this chapter, it is necessary to solve a concrete

problem

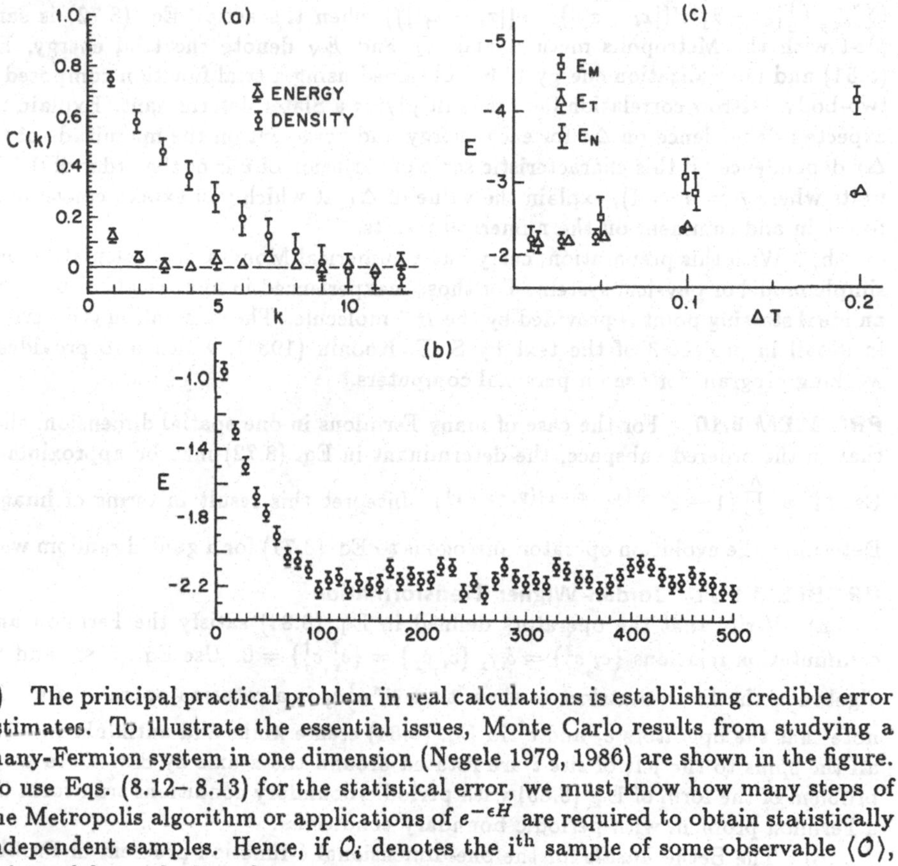

a) The principal practical problem in real calculations is establishing credible error estimates. To illustrate the essential issues, Monte Carlo results from studying a many-Fermion system in one dimension (Negele 1979, 1986) are shown in the figure. To use Eqs. (8.12– 8.13) for the statistical error, we must know how many steps of the Metropolis algorithm or applications of $e^{-\epsilon H}$ are required to obtain statistically independent samples. Hence, if $\mathcal{O}_i$ denotes the ith sample of some observable $\langle\mathcal{O}\rangle$, it is useful to calculate the autocorrelation function.

$$C(k) = \frac{\langle\langle \mathcal{O}_i \mathcal{O}_{i+k}\rangle\rangle - \langle\langle \mathcal{O}_i\rangle\rangle^2}{\langle\langle \mathcal{O}_i^2\rangle\rangle - \langle\langle \mathcal{O}_i\rangle\rangle^2}$$

where $\langle\langle \mathcal{O}_i \mathcal{O}_{i+k}\rangle\rangle \equiv \frac{1}{N}\sum_{i=1}^{N} \mathcal{O}_i \mathcal{O}_{i+k}$. Figure (a) shows the autocorrelation functions for the energy and density. Explain why these are different and how often to sample each for independent results. If a calculation had not yet equilibrated, how might this lack of equilibrium show up in the autocorrelation function? Suppose one uses Eq. (8.55) to calculate an observable by averaging over the complete time mesh. Are observables on adjacent time slices correlated? How would you take this into account in establishing errors?

b) Figure (b) shows the mean value of the energy obtained by starting with our initial configuration, sampling the energy and variance every $(m)^{\text{th}}$ update and averaging (n) such samples to obtain the energy and error bars shown. At what point has the energy equilibrated? Are the error bars past that point statistically sensible? Suggest several ways to check the equilibration and error estimates.

c) Figure (c) shows the energy as a function of time step size $\Delta\tau$ for three different calculations: E_M denotes the virial theorem average $\langle\sum_{i<j}\left(\frac{1}{2}|x_i - x_j|v'(|x_i - x_j|) + v(|x_i - x_j|)\right)\rangle$ when the action Eq. (8.72) is sampled with the Metropolis method and E_T and E_N denote the trial energy, Eq. (8.51) and normalization energy (8.61) obtained using a trial function composed of two-body Jastrow correlation factors multiplying a Slater determinant. Explain the expected dependence on $\Delta\tau$ for each energy and comment on the magnitude of the $\Delta\tau$ dependence. If this characteristic scale of variation of v is of the order of 0.2 (in units where $\hbar = m = 1$), explain the value of $\Delta\tau$ at which you expect convergence to set in and comment on the numerical results.

b) With this preparation, carry out a numerical Monte Carlo calculation on a simple model or physical system. For those inexperienced in numerical calculations, an ideal starting point is provided by the H^- molecule. The calculation is described in detail in project 8 of the text by S. E. Koonin (1985), which also provides a working program for use on personal computers.

PROBLEM 8.10 For the case of many Fermions in one spatial dimension, show that in the ordered subspace, the determinant in Eq. (8.72) may be approximated $\mathrm{Det}\, M \approx \prod_{i=2}^{N}\left(1 - e^{-\frac{m}{\epsilon}(x_i - x_{i-1})(y_i - y_{i-1})}\right)$. Intepret this result in terms of images. Determine the evolution operator analogous to Eq. (8.71) for a guided random walk.

PROBLEM 8.11 Jordan-Wigner Transformation

a) Verify that the operators defined in Eq. (8.84) satisfy the Fermion anti-commutation relations $\{c_i, c_j^\dagger\} = \delta_{ij}$, $\{c_i, c_j\} = \{c_i^\dagger, c_j^\dagger\} = 0$. Use Eq. (8.83) and the algebra of the Pauli matrices to show that $e^{i\pi a_j^\dagger a_j}$ is a rotation operator and thus note that the operators c_i and $c_i^\dagger$ in Eq. (8.84) create kinks and antikinks in which all the spins to the left of site i are rotated around the z axis by $\pm\pi$. Does a spin problem of the form of Eq. (8.85) with periodic boundary conditions transform into a Fermion problem with periodic boundary conditions?

b) The Bethe *ansatz* for the one-dimensional δ-function problem in Problem 1.9 was actually made for the Heisenberg spin Hamiltonian $H - J\sum_i S_i S_{i+1}$. Show that this may be rewritten in terms of Fermion operators as

$$H = -\frac{J}{2}\sum_i c_i^\dagger\left(c_{i+1} - 2c_i + c_{i-1}\right) - J\sum_i c_i^\dagger c_i c_{i+1}^\dagger c_{i+1}$$

and take the continuum limit to obtain the second quantized form of Eq. (1) of Problem 1.9. Utilize the possibility of rotating the spins on one sublattice to show that the ferromagnet corresponds to an attractive δ-function interaction and that the antiferromagnet corresponds to a repulsive δ-function.

REFERENCES

Abel, W. R., A. C. Anderson, and J. C. Wheatley, 1966, *Phys. Rev. Lett.* **17**, 74.

Abrikosov, A. A. and I. M. Khalatnikov, 1957, *Explt. Theoret. Phys.* (U.S.S.R.) **33**, 1154; *Soviet Physics JETP* **6**, 888 (1958).

Abrikosov, A. A., L. P. Gorkov, and I. E. Dzyaloshinski, 1963, *"Methods of Quantum Field Theory in Statistical Physics"*, (Prentice-Hall, Inc., New Jersey).

Achter, E. K., and L. Meyer, 1969, *Phys. Rev.* **188**, 291.

Alhassid, Y. and S. E. Koonin, 1981, *Phys. Rev.* **23**, 1590.

Anderson, J. B., 1975, *J. Chem. Phys.* **63**, 1499; 1976, *J. Chem. Phys.* **65**, 1421; 1980, *J. Chem. Phys.* **73**, 3897; 1981, *J. Chem. Phys.* **74**, 6307.

Anderson, P. W., 1958, *Phys. Rev.* **112**, 1900; 1963, *Phys. Rev.* **130**, 439.

Ashcroft, N. W., and N. D. Mermin, 1976, *Solid State Physics*, (Hold Rinehart and Winston, New York)

Aziz, R. A., V. P. S. Nain, J. S. Carley, W. L. Taylor, and T. T. McConville, 1979, *J. Chem. Phys.* **70**, 4330.

Baker, G. A., 1965, *Adv. Theoret. Phys.* **1**,1.

Ballan, R. and C. Bloch, 1974, *Ann. Phys.* (New York) **85**, 514.

Barber, M. N., 1983, *Phase Transitions and Critical Phenomena*, edited by C. Domb and J. L. Leibowicz (Academic Press, New York) vol. 8, p. 146.

Batrouni, G. G., G. R. Katz, A. S. Kronfeld, G. P. Lepage. B. Svetitsky, and K. G. Wilson, 1986, *Phys. Rev. D* **32**, 2736.

Baym, G. and S. A. Chin, 1976, *Nucl. Phys. A* **262**, 527.

Baym, G. and C. Pethick, 1976, in *The Physics of Liquid and Solid Helium*, Vol. 2, Bennemann and Ketterson, Eds. (John Wiley & Sons, New York).

Bell, J. S., and E. J. Squires, 1959, *Phys. Rev. Lett.* **3**, 96.

Bender, I., D. Gromes, H. Rothe, and K. Rothe, 1978, *Nucl. Phys. B* **36**, 259.

Bender, C. M. and T. T. Wu, 1969, *Phys. Rev.* **184**, 1231; 1971, *Phys. Rev. Lett.* **27**, 461; 1972, *Phys. Rev. D* **7**, 1620.

Berezin, F. A., 1965, *The Method of Second Quantization*, (Academic Press, New York).

Berry, M. V., 1966, *Proc. Phys. Soc.* (London) **89**, 479.

Bertsch, G. F., and T. T. S. Kuo, 1968, *Nucl. Phys.* **A112**, 204.

Bethe, H. A., 1931, *Z. Physik* **71**, 205.

Bethe, H. A., B. H. Brandow, and A. G. Petschek, 1963, *Phys. Rev.* **129**, 225.

Blaizot, J. P., and H. Orland, 1981, *Phys. Rev. C* **24**, 1740.

Bloch, C., and J. Horowitz, 1958, *Nucl. Phys.* **8**, 91–105.

Brandow, B. H., 1971, *Ann. Phys.* (New York) **64**, 21.

Brézin, E., J. C. LeGuillou, and J. Zinn-Justin, 1977, *Phys. Rev. D* **15**, 1544; 1558.

Brown, G. E., 1972, *Many Body Problems*, (North-Holland, Amsterdam).

Brueckner, K. A., and Sawada, 1957, *Phys. Rev.* **106**, 1117.

Brueckner, K. A., 1959, *The Many Body Problem*, edited by C. DeWitt (John Wiley and Sons, New York), p. 47.

Bogolvbov, N. N., 1947, *J. Phys. U.S.S.R.* **11**, 23.

Bohr, A., and B. R. Mottelson, 1969, *Nuclear Structure*, (W. A. Benjamin, New York).

Callaway, D. J. E., and A. Rahmann, 1982, *Phys. Rev. Lett.* **49**, 613.

Calogero, F., and A. Degasperis, 1975, *Phys. Rev.* **A11**, 265.

Carruthers, P. and M. M. Nieto, 1965, *Am. J. Phys.* **33**, 537.

Ceperley, D. M., and M. H. Kalos, 1979, *Monte Carlo Methods in Statistical Mechanics*, edited by K. Binder (Springer Verlag, New York).

Ceperley, D. M., and B. J. Alder, 1980, *Phys. Rev. Lett.* **45**, 566; 1981, *Physica B and C* (Amsterdam) **107**, 875.

Chin, S. A., J. W. Negele, and S. E. Koonin, 1984, *Ann. Phys.* **157**, 140.

Chin, S. A., O. S. van Roosmalen, E. A. Umland, and S. E. Koonin, 1985, *Phys. Rev.* **D31**, 3201.

Coleman, S., 1977, "The Uses of Instantons," in *The Why's of Subnuclear Physics*, A. Zichichi, ed. (Plenum Press, New York), p. 805.

Connor, J. N. L., and R. A. Marcus, 1971, *J. Chem. Phys.* **55**, 5636.

Creutz, M., 1983, *Phys. Rev. Lett.* **50**, 1411.

Creutz, M., 1983, *Quarks, Gluons, and Lattices*, (Cambridge University Press, Cambridge, U.K.)

Cowley, R. A., and A. D. B. Woods, 1971, *Can. J. Phys.* **49**, 177.

DeRaedt, H., and A. Legendijk, 1985, *Physics Reports*, **127**, 234.

Dickhoff, W. H., A. Faessler, H. Müther, and S. S. Wu, *Nucl. Phys. A* **405**, 534 (1983).

Dirac, P. A. M., 1933, *Physikalische Zeitschrift der Sowjetunion* **3**, 64.

Doniach, S., and E. H. Sondheimer, 1974, *Green's Functions for Solid State Physicists*, (W. A. Benjamin, Reading, MA).

Fantoni, S., B. L. Friman, and V. Pandharipande, 1981, *Phys. Lett. B* **104**, 89.

Feenberg, E., 1969, *Theory of Quantum Fluids* (Academic Press, New York).

Fetter, A. L., and J. D. Walecka, 1971, *Quantum Theory of Many-Particle Systems*, (McGraw-Hill, New York).

Feynman, R. P., 1948, *Rev. Mod. Phys.* **20**, 367; 1949, *Phys. Rev.* **76**, 769; 1950, *Phys. Rev.* **80**, 40.

Feynman, R. P., and A. R. Hibbs, 1965, *Quantum Mechanics and Path Integrals*, (McGraw-Hill, New York).

Forsythe, G. E., and R. A. Leibler, 1950, *MTAC* **4**, 127.

Frullani, S., and J. Mougey, 1984, *Advances in Nuclear Physics*, edited by J. W. Negele and E. Vogt (Plenum, New York), vol. 14.

Galitskii, V. M., 1958, *J. Exptl. Theoret. Phys.* (U.S.S.R.) **34**, 151; *Soviet Physics JETP* **7**, 104 (1958)

Gell-Mann, M., and F. E. Low, 1951, *Phys. Rev.* **84**, 350.

Gentile, G., 1940, *Nuovo Cimento* **17**, 493; 1942, *Nuovo Cimento* **19**, 109.

Ginzburg, V. L., 1960, *Soviet Phys. - Solid State* **2**, 1824.

Glauber, R., 1963, *Phys. Rev.* **131**, 2766.

Goldstone, J., 1957, *Proc. Roy Soc.* (London) **A239**, 267.

Goldstone, J., and K. Gottfried, 1959, *Nuovo Cimento Ser. X* **13**, 849.

Gordon, W. L., C. H. Shaw, and J. G. Daunt, 1958, *Phys. Chem. Solids* **5**, 117.

Graffi, S., V. Grecchi, and B. Simon, 1970, *Phys. Lett.* **32B**, 631.

Greywall, D. S., 1983, *Phys. Rev. B* **27**, 2747.

Gubernatis, J., *et al.* ed., 1985, Proceedings of "Frontiers of Quantum Monte Carlo" in *J. Stat. Phys.*, **43**, 729–1244.

Gutzwiller, M. C., 1967, *J. Math. Phys.* **8**, 1979

Hallock, R. B., 1972, *Phys. Rev.* **A5**, 320.

Handscomb, D. C., 1962, *Proc. Cambridge Philos. Soc.* **58**, 594.

Hardy, G. H., 1948, *Divergent Series* (University Press, Oxford, UK).

Heys, D. W., and D. R. Stump, 1983, *Phys. Rev. D.* **28**, 2067; 1985, *Nucl. Phys.* **B257**, 19.

Higgs, P. W., 1964, *Phys. Lett.* **12**, 132.

Hirsch, J. E., D. J. Scalapino, R. L. Sugar, and R. Blankenbecler, 1981, *Phys. Rev. Lett.* **47**, 1628.

Hirsch, J. E., R. L. Sugar, D. J. Scalapino, and R. Blankenbecler, 1982, *Phys. Rev. B* **26**, 5033.

Hirsch, J. E., 1983, *Phys. Rev. B* **28**, 4059; 1984, *Phys. Rev. B* **29**, 4159.

Horowitz, C., E. Moniz, and J. W. Negele, 1985, *Phys. Rev. D* **31**, 1689.

Huang, K., 1963, *Statistical Mechanics*, (Wiley & Sons, NY).

Hubbard, J., 1959, *Phys. Rev. Lett.* **3**, 77; 1963, *Proc. Roy. Soc.* **A276**, 238; 1964, *Proc. Roy. Soc.* **A277**, 237; 1964, *Proc. Roy. Soc.* **A281**, 401.

Henshaw, K. D. G., 1960, *Phys. Rev.* **119**, 9.

Hugenholtz, N., and D. Pines, 1959, *Phys. Rev.* **116**, 489.

Jeukenne, J. P., A. Lejeune, and C. Mahaux, 1976, *Phys. Rept.* **25**, 83.

Kadanoff, L. P., and G. Baym, 1962, *Quantum Statistical Mechanics*, (W. A. Benjamin, Menlo Park, CA).

Kalos, M. H., D. Levesque, and L. Verlet, 1974, *Phys. Rev. A* **9**, 2178.

Kalos, M. H., M. A. Lee, P. A. Whitlock, and G. V. Chester, 1981, *Phys. Rev. B* **24**, 115.

Kato, T., 1978, in *"Topics in Functional Analysis"*, edited by I. Gohberg and M. Kač (Academic Press, New York).

Kerman, A. K., S. Levit, and Troudet, 1983, *Ann. Phys.* (New York) **148**, 436.

Knuth, D. E., 1981, *The Art of Computer Programming*, (Addison-Wesley, Reading, MA), vol. 2.

Kogut, J., 1985, Proceedings of "Frontiers of Quantum Monte Carlo", in *Journal Statistical Physics*, in press.

Kohn, W., and J. M. Luttinger, 1960, *Phys. Rev.* **118**, 41.

Koltun, D., 1972, *Phys. Rev. Lett.* **28**, 182.

Kuti, J., and J. Polonyi, 1982, *Proc. XXI International Conf. on High Energy Physics*, (Paris).

Landau, L. D., 1956, *J. Exptl. Theoret. Phys.* (U.S.S.R.) **30**, 1058; 1957, *J. Exptl. Theoret. Phys.* (U.S.S.R.) **32**, 59; 1958, *J. Exptl. Theoret. Phys.* (U.S.S.R.) **35**, 97. (English translations: *Soviet Phys. JETP* **3**, 920 (1957); **5**, 101 (1957); **8**, 70 (1958).)

Landau, L. D., and E. M. Lifshitz, 1958, *Statistical Physics*, (Pergamon Press).

Langer, J., 1969, *Ann. Phys.* (New York) **54**, 258.

Lee, D. H., J. D. Joannopoulos, and J. W. Negele, 1984, *Phys. Rev. B* **30**, 1599.

Lee, M. A., K. E. Schmidt, and M. H. Kalos, 1981, *Phys. Rev. Lett.* **46**, 728.

Lee, T. D. and C. N. Yang, 1957, *Phys. Rev.* **105**, 1119.

Leggett, A. J., 1966, *Phys. Rev.* **147**, 119; 1968, *Ann. Phys.* (New York) **46**, 76.

LeGuillou, J. C., and J. Zinn-Justin, 1985, *Journal de Physique Letters* **46**, L137.

Lehmann, H., 1954, *Nuovo Cimento* **11**, 342.

Leiss, J. E. and R. E. Taylor, 1963, *Proc. Karlsruhe Photonuclear Conf.* (Data quoted by W. Czyz, *Phys. Rev.* **131**, 2141).

Levit, S., J. W. Negele, and Z. Paltiel, 1980a, *Phys. Rev. C* **21**, 1603; 1980b, *Phys. Rev. C* **22**, 1979.

Levit, S., *Phys. Rev. C.* **21**, 1594.

Lindhard, J., 1954, *Mat. Fys. Medd. Dan. Vid. Selsk.* **28**, no. 8.

Lipatov, L. N., 1976, *J.E.T.P. Lett.* **24**, 179; 1977, *J.E.T.P. Lett.* **25**, 116.

Loeffel, J. J., A. Martin, B. Simon, and A. S. Wightman, 1969, *Phys. Lett.* **30B**, 656.

Luttinger, J. M., and J. C. Ward, 1960, *Phys. Rev.* **115**, 1417.

Lyklema, J. W., 1982, *Phys. Rev. Lett.* **49**, 88.

Ma, S. K., 1976, *Modern Theory of Critical Phenomena*, (W. A. Benjamin, Reading, MA).

Martin, P., and J. Schwinger, 1959, *Phys. Rev.* **115**, 1342.

McGuire, J., 1965, *J. Math Phys.* **6**, 432.

Mermin, N. D., and H. Wagner, 1966, *Phys. Rev. Lett.* **17**, 1133.

Mermin, N. D., 1967, *Phys. Rev.* **159**, 161

Metropolis, N., A. Rosenbluth, M. Rosenbluth, A. Teller, and E. Teller, 1953, *J. Chem. Phys.* **21**, 1087.

Migdal, A. B., 1962, *Theory of Finite Fermi Systems and Applications to Atomic Nuclei* (Interscience, New York); 1978, *Rev. Mod. Phys.* **50**, 107.

Miller, W. H., 1970, *J. Chem. Phys.* **53**, 3578.

Moniz, E. J., 1969, *Phys. Rev.* **184**, 1154.

Moniz, E. J., I. Sick, R. R. Whitney, J. R. Ficenec, R. D. Kephart and W. P. Trower, 1971, *Phys. Rev. Lett.* **26**, 445.

Morse, P. M., and H. Feshbach, 1953, *Methods of Theoretical Physics*, (McGraw-Hill, New York).

Moskowitz, J. W., K. E. Schmidt, M. A. Lee, and M. H. Kalos, 1982, *J. Chem. Phys.* **77**, 349.

Negele, J. W., 1979, in *Lecture Notes in Physics*, Vol. 171, edited by K. Goeke and P. G. Reinhard (Springer Verlag, New York); 1982, *Rev. Mod. Phys.* **54**, 913; 1986, *Jour. Stat. Phys.* **43**, 991.

Negele, J. W., and K. Yazaki, 1981, *Phys. Rev. Lett.* **47**, 71.

Nohl, C. R., 1976, *Ann. Phys.* (New York) **96**, 234.

Onsager, L., 1944, *Phys. Rev.* **65**, 117.

Pollack, E. L., and D. M. Ceperley, 1984, *Phys. Rev. B* **30**, 2555.

Polonyi, J., and H. W. Wyld, 1983, *Phys. Lett.* **51**, 2257.

Polyakov, A. M., 1977, *Nucl. Phys.* **B121**, 429.

Pfeuty, P., and G. Toulouse, 1975, *Introduction to the Renormalization Group and to Critical Phenomena*, (J. Wiley & Sons, New York).

Press, W. H., B. P. Flannery, S. Teukolsky, and W. T. Vetterling, 1985, *Numerical Recipies*, University of Cambridge Press, Cambridge.

Reinhardt, H., 1978, *Nucl. Phys. A* **298**, 77; 1981, *Nucl. Phys. A* **367**, 269; 1982, *Nucl. Phys. A* **390**, 70.

Reynolds, P. J., D. M. Ceperley, B. J. Alder, and W. A. Lester, 1982, *J. Chem. Phys.* **77**, 1593.

Robkoff, H. N., D. A. Ewen and R. B. Hallock, 1979, *Phys. Rev. Lett.* **43**, 2006.

J. Schwinger, ed., 1953, *Selected Papers on Quantum Electrodynamics*, (Dover).

Samuel, S., 1980, *J. Math. Phys.* **21**, 2806.

Serot, B., S. E. Koonin, and J. W. Negele, 1983, *Phys. Rev. C* **28**, 1679.

Simon, B., 1979, *Functional Integrals and Quantum Mechanics*, (Academic Press, New York).

Stewart, G. R., 1984, *Rev. Mod. Phys.* **56**, 755.

Stirling, W. G., R. Scherm, P. A. Hilton, and R. A. Cowley, 1976, *J. Phys. C* **9**, 1643.

Stratonovich, R. L., 1957, *Dokl. Akad. Nauk S.S.S.R.* **115**, 1907. (English translation, 1958, *Sov. Phys. Dokl.* **2**, 416).

Sugiyama, G., and S. E. Koonin, 1986, *Ann. Phys.* (New York), **168**, 1.

Suzuki, M., 1976, *Prog. Theor. Phys.* **56**, 1454.

Svensson, E. C., V. F. Sears, A. B. D. Woods, and P. Martel, 1980, *Phys. Rev.B* **21**, 3638.

Thouless, D. J., 1972, *The Quantum Mechanics of Many-Body Systems*, (Academic Press, New York).

Trotter, H., 1959, *Proc. Am. Math. Soc.* **10**, 545.

Weinstein, J. J., and J. W. Negele, 1982, *Phys. Rev. Lett.*, **49**, 1016.

Wheatley, J. C., 1966, *Quantum Fluids*, D. F. Brewer, ed. (North-Holland, NY), p. 183; 1975, *Rev. Mod. Phys.* **47**, 415.

Whitlock, P. A., D. M. Ceperley, G. V. Chester, and M. H. Kalos, 1979, **Phys. Rev. B 19**, 5598.

Wick, G. C., 1950, *Phys. Rev.* **80**, 268.

Wiener, N., 1924, *Proc. Lond. Math. Soc.* **22**, 454; 1932, *J. Math. Phys.* **2**, 131.

Yoon, B., and J. W. Negele, 1977, *Phys. Rev.* **A16**, 1451.

Yang, C. N., 1967, *Phys. Lett.* **19**, 1313.

Zabolitzky, J. G., M. H. Kalos, 1981, *Nucl. Phys.* **A356**, 114.

Zabolitzky, J. G., K. E. Schmidt, and M. H. Kalos, 1982, *Phys. Rev.* *C* **25**, 1111.

Zinn-Justin, J., 1984, in *Recent Advances in Field Theory and Statistical Mechanics*, J. B. Zuber and R. Stora, eds. (North-Holland, Amsterdam).

INDEX

Printed in the United States
by Baker & Taylor Publisher Services